Principles and Practices of Molecular Properties: Theory, Modeling, and Simulations

Principles and Practices of Molecular Properties: Theory, Modeling, and Simulations

Patrick Norman
KTH Royal Institute of Technology, Stockholm, Sweden

Kenneth Ruud
University of Tromsø—The Arctic University of Norway, Norway

Trond Saue
CNRS/Université Toulouse III-Paul Sabatier, France

WILEY

This edition first published 2018

Registered Office(s)
John Wiley & Sons, Inc., 111 River Street, Hoboken, NJ 07030, USA
John Wiley & Sons Ltd, The Atrium, Southern Gate, Chichester, West Sussex, PO19 8SQ, UK

Editorial Office
9600 Garsington Road, Oxford, OX4 2DQ, UK

For details of our global editorial offices, customer services, and more information about Wiley products visit us at www.wiley.com.

Wiley also publishes its books in a variety of electronic formats and by print-on-demand. Some content that appears in standard print versions of this book may not be available in other formats.

Library of Congress Cataloging-in-Publication Data

Names: Norman, Patrick 1968- author. | Ruud, Kenneth, 1969- author. |
Saue, Trond, 1964- author.
Title: Principles and practices of molecular properties : theory, modeling and simulations / by Patrick Norman, Kenneth Ruud, Trond Saue.
Description: First edition. | Hoboken, NJ : John Wiley & Sons, 2018. |
Includes bibliographical references and index. |
Identifiers: LCCN 2017041617 (print) | LCCN 2017047781 (ebook) | ISBN 9781118794838 (pdf) | ISBN 9781118794814 (epub) | ISBN 9780470725627 (cloth)
Subjects: LCSH: Molecules–Models. | Chemistry, Theoretical Chemistry, Physical.
Classification: LCC QD480 (ebook) | LCC QD480 .N67 2018 (print) | DDC 539/.12–dc23
LC record available at https://lccn.loc.gov/2017041617

Cover Design: Wiley
Cover Image: Courtesy of Patrick Norman

Set in 10/12pt WarnockPro by SPi Global, Chennai, India

10 9 8 7 6 5 4 3 2 1

MIX
Paper from responsible sources
FSC www.fsc.org FSC® C013604

Contents

Preface *xi*

1 **Introduction** *1*

2 **Quantum Mechanics** *11*
2.1 Fundamentals *11*
2.1.1 Postulates of Quantum Mechanics *11*
2.1.2 Lagrangian and Hamiltonian Formalisms *11*
2.1.3 Wave Functions and Operators *18*
2.2 Time Evolution of Wave Functions *22*
2.3 Time Evolution of Expectation Values *25*
2.4 Variational Principle *27*
Further Reading *29*

3 **Particles and Fields** *31*
3.1 Microscopic Maxwell's Equations *32*
3.1.1 General Considerations *32*
3.1.2 The Stationary Case *34*
3.1.3 The General Case *38*
3.1.4 Electromagnetic Potentials and Gauge Freedom *39*
3.1.5 Electromagnetic Waves and Polarization *41*
3.1.6 Electrodynamics: Relativistic and Nonrelativistic Formulations *45*
3.2 Particles in Electromagnetic Fields *48*
3.2.1 The Classical Mechanical Hamiltonian *48*
3.2.2 The Quantum-Mechanical Hamiltonian *52*
3.3 Electric and Magnetic Multipoles *57*
3.3.1 Multipolar Gauge *57*
3.3.2 Multipole Expansions *59*
3.3.3 The Electric Dipole Approximation and Beyond *63*
3.3.4 Origin Dependence of Electric and Magnetic Multipoles *64*
3.3.5 Electric Multipoles *65*
3.3.5.1 General Versus Traceless Forms *65*
3.3.5.2 What We Can Learn from Symmetry *68*
3.3.6 Magnetic Multipoles *69*
3.3.7 Electric Dipole Radiation *70*

3.4 Macroscopic Maxwell's Equations *72*
3.4.1 Spatial Averaging *72*
3.4.2 Polarization and Magnetization *73*
3.4.3 Maxwell's Equations in Matter *77*
3.4.4 Constitutive Relations *79*
3.5 Linear Media *81*
3.5.1 Boundary Conditions *82*
3.5.2 Polarization in Linear Media *86*
3.5.3 Electromagnetic Waves in a Linear Medium *92*
3.5.4 Frequency Dependence of the Permittivity *96*
3.5.4.1 Kramers–Kronig Relations *97*
3.5.4.2 Relaxation in the Debye Model *98*
3.5.4.3 Resonances in the Lorentz Model *101*
3.5.4.4 Refraction and Absorption *105*
3.5.5 Rotational Averages *107*
3.5.6 A Note About Dimensions, Units, and Magnitudes *110*
Further Reading *111*

4 Symmetry *113*
4.1 Fundamentals *113*
4.1.1 Symmetry Operations and Groups *113*
4.1.2 Group Representation *117*
4.2 Time Symmetries *120*
4.3 Spatial Symmetries *125*
4.3.1 Spatial Inversion *125*
4.3.2 Rotations *127*
Further Reading *134*

5 Exact-State Response Theory *135*
5.1 Responses in Two-Level System *135*
5.2 Molecular Electric Properties *145*
5.3 Reference-State Parameterizations *151*
5.4 Equations of Motion *156*
5.4.1 Time Evolution of Projection Amplitudes *157*
5.4.2 Time Evolution of Rotation Amplitudes *159*
5.5 Response Functions *163*
5.5.1 First-Order Properties *166*
5.5.2 Second-Order Properties *166*
5.5.3 Third-Order Properties *169*
5.5.4 Fourth-Order Properties *174*
5.5.5 Higher-Order Properties *179*
5.6 Dispersion *179*
5.7 Oscillator Strength and Sum Rules *183*
5.8 Absorption *185*
5.9 Residue Analysis *190*
5.10 Relaxation *194*
5.10.1 Density Operator *195*

5.10.2 Liouville Equation *196*
5.10.3 Density Matrix from Perturbation Theory *200*
5.10.4 Linear Response Functions from the Density Matrix *201*
5.10.5 Nonlinear Response Functions from the Density Matrix *204*
5.10.6 Relaxation in Wave Function Theory *204*
5.10.7 Absorption Cross Section *207*
5.10.8 Einstein Coefficients *210*
Further Reading *211*

6 Electronic and Nuclear Contributions to Molecular Properties *213*
6.1 Born–Oppenheimer Approximation *213*
6.2 Separation of Response Functions *216*
6.3 Molecular Vibrations and Normal Coordinates *221*
6.4 Perturbation Theory for Vibrational Wave Functions *225*
6.5 Zero-Point Vibrational Contributions to Properties *227*
6.5.1 First-Order Anharmonic Contributions *227*
6.5.2 Importance of Zero-Point Vibrational Corrections *231*
6.5.3 Temperature Effects *234*
6.6 Pure Vibrational Contributions to Properties *235*
6.6.1 Perturbation Theory Approach *235*
6.6.2 Pure Vibrational Effects from an Analysis of the Electric-Field Dependence of the Molecular Geometry *238*
6.7 Adiabatic Vibronic Theory for Electronic Excitation Processes *244*
6.7.1 Franck–Condon Integrals *248*
6.7.2 Vibronic Effects in a Diatomic System *250*
6.7.3 Linear Coupling Model *252*
6.7.4 Herzberg–Teller Corrections and Vibronically Induced Transitions *252*
Further Reading *253*

7 Approximate Electronic State Response Theory *255*
7.1 Reference State Parameterizations *255*
7.1.1 Single Determinant *255*
7.1.2 Configuration Interaction *263*
7.1.3 Multiconfiguration Self-Consistent Field *266*
7.1.4 Coupled Cluster *268*
7.2 Equations of Motion *271*
7.2.1 Ehrenfest Theorem *271*
7.2.2 Quasi-Energy Derivatives *275*
7.3 Response Functions *276*
7.3.1 Single Determinant Approaches *276*
7.3.2 Configuration Interaction *281*
7.3.3 Multiconfiguration Self-Consistent Field *281*
7.3.4 Matrix Structure in the SCF, CI, and MCSCF Approximations *281*
7.3.5 Coupled Cluster *285*
7.4 Residue Analysis *288*
7.5 Relaxation *291*
Further Reading *293*

8 Response Functions and Spectroscopies *295*
8.1 Nuclear Interactions *296*
8.1.1 Nuclear Charge Distribution *296*
8.1.2 Hyperfine Structure *301*
8.1.2.1 Nuclear Magnetic Dipole Moment *301*
8.1.2.2 Nuclear Electric Quadrupole Moment *305*
8.2 Zeeman Interaction and Electron Paramagnetic Resonance *310*
8.3 Polarizabilities *317*
8.3.1 Linear Polarizability *317*
8.3.1.1 Weak Intermolecular Forces *321*
8.3.2 Nonlinear Polarizabilities *325*
8.4 Magnetizability *326*
8.4.1 The Origin Dependence of the Magnetizability *328*
8.4.2 Magnetizabilities from Magnetically Induced Currents *331*
8.4.3 Isotropic Magnetizabilities and Pascal's Rule *332*
8.5 Electronic Absorption and Emission Spectroscopies *335*
8.5.1 Visible and Ultraviolet Absorption *338*
8.5.2 Fluorescence Spectroscopy *343*
8.5.3 Phosphorescence *344*
8.5.4 Multiphoton Absorption *347*
8.5.4.1 Multiphoton Absorption Cross Sections *348*
8.5.4.2 Few-State Models for Two-Photon Absorption Cross Section *350*
8.5.4.3 General Multiphoton Absorption Processes *351*
8.5.5 X-ray Absorption *354*
8.5.5.1 Core-Excited States *355*
8.5.5.2 Field Polarization *358*
8.5.5.3 Static Exchange Approximation *360*
8.5.5.4 Complex or Damped Response Theory *362*
8.6 Birefringences and Dichroisms *364*
8.6.1 Natural Optical Activity *366*
8.6.2 Electronic Circular Dichroism *372*
8.6.3 Nonlinear Birefringences *375*
8.6.3.1 Magnetic Circular Dichroism *376*
8.6.3.2 Electric Field Gradient-Induced Birefringence *379*
8.7 Vibrational Spectroscopies *381*
8.7.1 Infrared Absorption *381*
8.7.1.1 Double-Harmonic Approximation *381*
8.7.1.2 Anharmonic Corrections *383*
8.7.2 Vibrational Circular Dichroism *384*
8.7.3 Raman Scattering *388*
8.7.3.1 Raman Scattering from a Classical Point of View *388*
8.7.3.2 Raman Scattering from a Quantum Mechanical Point of View *392*
8.7.4 Vibrational Raman Optical Activity *402*
8.8 Nuclear Magnetic Resonance *408*
8.8.1 The NMR Experiment *408*
8.8.2 NMR Parameters *413*
Further Reading *417*

Appendicies

A **Abbreviations** *419*

B **Units** *421*

C **Second Quantization** *423*
C.1 Creation and Annihilation Operators *423*
C.2 Fock Space *425*
C.3 The Number Operator *426*
C.4 The Electronic Hamiltonian on Second-Quantized Form *427*
C.5 Spin in Second Quantization *429*

D **Fourier Transforms** *431*

E **Operator Algebra** *435*

F **Spin Matrix Algebra** *439*

G **Angular Momentum Algebra** *441*

H **Variational Perturbation Theory** *445*

I **Two-Level Atom** *451*
I.1 Rabi Oscillations *452*
I.2 Time Dependent Perturbation Theory *454*
I.3 The Quasi-energy Approach *455*

Index *457*

Preface

This book has grown out of an ambition to collect the basic theoretical foundations needed in our daily work as computational chemists with a focus on calculating molecular spectroscopic properties. The result is of course a product strongly influenced by our past experiences and will, as such, provide both a subjective and selective view of the field. It is subjective in the ways we choose to present the subject matter and it is selective in the inclusion of some topics but leaving out others that by many may be deemed just as important. But it is our hope that we provide a basic understanding of key concepts so as to enable further reading of research articles providing either a more detailed treatment of theoretical specifics and algorithmic implementations, or covering any one of the several spectroscopies not mentioned in the book.

Every chapter in this book expresses a balance between *principles* and *practices*. A principle (such as the variational principle) can typically be expressed in an extremely compact manner and governs the practices in a wide range of applications (such as a whole category of methods in electronic structure theory). After an introductory chapter aimed at getting the reader into the right mind-set, the second chapter (on the topic of *quantum mechanics*) almost exclusively deals with principles. It sets the stage and introduces much of the notation used in the remainder of the book. This chapter, as well as this book, is best appreciated after having completed a dedicated course on quantum mechanics[1] at the level of the second cycle in higher education. The third chapter is devoted to *electrodynamics* and here we attempt not only to provide principles but also practices in a form adapted to its use in theoretical chemistry. It represents a core chapter of this book and it should be possible to study it even without any particular background in the field. The fourth chapter is devoted to the topic of *symmetry* and it focuses on principles and is hardly the material for a first time encounter. We expect that the reader is well familiar with the use of molecular point groups and one finds here merely a brief recapitulation of this topic.[2] However, we have tried to provide a novel perspective that takes the reader a bit further than the standard presentation, which will help prepare the reader to handle not only spin-free nonrelativistic wave functions, but also systems where the two-component (or four-component) nature of electronic wave functions must be considered. In the fifth chapter of this book, we reach the core of the theoretical exposé and become familiar with molecular *response properties*. This is intended to be a self-contained chapter and it is followed by the sixth chapter that explains the *separation of electronic and nuclear degrees of freedom* and discusses nuclear contributions to molecular properties. The seventh chapter provides a glimpse into the vast amount of work devoted to implementing the ideas of *response theory in approximate-state theory*. This chapter treads somewhere right in between principles and practices. It treats with certain detail some wave function models but leaves out much of

1 A recommended textbook is *Quantum Mechanics* by B. H. Bransden and C. J. Joachain.

2 A recommended reading is *Group Theory and Chemistry* by D. M. Bishop.

the background in electronic structure theory.[3] Our book is completed with the eighth chapter that illustrates the use of molecular response theory for the calculation of spectroscopic properties. We exemplify nuclear structure and spin properties, dispersive and absorptive electronic response properties, electronic birefringences and dichroisms, and vibrational spectroscopies. Needless to say, the list of examples in this final chapter could be made much longer, but we hope that our selection is diverse enough to provide compelling evidence for the impact and use of response theory in molecular modeling.

Before proceeding, let us note some notational conventions that will be used in this book. We will use the Système International d'unités (SI units). Vectors are typeset in boldface and operators are denoted by a hat, as for instance the molecular Hamiltonian $\hat{H}$. When referring to a specific matrix representation of an operator, we leave out the hat. We will indicate the real and imaginary parts of complex quantities by Re and Im, respectively. For compliance with the literature on electrodynamics, we will in Chapter 3 use **E** to denote the electric field. In the rest of the book, however, we will use **F** for the electric field in order to avoid confusion with the molecular energy that will be denoted by E. We will use implicit summation of repeated indices (the Einstein summation convention), such that

$$A_{\alpha\alpha} \Rightarrow \sum_{\alpha=x,y,z} A_{\alpha\alpha}.$$

Over the course of time spent in writing this book, we have discussed the material with a large number of people. We have organized several week-long courses where the manuscript has been used as teaching material and we have received numerous comments and corrections during these events. We are greatly indebted to all the enthusiastic students attending these schools on response theory, taking place in Chamonix and Luchon in France, as well as at Virginia Tech in the United States. The school at Virginia Tech has been co-organized with Daniel Crawford, who has provided excellent local arrangements and shared his expertise with us in a series of lectures. The U.S. variant of this school could never have been realized without his kind help. We have also benefited from the dedicated proof reading by several people, including Daniel Friese (Düsseldorf), Bin Gao (Tromsø), Michal Jaszuński (Warsaw), Marius Kadek (Tromsø), Nanna H. List (Stockholm), Roberto Di Remigio (Tromsø), and Magnus Ringholm (Tromsø).

A special thanks goes to the following persons for helpful discussions and exchange: Mats Aigner (Linköping), Radovan Bast (Tromsø), Hélène Bolvin (Toulouse), Richard A. Clarke (Surrey), David J. Griffiths (Reed), Trygve Helgaker (Oslo), John D. Jackson (Berkeley), Mathieu Linares (Stockholm), and Nanna H. List (Stockholm). We would also like to thank Barbara Helgaker for helping us iron out some language issues.

Things take longer time than one thinks, even when one takes into account that things take longer time than one thinks. Trond and Patrick would like to thank Hélène, Aurélien, and Sigurd and Dorthe, Jonathan, and Alexander, respectively, for their patience and understanding over the several years it has taken to complete this work.

Patrick Norman, Stockholm
Kenneth Ruud, Tromsø
Trond Saue, Toulouse

June, 2017

3 The most comprehensive treatment of today on this topic is found in *Molecular Electronic-Structure Theory* by T. Helgaker, J. Olsen, and P. Jørgensen.

1

Introduction

If you want to find the secrets of the universe, think in terms of energy, frequency and vibration.

Nicolas Tesla

This is a book about molecular properties, or to be more specific, molecular response properties. Response properties tell us about how molecules respond to electromagnetic fields. To understand these responses, we have to enter the microscopic world of atoms and molecules, governed by the laws of quantum mechanics. For that reason, the reader of this book can expect several intellectual challenges ranging from profound and conceptual cornerstones of quantum theory itself to trivial, yet mind-boggling, issues relating to the smallness of atomic sizes. Consider for instance the situation in which a collection of molecules are being exposed to the intense electric field of a laser, as illustrated in Figure 1.1. From a human perspective, the focal point of a laser is a dangerous place to be, but, from the atomic perspective, it is far less dramatic. In our example, there will be fewer photons than molecules, and, for instance, if the purpose is to protect the eye by efficient optical power limiting, only about every second molecule needs to absorb an incoming light quanta in order to reduce the energy in the transmitted light pulse to an eye-safe level. Furthermore, as strong as the electric field may appear to our eyes, to the individual electron it is several orders of magnitude smaller than the dominating forces exerted by the atomic nuclei and fellow electrons. To get an idea of magnitudes, one may note that the electric field below overhead power lines may reach 10^4 V m^{-1} and the maximum electric field strength possible in air without creating sparks is 3.0×10^6 V m^{-1}. In contrast, at the Bohr radius a_0 in the hydrogen atom, the electric field strength is 5.1×10^{11} V m^{-1}. This is a key point, namely, that we can expose molecules to fields that are strong enough so that we can detect the responses of their charges (nuclei and electrons) while at the same time the fields are weak enough to act as probes, not significantly perturbing the electronic and nuclear structure of the molecule.

Take a very simple example: What happens if a neutral atom (not even a molecule) is placed in a uniform electric field? An experimentalist will ask nature—that is, he or she may perform an experiment, where every macroscopic experiment relates to a very large number of probabilistic microscopic quantum events—by probing how the charge distribution of the atom is modified by the applied field. A theoretician will ask the wave function ψ. The quantum-mechanical equivalent to the outcome of the experiment is the expectation value

$$\langle \hat{\Omega} \rangle = \langle \psi | \hat{\Omega} | \psi \rangle, \tag{1.1}$$

where $\hat{\Omega}$ is the quantum-mechanical operator corresponding to the observable monitored by the experiment. Quantum mechanics is a probabilistic theory. The link between theory and

Principles and Practices of Molecular Properties: Theory, Modeling and Simulations, First Edition.
Patrick Norman, Kenneth Ruud, and Trond Saue.

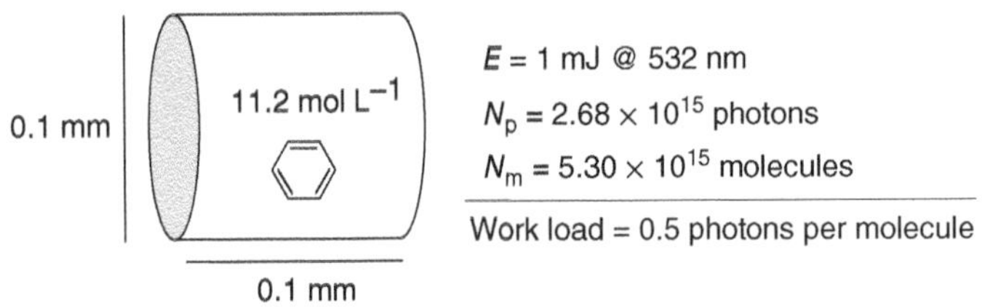

Figure 1.1 Liquid benzene in a small volume corresponding to the focal point of a laser operating at 532 nm and releasing pulses with an energy of 1 mJ.

experiment is made by considering a large number of systems prepared in the same state, prior to switching on the field. If we disregard measurement errors, then the possible outcomes of the individual quantum events are given by the eigenvalues $\{\omega_n\}$ of the operator $\hat{\Omega}$, defined by the eigenvalue equation

$$\hat{\Omega}\psi_n = \omega_n \psi_n. \tag{1.2}$$

Following the postulates of quantum mechanics, the operator $\hat{\Omega}$ is by necessity Hermitian, and the eigenvalues are thus real (corresponding to real-valued observables), and there is a probability $p_n = |\langle\psi_n|\psi\rangle|^2$ for the outcome ω_n in each of the single quantum events, leading to an expectation value that is

$$\langle\hat{\Omega}\rangle = \sum_n \omega_n \, p_n. \tag{1.3}$$

For example, indirect information about the charge distribution of the atom can be obtained from measurements of the electric dipole moment since the two quantities are connected through an expectation value of the form

$$\mu = \left\langle \psi \middle| -e \sum_{i=1}^{N_e} \mathbf{r}_i \middle| \psi \right\rangle = \int \mathbf{r}\rho(\mathbf{r}) \, d^3\mathbf{r}, \tag{1.4}$$

where N_e denotes the number of electrons and e is the elementary charge. However, the electronic charge density can in itself also be expressed as an expectation value

$$\rho(\mathbf{r}) = \left\langle \psi \middle| -e \sum_{i=1}^{N_e} \delta(\mathbf{r} - \mathbf{r}_i) \middle| \psi \right\rangle, \tag{1.5}$$

and it is possible to probe $\rho(\mathbf{r})$ in for instance X-ray diffraction experiments.

If the external electric field is weak compared to the internal atomic fields, we can expand the induced electronic charge density in a Taylor series with respect to field strength. In Figure 1.2, such a perturbation expansion is illustrated to fifth order for a neon atom. The electric field of strength F is applied along the vertical z-axis (directed upward in the figure) and will tend to pull the positive charge along the field and the negative charge in the opposite direction, resulting in an electronic charge density that can be expanded as

$$\rho(\mathbf{r}) = \rho^{(0)}(\mathbf{r}) + \rho^{(1)}(\mathbf{r})F + \frac{1}{2!}\rho^{(2)}(\mathbf{r})F^2 + \frac{1}{3!}\rho^{(3)}(\mathbf{r})F^3 + \frac{1}{4!}\rho^{(4)}(\mathbf{r})F^4 + \frac{1}{5!}\rho^{(5)}(\mathbf{r})F^5 + \cdots . \tag{1.6}$$

The zeroth-order density $\rho^{(0)}$ refers to that of neon in isolation and integrates to $-10e$. It follows from charge conservation that the higher-order densities all integrate to zero. The first-order density $\rho^{(1)}$ shows the charge separation of a dipole, and we then get more and more complicated

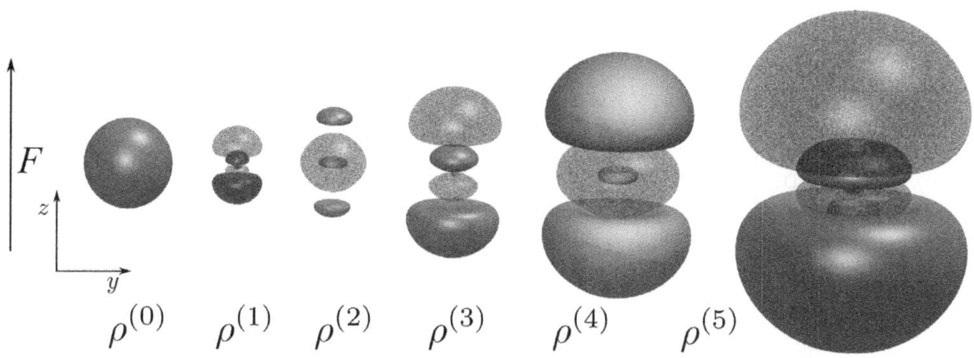

Figure 1.2 Electronic charge density of neon expanded in orders of the applied electric field *F*. Light and dark gray regions indicate positive and negative values, respectively.

structures with increasing order. It is also clear that the higher the order, the more diffuse the density becomes, and we can expect that an accurate description of higher-order responses put strong requirements on the wave function flexibility at large distances from the nucleus.

If we insert the expansion of the charge density into the expression for the dipole moment [Eq. (1.4)], even orders of the density will not contribute due to symmetry—this is a reflection of the fact that odd-order electric properties vanish in systems with a center of inversion. The resulting induced dipole moment, directed along the z-axis, becomes

$$\mu_z = \left[\int z\rho^{(1)}(\mathbf{r})\, d^3\mathbf{r}\right] F + \frac{1}{3!}\left[\int z\rho^{(3)}(\mathbf{r})\, d^3\mathbf{r}\right] F^3 + \frac{1}{5!}\left[\int z\rho^{(5)}(\mathbf{r})\, d^3\mathbf{r}\right] F^5 + \cdots . \quad (1.7)$$

This expression defines a series of proportionality constants between the induced dipole moment and powers of the field. The linear and cubic coupling constants are known as the electric dipole polarizability and second-order hyperpolarizability, and they are conventionally denoted by Greek letters α and γ, respectively.[1] It is the focus of this book to understand how these and other molecular properties can be determined by means of quantum-chemical calculations.

When judging the quality of quantum-chemical calculations, one typically considers the choice of method and basis set. These two quantities combined constitute a theoretical model chemistry, that is, a certain approximation level reaching toward the exact solution of the electronic wave function equation. There exist hierarchical series of basis sets that allow for systematic convergence toward the complete one-particle basis set limit, as indicated in Figure 1.3. An increase in the cardinal number of the basis set, from double- to triple-ζ and so forth, improves the description of the ground-state wave function, whereas levels of augmentation with diffuse functions in the basis set are particularly important for the description of the excited electronic states, and therefore also for many molecular properties. Likewise, in conventional wave function-based electronic structure theory, the configuration interaction (CI) and coupled cluster (CC) expansions provide systematic ways to reach the complete N-particle limit. Increased complexity of the theoretical model chemistry comes, however, at a sometimes staggering computational cost. In general, the computational cost scales as n^y, where the base n represents the size of the one-particle basis set, and therefore implicitly scales with the system size, and the exponent y is associated with a given electronic

1 Hyperpolarizability is a naming convention originally chosen on the basis of the presumed increase of the polarizability in the presence of a strong electric field, whereas hypopolarizability, a name not used any longer, was suggested for systems where there was a decrease.

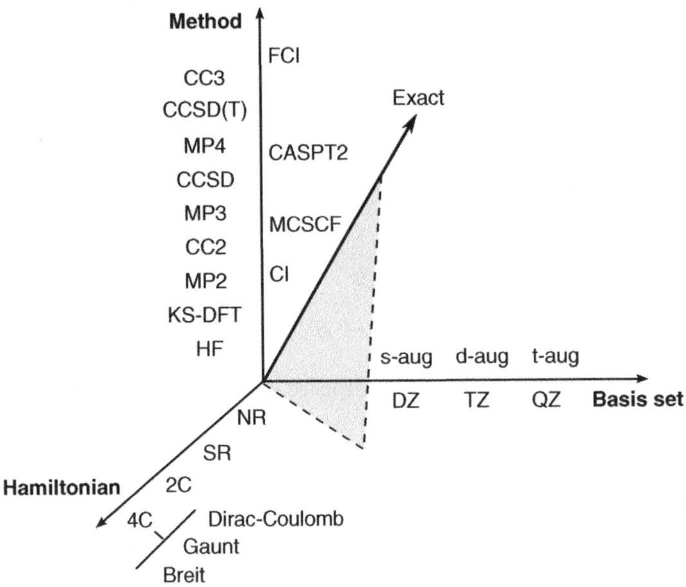

Figure 1.3 Hierarchy of quantum-chemical methods.

structure method. Starting from the Hartree–Fock (HF) method, which formally scales as n^4, each excitation level treated variationally (perturbatively) increases the exponent by two units (one unit). Accordingly, CC and CI models that include single and double (SD) excitations, CCSD and CISD, respectively, scale as n^6, but the CC expansion includes electron correlation in a more efficient manner than does CI and has other advantages such as size extensivity. Adding triple excitations perturbatively, as in CCSD(T), increases the exponent to seven. A great achievement of quantum chemistry has been to devise algorithms that significantly reduce these formal scalings.

Kohn–Sham density functional theory (KS-DFT) has become the most widely used method in quantum chemistry due to its efficient treatment of electron correlation at modest computational cost. It formally has the same scaling as HF theory as it also employs a single Slater determinant to describe the reference state of the fictitious noninteracting KS system, constrained to have the same electron density as the real interacting system. The similarity in the parametrization of the reference state has implications in the presentation of the time-dependent response approaches. Until the very final stages, we need not specify which of these two approaches we address, treating them instead in a uniform manner. We will present this formulation, common to time-dependent HF and KS-DFT theories, under the name self-consistent field (SCF) theory. At the end of the day, there is only one drawback that stands out as critical with the KS-DFT technique, and that is the lack of a systematic way to improve the exchange-correlation (XC) functional, which makes it impossible to provide a general ranking of DFT functionals. This leads to the necessity to benchmark the performance of different XC functionals, not only with respect to classes of molecular systems but also with respect to molecular properties.

Theoretical model chemistries are represented as points in the two-dimensional space spanned by the axis of electronic structure methods and that of the basis sets, as shown in Figure 1.3. Methods and basis sets should be chosen in a balanced manner. For instance, combining the HF method with a quadruple-ζ basis set is normally a waste of resources and time. On the other hand, in the 1980s, a widely used theoretical model chemistry was the CISD method combined with double-ζ basis sets. With increased computational power, however, it was shown numerically that such a combination of methods and basis sets provided reasonable results largely due to a fortuitous cancellation of errors—by increasing the quality of basis sets, the theoretical results moved away from the experimental ones, revealing the intrinsic error of the method. In general, triple-ζ basis sets or better should be used with CC and CI methods to provide an adequate description of the electron–electron cusp region. It should also be noted that standard basis sets are, in most cases, optimized according to energy criteria and that basis set requirements therefore may (and usually do) change when calculating molecular properties. This is easily understood from Figure 1.2 that demonstrates how the induced charge densities governing molecular properties become increasingly more diffuse as compared to the unperturbed density governing the energy.

Since the 1980s, it has become increasingly clear that a theoretical model chemistry providing a balanced description of *all* elements of the periodic table requires the inclusion of relativistic effects. Relativistic effects are normally associated with particles moving at speeds close to the speed of light c, which is the case for electrons in the vicinity of heavy nuclei. The kinetic energy of a 1s-electron in a hydrogen-like ion is

$$K = \frac{Z^2\hbar^2}{2m_e n^2 a_0^2}. \tag{1.8}$$

It is thus proportional to the square of the nuclear charge Z and ranges from 13.6 eV for hydrogen to about 10^5 eV for the heaviest elements with $Z \approx 100$. In Eq. (1.8), we have introduced the electron mass m_e, the Dirac constant $\hbar$, also known as the reduced Planck constant, and the quantum number n for the electron. For small kinetic energies, when

$$K = \frac{mv^2}{2}, \tag{1.9}$$

we thus expect a linear dependence between the electron speed and the nuclear charge Z, as illustrated in Figure 1.4(b). At high kinetic energies, on the other hand, this classical relation breaks down and we have instead

$$K = (\gamma - 1)\, mc^2, \tag{1.10}$$

where the Lorentz factor γ

$$\gamma = \frac{1}{\sqrt{1-(v/c)^2}}, \tag{1.11}$$

will increasingly deviate from its nonrelativistic limiting value of 1, as illustrated in Figure 1.4(a). In the region of gold (atomic number 79) in the periodic table, the speed of 1s-electrons reaches values of about 0.6c, corresponding to a Lorentz factor of about 1.15. The Lorentz factor appears in the expression for the relativistic mass increase $m = \gamma m_e$, and, since the Bohr radius is inversely proportional to the mass, we expect relativity to contract and stabilize orbitals.

One can of course argue, as did Dirac himself, that our focus on the 1s-orbitals is an extreme case for relativistic effects, and, because the 1s-electrons orbit very close to their respective nuclei, such relativistic effects will play an insignificant role in most of chemistry. However, relativistic effects propagate out to the valence through their modification of orbitals in the core region and, through the requirement of orthogonality, among all orbitals. An illustration of this aspect is given in Figure 1.4(c), in terms of the contraction of the valence ns-orbitals, which is seen to be particularly pronounced for the coinage metal atoms. The direct effect of relativity is to contract orbitals, but this contraction leads to increased screening of the nuclear charge, so there will be a competition between the direct and indirect effects of relativity. In practice, one finds that s- and p-orbitals tend to contract while d- and f-orbitals tend to expand.

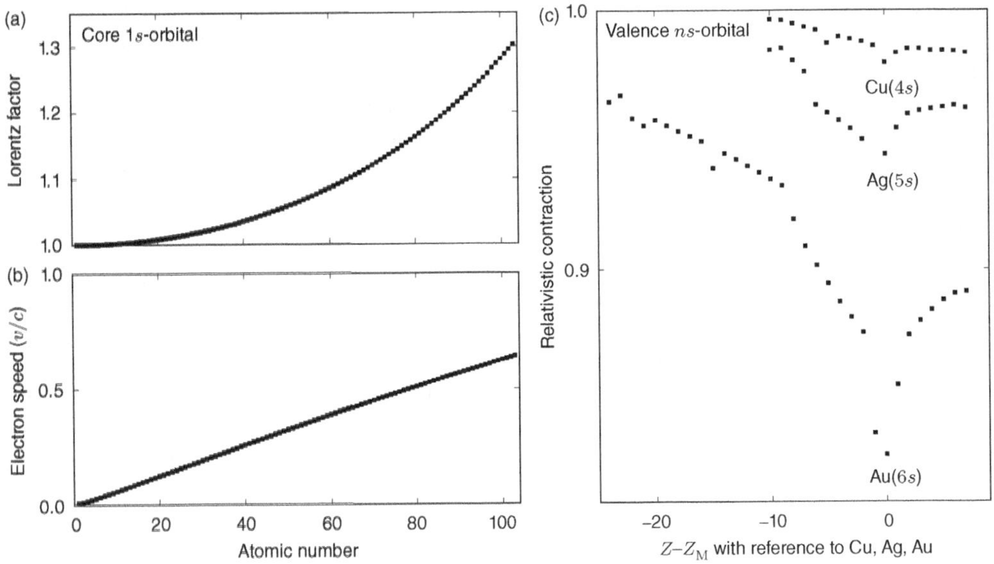

Figure 1.4 Speed of 1s-electrons relative to the speed of light (b) and the corresponding Lorentz factor (a) versus atomic number. Relativistic contractions (c), defined as ratio between relativistic and nonrelativistic HF radial expectation values, of valence *ns*-orbitals for elements of rows 4, 5, and 6 of the periodic table. Atomic numbers Z are given relative to atomic number Z_M of the coinage metal of each row (M=Cu, Ag, Au).

Relativistic effects associated with the relativistic mass increase of electrons discussed earlier are denoted scalar relativistic effects. A second relativistic effect is spin–orbit interaction. The name and typical operator form $\hat{h}_{SO} \sim \hat{\mathbf{s}} \cdot \hat{\mathbf{l}}$ suggests an interaction due to a coupling of the spin and angular momentum of the electron. However, this confuses the underlying physics that is simply magnetic induction. Spin–orbit coupling refers to the interaction of the electron spin with the magnetic fields generated by charged particles (nuclei or electrons) in relative motion. The orbital angular momentum operator $\hat{\mathbf{l}}$ appears in the operator describing the spin–orbit interaction induced by nuclei because it encodes the relative motion between the electron and the nucleus. Spin–orbit interactions constitute one of the several categories of intramolecular magnetic interactions that are unaccounted for in a nonrelativistic formulation, and they are often introduced in the nonrelativistic framework in an *ad hoc* manner, as they govern important phenomena in chemistry such as intersystem crossings and phosphorescence. Another example is given by electron paramagnetic resonance (EPR) spectroscopy where one of the parameters, the anisotropy of the molecular g tensor, is rooted in spin–orbit interactions. The spin–orbit interaction couples spin and spatial degrees of freedom, such that spin is no longer a good quantum number. Atomic orbitals will be characterized by their total angular momentum $\mathbf{j} = \mathbf{l} + \mathbf{s}$ with possible magnitude $j = l \pm 1/2$ and corresponding azimuthal quantum number $m_j = m_l + m_s$. As an example, the six degenerate p-orbitals split into two degenerate $p_{1/2}$- and four degenerate $p_{3/2}$-orbitals. The energy associated with the $p_{1/2}$-orbitals is lowered as a consequence of an optimum orientation of electron spin in space, see Figure 1.5(a). This flexibility for the direction of the spin to vary between different points in space is not accounted for in the nonrelativistic realm, where orbitals are either pure α- or pure β-spin orbitals. Interestingly, this is also the case for the relativistic $p_{3/2}$-orbital with $m_j = \pm 3/2$, as illustrated in Figure 1.5(b). The $m_j = 3/2$ orbital is pure α simply because β spin ($m_s = -1/2$) would require $m_l = 2$, which is not a valid quantum number for $l = 1$ angular functions.

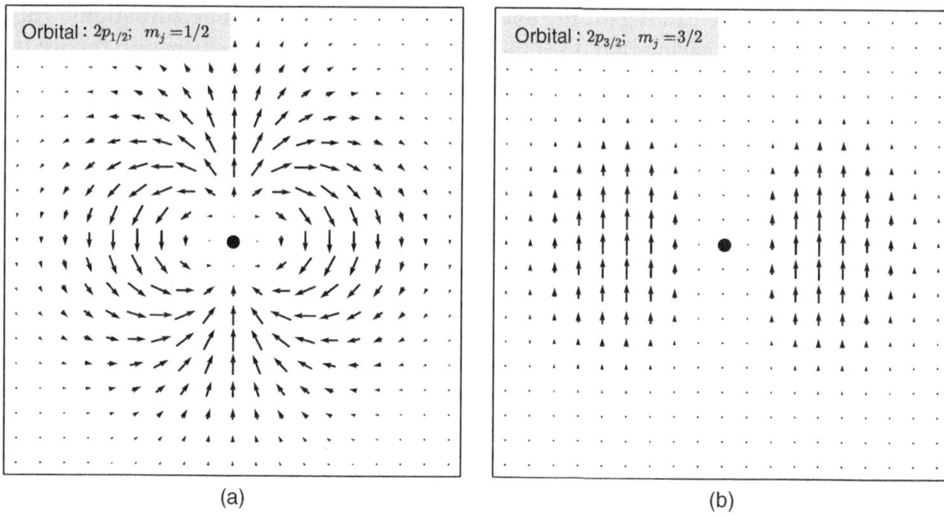

Figure 1.5 Hydrogen spin densities in the *xz*-plane for the (a) $2p_{1/2}(m_j = 1/2)$ and (b) $2p_{3/2}(m_j = 3/2)$ orbitals.

Dealing with elements from large parts of the periodic table thus corresponds to a situation in which conventional nonrelativistic calculations of the electronic structure may make little sense and relativistic modifications of the Hamiltonian are required, corresponding to the third axis in Figure 1.3. Scalar relativistic effects can be introduced without significant additional computational cost, whereas the inclusion of spin–orbit interactions requires Hamiltonians based on the four-component Dirac operator or two-component approximations to it. A simple way to introduce relativistic effects is to replace the relativistic core orbitals by effective core potentials, but this choice is only viable when is concerned with valence electron properties.

The choice of Hamiltonian is in principle one *independent* of the two other axes of our chart, which means that the conventional theoretical model chemistries known to us from the nonrelativistic realm can be extended to the relativistic domain. Another important consequence is that the choice of Hamiltonian only comes with a prefactor, independent of system size, to the overall computational cost.

At this point, some readers may feel that a lot of time has been spent discussing relativistic effects, especially if their focus is on molecular systems containing light atoms only. However, it should be kept in mind that a book about molecular properties is essentially about electromagnetic interactions, and electrodynamics—as elegantly summarized by Maxwell's equations—is a relativistic theory. In the nonrelativistic limit, electrodynamics scales down to electrostatics; that is, in this limit not only do the effects of the finite speed of interactions vanish, but also all magnetic interactions, including magnetic induction. We will demonstrate that nonrelativistic calculations of magnetic properties are made possible by using a relativistic coupling of particles to fields. This is a perfectly valid, pragmatic approach, but implies a mixing of theories with different transformation properties and makes the underlying physics less transparent than a fully relativistic formulation.

Most quantum-chemical calculations describe molecules alone in the universe and at 0 K. This may be a good approximation for dilute gases, since intermolecular interactions are often negligible unless explicitly addressed in the experimental studies. However, the scope of applications for the computational chemist is far wider, including molecules in liquid and solid phases, as well as at interfaces. With molecular response approaches, we can address cases where the properties of the macroscopic system is predominantly determined by the properties of the molecules, that is, one can regard the effects of the environment as perturbations. This is not, however, the same as saying that one employs the techniques of perturbation theory to account for these interactions; rather, self-consistent schemes are commonly adopted. Either way, the space is divided into regions treated at different levels of accuracy, where the core quantum region is large enough to encompass those parts of the electron density that give rise to the response signal in the experiment. The exterior of the core quantum region is treated by means of classical physics, either in terms of continuum dielectric medium models, such as the polarizable continuum model (PCM), or by the introduction of discrete charge multipole expansions representing the solvent molecules in molecular mechanics (MM) approaches. The latter of the two techniques can be rather straightforwardly combined with molecular dynamics (MD) simulations, be it classical or quantum MD, and it is a common procedure to extract snapshots from MD simulations and perform individual response theory calculations based on each of these snapshots. This is a way to sample the molecular configuration space under given experimental conditions of temperature, liquid, and chromophore densities, and the statistically averaged theoretical spectra represents the final result of the simulation. Figure 1.6 shows one such snapshot taken from simulations of luminescence properties of an oligothiophene chromophore used for optical probing of certain proteins. The partitioning of a solvent or a large molecular system into a target area treated accurately and with the remaining environment being treated in a more approximate manner is often referred to as focused models, and can be considered a computational equivalent of the chemical concept of active or functional regions in a complex molecular system.

This brief introduction aims to show that when the computational chemist needs to evaluate molecular response properties, a number of problems of computational and theoretical nature must be addressed, explicitly or at least assessed with respect to the expected relevance and importance. This book aims to be a valuable aid in this work, by introducing a response theory framework that is general enough to host the most commonly adopted model chemistries while leaving out most details of their implementations. The choice of basis set in the calculation is an obvious parameter that is not of much concern in the formulation of a theory, and, although less obvious, this holds largely true also for the choice of Hamiltonian. The reason for the latter simplification is that, before evaluating specific matrix elements, one can suppress the reference

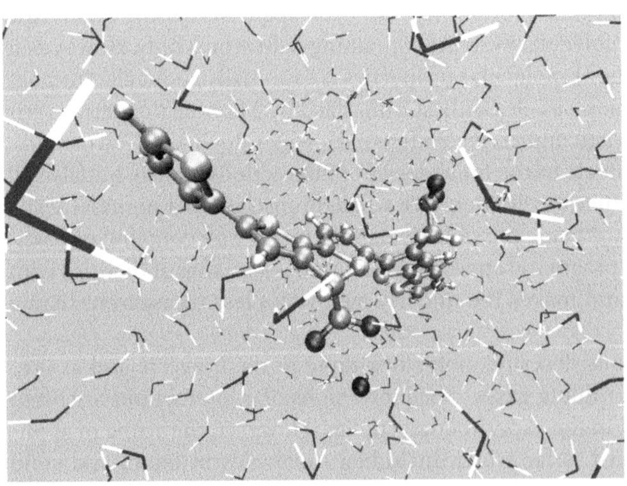

Figure 1.6 Anionic polythiophene acetic acid with sodium counterions in water solution.

to the explicit form of the Hamiltonian and write it in a general form, including summations of one-electron $\hat{h}(i)$ and two-electron $\hat{g}(i,j)$ terms—a form which is common to one-, two-, and four-component Hamiltonians. Even such a seemingly insurmountable complexity as that provided by the interactions with the surrounding environment can, from our perspective, be largely ignored. The key here is to define effective operators that will couple the external classical region to the core quantum region. Once this coupling is achieved, it can be viewed as a modification of the Hamiltonian that fits into our general framework. When it comes to the third axis in Figure 1.3—the electronic structure methods used in the model chemistries—the situation becomes substantially more complicated. In this case, care has to be taken already at the outset of the derivation of the response functions when choosing an equation of motion for the time evolution of the wave function. Although all formulations of exact-state wave mechanics give rise to identical time propagations of a given initial reference state, this is no longer true in approximate-state theories. We will discuss this issue at length and consider two fundamentally different formulations of time-dependent perturbation theory. These two formulations are referred to as the Ehrenfest and the quasi-energy approaches, respectively, and while the former is applicable only to variational electronic structure methods, the latter can also be applied to nonvariational methods.

2

Quantum Mechanics

Quantum mechanics, that mysterious, confusing discipline, which none of us really understands but which we know how to use. It works perfectly, as far as we can tell, in describing physical reality, but it is a 'counter-intuitive discipline', as social scientists would say. Quantum mechanics is not a theory, but rather a framework, within which we believe any correct theory must fit.

M. Gell-Mann (1980)

In this chapter we give a brief, and in many respects, an incomplete introduction to quantum mechanics, with a focus on the bare minimum of concepts that are central to the subsequent parts of the book. In addition to providing the necessary basis for the subsequent chapters, this chapter introduces much of the notation that will be used throughout the book.

2.1 Fundamentals

2.1.1 Postulates of Quantum Mechanics

The theory of quantum mechanics rests on a set of postulates as illustrated in Table 2.1 in the case of a single-particle system in one-dimensional (1D) motion. The precise knowledge of the system state and physical observables in the classical case has, in the quantum case, been turned into a situation where the values of observables can only be predicted probabilistically. The coordinate vectors $|x\rangle$ that are often used as a basis for the infinite-dimensional Hilbert space can be imagined (but not realized) as states of perfect particle localization $\langle x'|x\rangle = \delta(x - x')$, but in general the position of the quantum particle is described by the *probability density* (or simply the density)

$$n(x,t) = |\psi(x,t)|^2. \tag{2.3}$$

Changes in the density may be predictable and deterministic in accordance with Postulate D (the time-dependent Schrödinger equation), but they may also be indeterministic and associated with sudden "quantum jumps" to eigenstates of the operators $\hat{\Omega}$ in accordance with Postulate C.

2.1.2 Lagrangian and Hamiltonian Formalisms

From Postulate D, we see that the correspondence between classical and quantum mechanics goes through the Hamiltonian formalism. In order to better understand the construction of the Hermitian operators in quantum mechanics that are associated with physical observables, we

Principles and Practices of Molecular Properties: Theory, Modeling and Simulations, First Edition.
Patrick Norman, Kenneth Ruud, and Trond Saue.

Table 2.1 Postulates in classical and quantum mechanics.

	Classical mechanics	Quantum mechanics
State of system		
A	The state of a particle at any given time is given by the two variables $x(t)$ and $p(t)$, that is, a point in a 2D phase space.	The state of a particle is represented by a state vector $\|\psi(t)\rangle$ of unit norm in a Hilbert space. The projection on the basis $\|x\rangle$ gives the wave function $\langle x\|\psi(t)\rangle = \psi(x,t)$ in coordinate space.
Observables		
B	Every dynamical variable ω is a function of x and p: $\omega = \omega(x, p)$.	The independent variables x and p in classical mechanics are represented by the Hermitian operators $\hat{x}$ and $\hat{p}$ that obey the commutator relation $$[\hat{x}, \hat{p}] = i\hbar.$$ The operator that corresponds to the dynamical variable $\omega(x, p)$ is the Hermitian operator obtained from the substitution $$\hat{\Omega}(\hat{x}, \hat{p}) = \omega(x \to \hat{x}, p \to \hat{p}).$$
Measurements		
C	A measurement of the variable ω on a particle in a state given by x and p gives a value $\omega(x, p)$. After the measurement, the state is left unchanged.	A measurement of the variable, corresponding to operator $\hat{\Omega}$, on a particle in a normalized state $\|\psi\rangle$ results in an eigenvalue ω with probability $P(\omega) = \|\langle\omega\|\psi\rangle\|^2$. The state of the system is left in the associated eigenstate $\|\omega\rangle$ after the measurement.
Equations of motion		
D	The state variables have a time dependence given by Hamilton's equations $$\dot{x} = \frac{\partial H}{\partial p}; \quad \dot{p} = -\frac{\partial H}{\partial x}. \qquad (2.1)$$	The state vector $\|\psi(t)\rangle$ has a time dependence given by the time-dependent Schrödinger equation $$i\hbar\frac{\partial}{\partial t}\|\psi(t)\rangle = \hat{H}\|\psi(t)\rangle, \qquad (2.2)$$ where $\hat{H}(\hat{x}, \hat{p}, t) = H(x \to \hat{x}, p \to \hat{p}, t)$ is the quantum mechanical Hamiltonian and H is the classical Hamiltonian.

need to take a closer look at Hamiltonian mechanics. In fact, in order to get the complete picture of quantization, we shall start from *Lagrangian mechanics*. A fascinating aspect of Lagrangian mechanics is that it allows us to derive the equations of motion of a dynamical system of N particles from a single scalar function, the Lagrangian. Let $x \equiv x(t)$ collectively denote the $3N$ coordinates of the particles in the system—mapping out a point in the *configuration space* of the system—with corresponding velocities given by their total time derivatives $\dot{x} \equiv dx/dt$. In particular, let $x(t_a)$ and $x(t_b)$ be the particle coordinates at times t_a and t_b, respectively. We next introduce the *action* S defined as

$$S[x] = \int_{t_a}^{t_b} L(x, \dot{x}, t)dt, \qquad (2.4)$$

which is an example of a mathematical object known as a functional. In general, functionals map functions onto complex numbers, in contrast to functions that map sets into sets. In order to explore the possible values of the input function, a functional is often given as a definite integral over the function variables, that is

$$F[y] = \int_{x_1}^{x_2} f(y(x), y'(x), \ldots, x)\, dx. \tag{2.5}$$

The functional dependence of F on the function $y(x)$ is indicated by square brackets to separate it from arguments of a function that are kept inside parenthesis. In the calculus of variations, a central question is how the value of the functional depends on the variations of the function. For this purpose, the *first variation* of the functional F is defined as

$$\delta F = \lim_{\varepsilon \to 0} \frac{F[y + \varepsilon \delta y] - F[y]}{\varepsilon}, \tag{2.6}$$

where δy is an unspecified test function that can be viewed as pointing out a direction in the function space along which one takes an infinitesimal step by means of the multiplication with ε. A variation δy in the input function is connected to the variation of the functional through the functional derivative

$$\delta F = \int_{x_1}^{x_2} \left(\frac{\delta F}{\delta y(x)} \right) \delta y(x)\, dx. \tag{2.7}$$

This can be compared to the corresponding relation for functions, where an infinitesimal change in the argument is connected to the change in the function through the (function) derivative

$$dy = \left(\frac{dy}{dx} \right) dx. \tag{2.8}$$

In our applications, the function y is typically to be replaced by either the wave function ψ or the electron charge density ρ, and we will be scanning the respective function spaces in the search for elements that yield zero first-order variations. Since the function variations δy are arbitrary, this implies that the functional derivative itself is zero. In the present case, the functional under study is the action S, which is a functional of the system trajectory $x(t)$ and expressed as an integral over the *Lagrangian* from an initial time t_a to a final time t_b, as seen in Eq. (2.4).

The *principle of stationary action* states that the actual path taken by the system between times t_a and t_b is the one for which an infinitesimal change in the trajectories leaves the action unchanged (stationary), or, in other words, the first variation of the action vanishes

$$\delta S = \int_{t_a}^{t_b} \left[\left(\frac{\partial L}{\partial x} \right) \delta x(t) + \left(\frac{\partial L}{\partial \dot{x}} \right) \delta \dot{x}(t) \right] dt = 0. \tag{2.9}$$

To find the corresponding functional derivative of S with respect to the trajectory $x(t)$ in configuration space, we compare the above expression with the defining relation

$$\delta S = \int_{t_a}^{t_b} \frac{\delta S}{\delta x(t)} \delta x(t)\, dt. \tag{2.10}$$

We see that the first, but not the second term of Eq. (2.9), has the desired form. In order to bring the latter into the correct form, we carry out an integration by parts

$$\int_{t_a}^{t_b} \left(\frac{\partial L}{\partial \dot{x}} \right) \delta \dot{x}(t) dt = \left(\frac{\partial L}{\partial \dot{x}} \right) \delta x(t) \Big|_{t_a}^{t_b} - \int_{t_a}^{t_b} \left(\frac{d}{dt} \frac{\partial L}{\partial \dot{x}} \right) \delta x(t) dt. \tag{2.11}$$

We now impose that the start- and endpoints of the trajectories are known and fixed such that

$$\delta x(t_a) = \delta x(t_b) = 0. \tag{2.12}$$

Then the first term on the right-hand side of Eq. (2.11) vanishes and we obtain a form of δS that allows us to extract the functional derivative. Setting the functional derivative to zero for all coordinates x_i gives

$$\frac{\delta S}{\delta x_i(t)} = \frac{\partial L}{\partial x_i} - \frac{d}{dt}\left(\frac{\partial L}{\partial \dot{x}_i}\right) = 0; \quad i = 1, \ldots, 3N. \tag{2.13}$$

These equations of motion are known as the Euler–Lagrange equations. The condition $\delta S = 0$ implies that functional derivatives for individual coordinates are zero only to the extent that the coordinates are linearly independent. It may be that there are constraints on the motions of the particles; for instance, that our particles are constrained to move on the surface of a sphere, in which case this condition no longer holds. However, if there are k constraints on the form

$$f_\alpha(x, t) = 0; \quad \alpha = 1, \ldots, k, \tag{2.14}$$

then we may build them into our formalism by transforming to a set of $3N - k$ linearly independent *generalized coordinates* q_i, collectively denoted as q, which now defines our configuration space. In our derivations we will for simplicity ignore such constraints.

At this point it is important to note that the Lagrangian function is not given *a priori*, but should be chosen to give trajectories in accordance with the experiment. Building new Lagrangians therefore means building new physics. Generally, in the absence of dissipative forces such as friction, the Lagrangian has the form $L = T - U$ where T is the kinetic energy and $U = U(x, \dot{x}, t)$ is called the generalized potential. The kinetic energy is the Lagrangian of free particles, corresponding to the case $U = 0$. In the absence of any constraints, the kinetic energy is a function of velocities only and we may re-write the Euler–Lagrange equations, Eq. (2.13), in the form of Newton's second law as

$$F_i = \frac{d}{dt}\left(\frac{\partial T}{\partial \dot{x}_i}\right); \quad i = 1, \ldots, 3N, \tag{2.15}$$

where the generalized force F_i appears

$$F_i = -\frac{\partial U}{\partial x_i} + \frac{d}{dt}\left(\frac{\partial U}{\partial \dot{x}_i}\right). \tag{2.16}$$

Now let us consider the force $\mathbf{F}$ acting on a single particle. The work done by the force upon the particle along its trajectory between times t_1 and t_2 is given by a line integral

$$W = \int_{\mathbf{r}(t_1)}^{\mathbf{r}(t_2)} \mathbf{F} \cdot d\mathbf{r} = \int_{t_1}^{t_2} \mathbf{F} \cdot \mathbf{v}dt. \tag{2.17}$$

A first thing to note is that if we use Newton's second law $\mathbf{F} = m\mathbf{a}$, we can show that the total work along the trajectory is equal to the change in kinetic energy $T = \frac{1}{2}mv^2$ between the end-points of the trajectory, that is,

$$W = \int_{\mathbf{r}(t_1)}^{\mathbf{r}(t_2)} \mathbf{F} \cdot d\mathbf{r} = \int_{t_1}^{t_2} m\frac{d\mathbf{v}}{dt} \cdot \mathbf{v}dt = \frac{1}{2}m\int_{t_1}^{t_2} \frac{d(v^2)}{dt}dt = \int_{T(t_1)}^{T(t_2)} dT = T(t_2) - T(t_1). \tag{2.18}$$

Let us assume that the force is monogenic, that is, derivable from a single potential function $U(\mathbf{r}, \mathbf{v}, t)$, as in Eq. (2.16). We can then write the infinitesimal work as

$$\begin{aligned} dW = \mathbf{F} \cdot d\mathbf{r} &= \left[-\left(\frac{\partial U}{\partial \mathbf{r}}\right) + \frac{d}{dt}\left(\frac{\partial U}{\partial \mathbf{v}}\right)\right] \cdot d\mathbf{r} \\ &= -dU + \left(\frac{\partial U}{\partial \mathbf{v}}\right) \cdot d\mathbf{v} + \left(\frac{\partial U}{\partial t}\right) dt + \frac{d}{dt}\left(\frac{\partial U}{\partial \mathbf{v}}\right) \cdot d\mathbf{r}, \end{aligned} \tag{2.19}$$

where, in the final step, we have used the total differential

$$dU = \left(\frac{\partial U}{\partial \mathbf{r}}\right) \cdot d\mathbf{r} + \left(\frac{\partial U}{\partial \mathbf{v}}\right) \cdot d\mathbf{v} + \left(\frac{\partial U}{\partial t}\right) dt, \tag{2.20}$$

of the generalized potential. In many situations, but *not* in the presence of electromagnetic forces, the potential function has no explicit dependence on velocities and time. In such a case, the work along the trajectory is given by the value of the potential function at the endpoints, that is,

$$W = U(t_1) - U(t_2). \tag{2.21}$$

The work is therefore path independent and will for instance be zero around any closed path. Combining the above result with Eq. (2.19) we have

$$T(t_1) + U(t_1) = T(t_2) + U(t_2), \tag{2.22}$$

and the force is said to be *conservative.*

Returning to the general case, let us consider the total time derivative of the work. Starting from Eq. (2.19), we obtain

$$\frac{dW}{dt} = -\frac{dU}{dt} + \left(\frac{\partial U}{\partial \mathbf{v}}\right) \cdot \frac{d\mathbf{v}}{dt} + \frac{\partial U}{\partial t} + \frac{d}{dt}\left(\frac{\partial U}{\partial \mathbf{v}}\right) \cdot \frac{d\mathbf{r}}{dt} = -\frac{dU}{dt} + \frac{d}{dt}\left[\left(\frac{\partial U}{\partial \mathbf{v}}\right) \cdot \mathbf{v}\right] + \frac{\partial U}{\partial t}. \tag{2.23}$$

Keeping in mind that $dW = dT$, we can rearrange this to

$$\frac{d}{dt}\left[T + U - \left(\frac{\partial U}{\partial \mathbf{v}}\right) \cdot \mathbf{v}\right] = \frac{\partial U}{\partial t}. \tag{2.24}$$

We identify the expression in square brackets with energy

$$E = T + V; \quad V = U - \left(\frac{\partial U}{\partial \mathbf{v}}\right) \cdot \mathbf{v}, \tag{2.25}$$

where V is the potential energy.[1] Eq. (2.24) therefore demonstrates that the total energy is conserved if the general potential has no explicit time dependence.

Let us now turn to Hamiltonian mechanics. Again, a scalar function, this time the *Hamiltonian,* takes center stage. It can be obtained from the Lagrangian by a *Legendre transformation*

$$H(x, p, t) = \sum_{j=1}^{3N} \dot{x}_j p_j - L(x, \dot{x}, t). \tag{2.26}$$

A crucial difference between the Hamiltonian and the Lagrangian is that the dependence on velocity $\dot{x}$ in the latter is replaced by dependence on momentum p in the former, achieved through the Legendre transformation. We can also use the Legendre transformation to find the general definition of momentum: The Hamiltonian does not depend on velocity and so by taking the partial derivative with respect to velocity $\dot{x}$ on both sides of Eq. (2.26), we obtain

$$\frac{\partial H}{\partial \dot{x}_i} = 0 \quad \Rightarrow \quad p_i = \frac{\partial L}{\partial \dot{x}_i} = \frac{\partial T}{\partial \dot{x}_i} - \frac{\partial U}{\partial \dot{x}_i}. \tag{2.27}$$

If the generalized potential does not depend on velocity, that is, for a conservative system, the momentum reduces to its usual (nonrelativistic) form $p_i = m\dot{x}_i$, denoted by the *mechanical momentum.* However, as discussed in Chapter 3, magnetic forces depend on particle velocities, and hence the need to start from the correct Lagrangian to get a proper definition of momentum for use in the Hamiltonian, both in classical and quantum mechanics.

1 C. Lanczos. *The Variational Principles of Mechanics.* University of Toronto Press, 1949.

Hamilton's equations, which appear in Postulate D in Eq. (2.1), can be obtained from the Legendre transformation, Eq. (2.26), as well. Taking the partial derivative with respect to momentum we obtain the first equation. Taking the partial derivative with respect to coordinate x and using the Euler–Lagrange equations, Eq. (2.13), gives the second equation. Finally, taking the total time derivative of the Hamiltonian using the chain rule, we obtain

$$\frac{dH}{dt} = \left(\frac{\partial H}{\partial x}\right)\dot{x} + \left(\frac{\partial H}{\partial p}\right)\dot{p} + \frac{\partial H}{\partial t}. \tag{2.28}$$

By using Hamilton's equations, we find that the first two terms cancel. The final result is

$$\frac{dH}{dt} = \frac{\partial H}{\partial t} = -\frac{\partial L}{\partial t}, \tag{2.29}$$

where the final step is obtained by taking the explicit time derivative of the Legendre transformation. If the Lagrangian has no explicit time dependence, the above expression shows that the Hamiltonian is time independent and thus a constant of motion. In fact, we can obtain Eq. (2.29) directly by taking the total time derivative of the Lagrangian. For a single particle, we get

$$\frac{dL}{dt} = \left(\frac{\partial L}{\partial \mathbf{r}}\right)\cdot\frac{d\mathbf{r}}{dt} + \left(\frac{\partial L}{\partial \mathbf{v}}\right)\cdot\frac{d\mathbf{v}}{dt} + \frac{\partial L}{\partial t} = \frac{d}{dt}\left[\left(\frac{\partial L}{\partial \mathbf{v}}\right)\cdot\mathbf{v}\right] + \frac{\partial L}{\partial t}, \tag{2.30}$$

where the final step is obtained using the Euler–Lagrange equation, Eq. (2.13). By rearrangement, we obtain

$$\frac{d}{dt}\left[\mathbf{p}\cdot\mathbf{v} - L\right] = -\frac{\partial L}{\partial t}. \tag{2.31}$$

We see that the Legendre transformation leading to the Hamiltonian function drops out directly in the above expression. We can even go one step further and show that it is equivalent to the energy expression of Eq. (2.25) by writing the Legendre transformation, Eq. (2.26), as

$$H = T + V = \mathbf{v}\cdot\frac{\partial L}{\partial \mathbf{v}} - L \tag{2.32}$$

and carrying it out separately for the components T and U of the Lagrangian. From the kinetic energy, we get

$$\frac{\partial T}{\partial \mathbf{v}} - T = T, \tag{2.33}$$

and from the generalized potential we get directly the potential energy. The Hamiltonian can therefore be identified with the total energy,[2] which is conserved if there is no explicit time dependence in the generalized potential.

Let us illustrate these concepts with a very simple, albeit not very chemical, example, namely that of the free particle. We can deduce the nonrelativistic (NR) form of this Lagrangian from some simple considerations: (i) for a free particle there is no preferred moment in time (homogeneity of time), so the Lagrangian cannot depend explicitly on time, and (ii) there is no preferred point or direction in space (homogeneity and isotropy of space) so the Lagrangian cannot depend explicitly on the position or the velocity direction. Based upon these considerations, we write the Lagrangian as proportional to the square of the velocity (or speed) of the particle

$$L_0^{\mathrm{NR}} = T = \frac{1}{2}m\dot{x}^2, \tag{2.34}$$

2 In the case of constrained motions [Eq. (2.14)], Eq. (2.33) does not hold if there is an explicit time dependence in the functions f_α expressing the constraints. The Hamiltonian can then not be identified with energy.

where the constant of proportionality is fixed by experiment. Using Eq. (2.27), we find that the momentum of the free particle is $p = mx$. Next, we obtain the corresponding free-particle Hamiltonian by the Legendre transformation, Eq. (2.26),

$$H_0^{\mathrm{NR}} = m\dot{x}^2 - \frac{1}{2}m\dot{x}^2 = \frac{p^2}{2m}. \tag{2.35}$$

Note that the intermediate expression, referred to as the energy function, does not represent a valid Hamiltonian since it is expressed in terms of the velocity. Only after the velocity is substituted by the momentum does the Hamiltonian become a function of *position and momentum* as required. The velocity in terms of momentum is provided by the first of Hamilton's equations, whereas the second takes the form $\dot{p} = 0$, corresponding to Newton's first law. We again stress the fact that the momentum is defined from the Lagrangian and will therefore hinge upon a correct definition of this function. This point will turn out to be crucial when considering the introduction of electromagnetic fields, as will be discussed in Chapter 3.

The final quantum-mechanical Hamiltonian is obtained by replacing the position and momentum variables by their corresponding operators. These should be chosen such that the canonical commutation relation

$$\left[x_i, p_j\right] = i\hbar\delta_{ij}, \tag{2.36}$$

is obeyed. In the *coordinate representation*, we use the substitutions

$$x_i \to \hat{x}_i = x_i; \quad p_i \to \hat{p}_i = -i\hbar\frac{\partial}{\partial x_i}. \tag{2.37}$$

Another option is the *momentum representation*

$$x_i \to \hat{x}_i = i\hbar\frac{\partial}{\partial p_i}; \quad p_i \to \hat{p}_i = p_i, \tag{2.38}$$

but it is less useful for localized systems such as molecules.

The relativistic (R) free-particle case is more complicated and the appropriate form of the Lagrangian is less evident. In fact, since there is no longer a universal time, we should not single out a specific infinitesimal of time *dt* in the action [Eq. (2.4)] but rather use the Lorentz-invariant *proper* time τ, which is the time in the reference frame moving with the particle. The variation of the action becomes

$$\delta S = \delta\int_{\tau_a}^{\tau_b} \Lambda(x_\mu, v_\mu, \tau)\, d\tau = 0; \quad c\, d\tau = \sqrt{c^2dt^2 - dx^2}, \tag{2.39}$$

where the relativistic Lagrangian Λ is now a function of the four-position $x_\mu = (x, ict)$, four-velocity $v_\mu = \gamma\,(\dot{x}, ic)$, and the proper time τ. Following the same line of arguments as for the nonrelativistic free particle, it can be proposed that the relativistic Lagrangian should be proportional to the square of the four-velocity. In fact, the suitable form is $\Lambda = mv_\mu v_\mu = -mc^2$, which clearly is Lorentz invariant. We can go to a specific frame by noting that

$$c\, d\tau = c\, dt\gamma^{-1}, \tag{2.40}$$

where the Lorentz factor γ was defined in Eq. (1.11) and can be considered a diagnostic for relativistic effects. Relativistic effects will be important (and γ large) when the speed of the particle v is sizable compared to the speed of light, here denoted by c.

The action integral can now be re-expressed in its familiar form of Eq. (2.4) with the relativistic free-particle Lagrangian

$$L_0^{\mathrm{R}} = -mc^2\gamma^{-1}. \tag{2.41}$$

At first sight, Eqs. (2.34) and (2.41) look very different, but if we expand the inverse Lorentz factor to second order in $(v/c)^2$, we obtain

$$L_0^{\mathrm{R}} = -mc^2 + \frac{1}{2}mv^2 - \frac{1}{8}\frac{mv^4}{c^2}. \tag{2.42}$$

The first term, which does not contribute to the equation of motion since it is constant, is minus the rest mass, the second term is the nonrelativistic free-particle Lagrangian L_0^{NR}, Eq. (2.34), and the third term represents a relativistic correction known as the mass-velocity correction. Using Eq. (2.27), we obtain the relativistic momentum

$$p = \gamma mv, \tag{2.43}$$

which differs from the nonrelativistic definition by the Lorentz factor. The appearance of the Lorentz factor can be associated with a relativistic mass increase. A Legendre transformation of the Lagrangian in accordance with Eq. (2.26) gives

$$H_0^{\mathrm{R}} = \dot{x}p - L_0^{\mathrm{R}} = mc^2\gamma. \tag{2.44}$$

Before the identification of a valid Hamiltonian can be made, we must in this expression replace the velocity hidden in the Lorentz factor by the momentum. However, in order to carry out this replacement, we must proceed via a quadratic form and arrive at

$$\left(H_0^{\mathrm{R}}\right)^2 = (pc)^2 + (mc^2)^2 \quad \Rightarrow \quad H_0^{\mathrm{R}} = \pm\sqrt{m^2c^4 + c^2p^2}. \tag{2.45}$$

This is a problematic expression, for many reasons. One is that we face the possibility of free particles with negative sign. Another is that quantization is not straightforward because the momentum operator appears in the square root. We shall, however, postpone this discussion to Section 3.2.2.

2.1.3 Wave Functions and Operators

The wave function $\psi(x,t)$ associated with a single electron is, from a mathematical point of view, an element of a Hilbert space, which is a complete vector space equipped with a norm in the form of a scalar product. The probabilistic interpretation of the wave function implies that it should belong to the Hilbert space L^2 of square-integrable functions. More precisely, it can be enforced that

$$\|\psi\|^2 = \int_{\mathbb{R}^3} \psi^\dagger(x,t)\psi(x,t)d^3\mathbf{r} = 1; \quad \forall t. \tag{2.46}$$

Starting from the vector space $\mathbb{V}^{\mathrm{o}}$ associated with the scalar orbital part, we may successively add further degrees of freedom through a sequence of direct products

$$\mathbb{V} = \mathbb{V}^{\mathrm{o}} \otimes \mathbb{V}^{\mathrm{c}} \otimes \mathbb{V}^{\mathrm{s}}. \tag{2.47}$$

The electron spin is typically added by multiplying the spatial orbital by spin functions α and β, but we may also proceed in a more formal manner by using their representations

$$(\alpha, \beta) \rightarrow \mathbb{V}^{\mathrm{s}} = \left\{ \begin{pmatrix} 1 \\ 0 \end{pmatrix}, \begin{pmatrix} 0 \\ 1 \end{pmatrix} \right\}. \tag{2.48}$$

The corresponding representation of the spin operator $\hat{\mathbf{s}} = (\hat{s}_x, \hat{s}_y, \hat{s}_z)$ is given by $\hbar\sigma/2$, where the Pauli spin matrices appear

$$\sigma_x = \begin{pmatrix} 0 & 1 \\ 1 & 0 \end{pmatrix}; \quad \sigma_y = \begin{pmatrix} 0 & -i \\ i & 0 \end{pmatrix}; \quad \sigma_z = \begin{pmatrix} 1 & 0 \\ 0 & -1 \end{pmatrix}. \tag{2.49}$$

If we go all the way to the description of the electron provided by the Dirac equation, we also have to include a further two-dimensional part $\mathbb{V}^c$, now associated with the charge conjugation degrees of freedom (changing the sign of the charge but leaving other properties intact) because the Dirac equation describes both electrons and their antiparticles, the positrons. The four components of the Dirac wave function is constructed as

$$\psi^{L\alpha}(x,t)\otimes\begin{pmatrix}1\\0\end{pmatrix}\otimes\begin{pmatrix}1\\0\end{pmatrix};\quad \psi^{L\beta}(x,t)\otimes\begin{pmatrix}1\\0\end{pmatrix}\otimes\begin{pmatrix}0\\1\end{pmatrix};$$
$$\psi^{S\alpha}(x,t)\otimes\begin{pmatrix}0\\1\end{pmatrix}\otimes\begin{pmatrix}1\\0\end{pmatrix};\quad \psi^{S\beta}(x,t)\otimes\begin{pmatrix}0\\1\end{pmatrix}\otimes\begin{pmatrix}0\\1\end{pmatrix},\tag{2.50}$$

forming the general wave function as the sum of these four components to arrive at

$$\psi(x,t)=\begin{pmatrix}\psi^{L\alpha}\\ \psi^{L\beta}\\ \psi^{S\alpha}\\ \psi^{S\beta}\end{pmatrix}.\tag{2.51}$$

The label $s=\alpha,\beta$ of each component ψ^{cs} refers to spin. It is important to realize, though, that the label $c = L, S$, referring to large and small components, respectively, does not correspond directly to electronic or positronic degrees of freedom. The entire wave function is in fact either electronic or positronic, but related to that of its antiparticle through charge conjugation. Arriving at the full four-component form of the wave function, we note that the adopted ordering of vector spaces in Eq. (2.47), with $\mathbb{V}^c$ to the left of $\mathbb{V}^s$, leads to a collection of the two large (small) components in the upper (lower) half of the spinor.

In a two-component formalism, one suppresses the degrees of freedom associated with $\mathbb{V}^c$, and, in a one-component nonrelativistic formalism, one may also suppress the reference to $\mathbb{V}^s$ and instead introduce and manage electron spin in a more *ad hoc* manner, as described above. Because the nonrelativistic Hamiltonian is spin independent, one can always separate the spatial and spin parts of the wave function, but regardless of such simplifications in the representation and notation of the wave function, the true nature of the electronic wave function is described by Eq. (2.51), and we refer to such a wave function as a four-spinor.

The corresponding density is

$$n(x,t)=\psi^{\dagger}(x,t)\psi(x,t)=\sum_{c}^{L,S}\sum_{s}^{\alpha,\beta}|\psi^{cs}(x,t)|^2.\tag{2.52}$$

We will, however, often use the simplified notation of the density as introduced in Eq. (2.3) regardless of the number of components in the wave function.

We extract information that can be related to experiment from wave functions by forming integrals

$$\langle\psi_i|\hat{\Omega}|\psi_j\rangle=\langle\psi_i|\hat{\Omega}\psi_j\rangle.\tag{2.53}$$

In particular, expectation values ($i=j$) connect to observables, whereas the squared norm of transition moments ($i\neq j$) connects to spectral intensities. The right-hand side of Eq. (2.53) stresses that the operator $\hat{\Omega}$ acts to the right, on the ket. The *adjoint* operator $\hat{\Omega}^{\dagger}$ is defined to give the same result when acting on the bra, that is

$$\langle\hat{\Omega}^{\dagger}\psi_i|\psi_j\rangle=\langle\psi_j|\hat{\Omega}^{\dagger}|\psi_i\rangle^{*}=\langle\psi_i|\hat{\Omega}\psi_j\rangle.\tag{2.54}$$

In more informal usage one writes

$$\langle\psi|\hat{\Omega}=\langle\hat{\Omega}^{\dagger}\psi|,\tag{2.55}$$

Table 2.2 A selection of quantum mechanical operators.

Operator	Symbol	$\hat{\Omega}^{\mathrm{o}}$	$\hat{\Omega}^{\mathrm{c}}$	$\hat{\Omega}^{\mathrm{s}}$
Identity	$\hat{I}$	$\hat{I}$	I_2	I_2
Position	$\hat{x}$	x	I_2	I_2
Linear momentum	$\hat{p}_x$	$-i\hbar\partial/\partial x$	I_2	I_2
Orbital angular momentum	$\hat{l}_x$	$y\hat{p}_z - z\hat{p}_y$	I_2	I_2
Spin angular momentum	$\frac{\hbar}{2}\hat{\Sigma}_x$	$\hat{I}$	I_2	$\frac{\hbar}{2}\sigma_x$
Nonrelativistic velocity	$\hat{v}_x$	$\hat{p}_x/m$	I_2	I_2
Relativistic velocity	$c\hat{\alpha}_x$	$\hat{I}$	$\begin{pmatrix} 0 & 1 \\ 1 & 0 \end{pmatrix}$	$c\sigma_x$
Dirac β-matrix	$\hat{\beta}$	$\hat{I}$	$\begin{pmatrix} 1 & 0 \\ 0 & -1 \end{pmatrix}$	I_2

Operators relating to Cartesian components y and z are obtained by straightforward generalization.

that is, an operator becomes its adjoint when acting to the left.[3] Quantum-mechanical operators are selected to be self-adjoint ($\hat{\Omega}^{\dagger} = \hat{\Omega}$) such that expectation values are real. We usually refer to such operators as Hermitian, and we shall adhere to this usage, although there is a subtle difference between the two terms.[4]

Similar to what we discussed for wave functions, general electronic quantum-mechanical operators are written as a direct product between one part that acts in orbital space $\hat{\Omega}^{\mathrm{o}}$, one that acts in spin space $\hat{\Omega}^{\mathrm{s}}$, and one part associated with the charge conjugation degrees of freedom

$$\hat{\Omega} = \hat{\Omega}^{\mathrm{o}} \otimes \hat{\Omega}^{\mathrm{c}} \otimes \hat{\Omega}^{\mathrm{s}}. \tag{2.56}$$

The formation of a product operator means that we take the product in each individual operator space in accordance with

$$\hat{\Omega}\hat{\Lambda} = \hat{\Omega}^{\mathrm{o}}\hat{\Lambda}^{\mathrm{o}} \otimes \hat{\Omega}^{\mathrm{c}}\hat{\Lambda}^{\mathrm{c}} \otimes \hat{\Omega}^{\mathrm{s}}\hat{\Lambda}^{\mathrm{s}}. \tag{2.57}$$

A small selection of operators is given in Table 2.2. In this table, only one of the Cartesian components is provided for the vectorial operators, but other components follow suit.

With *pure* spatial (or spin) operators, we refer to situations when only $\hat{\Omega}^{\mathrm{s}}$ (or $\hat{\Omega}^{\mathrm{o}}$) is different from the identity operator—the identity operator itself is both a pure spatial and a pure spin operator. It is common practice to allow for simplified notations of operators, for example, to omit the symbol for the direct product or, for pure operators, to omit the identity operators. We shall use such conventions whenever there is no risk of confusion. An example of a pure orbital operator where the identity operators in $\mathbb{V}^{\mathrm{s}}$ and $\mathbb{V}^{\mathrm{c}}$ have been left out is given by the nonrelativistic Hamiltonian in the form given in Eq. (2.35). In dealing with pure operators,

3 From a mathematical point of view this is sloppy, since operators always act to the right. Furthermore, it may be that $\hat{\Omega}$ and its adjoint $\hat{\Omega}^{\dagger}$ have different domains; see next footnote.

4 The difference between Hermitian and self-adjoint operators arises from the fact that operators are mathematically defined not only by their actions, but also by their domain $\mathcal{D}(\hat{\Omega})$; that is, the function space they act on. Hermitian operators have the same action $\hat{\Omega}^{\dagger}\psi = \hat{\Omega}\psi$ for all functions $\psi \in \mathcal{D}(\hat{\Omega})$, but it may be that $\mathcal{D}(\hat{\Omega}^{\dagger}) \neq \mathcal{D}(\hat{\Omega})$, contrary to self-adjoint operators. This distinction becomes particularly important for unbounded operators, of which both the position and momentum operators are examples. Bounded operators satisfy $\|\hat{\Omega}\psi\| \leq M \|\psi\|$ for positive constant M and for all functions $\psi \in \mathcal{D}(\hat{\Omega})$. A good discussion of this and other subtle points has been discussed in F. Gieres. *Rep. Prog. Phys.*, 63:1893, 2000.

simplifications may be introduced in the notation, and it also makes it possible to separately treat symmetries in the vector and operator spaces. In the nonrelativistic domain, we do so for instance by the construction of spin-adapted wave functions and operators as well as by the use of symmetry-adapted spatial orbitals. In the relativistic domain, on the other hand, spin and spatial degrees of freedom are always coupled with the spin-orbit interactions and, as a consequence, it becomes impossible to separately treat spin and spatial symmetries.

The generalization of the vector space and the operators to a many-electron system is quite straightforward. For instance, the state vectors for two-electron systems are elements of a Hilbert space of the form

$$\mathbb{V} = \mathbb{V}^1 \otimes \mathbb{V}^2, \tag{2.58}$$

and the corresponding operators are given by

$$\hat{\Omega} = \hat{\Omega}'(1) \otimes \hat{\Omega}''(2). \tag{2.59}$$

For many-electron systems, we should keep in mind that electrons belong to the family of fermions and as such they obey the Pauli exclusion principle. This means that no two interacting electrons can be in the same quantum state, and the vector space $\mathbb{V}$ is therefore limited to include only antisymmetric two-particle vectors of the type

$$\psi(x_1, x_2) = \frac{1}{\sqrt{2}}\left[\psi_a(x_1) \otimes \psi_b(x_2) - \psi_b(x_1) \otimes \psi_a(x_2)\right] = \frac{1}{\sqrt{2}}\begin{vmatrix} \psi_a(x_1) & \psi_b(x_1) \\ \psi_a(x_2) & \psi_b(x_2) \end{vmatrix}. \tag{2.60}$$

In the last step we have made use of a determinant to write the antisymmetrized two-particle state, and we have also included an optional, explicit, normalization constant of $1/\sqrt{2}$ together with the assumption that the one-particle state vectors are individually normalized. Such a determinant is known as a Slater determinant in quantum chemistry, and one of the most notable properties of the Slater determinant is that when $a = b$, the wave function vanishes. The Slater determinant thus ensures that the Pauli principle is satisfied. It is clear that the set of all possible Slater determinants constructed from a complete set of one-particle states constitutes the natural basis of $\mathbb{V}$.

Another aspect to note—and which is a reminder in the handling of the vector as well as the operator spaces—is that there are vectors and operators that themselves may not be written as direct products. The general electronic wave function in Eq. (2.51) is one example as it is a sum of four direct product components, and for many-electron systems, a simple example is given by the coupled two-particle operator with a coordinate representation in terms of the Dirac delta function

$$\hat{\Omega} = \delta(x_1 - x_2). \tag{2.61}$$

Let us assume that $\{\psi_n\}$ forms a complete set of orthonormal one-particle functions. We then have

$$\delta(x_1 - x_2) = \langle x_2 | x_1 \rangle = \sum_n \langle x_2 | \psi_n \rangle \langle \psi_n | x_1 \rangle = \sum_n \psi_n^*(x_1)\psi_n(x_2), \tag{2.62}$$

which provides a resolution of $\hat{\Omega}$ into products of one-particle functions. Another and also very important example is given by an instantaneous Coulomb interaction operator. The Coulomb operator for a two-electron system takes the basic form

$$\hat{\Omega} = \frac{1}{|\mathbf{r}_1 - \mathbf{r}_2|}. \tag{2.63}$$

We may replace the orthogonality condition for the basis vectors with a condition of biorthogonality

$$\langle \psi_n | \phi_m \rangle = \delta_{nm}, \tag{2.64}$$

or, equivalently,

$$\sum_n \psi_n^*(\mathbf{r}_1)\phi_n(\mathbf{r}_2) = \delta(\mathbf{r}_1 - \mathbf{r}_2), \tag{2.65}$$

where the operator $\hat{\Omega}$ relates the pairs of biorthogonal vectors

$$\phi_m(\mathbf{r}) = \int \frac{\psi_m(\mathbf{r}')}{|\mathbf{r} - \mathbf{r}'|} d^3\mathbf{r}'. \tag{2.66}$$

We can then proceed to obtain a resolution of the Coulomb operator according to

$$\begin{aligned} \hat{\Omega} &= \sum_{n,m} \int\int \phi_n(\mathbf{r}_2)\psi_n^*(\mathbf{r}_2')\frac{1}{|\mathbf{r}_1' - \mathbf{r}_2'|}\psi_m(\mathbf{r}_1')\phi_m^*(\mathbf{r}_1)d^3\mathbf{r}_1' d^3\mathbf{r}_2' \\ &= \sum_n \phi_n^*(\mathbf{r}_1)\phi_n(\mathbf{r}_2), \end{aligned} \tag{2.67}$$

which provides a resolution of the Coulomb operator in a one-particle basis. Apart from providing a demonstration of how to decouple a two-particle interaction operator, it gives a glimpse into a family of techniques developed to achieve cost-efficient numerical treatments of two-electron integrals.

In this section, we have demonstrated how to construct wave functions and operators for many-electron systems. We emphasize that the foundations of quantum mechanics do not depend on the inclusion or exclusion of relativity, and to the extent possible, the aim has been to give a presentation that is valid in both situations. At the same time, it cannot be avoided that, at a certain level of detail, one must be specific about the choice of Hamiltonian (and other operators) and depending on that choice, there are particular and less general aspects to be explored and utilized in the calculations.

2.2 Time Evolution of Wave Functions

The time evolution of the state vector is given by the time-dependent Schrödinger equation, Postulate D, and is thus dictated by the Hamiltonian. Formally, we may describe the evolution of the wave function from some initial time t_0 by the introduction of a time evolution operator, or time propagator, according to

$$|\psi(t)\rangle = \hat{U}(t, t_0)|\psi(t_0)\rangle, \tag{2.68}$$

An equation of motion for the propagator is obtained by first inserting the above expression into the time-dependent Schrödinger equation. Since $|\psi(t_0)\rangle$ is independent of the time variable t, we obtain a resulting operator equation

$$i\hbar\frac{\partial}{\partial t}\hat{U}(t, t_0) = \hat{H}\hat{U}(t, t_0), \tag{2.69}$$

which is equivalent to the time-dependent Schrödinger equation. Expressions for the time evolution operator are now obtained by integrating this equation. Alternatively, we can proceed via an infinitesimal time evolution operator. For an infinitesimal time interval dt, we may Taylor expand the wave function to first order and use the time-dependent Schrödinger equation to arrive at

$$|\psi(t_0 + dt)\rangle = |\psi(t_0)\rangle + dt\frac{\partial}{\partial t}|\psi(t)\rangle\bigg|_{t=t_0} = |\psi(t_0)\rangle - \frac{i}{\hbar}dt\hat{H}(t_0)|\psi(t_0)\rangle, \tag{2.70}$$

which shows that the infinitesimal time evolution operator is

$$\hat{U}(t_0 + dt, t_0) = \hat{I} - \frac{i}{\hbar}dt\hat{H}(t_0). \tag{2.71}$$

Since we only retain the infinitesimal interval dt to first order and since the Hamiltonian is Hermitian, it follows that the infinitesimal time evolution operator is unitary

$$\hat{U}^\dagger(t_0 + dt, t_0)\hat{U}(t_0 + dt, t_0) = \left(\hat{I} + \frac{i}{\hbar}dt\hat{H}^\dagger(t_0)\right)\left(\hat{I} - \frac{i}{\hbar}dt\hat{H}(t_0)\right) = \hat{I}. \tag{2.72}$$

For a finite time interval, we may express the time evolution operator as a product of infinitesimal time evolution operators by dividing the time interval into sufficiently small pieces

$$t_k = t_0 + k\,\frac{t - t_0}{n}; \quad k = 0, \ldots, n, \tag{2.73}$$

which yields a time propagator that reads as

$$\hat{U}(t, t_0) = \lim_{n\to\infty}\prod_{k=0}^{n-1}\hat{U}(t_{k+1}, t_k) = \lim_{n\to\infty}\prod_{k=0}^{n-1}\left[1 - \frac{i}{\hbar}\frac{(t - t_0)}{n}\hat{H}(t_k)\right]. \tag{2.74}$$

This product form of individually unitary operators shows that the time evolution for a finite interval is unitary as well. When the Hamiltonian is time independent, the time evolution operator attains a simple exponential form

$$\hat{U}(t, t_0) = \lim_{n\to\infty}\left[1 - \frac{i}{\hbar}\frac{(t - t_0)}{n}\hat{H}\right]^n = e^{-i\hat{H}(t-t_0)/\hbar}. \tag{2.75}$$

If the initial state happens to be an eigenstate of the Hamiltonian, the time dependence becomes an overall phase factor of the wave function. Let us assume that at time t_0, the wave function is an eigenstate of $\hat{H}$ denoted by ψ_n with eigenvalue E_n. From Eq. (2.68), we get

$$\psi(x, t) = \psi_n(x)e^{-iE_n(t-t_0)/\hbar}, \tag{2.76}$$

and the particle density, Eq. (2.3), is thus in this case time independent

$$n(x) = |\psi_n(x)|^2. \tag{2.77}$$

For this reason, the eigenstates of the Hamiltonian are also referred to as the *stationary* states of the system and are characterized by being separable in space and time. It is important to note that even for a system in a stationary state there can be an associated nonzero linear momentum as illustrated by the case of the nonrelativistic free particle. The stationary states can in this case be written as

$$\psi_p(x, t) = e^{ipx/\hbar}e^{-iE_p(t-t_0)/\hbar}, \tag{2.78}$$

and the associated linear momentum is given by

$$\langle\hat{p}\rangle = p. \tag{2.79}$$

This corresponds to particle motion at a speed of $v = p/m$ for the stationary state. For a molecular system with bound electrons, a nonzero linear momentum may be found in orbitally degenerate states—such as the p-states in hydrogen, or the E-states in benzene—but the current density is such that the particle flow into a probing infinitesimal volume ΔV equals the flow out of it.

In general, the initial state will not be an eigenstate of the Hamiltonian, and consequently the density of a system will be time dependent, even when its Hamiltonian is not. This can be illustrated by the nonrelativistic harmonic oscillator with a displaced Gaussian wave packet serving as the initial state at time $t_0 = 0$

$$\psi(x,0) = \left(\frac{m\omega}{\pi\hbar}\right)^{1/4} e^{-m\omega(x-x_0)^2/2\hbar}, \tag{2.80}$$

centered about $x = x_0$, and the initial wave packet corresponds to a translated ground-state harmonic oscillator wave function. We follow the propagation of the generally complex wave packet ψ and its amplitude $|\psi|$ in Figure 2.1. At the initial time, the phase of the wave function is independent of the spatial coordinate x—in fact the phase is equal to zero since the initial state is real—and the expectation value of the Hermitian, but imaginary linear momentum operator, must therefore vanish. In the initial state, $\langle \hat{p} \rangle = 0$ and $\langle \hat{x} \rangle = x_0$, corresponding to a classical turning point of the harmonic oscillator, as illustrated in the lowest curve plot in Figure 2.1. At a later time, the wave function is found by application of Eq. (2.68). After a short while, at $t = 0.002T$ ($T = 2\pi/\omega$), it will take the form illustrated in the second curve plot. Here it is visible that we plot not only the amplitude but also the complex value of the wave function. The z and y directions are used for plotting the real and imaginary parts of $\psi(x,t)$, respectively, and since we adopt a right-handed coordinate system with a z-axis directed upward, the positive y axis points into the plane of the paper. In this second curve plot, the phase starts to become negative for all values of x.

At a time equal to one quarter of the period time ($t = T/4$), the phase changes rapidly in space corresponding to a maximum in the linear momentum, and we may think of the complex wave function in terms of a corkscrew drilling its way through space. By the time the wave packet

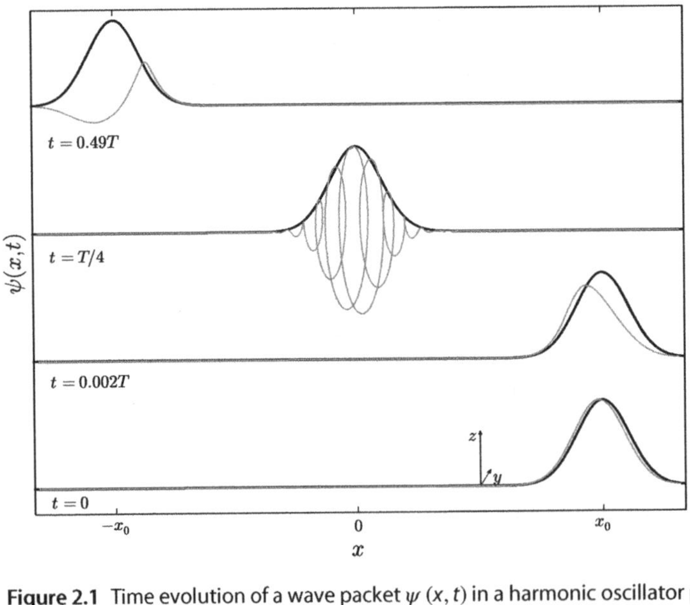

Figure 2.1 Time evolution of a wave packet $\psi(x,t)$ in a harmonic oscillator potential. The black line indicates the amplitude $|\psi|$ of the generally complex wave function. The full wave packet (gray line) has been traced in the complex plane normal to the x-axis; the orientation of the y imaginary axis and z real axis is indicated in the lower curve plot.

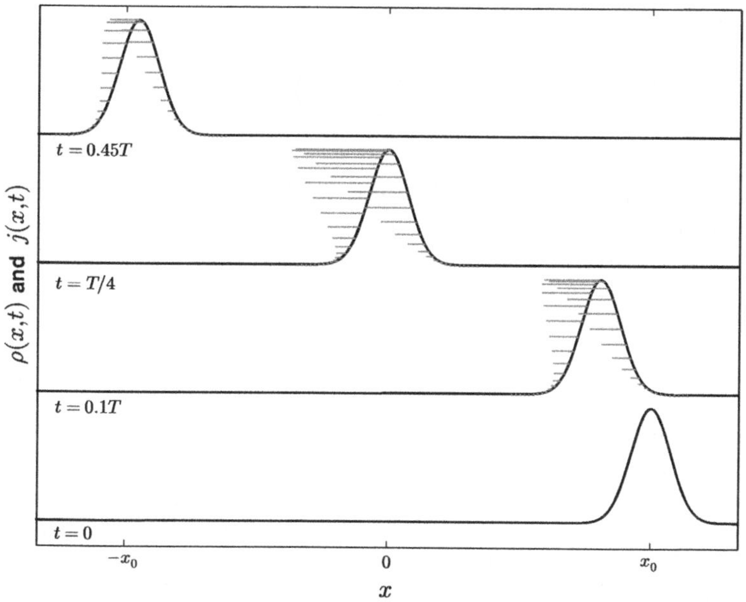

Figure 2.2 Time dependence of the probability current density for the harmonic oscillator.

reaches the other classical turning point, at time $t = T/2$, the phase again becomes independent of the spatial coordinate so that the linear momentum will be zero. The uppermost curve plot is a snapshot taken right before the turning point. This example illustrates that although the changes in the density function are not very dramatic—the density performs an oscillatory motion with a shape that is constant in time—the time and spatial dependencies of the phase function are all the more violent.

Associated with the time-dependent density there is a nonzero probability current density which is

$$j(x,t) = \mathrm{Re}\left[\psi^*(x,t)\,\frac{\hat{p}}{m}\,\psi(x,t)\right]. \tag{2.81}$$

For the displaced harmonic oscillator, $j(x,t)$ is illustrated together with the density in Figure 2.2. Different points on the density curve are shown by lines pointing in the direction of the current and with lengths proportional to the size of the current. At the two turning points the current density vanishes since the linear momentum vanishes.

2.3 Time Evolution of Expectation Values

It is interesting to note molecular responses to electromagnetic fields and the time evolution of expectation values, since they correspond to observables such as electric and magnetic moments. The direct approach to this problem is to determine the time dependence of the reference wave function and then *subsequently* form the expectation value to determine the property. However, in a situation where the Hamiltonian depends on time, the direct integration of the Schrödinger equation cannot be carried out to form a simple propagator expression as in Eq. (2.68). Instead, we can establish an equation of motion for the expectation value itself, and from this equation determine the property of interest. The time derivative of an expectation value takes the form

$$\frac{\partial}{\partial t}\langle\hat{\Omega}\rangle = \left\langle\frac{\partial\psi}{\partial t}|\hat{\Omega}|\psi(t)\right\rangle + \left\langle\psi(t)|\hat{\Omega}|\frac{\partial\psi}{\partial t}\right\rangle + \left\langle\psi(t)|\frac{\partial\hat{\Omega}}{\partial t}|\psi(t)\right\rangle = -\frac{i}{\hbar}\langle[\hat{\Omega},\hat{H}]\rangle + \left\langle\frac{\partial\hat{\Omega}}{\partial t}\right\rangle, \tag{2.82}$$

where, in the last step, the substitution of the time derivative of the wave function is made with use of the time-dependent Schrödinger equation. This equation is known as the Ehrenfest theorem and it is equivalent to the Schrödinger equation in the sense that it provides neither more nor less information. As an example, we see that, for a system described by a time-independent Hamiltonian, we have

$$\frac{\partial}{\partial t}\langle\hat{H}\rangle = 0. \tag{2.83}$$

To the extent that the Hamiltonian can be identified with the energy as discussed in connection with Eq. (2.33), the above result demonstrates the conservation of energy of the quantum-mechanical system. An example of paramount importance to us is given by molecular systems subjected to external static electromagnetic fields, leading to well-defined, field-dependent energies that can be used as starting points to define molecular properties.

Another important application of the Ehrenfest theorem is provided by $\hat{\Omega} = \hat{I}$ that leads to

$$\frac{\partial}{\partial t}\langle\psi(t)|\psi(t)\rangle = 0, \tag{2.84}$$

showing that the time-dependent Schrödinger equation conserves the norm of the wave function. This implies that

$$\langle\psi(t)|\psi(t)\rangle = \langle\psi(t_0)|\hat{U}^\dagger(t,t_0)\hat{U}(t,t_0)|\psi(t_0)\rangle = \langle\psi(t_0)|\psi(t_0)\rangle, \tag{2.85}$$

which is another demonstration of the fact that the time-evolution operator is unitary.

If we insert the time evolution operator into the expectation value expression for a general operator, we can shift the time dependence of the wave function to the operator

$$\langle\hat{U}(t,t_0)\psi(t_0)|\;\hat{\Omega}\;|\hat{U}(t,t_0)\psi(t_0)\rangle = \langle\psi(t_0)|\hat{\Omega}_{\mathrm{H}}(t)|\psi(t_0)\rangle, \tag{2.86}$$

where

$$\hat{\Omega}_{\mathrm{H}}(t) = \hat{U}^\dagger(t,t_0)\;\hat{\Omega}\;\hat{U}(t,t_0). \tag{2.87}$$

This corresponds to going from the Schrödinger picture to the Heisenberg picture, a transformation which is indicated by a subscript H on the now generally time-dependent operator. Analogous to the Ehrenfest theorem, we can obtain an equation of motion for operators in the Heisenberg picture by using the equation for the time evolution operator Eq. (2.69). The resulting Heisenberg equation of motion takes the form

$$\frac{d\hat{\Omega}_{\mathrm{H}}(t)}{dt} = -\frac{i}{\hbar}\left[\hat{\Omega},\hat{H}\right]_{\mathrm{H}} + \left(\frac{\partial\hat{\Omega}}{\partial t}\right)_{\mathrm{H}}. \tag{2.88}$$

Choosing $\hat{\Omega}$ to be the position operator $\mathbf{r}$ leads to

$$\frac{d\mathbf{r}_{\mathrm{H}}}{dt} = \frac{\hat{\mathbf{p}}_{\mathrm{H}}}{m}. \tag{2.89}$$

In the nonrelativistic domain, this equation suggests that the velocity operator $\mathbf{v}_{\mathrm{H}}$ is the momentum operator divided by the particle mass, which is reasonable. The relativistic result is more intriguing. The velocity operator, defined in Table 2.2, in the Heisenberg picture is denoted by

$\hat{\mathbf{v}}_{\mathrm{H}}^{\mathrm{R}}$ and becomes $c\boldsymbol{\alpha}_{\mathrm{H}}$. Since the square of any of the Dirac $\boldsymbol{\alpha}$ matrices is the unit matrix, this suggests that each component of the particle velocity is on the order of the speed of light c, such that the speed of the particle is beyond c. We must remember, however, that we cannot independently probe the three velocity components in any conceivable experiment because the three Dirac $\boldsymbol{\alpha}$ matrices fail to commute. The curious form of the relativistic velocity operator is a manifestation of the particle motion proposed by Schrödinger, known as *Zitterbewegung*, that is, the highly oscillatory motion due to the interference of positive and negative energy states that is superimposed on the average trajectory of the relativistic electron.

2.4 Variational Principle

A fundamental property of the eigenfunctions $|\omega_n\rangle$ of a Hermitian operator $\hat{\Omega}$ is that they form a complete set and can be used for the formation of an orthonormal basis of our wave function space. The eigenstates and eigenvalues of Hermitian operators are of particular interest to us because they appear in connection with measurements as described in Postulate C in Table 2.1. An arbitrary wave function can be expanded in the basis of eigenstates

$$|\psi\rangle = \sum_n c_n \, |\omega_n\rangle, \tag{2.90}$$

where the expansion coefficient c_n is to be regarded as a projection coefficient of $|\psi\rangle$ onto the basis vector $|\omega_n\rangle$, that is

$$c_n = \langle\omega_n|\psi\rangle. \tag{2.91}$$

The change from one orthonormal basis $|n\rangle$ to another $|\omega_n\rangle$ represents a unitary transformation according to

$$|\omega_n\rangle = \hat{U}|n\rangle = \sum_k |k\rangle\langle k|\hat{U}|n\rangle = \sum_k |k\rangle U_{kn}, \tag{2.92}$$

and so, if the matrix representation of the operator $\hat{\Omega}$ is Ω in the basis $|n\rangle$, then we can diagonalize this matrix by means of the unitary transformation matrix U

$$D = U^{\dagger}\Omega U. \tag{2.93}$$

The matrix D now contains the eigenvalues of Ω as diagonal elements. This result provides us with a convenient way to evaluate the function of a matrix $f(\Omega)$ by employing $|\omega_n\rangle$ as an auxiliary basis. The explicit formula becomes

$$f(\Omega) = Uf(D)U^{\dagger}, \tag{2.94}$$

where $f(D)$ is a diagonal matrix with diagonal elements that are $f(\omega_n)$. A closely related result is that

$$f(\hat{\Omega})|\omega_n\rangle = f(\omega_n)|\omega_n\rangle. \tag{2.95}$$

If one could be granted access to the eigenfunctions (and eigenvalues) of a single Hermitian operator, the first choice would be to pick those of the Hamiltonian. Already from the eigenvalues, or, to be more precise, the differences between them, it is possible to compare the theoretical results to spectral peak positions arising from any of the several different available absorption spectroscopies. Additionally, the energy basis also stands out as one in which certain technical calculations can be performed with ease. Let us for instance assume that the wave function at time t_0 is known and that we are interested in its time propagation. With the use of an expansion of $\psi(t_0)$ in the energy basis, the result can immediately be written as

$$|\psi(t)\rangle = \hat{U}(t,t_0)\sum_n c_n(t_0)\,|\psi_n\rangle = \sum_n c_n(t_0)\,|\psi_n\rangle e^{-iE_nt/\hbar}, \tag{2.96}$$

where the time propagator, being a function of the Hamiltonian, acts on the energy eigenstates in analogy with Eq. (2.95).

As attractive as these results may be, they represent a detour from our destination in this section since the true eigenstates of the Hamiltonian are rarely accessible to us in practical calculations. But the mere fact that the energy basis does exist and that an arbitrary ket vector can *in principle* be expanded in this basis allows for an important conclusion with practical implications in approximate-state theory. Any wave function in approximate-state theory ψ^{A} can also be expanded in the energy basis according to

$$|\psi^{\mathrm{A}}\rangle = \sum_n c_n^{\mathrm{A}}|\psi_n\rangle. \tag{2.97}$$

The energy E^{A} associated with the wave function ψ^{A} is given by the expectation value of the Hamiltonian. For this expectation value, we conclude that

$$E^{A} = \langle\psi^{\mathrm{A}}|\,\hat{H}\,|\psi^{\mathrm{A}}\rangle = \sum_n |c_n^{\mathrm{A}}|^2 E_n \geq E_0, \tag{2.98}$$

where E_0 is the lowest eigenvalue of the Hamiltonian, corresponding to the ground state.[5] This result is known as the *variational principle*, and it is often used as a measure of the quality of approximate-state wave functions. The wave function ψ^{A} includes a set of parameters, such as for example, molecular orbital coefficients, and the best values of these parameters are defined as those minimizing the energy E^{A}. This approach is directly applicable to the optimization of ground-state wave functions, but with some constraints it can also be adopted for the optimization of wave functions representing excited states.

In Eq. (2.98), it is natural to consider E^{A} as a mapping of wave functions to real numbers. This turns out to represent an example of a functional as introduced in Eq. (2.5). With respect to wave functions, the simplest functional of interest to us is the norm of the wave function

$$N[\psi] = \langle\psi|\psi\rangle. \tag{2.99}$$

This functional is special in the sense that it maps *all* wave functions onto the real number 1, and it is clear that the first variation of N vanishes for all allowed ψ. We can also formulate this as a requirement on the test function $\delta\psi$ (or variations in ψ) that results in

$$\delta N[\psi] = \langle\delta\psi|\psi\rangle + \langle\psi|\delta\psi\rangle = 0. \tag{2.100}$$

This means that, when we move from ψ to $\psi + \epsilon\delta\psi$ in function space, we are restricted to move in directions given by

$$|\delta\psi\rangle = |\delta\psi^{\perp}\rangle + i\eta|\psi\rangle; \quad \eta \in \mathcal{R}, \tag{2.101}$$

where $\delta\psi^{\perp}$ is strictly orthogonal to ψ.

Next we consider the stationary states in quantum mechanics which, as we have seen, are separable in space and time variables. The spatial parts of these particular solutions to the Schrödinger equation fulfill the eigenvalue equation of the Hamiltonian

$$\hat{H}\psi_n(x) = E_n\psi_n(x), \tag{2.102}$$

5 This assumes that the Hamiltonian is bounded from below, which is true for the nonrelativistic, but not for the relativistic, case.

often referred to as the time-independent Schrödinger equation. We can isolate the energy E_n in Eq. (2.102) by the multiplication with a bra vector

$$\langle \psi_n | \hat{H} | \psi_n \rangle = E_n. \tag{2.103}$$

It is clear that the multiplication with a vector belonging to the orthogonal complement yields

$$\langle \psi_n^{\perp} | \hat{H} | \psi_n \rangle = 0. \tag{2.104}$$

For allowed variations of the form given in Eq. (2.101), by using Eq. (2.104) we can conclude that the first variation in the energy functional vanishes

$$\delta E_n = \langle \delta\psi_n | \hat{H} | \psi_n \rangle + \langle \psi_n | \hat{H} | \delta\psi_n \rangle = 0. \tag{2.105}$$

All eigenstates of the Hamiltonian are characterized by having a zero electronic gradient.

In many cases, the Hamiltonian carries a dependence on a parameter F (such as an external static field amplitude), and we can trace the dependence of the eigenfunctions of the Hamiltonian on this parameter. Due to the normalization condition, it is clear that

$$\frac{d}{dF}\langle \psi_n(F) | \psi_n(F) \rangle = \left\langle \frac{d\psi_n}{dF} | \psi_n(F) \right\rangle + \left\langle \psi_n(F) | \frac{d\psi_n}{dF} \right\rangle = 0, \tag{2.106}$$

which tells us that $\delta\psi_n = d\psi_n/dF$ is an allowed variation of the wave function in accordance with Eq. (2.100). The derivative of the energy with respect to the parameter F thereby becomes

$$\begin{aligned} \frac{dE_n}{dF} &= \left\langle \frac{d\psi_n}{dF} | \hat{H} | \psi_n \right\rangle + \left\langle \psi_n | \hat{H} | \frac{d\psi_n}{dF} \right\rangle + \left\langle \psi_n \left| \frac{\partial \hat{H}}{\partial F} \right| \psi_n \right\rangle \\ &= \delta E_n + \left\langle \psi_n \left| \frac{\partial \hat{H}}{\partial F} \right| \psi_n \right\rangle \\ &= \left\langle \psi_n \left| \frac{\partial \hat{H}}{\partial F} \right| \psi_n \right\rangle, \end{aligned} \tag{2.107}$$

where, the last step is based on the fact that the first-order variation in the energy vanishes for all eigenfunctions of the Hamiltonian [see Eq. (2.105)]. The resulting equation is known as the *Hellmann–Feynman theorem*, and it holds as long as the wave function is variationally optimized with respect to the Hamiltonian. As can be seen in Chapter 5, this theorem is highly relevant, since it provides a bridge between energy derivatives and molecular properties, through the left- and right-hand sides of the equation, respectively.

Further Reading

1 J. E. Baggott. *Beyond Measure: Modern Physics, Philosophy, and the Meaning of Quantum Theory*. Oxford University Press, 2004.

2 B. H. Bransden and C. J. Joachain. *Quantum Mechanics*. Pearson Education, 2000.

3 C. Cohen-Tannoudji, B. Diu, and F. Laloë. *Quantum Mechanics*. Wiley, 1977.

4 K. G. Dyall and K. Fægri Jr. *Introduction to Relativistic Quantum Chemistry*. Oxford University Press, 2007.

5 W. Greiner. *Relativistic Quantum Mechanics*, volume 3. Springer, 1990.

6 E. Merzbacher. *Quantum Mechanics*. Wiley, 1998.

7 D. J. Tannor. *Introduction to Quantum Mechanics*. University Science Books, 2007.

3

Particles and Fields

From a long view of the history of mankind, seen from, say, ten thousand years from now, there can be little doubt that the most significant event of the nineteenth century will be judged as Maxwell's discovery of the laws of electrodynamics.

R. Feynman

The overall theme of this book is the behavior of molecules when subject to combinations of electromagnetic fields. We therefore need to know how to introduce external electromagnetic fields into our molecular or electronic Hamiltonian. The same procedure can also be used to introduce internal fields not normally taken into account in our unperturbed Hamiltonian, such as those associated with nuclear magnetic moments.

For a system of particles and fields we can generally write the total Hamiltonian as

$$\hat{H} = \hat{H}_{\mathrm{p}} + \hat{H}_{\mathrm{int}} + \hat{H}_{\mathrm{f}}, \quad (3.1)$$

where $\hat{H}_{\mathrm{p}}$, $\hat{H}_{\mathrm{f}}$, and $\hat{H}_{\mathrm{int}}$ describe particles, fields, and their interactions, respectively. In practice one never employs the full Hamiltonian, containing all degrees of freedom associated with both particles and fields. Usually, the external fields are considered fixed and enter the Hamiltonian as parameters rather than as variables. This means that the fields do not react to the particle motion so that the field term $\hat{H}_{\mathrm{f}}$ becomes a constant and contributes a mere shift of the energy that can be neglected. In the framework of classical mechanics, such a constant term will disappear from the equations of motion because they only involve derivatives of the Hamiltonian with respect to the variables (see Postulate D in Chapter 2). It is also possible to go to the other extreme and fix the particle motion, such that the particle term $\hat{H}_{\mathrm{p}}$ disappears from the equations of motion. The particles are now simply sources of electromagnetic fields and the resulting equations of motion in classical electrodynamics are Maxwell's equations.

The latter view is often the perspective of the experimentalist, who will probe a molecular sample with electromagnetic fields and monitor their changes. A simple example is optical activity where the experimentalist observes the rotation of the plane of polarization as light passes through the sample. We thus have, on the one hand, the theoretician studying the responses of the molecular charge density and, on the other hand, the experimentalist studying the responses of the electromagnetic fields. A further distinction to keep in mind is that the theoretician starts from a microscopic description of the system, whereas the experimentalist necessarily, due to the size of the measuring apparatus, works at a macroscopic level. Connecting theory and experiment therefore means that the response of molecules and fields as well as the microscopic and macroscopic descriptions of phenomena must be reconciled.

Principles and Practices of Molecular Properties: Theory, Modeling and Simulations, First Edition.
Patrick Norman, Kenneth Ruud, and Trond Saue.

3.1 Microscopic Maxwell's Equations

3.1.1 General Considerations

Classical electrodynamics is summarized by the celebrated Maxwell's equations. These are first-order differential equations consisting of a homogeneous and an inhomogeneous pair. The homogeneous pair, with right-hand sides being equal to zero, reads as

$$\nabla \cdot \mathbf{B} = 0, \tag{3.2}$$

$$\nabla \times \mathbf{E} + \partial_t \mathbf{B} = \mathbf{0}, \tag{3.3}$$

and thus involves only the electric $\mathbf{E}$ and magnetic $\mathbf{B}$ fields. In the above equations, ∂_t represents the derivative with respect to time, while ∇ is the gradient, or del, operator, which in Cartesian coordinates takes the form

$$\nabla = \partial_x \mathbf{e}_x + \partial_y \mathbf{e}_y + \partial_z \mathbf{e}_z = \left(\frac{\partial}{\partial x}, \frac{\partial}{\partial y}, \frac{\partial}{\partial z} \right). \tag{3.4}$$

The inhomogeneous pair, with right-hand sides different from zero, is given by

$$\nabla \cdot \mathbf{E} = \rho / \varepsilon_0, \tag{3.5}$$

$$\nabla \times \mathbf{B} - \frac{1}{c^2} \partial_t \mathbf{E} = \mu_0 \mathbf{j}, \tag{3.6}$$

in which appear the electric and magnetic constants ε_0 and μ_0, respectively, also known as the permittivity and permeability of free space. These constants are connected to the vacuum speed of light c through the relation

$$\mu_0 \varepsilon_0 = \frac{1}{c^2}. \tag{3.7}$$

The inhomogeneous pair adds the sources of the fields, that is, the charge density ρ and current density $\mathbf{j}$. For a classical point charge with charge q, as depicted in Figure 3.1, the charge density reduces to charge times a Dirac delta function fixing the instantaneous position to the trajectory $\mathbf{r}(t)$ of the particle whereas the current density is charge times the instantaneous velocity, or, equivalently,

$$\rho(\mathbf{r}', t) = q\, \delta(\mathbf{r}' - \mathbf{r}(t)); \quad \mathbf{j}(\mathbf{r}', t) = q\dot{\mathbf{r}}(t)\, \delta(\mathbf{r}' - \mathbf{r}(t)). \tag{3.8}$$

Maxwell's equations determine the electric and magnetic fields given a set of sources. Equation (3.5) is known as Gauss' law, and by analogy Eq. (3.2) is sometimes called Gauss' law of magnetism. Equation (3.3) corresponds to Faraday's law and describes electromagnetic induction, that is, the generation of an electric field by a magnetic field varying in time.

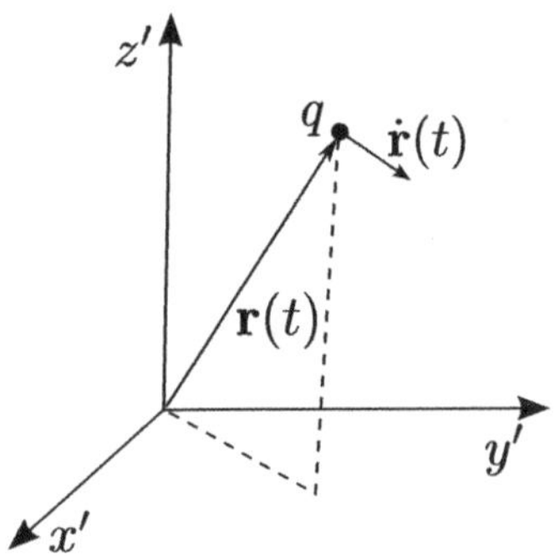

Figure 3.1 A classical point charge in motion.

Equation (3.6) is known as the Ampère–Maxwell equation and enounces that a magnetic field can be generated from currents (Ampère's law) and time-varying electric fields. The latter effect, pointed out by Maxwell, is important and leads to a relation between the charge and current density as follows:

For a general vector field **F**, the divergence of its curl vanishes, or in other words,

$$\nabla \cdot (\nabla \times \mathbf{F}) = 0. \tag{3.9}$$

By taking the divergence of Eq. (3.6), we are thereby led to conclude

$$-\frac{1}{c^2}\partial_t(\nabla \cdot \mathbf{E}) = \mu_0(\nabla \cdot \mathbf{j}), \tag{3.10}$$

and, with further use of Eqs. (3.5) and (3.7), we arrive at

$$(\nabla \cdot \mathbf{j}) + \partial_t \rho = 0. \tag{3.11}$$

This important relation is known as the *continuity equation*, and it expresses the conservation of charge, as will become clear by the consideration to follow.

The integral of the divergence of a vector function **F** over some volume V is related to its *flux* through the surface S enclosing the volume according to the *divergence theorem*[1]

$$\int_V (\nabla \cdot \mathbf{F}) dV = \int_S (\mathbf{F} \cdot \mathbf{n}) dS. \tag{3.12}$$

On the right-hand side of Eq. (3.12), **n** is a unit vector normal to the surface that, by convention, points outward from the surface at each point and dS is an infinitesimal surface element. The right-hand integral thus represents a surface summation of the component of the vector function **F** perpendicular to the surface, which in turn encloses the entire volume. An example of the use of the divergence theorem is to recast the continuity equation, Eq. (3.11), in integral form

$$\int_V (\nabla \cdot \mathbf{j}) dV = \int_S (\mathbf{j} \cdot \mathbf{n}) dS = -\int_V \partial_t \rho \, dV = -\partial_t Q. \tag{3.13}$$

The surface integral represents the charge flowing out of the enclosed volume and it is equal to the negative time derivative of the total charge Q inside the volume, showing that the continuity equation expresses charge conservation.

Correspondingly, by the *curl theorem*[2], the integral of the outward component of the curl of a vector function over some surface S is related to its *circulation* around a curve C enclosing the surface

$$\int_S (\nabla \times \mathbf{F}) \cdot \mathbf{n} \, dS = \oint_C (\mathbf{F} \cdot \mathbf{t}) dl. \tag{3.14}$$

On the right-hand side of Eq. (3.14), **t** is a unit vector tangential to the curve at each point and dl is an infinitesimal line element such that the line integral samples the tangential component of the vector function along the curve. If the normal vector **n** points toward the reader, the line integration follows the tangential vector **t** in a counterclockwise sense around the curve.

We can derive a specific form of the divergence theorem by considering the vector function

$$\mathbf{F} = f\mathbf{m}, \tag{3.15}$$

where **m** is a constant vector. Its divergence is

$$\nabla \cdot \mathbf{F} = \mathbf{m} \cdot \nabla f. \tag{3.16}$$

---

1 Other names are Gauss' theorem, Ostrogradsky's theorem, and Gauss–Ostrogradsky theorem.

2 Also known as Stokes' theorem.

From the divergence theorem it then follows that

$$\int_V \nabla f \, dV = \int_S f\mathbf{n} \, dS, \tag{3.17}$$

which is useful when integration by parts is to be done in vector calculus. For a product function with a vanishing surface integral, we have

$$0 = \int_V \nabla(fg) \, dV = \int_V g\nabla f \, dV + \int_V f\nabla g \, dV. \tag{3.18}$$

We can alternatively define the divergence of a vector function $\mathbf{F}$ in a point by a limiting process in which the surrounding volume in Eq. (3.12) tends toward zero, that is

$$\nabla \cdot \mathbf{F} = \lim_{\Delta V \to 0} \frac{1}{\Delta V} \int_S (\mathbf{F} \cdot \mathbf{n}) dS. \tag{3.19}$$

In the same manner, we can define the curl of a vector function $\mathbf{F}$ in a point and in the direction normal to some surface by a limiting process in which the surface area in Eq. (3.14) tends toward zero, that is

$$(\nabla \times \mathbf{F}) \cdot \mathbf{n} = \lim_{\Delta S \to 0} \frac{1}{\Delta S} \oint_C (\mathbf{F} \cdot \mathbf{t}) dl. \tag{3.20}$$

These limiting values are independent of the shape of the corresponding volumes and surfaces.

3.1.2 The Stationary Case

Let us now explore the physical content of Maxwell's equations. We begin by a treatment of the *stationary case* in which the electric and magnetic fields are time independent. Under such circumstances, Maxwell's equations simplify to

$$\begin{aligned} \nabla \cdot \mathbf{B} &= 0; \quad \nabla \cdot \mathbf{E} = \rho/\varepsilon_0; \\ \nabla \times \mathbf{E} &= \mathbf{0}; \quad \nabla \times \mathbf{B} = \mu_0 \mathbf{j}. \end{aligned} \tag{3.21}$$

This restriction on the electric and magnetic fields also implies restrictions on the sources: Taking the divergence of the last relation, we obtain $\nabla \cdot \mathbf{j} = 0$, which corresponds to *steady* currents and, from the continuity equation, time-independent charge densities. In other words, steady current implies that there is no net transport of charge. As a further restriction, if there exist no currents ($\mathbf{j} = \mathbf{0}$), we obtain the *static case.*

The left-hand sides of the stationary Maxwell equations involve the divergence and curl of the electric and magnetic fields. Let us therefore first look at these vector operations in more detail. We note that the sources determine directly the curl and divergence of the fields, but not the fields themselves. *Boundary conditions* must be added to uniquely determine the fields for a given set of sources. The conventional boundary condition is to let $\mathbf{E}$ and $\mathbf{B}$ go to zero at infinite distance from the sources.

As an example, consider the case where there are no sources. Equation (3.21) shows that the divergence and the curl of the electric and magnetic fields are zero everywhere in space. But both vector fields $\mathbf{F}_1 = (0, 0, 0)$ and $\mathbf{F}_2 = (yz, zx, xy)$ have zero divergence and curl and it is not until we impose the boundary condition that we can conclude that $\mathbf{F}_1$ is the only acceptable solution of the two. Can there be additional acceptable solutions? In order to address this question, we employ the vector field identity

$$\nabla^2 \mathbf{F} = \nabla(\nabla \cdot \mathbf{F}) - \nabla \times (\nabla \times \mathbf{F}), \tag{3.22}$$

which connects the divergence and the curl of a vector function. If both are zero, then each component of $\mathbf{F}$ satisfies *Laplace's equation*

$$\nabla^2 F_i(\mathbf{r}, t) = 0. \tag{3.23}$$

Using the divergence theorem, Eq. (3.12), we find that this implies

$$\int_V \nabla^2 F_i dV = \int_V \boldsymbol{\nabla} \cdot (\boldsymbol{\nabla} F_i)\, dV = \int_S (\boldsymbol{\nabla} F_i \cdot \mathbf{n}) dS = 0, \tag{3.24}$$

that is, integrating the gradient of F_i over the boundary surface of any volume gives a vanishing result. Suppose now that F_i has a maximum (minimum) at some point $\mathbf{r}_1$, that is $\boldsymbol{\nabla} F_i(\mathbf{r}_1) = \mathbf{0}$. Then, by definition, the normal derivative of F_i on a small sphere surrounding $\mathbf{r}_1$ must be negative (positive) everywhere, and Eq. (3.24) cannot hold true. This means that F_i can have no local maxima and minima. If we now require F_i to go to zero at infinity, then it is concluded that F_i has to be zero everywhere. Hence, there are no acceptable solutions other than $\mathbf{F}_1 = (0, 0, 0)$ in the absense of sources.

In the general case of nonzero charges and currents, we benefit from the decomposition of vector fields made in Eq. (3.22), that is,

$$\nabla^2 \mathbf{F} = \boldsymbol{\nabla} D - \boldsymbol{\nabla} \times \mathbf{C}, \tag{3.25}$$

where D and $\mathbf{C}$ represent the divergence and curl of $\mathbf{F}$. We are not going to solve this second-order partial differential equation directly. Rather, we recognize that each component of the vector function $\mathbf{F}$ satisfies *Poisson's equation*

$$\nabla^2 F_i(\mathbf{r}, t) = f(\mathbf{r}, t) \tag{3.26}$$

with solutions

$$F_i(\mathbf{r}_1, t) = -\frac{1}{4\pi} \int \frac{f(\mathbf{r}_2, t)}{r_{12}} d^3\mathbf{r}_2. \tag{3.27}$$

We thus obtain

$$\mathbf{F}(\mathbf{r}_1, t) = -\frac{1}{4\pi} \int \frac{\boldsymbol{\nabla}_2 D(\mathbf{r}_2, t)}{r_{12}} d^3\mathbf{r}_2 + \frac{1}{4\pi} \int \frac{\boldsymbol{\nabla}_2 \times \mathbf{C}(\mathbf{r}_2, t)}{r_{12}} d^3\mathbf{r}_2. \tag{3.28}$$

Next, we use Eq. (3.18) to carry out integrations by parts and arrive at

$$\mathbf{F}(\mathbf{r}_1, t) = \frac{1}{4\pi} \int D(\mathbf{r}_2, t) \boldsymbol{\nabla}_2 \frac{1}{r_{12}} d^3\mathbf{r}_2 + \frac{1}{4\pi} \int \mathbf{C}(\mathbf{r}_2, t) \times \boldsymbol{\nabla}_2 \frac{1}{r_{12}} d^3\mathbf{r}_2. \tag{3.29}$$

Finally, we use the relation

$$\boldsymbol{\nabla}_2 \frac{1}{r_{12}} = -\boldsymbol{\nabla}_1 \frac{1}{r_{12}}, \tag{3.30}$$

to pull the gradient operators outside the integral sign to give

$$\mathbf{F}(\mathbf{r}) = -\boldsymbol{\nabla} s(\mathbf{r}) + \boldsymbol{\nabla} \times \mathbf{v}(\mathbf{r}). \tag{3.31}$$

The scalar function s and the vector function $\mathbf{v}$ can be constructed from the divergence and curl of $\mathbf{F}$ according to

$$s(\mathbf{r}_1) = \frac{1}{4\pi} \int \frac{\boldsymbol{\nabla}_2 \cdot \mathbf{F}(\mathbf{r}_2)}{r_{12}} d^3\mathbf{r}_2; \quad \mathbf{v}(\mathbf{r}_1) = \frac{1}{4\pi} \int \frac{\boldsymbol{\nabla}_2 \times \mathbf{F}(\mathbf{r}_2)}{r_{12}} d^3\mathbf{r}_2. \tag{3.32}$$

The two last equations constitute the *Helmholtz theorem*, and guarantees the uniqueness of solutions to the stationary Maxwell equations. It should be noted, however, that it comes with two restrictions: First, as $r \to \infty$, the divergence and curl of $\mathbf{F}$ must go to zero faster than r^{-2}, or the integrals in Eq. (3.32) diverge. The second restriction is that the general solution $\mathbf{F}$ is obtained from Poisson's equation (Eq. (3.26)). To the solution in Eq. (3.31) we can always add the solution of the homogeneous equation, that is, Laplace's equation, but as we have seen earlier, this solution is simply equal to the null vector as long as we insist that $\mathbf{F}$ tends to zero as $r \to \infty$.

Using Eqs. (3.31) and (3.32), it is straightforward to write down the general solutions for the electric and magnetic fields in the stationary case. The electric field becomes

$$\mathbf{E}(\mathbf{r}_1) = \frac{1}{4\pi\varepsilon_0}\int \frac{\mathbf{r}_{12}\rho(\mathbf{r}_2)}{r_{12}^3} d^3\mathbf{r}_2, \tag{3.33}$$

which is known as Coulomb's law in electrostatics. For the magnetic field, we obtain the solution

$$\mathbf{B}(\mathbf{r}_1) = -\frac{\mu_0}{4\pi}\int \frac{(\mathbf{r}_{12} \times \mathbf{j}(\mathbf{r}_2))}{r_{12}^3} d^3\mathbf{r}_2, \tag{3.34}$$

which is the fundamental law of magnetostatics known as the Biot–Savart law.

An important consequence of the Helmholtz theorem in Eq. (3.31) is that it provides a decomposition of *any* vector function that goes to zero faster than $1/r$ when $r \to \infty$. The first term, $\mathbf{F}_{\parallel} = -\boldsymbol{\nabla} s(\mathbf{r})$, and the second term, $\mathbf{F}_{\perp} = \boldsymbol{\nabla} \times \mathbf{v}(\mathbf{r})$, in Eq. (3.31) are referred to as the *irrotational* and the *solenoidal* parts, respectively. The former has zero curl, $\boldsymbol{\nabla} \times \mathbf{F}_{\parallel} = \mathbf{0}$, whereas the latter has zero divergence, $\boldsymbol{\nabla} \cdot \mathbf{F}_{\perp} = 0$. From Eq. (3.21), we see that in the stationary case the electric and magnetic fields are *irrotational* and *solenoidal* vector functions, respectively. In fact, from Eq. (3.2) we see that the magnetic field is solenoidal in the time-dependent case as well.[3]

As we have seen in Eqs. (3.19) and (3.20), the divergence and curl of a vector field in a point probes the flux and circulation of the function in the near vicinity of that point. Let us now try to get an idea about what the irrotational and solenoidal parts of a vector function look like. First we note that any constant addition to the vector field will contribute neither to the curl nor the divergence, so, without loss of generality, we can globally subtract the value at the reference point $\mathbf{r}_1$ to obtain a vector function for which $\mathbf{F}(\mathbf{r}_1) = \mathbf{0}$. Furthermore, since the limiting value in Eq. (3.19) is independent of the shape of the volume, the divergence of a vector function can be obtained from consideration of the ratio of flux through a sphere and the volume of the sphere. Likewise, for the curl in Eq. (3.20), we can consider the ratio of circulation around a circle in the xy-plane and the surface of this circle. For convenience, we therefore introduce a spherical coordinate system centered at the reference point. Divergence and curl in spherical coordinates are constructed to comply with the limiting processes of Eqs. (3.19) and (3.20), respectively, and are given by

$$\boldsymbol{\nabla} \cdot \mathbf{F} = \frac{1}{r^2 \sin\theta}\left[\sin\theta \frac{\partial}{\partial r}(r^2 F_r) + r\frac{\partial}{\partial \theta}(\sin\theta F_\theta) + r\frac{\partial F_\phi}{\partial \phi}\right],$$

$$\begin{aligned}\boldsymbol{\nabla} \times \mathbf{F} = \frac{1}{r\sin\theta}\left\{\boldsymbol{e}_r\left(\frac{\partial}{\partial\theta}(\sin\theta F_\phi) - \frac{\partial F_\theta}{\partial\phi}\right) + \boldsymbol{e}_\theta\left(\frac{\partial F_r}{\partial\phi} - \sin\theta\frac{\partial}{\partial r}(rF_\phi)\right)\right.\\ \left. + \sin\theta\boldsymbol{e}_\phi\left(\frac{\partial}{\partial r}(rF_\theta) - \frac{\partial F_r}{\partial\theta}\right)\right\}.\end{aligned}$$

The unit vectors appearing in the expression for the curl can be introduced in the following manner: an infinitesimal displacement is given by

$$d\mathbf{r} = \mathbf{e}_r dr + r\mathbf{e}_\theta d\theta + r\sin\theta\mathbf{e}_\phi d\phi, \tag{3.35}$$

3 An amusing illustration of the combined misuse of the divergence and curl theorems to make magnetic fields vanish has been given in G. B. Arfken. *Am. J. Phys.*, 27:608, 1959.

where

$$\begin{aligned} \mathbf{e}_r &= \mathbf{e}_x \sin\theta\cos\phi + \mathbf{e}_y \sin\theta\sin\phi + \mathbf{e}_z\cos\theta, \\ \mathbf{e}_\theta &= \mathbf{e}_x \cos\theta\cos\phi + \mathbf{e}_y\cos\theta\sin\phi - \mathbf{e}_z\sin\theta, \\ \mathbf{e}_\phi &= -\mathbf{e}_x\sin\phi + \mathbf{e}_y\cos\phi. \end{aligned} \tag{3.36}$$

It should be noted that, contrary to the Cartesian unit vectors $\mathbf{e}_x$, $\mathbf{e}_y$, and $\mathbf{e}_z$, the unit vectors in the spherical coordinate system change direction at every point in space. In the xy-plane, where $\theta = \pi/2$, their relations simplify to

$$\begin{aligned} \mathbf{e}_r &= \mathbf{e}_x\cos\phi + \mathbf{e}_y\sin\phi, \\ \mathbf{e}_\theta &= -\mathbf{e}_z, \\ \mathbf{e}_\phi &= -\mathbf{e}_x\sin\phi + \mathbf{e}_y\cos\phi, \end{aligned} \tag{3.37}$$

and, in this plane, vectors $\mathbf{e}_r$ and $\mathbf{e}_\phi$ correspond to the normal and tangential unit vectors, respectively, of the circle, see Figure 3.2.

As an illustrative example, let us consider a vector function

$$\mathbf{F}(\mathbf{r}) = F_r(r)\mathbf{e}_r + F_\phi(r)\mathbf{e}_\phi,$$

for which we impose that $r^2F_r(r) \to 0$ as $r \to \infty$ to comply with the Helmholtz theorem. This field, along a cirle of radius r in the xy-plane, is illustrated in Figure 3.2(a). Note that we have made the reasonable simplification of choosing radial and angular components, that is, F_r and F_ϕ, that are constant along the circle and thus functions of the radius r only. They are depicted in Figure 3.2(b) and (c), respectively. We see that the components F_r and F_ϕ describe flux and circulation of the vector function, respectively, and correspond to the irrotational and solenoidal fields, that is

$$\mathbf{F}_\parallel = F_r\mathbf{e}_r; \quad \mathbf{F}_\perp = F_\phi\mathbf{e}_\phi.$$

Indeed, we find that

$$\nabla\cdot\mathbf{F} = \frac{1}{r^2}\frac{\partial}{\partial r}(r^2F_r); \quad \nabla\times\mathbf{F} = \frac{1}{r\sin\theta}\left\{\mathbf{e}_r\cos\theta F_\phi - \mathbf{e}_\theta\sin\theta\frac{\partial}{\partial r}(rF_\phi)\right\} \overset{\theta=\frac{\pi}{2}}{\to} \frac{1}{r}\frac{\partial}{\partial r}(rF_\phi)\mathbf{e}_z. \tag{3.38}$$

From the preceding discussion we can easily understand this decomposition. Since the radial component F_r is constant on any sphere centered at the origin, integrating its value on the sphere clearly gives a nonzero result and $F_r\mathbf{e}_r$ is clearly not solenoidal. In the same manner, we

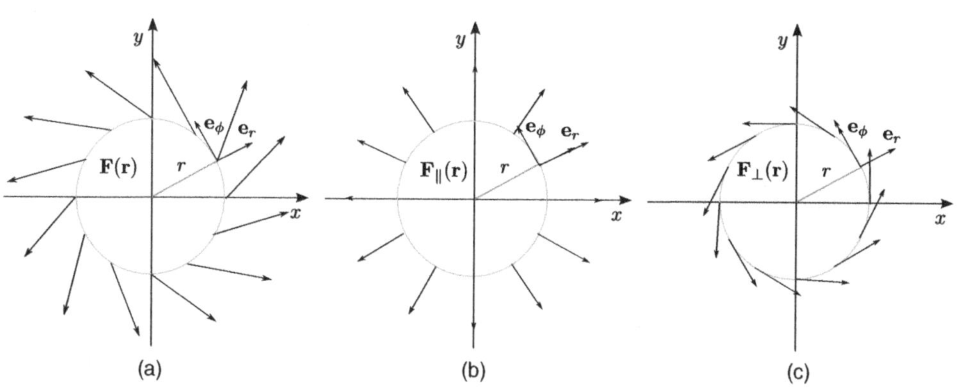

Figure 3.2 Decomposition of vector field $\mathbf{F}(\mathbf{r}) = F_r(r)\mathbf{e}_r + F_\phi(r)\mathbf{e}_\phi$ into irrotational $\mathbf{F}_\parallel(\mathbf{r})$ and solenoidal $\mathbf{F}_\perp(\mathbf{r})$ parts.

can easily understand that $F_\phi \mathbf{e}_\phi$ cannot be irrotational in the xy-plane since its integration along any circle centered at the origin gives a nonzero result.

3.1.3 The General Case

It is possible to give general solutions for the electric and magnetic fields in terms of their sources. Taking the curl of Eq. (3.3) and using Eqs. (3.5) and (3.6) as well as the vector relation Eq. (3.22), we obtain

$$\Box^2 \mathbf{E} = \nabla \rho/\varepsilon_0 + \mu_0 \partial_t \mathbf{j}, \tag{3.39}$$

where we have introduced the *d'Alembertian operator*

$$\Box^2 = \nabla^2 - \frac{1}{c^2}\partial_t^2. \tag{3.40}$$

Similar manipulation of Eq. (3.6) gives

$$\Box^2 \mathbf{B} = -\mu_0(\nabla \times \mathbf{j}). \tag{3.41}$$

For each component of the electric and magnetic fields, the above equations take the form of an *inhomogeneous electromagnetic wave equation* according to

$$\Box^2 \Psi(\mathbf{r}, t) = f(\mathbf{r}, t), \tag{3.42}$$

where $f(\mathbf{r}, t)$ can be considered a source function. The solution to this wave equation can be written as

$$\Psi(\mathbf{r}_1, t) = -\frac{1}{4\pi}\int \frac{f(\mathbf{r}_2, t_r)}{r_{12}} d^3\mathbf{r}_2, \tag{3.43}$$

where appears the *retarded* time defined as

$$t_r = t - \frac{r_{12}}{c}. \tag{3.44}$$

This is where relativity shows its teeth. Relativity imposes a "speed limit" c on any interaction such that there can be no instantaneous interactions. We can think of electromagnetic interactions between particles as an exchange of virtual photons. These photons travel at the finite speed of light and so there is a delay between the emission and reception of this signal. By the same token, when you are looking at a starry sky you are not only looking into space but also back in time. The effects of retardation complicate the description of electromagnetic interactions enormously, so that a complete description of a system of charged particles requires not only their instantanous positions and velocities but also their past history.

Using Eq. (3.43), we obtain the following general solutions for the electric field

$$\mathbf{E}(\mathbf{r}_1, t) = -\frac{1}{4\pi}\int \frac{[\nabla_2 \rho(\mathbf{r}_2, t)/\varepsilon_0 + \mu_0 \partial_t \mathbf{j}(\mathbf{r}_2, t)]_{t=t_r}}{r_{12}} d^3\mathbf{r}_2, \tag{3.45}$$

where the quantity in brackets is to be evaluated at the retarded time. By performing an integration by parts, we may transform Eq. (3.45) to a form closer to Coulomb's law in Eq. (3.33). But in doing so, some care is called for when considering the first term in the numerator of the integrand. The replacement $t = t_r$ is to be made *after* the gradient operator has acted, or otherwise there would appear a contribution to the gradient of the density from the time retardation. Before carrying out the integration by parts, we reverse these two operations by means of the relation

$$[\nabla_2 \rho(\mathbf{r}_2, t)]_{t=t_r} = \nabla_2 \rho(\mathbf{r}_2, t_r) - \frac{1}{c}\partial_t \rho(\mathbf{r}_2, t_r)\frac{\mathbf{r}_{12}}{r_{12}}, \tag{3.46}$$

to arrive at

$$\mathbf{E}(\mathbf{r}_1, t) = \frac{1}{4\pi} \int \left[\frac{\mathbf{r}_{12}\rho(\mathbf{r}_2, t_r)}{\varepsilon_0 r_{12}^3} + \frac{\mathbf{r}_{12}\partial_t\rho(\mathbf{r}_2, t_r)}{c\varepsilon_0 r_{12}^2} - \frac{\mu_0 \partial_t \mathbf{j}(\mathbf{r}_2, t_r)}{r_{12}} \right] d^3\mathbf{r}_2. \tag{3.47}$$

By the same procedure, we obtain for the magnetic field

$$\mathbf{B}(\mathbf{r}_1, t) = -\frac{\mu_0}{4\pi} \int \left[\frac{\mathbf{r}_{12} \times \mathbf{j}(\mathbf{r}_2, t_r)}{r_{12}^3} + \frac{\mathbf{r}_{12} \times \partial_t \mathbf{j}(\mathbf{r}_2, t_r)}{c r_{12}^2} \right] d^3\mathbf{r}_2. \tag{3.48}$$

These rather impressive-looking expressions for the electric and magnetic fields are in the literature referred to as *Jefimenko's equations*. Just as in the stationary case, we can add the solutions of the homogeneous equation (no sources) to the above. In the time-dependent case, these are electromagnetic waves that are studied in considerable detail in Section 3.1.5. We finally note that if we ignore retardation and restrict ourselves to time-independent sources, Jefimenko's equations reduce to the (generalized) Coulomb and Biot–Savart laws, Eqs. (3.33) and (3.34), respectively, as expected.

3.1.4 Electromagnetic Potentials and Gauge Freedom

Although we now have expressions for electric and magnetic fields for arbitrary sources, that is Eqs. (3.47) and (3.48), they are in practice of limited use since one soon runs into integrals that are difficult to evaluate. An alternative route to the solution of Maxwell's equations is to exploit the decomposition of electric and magnetic fields into irrotational and solenoidal components suggested by the Helmholtz theorem in Eq. (3.31). From this point of view, Eq. (3.2) states that the magnetic field is purely solenoidal (no magnetic charge) and can therefore be written as the curl of some vector function $\mathbf{A}$ that is called the *vector potential*

$$\mathbf{B} = \nabla \times \mathbf{A}. \tag{3.49}$$

Likewise, when the magnetic field is absent or time independent, Eq. (3.3) shows that the electric field is purely irrotational. In general, by inserting expression Eq. (3.49) into Eq. (3.3), we find that the curl of $\mathbf{E} + \partial_t \mathbf{A}$ is zero. This quantity can therefore be expressed as the gradient of a scalar function, leading to the expression

$$\mathbf{E} = -\nabla\phi - \partial_t \mathbf{A}, \tag{3.50}$$

where ϕ is the *scalar potential*. The electric field can be split into an irrotational and a solenoidal component according to

$$\mathbf{E}_{\parallel} = -\nabla\phi - \partial_t \mathbf{A}_{\parallel}; \quad \mathbf{E}_{\perp} = -\partial_t \mathbf{A}_{\perp}. \tag{3.51}$$

The benefit of writing the electric and magnetic fields in terms of the vector and scalar potentials is that we are able to solve the homogeneous pair of Maxwell's equations, Eqs. (3.2) and (3.3), and can focus on solving the inhomogeneous pair, Eqs. (3.5) and (3.6). In terms of potentials, these can be written as

$$\nabla^2\phi + \partial_t(\nabla \cdot \mathbf{A}) = -\rho/\varepsilon_0, \tag{3.52}$$

$$\Box^2\mathbf{A} - \nabla\left[(\nabla \cdot \mathbf{A}) + \frac{1}{c^2}\partial_t\phi\right] = -\mu_0\mathbf{j}. \tag{3.53}$$

However, the introduction of electromagnetic potentials is not unambiguous but leads to *gauge freedom*, that is, there are infinitely many choices of electromagnetic potentials that give identical electric and magnetic fields. To see this, we again look at Eq. (3.49). The magnetic field is purely solenoidal and can therefore be written as the curl of a vector potential. However, a modification of the *irrotational* part of the vector potential leaves the magnetic field unchanged. Such a modification can be written as

$$\mathbf{A}_{\parallel} \rightarrow \mathbf{A}'_{\parallel} = \mathbf{A}_{\parallel} + \boldsymbol{\nabla}\chi, \tag{3.54}$$

where we have introduced the *gauge function* $\chi \equiv \chi(\mathbf{r}, t)$. However, the irrotational component of the vector potential also contributes to the irrotational component of the electric field as seen in Eq. (3.51). In order to leave the electric field unchanged, a concurrent modification of the scalar potential is required in accordance with

$$\phi \quad \rightarrow \quad \phi' = \phi - \partial_t \chi. \tag{3.55}$$

The above transformations, as expressed in Eqs. (3.54) and (3.55), constitute a *gauge transformation*.

A Lorentz-invariant choice of gauge is the Lorentz gauge

$$(\boldsymbol{\nabla} \cdot \mathbf{A}) + \frac{1}{c^2}\partial_t \phi = 0, \tag{3.56}$$

which reduces Maxwell's equations to

$$\Box^2 \mathbf{A} = -\mu_0 \mathbf{j}; \quad \Box^2 \phi = -\rho/\varepsilon_0. \tag{3.57}$$

These equations have the form of the inhomogeneous electromagnetic wave equation in Eq. (3.42) and, in the Lorentz gauge, we can thus immediately write down the expressions for the electromagnetic potentials as

$$\phi(\mathbf{r}_1, t) = \frac{1}{4\pi\varepsilon_0}\int \frac{\rho(\mathbf{r}_2, t_r)}{r_{12}} d^3\mathbf{r}_2; \quad \mathbf{A}(\mathbf{r}_1, t) = \frac{\mu_0}{4\pi}\int \frac{\mathbf{j}(\mathbf{r}_2, t_r)}{r_{12}} d^3\mathbf{r}_2. \tag{3.58}$$

These potentials contain retarded time reflecting the finite speed of electromagnetic interactions, such that the integration over space is also an integration back in time.

Another commonly adopted choice of gauge is the Coulomb gauge

$$\boldsymbol{\nabla} \cdot \mathbf{A} = 0, \tag{3.59}$$

which sets the irrotational part of the vector potential to zero. Maxwell's equations then reduce to

$$\nabla^2 \phi = -\rho/\varepsilon_0, \tag{3.60}$$

$$\Box^2 \mathbf{A} - \boldsymbol{\nabla}\frac{1}{c^2}\partial_t \phi = -\mu_0 \mathbf{j}. \tag{3.61}$$

The first equation is the Poisson equation, see also Eq. (3.26), with solution

$$\phi(\mathbf{r}_1, t) = \frac{1}{4\pi\varepsilon_0}\int \frac{\rho(\mathbf{r}_2, t)}{r_{12}} d^3\mathbf{r}_2. \tag{3.62}$$

In contrast to Eq. (3.58), the above expression for the scalar potential contains instantaneous and not retarded time. This seems to violate the fundamental tenet of special relativity, that instantaneous interactions are not allowed. However, it should be kept in mind that the scalar potential can be considered an auxilliary quantity. All effects of retardations are, in this case, found in Eq. (3.61) involving the vector potential. Furthermore, we may split this equation into its irrotational and solenoidal components, as we in fact already did in Section 3.1.1 when we derived the continuity equation. The solenoidal part becomes

$$\Box^2 \mathbf{A}_{\perp} = -\mu_0 j_{\perp}, \tag{3.63}$$

with solution

$$\mathbf{A}_\perp(\mathbf{r}_1,t)=\frac{\mu_0}{4\pi}\int\frac{\mathbf{j}_\perp(\mathbf{r}_2,t_r)}{r_{12}}d^3\mathbf{r}_2, \tag{3.64}$$

where retarded time does appear.

It should be noted that Lorentz and Coulomb gauge refer to classes of electromagnetic potentials. Consider for instance a vector potential **A** that fulfills the Coulomb gauge condition in Eq. (3.59). Provided that a gauge function χ satisfies the Laplace equation, Eq. (3.23), then, for the transformed vector potential $\mathbf{A}'$, we have

$$\boldsymbol{\nabla}\cdot\mathbf{A}'=\boldsymbol{\nabla}\cdot\mathbf{A}+\nabla^2\chi=0, \tag{3.65}$$

which shows that it will also fulfill the Coulomb gauge. In the same manner, one can show that the Lorentz gauge is conserved under gauge transformations for which the gauge function satisfies the *electromagnetic wave equation*, that is,

$$\Box^2\chi=0. \tag{3.66}$$

3.1.5 Electromagnetic Waves and Polarization

In this section, we consider electromagnetic waves that are solutions to Maxwell's equations in the absence of sources. In fact, this is an idealized situation; we are rather considering the situation where the sources are sufficiently remote that we can set them to zero. Here the alert reader might protest, because in Section 3.1.2 we pointed out the importance of boundary conditions and even stated that the fields should go to zero at sufficient distance from the sources. However, this concerns the solutions to the *inhomogeneous* differential equations. To those solutions we can always add, as we pointed out in Section 3.1.4, the solutions of the homogeneous problem, and those are the solutions that we consider now.

The homogeneous versions of Eqs. (3.39) and (3.41) are

$$\Box^2\mathbf{E}=\mathbf{0};\quad \Box^2\mathbf{B}=\mathbf{0}. \tag{3.67}$$

Interestingly, as seen from Eq. (3.39), we get the same equation for the electric field if the charge density is homogeneous over all space and the current density is time independent. Likewise, as seen from Eq. (3.41), we get the same equation for the magnetic field if the current density is purely irrotational. From Eqs. (3.2) and (3.5), we see that in the absence of sources, both the electric and magnetic fields have zero divergence and are therefore purely solenoidal. There is thus no need to introduce a scalar potential, and we can do with a purely transversal vector potential. The latter is precisely what the Coulomb gauge gives us, and in this gauge the source-free Maxwell equations reduce to

$$\Box^2\mathbf{A}=\mathbf{0}. \tag{3.68}$$

In the absence of sources, the components of the electric field, of the magnetic field, and of the vector potential all satisfy the electromagnetic wave equation

$$\Box^2\Psi(\mathbf{r},t)=\left[\nabla^2-\frac{1}{c^2}\partial_t^2\right]\Psi(\mathbf{r},t)=0. \tag{3.69}$$

A real solution to this equation is provided by a cosine function of the form

$$\Psi(\mathbf{r},t)=\Psi_0\cos(\mathbf{k}\cdot\mathbf{r}-\omega t+\delta), \tag{3.70}$$

where appears the wave vector **k**, the angular frequency ω, and the phase constant δ. By manipulation of the phase constant or by taking a derivative, the cosine function is converted into a sine function. Mathematically, it is therefore much more convenient to work with complex exponentials

$$\Psi(\mathbf{r},t) = \Psi_0 \exp[i(\mathbf{k}\cdot\mathbf{r} - \omega t + \delta)], \tag{3.71}$$

in particular, since the phase constant can be factorized out while still retaining the same exponential form. Feeding this solution back into the wave equation, Eq. (3.69), we find that the amplitude of the wave vector is related to the angular frequency and the speed of light according to

$$\Box^2\Psi(\mathbf{r},t) = \left(-k^2 + \frac{\omega^2}{c^2}\right)\Psi(\mathbf{r},t) = 0 \quad \Rightarrow \quad k = \frac{\omega}{c}. \tag{3.72}$$

Therefore, the functions in Eqs. (3.70) and (3.71) are completely specified by giving the wave vector $\mathbf{k}$ and the phase constant δ. The quantity $(\mathbf{k}\cdot\mathbf{r} - \omega t + \delta)$ is the *phase* of the wave function. The set of points in space giving the same phase at a particular instant of time constitute a surface that is called the *wavefront*. In our particular case, the wavefront is a plane. This is easily seen by splitting the position vector into two parts $\mathbf{r} = \mathbf{r}_\parallel + \mathbf{r}_\perp$, where $\mathbf{r}_\parallel$ and $\mathbf{r}_\perp$ are parallel and perpendicular to the wave vector $\mathbf{k}$, respectively. We thus have $\mathbf{k}\cdot\mathbf{r} = \mathbf{k}\cdot\mathbf{r}_\parallel$ such that fixing $\mathbf{r}_\parallel$ fixes the value of $\mathbf{k}\cdot\mathbf{r}$ and thus the phase at a given instant of time. The perpendicular component $\mathbf{r}_\perp$ can be chosen freely, which means that a fixed value of $\mathbf{k}\cdot\mathbf{r}$ defines a plane perpendicular to $\mathbf{k}$. Using the dispersion relation, the rightmost relation in Eq. (3.72), we can express the phase as

$$\mathbf{k}\cdot\mathbf{r} - \omega t + \delta = \pm\frac{\omega}{c}(r_\parallel \mp ct) + \delta, \tag{3.73}$$

where the plus and minus signs refer to $\mathbf{k}$ and $\mathbf{r}_\parallel$ being parallel and antiparallel, respectively. The phase is invariant under the combined substitution $r_\parallel \to r_\parallel \pm c\tau$ and $t \to t + \tau$ where τ is some arbitrary time interval. This shows that the plane wave propagates at the speed c in the direction of the wave vector $\mathbf{k}$. The role of the phase constant δ can be understood as follows. Suppose that we start with $\delta = 0$ in the expression above. Now we add a constant δ. The plane that was defined by a particular value of the phase at some instant in time is then moved δ/ω forward in time or δ/k backward in space.

From the periodicity of the exponential function it follows that the value of $\Psi(\mathbf{r},t)$ is conserved by adding $m2\pi$, that is, an integer multiple of 2π, to the phase. This implies that planes with identical values of $\Psi(\mathbf{r},t)$ are separated by a *period* $T = 2\pi/\omega$ in time and *wavelength* $\lambda = cT$ in space. The *frequency* $\nu = T^{-1}$ is the number of oscillations per unit time. In spectroscopy, one often refers to the *wavenumber*, which is defined as the inverse wavelength, that is, $\tilde{\nu} = \lambda^{-1}$.

The electric and magnetic components of a *monochromatic* (single-frequency) electromagnetic wave are written as

$$\mathbf{E} = \tilde{\mathbf{E}}_0 \exp[i(\mathbf{k}\cdot\mathbf{r} - \omega t)]; \quad \mathbf{B} = \tilde{\mathbf{B}}_0 \exp[i(\mathbf{k}\cdot\mathbf{r} - \omega t)], \tag{3.74}$$

where it is to be understood that the physical fields are the *real* parts of $\mathbf{E}$ and $\mathbf{B}$. The phase constants have been incorporated into complex amplitudes, for example, $\tilde{E}_{0;x} = E_{0;x}\exp(i\delta_x)$. Both fields are solenoidal, which implies that they are perpendicular to the wave vector

$$\mathbf{k}\cdot\mathbf{E} = 0; \quad \mathbf{k}\cdot\mathbf{B} = 0, \tag{3.75}$$

and they are connected through Maxwell's equations. Inserting Eq. (3.74) into Eq. (3.3) shows that $\mathbf{k}$, $\mathbf{E}$, and $\mathbf{B}$ are mutually perpendicular, forming a right-handed axis system with

$$\mathbf{k}\times\mathbf{E} = \omega\mathbf{B} \quad \Rightarrow \quad E = cB. \tag{3.76}$$

This means that if the wave vector is pointing along the z-axis and we orient our coordinate system such that the electric field vector points along the x-axis, then the magnetic field vector is pointing along the y-axis. The above relation also shows that the amplitude of the electric field is a factor c larger than the amplitude of the magnetic field.

An important property of electromagnetic waves is their *polarization*. Consider an electromagnetic field propagating along the z-direction, that is, $\mathbf{k} = k\mathbf{e}_z$. The most general form of the electric field is then

$$\begin{aligned}\mathbf{E} &= \mathrm{Re}\{(\tilde{E}_{0;x}\mathbf{e}_x + \tilde{E}_{0;y}\mathbf{e}_y)\exp[i(kz - \omega t)]\}\\ &= E_{0;x}\cos(kz - \omega t + \delta_x)\mathbf{e}_x + E_{0;y}\cos(kz - \omega t + \delta_y)\mathbf{e}_y.\end{aligned} \tag{3.77}$$

In the simplest case, the phase constants in the x- and y-directions are equal; that is, $\delta_x = \delta_y = \delta$, so that the physical wave can be rewritten as

$$\mathbf{E}_{\mathrm{lp}} = (E_{0;x}\mathbf{e}_x + E_{0;y}\mathbf{e}_y)\mathrm{Re}\{\exp[i(kz - \omega t + \delta)]\} \tag{3.78}$$

With the aid of Figure 3.3, we can write the wave as

$$\mathbf{E}_{\mathrm{lp}} = E_0(\cos\theta\mathbf{e}_x + \sin\theta\mathbf{e}_y)\cos(kz - \omega t + \delta), \tag{3.79}$$

showing that it corresponds to an oscillating electric field vector of amplitude E_0 at a fixed angle θ with respect to the x-axis. From Eq. (3.76), it follows that the corresponding magnetic field vector is given by

$$\mathbf{B}_{\mathrm{lp}} = (-E_y/c, E_x/c, 0), \tag{3.80}$$

and that it makes an angle $\theta + \pi/2$ with the x-axis. This particular configuration corresponds to *linearly polarized (lp) light*. More generally, linearly polarized light corresponds to a phase difference, which is a multiple of π, that is, $\delta_y - \delta_x = m\pi$.

Consider next the case when we add $\pi/2$ to the phase difference, that is, $\delta_y - \delta_x = \frac{\pi}{2} + m\pi$. The electric field can then be written as

$$\mathbf{E}_{\mathrm{ep}} = E_{0;x}\cos(kz - \omega t + \delta_x)\mathbf{e}_x - (-1)^m E_{0;y}\sin(kz - \omega t + \delta_x)\mathbf{e}_y, \tag{3.81}$$

and it has a time-dependent orientation. We straightforwardly obtain the relation

$$\frac{E_x^2}{E_{0;x}^2} + \frac{E_y^2}{E_{0;y}^2} = 1, \tag{3.82}$$

which shows that in a given plane, for example, $z = 0$, the tip of the electric field vector traces out an ellipse. This is a particular case of *elliptic polarization (ep)*. When $E_{0;y} > E_{0;x}$ the semimajor and semiminor axes of the ellipse are given by $E_{0;y}$ and $E_{0;x}$, respectively. When $E_{0;y} < E_{0;x}$ the roles are reversed.

An important special case of elliptic polarization is when $E_{0;x} = E_{0;y} = E_0$. The tip of the electric (and magnetic) field vectors then trace a circle and we say that the electromagnetic wave is

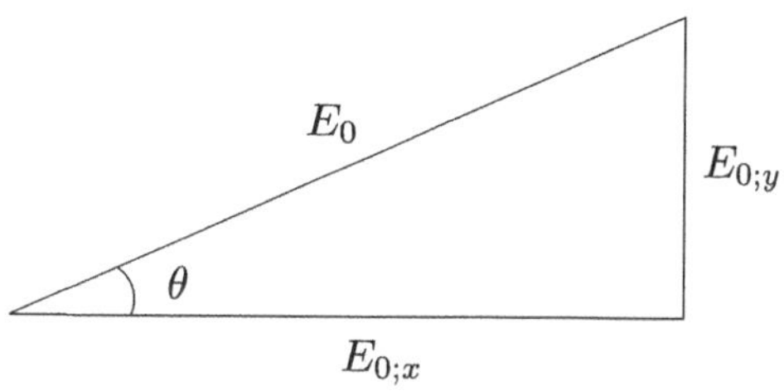

Figure 3.3 Linearly polarized light oscillating with an amplitude E_0 at an angle θ with the x-axis.

circularly polarized (cp). The physical wave now takes the form

$$\begin{aligned}\mathbf{E}_{cp}(z,t) &= \text{Re}\{E_0(\mathbf{e}_x + (-1)^m i\mathbf{e}_y)\exp[i(kz-\omega t+\delta_x)]\} \\ &= E_0[\cos(kz-\omega t+\delta_x)\mathbf{e}_x - (-1)^m \sin(kz-\omega t+\delta_x)\mathbf{e}_y],\end{aligned} \tag{3.83}$$

where appears complex unit vectors

$$\mathbf{e}_\pm = \mathbf{e}_x \pm i\mathbf{e}_y, \tag{3.84}$$

where the sign depends on whether the integer m is even (+) or odd (−). For later use we note that if we rotate the Cartesian units vectors $\mathbf{e}_x$ and $\mathbf{e}_y$ by an angle ϕ, the complex unit vectors $\mathbf{e}_\pm$ turn into

$$\mathbf{e}'_\pm = e^{\mp i\phi}\mathbf{e}_\pm. \tag{3.85}$$

Let us investigate the case where m is odd. Let us first consider the curve traced by the tip of the electric field vector along the z-axis for a given instant of time. For simplicity we choose $t = \delta_x/\omega$ and get

$$\mathbf{E}_{cp}(z, t=\delta_x/\omega) = E_0[\cos(kz)\mathbf{e}_x + \sin(kz)\mathbf{e}_y] \quad (m \text{ odd}). \tag{3.86}$$

When $z = 0$ the field vector is along the x-axis and will next move toward the positive end of the y-axis. From this we see that a plot of the field vector as a function of z at a given moment in time gives a right-handed helix. If we instead fix z and trace the electric field vector as a function of time we obtain a *left*-handed helix. There is in fact two opposite conventions of handedness for circularly polarized light in use depending on whether one follows the field vector in time or in space.[4] In this book we will, following the IUPAC recommendation,[5] employ the handedness convention based on the spatial configuration. This convention implies that we will call a circularly polarized electromagnetic wave right-handed if it traces a right-handed helix when frozen in time. Furthermore, when we look directly toward the circularly polarized electromagnetic wave coming toward us, the field vector traces a circle in the clockwise direction. Pointing our right thumb *in the direction of the source,* our fingers then curl in the direction of rotation of the field vector. In other words, in this convention the polarization is right-handed (odd m or $\mathbf{e}_-$) from the point of view of the receiver. The opposite case is illustrated in Figure 3.4, showing a left-handed circularly polarized wave. If we now point our right thumb *in the direction of propagation,* our fingers now curl in the direction of rotation of the field vector. The handedness convention based on temporal convention is accordingly from the point of view of the

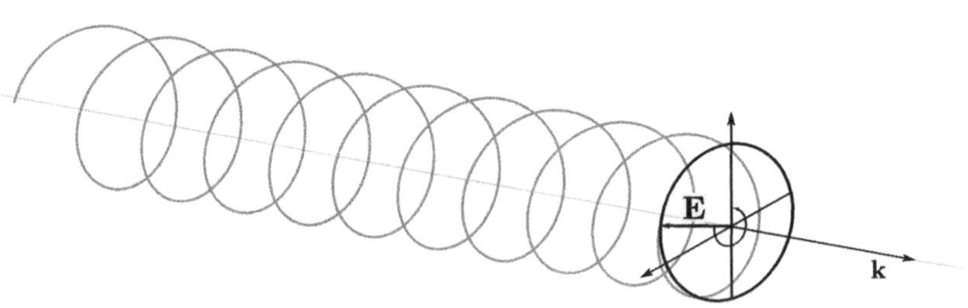

Figure 3.4 Left-handed circularly polarized wave with wave vector **k**. The electric field vector is illustrated at a certain coordinate z along the direction of propagation (at a certain instant of time) and the direction of rotation (here anticlockwise) of this vector as time passes is depicted.

---

4 See D. S. Kliger, J. W. Lewin, and C. E. Randall. *Polarized Light in Optics and Spectroscopy. Academic Press, San Diego, CA,* 1990.
5 S. E. Braslavsky. *Pure Appl. Chem.,* 79:293–465, 2009.

source. Note that this convention agrees with the definition of helicity in quantum mechanics where a particle is called right-handed if its spin vector is parallel to the direction of motion. The quantum of the electromagnetic field, the photon, carries spin $s = 1$. The two components $m_s = +1$ and $m_s = -1$ correspond to left-handed and right-handed photons, respectively, or equivalently, to positive and negative helicity. The component $m_s = 0$ is absent since electromagnetic waves are purely transversal.

In Eq. (3.77), we represented the electric field vector in terms of Cartesian unit vectors $\mathbf{e}_x$ and $\mathbf{e}_y$. However, as we have also seen, a completely equivalent and sometimes more advantageous representation is provided by the complex vectors $\mathbf{e}_\pm$ in accordance with

$$\mathbf{E} = \mathrm{Re}\{(\tilde{E}_{0;+}\mathbf{e}_+ + \tilde{E}_{0;-}\mathbf{e}_-)\exp[i(kz - \omega t)]\}, \tag{3.87}$$

where

$$\tilde{E}_{0;\pm} = E_{0;\pm}e^{i\delta_\pm} = \frac{1}{2}(\tilde{E}_{0;x} \mp i\tilde{E}_{0;y}). \tag{3.88}$$

In this representation, a linearly polarized wave of amplitude E_0 at an angle θ with respect to the x-axis is given by

$$\begin{aligned} \mathbf{E}_{\mathrm{lp}} &= \frac{1}{2}E_0\mathrm{Re}\{\mathbf{e}_- \exp[i(kz - \omega t + \delta + \theta)]\} \\ &+ \frac{1}{2}E_0\mathrm{Re}\{\mathbf{e}_+ \exp[i(kz - \omega t + \delta - \theta)]\}, \end{aligned} \tag{3.89}$$

which shows that a linearly polarized wave can be decomposed into two circularly polarized waves of opposite handedness. This result is important because it shows that if the two circular components of a linearly polarized electromagnetic wave propagate at different speeds in a medium, a rotation of the plane of polarization (optical activity) will be observed. A left- and right-handed circularly polarized electromagnetic wave, according to the convention we employ in this book, is obtained by setting $\tilde{E}_{0;-}$ and $\tilde{E}_{0;+}$, respectively, to zero. An elliptically polarized wave with the ellipse aligned along the main Cartesian axes is obtained by setting $\delta_- = \delta_+$. Finally, let us consider the most general case with a phase difference $\alpha = \delta_- - \delta_+$. We can then rework Eq. (3.87) into the form

$$\mathbf{E} = \mathrm{Re}\{(E_{0;+}e^{-i\alpha/2}\mathbf{e}_+ + E_{0;-}e^{+i\alpha/2}\mathbf{e}_-)\exp[i(kz - \omega t + \bar{\delta})]\}, \tag{3.90}$$

where $\bar{\delta} = (\delta_+ + \delta_-)/2$ is the mean phase. Using Eq. (3.85), we can see that this corresponds to an elliptically polarized electromagnetic wave with the polarization ellipse rotated by an angle $\alpha/2$ with respect to the x-axis.

3.1.6 Electrodynamics: Relativistic and Nonrelativistic Formulations

Maxwell's equations can be traced back to an 1861 paper by Maxwell, some 40 years before Einstein formulated the special theory of relativity in 1905; yet, these equations are inherently relativistic and need no modifications. The work of Einstein is based on the principle of relativity (Galileo Galilei, 1631) which states that the laws of physics are the same in all inertial frames, that is, reference frames moving with constant relative velocities. Einstein, however, added the postulate that the speed of light is the same in all inertial frames, thus creating the special theory of relativity.[6] This is contrary to our everyday experience with velocities. We certainly do not expect to measure the same relative speed of a passing car when we stand on the side of the

6 Later Einstein formulated a *general* theory allowing for acceleration between reference frames.

road or drive another car. The constancy of the speed of light in all inertial frames requires both position and time to change between inertial frames, and the Galilean transformation of these quantities, known from Newtonian mechanics, is accordingly replaced by the Lorentz transformation: If an inertial frame K' moves with velocity $\mathbf{v}$ with respect to the inertial frame K and their origins coincide at $t' = t = 0$, then their space and time coordinates are related by

$$\mathbf{r}' = \mathbf{r} + (\gamma - 1)\frac{(\mathbf{r}\cdot\mathbf{v})\mathbf{v}}{v^2} - \gamma\mathbf{v}t; \quad t' = \gamma\left(t - \frac{\mathbf{r}\cdot\mathbf{v}}{c^2}\right), \tag{3.91}$$

where we again meet the Lorentz factor, Eq. (1.11). Maxwell's equations indeed obey the Lorentz transformation. The revolutionary nature of Einstein's theory was to propose that *all* laws of physics should obey the Lorentz transformation.

Carrying out a Lorentz transformation of a physical law can be a tedious exercise. There is fortunately a simpler way to reveal the relativistic nature of a law, namely to express it on Lorentz covariant form, that is, in terms of quantities that all transform in the same way under a Lorentz transformation. An example of such quantities are *4-vectors*. The 4-position

$$r_\mu = (\mathbf{r}, ict), \tag{3.92}$$

is a vector of position $\mathbf{r}$ and time t that are connected by Lorentz transformations.[7] In fact, the Lorentz transformation is constructed such that it conserves the norm $r_\mu r_\mu = r^2 - c^2t^2$ of the 4-position, which ensures constant speed of light in all inertial frames. Note that the time part of the 4-position is multiplied with the speed of light such that it has the dimension of length. Some other 4-vectors are the 4-momentum, 4-potential, 4-current, and 4-gradient for which the respective expressions read

$$p_\mu = (\mathbf{p}, iE/c), \tag{3.93}$$

$$A_\mu = (\mathbf{A}, i\phi/c), \tag{3.94}$$

$$j_\mu = (\mathbf{j}, ic\rho), \tag{3.95}$$

$$\partial_\mu = (\boldsymbol{\nabla}, -i\partial_t/c). \tag{3.96}$$

The quantities appearing in these vectors transform as the corresponding quantities in the 4-position. For instance, by direct comparison we see that the scalar potential ϕ transforms as c^2t. This has the interesting consequence that, when we Lorentz transform a scalar potential to some other inertial frame, a nonzero vector potential will appear. This phenomenon is magnetic induction, a truly relativistic phenomenon, as we shall see in the following text.

Let us now explore the relativistic nature of electrodynamics. We shall see that the introduction of 4-vectors combined with the Einstein summation convention leads to a very compact formulation of electrodynamics. Note, however, that when 4-vectors are involved the summation goes over the *four* components of the vector. As a warm-up exercise, we note that the continuity equation in Eq. (3.11) can be compactly expressed in terms of 4-vectors as

$$\partial_\mu j_\mu = 0. \tag{3.97}$$

Likewise, we find that the d'Alembertian operator in Eq. (3.40) is simply

$$\Box^2 = \partial_\mu \partial_\mu. \tag{3.98}$$

7 We employ imaginary i in the time-like part of 4-vectors, rather than to introduce a metric, and will therefore not make a distinction between covariant and contravariant quantities. This simplification is not possible in the general theory of relativity where the metric is a function of space-time.

Gauge transformations, as expressed in Eqs. (3.54) and (3.55), can be written on a manifestly Lorentz-invariant form according to

$$A_\mu \quad \rightarrow \quad A'_\mu = A_\mu + \partial_\mu \chi, \tag{3.99}$$

as is also the case for the Lorentz gauge condition in Eq. (3.56), which takes the form

$$\partial_\mu A_\mu = 0. \tag{3.100}$$

In terms of scalar and vector potentials, the inhomogeneous Maxwell's equations are expressed by Eqs. (3.52) and (3.53). However, Eq. (3.52) can be reworked to

$$\Box^2 \phi + \partial_t \left[(\nabla \cdot \mathbf{A}) + \frac{1}{c^2} \partial_t \phi \right] = -\rho/\varepsilon_0. \tag{3.101}$$

Using Eq. (3.7), we can then combine Eqs. (3.101) and (3.53) into a *single* equation written in Lorentz covariant form according to

$$\Box^2 A_\mu - \partial_\mu (\partial_\nu A_\nu) = -\frac{1}{\varepsilon_0 c^2} j_\mu, \tag{3.102}$$

where, following our convention, there is an implicit summation over ν, but not over μ. Note that, if we invoke the Lorentz gauge condition given in Eq. (3.100), the second term disappears and we obtain Eq. (3.57). This is not only an extremely compact version of Maxwell's equations, but it also shows the manifestly relativistic nature of the fundamental equations of electrodynamics.

Curiously, in the early days, Maxwell's equations were considered as *nonrelativistic* since their transformation between inertial frames did not obey Galilean invariance. This naturally leads us to the question: What is the nonrelativistic limit of Maxwell's equations? We cannot get an experimental answer since we only have a relativistic world at our disposal. Theoretically we obtain the nonrelativistic limit by letting the speed of light c tend to infinity, but this procedure is somewhat complicated due to the fact that the speed of light appears in different ways according to the electromagnetic units in which Maxwell's equations are expressed. By using Eq. (3.7) to eliminate the magnetic constant μ_0 everywhere in Maxwell's equations and then taking the limit $c \rightarrow \infty$, we obtain

$$\nabla \cdot \mathbf{B} = 0, \tag{3.103}$$

$$\nabla \times \mathbf{E} + \partial_t \mathbf{B} = \mathbf{0}, \tag{3.104}$$

$$\nabla \cdot \mathbf{E} = \rho/\varepsilon_0, \tag{3.105}$$

$$\nabla \times \mathbf{B} = \mathbf{0}. \tag{3.106}$$

We see that the divergence *and* the curl of the magnetic field are zero everywhere in space. If we combine this information with the boundary condition that the magnetic field $\mathbf{B}$ should go to zero at infinity, we are forced to conclude that the magnetic field itself is zero everywhere in space. The electric field is then purely irrotational and can be represented in terms of the scalar potential only. There is no vector potential and thus no gauge freedom. In fact our result suggests that, in the nonrelativistic limit of electrodynamics, there are no magnetic interactions and no effects of retardation. Electrodynamics reduces to electrostatics. Our result also explains why the Coulomb gauge holds interest both in a relativistic and nonrelativistic framework since the Coulomb gauge singles out the instantaneous Coulomb interaction which is the most important interaction in chemistry. All retardation effects as well as magnetic interactions are to be considered as corrections of a perturbation expansion of the total interaction in $1/c^2$ (cf. Eq. (3.64)).

3.2 Particles in Electromagnetic Fields

3.2.1 The Classical Mechanical Hamiltonian

Let us now put everything together and consider a system of particles and fields, such as a molecule in a laser field. In complete analogy with Eq. (3.1), we may write the total Lagrangian as

$$L = L_{\mathrm{p}} + L_{\mathrm{int}} + L_{\mathrm{f}}, \tag{3.107}$$

where L_{p} and L_{f} are the separate Lagrangians for the particles and fields, respectively, and L_{int} describes their interactions. We have discussed the Lagrangian for particles, for both the relativistic and nonrelativistic cases, in Chapter 2, but we have yet to consider the Lagrangian for electromagnetic fields as well as their interactions with particles.

A fundamental difference between particles and fields is that particles are discrete quantities, whereas fields are continuous quantities. The fundamental quantity in the Lagrangian formalism for continuous systems is the Lagrangian *density* $\mathcal{L}$, which integrates to the Lagrangian

$$L = \int \mathcal{L} d^3\mathbf{r}. \tag{3.108}$$

For an electromagnetic field, the Lagrangian density is a function of the scalar and vector potential and their space and time derivatives

$$\mathcal{L}(A_\alpha, \partial_\beta A_\alpha) = j_\alpha A_\alpha - \frac{1}{4}\varepsilon_0 c^2 F_{\alpha\beta} F_{\alpha\beta}, \tag{3.109}$$

where we have introduced the *electromagnetic tensor*

$$F_{\alpha\beta} = \partial_\alpha A_\beta - \partial_\beta A_\alpha. \tag{3.110}$$

Maxwell's equations, as given in Eq. (3.102), are obtained from the Euler–Lagrange equations for continuous systems

$$\frac{\partial \mathcal{L}}{\partial A_\alpha} - \partial_\beta \left(\frac{\partial \mathcal{L}}{\partial_\beta A_\alpha} \right) = 0, \tag{3.111}$$

which is to be compared with the equivalent equations for discrete systems given in Eq. (2.13). In the above formulas, we have employed the compact 4-vector notation introduced in Section 3.1.6. Translating to a perhaps more familiar notation, the elements of the anti-symmetric electromagnetic field tensor are seen to be components of the electric and magnetic fields

$$F = \begin{bmatrix} 0 & B_z & -B_y & -iE_x/c \\ -B_z & 0 & B_x & -iE_y/c \\ B_y & -B_x & 0 & -iE_z/c \\ iE_x/c & iE_y/c & iE_z/c & 0 \end{bmatrix}, \tag{3.112}$$

whereas the Lagrangian density for the isolated electromagnetic field becomes

$$\mathcal{L}_{\mathrm{f}} = \frac{1}{2}\varepsilon_0 c^2 (E^2 - c^2 B^2). \tag{3.113}$$

The Lagrangian describing the interactions between particles and fields is given by

$$L_{\mathrm{int}} = \int j_\mu A_\mu d^3\mathbf{r}' = \int \left[\mathbf{j}(\mathbf{r}', t) \cdot \mathbf{A}(\mathbf{r}', t) - \rho(\mathbf{r}', t)\phi(\mathbf{r}', t) \right] d^3\mathbf{r}', \tag{3.114}$$

which makes it clear that the current density couples to the vector potential whereas the charge density couples to the scalar potential. For discrete systems of point-localized particles, densities and currents are expressed in terms of Dirac delta functions and the above spatial integration then leaves us with a Lagrangian that becomes a function of the particle coordinates. For continuous systems, on the other hand, the Lagrangian becomes a mere scalar function of time.

At the beginning of this chapter, we pointed out that, in principle, the equations of motion for the combined system of particles and fields, described by the Lagrangian in Eq. (3.107), should be solved self-consistently, but that in practice this is seldom done. Maxwell's equations are equations of motion for the fields obtained by fixing the particles as sources for the fields. Let us now instead consider the equations of motion for particles in applied external electromagnetic fields. Let us start very simply by restricting our attention to a single point charge. Using Eq. (3.8), the interaction Lagrangian reduces to

$$L_{\text{int}} = q\mathbf{v}(t) \cdot \mathbf{A}(\mathbf{r}, t) - q\phi(\mathbf{r}, t), \tag{3.115}$$

where $\mathbf{r} = \mathbf{r}(t)$ refers to the trajectory of the particle. The generalized, or canonical, momentum is

$$\mathbf{p} = \frac{\partial L}{\partial \mathbf{v}} = \boldsymbol{\pi} + q\mathbf{A}; \quad \boldsymbol{\pi} = \frac{\partial L_{\text{p}}}{\partial \mathbf{v}}, \tag{3.116}$$

where we have introduced the *mechanical momentum* $\boldsymbol{\pi}$ as the momentum obtained from the free-particle Lagrangian L_{p} (see Section 2.1), be it relativistic or not. The Hamiltonian, coupling the particle to the external fields, is obtained from the Lagrangian and becomes

$$\hat{H} = \frac{\partial L}{\partial \mathbf{v}} \cdot \mathbf{v} - L = (\boldsymbol{\pi} \cdot \mathbf{v} - L_{\text{p}}) + q\phi(\mathbf{r}, t). \tag{3.117}$$

In a final subsequent step, one substitutes velocities and mechanical momenta with the generalized momenta and coordinates that are the proper variables of the Hamiltonian. The explicit form of the particle Lagrangian depends on whether we adopt a nonrelativistic or relativistic framework, but, for both cases, Eqs. (3.116) and (3.117) give us the recipe for introducing external fields into the Hamiltonian by the substitutions

$$p_\mu \rightarrow p_\mu - qA_\mu, \tag{3.118}$$

as expressed in terms of the 4-momentum and 4-potential given in Eqs. (3.93) and (3.94). Note that we have identified the (free-particle) Hamiltonian with energy, which allows us to express the minimal substitutions in a compact Lorentz-invariant form using the 4-momentum, which shows its *relativistic* origin. As mentioned, however, we have not specified whether the starting free-particle Lagrangian is the nonrelativistic or the relativistic one. In practice, the relativistic coupling to the field is employed *both* in a relativistic and a nonrelativistic framework. In the latter case, this may be formally inconsistent, but it is a perfectly pragmatic solution. The only internal property of the particle that appears in Eq. (3.118) is its charge. This is the *principle of minimal electromagnetic coupling* and the above equations are referred to as the minimal substitutions.

In the nonrelativistic realm, we have

$$L_{\text{p}} = \frac{m\mathbf{v} \cdot \mathbf{v}}{2}; \quad \boldsymbol{\pi} = m\mathbf{v}, \tag{3.119}$$

which results in the Hamiltonian

$$\hat{H} = \frac{(\mathbf{p} - q\mathbf{A}) \cdot (\mathbf{p} - q\mathbf{A})}{2m} + q\phi(\mathbf{r}, t). \tag{3.120}$$

Let us now derive the force acting on a point charge in an electromagnetic field. From Newtonian mechanics, we know that

$$\mathbf{F} = m\mathbf{a}, \tag{3.121}$$

which tells us that the force **F** is the total time derivative of the *mechanical* momentum

$$\mathbf{F} = \frac{d\boldsymbol{\pi}}{dt} = \boldsymbol{\nabla}L - q\frac{d\mathbf{A}}{dt} = \boldsymbol{\nabla}L_{\text{int}} - q\frac{d\mathbf{A}}{dt}. \tag{3.122}$$

The right-hand side of the formula is obtained using the Euler–Lagrange equation in Eq. (2.13) and the fact that the particle Lagrangian is independent of position. Taking the total time derivative requires a bit of care since the position **r** appearing in Eq. (3.115) refers to the time-dependent trajectory of the particle. With this in mind we obtain, admittedly with some manipulation,

$$\begin{aligned}\mathbf{F} &= q\boldsymbol{\nabla}(\mathbf{v}(t)\cdot\mathbf{A}) - q\boldsymbol{\nabla}\phi - q\frac{\partial\mathbf{A}}{\partial t} - \left(\frac{d\mathbf{r}}{dt}\cdot\boldsymbol{\nabla}\right)\mathbf{A} \\ &= q\left[-\boldsymbol{\nabla}\phi - \frac{\partial\mathbf{A}}{\partial t} + \boldsymbol{\nabla}\times\mathbf{A}\right] \\ &= q\mathbf{E} + q(\mathbf{v}\times\mathbf{B}),\end{aligned} \tag{3.123}$$

where, in the final step, we made use of Eqs. (3.49) and (3.50). This is known as the *Lorentz force* and its expression shows that the electric force is parallel or antiparallel to the electric field, depending on the sign of the charge of the particle. The magnetic force, on the other hand, is not only perpendicular to the magnetic field, it is also perpendicular to the velocity (or current $\mathbf{j} = q\mathbf{v}$) of the particle. The latter implies that magnetic forces do not achieve any work. Work is generally given as a line integral along the trajectory $\mathbf{r}(t)$ of the particle

$$W = \int_{\mathbf{r}(t_a)}^{\mathbf{r}(t_b)} \mathbf{F}\cdot d\mathbf{r} = \int_{t_a}^{t_b} (\mathbf{F}\cdot\mathbf{v})dt; \quad \mathbf{v}(t) = \frac{d\mathbf{r}}{dt}. \tag{3.124}$$

Inserting the Lorentz force into the general expression, we only get a contribution from the electric field

$$W = \int_{t_a}^{t_b} q(\mathbf{E}\cdot\mathbf{v})dt. \tag{3.125}$$

Returning to Eq. (3.117), we see that when particles interact with electromagnetic fields, it is only the coupling to the scalar potential that enters the Hamiltonian. The replacement of canonical momentum by mechanical momentum is the *removal* of the magnetic contribution to the former. This is indeed an astonishing result that seems to be in contradiction with our everyday experience with magnetic forces. However, closer scrutiny of such situations reveals that electric forces are actually doing the work. Take for instance the application of a magnetic field perpendicular to a current-carrying wire. The current consists of mobile electrons that now experiences a force in a direction perpendicular to both the wire and the applied magnetic field. The resulting charge displacement sets up an electric field such that the stationary positively charged cores (that is, the atomic nuclei and the core electrons) and thus the wire, follow the electrons as illustrated in Figure 3.5.

The intimate connection between electric and magnetic fields is seen in Maxwell's equations and had actually motivated Einstein to develop his theory of special relativity. It also comes into play if we look at the instantaneous power, that is, the rate at which mechanical work is performed on particles by the electromagnetic field. Starting from Eq. (3.125), but generalizing

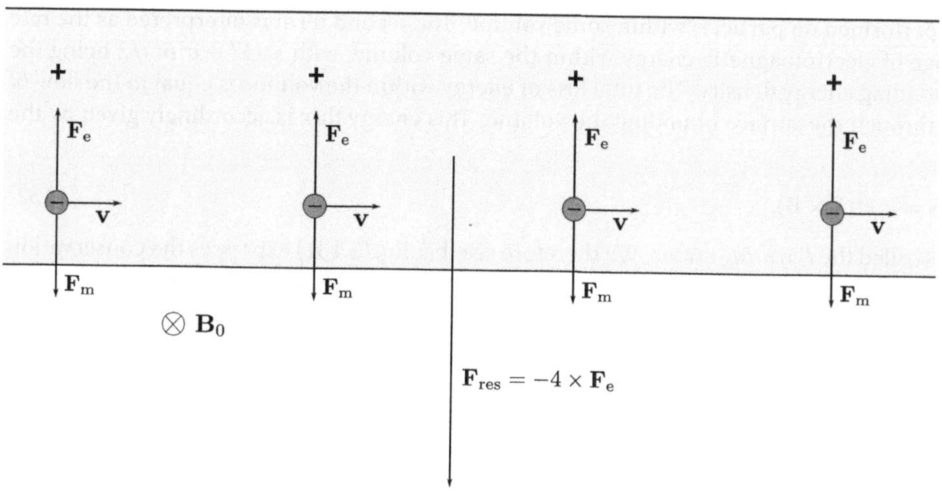

Figure 3.5 Conducting electrons in a wire placed in a uniform, static, magnetic field $\mathbf{B}_0$ with inward direction. Electrons travel at velocity **v**, giving rise to a magnetic Lorentz force $\mathbf{F}_m$ that in turn causes a displacement of the electronic charge density. The net charge inhomogeneity gives rise to an electric force $\mathbf{F}_e = -\mathbf{F}_m$ on the electrons and a resulting counterforce $\mathbf{F}_{res}$ on the wire.

from a single moving charge to a current density, we get

$$\frac{dW}{dt} = \int (\mathbf{E} \cdot \mathbf{j}) d^3\mathbf{r}. \tag{3.126}$$

We next use Maxwell's equation in Eq. (3.6) to replace the source term **j** by the fields themselves

$$\frac{dW}{dt} = \varepsilon_0 \int [c^2\mathbf{E} \cdot (\nabla \times \mathbf{B}) - \mathbf{E} \cdot \partial_t \mathbf{E}] d^3\mathbf{r} - \varepsilon_0 \int \left[c^2\mathbf{E} \cdot (\nabla \times \mathbf{B}) - \frac{1}{2}\partial_t E^2\right] d^3\mathbf{r}. \tag{3.127}$$

Note how, in the final expression, we have shifted the time derivative of the second term to the left. We can treat the spatial derivative of the first term in an analogous manner using the relation

$$\nabla \cdot (\mathbf{E} \times \mathbf{B}) = \mathbf{B} \cdot (\nabla \times \mathbf{E}) - \mathbf{E} \cdot (\nabla \times \mathbf{B}), \tag{3.128}$$

to obtain

$$\frac{dW}{dt} = \varepsilon_0 \int \left[c^2\mathbf{B} \cdot (\nabla \times \mathbf{E}) - c^2\nabla \cdot (\mathbf{E} \times \mathbf{B}) - \frac{1}{2}\partial_t E^2\right] d^3\mathbf{r}. \tag{3.129}$$

Using Faraday's law, Eq. (3.3), and rearranging terms, we obtain the following expression for the instantaneous power

$$\frac{dW}{dt} = -\frac{1}{2}\varepsilon_0 \int \partial_t (E^2 + c^2B^2) d^3\mathbf{r} - \varepsilon_0 c^2 \int \nabla \cdot (\mathbf{E} \times \mathbf{B}) d^3\mathbf{r}. \tag{3.130}$$

Finally, we apply the divergence theorem, Eq. (3.12), on the last term and rearrange terms to obtain *Poynting's theorem*

$$\frac{dW}{dt} + \frac{d}{dt}\int_V \frac{1}{2}\varepsilon_0 (E^2 + c^2B^2) d^3\mathbf{r} = -\int_S \varepsilon_0 c^2 (\mathbf{E} \times \mathbf{B}) \cdot \mathbf{n}\, dS. \tag{3.131}$$

Note that we have pulled a time derivative outside an integration sign and hence assume that the integration volume does not change over time. We can interpret this expression as follows: from Eq. (3.126), we see that the first term on the left side represents the rate of change of mechanical energy performed on particles within some volume. The second term is interpreted as the rate of change of electromagnetic energy within the same volume, with $\varepsilon_0(E^2 + c^2B^2)/2$ being the corresponding energy density. The total loss of energy within the volume is equal to the flow of energy through the surface bounding the volume. This energy flux is accordingly given by the vector

$$\mathbf{S} = \varepsilon_0 c^2(\mathbf{E} \times \mathbf{B}), \tag{3.132}$$

which is called the *Poynting vector*. We therefore see that Eq. (3.131) expresses the conservation of energy of a system of particles and fields and has a form analogous to the continuity equation in Eq. (3.11).

For an electromagnetic wave we find, using Eqs. (3.75) and (3.76), that the Poynting vector is

$$\mathbf{S} = \frac{1}{\omega}\varepsilon_0 c^2 E^2 \mathbf{k} = \varepsilon_0 c E^2 \mathbf{e}_k. \tag{3.133}$$

We see that the energy flows in the direction of the wave. The *intensity* is the energy flux averaged over a period of the wave. For a monochromatic electromagnetic wave, as given in Eq. (3.74), we get an intensity

$$I = \langle S \rangle_T = \varepsilon_0 c \frac{1}{T}\int_{-\frac{T}{2}}^{+\frac{T}{2}} E^2 dt = \varepsilon_0 c E_0^2 \frac{1}{T}\int_{-\frac{T}{2}}^{+\frac{T}{2}} \cos^2(\mathbf{k}\cdot\mathbf{r} - \omega t + \delta) dt = \frac{1}{2}\varepsilon_0 c E_0^2. \tag{3.134}$$

The final step is obtained by noting that the time average of $\cos^2\theta$ must be the same as $\sin^2\theta$. From the relation $\cos^2\theta + \sin^2\theta = 1$, we therefore directly obtain the result $\langle\cos^2\theta\rangle_T = \langle\sin^2\theta\rangle_T = 1/2$. It should also be noted that we have arrived at this result by employing the real, physical form of the fields. In terms of complex fields, the same result is obtained by writing

$$\langle \mathbf{S} \rangle_T = \frac{1}{2}\varepsilon_0 c^2 \,\mathrm{Re}(\mathbf{E} \times \mathbf{B}^*). \tag{3.135}$$

3.2.2 The Quantum-Mechanical Hamiltonian

In this book, we are concerned with quantum-chemical calculations of molecular properties, relevant to the response of molecules to electromagnetic fields. We will limit our attention to weak fields such that the molecular response can be calculated by perturbation theory. In this section, we therefore consider the choice of the zeroth-order quantum-mechanical Hamiltonian and how to extend it to include the interaction of the molecule with the fields. For the moment, we focus on the electronic Hamiltonian and postpone a discussion of the full molecular Hamiltonian to Chapter 6.

Within the Born–Oppenheimer approximation, the electronic Hamiltonian, relativistic or not, can be represented by the *generic* form

$$\hat{H} = \sum_i \hat{h}(i) + \frac{1}{2}\sum_{i \neq j} \hat{g}(i,j) + V_{\mathrm{NN}}, \tag{3.136}$$

where V_{NN} is the classical repulsion of fixed nuclei

$$V_{\mathrm{NN}} = \frac{1}{2}\sum_{A \neq B} \frac{Z_A Z_B e^2}{4\pi\varepsilon_0 R_{AB}}. \tag{3.137}$$

Furthermore, $\hat{h}(i)$ is the one-electron Hamiltonian and $\hat{g}(i,j)$ represents the interaction between electrons i and j—with a factor of one half to correct for the double counting of the interactions.

As pointed out in the Chapter 1, this generic form is generally sufficient to derive the working equations for quantum-chemical methods.

The zeroth-order one-electron Hamiltonian can be further divided into two terms

$$\hat{h} = \hat{T} + \hat{V}, \tag{3.138}$$

where $\hat{T}$ is the free-particle Hamiltonian and $\hat{V}$ describes the interaction of the electron with the electrostatic potential of the clamped nuclei. Other interactions, such as those arising from nuclear spin, are excluded since they are much weaker and may break symmetries. This implies that the fields to be added to the zeroth-order Hamiltonian in response calculations may be both external and internal.

The most accurate one-electron Hamiltonian that will be considered in this book is the four-component Dirac Hamiltonian

$$\hat{h}_{\mathrm{D}} = \beta m_{\mathrm{e}} c^2 + c(\boldsymbol{\alpha} \cdot \hat{\mathbf{p}}) + \hat{V}. \tag{3.139}$$

It is a 4×4 matrix operator due to the presence of the Dirac matrices

$$\boldsymbol{\alpha} = \begin{bmatrix} 0_2 & \boldsymbol{\sigma} \\ \boldsymbol{\sigma} & 0_2 \end{bmatrix}; \quad \beta = \begin{bmatrix} I_2 & 0_2 \\ 0_2 & -I_2 \end{bmatrix}, \tag{3.140}$$

where appear the Pauli spin matrices, Eq. (2.49), and the 2×2 identity and zero matrices denoted by I_2 and 0_2, respectively. After subtraction of the electron rest mass $m_{\mathrm{e}}c^2$ to align the relativistic and nonrelativistic energy scales, the corresponding time-independent wave equation reads

$$\begin{bmatrix} \hat{V} & c(\boldsymbol{\sigma} \cdot \hat{\mathbf{p}}) \\ c(\boldsymbol{\sigma} \cdot \hat{\mathbf{p}}) & \hat{V} - 2m_{\mathrm{e}}c^2 \end{bmatrix} \begin{bmatrix} \psi^{\mathrm{L}} \\ \psi^{\mathrm{S}} \end{bmatrix} = \begin{bmatrix} \psi^{\mathrm{L}} \\ \psi^{\mathrm{S}} \end{bmatrix} E, \tag{3.141}$$

where the large ψ^{L} and small ψ^{S} components are two-component functions

$$\psi^{X} = \begin{bmatrix} \psi^{X\alpha} \\ \psi^{X\beta} \end{bmatrix}; \quad X = \mathrm{L}, \mathrm{S}. \tag{3.142}$$

The most important observation to be made when comparing the well-known nonrelativistic and the above relavistic form of the Hamiltonian for the single-electron system is that in the latter case, the dependence on the momentum operator is linear (as opposed to quadratic in the nonrelativistic case). This is a necessary requirement for a relativistic Hamiltonian since the time differentiation in the Schrödinger equation is linear and a Lorentz-invariant theory requires differentiation with respect to time and spatial variables to be done on an equal footing. A troublesome feature of the Dirac equation is the presence of continuum solutions of negative energy, separated by an energy gap of about $2m_{\mathrm{e}}c^2$ from the positive-energy solutions. In classical physics, these solutions can be discarded as unphysical since the theory does not allow discontinuous energy changes. In quantum theory, however, there is always a finite probabililty that an electron would descend down into the negative-energy continuum, thus liberating an infinite amount of energy on its way down the spectrum and making matter unstable (the hydrogen atom would disintegrate in about a nanosecond). Dirac solved this problem by postulating that all negative-energy solutions are occupied, but not observable, thus blocking further electron occupation by the Pauli principle. A remarkable consequence of this daring proposal was the discovery of positrons, the antiparticles of electrons. In applications in chemistry, however, the negative-energy solutions are not of great interest and, as a consequence, various two-component relativistic Hamiltonians have been developed. Examples of such Hamiltonians are the exact two-component Hamiltonian (X2C) and approximations to it, for instance the Douglas–Kroll–Hess (DKH) and the zeroth-order regular approximation

(ZORA) Hamiltonians. They reproduce in an exact or approximate manner the positive-energy spectrum of the original Dirac equation, but only for the potential defining the decoupling of the positive- and negative-energy solutions.

The most celebrated feature of the Dirac equation is that it, in a natural manner, introduces the electron spin degree of freedom. What is less recognized, however, is the possibility to argue for the presence of electron spin also in the nonrelativistic framework. The electron spin is, from this point of view, also present in the one-electron Hamiltonian, but it is "hidden" in the absence of external magnetic fields. This can be seen from a consideration of the nonrelativistic limit of the Dirac equation, Eq. (3.141), which we write in a more compact manner as

$$\hat{h}_{\mathrm{D}}\psi = \psi E. \tag{3.143}$$

The nonrelativistic limit is normally obtained by setting the speed of light c to infinity, but this recipe would lead to an "explosion" of the Dirac Hamiltonian $\hat{h}_{\mathrm{D}}$ since it contains terms linear and quadratic in c. The solution is to carry out a nonunitary transformation

$$\psi = W\tilde{\psi}, \tag{3.144}$$

which changes the metric of the Dirac equation in accordance with

$$W^{\dagger}\hat{h}_{\mathrm{D}}W\tilde{\psi} = W^{\dagger}W\tilde{\psi}E. \tag{3.145}$$

Specificially, we select the transformation

$$\begin{bmatrix} \psi^{\mathrm{L}} \\ \psi^{\mathrm{S}} \end{bmatrix} = \begin{bmatrix} I_2 & 0_2 \\ 0_2 & I_2 c^{-1} \end{bmatrix} \begin{bmatrix} \psi^{\mathrm{L}} \\ \tilde{\psi}^{\mathrm{S}} \end{bmatrix}, \tag{3.146}$$

which gives a transformed Dirac equation that reads

$$\begin{bmatrix} \hat{V} & (\boldsymbol{\sigma}\cdot\hat{\mathbf{p}}) \\ (\boldsymbol{\sigma}\cdot\hat{\mathbf{p}}) & \hat{V}/c^2 - 2m_{\mathrm{e}} \end{bmatrix} \begin{bmatrix} \psi^{\mathrm{L}} \\ \tilde{\psi}^{\mathrm{S}} \end{bmatrix} = \begin{bmatrix} I_2 & 0_2 \\ 0_2 & I_2 c^{-2} \end{bmatrix} \begin{bmatrix} \psi^{\mathrm{L}} \\ \tilde{\psi}^{\mathrm{S}} \end{bmatrix} E. \tag{3.147}$$

This equation certainly looks more promising, since the speed of light now only appears in denominators, but caution still has to be exercised when letting c go to infinity in order to reach the nonrelativistic limit, because when the nuclei are represented by point charges, the resulting electrostatic potential has singularities at the nuclear positions and thus the condition $V \ll 2m_{\mathrm{e}}c^2$ does not hold at all points in space. This difficulty can be circumvented by treating the nuclei as extended charges, which in any case is a physically more correct model. Another complication is that the energies of the negative-energy solutions are on the order of $-2m_{\mathrm{e}}c^2$ and so $|E| \ll 2m_{\mathrm{e}}c^2$ does not hold for those. Therefore, the nonrelativistic limit will be considered only for the positive-energy solutions. With these premises, we arrive at an equation known as the Lévy-Leblond equation

$$\begin{bmatrix} \hat{V} & (\boldsymbol{\sigma}\cdot\hat{\mathbf{p}}) \\ (\boldsymbol{\sigma}\cdot\hat{\mathbf{p}}) & -2m \end{bmatrix} \begin{bmatrix} \psi^{\mathrm{L}} \\ \tilde{\psi}^{\mathrm{S}} \end{bmatrix} = \begin{bmatrix} I_2 & 0_2 \\ 0_2 & 0_2 \end{bmatrix} \begin{bmatrix} \psi^{\mathrm{L}} \\ \tilde{\psi}^{\mathrm{S}} \end{bmatrix} E. \tag{3.148}$$

This is a four-component *nonrelativistic* equation, which has a positive-energy spectrum that matches that of the standard one-component nonrelativistic Schrödinger equation. It is noteworthy, however, that in the calculation of magnetic properties, the four-component variant has formal and computational advantages. The Schrödinger equation can be recovered from the Lévy-Leblond equation by first writing the latter as two coupled equations and then eliminating the small component $\tilde{\psi}^{\mathrm{S}}$. We thereby obtain an equation involving the large component only

$$\left[\frac{(\boldsymbol{\sigma}\cdot\hat{\mathbf{p}})(\boldsymbol{\sigma}\cdot\hat{\mathbf{p}})}{2m} + \hat{V}\right]\psi^{\mathrm{L}} = \psi^{\mathrm{L}}E, \tag{3.149}$$

which we recognize as the standard Schrödinger equation by employing the Dirac identity [see Eq. (F.2)]

$$(\boldsymbol{\sigma}\cdot\hat{\mathbf{p}})(\boldsymbol{\sigma}\cdot\hat{\mathbf{p}}) = \hat{p}^2. \tag{3.150}$$

Note, however, that in Eq. (3.149) the wave function ψ^{L} is a two-component quantity and, as we shall shortly see, the choice between the left- and right-hand side of Eq. (3.150) does make a difference in the presence of an external magnetic field.

Just as in the classical case, electromagnetic fields are introduced into the quantum-mechanical Hamiltonian through the principle of minimal electromagnetic coupling, Eq. (3.118), and via the potentials rather than the fields themselves. Starting from the free-particle Dirac Hamiltonian ($\hat{V} = 0$), we obtain

$$\hat{h}_{\mathrm{D}} = \beta m_{\mathrm{e}}c^2 + c(\boldsymbol{\alpha}\cdot\hat{\mathbf{p}}) - e\phi + ec(\boldsymbol{\alpha}\cdot\mathbf{A}), \tag{3.151}$$

whereas the corresponding extension of the Schrödinger equation gives

$$\hat{h} = \frac{\hat{\pi}^2}{2m_{\mathrm{e}}} - e\phi; \quad \hat{\boldsymbol{\pi}} = \hat{\mathbf{p}} + e\mathbf{A}. \tag{3.152}$$

Further manipulation gives

$$\hat{h} = \frac{\hat{p}^2}{2m_{\mathrm{e}}} + \frac{e}{2m_{\mathrm{e}}}(\hat{\mathbf{p}}\cdot\mathbf{A} + \mathbf{A}\cdot\hat{\mathbf{p}}) + \frac{e^2A^2}{2m_{\mathrm{e}}} - e\phi. \tag{3.153}$$

In the Coulomb gauge ($\boldsymbol{\nabla}\cdot\mathbf{A} = 0$), we can simplify this equation even further as

$$\hat{h} = \frac{\hat{p}^2}{2m_{\mathrm{e}}} + \frac{e}{m_{\mathrm{e}}}\mathbf{A}\cdot\hat{\mathbf{p}} + \frac{e^2A^2}{2m_{\mathrm{e}}} - e\phi \quad \text{(Coulomb gauge)}. \tag{3.154}$$

Spin interactions are absent from Eq. (3.153). However, if we consider spin as "hidden" in the nonrelativistic free-particle Hamiltonian and use the left-hand side of Eq. (3.150), we get

$$\hat{h} = \frac{(\boldsymbol{\sigma}\cdot\hat{\boldsymbol{\pi}})^2}{2m_{\mathrm{e}}} - e\phi. \tag{3.155}$$

If we again use the Dirac identity [Eq. (F.2)], we get two terms

$$(\boldsymbol{\sigma}\cdot\hat{\boldsymbol{\pi}})(\boldsymbol{\sigma}\cdot\hat{\boldsymbol{\pi}}) = \hat{\pi}^2 + i\boldsymbol{\sigma}\cdot(\hat{\boldsymbol{\pi}}\times\hat{\boldsymbol{\pi}}). \tag{3.156}$$

The first one is already obtained using the right-hand side of Eq. (3.150). The second term appears to be zero since it is the vector product of a vector with itself. However, $\hat{\boldsymbol{\pi}}$ is a vector *operator* and the zero result is only valid when all components of the operator commute, which is not the case here. In fact, proper manipulation gives

$$i\boldsymbol{\sigma}\cdot(\hat{\boldsymbol{\pi}}\times\hat{\boldsymbol{\pi}}) = i\boldsymbol{\sigma}\cdot(\hat{\mathbf{p}}\times e\mathbf{A} + e\mathbf{A}\times\hat{\mathbf{p}}) = e\hbar\boldsymbol{\sigma}\cdot(\boldsymbol{\nabla}\times\mathbf{A}), \tag{3.157}$$

which can be recognized as the interaction between spin and an external magnetic field. The full Hamiltonian now reads

$$\hat{h} = \frac{\hat{p}^2}{2m_{\mathrm{e}}} + \frac{e}{2m_{\mathrm{e}}}(\hat{\mathbf{p}}\cdot\mathbf{A} + \mathbf{A}\cdot\hat{\mathbf{p}}) + \frac{e^2A^2}{2m_{\mathrm{e}}} + \frac{e\hbar}{2m_{\mathrm{e}}}\boldsymbol{\sigma}\cdot\mathbf{B} - e\phi. \tag{3.158}$$

The development of the full Hamiltonian, Eq. (3.158), from the nonrelativistic free-particle Hamiltonian is a fine demonstration of the economy of information in the laws of Nature. The specification of the free-particle Hamiltonian only requires input about the particle mass, not about its possible charge and spin, since this is not needed to describe the dynamics of a free particle. The particle charge only comes into play when there are electromagnetic fields to interact with and the spin only when there are magnetic fields present. Another point to note is that both the relativistic and the nonrelativistic Hamiltonians can accommodate spin, which seems to indicate that spin is not a relativistic phenomenon. However, we should keep in mind that we have used a manifestly relativistic coupling of particles and fields to derive Eq. (3.158) and, following the discussion in Section 3.1.6, magnetic fields cannot be generated in a purely nonrelativistic world. This leads us to conclude that spin is indeed a relativistic effect.

Let us now consider the two-electron interaction. One way to construct the operator is to consider the coupling of one electron to the fields generated by a second electron. This means that the generic form of the two-electron operator would be

$$\hat{g}(1,2) = q_1\phi_2(\mathbf{r}_1) - q_1\mathbf{v}_1 \cdot \mathbf{A}_2(\mathbf{r}_1). \tag{3.159}$$

If we choose Coulomb gauge, then the scalar and vector potentials are given by Eqs. (3.62) and (3.64), respectively. In the nonrelativistic case, the vector potential vanishes as electrodynamics reduces to electrostatics and the two-electron interaction is given by the instantaneous Coulomb interaction

$$\hat{g}^{\mathrm{C}}(1,2) = \frac{e^2}{4\pi\varepsilon_0 r_{12}}. \tag{3.160}$$

In the relativistic case, things get more complicated since we now have to take into account both the magnetic interaction and the effect of retardation. In fact, no simple expression is available. However, if we ignore retardation and use the relativistic velocity operator $c\boldsymbol{\alpha}$ (see Section 2.3), the Coulomb operator is supplemented by the Breit operator

$$\hat{g}^{\mathrm{B}}(1,2) = -\frac{e^2}{8\pi\varepsilon_0 c^2}\left[\frac{c\boldsymbol{\alpha}_1 \cdot c\boldsymbol{\alpha}_2}{r_{12}} + \frac{(c\boldsymbol{\alpha}_1 \cdot \mathbf{r}_{12})(c\boldsymbol{\alpha}_2 \cdot \mathbf{r}_{12})}{r_{12}^3}\right], \tag{3.161}$$

where two terms occur because we are subtracting the irrotational current from the total one to obtain the solenoidal part [see Eq. (3.64)].

Let us now consider the charge and current density of a single electron. The expressions we derive can be generalized to many-electron systems, but the final form will then depend on the wave-function model chosen. The expectation value of the interaction Hamiltonian, that is, the part of the Hamiltonian which describes the coupling of particles and fields, is

$$\langle \hat{h}_{\mathrm{int}} \rangle = \int [\rho(\mathbf{r},t)\phi(\mathbf{r},t) - \mathbf{j}(\mathbf{r},t) \cdot \mathbf{A}(\mathbf{r},t)]d^3\mathbf{r}. \tag{3.162}$$

This suggests that we can extract quantum-mechanical expressions for the charge and current density through functional derivatives

$$\rho(\mathbf{r},t) = \frac{\delta\langle \hat{h}_{\mathrm{int}} \rangle}{\delta\phi(\mathbf{r},t)}; \quad \mathbf{j}(\mathbf{r},t) = -\frac{\delta\langle \hat{h}_{\mathrm{int}} \rangle}{\mathbf{A}(\mathbf{r},t)}. \tag{3.163}$$

In the relativistic case the expectation value of the interaction Hamiltonian is

$$\langle \hat{h}^{\mathrm{R}}_{\mathrm{int}} \rangle = \int \psi^\dagger(\mathbf{r},t)[-e\phi + ec(\boldsymbol{\alpha} \cdot \mathbf{A})]\psi(\mathbf{r},t)d^3\mathbf{r}, \tag{3.164}$$

and we straightforwardly obtain

$$\rho^{\mathrm{R}}(\mathbf{r},t) = -e\psi^\dagger\psi; \quad \mathbf{j}^{\mathrm{R}}(\mathbf{r},t) = -e\psi^\dagger c\boldsymbol{\alpha}\psi. \tag{3.165}$$

These expressions make physical sense: the charge density is the charge of the electron times its probability distribution, whereas the current density is the charge of the electron times its velocity distribution.

In the nonrelativistic case, the expectation value of the interaction Hamiltonian is

$$\langle \hat{h}^{\text{NR}}_{\text{int}} \rangle = \int \psi^{\dagger}(\mathbf{r},t)\left[-e\phi + \frac{e}{2m_{\text{e}}}(\hat{\mathbf{p}}\cdot\mathbf{A} + \mathbf{A}\cdot\hat{\mathbf{p}}) + \frac{e^2A^2}{2m_{\text{e}}} + \frac{e\hbar}{2m_{\text{e}}}\boldsymbol{\sigma}\cdot\mathbf{B}\right]\psi(\mathbf{r},t)d^3\mathbf{r}. \quad (3.166)$$

After a bit more work, involving partial integration for the current density, we obtain

$$\rho^{\text{NR}}(\mathbf{r},t) = -e\psi^{\dagger}\psi, \quad (3.167)$$

$$\mathbf{j}^{\text{NR}}(\mathbf{r},t) = -\frac{e}{2m_{\text{e}}}[\psi^{\dagger}\hat{\mathbf{p}}\psi - \psi^{T}\hat{\mathbf{p}}\psi^{*}] - \frac{e^2}{m_{\text{e}}}\psi^{\dagger}\mathbf{A}\psi - \frac{e}{2m_{\text{e}}}\boldsymbol{\nabla}\times[\psi^{\dagger}\boldsymbol{\sigma}\psi]. \quad (3.168)$$

While the nonrelativistic and relativistic charge densities are identical, the expressions for the current density are quite different. The nonrelativistic expression for current density separates the orbital and spin currents. The first contribution is constructed from the nonrelativistic velocity operator, which is mechanical momentum, Eq. (3.116), divided by electron mass, and thus refers to a velocity distribution. The orbital contributions are called the paramagnetic and the diamagnetic current densities, respectively. The spin current contribution is in the form of the curl of spin magnetization and has no classical counterpart, although we shall meet a similar expression later on in this chapter (see the second term of Eq. (3.297)).

3.3 Electric and Magnetic Multipoles

3.3.1 Multipolar Gauge

In Section 3.2, we saw that electromagnetic fields are introduced into the Hamiltonian through the corresponding *potentials*. In this section, we shall see that it is possible to express the interaction Hamiltonian directly in terms of the electromagnetic fields, although without giving up the gauge freedom. We will consider the so-called *multipolar gauge* that leads directly to the definition of electric and magnetic multipoles, a very useful concept for the description of the interaction of light with matter. In fact, most molecular properties can be interpreted as the modification of electric and magnetic multipoles by external fields.

We start by considering a Taylor expansion of the scalar and vector potentials about a reference point, which, for simplicity, is set equal to the origin. For the scalar potential, we obtain

$$\begin{aligned}\phi(\mathbf{r},t) &= \phi^{[0]} + r_i\phi^{[1]}_i + \frac{1}{2}r_ir_j\phi^{[2]}_{ij} + \cdots \\ &= \sum_{n=0}^{\infty}\frac{1}{n!}r_{j_1}r_{j_2}\ldots r_{j_n}\phi^{[n]}_{j_1j_2\cdots j_n}; \quad \phi^{[n]}_{j_1j_2\cdots j_n} = \left.\frac{\partial^n\phi}{\partial r_{j_1}\partial r_{j_2}\cdots\partial r_{j_n}}\right|_{\mathbf{0}}.\end{aligned} \quad (3.169)$$

Likewise, for the three components of the vector potential, we obtain

$$A_i(\mathbf{r},t) = A^{[0]}_i + r_jA^{[1]}_{i;j} + \cdots = \sum_{n=0}^{\infty}\frac{1}{n!}r_{j_1}r_{j_2}\cdots r_{j_n}A^{[n]}_{i;j_1j_2\cdots j_n}. \quad (3.170)$$

We remind the reader that we make use of the Einstein summation convention.

The second term of the expansion of the scalar potential contains the gradient of the scalar potential in the chosen expansion point. This gradient also appears in the expression for the electric field in terms of the potentials, Eq. (3.50). Using this latter expression, we can rework our Taylor expansion to obtain

$$\begin{aligned}\phi(\mathbf{r},t) = {} & \phi^{[0]} - r_iE^{[0]}_i - \tfrac{1}{2}r_ir_jE^{[1]}_{i;j} - \cdots \\ & -r_i\partial_tA^{[0]}_i - \tfrac{1}{2}r_ir_j\partial_tA^{[1]}_{i;j} - \cdots .\end{aligned} \quad (3.171)$$

We see that we have eliminated all reference to the scalar potential, except for its value at the expansion point, but at the cost of introducing time derivatives of the vector potential and its spatial derivatives at the expansion point. However, we can get rid of the terms involving the vector potential by a gauge transformation using the gauge function

$$\chi(t) = -\sum_{n=0}^{\infty} \frac{1}{(n+1)!} r_{j_1} r_{j_2} \cdots r_{j_n} \left(\mathbf{r} \cdot \mathbf{A}^{[n]}_{j_1 j_2 \cdots j_n} \right). \tag{3.172}$$

A more compact way of writing the above expression is

$$\chi(t) = -\int_0^1 \mathbf{r} \cdot \mathbf{A}(u\mathbf{r}, t)\, du, \tag{3.173}$$

which can be seen by Taylor-expanding $\mathbf{A}(u\mathbf{r}, t)$ about $\mathbf{r} = \mathbf{0}$. For the corresponding gauge transformation of the vector potential, we need the gradient of the gauge function, which can be expressed as

$$-\nabla_i \chi = A^{[0]}_i + \frac{1}{2} r_j A^{[1]}_{i;j} + \frac{1}{2} r_j A^{[1]}_{j;i} + \frac{1}{6} r_j r_k A^{[2]}_{i;jk} + \frac{1}{3} r_j r_k A^{[2]}_{j;ki} + \cdots, \tag{3.174}$$

or, more compactly[8]

$$\nabla_i \chi = \int_0^1 \left[A_i + u r_j \nabla_i A_j \right] (u\mathbf{r}, t)\, du. \tag{3.175}$$

We now carry out the gauge transformation of the vector potential, which gives

$$A'_i = \frac{1}{2} r_j \left(A^{[1]}_{i;j} - A^{[1]}_{j;i} \right) + \frac{1}{3} r_j r_k \left(A^{[2]}_{i;jk} - A^{[2]}_{j;ik} \right) + \cdots. \tag{3.176}$$

Consider now a specific component of the transformed vector potential, for instance

$$\begin{aligned} A'_x &= \frac{1}{2} \left\{ x \left(A^{[1]}_{x;x} - A^{[1]}_{x;x} \right) + y \left(A^{[1]}_{x;y} - A^{[1]}_{y;x} \right) + z \left(A^{[1]}_{x;z} - A^{[1]}_{z;x} \right) \right\} + \cdots \\ &= -\frac{1}{2} \left\{ y B^{[0]}_z - z B^{[0]}_y \right\} + \cdots \\ &= -\frac{1}{2} \left(\mathbf{r} \times \mathbf{B}^{[0]} \right)_x + \cdots. \end{aligned}$$

The vector potential can therefore be entirely rewritten in terms of the value of the magnetic field and its derivatives at the expansion point. After gauge transformation, the potentials are thereby given as

$$\begin{aligned} \phi(\mathbf{r}, t) &= \phi^{[0]} - \sum_{n=0}^{\infty} \tfrac{1}{(n+1)!} r_{j_1} r_{j_2} \cdots r_{j_n} \left(\mathbf{r} \cdot \mathbf{E}^{[n]}_{j_1 j_2 \cdots j_n} \right), \\ \mathbf{A}(\mathbf{r}, t) &= -\sum_{n=0}^{\infty} \tfrac{n+1}{(n+2)!} r_{j_1} r_{j_2} \cdots r_{j_n} \left(\mathbf{r} \times \mathbf{B}^{[n]}_{j_1 j_2 \cdots j_n} \right), \end{aligned} \tag{3.177}$$

or, in more compact form

$$\phi(\mathbf{r}, t) = \phi^{[0]} - \int_0^1 \mathbf{r} \cdot \mathbf{E}(u\mathbf{r}, t) du; \quad \mathbf{A}(\mathbf{r}, t) = \int_0^1 u\mathbf{r} \times \mathbf{B}(u\mathbf{r}, t) du. \tag{3.178}$$

What happened to gauge freedom? The answer is that it is still there and now corresponds to our choice of expansion point.

---

8 For aficionados: The contributions to Eq. (3.174) from the second term are obtained from a Taylor expansion of $\nabla_i A_j$. The gradient operator ∇_i refers to the coordinate $u\mathbf{r}$ rather than $\mathbf{r}$ so that we get an additional factor u in the second term.

To get some insight into the multipolar gauge, let us consider two simple examples:

1) In the case of a uniform electric field (no spatial derivatives), we get

$$\phi(\mathbf{r},t) = -\mathbf{r}\cdot\mathbf{E}^{[0]};\quad \mathbf{A}(\mathbf{r},t) = \mathbf{0}. \tag{3.179}$$

2) In the case of a uniform magnetic field, we get

$$\phi(\mathbf{r},t) = 0;\quad \mathbf{A}(\mathbf{r},t) = \frac{1}{2}\left(\mathbf{B}^{[0]}\times\mathbf{r}\right). \tag{3.180}$$

Note in the latter case, however, that, if the magnetic field is time dependent, we get a nonuniform electric field as well according to

$$\mathbf{E} = -\boldsymbol{\nabla}\phi - \partial_t\mathbf{A} = \frac{1}{2}\left(\mathbf{r}\times\partial_t\mathbf{B}^{[0]}\right). \tag{3.181}$$

This once again shows the intimate connection between electric and magnetic fields. Note that the value of the electric field at the expansion point is zero, and so is any derivative higher than first order. However, the nondiagonal components of the electric field gradient $E^{[1]}_{i;j}$ are nonzero at the expansion point. Using the Levi-Civita symbol, we can write them as

$$E^{[1]}_{i;j} = -\frac{1}{2}\epsilon_{ijk}\partial_t B^{[0]}_k. \tag{3.182}$$

Should there not be contributions to the scalar potential of Eq. (3.177)? In fact, the answer is no since the first-order term in question can be expressed as

$$-\frac{1}{2}r_j r_i E^{[1]}_{i;j} = \frac{1}{4}\epsilon_{ijk}r_j r_i \partial_t B^{[0]}_k = \frac{1}{4}(\mathbf{r}\times\mathbf{r})\cdot\partial_t\mathbf{B}^{[0]}, \tag{3.183}$$

and is therefore strictly zero.

The two specific sets of potentials given earlier both satisfy the Coulomb gauge, but this does not hold in general since

$$\boldsymbol{\nabla}\cdot\mathbf{A} = -\int_0^1 u^2\mathbf{r}\cdot[\boldsymbol{\nabla}\times\mathbf{B}](u\mathbf{r},t)\,du. \tag{3.184}$$

Comparing with Eq. (3.6), we see that Coulomb gauge is only fulfilled if the electric field is static and there are no external currents. On the other hand, one can easily show that the multipolar gauge satisfies

$$\mathbf{r}\cdot\mathbf{A} = 0. \tag{3.185}$$

3.3.2 Multipole Expansions

A very interesting feature of the multipolar gauge is seen when inserting the scalar and vector potentials in Eq. (3.177) into the expression for the expectation value of the interaction Hamiltonian in Eq. (3.162). We then get

$$\langle \hat{H}_{\text{int}}\rangle = Q^{[0]}\phi^{[0]} - \sum_{n=1}^{\infty}\frac{1}{n!}\mathbf{Q}^{[n]}_{j_1\cdots j_{n-1}}\cdot\mathbf{E}^{[n-1]}_{j_1\cdots j_{n-1}} - \sum_{n=1}^{\infty}\frac{1}{n!}\mathbf{m}^{[n]}_{j_1\cdots j_{n-1}}\cdot\mathbf{B}^{[n-1]}_{j_1\cdots j_{n-1}}, \tag{3.186}$$

where the first term is the interaction of the total charge with the value of the scalar potential at the expansion point. The total charge corresponds to the electric monopole. The second term corresponds to interactions with more elaborate charge distributions, represented by electric multipoles

$$Q^{[n]}_{j_1\cdots j_n} = \int r_{j_1}r_{j_2}\cdots r_{j_n}\rho(\mathbf{r},t)d^3\mathbf{r}. \tag{3.187}$$

To lowest order, we have the electric dipole moment

$$\boldsymbol{\mu} = Q_i^{[1]}\mathbf{e}_i = \int \mathbf{r}\rho(\mathbf{r}, t)d^3\mathbf{r}, \tag{3.188}$$

which couples to the electric field at the expansion point. To the next order, we have the electric quadrupole moment

$$Q_{ij}^{[2]} = \int r_i r_j \rho(\mathbf{r}, t)d^3\mathbf{r}, \tag{3.189}$$

which interacts with the electric field gradient at the expansion point. Higher electric multipoles probe further inhomogeneities of the electric field.

The third and final term in Eq. (3.186) is the interaction of magnetic multipoles given by

$$m_{i;j_1\cdots j_{n-1}}^{[n]} = \frac{n}{n+1}\int r_{j_1} r_{j_2}\cdots r_{j_{n-1}}[\mathbf{r}\times\mathbf{j}(\mathbf{r}, t)]_i d^3\mathbf{r}, \tag{3.190}$$

which involves the magnetic field and its derivatives. There is no magnetic monopole due to the purely transversal nature of magnetic fields. To the lowest order, we therefore have the magnetic dipole moment

$$\mathbf{m}^{[1]} = \frac{1}{2}\int [\mathbf{r}\times\mathbf{j}(\mathbf{r}, t)]d^3\mathbf{r}, \tag{3.191}$$

which couples to the magnetic field at the expansion point, whereas higher magnetic multipoles probe the inhomogeneities of the magnetic field.

Electric multipoles can be constructed from point charges, hence their name. An electric dipole is created by two charges $+q$ and $-q$ separated by distance vector $\boldsymbol{\delta}$. A static *point electric dipole* is obtained by taking the limit $\delta \to 0$, keeping constant the quantity $\boldsymbol{\mu} = q\boldsymbol{\delta}$. To see how this works, let us construct the scalar potential of a point dipole. The potential of two point charges is

$$\phi(\mathbf{r}) = \frac{1}{4\pi\varepsilon_0}\left[\frac{q_A}{|\mathbf{r}-\mathbf{r}_A|} + \frac{q_B}{|\mathbf{r}-\mathbf{r}_B|}\right] = \frac{1}{4\pi\varepsilon_0}\left[\frac{Q}{r} + \frac{\boldsymbol{\mu}\cdot\mathbf{r}}{r^3} + \cdots\right], \tag{3.192}$$

where appears the electric monopole $Q = q_A + q_B$ and dipole $\boldsymbol{\mu} = q_A\mathbf{r}_A + q_B\mathbf{r}_B$—this setup is illustrated in Figure 3.6. By setting $q = q_A = -q_B$ and $\boldsymbol{\delta} = \mathbf{r}_{AB}$ and carrying out the appropriate limiting process, we obtain

$$\phi_{\mathrm{pE1}}(\mathbf{r}) = \frac{\boldsymbol{\mu}\cdot\mathbf{r}}{4\pi\varepsilon_0 r^3}. \tag{3.193}$$

The name multipole is less appropriate in the magnetic case than in the electric case since magnetic multipoles are made from current loops rather than point charges. A *point magnetic*

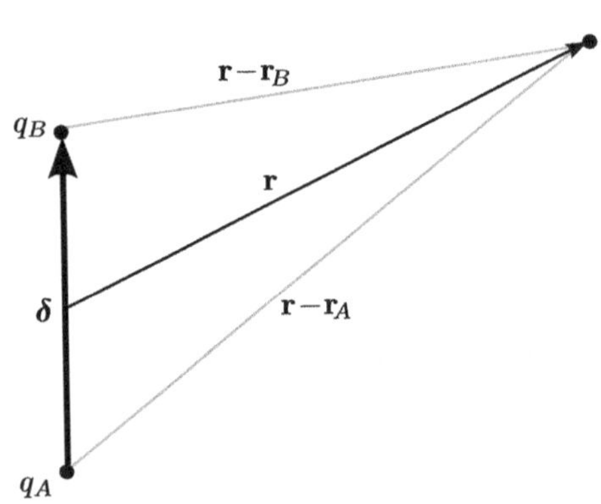

Figure 3.6 Electric dipole formed by point charges q_A and q_B separated by the distance vector $\boldsymbol{\delta}$. An observer point $\mathbf{r}$ is indicated at which the scalar potential $\phi(\mathbf{r})$ is given by Eq. (3.192).

Figure 3.7 Magnetic dipole generated by current loop.

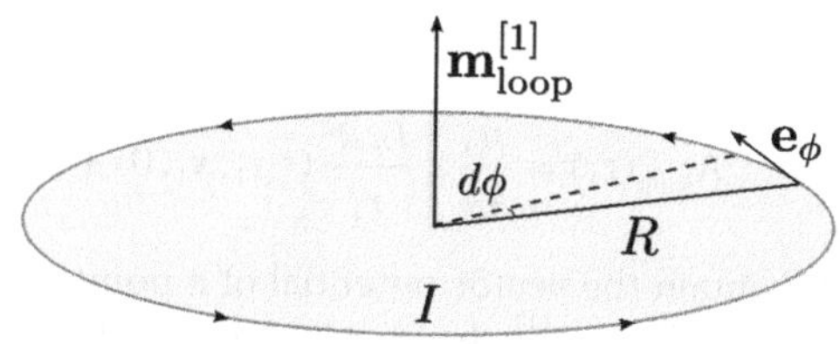

dipole is generated from a current loop, letting the coil area go to zero. This is mathematically a bit more challenging, but it is nevertheless instructive to analyze this in some detail: We consider a loop of radius R centered in the xy-plane in which a current I circulates in the counterclockwise direction, as shown in Figure 3.7. We start by calculating the magnetic dipole moment, Eq. (3.191). In this case, it reduces to a line integral

$$\mathbf{m}^{[1]}_{\text{loop}} = \frac{1}{2}\int (\mathbf{r}\times I d\mathbf{l}), \tag{3.194}$$

following the current in the loop. It will be convenient to use spherical coordinates (r, θ, ϕ). The position vector $\mathbf{r}$ follows the loop and has $r = R$, $\theta = \pi/2$, and all allowed values of ϕ. It may be expressed as

$$\mathbf{r} = (R\cos\phi, R\sin\phi, 0). \tag{3.195}$$

A general infinitesimal displacement is given by Eq. (3.35), but in this case the displacement is restricted to the loop and we may express it as

$$d\mathbf{l} = R\mathbf{e}_\phi d\phi = R(-\sin\phi, \cos\phi, 0)d\phi. \tag{3.196}$$

Upon integration we therefore obtain

$$\mathbf{m}^{[1]}_{\text{loop}} = I\pi R^2\mathbf{e}_z. \tag{3.197}$$

The magnetic dipole is accordingly seen to lie along the z-axis with a magnitude equal to the current times the loop area.

Next we consider the vector potential associated with the current loop. In the stationary case, when currents are steady and thus purely solenoidal, the general form of the vector potential is

$$\mathbf{A}(\mathbf{r}_1) = \frac{\mu_0}{4\pi}\int \frac{\mathbf{j}(\mathbf{r}_2)}{r_{12}} d^3\mathbf{r}_2. \tag{3.198}$$

It can be easily verified that this vector potential generates a magnetic field of the form of the Biot–Savart law, Eq. (3.34). In the present case, the volume integral reduces to a line integral

$$\mathbf{A}_{\text{loop}}(\mathbf{r}_1) = \frac{\mu_0}{4\pi}\int \frac{Id\mathbf{l}_2}{r_{12}} = \frac{\mu_0}{4\pi}\int \frac{I}{\sqrt{r_1^2 + R^2 - 2r_1 R\sin\theta_1\cos(\phi_1 - \phi_2)}} d\mathbf{l}_2. \tag{3.199}$$

Looking at the expression for the infinitesimal line segment, Eq. (3.196), we can see that the z-component of the vector potential is strictly zero. We now expand the integrand about $R = 0$. This gives

$$\mathbf{A}_{\text{loop}}(\mathbf{r}_1) = \frac{\mu_0}{4\pi}\left[\frac{1}{r_1}\int Id\mathbf{l}_2 + \frac{IR\sin\theta_1}{r_1^2}\int \cos(\phi_1 - \phi_2)d\mathbf{l}_2 \right. \tag{3.200}$$
$$\left. + \frac{IR^2}{r_1^3}\int (3\sin^2\theta_1\cos^2(\phi_1 - \phi_2) - 1)d\mathbf{l}_2 + \cdots\right].$$

The first term is an integral over the current and vanishes. Further manipulation gives

$$\mathbf{A}_{\text{loop}}(\mathbf{r}_1) = \frac{\mu_0}{4\pi}\left[\frac{I\pi R^2}{r_1^3}(-y_1, x_1, 0) + O(R^3)\right] = \frac{\mu_0}{4\pi}\left[\frac{\mathbf{m}^{[1]}_{\text{loop}} \times \mathbf{r}_1}{r_1^3} + O(R^3)\right]. \tag{3.201}$$

We obtain the vector potential of a point magnetic dipole by letting $R \to 0$ while keeping the magnitude of $\mathbf{m}^{[1]}_{\text{loop}}$ fixed. All higher-order terms then vanish and we are left with

$$\mathbf{A}_{\text{pM1}}(\mathbf{r}) = \frac{\mu_0}{4\pi}\frac{\mathbf{m}^{[1]} \times \mathbf{r}}{r^3}, \tag{3.202}$$

where $\mathbf{m}^{[1]}$ refers to the resulting point magnetic dipole.

It is instructive to compare the electric field of a point electric dipole with the magnetic field of a point magnetic dipole. In the former case we have, starting with Eq. (3.193),

$$\mathbf{E}_{\text{pE1}} = -\boldsymbol{\nabla}\phi_{\text{pE1}} = -\frac{\boldsymbol{\mu}}{3\varepsilon_0}\delta(\mathbf{r}) + \frac{1}{4\pi\varepsilon_0}\left[\frac{3\mathbf{r}(\mathbf{r}\cdot\boldsymbol{\mu}) - \boldsymbol{\mu}r^2}{r^5}\right], \tag{3.203}$$

whereas, in the latter case, one has, starting from Eq. (3.202)

$$\mathbf{B}_{\text{pM1}} = \boldsymbol{\nabla} \times \mathbf{A}_{\text{pM1}} = \frac{1}{c^2}\left(\frac{2\mathbf{m}}{3\varepsilon_0}\delta(\mathbf{r}) + \frac{1}{4\pi\varepsilon_0}\left[\frac{3\mathbf{r}(\mathbf{r}\cdot\mathbf{m}) - \mathbf{m}r^2}{r^5}\right]\right). \tag{3.204}$$

It can be seen that the fields have exactly the same form, except for a single point, namely, the position of the dipole, and which can be explained from the different limiting processes employed for their generation: the point electric dipole is generated from two point charges, whereas the point magnetic is generated from a current loop. An illustration of the fields due to electric and magnetic dipoles is given in Figure 3.8. If we orient the dipole along the z-axis, the common part of the dipole field, that is, excluding the origin, becomes proportional to the vector function

$$\mathbf{F}(\mathbf{r}) = \frac{(3xz, 3yz, 3z^2 - r^2)}{r^5} = \frac{2\cos\theta\mathbf{e}_r + \sin\theta\mathbf{e}_\theta}{r^3}. \tag{3.205}$$

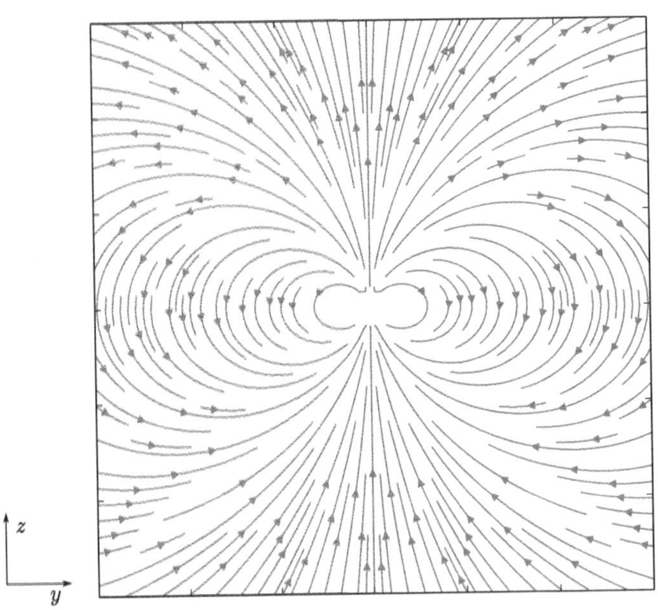

Figure 3.8 Field of an electric or magnetic dipole aligned with the z-axis.

From the final term on the right-hand side, we see that there is no azimuthal component F_ϕ, which implies axial symmetry as clearly seen in Figure 3.8. We also see that the field has an r^{-3} dependence with respect to the origin of the dipole. From the first term on the right-hand side, we see that the x- and y-components change sign upon the transformation $z \rightarrow -z$, whereas the z-component is unchanged, as is clear from Figure 3.8.

3.3.3 The Electric Dipole Approximation and Beyond

We have seen that the multipolar gauge leads to an ordering of interactions according to the inhomogeneities of the electric and magnetic fields and this leads to a hierarchy of very useful approximations. So far, we have considered general electromagnetic fields and the summation in Eq. (3.186) is accordingly infinite. Let us now consider the important example of the interaction of a molecule with a monochromatic electromagnetic wave as given in Eq. (3.74). Truncations in Eq. (3.186) may then be done based on the wave vector k in Eq. (3.72) as order parameter, and since we in Eq. (3.76) have seen that

$$kE = \omega B, \tag{3.206}$$

we can conclude that a truncation at order n in the electric field is to be accompanied by a truncation at order $n-1$ in the magnetic field. To zeroth order, we thus only get a contribution from the electric field

$$k^0 : -\mathbf{Q}^{[1]} \cdot \mathbf{E}^{[0]}. \tag{3.207}$$

This approximate truncation of the full interaction is the *electric dipole approximation*. Physically, it corresponds to a situation where the wavelength of the radiation is much longer than the molecular dimensions such that the molecule effectively only sees a uniform electric field rather than the full electromagnetic wave. This approximation is therefore reasonable for small- and medium-sized molecules exposed to electromagnetic radiation with wavelengths in the ultraviolet region, or longer. For nanosized systems or in the wavelength region of soft X-rays, and shorter, the validity of the electric dipole approximation is reduced, as depicted in Figure 3.9. In this figure, we illustrate a monochromatic plane wave propagating along the z-axis with an electric field polarized along the x-axis in accordance with

$$\mathbf{E}(\mathbf{r}, t) = E_0 \cos(kz - \omega t)\, \mathbf{e}_x. \tag{3.208}$$

In the upper panel, we consider the case of an electromagnetic wave in the near ultraviolet region with a wavelength of 200 nm, and we plot the electric field strength at times corresponding to ωt equal to 0 and $\pi/2$. The variation of the field strength along the direction of propagation is shown for a spatial region of 2 nm. For comparison, the size of a medium-sized molecule such as the fullerene C_{60} is depicted by a horizontal bar. At time zero, the gradient of the electric field at the coordinate origin, that is, $\mathbf{E}^{[1]}$, is zero and there is no visible spatial variation of the electric field. At $\omega t = \pi/2$, the electric field has become a sine function and $E^{[1]}_{x;z}$ is at its maximum and has the value $E_0 k$. One can now discern a clear spatial variation in the electric field along the z-axis. In the lower panel, the same presentation is done, but for a wavelength of 4.28 nm, corresponding to a photon energy of 290 eV, which refers approximately to energies at the carbon K-edge in X-ray spectroscopies. At time zero, $\mathbf{E}^{[1]}$ is still equal to zero of course, but there is now a very strong contribution from the second-order derivative $E^{[2]}_{x;zz}$ that is equal to $-E_0 k^2$. For this short wavelength, the accuracy of the electric dipole approximation is not obvious and should be assessed for the spectroscopy at hand. For X-ray absorption spectroscopies, the K-edge regions of second-row elements are addressed in the electric dipole approximation.

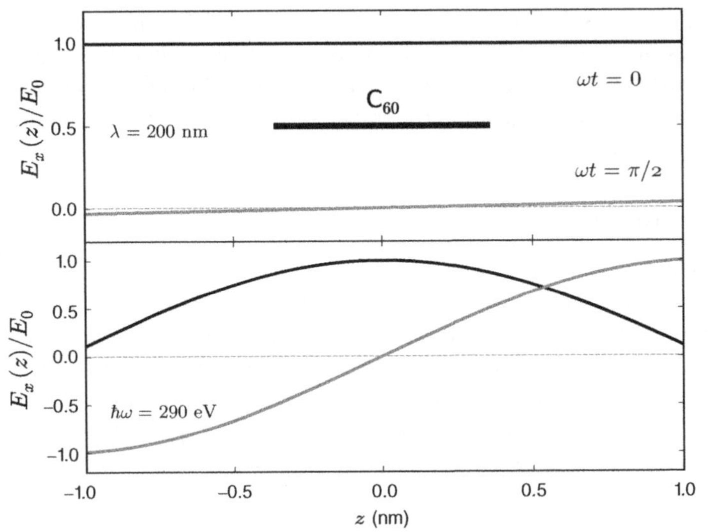

Figure 3.9 Electric field strength of a monochromatic plane wave along the direction of propagation and at two separate times corresponding to ωt equal to 0 (solid black line) and $\pi/2$ (solid gray line). The upper and lower panels refer to wavelengths in the near ultraviolet and soft X-ray regions, respectively. The horizontal bar depicts the size of the C_{60} fullerene.

The next level of approximation is to include also the linear term in the expansion of the vector potential

$$k^1 : -\frac{1}{2}\mathbf{Q}_j^{[2]} \cdot \mathbf{E}_j^{[1]} - \mathbf{m}^{[1]} \cdot \mathbf{B}^{[0]}, \tag{3.209}$$

which leads to the electric quadrupole–magnetic dipole approximation, invoking the electric field gradient and the magnetic field at the expansion point.

3.3.4 Origin Dependence of Electric and Magnetic Multipoles

To be experimentally observable, the electric and magnetic multipoles must be origin independent. Generally, one finds that this holds true only for the lowest-order nonzero multipole. Consider for instance the effect on the electric dipole moment caused by changing the origin to some point $\mathbf{G}$. The replacement $\mathbf{r} \rightarrow \mathbf{r} - \mathbf{G}$ gives

$$\boldsymbol{\mu}_{\mathrm{R}} = \int (\mathbf{r} - \mathbf{G})\rho(\mathbf{r}, t)d^3\mathbf{r} = \int \mathbf{r}\rho(\mathbf{r}, t)d^3\mathbf{r} - \mathbf{G}\int \rho(\mathbf{r}, t)d^3\mathbf{r} = \boldsymbol{\mu}_0 - \mathbf{G}Q^{[0]}, \tag{3.210}$$

which tells us that the electric dipole moment is origin independent only if the monopole, that is the total charge, is zero. By performing the same analysis for the magnetic dipole moment, we obtain

$$\mathbf{m}_{\mathrm{R}}^{[1]} = \frac{1}{2}\int [(\mathbf{r} - \mathbf{G}) \times \mathbf{j}(\mathbf{r}, t)]d^3\mathbf{r} = \frac{1}{2}\int [\mathbf{r} \times \mathbf{j}(\mathbf{r}, t)]d^3\mathbf{r} - \mathbf{G} \times \frac{1}{2}\int \mathbf{j}(\mathbf{r}, t)d^3\mathbf{r}, \tag{3.211}$$

which shows that a magnetic dipole is origin independent only if the total current is zero.

An important point to note is that a change of multipole origin corresponds to a change of origin of the fields and therefore constitutes a gauge transformation. As an illustration, consider a uniform, time-dependent magnetic field. In multipolar gauge, with the expansion point set to $\mathbf{r} = \mathbf{0}$, the scalar and vector potentials are given by Eq. (3.180), and we have already seen that the time-dependent uniform magnetic field is accompanied by a time-dependent nonuniform electric field, given by Eq. (3.181). In fact, this follows from Faraday's law, Eq. (3.3).

Let us now change the expansion point of such a uniform magnetic field to position **G**. We therefore write the vector potential as

$$\mathbf{A}_G(\mathbf{r},t) = \frac{1}{2}\mathbf{B}(t) \times \mathbf{r}_G; \quad \mathbf{r}_G = \mathbf{r} - \mathbf{G}. \tag{3.212}$$

The new and the old vector potentials are related as

$$\mathbf{A}_G(\mathbf{r},t) = \frac{1}{2}\mathbf{B}(t) \times \mathbf{r} - \frac{1}{2}\mathbf{B}(t) \times \mathbf{G} = \mathbf{A}_0(\mathbf{r},t) - \mathbf{A}_0(\mathbf{G},t), \tag{3.213}$$

which can be expressed as a gauge transformation

$$\mathbf{A}_G(\mathbf{r},t) = \mathbf{A}_0(\mathbf{r},t) - \nabla\chi; \quad \chi(\mathbf{r},t) = \mathbf{r} \cdot \mathbf{A}_0(\mathbf{G},t). \tag{3.214}$$

Since the gauge function is manifestly time dependent, there will be a corresponding transformation of the scalar potential

$$\phi_G(\mathbf{r},t) = \phi_0(\mathbf{r},t) + \partial_t\chi(\mathbf{r},t) = \frac{1}{2}\mathbf{r} \cdot (\partial_t\mathbf{B} \times \mathbf{G}). \tag{3.215}$$

The resulting electric and magnetic fields

$$\mathbf{E}(\mathbf{r},t) = \frac{1}{2}(\mathbf{r} \times \partial_t\mathbf{B}); \quad \mathbf{B}(\mathbf{r},t) = \mathbf{B}, \tag{3.216}$$

are unchanged, as they should be.

3.3.5 Electric Multipoles

3.3.5.1 General Versus Traceless Forms

The set of electric multipoles of a given order n is given by

$$Q^{[n]} = \int x^i y^j z^k \rho(\mathbf{r},t) d^3\mathbf{r}; \quad \forall\,(i+j+k) = n, \tag{3.217}$$

and is seen to be symmetric in all n indices. The number of independent electric multipoles of order n is therefore the same as the number of Cartesian Gaussians of the same order. To find this number we note that the triples (i,j,k) can be inserted into a square matrix of dimension $(n+1)$. For instance, for $n = 2$ we find

$$\begin{matrix} (0,0,2) & (0,1,1) & (0,2,0) \\ (1,0,1) & (1,1,0) & \\ (2,0,0) & & \end{matrix} \tag{3.218}$$

The indices i and j determine the position in the matrix, whereas the index k follows from the condition $(i+j+k) = n$. The number of elements that we fill in corresponds to one triangle and the diagonal, from which we conclude that there are $(n+2)(n+1)/2$ independent electric multipoles of order n.

Let us now look at the interaction of electric multipoles with electric fields. In Section 3.3.2, we found that the nth-order multipole $\mathbf{Q}^{[n]}$ interacts with the $(n-1)$th-order derivative of the electric field $\mathbf{E}^{[n-1]}$, that is

$$\langle \hat{H}_{En} \rangle = -\frac{1}{n!} Q^{[n]}_{j_1 j_2 \cdots j_n} E^{[n-1]}_{j_1; j_2 \cdots j_n}. \tag{3.219}$$

In contrast to the electric multipole, the electric field derivative

$$E^{[n-1]}_{j_1; j_2 \cdots j_n} = \left.\frac{\partial^{n-1} E_{j_1}}{\partial r_{j_2} \cdots \partial r_{j_n}}\right|_0, \tag{3.220}$$

is symmetric in all but one index, the one which indicates the component of the electric field. However, in the stationary case, where the electric field is irrotational (see Section 3.1.2), the electric field derivative is symmetric in *all* indices since we have

$$\mathbf{E} = -\nabla\phi \quad \Rightarrow \quad E^{[n-1]}_{j_1;j_2\cdots j_n} = -\phi^{[n]}_{j_1\cdots j_n}. \tag{3.221}$$

The complete electric multipole interaction can therefore in the stationary case be expressed as

$$\langle \hat{H}_E \rangle = \sum_{n=0}^{\infty} \frac{1}{n!} Q^{[n]}_{j_1\ldots j_n} \phi^{[n]}_{j_1\ldots j_n}. \tag{3.222}$$

We obtain the same expression by starting from

$$\langle \hat{H}_{\text{int}} \rangle = \int \rho(\mathbf{r})\phi(\mathbf{r})d^3\mathbf{r}; \quad \phi(\mathbf{r}) = \frac{1}{4\pi\varepsilon_0} \int \frac{\rho'(\mathbf{r}')}{|\mathbf{r}-\mathbf{r}'|} d^3\mathbf{r}', \tag{3.223}$$

and then Taylor expanding the scalar potential $\phi(\mathbf{r})$ about the origin. We shall explore the difference between the stationary and the general case in the following text.

First, we note that terms in the interaction may vanish if the electric field is generated by some localized charge distribution and the expansion point $\mathbf{r} = \mathbf{0}$ of the electric field is *outside* this charge distribution, because we then have

$$[\nabla \cdot \mathbf{E}]_{\mathbf{r}=\mathbf{0}} = E^{[1]}_{i;i} = [-\nabla^2\phi - \partial_t(\nabla \cdot \mathbf{A})]_{\mathbf{r}=\mathbf{0}} = \rho^{[0]}/\varepsilon_0 = 0. \tag{3.224}$$

In the stationary case, we obtain the stronger condition that the scalar potential should fulfill Laplace's equation at the expansion point

$$[-\nabla^2\phi]_{\mathbf{r}=\mathbf{0}} = -\phi^{[2]}_{ii} = \rho^{[0]}/\varepsilon_0 = 0. \tag{3.225}$$

This will also hold true in Coulomb gauge ($\nabla \cdot \mathbf{A} = 0$), but we have seen above that the multipole gauge and the Coulomb gauge only coincide for time-independent electric fields and no external currents (the static case). These restrictions can be built into electric multipoles by introducing *traceless* multipoles that refer to forms in which tracing over any two indices gives zero. To see how this works in practice, we shall consider a specific example.

Let us consider the interaction of the electric quadrupole with the electric field gradient in the stationary case

$$\begin{aligned}\langle \hat{H}_{E2} \rangle &= \frac{1}{2} Q^{[2]}_{ij} \phi^{[2]}_{ij} \\ &= \frac{1}{2}\left[Q^{[2]}_{xx}\phi^{[2]}_{xx} + Q^{[2]}_{yy}\phi^{[2]}_{yy} + Q^{[2]}_{zz}\phi^{[2]}_{zz} + Q^{[2]}_{xy}\phi^{[2]}_{xy} + \cdots\right].\end{aligned} \tag{3.226}$$

Let us have a look at the first three terms. We note that $Q^{[2]}_{ij}$ and $\phi^{[2]}_{ij}$ are two-index quantities and can be considered as elements of separate matrices. The sum $\phi^{[2]}_{ii}$ of the diagonal elements of the latter matrix (i.e., the trace) is zero in the stationary case, as shown by Eq. (3.225). However, in the expression for the interaction energy, Eq. (3.226), the diagonal elements $\phi^{[2]}_{ii}$ appear in a *weighted* sum. If the diagonal terms $Q^{[2]}_{ii}$ of the electric quadrupole matrix $Q^{[2]}$ were identical, the first three terms of Eq. (3.226) would be zero, but this is generally not the case. However, we can subtract from each diagonal element of $Q^{[2]}$ their average value $\mathrm{Tr}\{Q^{[2]}\}/3$ to build in the trace condition Eq. (3.225). Introducing the traceless electric quadrupole moment

$$\Theta^{[2]}_{ij} = \frac{3}{2}\left(Q^{[2]}_{ij} - \frac{1}{3}Q^{[2]}_{kk}\delta_{ij}\right), \tag{3.227}$$

where the prefactor comes from convention, we can rewrite the above interaction energy as

$$\langle \hat{H}_{E2} \rangle = -\frac{1}{2} Q^{[2]}_{ij} E^{[1]}_{i;j} = -\frac{1}{3}\Theta_{ij} E^{[1]}_{i;j} - \frac{1}{6} Q^{[2]}_{kk} E^{[1]}_{i;i}. \tag{3.228}$$

The second term on the right-hand side of the quadrupole coupling is known as the Poisson term since it contains the divergence of the electric field at the expansion point. It is strictly zero if the charge distribution generating the electric field is outside the expansion point.

A general formula for traceless electric multipoles to any order n can be generated by looking in more detail at the Taylor expansion of the scalar potential. In the quadrupole coupling appears

$$\phi_{ij}^{[2]} = \frac{1}{4\pi\varepsilon_0}\int \frac{(3r_ir_j - \delta_{ij}r^2)\rho(\mathbf{r})}{r^5}d^3\mathbf{r} - \frac{1}{3\varepsilon_0}\rho^{[0]}\delta_{ij}, \tag{3.229}$$

where we have used the formula

$$\nabla_i\nabla_j r^{-1} = 3r_ir_jr^{-5} - \delta_{ij}r^{-3} - \frac{4\pi}{3}\delta(\mathbf{r})\delta_{ij}. \tag{3.230}$$

In the first term under the integral sign, we see a form corresponding to that of the traceless electric quadrupole, Eq. (3.227), multiplied by r^{-5} and with the missing factor 1/2 coming from the prefactor of the Taylor expansion. If we trace over the second derivative matrix $\phi_{ij}^{[2]}$ of the scalar potential, the term corresponding to the traceless electric quadrupole vanishes, showing that it is indeed traceless, and we obtain the Poisson equation, Eq. (3.60).

Traceless electric multipoles to any order n can be generated from the formula

$$\Theta_{j_1\cdots j_n}^{[n]} = \int \rho(\mathbf{r})\frac{(-1)^n}{n!}r^{2n+1}\left\{\Pi_{k=1}^{n}\nabla_{j_k}\right\}\frac{1}{r}d^3\mathbf{r}, \tag{3.231}$$

where we momentarily assume $\rho^{[0]} = 0$ in order to not get contributions associated with Dirac delta functions. For the traceless octupole and hexadecapole moments, we then obtain

$$\begin{aligned}
\Theta_{ijk}^{[3]} &= \tfrac{1}{2}\left(5Q_{ijk}^{[3]} - Q_{ill}^{[3]}\delta_{jk} - Q_{jll}^{[3]}\delta_{ik} - Q_{kll}^{[3]}\delta_{ij}\right), \\
\Theta_{ijkl}^{[4]} &= \tfrac{1}{8}\Big(35Q_{ijkl}^{[4]} - 5\delta_{ij}Q_{klmm}^{[4]} - 5\delta_{ik}Q_{jlmm}^{[4]} - 5\delta_{il}Q_{jkmm}^{[4]} - 5\delta_{jk}Q_{ilmm}^{[4]} \\
&\quad -5\delta_{jl}Q_{ikmm}^{[4]} - 5\delta_{kl}Q_{ilmm}^{[4]} \\
&\quad + \left(\delta_{ij}\delta_{kl} + \delta_{ik}\delta_{jl} + \delta_{il}\delta_{jk}\right)Q_{mmnn}^{[4]}\Big).
\end{aligned} \tag{3.232}$$

Inspection of these forms show that they are constructed from the original quadrupole moments with the constraints subtracted. The number of constraints for an electric multipole of order n is equal to the number of index pairs, that is, $\frac{1}{2}n(n-1)$. The effective number of traceless multipoles of order n is therefore $\frac{1}{2}(n+2)(n+1) - \frac{1}{2}n(n-1) = (2n+1)$.

The number of traceless and general electric multipoles of order n corresponds exactly to the number of Cartesian and spherical Gaussian-type orbitals, respectively, of order ℓ. Indeed, an electric multipole expansion with the same number of degrees of freedom can be carried out in spherical harmonics using the addition theorem

$$\frac{1}{|\mathbf{r}-\mathbf{r}'|} = \sum_{\ell=0}^{\infty}\sum_{m=-\ell}^{+\ell}\frac{4\pi}{2\ell+1}\frac{r_<^{\ell}}{r_>^{\ell+1}}Y_{\ell m}^*\left(\frac{\mathbf{r}}{r}\right)Y_{\ell m}\left(\frac{\mathbf{r}'}{r'}\right), \tag{3.233}$$

where $r_> = \max(r, r')$ and $r_< = \min(r, r')$. Inserting this into the interaction energy we obtain, assuming remote sources, that is, $r' = r_>$

$$\langle\hat{H}_{\text{int}}\rangle = \int \rho(\mathbf{r})\phi(\mathbf{r})d^3\mathbf{r} = \frac{1}{4\pi\varepsilon_0}\int\int \frac{\rho(\mathbf{r})\rho'(\mathbf{r}')}{|\mathbf{r}-\mathbf{r}'|}d^3\mathbf{r} = \sum_{\ell=0}^{\infty}\sum_{m=-\ell}^{\ell} q_{\ell m}E_{\ell m}, \tag{3.234}$$

with spherical electric multipoles

$$q_{\ell m} = \int \hat{q}_{\ell m}\rho(\mathbf{r})d^3\mathbf{r} = \int r^{\ell}Y_{\ell m}^*(\theta,\phi)\rho(\mathbf{r})d^3\mathbf{r}, \tag{3.235}$$

and spherical components of the electric field

$$E_{\ell m} = \frac{1}{4\pi\varepsilon_0}\frac{4\pi}{2\ell+1}\int r^{-(\ell+1)}Y_{\ell m}(\theta,\phi)\rho'(\mathbf{r})d^3\mathbf{r}. \tag{3.236}$$

In the stationary case, the traceless and spherical multipoles are equivalent to the general ones, but what about their validity in the general time-dependent case? We have already found that the traceless quadrupole coupling is generally valid to the extent that the generating charge distribution is zero at the expansion point. For octupole and hexadecapole moments, we find the following interaction energies

$$\begin{aligned} -\frac{1}{6}Q^{[3]}_{ijk}E^{[2]}_{i;jk} &= -\frac{1}{15}\Theta^{[3]}_{ijk}E^{[2]}_{i;jk} - \frac{1}{15}Q^{[3]}_{ill}E^{[2]}_{i;jj} - \frac{1}{30}Q^{[3]}_{kll}E^{[2]}_{i;ik}, \\ -\frac{1}{24}Q^{[4]}_{ijkl}E^{[3]}_{ijkl} &= -\frac{1}{105}\Theta^{[4]}_{ijkl}E^{[3]}_{i;jkl} - \frac{1}{56}Q^{[4]}_{ilmm}E^{[3]}_{i;jjl} - \frac{1}{56}Q^{[4]}_{klmm}E^{[3]}_{i;ikl} + \frac{1}{280}Q^{[4]}_{mmnn}E^{[4]}_{i;ikk}. \end{aligned} \tag{3.237}$$

If we again assume that the charge distribution generating the electric field is outside the expansion point, we find that the third term of the octupole coupling and the third and fourth terms of the hexadecapole coupling are zero. This leaves some terms involving general electric multipoles that are not zero, which means that if one wants to introduce electric multipoles that can be applied in both static and dynamic situations, the general definition Eq. (3.187) is to be preferred over the traceless forms, Eq. (3.231).

Before concluding this section, we include a useful expansion of the electric field in the static case. Consider the electrostatic potential of a local charge distribution centered around $\boldsymbol{R}_A$

$$\phi_{\mathrm{A}}(\boldsymbol{r}) = \frac{1}{4\pi\varepsilon_0}\int \frac{\rho_{\mathrm{A}}(\mathbf{r}')}{|\boldsymbol{r}-\boldsymbol{r}'|}d^3\boldsymbol{r}'. \tag{3.238}$$

For an observer point $\boldsymbol{r} \neq \boldsymbol{R}_{\mathrm{A}}$ we expand $f(\boldsymbol{r}') = \frac{1}{|\boldsymbol{r}-\boldsymbol{r}'|}$ about $\boldsymbol{r}' = \boldsymbol{R}_{\mathrm{A}}$ to give

$$\begin{aligned} \phi_{\mathrm{A}}(\boldsymbol{r}) = {} & \frac{Q^{[0]}_{\mathrm{A}}}{4\pi\varepsilon_0}\left[\frac{1}{|\boldsymbol{r}-\boldsymbol{R}_{\mathrm{A}}|}\right] + \frac{Q^{[1]}_{\mathrm{A};i}}{4\pi\varepsilon_0}\left[\frac{(\boldsymbol{r}-\boldsymbol{R}_{\mathrm{A}})_i}{|\boldsymbol{r}-\boldsymbol{R}_{\mathrm{A}}|^3}\right] \\ & + \frac{Q^{[2]}_{\mathrm{A};ij}}{4\pi\varepsilon_0}\left[\frac{3(\boldsymbol{r}-\boldsymbol{R}_{\mathrm{A}})_i(\boldsymbol{r}-\boldsymbol{R}_{\mathrm{A}})_j - \delta_{ij}|\boldsymbol{r}-\boldsymbol{R}_{\mathrm{A}}|^2}{|\boldsymbol{r}-\boldsymbol{R}_{\mathrm{A}}|^5}\right] + \cdots \end{aligned} \tag{3.239}$$

The corresponding field is

$$\begin{aligned} \boldsymbol{E}_{\mathrm{A};i}(\boldsymbol{r}) = -\nabla_i\phi_{\mathrm{A}}(\boldsymbol{r}) = {} & \frac{Q^{[0]}_{\mathrm{A}}}{4\pi\varepsilon_0}\left[\frac{(\boldsymbol{r}-\boldsymbol{R}_{\mathrm{A}})_i}{|\boldsymbol{r}-\boldsymbol{R}_{\mathrm{A}}|^3}\right] \\ & + \frac{Q^{[1]}_{\mathrm{A};j}}{4\pi\varepsilon_0}\left[\frac{3(\boldsymbol{r}-\boldsymbol{R}_{\mathrm{A}})_i(\boldsymbol{r}-\boldsymbol{R}_{\mathrm{A}})_j - \delta_{ij}|\boldsymbol{r}-\boldsymbol{R}_{\mathrm{A}}|^2}{|\boldsymbol{r}-\boldsymbol{R}_{\mathrm{A}}|^5}\right] + \cdots. \end{aligned} \tag{3.240}$$

We see that we have an expansion of the field in terms of electric multipoles of the source, Eq. (3.217), multiplied with terms involving the observer point that have the traceless forms.

3.3.5.2 What We Can Learn from Symmetry

We can use symmetry to determine the possible nonzero multipoles since we know from group theory[9] that this requires a totally symmetric integrand. As an important example, consider a system described by a Hamiltonian that commutes with the inversion operator $\hat{P}$ such that the eigenfunctions have well-defined parity

$$[\hat{H},\hat{P}] = 0 \quad \Rightarrow \quad \hat{P}\psi(\mathbf{r}) = \pm\psi(\mathbf{r}). \tag{3.241}$$

9 A crash course is given in the next chapter.

The density is therefore *gerade* (totally symmetric) under inversion

$$\rho(\mathbf{r}) = \rho(-\mathbf{r}). \tag{3.242}$$

On the other hand, the parity of the coordinate product $x^i y^j z^k$ is determined by the order of the multipole

$$\hat{P}x^i y^j z^k = (-1)^n x^i y^j z^k; \quad n = i + j + k. \tag{3.243}$$

Under such conditions, odd-order electric multipoles, such as the electric dipole moment of an isolated atom, are strictly zero because the total integrand is *ungerade*. To the extent that nuclear forces are parity conserving, this also holds for an atomic nucleus.

A slightly more complicated example is atoms (and other systems) with well-defined angular momentum (the reader may consult Appendix G for more on angular momentum and its algebra). Let the state function be expressed as $|\eta LM\rangle$ where

$$\hat{L}^2|\eta LM\rangle = \hbar^2 L(L+1)|\eta LM\rangle; \quad \hat{L}_z|\eta LM\rangle = \hbar M|\eta LM\rangle, \tag{3.244}$$

and η collects other state-specific information. Consider now the action of the spherical multipole operator $\hat{q}_{lm}$ from Eq. (3.235) on such a state function. Since such state functions form a complete set, we can express the result as

$$\hat{q}_{\ell m}|\eta LM\rangle = \sum_{\ell'=|\ell-L|}^{\ell+L} C_{\ell'}|\eta\ell'(M+m)\rangle. \tag{3.245}$$

The restriction of the summation on the right-hand side follows from the coupling of angular momenta l and L (Clebsch–Gordan series), which also implies that the resulting azimuthal quantum number should be $m' = m + M$. From the orthogonality of the state functions, it now follows that the integral

$$\langle\eta' L' M'|\hat{q}_{\ell m}|\eta LM\rangle = \sum_{\ell'=|\ell-L|}^{\ell+l} C_{\ell'}\langle\eta' L' M'|\eta\ell'(M+m)\rangle, \tag{3.246}$$

is nonzero only for

$$L' = |\ell - L|, \ldots, \ell + L; \quad M' = M + m, \tag{3.247}$$

and that the state $|\eta LM\rangle$ has at most an electric multipole moment of order $2L$.

3.3.6 Magnetic Multipoles

In this section, we will take a closer look at the magnetic multipoles defined in Eq. (3.190), which shows that magnetic multipoles are symmetric in all but the first index. The set of magnetic multipoles of a given order n is given by

$$\mathbf{m}^{[n]} = \int x^i y^j z^k [\mathbf{r} \times \mathbf{j}(\mathbf{r}, t)] d^3\mathbf{r}; \quad \forall\, (i + j + k) = n - 1, \tag{3.248}$$

which suggests that the number of independent magnetic multipoles of order n is the number of independent electric multipoles of order $(n-1)$ times three, that is, $3n(n+1)/2$. However, in contrast to electric fields, a trace condition on magnetic fields applies to any point in space, namely that the divergence of a magnetic field is zero. Indeed, we find that if we set the final index i in Eq. (3.190) equal to any other index of a magnetic multipole and trace over the index pair, we do get zero. Magnetic multipoles are therefore inherently traceless. The additional $(n-1)$ constraints imply that there are only $(3n^2 + n + 2)/2$ independent magnetic multipoles. The number of degrees of freedom in the magnetic multipole expansion can be further reduced by considering symmetry. For instance, since the classical current is charge times velocity, which transforms as the coordinates, we expect *even* magnetic multipoles to vanish in systems with well-defined parity. Other symmetries, such as time reversal symmetry,[10] may further reduce the effective number of magnetic multipoles.

For electric multipoles, we found additional constraints when assuming that the interacting electric field was generated by a local charge distribution being zero at the multipole expansion point. For magnetic multipoles we may consider the case where the magnetic field is generated by a local current distribution. In the stationary case, the curl of the resulting magnetic field is then zero in all points outside the source current distribution

$$\boldsymbol{\nabla} \times \mathbf{B} \mid_{\mathbf{r}=\mathbf{0}} = \mu_0\, \mathbf{j}^{[0]} = \mathbf{0}. \tag{3.249}$$

As a specific example, consider the magnetic quadrupole interaction

$$\langle \hat{H}_{\text{int}} \rangle = -\frac{1}{2} m^{[2]}_{i;j} B^{[1]}_{i;j}; \quad m^{[2]}_{i;j} = \frac{2}{3} \int r_j (\mathbf{r} \times \mathbf{j})_i d^3\mathbf{r}; \quad B^{[1]}_{i;j} = \nabla_j B_i \mid_{\mathbf{r}=\mathbf{0}}, \tag{3.250}$$

for which the curl condition in Eq. (3.249) connects two elements of the magnetic field gradient, for instance

$$\nabla_x B_y - \nabla_y B_x = 0.$$

The corresponding antisymmetric combination $m^{[2]}_{y;x} - m^{[2]}_{x;y}$ of the magnetic quadrupole is therefore redundant and can be projected out, similar to what was done for the electric quadrupole coupling. The nonredundant combinations of magnetic quadrupole elements away from the sources are thus the *symmetric* ones.

3.3.7 Electric Dipole Radiation

In cases when electric dipoles are time dependent, they will be the sources of radiation, much like microscopic antennas. Let us consider a time-dependent point electric dipole described by the electric dipole density

$$\mathbf{p}(\mathbf{r}, t) = \boldsymbol{\mu} e^{-i\omega t} \delta(\mathbf{r}), \tag{3.251}$$

where, just as before, it is understood that the complex notation is used for mathematical convenience and that it is the real part that corresponds to the physical dipole moment. Let us further assume that the dipole is oriented along the z-axis so that $\boldsymbol{\mu} = \mu \mathbf{e}_z$. The point electric dipole corresponds to a current density

$$\mathbf{j}(\mathbf{r}, t) = \frac{\partial \mathbf{p}(\mathbf{r}, t)}{\partial t} = -i\omega \boldsymbol{\mu} e^{-i\omega t} \delta(\mathbf{r}), \tag{3.252}$$

which, in the Lorentz gauge, Eq. (3.58), gives rise to a retarded vector potential

$$\mathbf{A}(\mathbf{r}, t) = \frac{-i\omega\mu_0}{4\pi r} \boldsymbol{\mu} e^{-i\omega(t-r/c)}. \tag{3.253}$$

The magnetic field then becomes

$$\mathbf{B}(\mathbf{r}, t) = \boldsymbol{\nabla} \times \mathbf{A}(\mathbf{r}, t) = \frac{\omega^2}{4\pi\varepsilon_0 c^3} \frac{1}{r^2} (\mathbf{r} \times \boldsymbol{\mu}) \left(1 + \frac{ic}{\omega r}\right) e^{-i\omega(t-r/c)}, \tag{3.254}$$

where we have used the vector operation

$$(\boldsymbol{\nabla} \times [\mathbf{a} f(r)])_i = \varepsilon_{ijk} \partial_j f(r) a_k = \varepsilon_{ijk} r_j a_k \frac{1}{r} \frac{\partial f(r)}{\partial r}, \tag{3.255}$$

10 These symmetries will be further discussed in Section 4.2.

or, equivalently,

$$\nabla \times [\mathbf{a}f(r)] = (\mathbf{r} \times \mathbf{a})\frac{1}{r}\frac{\partial f(r)}{\partial r}. \tag{3.256}$$

From the Lorentz gauge condition, Eq. (3.56), we have

$$\partial_t \phi = -c^2 \nabla \cdot \mathbf{A}(\mathbf{r}, t) = \frac{-\omega^2}{4\pi\varepsilon_0 c r^2}(\mathbf{r} \cdot \boldsymbol{\mu})\left(1 + \frac{ic}{\omega r}\right) e^{-i\omega(t-r/c)}, \tag{3.257}$$

where we have used

$$\nabla \cdot [\mathbf{a}f(r)] = (\mathbf{r} \cdot \mathbf{a})\frac{1}{r}\frac{\partial f(r)}{\partial r}. \tag{3.258}$$

By means of time integration, we find that the retarded scalar potential is

$$\phi(\mathbf{r}, t) = \frac{-i\omega}{4\pi\varepsilon_0 c}\frac{1}{r^2}(\mathbf{r} \cdot \boldsymbol{\mu})\left(1 + \frac{ic}{\omega r}\right) e^{-i\omega(t-r/c)}, \tag{3.259}$$

where the arbitrary constant in the primitive function has been set to zero so that the scalar potential tends to zero at large r. In order to determine the electric field, we need

$$-\partial_t \mathbf{A}(\mathbf{r}, t) = \frac{\omega^2}{4\pi\varepsilon_0 c^2 r}\boldsymbol{\mu}\, e^{-i\omega(t-r/c)} \tag{3.260}$$

and

$$\begin{aligned} -\nabla\phi = \frac{i\omega}{4\pi\varepsilon_0 c}&\left[-2\frac{\mathbf{r}}{r^4}(\mathbf{r} \cdot \boldsymbol{\mu})\left(1 + \frac{ic}{\omega r}\right) e^{-i\omega(t-r/c)}\right. \\ &+\frac{1}{r^2}\boldsymbol{\mu}\left(1 + \frac{ic}{\omega r}\right) e^{-i\omega(t-r/c)} \\ &+\frac{1}{r^2}(\mathbf{r} \cdot \boldsymbol{\mu})\frac{-ic\mathbf{r}}{\omega r^3}e^{-i\omega(t-r/c)} \\ &\left.+\frac{1}{r^2}(\mathbf{r} \cdot \boldsymbol{\mu})\left(1 + \frac{ic}{\omega r}\right)\left(\frac{i\omega\mathbf{r}}{cr}\right) e^{-i\omega(t-r/c)}\right], \end{aligned} \tag{3.261}$$

where we have made repeated use of

$$\nabla f(r) = \frac{\mathbf{r}}{r}\frac{\partial f(r)}{\partial r}. \tag{3.262}$$

After collecting terms, the electric field from an oscillating electric dipole is found to be

$$\begin{aligned} \mathbf{E}(\mathbf{r}, t) &= -\nabla\phi - \partial_t \mathbf{A} \\ &= \frac{\omega^2}{4\pi\varepsilon_0 c^2 r}\left(\boldsymbol{\mu} - \frac{(\mathbf{r} \cdot \boldsymbol{\mu})\mathbf{r}}{r^2}\right) e^{-i\omega(t-r/c)} \\ &+\frac{1}{4\pi\varepsilon_0}\left(\frac{1}{r^3} - \frac{i\omega}{cr^2}\right)\left(3\frac{(\mathbf{r} \cdot \boldsymbol{\mu})\mathbf{r}}{r^2} - \boldsymbol{\mu}\right) e^{-i\omega(t-r/c)}. \end{aligned} \tag{3.263}$$

The far-field limit of these electromagnetic fields is important to us, since these fields correspond to an experiment where photons scattered from a sample are collected at a macroscopic distance away from the sources (molecules). For large r, the leading terms in the fields are

$$\mathbf{E}(\mathbf{r}, t) = \frac{\omega^2}{4\pi\varepsilon_0 c^2 r}\left(\boldsymbol{\mu} - \frac{(\mathbf{r} \cdot \boldsymbol{\mu})\mathbf{r}}{r^2}\right) e^{-i\omega(t-r/c)}, \tag{3.264}$$

$$\mathbf{B}(\mathbf{r}, t) = \frac{\omega^2}{4\pi\varepsilon_0 c^3 r^2}(\mathbf{r} \times \boldsymbol{\mu}) e^{-i\omega(t-r/c)}. \tag{3.265}$$

The associated time-averaged Poynting vector (or intensity), defined in Eq. (3.135) for complex fields, becomes

$$\begin{aligned}\langle \mathbf{S}\rangle_{\mathrm{T}} &= \frac{1}{2}\varepsilon_0 c^2\ \mathrm{Re}(\mathbf{E}\times\mathbf{B}^*) \\ &= \frac{\omega^4\mu^2}{32\pi^2\varepsilon_0 c^3}\frac{\mathbf{r}}{r^3}\sin^2\theta,\end{aligned} \tag{3.266}$$

where θ is the angle between the induced molecular dipole $\boldsymbol{\mu}$ and the direction of observation $\mathbf{r}$. This scattering intensity is the microscopic origin of the observed photons in both Rayleigh and Raman scattering processes, to be discussed in Section 8.7.3.

The near-field limit, on the other hand, is defined as the region where

$$r \ll \frac{c}{\omega}.$$

To give an example, for UV radiation with a wavelength of 200 nm, we get $c/\omega \approx 1400\ a_0$, which means that the near-field region in this case encloses all molecular sites considered in typical environment models. In this limit, the leading terms of the electromagnetic fields are given by

$$\mathbf{E}(\mathbf{r},t) = \frac{1}{4\pi\varepsilon_0}\frac{1}{r^3}\left(3\frac{(\mathbf{r}\cdot\boldsymbol{\mu})\mathbf{r}}{r^2} - \boldsymbol{\mu}\right)e^{-i\omega t}, \tag{3.267}$$

$$\mathbf{B}(\mathbf{r},t) = \frac{i\omega}{4\pi\varepsilon_0 c^2}\frac{1}{r^3}(\mathbf{r}\times\boldsymbol{\mu})e^{-i\omega t}. \tag{3.268}$$

There are a few interesting observations to be made from these equations. First, we note that the time-dependent electric field associated with an oscillating electric dipole in the near-field limit is identical to what one would have obtained by multiplying the electric field due to the *static* electric dipole in Eq. (3.203) (less the Dirac-delta term) with an exponential $\exp(-i\omega t)$. Second, the magnetic field is out of phase by an angle $\pi/2$ with respect to the electric field. This is due to the fact that a maxima for the electric and magnetic fields occur in time at points where the dipole and current densities reach maxima, respectively, and these are out of phase, as seen by comparing Eqs. (3.251) and (3.252). Third, the magnetic field is much weaker than the electric field (by a factor of $\omega r/c^2$) and can typically be ignored in the construction of intermolecular interaction models.

3.4 Macroscopic Maxwell's Equations

3.4.1 Spatial Averaging

So far, we have been considering electrodynamics at the level of individual molecules, corresponding to the microscopic description typically employed by a theoretical chemist. Let us now turn to the macroscopic description typically offered by experiment.

One of the most striking visualizations of magnetic field lines is obtained by sprinkling iron filings onto a piece of paper held above a magnet (cf. Figure 3.10). Although an intricate pattern appears, it hides the full details of the magnetic field at the length scale of molecules because iron filings are macroscopic objects. When measuring electromagnetic fields in matter, as in the example with the iron filings, the size of the experimental probe is several orders of magnitude larger than molecular dimensions, and it can therefore only provide smoothed representations of the fields (and sources). In theory, we mimic macroscopic fields by performing spatial averaging of the microscopic counterparts.

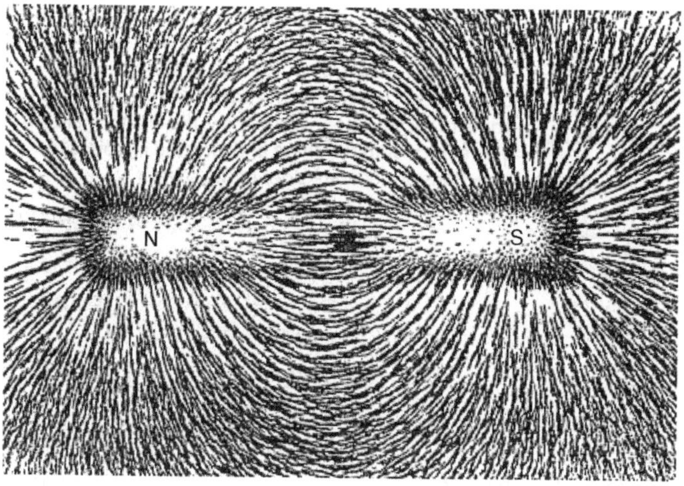

Figure 3.10 Visualization of magnetic field lines using iron fillings sprinkled on a sheet of paper held over a bar magnet. (From Wikipedia, the free encyclopedia).

Let $\mathbf{F}(\mathbf{r}, t)$ be a vector function, corresponding for instance to the microscopic electric or magnetic field, and $w(\mathbf{r})$ is a suitable weighting function normalized to unity and carrying units of inverse volume. It is reasonable to insist that the weighting function is isotropic and positive. The spatially averaged vector function $\overline{\mathbf{F}}$ is then defined as

$$\overline{\mathbf{F}}(\mathbf{r}, t) = \langle \mathbf{F} \rangle = \int w(\mathbf{r}' - \mathbf{r})\mathbf{F}(\mathbf{r}', t)\, d^3\mathbf{r}'; \quad \int w(\mathbf{r})\, d^3\mathbf{r} = 1. \tag{3.269}$$

It is easy to see that time derivatives of spatial averages equal spatial averages of time derivatives, but it is perhaps less obvious that this holds also for spatial derivatives. This becomes clear, however, after a simple variable substitution

$$\mathbf{u} = \mathbf{r}' - \mathbf{r}, \quad d^3\mathbf{u} = d^3\mathbf{r}' \quad \rightarrow \quad \overline{\mathbf{F}}(\mathbf{r}, t) = \int w(\mathbf{u})\mathbf{F}(\mathbf{u} + \mathbf{r}, t)\, d^3\mathbf{u}. \tag{3.270}$$

The implication of this is that Maxwell's equations given in Eqs. (3.2)–(3.6) also hold when the microscopic fields and sources are replaced by their macroscopic, spatially averaged, counterparts. However, since spatial averaging manifestly treats space and time on a different footing, it is at best questionable whether the resulting macroscopic Maxwell's equations are Lorentz invariant.

3.4.2 Polarization and Magnetization

When applying external electromagnetic fields, the charge and current densities of molecules are generally modified. In order to investigate how the *response* of matter contributes to total electromagnetic fields, let us start by a simple electrostatic example that at the same time demonstrates the use of *Gaussian surfaces* to calculate electric fields. We shall consider a capacitor (also known as a condensator) consisting of two oppositely charged parallel plates some distance d apart. If the distance d is small compared to the dimensions of the plates, then they can be thought of as infinite charged sheets.

To calculate the electric field generated by such a capacitor, we first consider one of the sheets and draw a cylinder bisected by the sheet, as illustrated in Figure 3.11. Using the divergence

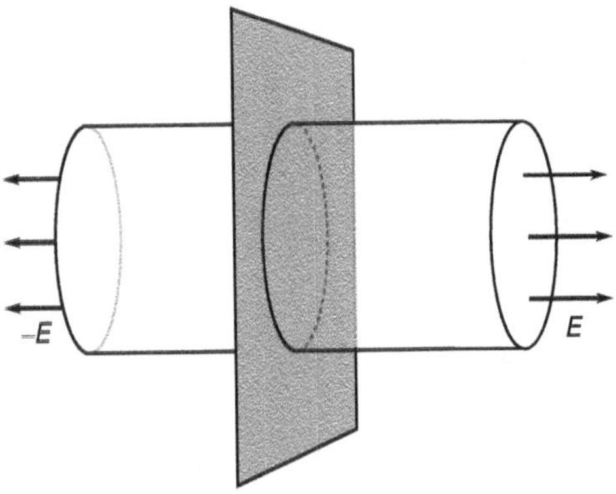

Figure 3.11 Gaussian cylinder surface bisected by an infinite flat sheet with a constant and homogeneous surface charge density σ. The area of the cylinder cross-section is A so that the enclosed charge Q_{encl} is equal to σA. Because of symmetry, the electric field $\mathbf{E}$ is perpendicular to the sheet and directed in opposite directions on separate sides of the sheet.

theorem, Eq. (3.12), we find that

$$\varepsilon_0 \int_V \nabla \cdot \mathbf{E}\, dV = \varepsilon_0 \int_S \mathbf{E} \cdot \mathbf{n}\, dS = \int_V \rho\, dV = Q_{\text{encl}}, \tag{3.271}$$

which means that the total flux of electric field out of the cylinder surface is equal to the total charge Q_{encl} contained inside. The enclosed charge is equal to the surface charge density σ times the area A of the base of the cylinder. Since any parallel component of the electric field must cancel due to symmetry, we find that only the component normal to the sheet is nonzero. Symmetry also tells us that the electric field is constant at the top and bottom surface of the cylinder, so we may write

$$2\varepsilon_0 E \int dA = \sigma A \quad \Rightarrow \quad E = \frac{\sigma}{2\varepsilon_0}. \tag{3.272}$$

We now bring together two oppositely charged sheets to form a capacitor as illustrated in Figure 3.12. Outside the plates the fields cancel, whereas between the plates the electric field is constant and adds up to

$$\mathbf{E} = \frac{\sigma}{\varepsilon_0}\mathbf{n}, \tag{3.273}$$

where the unit vector $\mathbf{n}$ points from the positive to the negative plate.

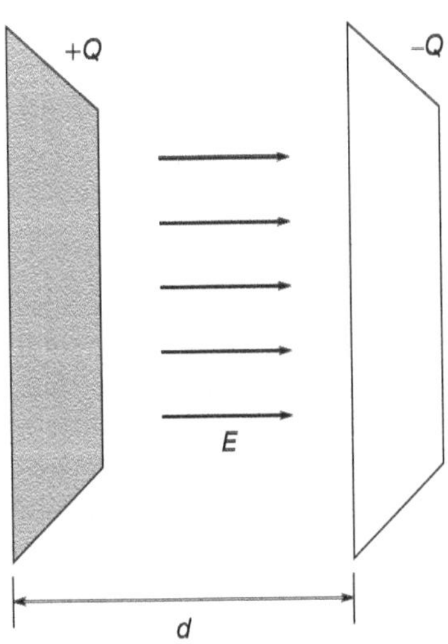

Figure 3.12 Electric field due to a parallel-plate capacitor. The electric field $\mathbf{E}$ inside the capacitor is constant and homogeneous, whereas outside it is zero.

We can measure the electric field between the plates using a voltmeter, since the potential difference between the plates is given by

$$\Delta\phi = Ed. \tag{3.274}$$

If we insert a dielectric material between the plates, the voltage, and thus the electric field, changes and becomes

$$\mathbf{E} = \frac{\sigma}{\varepsilon}\mathbf{n}, \tag{3.275}$$

where ε is the *permittivity* of the material. The permittivity is related to the electric constant ε_0—that is, the vacuum permittivity—by a factor ε_r known as the *relative permittivity*

$$\varepsilon = \varepsilon_0\varepsilon_r. \tag{3.276}$$

This reduction of the electric field by a factor ε_r corresponds to an apparent reduction of the surface charge. We say apparent because there has not occured any charge transfer between the dielectric medium and the metal plates, but the charge distribution in the dielectric material has become re-organized due to the electric field such that it partially cancels the effect of the charges on the plates. The difference in surface charge in the absence and the presence of the dielectric medium is

$$\mathbf{P} = \sigma\left(1 - \frac{1}{\varepsilon_r}\right)\mathbf{n} = \varepsilon_0(\varepsilon_r - 1)\mathbf{E} = \varepsilon_0\chi_e\mathbf{E}, \tag{3.277}$$

where $\mathbf{P}$ is called the *polarization*. The quantity $(\varepsilon_r - 1)$ is the *electric susceptibility* χ_e and is accordingly zero for the vacuum. The original surface density σ, due to "true" charges, times the unit normal vector $\mathbf{n}$ is called the *electric displacement* $\mathbf{D}$ and is given by

$$\mathbf{D} = \varepsilon_0\mathbf{E} + \mathbf{P}. \tag{3.278}$$

In this particular case, the polarization is proportional to the electric field such that $\mathbf{E}$ and $\mathbf{P}$, and therefore also $\mathbf{D}$, are all parallel, but this does not generally hold true.

In the same manner as matter may be polarized by electric fields, it can be magnetized by magnetic fields. To illustrate this and at the same time demonstrate the use of Ampèrian loops to calculate magnetic fields, we shall consider a toroidal solenoid consisting of a conducting wire tightly wound around an iron core as illustrated in Figure 3.13.

First, we consider the magnetic field generated by a steady current I passing through the toroidal coil *in the absence of the iron core.* For this purpose, we shall employ the

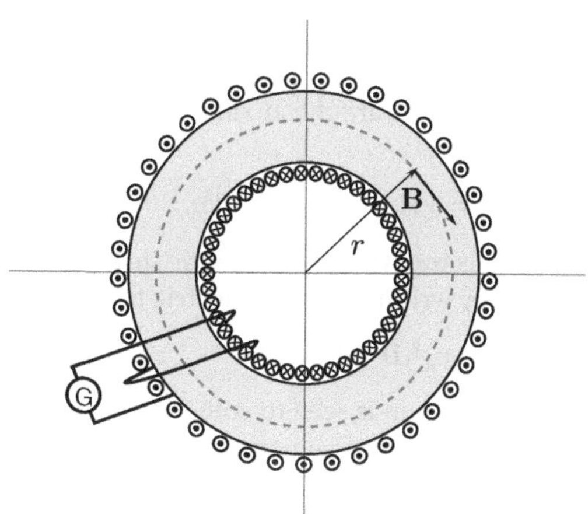

Figure 3.13 Rowland ring in terms of an iron toroid that is magnetized by a tightly wound coil. A sampling coil with a small number of windings measures the induced magnetic flux. G indicates a galvanometer. The direction of the magnetic field **B** in the figure corresponds to current entering/leaving the page in the inner/outer ring.

Ampère–Maxwell equation, Eq. (3.6). We saw in the stationary case in Section 3.1.2 that a time-independent electric field implies steady currents. However, the converse is not true, since Eq. (3.10) shows that a steady current implies that there is no irrotational electric field in the system whereas it does not exclude solenoidal electric fields. This effect is quite small, though, and will be ignored in the following, effectively reducing the Ampère–Maxwell equation to Ampère's law. Using the curl theorem in Eq. (3.14), we obtain

$$\int_S (\nabla \times \mathbf{B}) \cdot \mathbf{n}\, dS = \oint_C (\mathbf{B} \cdot \mathbf{t})\, dl = \mu_0 \int_S (\mathbf{j} \cdot \mathbf{n})\, dS = \mu_0 I_{\text{encl}}, \tag{3.279}$$

which shows that the line integral of the magnetic field along a closed loop is equal to the current I_{encl} enclosed by the loop. To calculate the magnetic field generated by the toroid coil we will therefore draw a sequence of circles, denoted as Ampèrian loops, of different radius r centered in the plane of the toroid. The unit vector $\mathbf{n}$ is then normal to the toroidal plane, whereas the vector $\mathbf{t}$ is tangential to the circle in the counterclockwise direction. From symmetry arguments, we find that the tangential component of the magnetic field, if nonzero, must be constant along such a circle, giving the general expression

$$\mathbf{B} = \mu_0 \frac{I_{\text{encl}}}{2\pi r}\mathbf{e}_\phi = \begin{cases} \mu_0 \dfrac{NI}{2\pi r}\mathbf{e}_\phi, & \text{(inside toroid)} \\ \mathbf{0}, & \text{(outside toroid)} \end{cases} \tag{3.280}$$

where N is the number of windings of the coil. For a radius r smaller than the toroid, the enclosed current is zero. For a radius r larger than the toroid, the enclosed current contains currents of opposite sign, exactly cancelling to zero.

We can measure the magnetic field inside the toroid by adding a secondary coil to the toroidal coil, forming a *Rowland ring*. If we turn the current on or off, we have a time-varying field inside the toroid and a current is induced in the secondary coil. By connecting the secondary coil to a ballistic galvanometer[11] we can recover the value of the total magnetic field inside the toroid. If we equip our toroidal coil by an iron core, there will be a dramatic increase by a factor μ_r (*relative permeability*) in the total magnetic field $\mathbf{B}$ inside the toroid according to

$$\mathbf{B} = \mu \frac{NI}{2\pi r}\mathbf{e}_\phi, \quad \mu = \mu_r \mu_0, \tag{3.281}$$

where μ is known as the permeability. Analogous to the case of polarization, we may split the total magnetic field $\mathbf{B}$ into a controllable applied part $\mu_0\mathbf{H}$, corresponding to the magnetic field in the absence of the iron coil, and a contribution $\mu_0\mathbf{M}$ reflecting the response of the medium, giving

$$\mathbf{B} = \mu_0(\mathbf{H} + \mathbf{M}), \tag{3.282}$$

where $\mathbf{H}$ and $\mathbf{M}$ are the auxilliary field and the magnetization, respectively. In this particular case, we have

$$\mathbf{M} = \chi\mathbf{H}; \quad \chi = (\mu_r - 1), \tag{3.283}$$

where χ is the (volume) *magnetic susceptibility*. The analogy is not complete, however, as can be seen by writing the total electric field as

$$\mathbf{E} = \varepsilon_0^{-1}(\mathbf{D} - \mathbf{P}). \tag{3.284}$$

The polarization is seen to come with a negative sign since it tends to reduce the applied electric field, whereas magnetization tends to strengthen the applied magnetic fields. The latter

11 A ballistic galvanometer is a galvanometer for measuring transient currents.

is not true in general, however. We will be concerned with molecules for which the magnetic susceptibilty χ is small in magnitude (typically less than 0.001) but sometimes positive (paramagnetic systems) and at other times negative (diamagnetic systems). The above iron core is an example of a ferromagnetic solid for which the magnetic susceptibility is positive and very large (for iron $\chi \approx 3000$).

3.4.3 Maxwell's Equations in Matter

Let us now attempt to establish connections between macroscopic polarization and magnetization to properties of individual molecules in matter. For this purpose, it will be convenient to introduce a distinction between free and bound charge and current densities according to

$$\rho = \rho_{\text{free}} + \rho_{\text{bound}}; \quad \mathbf{j} = \mathbf{j}_{\text{free}} + \mathbf{j}_{\text{bound}}. \tag{3.285}$$

In the case of the parallel-plate capacitor that we investigated in Section 3.4.2, the free charge density corresponds to the charges deposited on the capacitor plates. In metals and other conductors, part of the electron density is highly mobile (associated with electrons in conduction bands) and would also be included in the free charge density. In a dielectric medium, on the other hand, all electrons are bound to individual molecules (or atoms) and we may partition the bound charge density into molecular volumes according to

$$\rho_{\text{bound}}(\mathbf{r}, t) = \sum_M \rho_M(\mathbf{r} - \mathbf{R}_M, t), \tag{3.286}$$

where $\mathbf{R}_M$ denotes an associated molecular origin inside the volume. These molecular charge densities comprise both electrons and nuclei. The spatially averaged bound charge density is then

$$\begin{aligned}\overline{\rho}_{\text{bound}}(\mathbf{r}, t) &= \sum_M \int w(\mathbf{r}' - \mathbf{r})\rho_M(\mathbf{r}' - \mathbf{R}_M, t)\, d^3\mathbf{r}' \\ &= \sum_M \int w(\mathbf{r}_M + \mathbf{R}_M - \mathbf{r})\rho_M(\mathbf{r}_M, t)\, d^3\mathbf{r}_M,\end{aligned} \tag{3.287}$$

where we in the final expression have switched variables from the general coordinate $\mathbf{r}'$ to the internal coordinates $\mathbf{r}_M = \mathbf{r}' - \mathbf{R}_M$ of each molecular charge distribution, as illustrated in Figure 3.14.

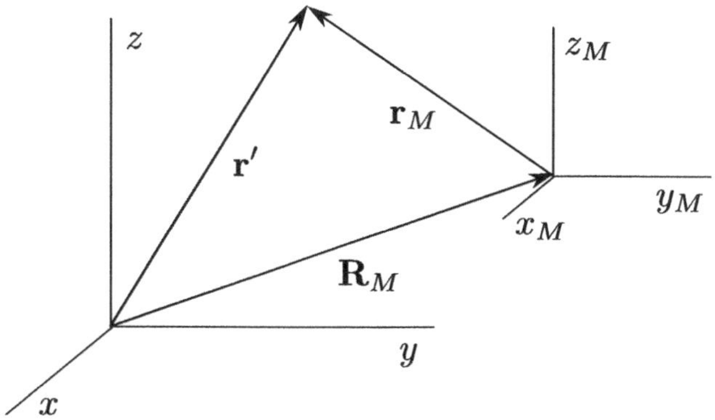

Figure 3.14 Spatial averaging of the distribution of bound charges is carried out over the dummy integration variable $\mathbf{r}'$ to obtain $\overline{\rho}_{\text{bound}}(\mathbf{r}, t)$, and it introduces, for each molecule M, a frame of reference with internal coordinates $\mathbf{r}_M$. The molecular charge density is $\rho_M(\mathbf{r}_M, t)$. The weight function depends on the distance between the observer point $\mathbf{r}$ and $\mathbf{r}'$.

Next, we perform a Taylor expansion of the weight function with respect to the integration variable around the point $\mathbf{r}_M = \mathbf{0}$ to arrive at

$$\begin{aligned}\bar{\rho}_{\text{bound}}(\mathbf{r},t) = &\sum_M w(\mathbf{R}_M - \mathbf{r}) \int \rho_M(\mathbf{r}_M,t)\, d^3\mathbf{r}_M \\ &-\boldsymbol{\nabla}\cdot \sum_M w(\mathbf{R}_M - \mathbf{r}) \int \mathbf{r}_M \rho_M(\mathbf{r}_M,t)\, d^3\mathbf{r}_M \\ &+\frac{1}{2}\sum_M \int [(\mathbf{r}_M\cdot\boldsymbol{\nabla})(\mathbf{r}_M\cdot\boldsymbol{\nabla}) w(\mathbf{R}_M - \mathbf{r})]\rho_M(\mathbf{r}_M,t)\, d^3\mathbf{r}_M - \cdots,\end{aligned} \tag{3.288}$$

where we have used that

$$\boldsymbol{\nabla}_M w = -\boldsymbol{\nabla} w. \tag{3.289}$$

The gradient operator $\boldsymbol{\nabla}$ refers to differentiation with respect to coordinate $\mathbf{r}$, independent of $\mathbf{r}_M$, and can therefore be taken outside the integral. By introducing molecular electric multipoles with respect to each molecular origin $\mathbf{R}_M$ as

$$Q^{[n]}_{M;j_1\cdots j_n} = \int r_{M;j_1} r_{M;j_2} \cdots r_{M;j_n} \rho_M(\mathbf{r}_M,t)\, d^3\mathbf{r}_M \tag{3.290}$$

and treating them as spatially averaged point electric multipoles

$$\overline{Q}^{[n]}_M(\mathbf{r}) = \langle Q^{[n]}_M \delta(\mathbf{r} - \mathbf{R}_M)\rangle, \tag{3.291}$$

the smoothed, bound charge density can be written as

$$\bar{\rho}_{\text{bound}}(\mathbf{r},t) = \rho^{\text{ion}} - \boldsymbol{\nabla}\cdot\mathbf{P}(\mathbf{r},t). \tag{3.292}$$

The first term is the spatially averaged molecular total charge

$$\rho^{\text{ion}}(\mathbf{r},t) = \sum_M \overline{Q}^{[0]}_M(\mathbf{r},t), \tag{3.293}$$

which in most cases is zero. The second term is the *macroscopic polarization* with components

$$P_i(\mathbf{r},t) = \sum_M \overline{Q}^{[1]}_{M;i}(\mathbf{r},t) + \frac{1}{2}\nabla_j \sum_M \overline{Q}^{[2]}_{M;ij}(\mathbf{r},t) + \cdots. \tag{3.294}$$

It is composed of the macroscopic electric dipole density as well as derivatives of higher macroscopic electric multipole densities, where the latter for most purposes can be ignored. Inserting into the macroscopic Maxwell's equations, we obtain

$$\boldsymbol{\nabla}\cdot\mathbf{D} = \rho_{\text{free}} \tag{3.295}$$

expressed in terms of the displacement, Eq. (3.278).

Proceeding in the same manner with the bound current density, one would be tempted to write it as a sum of molecular current densities $\mathbf{j}_M$. However, each molecular current density is charge times a velocity distribution where the velocities are relative to the molecular origins $\mathbf{R}_M$. The molecular current densities must therefore be supplemented with terms describing the displacement of the molecular volumes with respect to a *common* reference point, that is, the origin. The smoothed current density is thus

$$\bar{\mathbf{j}}_{\text{bound}}(\mathbf{r},t) = \sum_M \int d^3\mathbf{r}_M [\mathbf{V}_M \rho_M(\mathbf{r}_M,t) + \mathbf{j}_M(\mathbf{r}_M,t)] w(\mathbf{R}_M + \mathbf{r}_M - \mathbf{r}), \tag{3.296}$$

where $\mathbf{V}_M = \partial_t \mathbf{R}_M$. After a bit of work, involving use of the continuity equation, Eq. (3.11), we arrive at the following expression for the macroscopic bound current density

$$\bar{\mathbf{j}}_{\text{bound}}(\mathbf{r}, t) = \partial_t \mathbf{P} + (\boldsymbol{\nabla} \times \mathbf{M}) + \mathbf{j}^{\text{ion}} + \mathbf{j}^{\text{Röntgen}}. \tag{3.297}$$

The first term is the time derivative of the macroscopic polarization, Eq. (3.294). The second term is the curl of the magnetic equivalent, that is, the *macroscopic magnetization*

$$\mathbf{M}(\mathbf{r}, t) = \left\langle \sum_M \mathbf{m}_M^{[1]} \delta(\mathbf{R}_M - \mathbf{r}) \right\rangle + \cdots , \tag{3.298}$$

comprising the macroscopic magnetic dipole density as well as the derivatives of higher-order magnetic multipoles that will be ignored in the following. The third term comprises the spatially averaged currents due to total molecular charge

$$\mathbf{j}^{\text{ion}} = \left\langle \sum_M \mathbf{V}_M Q_M^{[0]} \delta(\mathbf{R}_M - \mathbf{r}) \right\rangle, \tag{3.299}$$

and is the current equivalent of ρ^{ion} in Eq. (3.292). We shall ignore it by assuming neutral molecules in our sample. The final term is the *Röntgen current*

$$\mathbf{j}^{\text{Röntgen}} = \boldsymbol{\nabla} \times \left[\left\langle \sum_M Q_{M;i}^{[1]} \delta(\mathbf{R}_M - \mathbf{r}) \right\rangle \times \mathbf{V}_M + \cdots \right], \tag{3.300}$$

which couples the overall velocity of the molecules to polarization. We shall ignore it by assuming that the center-of-mass velocity of the molecules in our sample relative to the laboratory frame is small.

Inserting the above expression for the macroscopic bound current density into the macroscopic equivalent of Maxwell's equation, Eq. (3.6), we obtain

$$\boldsymbol{\nabla} \times \mathbf{H} - \partial_t \mathbf{D} = \mathbf{j}_{\text{free}}, \tag{3.301}$$

coupling the auxilliary fields **H** and **D** and the macroscopic free current density $\mathbf{j}_{\text{free}}$.

We have now achieved the macroscopic equivalents of the microscopic Maxwell equations of Section 3.1.1. They can be summarized as

$$\begin{array}{llll} \boldsymbol{\nabla} \cdot \mathbf{B} &= 0 \ \text{(a)}; & \boldsymbol{\nabla} \cdot \mathbf{D} &= \rho_{\text{free}} \ \text{(c)} \\ \boldsymbol{\nabla} \times \mathbf{E} + \partial_t \mathbf{B} &= \mathbf{0} \ \text{(b)}; & \boldsymbol{\nabla} \times \mathbf{H} - \partial_t \mathbf{D} &= \mathbf{j}_{\text{free}} \ \text{(d)}. \end{array} \tag{3.302}$$

Although the homogeneous equations look exactly the same as before, it should be kept in mind that they now involve the spatially averaged macroscopic electric and magnetic fields. Also, the definitions of polarization **P** and magnetization **M**, and thereby the auxilliary fields **D** and **H**, depend on the chosen separation of free and bound charge/current density. Note also that the expressions for bound charge and current densities (Eqs. (3.292) and (3.297)), sample only the *irrotational* component of the polarization and the *solenoidal* component of magnetization, thus leaving some gauge freedom in their definition.

3.4.4 Constitutive Relations

We have seen that by separating the sources into free and bound parts we obtain equations for the auxilliary fields $\mathbf{D}/\varepsilon_0$ and $\mu_0\mathbf{H}$ in terms of the free charge and current densities, which are typically the controllable parts in our experiment; for instance, a beam of photons or spin-polarized neutrons passing through our sample. However, the homogeneous equations still involve the full electric and magnetic fields **E** and **B** containing the contribution of the bound sources through the polarization **P** and magnetization **M**. In order to solve Maxwell's equations in matter, Eq. (3.302), we need to connect the auxilliary fields with the total fields. Formally, as we have seen, the connections are

$$\mathbf{D} = \varepsilon_0\mathbf{E} + \mathbf{P}; \quad \mathbf{H} = \mu_0^{-1}\mathbf{B} - \mathbf{M}, \tag{3.303}$$

but we now have to take into account that the polarization **P** and magnetization **M** arise not just from the *permanent* (molecular) electric and magnetic multipoles associated with the bound charge and current density of the medium. Rather, they may have significant, and even dominant, contributions from multipole moments *induced* by the applied fields, and, to make things really complicated, in order to calculate the induced multipole moments it is not enough to know the applied fields; we need to know the *total* electric and magnetic fields. This leads to *constitutive relations* for the auxilliary fields that are typically complicated functionals of the total fields

$$\mathbf{D} \equiv \varepsilon_0\mathbf{E} + \mathbf{P}[\mathbf{E}, \mathbf{B}]; \quad \mathbf{H} \equiv \mu_0^{-1}\mathbf{B} - \mathbf{M}[\mathbf{E}, \mathbf{B}]. \tag{3.304}$$

The form of these equations is somewhat misleading, giving the impression that the auxiliary fields **D** and **H** arise in response to the total fields **E** and **B**, whereas rather the opposite is the case. The form of the constitutive relations arise precisely because the induced polarization and magnetization depend on the *total* fields, and not only on the applied ones.

As an example, we may consider the measurement of magnetization in a ferromagnetic material. Our experimental setup will be a sophisticated version of the Rowland ring that we discussed in Section 3.4.2. For continuous magnetic induction we connect our primary coil to an AC power supply. The total magnetic field **B** and the applied magnetic field $\mu_0 H$ can be found from Eqs. (3.302b) and (3.302d), respectively. In practice, we find $\mu_0 H$ by measuring the current (in fact, voltage) of the primary coil and the total magnetic field **B** by time integration of the induced voltage in the secondary coil (using a operational amplifier integrator). If we start from unmagnetized iron, then the total magnetic field B as a function of the applied magnetic field $\mu_0 H$ first traces the curve 1 → 2 as sketched in Figure 3.15 and which is far from the linear relation suggested by Eq. (3.283). Next, for each cycle of alternating current, the curve 2 → 3 → 4 → 5 is traced, showing that the total field B is a multivalued function of the applied field $\mu_0 H$, depending on the history of the magnetization. The resulting *hysteresis* curve arises from the modification of magnetic domains inside the material.

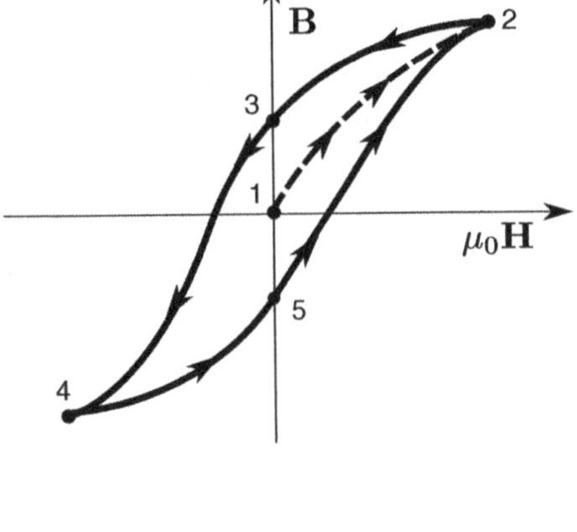

Figure 3.15 Magnetic hysteresis curve.

3.5 Linear Media

When the total electric and magnetic fields **E** and **B** are not too strong, it is reasonable to assume that they are connected in a linear fashion to the auxiliary fields **D** and **H**. It should be noted, though, that this implies that not only the applied field, but also the response of the medium, is weak, an assumption that clearly does not hold for ferromagnetic materials. The most general form of the constitutive relations for linear media is

$$D_i(\mathbf{r},t) = \sum_j \int\int [A_{ij}(\mathbf{r},\mathbf{r}',t,t')E_j(\mathbf{r}',t') + T_{ij}(\mathbf{r},\mathbf{r}',t,t')B_j(\mathbf{r}',t')]d^3\mathbf{r}'dt', \tag{3.305}$$

$$H_i(\mathbf{r},t) = \sum_j \int\int [U_{ij}(\mathbf{r},\mathbf{r}',t,t')E_j(\mathbf{r}',t') + X_{ij}(\mathbf{r},\mathbf{r}',t,t')B_j(\mathbf{r}',t')]d^3\mathbf{r}'dt', \tag{3.306}$$

where the *material constants* A, T, U, and X appear. They are written as functions of two points in time and space. This allows for nonlocality in both space and time, that is, the polarization and magnetization at a given point in space and time may not only depend on the value of the electric and magnetic fields in the same point, but also on the neighboring points in space as well as (previous) times. However, unless the properties of the material evolves in time, the two times can be replaced by the time difference. Likewise, if the material is homogeneous in space on a macroscopic scale, we may replace the two space points by their difference. The constitutive relations then become

$$D_i(\mathbf{r},t) = \sum_j \int\int [A_{ij}(\mathbf{r}-\mathbf{r}',t-t')E_j(\mathbf{r}',t') + T_{ij}(\mathbf{r}-\mathbf{r}',t-t')B_j(\mathbf{r}',t')]d^3\mathbf{r}'dt', \tag{3.307}$$

$$H_i(\mathbf{r},t) = \sum_j \int\int [U_{ij}(\mathbf{r}-\mathbf{r}',t-t')E_j(\mathbf{r}',t') + X_{ij}(\mathbf{r}-\mathbf{r}',t-t')B_j(\mathbf{r}',t')]d^3\mathbf{r}'dt'. \tag{3.308}$$

These expressions can now be recognized as convolutions, and a Fourier transform in space and time gives

$$D_i(\mathbf{k},\omega) = \sum_j [A_{ij}(\mathbf{k},\omega)E_j(\mathbf{k},\omega) + T_{ij}(\mathbf{k},\omega)B_j(\mathbf{k},\omega)], \tag{3.309}$$

$$H_i(\mathbf{k},\omega) = \sum_j [U_{ij}(\mathbf{k},\omega)E_j(\mathbf{k},\omega) + X_{ij}(\mathbf{k},\omega)B_j(\mathbf{k},\omega)]. \tag{3.310}$$

Spatial dispersion, or the dependence of the material constants on the wave vector **k**, is determined by the ratio between characteristic lengths in the medium, such as molecular dimensions in a solution, to the wave length of the fields. We can conclude that spatial dispersion can be ignored to the extent that the electric dipole approximation (see Section 3.3.2) holds. Temporal dispersion, often simply referred to as dispersion, will depend on the ratio between the time scale of relaxation/resonance processes in the medium to the time scale, or inverse frequency, of the fields. The range of energies in a molecule is very wide: from rotations in the microwave domain, through vibrations in the infrared, electronic excitations in the visible, ultraviolet, and X-ray domains to nuclear excitations in the gamma region. Overlap with the energies, or frequencies, associated with the field is therefore quite likely, so that temporal dispersion can in general not be ignored. Relaxation processes may furthermore introduce phase differences between the total and auxiliary fields, which are conveniently taken into account by expressing the material constants as complex quantities.

One should also note that the material constants in general are 3×3 matrices, reflecting the fact that the response of the medium may depend on the relative orientation with respect to the applied fields, leading to effects of *anisotropy* such as birefringence and dichroism, which we shall discuss in Section 8.6.

In the following text, we shall mostly be concerned with *simple linear dielectrics*, where we have

$$\mathbf{D} = \varepsilon_0 \varepsilon_r \mathbf{E} \quad \Rightarrow \quad \mathbf{P} = \varepsilon_0(\varepsilon_r - 1)\mathbf{E}. \tag{3.311}$$

3.5.1 Boundary Conditions

In order to treat electric and magnetic fields in different media, we have to connect fields through interfaces. The boundary conditions follow directly from the macroscopic Maxwell equations, Eq. (3.302). They involve the divergence and curl of the fields, but with the help of the divergence and curl theorems, Eqs. (3.12) and (3.14), we can directly address the normal and tangential components, respectively, of the fields on both sides of the interface.

Consider the interface between medium A and B. At each point on the interface we introduce a normal vector $\mathbf{n}_{AB}$ pointing from B into A. To connect the normal components of field vectors across the interface, we start by drawing a small flat cylinder ("pillbox") with axis parallel to $\mathbf{n}_{AB}$, of surface dA and height dh straddling the interface, as shown in Figure 3.16(a). From the divergence theorem we obtain

$$\int_V \rho_f dV = \int_V (\nabla \cdot \mathbf{D}) dV = \int_S (\mathbf{D} \cdot \mathbf{n}) \; dS, \tag{3.312}$$

where S represents the surface enclosing the volume V of the pillbox and $\mathbf{n}$ a unit vector in each point directed outward normal to the surface. Let us now consider what happens as the cylinder is squashed into the surface from both sides. As we compress the height of the pillbox, the contributions of $\mathbf{D}$ parallel to the surface are eliminated. For a sufficiently small surface of the pillbox we can also take the normal component of $\mathbf{D}$ to be constant over the surface dA. The total charge in the pillbox is $dq = \rho_f dh dA$ and goes to zero unless there is a singularity of ρ_f at the interface, giving a nonzero surface *free* charge density $\sigma_f = \rho_f dh$. The result of the limiting process is therefore

$$\mathbf{n}_{AB} \cdot (\mathbf{D}_A - \mathbf{D}_B) = \sigma_f. \tag{3.313}$$

Following the same argument, we obtain for the magnetic field

$$\mathbf{n}_{AB} \cdot (\mathbf{B}_A - \mathbf{B}_B) = 0. \tag{3.314}$$

To connect the tangential components of the field vectors across the interface, we imagine a long razorblade cutting into the interface, as shown Figure 3.16(b). In more mathematical

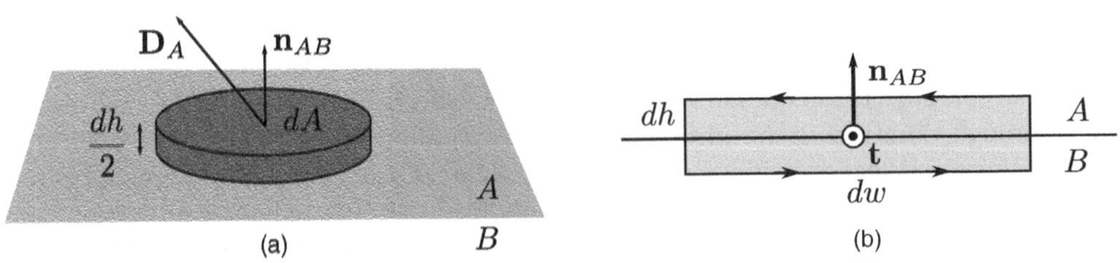

Figure 3.16 Surface boundary between media A and B. (a) A cylinder with area *dA* and height *dh* straddles the interface, with the cylinder axis being parallel with the normal $\mathbf{n}_{AB}$ of the boundary surface. (b) A rectangle with width *dw* and heigth *dh* is placed across the surface, oriented so that the surface tangent **t** is a normal to the rectangle.

terms, we consider a small rectangle of height dh and width dw ($dh \ll dw$) with orientation defined by the normal vector $\mathbf{t}$ tangential to the interface. From the curl theorem, we obtain

$$\int_S (\mathbf{j}_f + \partial_t \mathbf{D}) \cdot \mathbf{t}\, dS = \int_S (\nabla \times \mathbf{H}) \cdot \mathbf{t}\, dS = \oint_C \mathbf{H} \cdot d\mathbf{l}, \tag{3.315}$$

where the final term is a line integral around the edges of the rectangle. As the rectangle is compressed into a single line along the interface, the components of $\mathbf{H}$ perpendicular to the surface vanish. The total flux of current through the rectangle will vanish as well unless there is a singularity in the tangential free current density corresponding to a nonzero *free* surface current density $\mathbf{K}_\text{f} = \mathbf{j}_\text{f} dh$. The displacement $\mathbf{D}$ is assumed finite on the interface and so the time derivative of its flux through the surface vanishes. The result of the limiting process is therefore

$$\mathbf{K}_\text{f} \cdot \mathbf{t} = (\mathbf{H}_A - \mathbf{H}_B) \cdot (\mathbf{t} \times \mathbf{n}_{AB}) = \{\mathbf{n}_{AB} \times (\mathbf{H}_A - \mathbf{H}_B)\} \cdot \mathbf{t}. \tag{3.316}$$

Since the orientation of $\mathbf{t}$ on the interface is arbitrary, we can write

$$\mathbf{n}_{AB} \times (\mathbf{H}_A - \mathbf{H}_B) = \mathbf{K}_\text{f}, \tag{3.317}$$

where it is understood that the surface current is oriented in the plane of the interface. Following the same line of reasoning, we obtain for the electric field

$$\mathbf{n}_{AB} \times (\mathbf{E}_A - \mathbf{E}_B) = \mathbf{0}. \tag{3.318}$$

Instead of starting from the Maxwell equations involving the auxiliary fields $\mathbf{D}$ and $\mathbf{H}$, we can start from the macroscopic equivalent of Eqs. (3.5) and (3.6). We then obtain the boundary conditions

$$\mathbf{n}_{AB} \cdot (\mathbf{E}_A - \mathbf{E}_B) = \sigma/\varepsilon_0, \tag{3.319}$$
$$\mathbf{n}_{AB} \times (\mathbf{B}_A - \mathbf{B}_B) = \mu_0 \mathbf{K}, \tag{3.320}$$

where σ and $\mathbf{K}$ are the *total* surface charge and current densities, respectively. If we combine the above boundary conditions with the definitions of the auxiliary fields in Eqs. (3.278) and (3.282), we find the relations

$$\mathbf{n}_{AB} \cdot (\mathbf{P}_A - \mathbf{P}_B) = -\sigma_\text{b}, \tag{3.321}$$
$$\mathbf{n}_{AB} \times (\mathbf{M}_A - \mathbf{M}_B) = \mathbf{K}_\text{b}, \tag{3.322}$$

where σ_b and $\mathbf{K}_\text{b}$ are the surface *bound* charge and current densities, respectively.

As an example of the use of the boundary conditions, consider a dielectric sphere of radius a and relative permittivity ε_A placed inside some other dielectric material of relative permittivity ε_B—this setup is illustrated in Figure 3.17. The aim is to investigate the fields in our system when applying an external uniform electric field. This is a purely electrostatic problem and so we can work with scalar potentials only. There are no free charges in our system, including surface charges, and so the Laplace equation

$$\nabla^2 \phi(\mathbf{r}) = 0, \tag{3.323}$$

is valid all over space. We shall work in spherical coordinates and put the origin at the center of the sphere. The boundary condition Eq. (3.313) translates into

$$\mathbf{n}_{AB} \cdot (\mathbf{D}_A - \mathbf{D}_B) = -\varepsilon_0 \mathbf{n}_{AB} \cdot (\varepsilon_A \nabla \phi_A - \varepsilon_B \nabla \phi_B) = 0, \tag{3.324}$$

which in the present case is equivalent to

$$\varepsilon_A \left.\frac{\partial \phi_A}{\partial r}\right|_{r=a} = \varepsilon_B \left.\frac{\partial \phi_B}{\partial r}\right|_{r=a}. \tag{3.325}$$

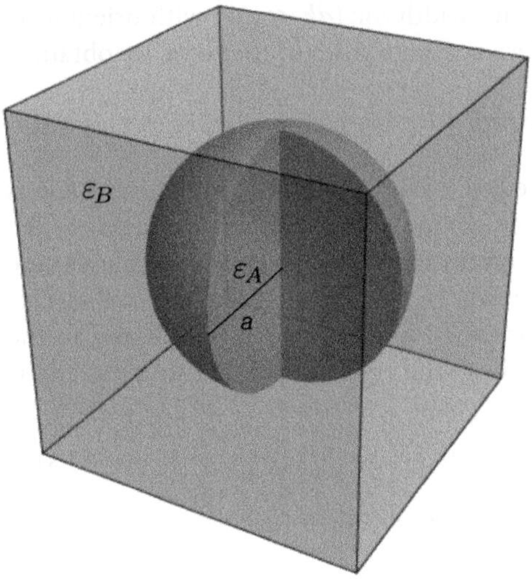

Figure 3.17 Dielectric sphere in dielectric medium.

In order for the derivatives to exist, the scalar potential must be continuous over the boundary

$$\phi_A|_{r=a} = \phi_B|_{r=a}. \tag{3.326}$$

The relation

$$\hat{\mathbf{l}} \cdot \hat{\mathbf{l}} = (\mathbf{r} \times \hat{\mathbf{p}}) \cdot (\mathbf{r} \times \hat{\mathbf{p}}) = (\mathbf{r}.\boldsymbol{\nabla})(\mathbf{r}.\boldsymbol{\nabla}) + (\mathbf{r}.\boldsymbol{\nabla}) - r^2\nabla^2, \tag{3.327}$$

allows us to write the Laplace equation as

$$\frac{1}{r^2}\left[r\frac{\partial}{\partial r}\left(r\frac{\partial}{\partial r}+1\right) - \hat{\mathbf{l}}^2\right]\phi(\mathbf{r}) = 0, \tag{3.328}$$

and perform a separation of radial and angular variables. The spherical harmonics $Y_{lm}(\theta, \phi)$ are eigenfunctions of the squared orbital angular momentum operator with eigenvalues $l(l+1)$, but, in the present case, we have axial symmetry. We can therefore, without loss of generality, align the z-axis with the applied field, set $m = 0$ and work with the Legendre polynomials $P_l(\cos\theta)$. The purely radial problem now reads

$$\left[r\frac{\partial}{\partial r}\left(r\frac{\partial}{\partial r}+1\right) - l(l+1)\right]R(r) = 0. \tag{3.329}$$

Setting the radial solutions $R(r) = r^n$ (integer n), we obtain

$$n(n+1) = l(l+1), \tag{3.330}$$

with two solutions, $n = l$ and $n = -(l+1)$. The general solution of the Laplace equation in the case of axial symmetry can thereby be expressed as

$$\phi(\mathbf{r}) = \sum_{l=0}^{\infty}\left(A_l r^l + \frac{B_l}{r^{l+1}}\right)P_l(\cos\theta). \tag{3.331}$$

We now use the boundary conditions to find the coefficients A_l and B_l inside and outside the sphere.

Let $\mathbf{E}_0$ be the applied uniform field in medium B in the absence or sufficiently far from medium A. In other words, we impose

$$\lim_{r\to\infty}\phi_B(\mathbf{r}) = -\mathbf{E}_0 \cdot \mathbf{r} = -E_0 r\cos\theta, \tag{3.332}$$

where we recognize that $\cos\theta$ is the Legendre polynomial for $l = 1$. From the fact that the Legendre polynomials form an orthonormal set, we conclude that outside the sphere we only get contributions from $l = 1$ and that the coefficient $A_1^{\text{out}} = -E_0$, whereas the coefficient B_1^{out} so far is undetermined. From the continuity of the scalar potential across the boundary between dielectrics A and B, Eq. (3.326), we find that inside the sphere, $l = 1$ as well. We furthermore have $B_1^{\text{in}} = 0$ to avoid a singularity at the origin. At the boundary, we therefore have

$$A_1^{\text{in}} a = -E_0 a + \frac{B_1^{\text{out}}}{a^2} \quad \Rightarrow \quad B_1^{\text{out}} = a^3(A_1^{\text{in}} + E_0). \tag{3.333}$$

Connecting in the same manner the normal derivatives across the boundary, Eq. (3.325), we have

$$\varepsilon_A A_1^{\text{in}} = \varepsilon_B \left(-E_0 - \frac{2B_1^{\text{out}}}{a^3} \right). \tag{3.334}$$

Solving for the two unknowns, we obtain the general solutions

$$\phi_A(\mathbf{r}) = -\left[\frac{3\varepsilon_B}{\varepsilon_A + 2\varepsilon_B} \right] E_0 r \cos\theta; \quad \phi_B(\mathbf{r}) = -\left[1 - \left(\frac{\varepsilon_A - \varepsilon_B}{\varepsilon_A + 2\varepsilon_B} \right) \frac{a^3}{r^3} \right] E_0 r \cos\theta. \tag{3.335}$$

Note that if $\varepsilon_A = \varepsilon_B$ we have a uniform field $\mathbf{E}_0$ all over space.

Let us now consider the special case of a *dielectric sphere of relative permittivity* ε_r *in a uniform electric field* $\mathbf{E}_0$. We then set $\varepsilon_A = \varepsilon_r$ and $\varepsilon_B = 1$, leading to

$$\phi^{\text{in}}(\mathbf{r}) = -\left[\frac{3}{\varepsilon_r + 2} \right] E_0 r \cos\theta; \quad \phi^{\text{out}}(\mathbf{r}) = -\left[1 - \left(\frac{\varepsilon_r - 1}{\varepsilon_r + 2} \right) \frac{a^3}{r^3} \right] E_0 r \cos\theta. \tag{3.336}$$

The electric field and polarization inside the sphere is accordingly

$$\mathbf{E}^{\text{in}} = \left(\frac{3}{\varepsilon_r + 2} \right) \mathbf{E}_0 = \mathbf{E}_0 - \left(\frac{\varepsilon_r - 1}{\varepsilon_r + 2} \right) \mathbf{E}_0, \tag{3.337}$$

$$\mathbf{P}^{\text{in}} = \varepsilon_0(\varepsilon_r - 1)\mathbf{E}^{\text{in}} = 3\varepsilon_0 \left(\frac{\varepsilon_r - 1}{\varepsilon_r + 2} \right) \mathbf{E}_0. \tag{3.338}$$

We see that inside the dielectric sphere the applied field is indeed reduced by polarization. On the other hand, the electric displacement, $\mathbf{D}^{\text{in}} = \varepsilon_0\varepsilon_r\mathbf{E}^{\text{in}}$, in this case, does not correspond to the controllable applied field. The corresponding fields outside the sphere are

$$\mathbf{E}^{\text{out}} = \mathbf{E}_0 + E_0 \left(\frac{\varepsilon_r - 1}{\varepsilon_r + 2} \right) \frac{a^3}{r^5} (3xz, 3yz, 3z^2 - r^2), \tag{3.339}$$

$$\mathbf{P}^{\text{out}} = \mathbf{0}. \tag{3.340}$$

Note that the outside polarization is zero because $\varepsilon_B = 1$. The charge density associated with the bound charge is zero since the divergence of the polarization is zero everywhere, see Eq. (3.292). However, there is a discontinuity in the polarization at the interface, which reflects a surface-bound charge density

$$\sigma_b = \boldsymbol{e}_r \cdot (\mathbf{P}^{\text{in}} - \mathbf{P}^{\text{out}}) = 3\varepsilon_0 \left(\frac{\varepsilon_r - 1}{\varepsilon_r + 2} \right) E_0 \cos\theta. \tag{3.341}$$

Let us now try to solve the same problem, but without making a distinction between free and bound charges. We again use spherical coordinates and exploit axial symmetry. The inside and outside potentials have the form

$$\phi^{\text{in}}(\mathbf{r}) = A_1^{\text{in}} r \cos\theta; \quad \phi^{\text{out}}(\mathbf{r}) = \left(-E_0 r \cos\theta + \frac{B_1^{\text{out}}}{r^2} \right) \cos\theta. \tag{3.342}$$

Continuity of the scalar potential across the boundary leads again to Eq. (3.333), but due to the surface-bound charge density, the normal derivatives across the boundary are connected as

$$\varepsilon_0\left(\frac{\partial\phi^{\text{in}}}{\partial r}-\frac{\partial\phi^{\text{out}}}{\partial r}\right)_{r=a}=\sigma_b=3\varepsilon_0\left(\frac{\varepsilon_r-1}{\varepsilon_r+2}\right)E_0\cos\theta. \quad (3.343)$$

Proceeding as above, we arrive at the same potentials as before.

3.5.2 Polarization in Linear Media

We now arrive at the heart of the matter. In this section, we shall investigate how the macroscopic polarization of a linear dielectric is related to microscopic molecular properties. Let us first consider a linear medium consisting of molecules without a permanent electric dipole moment. Polarization then arises primarily from *induced* electric dipole moments and can be expressed as

$$\mathbf{P}(\mathbf{r},t)=\sum_M\left\langle \mathbf{Q}_M^{[1]}\delta(\mathbf{r}-\mathbf{M})\right\rangle=\sum_M\alpha_M\langle\mathbf{E}_i\delta(\mathbf{r}-\mathbf{M})\rangle=\sum_M\alpha_M\overline{\mathbf{E}}_i, \quad (3.344)$$

where appears the (isotropic) dipole polarizability α_M of molecule M. There is, however, a catch, indicated by the subscript i in the above equation: in order to calculate the induced electric dipole moment we need to know the electric field "felt" by the molecule, which is the total electric field *minus* the electric field associated with the molecule itself. This quantity is in the literature called the *internal* field.

We shall consider three different ways of connecting the microscopic polarizability to the macroscopic polarizability, each of which sheds a different light on this connection. For simplicity, we will assume that our medium consists of identical molecules, each occupying a spherical cavity of radius R inside the medium. This means that the number density $\mathcal{N}$ of molecules satisfies

$$\mathcal{N}\frac{4\pi}{3}R^3=1. \quad (3.345)$$

One way to calculate the internal field is to cut out the part of the medium corresponding to this cavity and then calculate the electric field inside this medium due to the surrounding medium. Notice that this is different from our previous calculation of the electric field inside a spherical cavity since we now keep the overall polarization frozen, that is, it is not allowed to adapt to this physical situation. Rather than directly calculating this electric field, it is simpler to calculate the electric field $\mathbf{E}_{\text{plug}}$ associated with the spherical volume removed from the medium and then subtract it from the total field (see Figure 3.18), that is

$$\overline{\mathbf{E}}_i=\mathbf{E}-\mathbf{E}_{\text{plug}}. \quad (3.346)$$

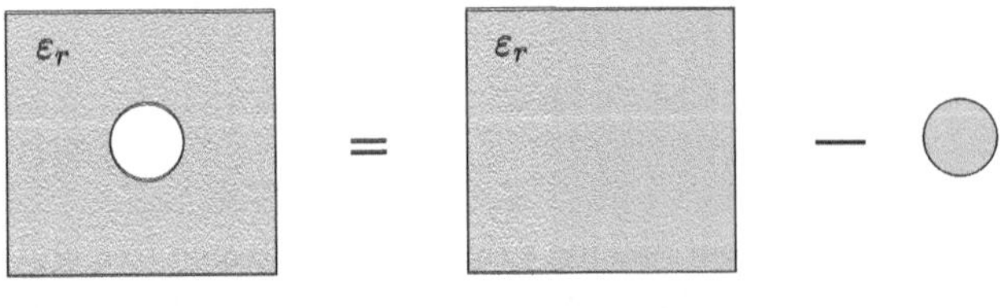

Figure 3.18 The internal field $\mathbf{E}_i$ can be calculated as the difference between the total field inside the dielectric and the field associated with the molecular volume.

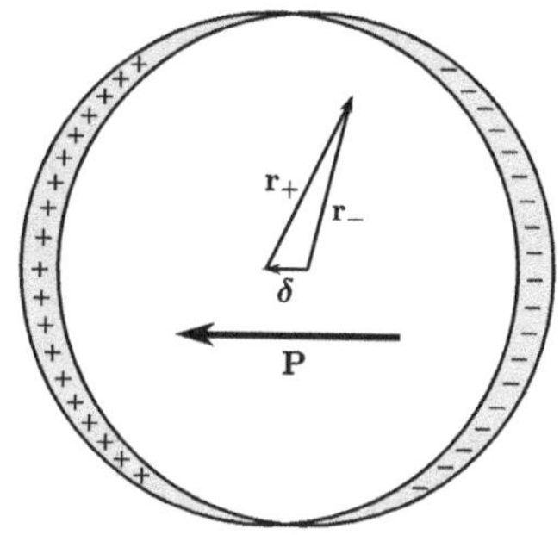

Figure 3.19 The field $\mathbf{E}_{plug}$ of a spherical plug of frozen polarization can be calculated by considering two uniformly charged spheres.

The polarization inside the sphere is homogeneous and there is accordingly no bound charge density, according to Eq. (3.292). At the boundary of the sphere there is clearly a discontinuity since the polarization is zero outside, and thus an associated surface bound charge density (see Eq. (3.321)). We can mimic this situation by considering two identical spheres, uniformly charged with density ρ, but of opposite sign (see Figure 3.19). If the two spheres are fully overlapping, the total charge density is zero all over space. Polarization is induced by the introduction of a slight displacement $\boldsymbol{\delta}$, giving

$$\mathbf{P} = \rho\boldsymbol{\delta}. \tag{3.347}$$

In the bulk of the sphere, the total charge density is still zero, but an inhomogeneous surface charge appears. To calculate the total field inside the combined sphere, we consider each sphere separately. We can use Gaussian surfaces to calculate the associated electric field. Outside the sphere the electric field is

$$\mathbf{E}^{out}(\mathbf{r}) = \frac{Q\mathbf{r}}{4\pi\varepsilon_0 r^3}; \quad Q = \frac{4\pi}{3}R^3\rho. \tag{3.348}$$

In this case, the Gaussian surface enclosed all the charge Q. For an interior point, the electric field has the same form, but the enclosed charge is reduced

$$\mathbf{E}^{in}(\mathbf{r}) = \frac{Q_r\mathbf{r}}{4\pi\varepsilon_0 r^3}; \quad Q_r = \frac{4\pi}{3}r^3\rho, \tag{3.349}$$

so that

$$\mathbf{E}^{in}(\mathbf{r}) = \frac{\rho\mathbf{r}}{3\varepsilon_0}. \tag{3.350}$$

The electric field inside the dielectric spherical volume is thus

$$\mathbf{E}_{plug}(\mathbf{r}) = \frac{\rho_+\mathbf{r}_+}{3\varepsilon_0} + \frac{\rho_-\mathbf{r}_-}{3\varepsilon_0}, \tag{3.351}$$

where $\mathbf{r}_\pm$ is the distance to the observer point with respect to the centers of the two spheres. We obtain

$$\mathbf{E}_{plug}(\mathbf{r}) = \frac{\rho}{3\varepsilon_0}(\mathbf{r}_+ - \mathbf{r}_-) = \frac{-\rho\boldsymbol{\delta}}{3\varepsilon_0} = -\frac{\mathbf{P}}{3\varepsilon_0}, \tag{3.352}$$

showing that the total field inside the sphere is uniform. The internal field is accordingly

$$\overline{\mathbf{E}}_i = \frac{\varepsilon_r + 2}{3}\mathbf{E}. \tag{3.353}$$

For a nonpolar dielectric, we thereby find

$$\mathbf{P} = \mathcal{N}\alpha\overline{\mathbf{E}}_i = \mathcal{N}\alpha\frac{\varepsilon_\mathrm{r}+2}{3}\mathbf{E} = \varepsilon_0(\varepsilon_\mathrm{r}-1)\mathbf{E}, \tag{3.354}$$

which upon rearrangement gives

$$\frac{\mathcal{N}\alpha}{3\varepsilon_0} = \frac{\varepsilon_\mathrm{r}-1}{\varepsilon_\mathrm{r}+2}, \tag{3.355}$$

known as the Clausius–Mossotti relation, which relates the microscopic polarizability and the macroscopic relative permittivity.

Another way to obtain the same relation is to take a macroscopic spherical piece of our medium and place it in a uniform electric field $\mathbf{E}_0$, where the subscript 0 emphasizes that this is the electric field in vacuum. We have already calculated the electric field in the medium (see Eq. (3.337))

$$\mathbf{E} = \frac{3}{\varepsilon_\mathrm{r}+2}\mathbf{E}_0, \tag{3.356}$$

such that we directly obtain

$$\mathbf{P} = 3\varepsilon_0\left(\frac{\varepsilon_\mathrm{r}-1}{\varepsilon_\mathrm{r}+2}\right)\mathbf{E}_0 = \mathcal{N}\alpha\mathbf{E}_i. \tag{3.357}$$

Here we can basically read off the Clausius–Mossotti relation, Eq. (3.355), but only if the internal field is equal to the external field $\mathbf{E}_0$ in vacuum. By direct calculation, we find that this is in fact true

$$\overline{\mathbf{E}}_i = \mathbf{E} + \frac{\mathbf{P}}{3\varepsilon_0} = \frac{3}{\varepsilon_\mathrm{r}+2}\mathbf{E}_0 + \frac{\varepsilon_\mathrm{r}-1}{\varepsilon_\mathrm{r}+2}\mathbf{E}_0 = \mathbf{E}_0, \tag{3.358}$$

a result we could also have obtained by simply expressing Eq. (3.356) in terms of $\mathbf{E}_0$ and comparing with Eq. (3.353). Thinking about this, it does makes sense, because the internal field is the electric field in the absence of the dielectric medium, and that is the corresponding electric field in vacuum.

A final, somewhat more complicated, but also more powerful approach is to introduce the notion of a *reaction field*. In our first approach, as shown in Figure 3.19, we calculated the internal field $\overline{\mathbf{E}}_i$ by carving out a spherical cavity from a dielectric medium. The internal field is then the electric field inside this cavity when the surrounding medium is not allowed to react. Let us see what happens if we relax this condition. It corresponds to setting $\varepsilon_A = 1$ and $\varepsilon_B = \varepsilon_\mathrm{r}$ in the general solution in Eq. (3.335). We find that the electric field inside the spherical cavity is

$$\mathbf{E}_\mathrm{cav} = \left(\frac{3\varepsilon_\mathrm{r}}{2\varepsilon_\mathrm{r}+1}\right)\mathbf{E}. \tag{3.359}$$

It is clearly different from the internal field, where polarization is frozen. The difference arises from the fact that in our case the spherical cavity is not empty, but contains an induced dipole moment. Let us calculate the fields associated with the induced dipole. We have already seen that the scalar potential of an isolated point dipole, induced or not, is given by Eq. (3.193). Now we place the dipole at the center of a spherical cavity of radius R inside a linear medium of relative permittivity ε_r. Without loss of generality we will align the dipole with the z-axis. The general form of the scalar potential is then given by the expansion Eq. (3.331). If we probe the scalar potential close to the center of the cavity we should recover the scalar potential of an isolated dipole, so we impose the boundary condition

$$\lim_{r \to 0} \phi(\mathbf{r}) = \frac{\mu \cos\theta}{4\pi\varepsilon_0 r^2}. \tag{3.360}$$

From the orthonormality of the Legendre functions it then follows that only the term $l = 1$ contributes to the scalar potential inside the cavity. From continuity across the boundary, the same conclusion also holds outside the cavity. The scalar potential therefore has the general form

$$\phi^{\text{in}}(\mathbf{r}) = \left[Ar + \frac{\mu}{4\pi\varepsilon_0 r^2} \right] \cos\theta, \tag{3.361}$$

$$\phi^{\text{out}}(\mathbf{r}) = \frac{B}{r^2} \cos\theta. \tag{3.362}$$

It should be noted that the term of the potential linear in the radial coordinate is deleted for the outside potential to avoid that it blows up in the limit $r \to \infty$. The boundary conditions in Eqs. (3.325) and (3.326) then lead to the equations

$$A - \frac{3\mu}{4\pi\varepsilon_0 R^3} = -\frac{3B}{R^3}\varepsilon_r; \quad AR + \frac{\mu}{4\pi\varepsilon_0 R^2} = \frac{B}{R^2}. \tag{3.363}$$

Solving this system of equations leads to the final expressions

$$\phi^{\text{in}}(\mathbf{r}) = \left[-\frac{2(\varepsilon_r - 1)}{2\varepsilon_r + 1} \frac{\mu}{4\pi\varepsilon_0 R^3} r + \frac{\mu}{4\pi\varepsilon_0 r^2} \right] \cos\theta, \tag{3.364}$$

$$\phi^{\text{out}}(\mathbf{r}) = \frac{3}{(2\varepsilon_r + 1)} \frac{\mu}{4\pi\varepsilon_0 r^2} \cos\theta. \tag{3.365}$$

We note that the outside potential corresponds to that of the dipole screened by a factor $3/(2\varepsilon_r + 1)$. Inside the cavity we see the potential of an isolated dipole, but also an additional contribution, which is the potential of the reaction field due to the reaction of the surrounding medium and which is proportional to the dipole

$$\mathbf{R} = f\boldsymbol{\mu}, \tag{3.366}$$

where the factor of proportionality is given by

$$f = \frac{2(\varepsilon_r - 1)}{2\varepsilon_r + 1} \frac{1}{4\pi\varepsilon_0 R^3} = \frac{2(\varepsilon_r - 1)}{2\varepsilon_r + 1} \frac{\mathcal{N}}{3\varepsilon_0}. \tag{3.367}$$

The final form is obtained using Eq. (3.345). The internal field is the sum of all fields inside the cavity, except that of the isolated dipole, that is

$$\overline{\mathbf{E}}_i = \mathbf{E}_{\text{cav}} + \mathbf{R}. \tag{3.368}$$

Knowing from our second approach that $\overline{\mathbf{E}}_i = \mathbf{E}_0$, we can instead write

$$\mathbf{E}_0 = \frac{3\varepsilon_r}{2\varepsilon_r + 1}\mathbf{E} + \mathbf{R} = \frac{3\varepsilon_r}{2\varepsilon_r + 1} \frac{3}{\varepsilon_r + 2}\mathbf{E}_0 + \mathbf{R}, \tag{3.369}$$

which leads to

$$\mathbf{R} = \frac{2(\varepsilon_r - 1)^2}{(2\varepsilon_r + 1)(\varepsilon_r + 2)}\mathbf{E}_0 = \frac{2(\varepsilon_r - 1)}{(2\varepsilon_r + 1)} \frac{\mathcal{N}\alpha}{3\varepsilon_0}\mathbf{E}_0, \tag{3.370}$$

and from which we obtain the Clausius–Mossotti relation, Eq. (3.355), by elimination of common factors.

Let us now consider what happens if the molecules carry permanent dipoles $\boldsymbol{\mu}_0$. The polarization will now clearly depend on the orientation of these dipoles in the medium. We know that the energy of interaction between a dipole $\boldsymbol{\mu}_0$ and an electric field $\mathbf{E}_d$ is

$$U(\theta) = -\boldsymbol{\mu}_0 \cdot \mathbf{E}_\mathrm{d} = -\mu_0 E_\mathrm{d} \cos\theta, \tag{3.371}$$

which favors a parallel alignment. We can extract the average orientation of dipoles in our sample from the Boltzmann ensemble average of the interaction energy, which in this case takes the form

$$\langle U(\theta)\rangle = \frac{N_B}{D_B} = -\mu_0 E_\mathrm{d}\langle\cos\theta\rangle = \frac{\int_0^\pi U(\theta)\exp[-\beta U(\theta)]p(\theta)d\theta}{\int_0^\pi \exp[-\beta U(\theta)]p(\theta)d\theta}; \quad \beta = \frac{1}{kT}, \tag{3.372}$$

and where $p(\theta)d\theta$ is the probability of finding the dipole oriented at an angle θ with respect to the electric field. Notice that the form of the interaction does not favor any particular orientation of the component of the dipole perpendicular to the electric field, and so this component averages to zero. This means that the polarization is given by the component of the dipole moments oriented along the electric field, that is

$$\mathbf{P} = \mathcal{N}\mu_0\langle\cos\theta\rangle\frac{\mathbf{E}_\mathrm{d}}{E_\mathrm{d}}. \tag{3.373}$$

Once again the devil is in the details, in this case in the subscript d of $\mathbf{E}_\mathrm{d}$. An important point to note is that we should include in the ensemble average only the contributions to the interaction energy that actually has an orientation dependence. Let us investigate what they actually are. The full interaction energy is given by

$$U = -\boldsymbol{\mu}_0 \cdot \overline{\mathbf{E}}_i = -\boldsymbol{\mu}_0 \cdot (\mathbf{E}_\mathrm{cav} + \mathbf{R}) = -\boldsymbol{\mu}_0 \cdot (\mathbf{E}_\mathrm{cav} + f\boldsymbol{\mu}), \tag{3.374}$$

where we have used Eqs. (3.366) and (3.368). The total dipole $\boldsymbol{\mu}$ is

$$\boldsymbol{\mu} = \boldsymbol{\mu}_0 + \alpha\overline{\mathbf{E}}_i = \boldsymbol{\mu}_0 + \alpha(\mathbf{E}_\mathrm{cav} + f\boldsymbol{\mu}). \tag{3.375}$$

Re-arranging, we find

$$\boldsymbol{\mu} = \frac{1}{1-f\alpha}\boldsymbol{\mu}_0 + \frac{\alpha}{1-f\alpha}\mathbf{E}_\mathrm{cav}. \tag{3.376}$$

Inserting into Eq. (3.374), we find

$$U = -\boldsymbol{\mu}_0 \cdot \left(\mathbf{E}_\mathrm{cav} + \frac{f}{1-f\alpha}\boldsymbol{\mu}_0 + \frac{f\alpha}{1-f\alpha}\mathbf{E}_\mathrm{cav}\right) = -\frac{1}{1-f\alpha}\mu_0^2 - \frac{1}{1-f\alpha}\boldsymbol{\mu}_0 \cdot \mathbf{E}_\mathrm{cav}. \tag{3.377}$$

Only the second term has a dependence on θ and can be thought of as the interaction of the permanent electric dipole with the directional component

$$\mathbf{E}_\mathrm{d} = \frac{1}{1-f\alpha}\mathbf{E}_\mathrm{cav} = \frac{1}{1-f\alpha}\left(\frac{3\varepsilon_\mathrm{r}}{2\varepsilon_\mathrm{r}+1}\right)\mathbf{E}, \tag{3.378}$$

of the electric field, hence the subscript d. Another way to arrive at the same result is to consider the torque

$$\mathbf{N} = \boldsymbol{\mu}_0 \times \overline{\mathbf{E}}_i, \tag{3.379}$$

which induces orientation of the permanent dipole in the presence of the internal field. Only the components of $\overline{\mathbf{E}}_i$ which are not parallel to $\boldsymbol{\mu}_0$, corresponding to $\mathbf{E}_\mathrm{d}$, will contribute to this torque.

Let us now proceed to the calculation of the ensemble average. We first calculate the probability $p(\theta)d\theta$. Consider a sphere of radius R. The probability $p(\theta)d\theta$ is the ratio between the area of a ribbon between θ and $\theta + d\theta$ and the total area $4\pi R^2$ of the sphere. The area of the ribbon is

$$\int_0^{2\pi}\int_\theta^{\theta+d\theta} R^2 \sin\theta d\theta d\phi = 2\pi R^2 \sin\theta d\theta, \tag{3.380}$$

so the probability is

$$p(\theta)d\theta = \frac{1}{2}\sin\theta d\theta. \tag{3.381}$$

We next calculate the denominator of the ensemble average

$$D_B = \frac{1}{2}\int_0^\pi \exp[-\beta U(\theta)]\sin\theta d\theta = \frac{1}{2\mu_0 E_\mathrm{d}}\int_{U(0)}^{U(\pi)} \exp[-\beta U'(\theta)]dU' = \frac{1}{x}\sinh x, \tag{3.382}$$

where

$$x = \beta\mu_0 E_\mathrm{d}.$$

When calculating the numerator it helps to note that it is a simple derivative of the denominator

$$N_B = \frac{1}{2}\int_0^\pi U(\theta)\exp[-\beta U(\theta)]\sin\theta d\theta = -\frac{\partial}{\partial\beta}D_B = \mu_0 E_\mathrm{d}\left(\frac{1}{x^2}\sinh x - \frac{1}{x}\cosh x\right). \tag{3.383}$$

The orientational average then becomes

$$\langle\cos\theta\rangle = L(x), \tag{3.384}$$

where we have introduced the *Langevin function*

$$L(x) = \coth x - \frac{1}{x}. \tag{3.385}$$

The Taylor expansion of $\coth x$ is

$$\coth x = \frac{1}{x} + \frac{x}{3} - \frac{x^3}{45} + \frac{2x^5}{945} + \cdots, \tag{3.386}$$

so that for small x we have

$$L(x) \approx \frac{x}{3} \quad\Rightarrow\quad \langle\cos\theta\rangle = \frac{\mu_0 E_\mathrm{d}}{3kT}. \tag{3.387}$$

The polarization then becomes

$$\mathbf{P} = \mathcal{N}\frac{\mu_0^2}{3kT}\mathbf{E}_\mathrm{d}. \tag{3.388}$$

Notice that it is only within this linear regime that the description of the medium is linear. Figure 3.20 displays the Langevin function and its linear approximation together with a tabulation of some representative values. For $x = 0.1$, the error of the linear approximation is negligible, yet it typically corresponds to quite strong directing electric fields. As an example, we can consider the water molecule, whose dipole moment in gas phase is 1.85 Debye or 6.17×10^{-30} C m. At room temperature, a value $x = 0.1$ corresponds to an electric field strength of $6.6 \times 10^7\,\mathrm{V\,m^{-1}}$, which is indeed strong, keeping in mind that an electric field of $3 \times 10^6\,\mathrm{V\,m^{-1}}$ will create sparks in air. On the other hand, we see that the field causes very little alignment of water molecules.

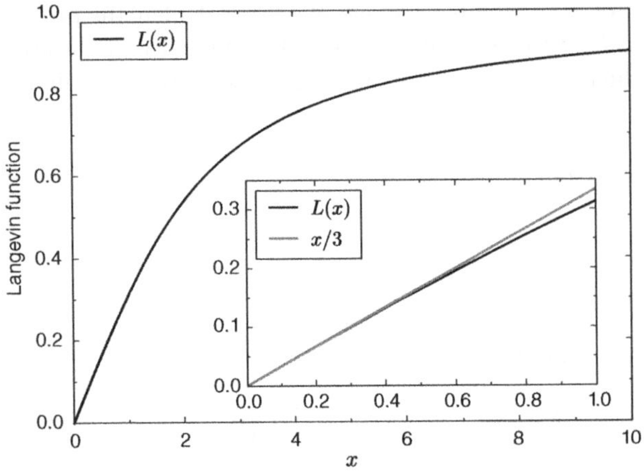

x	$L(x)$	$x/3$	Deviation (%)
0.1	0.0333	0.0333	0.07
0.2	0.0665	0.0667	0.27
0.5	0.1640	0.1667	1.65
1.0	0.3130	0.3333	6.48

Figure 3.20 The Langevin function, Eq. (3.385), and its linear approximation $x/3$. Selected values are tabulated to the right.

For molecular properties that have nonlinear dependencies on the external fields, one generally finds that the orientational effects can be expressed in a power series of the inverse temperature

$$A_0 + \frac{1}{T}A_i + \frac{1}{T^2}A_2 + \cdots . \tag{3.389}$$

By fitting observable quantities measured at different temperatures to this power series, we can extract the pure electronic contributions themselves from the temperature-independent term A_0 as well as the lower-order multipole moments or response properties from the coefficients A_1, $\mathrm{A}_2, \ldots$. Such nonlinear temperature dependencies are particularly common for nonlinear hyperpolarizabilities or birefringences, as we will briefly touch upon in Sections 8.3.2 and 8.6.

3.5.3 Electromagnetic Waves in a Linear Medium

In this section, we will derive equations for electromagnetic waves in linear media, and then study their propagation as well as their attenuation due to absorption. The macroscopic Maxwell equations are given in Eq. (3.302). In order to allow for frequency-dependent permittivity ε and permeability μ, we first perform a Fourier decomposition. We then obtain

$$\begin{aligned} \nabla\cdot\mathbf{D}(\mathbf{r},\omega) &= \rho_f(\mathbf{r},\omega); \quad \nabla\times\mathbf{E}(\mathbf{r},\omega) - i\omega\mathbf{B}(\mathbf{r},\omega) = 0; \\ \nabla\times\mathbf{H}(\mathbf{r},\omega) + i\omega\mathbf{D}(\mathbf{r},\omega) &= \mathbf{j}_f(\mathbf{r},\omega); \quad \nabla\cdot\mathbf{B}(\mathbf{r},\omega) = 0. \end{aligned} \tag{3.390}$$

We now consider a case where

$$\mathbf{B}(\mathbf{r},\omega) = \mu(\omega)\mathbf{H}(\mathbf{r},\omega); \quad \mathbf{D}(\mathbf{r},\omega) = \varepsilon(\omega)\mathbf{E}(\mathbf{r},\omega), \tag{3.391}$$

and insert these expressions into Eq. (3.375) to arrive at

$$\begin{aligned} \nabla\cdot\mathbf{E}(\mathbf{r},\omega) = \rho_f(\mathbf{r},\omega)/\varepsilon(\omega); &\quad \nabla\times\mathbf{E}(\mathbf{r},\omega) - i\omega\mathbf{B}(\mathbf{r},\omega) = 0; \\ \nabla\times\mathbf{B}(\mathbf{r},\omega) + i\omega\mu(\omega)\varepsilon(\omega)\mathbf{E}(\mathbf{r},\omega) = \mu(\omega)\mathbf{j}_f(\mathbf{r},\omega); &\quad \nabla\cdot\mathbf{B}(\mathbf{r},\omega) = 0. \end{aligned} \tag{3.392}$$

From this result and by following the approach in Section 3.1.3, we may extract the wave equations

$$\left[\nabla^2 + \frac{\omega^2}{c^2}\mu_r(\omega)\varepsilon_r(\omega)\right] \mathbf{E}(\mathbf{r},\omega) = \boldsymbol{\nabla}\rho_f(\mathbf{r},\omega)/\varepsilon(\omega) - i\omega\mu(\omega)\mathbf{j}_f(\mathbf{r},\omega),$$

$$\left[\nabla^2 + \frac{\omega^2}{c^2}\mu_r(\omega)\varepsilon_r(\omega)\right] \mathbf{B}(\mathbf{r},\omega) = -\mu(\omega)(\boldsymbol{\nabla}\times\mathbf{j}_f(\mathbf{r},\omega)).$$

In the absence of free sources, we get equations for electromagnetic waves propagating in the medium. Plane wave solutions

$$\mathbf{E}(\mathbf{r},\omega) = \mathbf{E}_0 \exp[i(\mathbf{k}\cdot\mathbf{r})]; \quad \mathbf{B}(\mathbf{r},\omega) = \mathbf{B}_0 \exp[i(\mathbf{k}\cdot\mathbf{r})], \tag{3.393}$$

are obtained with the condition

$$k^2(\omega) = \frac{\omega^2}{c^2}\mu_r(\omega)\varepsilon_r(\omega). \tag{3.394}$$

We introduce the frequency-dependent refractive index $n(\omega)$ by the relation

$$n^2(\omega) = \mu_r(\omega)\varepsilon_r(\omega). \tag{3.395}$$

Comparing with Section 3.1.5 we see that each frequency component of the wave has its own velocity of propagation in the medium

$$\mathbf{v}(\omega) = \frac{c}{n(\omega)}\mathbf{e}_k, \tag{3.396}$$

determined by the frequency-dependent refractive index. A subtle distinction needs to be made: the above velocity is the *phase velocity* $\mathbf{v}_\mathrm{p}$ of the wave. If the frequency-components of the total wave are concentrated around a mean frequency $\overline{\omega}$, it makes sense to also define a *group velocity*

$$\mathbf{v}_\mathrm{g} = \left[\left|\frac{\partial\mathbf{k}}{\partial\omega}\right|^{-1}\right]_{\omega=\overline{\omega}}, \tag{3.397}$$

which describes the speed at which the overall envelope of the resulting wave packet travels. Furthermore, there is a quantity called the *signal velocity* which describes the speed at which an electromagnetic wave carries information. It is usually equal to the group velocity, but in specially designed media it is possible to achieve group velocities beyond the speed of light and the signal velocity.

Let us now investigate the energy transport associated with an electromagnetic wave propagating through a linear medium. Following the approach used in Section 3.2.1, we start by considering the mechanical work associated with the *free* current in some medium

$$\frac{dW}{dt} = \int (\mathbf{E}\cdot\mathbf{j}_\mathrm{f})\, d^3\mathbf{r}. \tag{3.398}$$

Using Maxwell's macroscopic laws, Eq. (3.302), we obtain

$$\frac{dW}{dt} = \int \mathbf{E}\cdot(\boldsymbol{\nabla}\times\mathbf{H} - \partial_t\mathbf{D})d^3\mathbf{r} = \int [(\mathbf{H}\cdot(\boldsymbol{\nabla}\times\mathbf{E}) - \boldsymbol{\nabla}\cdot(\mathbf{E}\times\mathbf{H})) - \mathbf{E}\cdot\partial_t\mathbf{D}]d^3\mathbf{r}, \tag{3.399}$$

which can be further worked into

$$\frac{dW}{dt} + \int [\mathbf{E}\cdot\partial_t\mathbf{D} + \mathbf{H}\cdot\partial_t\mathbf{B}]d^3\mathbf{r} = -\int \boldsymbol{\nabla}\cdot(\mathbf{E}\times\mathbf{H})\, d^3\mathbf{r}. \tag{3.400}$$

In the general case, there is no simple way to proceed since the connection between electric and magnetic fields and the auxilliary fields **D** and **H** may be complicated, as illustrated in Figure 3.15. In the special case of simple linear media, Eq. (3.391), we obtain

$$\frac{dW}{dt} + \frac{1}{2}\frac{d}{dt}\int\left(\varepsilon E^2 + \frac{1}{\mu}B^2\right)d^3\mathbf{r} = -\int \nabla\cdot(\mathbf{E}\times\mathbf{H})d^3\mathbf{r} = -\int(\mathbf{S}\cdot\mathbf{n})\,dS, \tag{3.401}$$

where appears the Poynting vector

$$\mathbf{S} = \frac{1}{\mu}(\mathbf{E}\times\mathbf{B}), \tag{3.402}$$

generalized to linear media.

Consider now a nonmagnetic medium ($\mu_r = 1$) with a complex frequency-dependent relative permittivity written in the form

$$\varepsilon_{\rm r} = \varepsilon_1 + i\varepsilon_2. \tag{3.403}$$

The wave vector then becomes complex with magnitude $\tilde{k} = k + i\kappa$, and the electromagnetic wave *decays* inside the material according to

$$\mathbf{E}(\mathbf{r},\omega) = \mathbf{E}_0 \exp[i(\mathbf{k}\cdot\mathbf{r})]\exp[-\boldsymbol{\kappa}\cdot\mathbf{r}],$$
$$\mathbf{B}(\mathbf{r},\omega) = \mathbf{B}_0 \exp[i(\mathbf{k}\cdot\mathbf{r})]\exp[-\boldsymbol{\kappa}\cdot\mathbf{r}].$$

We can connect the real and imaginary parts of the wave vector to the corresponding components of the relative permittivity using Eq. (3.394)

$$k^2 - \kappa^2 = \frac{\omega^2}{c^2}\varepsilon_1; \quad 2k\kappa = \frac{\omega^2}{c^2}\varepsilon_2. \tag{3.404}$$

With a bit more work, we find the relations

$$k = \frac{\omega}{c}\sqrt{\frac{1}{2}(|\varepsilon_{\rm r}| + \varepsilon_1)}; \quad \kappa = \frac{\omega}{c}\sqrt{\frac{1}{2}(|\varepsilon_{\rm r}| - \varepsilon_1)}. \tag{3.405}$$

The intensity associated with frequency ω of the wave is given by

$$I(\omega) = \langle S\rangle_T; \quad \langle\mathbf{S}\rangle_T = \frac{1}{2\mu}\mathrm{Re}(\mathbf{E}\times\mathbf{B}^*), \tag{3.406}$$

where we have used Eq. (3.135). The intensity is position dependent; the intensities at two different points $\mathbf{r}_1$ and $\mathbf{r}_2$ are related by

$$I_1 = I_2 e^{-2\boldsymbol{\kappa}\cdot(\mathbf{r}_1-\mathbf{r}_2)}. \tag{3.407}$$

This leads to the relation

$$-\ln\frac{I}{I_0} = \alpha l, \tag{3.408}$$

where I and I_0 are the intensity of transmitted and incident radiation and l is the path length through the sample. The proportionality constant $\alpha = 2\kappa$ is called the *absorption coefficient* and is proportional to the number density $\mathcal{N}$ of the absorbing molecules in the sample

$$\alpha = \mathcal{N}\sigma. \tag{3.409}$$

Defining *absorbance* as

$$A = -\log_{10}\frac{I}{I_0}, \tag{3.410}$$

we combine Eqs. (3.408) and (3.409) to express the empirical Beer–Lambert law[12]

$$A = \varepsilon c l \quad \text{or} \quad I = I_0 e^{-\mathcal{N}\sigma l} = I_0 \times 10^{-\varepsilon c l}, \tag{3.411}$$

where appears the molar concentration c (the number density is $\mathcal{N} = N_A c$) of the absorbing molecules and the molar (decadic) absorption coefficient

$$\varepsilon = \frac{N_A \sigma}{\ln 10} = \frac{\alpha}{c \ln 10}. \tag{3.412}$$

The absorption cross-section σ expresses the probability of absorption and has the dimension of area. To get a feeling for the physical meaning of an absorption cross section, we can think in classical terms about an absorption experiment where a beam of projectiles (photons) is sent through some medium as illustrated in Figure 3.21. We can think of the medium as a sequence of slabs of surface A and thickness dl in which target particles of cross section σ are embedded, and impose that a projectile is absorbed if it hits a target particle. The probability of absorption of a single projectile is therefore given by the fraction of the total surface covered by target particles. The probability is accordingly $(\mathcal{N} A dl)\sigma / A = \mathcal{N}\sigma dl$, where $\mathcal{N}$ is the target number density. The total number of projectiles hitting the surface in some time interval Δt is $JA\Delta t$, where J is the projectile particle flux. The change in projectile particle flux upon passage through the slab is thereby $dJ = -J\mathcal{N}\sigma dl$. We convert from photon flux to energy flux (intensity) by $I = \hbar\omega J$ and obtain the Beer–Lambert law in Eq. (3.411).

An alternative derivation of the Beer–Lambert law proceeds as follows[13]: Consider a straight tube of length l and (target particle) cross-section σ going through the medium. The transmittance I/I_0 measures the fraction of photons going through the medium without being absorbed and therefore corresponds to the probability that a single photon goes through the tube without being absorbed. Owing to the high speed of light we can ignore all molecular motion and

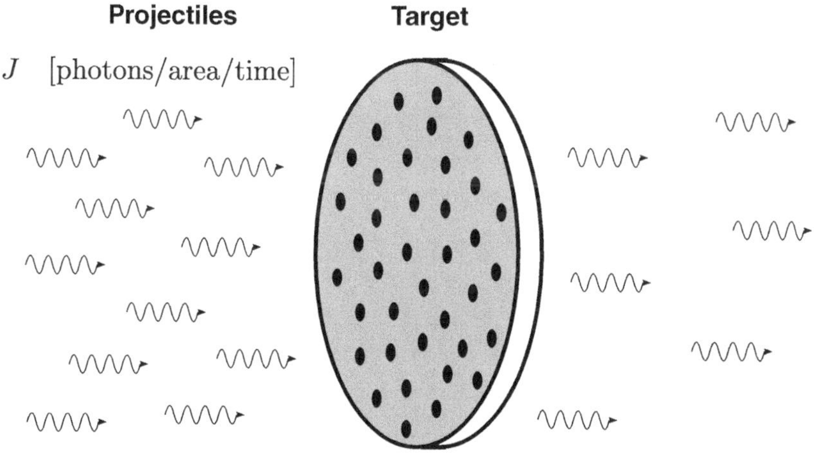

Figure 3.21 Projectiles (photons) with flux J are incident on a target medium slab of thickness dl and area A containing particles with individual cross section σ. As a result of the interaction with the medium, the photon flux is reduced upon passage.

12 A more appropriate name is perhaps the Beer–Lambert–Bouguer law. Pierre Bouguer (1698–1758) was apparently the first to suggest that absorbance is proportional to optical path length, a finding later reported by Johann Heinrich Lambert (1728–1777). August Beer (1825–1863) proposed that absorbance was proportional to concentration.

13 See M. N. Berberan-Santos. *J. Chem. Educ.*, 67:757, 1990.

consider the probability of all molecules being outside the tube at a given instant. If there are N particles in the total volume V, then this probability is given as

$$\frac{I}{I_0} = \left(1 - \frac{v}{V}\right)^N. \tag{3.413}$$

Since the tube volume $v = \sigma l$ is very much smaller than the total volume, we can write the probability (transmittance) as

$$\frac{I}{I_0} = e^{-\frac{N}{V}v} = e^{-\mathcal{N}\sigma l}, \tag{3.414}$$

which again is the Beer–Lambert law.

3.5.4 Frequency Dependence of the Permittivity

In this section, we will look at the dependency of the permittivity on the frequency in linear media, where the relation

$$P(\omega) = \varepsilon_0[\varepsilon_r(\omega) - 1]E(\omega), \tag{3.415}$$

holds. The relative permittivity here is complex as defined in Eq. (3.403). In the following text, we shall use liquid water as an example. In Figure 3.22, we show the real and imaginary parts of the relative permittivity of liquid water at 25° C for a very sizable portion of the electromagnetic spectrum. The solid lines are based on the compilation by Segelstein,[14]

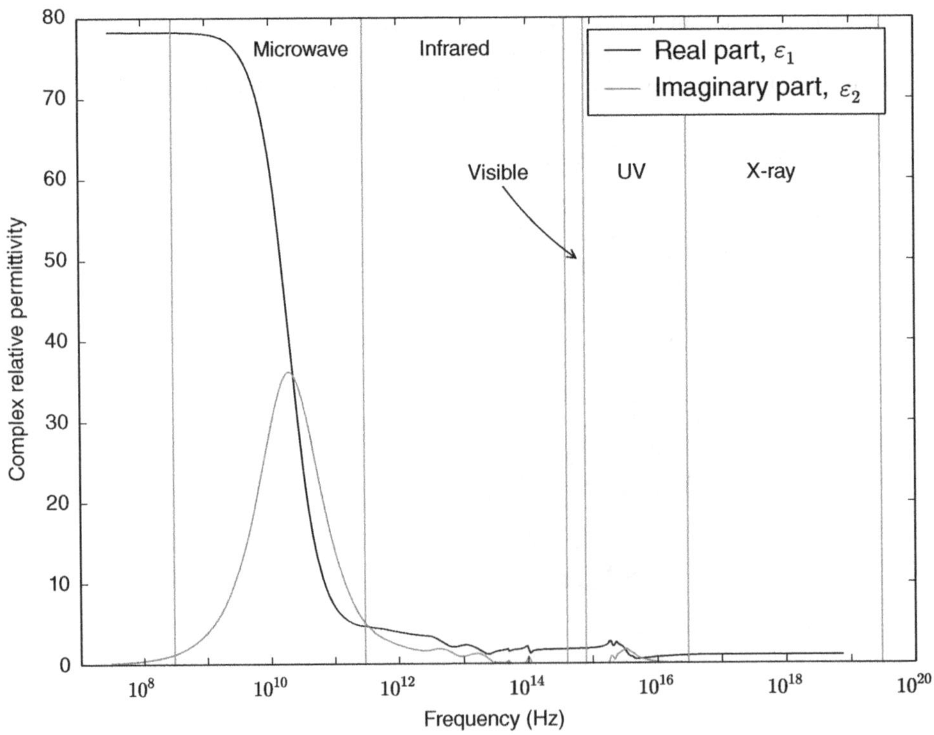

Figure 3.22 Complex relative permittivity of liquid water at 25 °C as a function of frequency on a logarithmic scale.

14 D. J. Segelstein. *The complex refractive index of water*. M.S. thesis, University of Missouri–Kansas City, 1981.

whereas the extension into the X-ray domain, represented by dashed lines, is generated from atomic scattering factors.[15] The real part ε_1 has an overall inverted sigmoidal shape, starting off from around 80 and then decreasing quickly through the microwave region. One clearly sees some structure (wiggles) in the infrared through ultraviolet region, but ε_1 finally converges to the vacuum value $\varepsilon_r = 1$ in the X-ray region. The imaginary part ε_2 starts off and ends at zero, but shows a pronounced maximum at the inflection point of ε_1 in the microwave region and wiggles at higher frequencies.

3.5.4.1 Kramers–Kronig Relations

The real and imaginary parts of the complex permittivity are not independent. Instead, they are connected through the Kramers–Kronig relations. We can show this in the following manner: the time-dependent polarization can be expressed in terms of its inverse Fourier transform

$$P(t) = \int_{-\infty}^{+\infty} P(\omega)e^{-i\omega t}d\omega = \varepsilon_0 \int_{-\infty}^{+\infty} [\varepsilon_r(\omega) - 1]E(\omega)e^{-i\omega t}d\omega, \tag{3.416}$$

where we have used Eq. (3.415). The time-dependent polarization is seen to be the inverse Fourier transform of a product of frequency-dependent functions and can therefore be expressed as a convolution of their Fourier transform partners, as discussed in Appendix D. Specifically, we have

$$P(t) = \frac{\varepsilon_0}{2\pi} \int_{-\infty}^{+\infty} E(t')G(t - t')dt', \tag{3.417}$$

where appears the response function

$$G(t) = \int_{-\infty}^{+\infty} [\varepsilon_r(\omega) - 1]e^{-i\omega t}d\omega. \tag{3.418}$$

Let us assume that the electric field is turned on at $t = 0$. A crucial observation is then that the response function has to be zero for $t < 0$, and, since relativity does not allow instantaneous interactions, also for $t = 0$. It is this condition of causality that is at the origin of the Kramers–Kronig relations, as we shall now demonstrate.

We may write that inverse transform of Eq. (3.418) as

$$\varepsilon_r(\omega) - 1 = \int_{-\infty}^{+\infty} G(t)e^{+i\omega t}dt = \int_{-\infty}^{+\infty} G(t)\cos\omega t dt + i\int_{-\infty}^{+\infty} G(t)\sin\omega t dt, \tag{3.419}$$

where, in the final expression we have expanded the complex exponential. The cosine and sine functions are even and odd functions, respectively, with respect to the time argument. Generally, even and odd functions are defined by

$$f_{\text{even}}(t) = f_{\text{even}}(-t); \quad f_{\text{odd}}(t) = -f_{\text{odd}}(-t). \tag{3.420}$$

A second crucial observation is that since the integration domain of the Fourier transform is symmetric about $t = 0$, the total integrand has to be *even* in order to have a nonzero integral. This allows us to decompose the Fourier transform into real and imaginary parts

$$\varepsilon_1(\omega) - 1 = \int_{-\infty}^{+\infty} G_{\text{even}}(t)\cos\omega t dt; \quad \varepsilon_2(\omega) = \int_{-\infty}^{+\infty} G_{\text{odd}}(t)\sin\omega t dt. \tag{3.421}$$

Let us now consider the construction of the even and odd parts of the response function $G(t)$. The general formulas are

$$f_{\text{even}}(t) = \frac{1}{2}(f(t) + f(-t)); \quad f_{\text{odd}}(t) = \frac{1}{2}(f(t) - f(-t)). \tag{3.422}$$

15 See B. L. Henke, E. M. Gullikson, and J. C. Davis. *At. Data Nucl. Data Tables*, 54:181, 1993.

Owing to the causal nature of the response function, we find that the sign of the odd part depends on the sign of the time argument. We can express this fact by using the sign function sgn(t) defined in Eq. (D.13) as follows

$$G_{\text{even}}(t) = \frac{1}{2}G(t); \quad G_{\text{odd}}(t) = \text{sgn}(t)\frac{1}{2}G(t) = \text{sgn}(t)G_{\text{even}}(t). \tag{3.423}$$

These expressions imply that the Fourier transform of the odd part of the response function is the Fourier transform of a product of the sign function and the even part of the response function, which can therefore be expressed as a convolution in the frequency domain (Eq. (D.20)), that is

$$\mathcal{F}[G_{\text{odd}}(t)](\omega) = -i\varepsilon_2(\omega) = \int_{-\infty}^{+\infty} \mathcal{F}[\text{sgn}](\omega - \omega')\mathcal{F}[G_{\text{even}}](\omega')d\omega'. \tag{3.424}$$

Since we can equivalently write $G_{\text{even}}(t) = \text{sgn}(t)G_{\text{odd}}(t)$, we similarly obtain

$$\mathcal{F}[G_{\text{even}}(t)](\omega) = \varepsilon_1(\omega) = \int_{-\infty}^{+\infty} \mathcal{F}[\text{sgn}](\omega - \omega')\mathcal{F}[G_{\text{odd}}](\omega')d\omega'. \tag{3.425}$$

The Fourier transform of the sign function is given in Eq. (D.16) and the final result is thus

$$\varepsilon_1(\omega) = 1 - \frac{1}{\pi}\int_{-\infty}^{+\infty} \frac{\varepsilon_2(\omega')}{\omega - \omega'}d\omega'; \quad \varepsilon_2(\omega) = \frac{1}{\pi}\int_{-\infty}^{+\infty} \frac{\varepsilon_1(\omega') - 1}{\omega - \omega'}d\omega', \tag{3.426}$$

which is known as the Kramers–Kronig relations. These relations can for instance provide a useful check on experimental data.

3.5.4.2 Relaxation in the Debye Model

The relative permittivity is expressed as a complex quantity when there is a phase lag ϕ between the electric displacement and the total electric field. If we consider a monochromatic electric field

$$E(t) = E_0 e^{-i\omega t}, \tag{3.427}$$

where we again make use of the complex representation of the electric field and it is understood that it is the real part that corresponds to the physical field. With the use of the complex permittivity in Eq. (3.403), we may write the corresponding electric displacement as

$$D(t) = \varepsilon_0\varepsilon_r E(t) = D_0 e^{-i(\omega t - \phi)}, \tag{3.428}$$

which demonstrates that the imaginary part of the permittivity causes a phase lag in the electric displacement with respect to the electric field. We also note that we are allowed to write the electric displacement and the total electric field as scalars since they have the same direction. The underlying physical explanation for this phase lag is connected with absorption as we have already seen that the imaginary part of the relative permittivity is associated with absorption, and therefore energy loss, in a dielectric medium.

One mechanism of phase lag is *relaxation*, whereby polarization builds up with a delay when an electric field is turned on, and also does not go away instantaneously when the electric field is turned off. Suppose that at time t_0 we turn on a constant electric field E_0. We may then write the polarization at some later time t as

$$P(t) = a(t - t_0)E_0. \tag{3.429}$$

In order to establish causality, the relaxation function $a(t)$ must clearly be zero for negative time arguments. One simple form is

$$a(t) = a_0[1 - e^{-t/\tau}]; \quad t > 0, \tag{3.430}$$

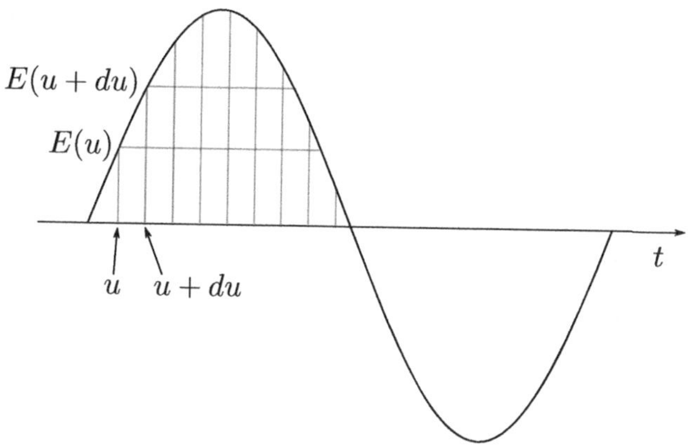

Figure 3.23 A decomposition of a time-dependent electric field into a sum of static fields.

where τ is the *relaxation time* and a_0 a constant of proportionality. We are interested in time-dependent rather than static fields. However, we can write a time-dependent electric field as a sum of static fields with different durations. For instance, at time $u + du$ we have

$$E(u + du) = E(u) + \frac{dE(u)}{du}du, \tag{3.431}$$

as illustrated in Figure 3.23. Using the superposition principle we can therefore write the total polarization as

$$P(t) = a(t - t_0)E(t_0) + \int_{t_0^+}^{t} a(t-u)\frac{dE(u)}{du}du, \tag{3.432}$$

where the lower bound t_0^+ means that integration starts immediately after the field is switched on, to avoid an ill-defined derivative. We next carry out integration by parts to obtain

$$P(t) = a(0)E(t) - \int_{t_0^+}^{t} E(u)\frac{da(t-u)}{du}du. \tag{3.433}$$

Comparing with Eq. (3.417), we see that $\varepsilon_0 G(t) = 2\pi da(t)/dt$. A nonzero $a(0)$ in the first term would imply an instantaneous onset of polarization, which is not only forbidden by relativity but also impossible to achieve in practice. However, it will be useful to single out a component of polarization, associated with vibrational and electronic degrees of freedom, which follows the electric field (almost) perfectly and which we denote

$$P_\infty(t) = \varepsilon_0(\varepsilon_\infty - 1)E(t). \tag{3.434}$$

Following Debye, we invoke the relaxation function Eq. (3.430) for rotational degrees of freedom. The relaxation part of the total polarization can then be expressed as

$$P_{\rm rel}(t) = \frac{a_0 E_0}{\tau}e^{-t/\tau}\int_{t_0^+}^{t} e^{(1/\tau - i\omega)u}du = \frac{a_0 E_0}{1 - i\omega\tau}e^{-t/\tau}\left[e^{(1/\tau - i\omega)u}\right]_{t_0^+}^{t}. \tag{3.435}$$

The electric field is suddenly switched on at t_0. If we consider times t sufficiently distant so that there is no longer any memory of the onset of the field, then the relaxation part of the polarization attains the form

$$P_{\rm rel}(t) = \frac{a_0}{1 - i\omega\tau}E(t). \tag{3.436}$$

From the form of the total polarization in linear media, Eq. (3.311), we can make the identification

$$\varepsilon_r = \frac{a_0}{\varepsilon_0(1 - i\omega\tau)} + \varepsilon_\infty. \tag{3.437}$$

We determine the constant a_0 by considering the static limit

$$\varepsilon_s \equiv \lim_{\omega\to 0} \varepsilon_r(\omega) = \frac{a_0}{\varepsilon_0} + \varepsilon_\infty; \quad \Rightarrow \quad a_0 = \varepsilon_0(\varepsilon_s - \varepsilon_\infty). \tag{3.438}$$

Using Eq. (3.437), we can separate the real and imaginary parts of the relative permittivity

$$\varepsilon_1 = \varepsilon_\infty + \frac{\varepsilon_s - \varepsilon_\infty}{1 + \omega^2\tau^2}; \quad \varepsilon_2 = \frac{\omega\tau(\varepsilon_s - \varepsilon_\infty)}{1 + \omega^2\tau^2}. \tag{3.439}$$

These expressions correspond basically to what we see in Figure 3.22. In particular, we see the decrease of the real part ε_1 from a static value ε_s that, in the figure, can be read off in the low-frequency limit as being equal to 78.3. By taking derivatives of ε_2 in Eq. (3.439), we find that it goes through a maximum at $\omega_{max} = 2\pi\nu_{max} = 1/\tau$ where it attains the value $\varepsilon_{2;max} = (\varepsilon_s - \varepsilon_\infty)/2$. For liquid water at 25 °C, this maximum occurs around $10^{10.3}$ Hz = 20 GHz. This corresponds to a relaxation time of τ of around 8 ps, which is associated with rotational relaxation within the hydrogen-bonded network of water molecules. From the experimental data, we find that the maximum value is equal to $\varepsilon_{2;max} = 36.1$, and we can infer a high-frequency limit ε_∞ of 6.0. In other words, setting ε_∞ to this value will cause the maximum of ε_2 in the Debye model to agree with the experimental one.

In Figure 3.24, we plot the imaginary part of the relative permittivity, ε_2, in the microwave and infrared regions as obtained by experiment and the Debye model, respectively. The agreement is remarkable over a large frequency range. However, when entering the THz domain the Debye

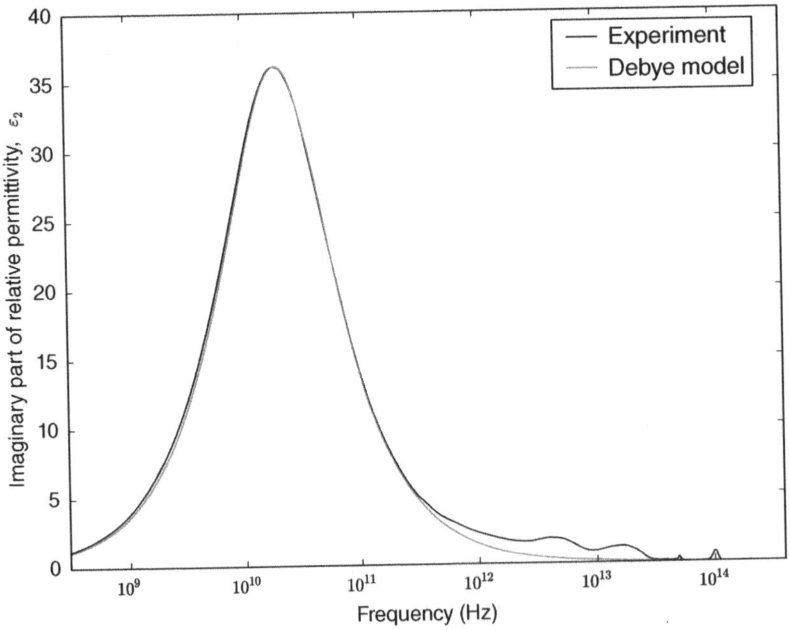

Figure 3.24 Imaginary part ε_2 of the complex permittivity of liquid water at 25 °C as a function of frequency on a logarithmic scale. The displayed frequencies are in the infrared and microwave regions of the spectrum.

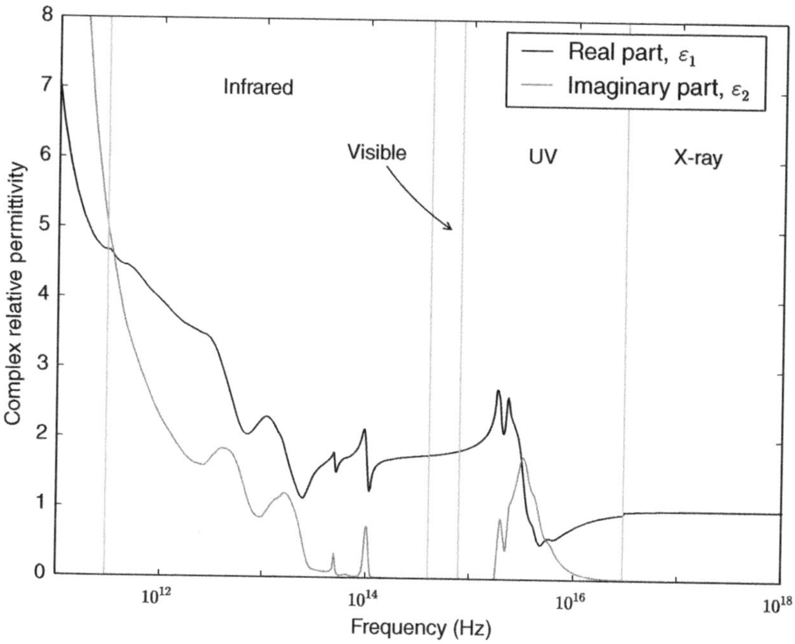

Figure 3.25 Complex relative permittivity of liquid water at 25 °C as a function of frequency on a logarithmic scale. Compared to that shown in Figure 3.22, the infrared to X-ray domain has been enlarged.

curve falls off too quickly, which may indicate that further relaxation mechanisms are at play.[16] Further, the experimental curve shows wiggles that are not taken into account by the Debye model. These wiggles, shown in more detail in Figure 3.25, can be attributed to resonances associated with vibrational and electronic degrees of freedom. Figure 3.25 also shows that the real permittivty ε_1 has a rather constant value of around 1.8 in the visible region. The *optical permittivity* is typically chosen as the value of ε_1 at 25 °C and a frequency corresponding to the sodium D-line (589.29 nm).

3.5.4.3 Resonances in the Lorentz Model

Following Lorentz, we may model the resonant part of the spectrum by charges undergoing forced oscillations about their equilibrium positions due to an applied electric field. In the *Lorentz model*, electrons are bound to nuclei by springs such that there is a restoring force $\mathbf{F} = -k\mathbf{r}$ upon displacements $\mathbf{r}$ from the equilibrium position. In the presence of an external electric field, the equation of motion is

$$m_e \frac{d^2\mathbf{r}}{dt^2} + m_e\omega_0^2\mathbf{r} = -e\mathbf{E}(t); \quad \omega_0 = \sqrt{\frac{k}{m_e}}, \tag{3.440}$$

where ω_0 is the eigenfrequency of the oscillator. By Fourier transformation, we obtain for each frequency component

$$m_e\mathbf{r}(\omega)(-\omega^2 + \omega_0^2) = -e\mathbf{E}(\omega), \tag{3.441}$$

16 See W. J. Ellison. *J. Phys. Chem. Ref. Data*, 36:1, 2007, and I. Popov, P. B. Ishai, A. Khamzin, and Y. Feldman. *Phys. Chem. Chem. Phys.*, 18:13941, 2016.

so that the induced dipole moment per electron is

$$\boldsymbol{\mu}(\omega) = -e\mathbf{r}(\omega) = \alpha(\omega)\mathbf{E}(\omega); \quad \alpha(\omega) = \frac{e^2}{m_e}\frac{1}{\omega_0^2 - \omega^2}, \tag{3.442}$$

where $\alpha(\omega)$ is the electric dipole polarizability associated with the single electron. In the time domain it reads

$$\alpha(t) = \frac{\pi e^2}{m_e \omega_0} \sin \omega_0 t, \quad t > 0,$$

and from Eq. (D. 21) we see that the corresponding time-dependent induced dipole moment $\mu(t)$ will be the convolution of $\alpha(t)$ with $\mathbf{E}(t)$.

We can generalize this result to a collection of independent oscillators by writing the total dipole polarizability as

$$\alpha(\omega) = \frac{e^2}{m_e}\sum_i f_i \frac{1}{\omega_i^2 - \omega^2}, \tag{3.443}$$

where f_i is the number of oscillators with associated frequency ω_i. Using Eq. (3.354), but ignoring the difference between the internal field $\mathbf{E}_i$ and the total field $\mathbf{E}$, we convert this result into a macroscopic expression for the polarization from which we extract

$$\varepsilon_r(\omega) = 1 + \frac{\mathcal{N} e^2}{m_e \varepsilon_0}\sum_i f_i \frac{1}{\omega_i^2 - \omega^2}. \tag{3.444}$$

However, this expression is problematic in that the permittivity is real at all frequencies and in addition has infinite values under conditions of resonance when $\omega = \omega_i$. A more realistic model is obtained by the introduction of a phenomenological damping factor γ as follows

$$m_e \frac{d^2\mathbf{r}}{dt^2} + 2m_e\gamma\frac{d\mathbf{r}}{dt} + m_e\omega_0^2\mathbf{r} = -e\mathbf{E}(t). \tag{3.445}$$

As in Eq. (3.344), we can convert this microscopic equation for the electron displacement into a macroscopic equation for the polarization $\mathbf{P}(t)$ and the macroscopic electric field $\overline{\mathbf{E}}(t)$. Since these two vector quantities will be parallel, we will from now on consider them to be scalar functions. We thereby get

$$\frac{d^2P}{dt^2} + 2\gamma\frac{dP}{dt} + \omega_0^2 P = \frac{\mathcal{N} e^2}{m_e}\overline{E}(t), \tag{3.446}$$

with solution

$$P(\omega) = \frac{\mathcal{N} e^2}{m_e}\frac{1}{(\omega_0^2 - \omega^2 - i2\gamma\omega)}; \Rightarrow P(t) = -\pi i \frac{\mathcal{N} e^2}{m_e \varepsilon_0}\frac{e^{-\gamma t}e^{+i\sqrt{\omega_0^2 - \gamma^2}t}}{\sqrt{\omega_0^2 - \gamma^2}}, \quad t > 0$$

Polarization in the Lorentz model is generated from damped harmonic oscillators driven by the external electric field. The damping factor γ has the dimension of frequency and its inverse $\tau = 1/\gamma$ represents a relaxation time at which the amplitude has been reduced to $1/e$ of its original value, where e is here Euler's number and not the elementary charge.

In the above context, it is interesting to note that using the Leibniz integral rule, we can convert Eq. (3.435) into a first-order differential equation

$$\frac{dP_{\text{rel}}}{dt} + \frac{1}{\tau}P_{\text{rel}} = \frac{a_0}{\tau}E(t). \tag{3.447}$$

We thereby see that polarization in the Debye model arises from an overdamped harmonic oscillator ($\gamma > \omega_0$), meaning that the damping is so strong that the system never enters oscillatory motion, but decays exponentially to the equilibrium position.

If we generalize the Lorentz model to a collection of independent oscillators, we obtain the following expression for the complex permittivity

$$\varepsilon_r(\omega) = 1 + \frac{\mathcal{N} e^2}{m_e \varepsilon_0} \sum_i f_i \frac{1}{\omega_i^2 - \omega^2 - i2\gamma_i\omega}. \tag{3.448}$$

The real and imaginary parts are

$$\varepsilon_1(\omega) = 1 + \frac{\mathcal{N} e^2}{m_e \varepsilon_0} \sum_i f_i \frac{\omega_i^2 - \omega^2}{(\omega_i^2 - \omega^2)^2 + (2\gamma_i\omega)^2}, \tag{3.449}$$

$$\varepsilon_2(\omega) = \frac{\mathcal{N} e^2}{m_e \varepsilon_0} \sum_i f_i \frac{2\gamma_i\omega}{(\omega_i^2 - \omega^2)^2 + (2\gamma_i\omega)^2}. \tag{3.450}$$

A first thing to note is that in the high-frequency limit, beyond all resonances, the complex permittivity behaves as

$$\varepsilon_1(\omega) = 1 - \frac{\omega_p^2}{\omega^2}; \quad \varepsilon_2(\omega) = \frac{2\mathcal{N} e^2}{m_e \varepsilon_0 \omega^3} \sum_i f_i \gamma_i; \quad \omega_p^2 = \frac{\mathcal{N} e^2}{m_e \varepsilon_0} \sum_i f_i \approx \frac{\mathcal{N} e^2 N_e}{m_e \varepsilon_0}, \tag{3.451}$$

where appears the plasma (angular) frequency ω_p, associated with collective excitations (Langmuir waves) of all electrons.

In the final expression, we can for example identify the sum over all oscillators with the number of electrons (N_e=10) in the water molecule. Starting from the number density $\mathcal{N}$ = 55.3 mol l^{-1} at 25 °C, we find that $\omega_p^2 = 1.06 \times 10^{33}$ rad^2 s^{-2}, which translates into a plasma frequency $\nu_p = 5.18 \times 10^{15}$ Hz in the extreme UV region. Inspecting the last 10 data points of the Segelstein compilation, we indeed find that $1 - \varepsilon_1$ and ε_2 approximately falls off as ω^{-2} and ω^{-3} with proportionality constants 0.95×10^{33} rad^2 s^{-2} and 2.59×10^{49} rad^3 s^{-3}, respectively.

Let us now look at the shape of the real and imaginary parts of the complex permittivity in the vicinity of a specific resonance ω_k. We first of all assume that resonances are sufficiently far apart so that we can ignore contributions from other resonances. Next, introducing $\Delta\omega_k = \omega_k - \omega$, we can write

$$\omega_k^2 - \omega^2 = \Delta\omega_k(2\omega + \Delta\omega_k) \approx 2\omega\Delta\omega_k, \tag{3.452}$$

where the final expression comes from the assumption that $\Delta\omega_k$ is very small, that is, we are in the vicinity of ω_k. This allows us to write

$$\varepsilon_1(\omega) = 1 + A_k \left[\frac{\omega_k - \omega}{(\omega_k - \omega)^2 + \gamma_k^2} \right], \tag{3.453}$$

$$\varepsilon_2(\omega) = A_k \left[\frac{\gamma_k}{(\omega_k - \omega)^2 + \gamma_k^2} \right], \tag{3.454}$$

where

$$A_k = \frac{\mathcal{N} e^2 f_k}{2m\varepsilon_0\omega_k}.$$

In the expression for A_k, we have set $\omega = \omega_k$, making $\varepsilon_1 - 1$ and ε_2 antisymmetric and symmetric in $\omega - \omega_k$, respectively, which are relations that do not hold for the full expressions in

Eqs. (3.445) and (3.450). The imaginary part ε_2 can alternatively be written as

$$\varepsilon_2(\omega) = A_k \pi f(\omega; \omega_k, \gamma_k); \; f(\omega; \omega_k, \gamma_k) = \frac{1}{\pi}\left[\frac{\gamma_k}{(\omega_k - \omega)^2 + \gamma_k^2}\right], \tag{3.455}$$

where $f(\omega; \omega_k, \gamma_k)$ is known as a Cauchy distribution, giving a Lorentzian lineshape. The maximum of ε_2 occurs at resonance, where we have

$$\varepsilon_1(\omega_k) = 1; \; \varepsilon_2(\omega_k) = -\frac{d\varepsilon_1(\omega_k)}{d\omega} = \frac{A_k}{\gamma_k}. \tag{3.456}$$

The extreme values of ε_1 occurs at $\omega = \omega_k \pm \gamma_k$. At these values of the frequency, we have

$$\varepsilon_1(\omega_k \pm \gamma_k) = 1 \mp \frac{A_k}{2\gamma_k}; \; \varepsilon_2(\omega_k \pm \gamma_k) = \frac{A_k}{2\gamma_k}, \tag{3.457}$$

which also shows that the full width at half maximum (FWHM) of ε_2 is equal to $2\gamma_k$. For later use we also note two properties of the Cauchy distribution: (i) it is normalized to one, that is

$$\int_{-\infty}^{\infty} f(\omega; \omega_k, \gamma)\, d\omega = 1, \tag{3.458}$$

and (ii) it provides a representation of the Dirac-delta function

$$\lim_{\gamma \to 0} f(\omega; \omega_k, \gamma) = \delta(\omega_k - \omega). \tag{3.459}$$

In Figure 3.26, we plot the complex permittivity of liquid water at 25 °C around the first electronic excitation, occurring at $\nu \approx 1.97$ PHz and therefore blue-shifted with respect to the corresponding $^1B_1 \leftarrow {}^1A_1$ transition at 1.80 PHz in gaseous water. We have also plotted the complex permittivity from the Lorentz model using parameters extracted from the experimental data. The fit is less straightforward than for the Debye model since the resonances are not isolated. The resonance frequency ω_k is extracted from the peak of ε_2. The positions $\omega = \omega_k \pm \gamma_k$ have been located from the extreme values of ε_2 around the resonance and from the corresponding line broadening we find a value for the damping term equal to $\gamma_k \approx 0.1 \times 10^{15}\text{s}^{-1}$, corresponding to a relaxation time in the order of 5 fs.

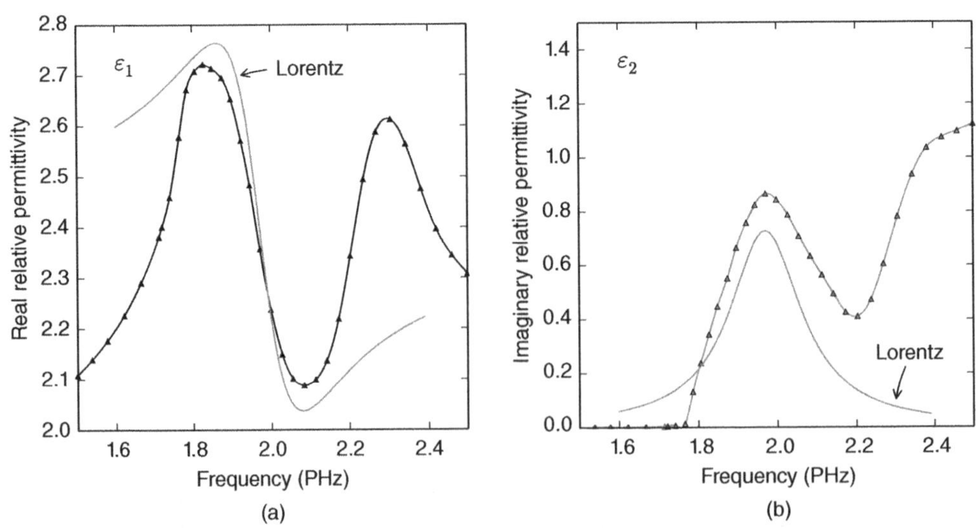

Figure 3.26 Complex permittivity of liquid water at 25 °C as a function of frequency around the first electronic excitation, found in the UV region. Results from the Lorentz model are compared with experimental data.

The Lorentz model also works quite well for vibrational resonances. For the prominent ε_2 peak around 101 THz, corresponding to the asymmetric stretch ν_3, we identify a value of the damping term from the line broadening $\gamma \approx 7 \times 10^{12}$ s^{-1}, corresponding to a relaxation time of about 70 fs.

3.5.4.4 Refraction and Absorption

It should be kept in mind that although the imaginary part ε_2 of the complex permittivity is associated with energy loss, it does not constitute an absorption spectrum. In Figure 3.27, we give a log–log plot of the absorption coefficient $\alpha = 2\kappa$ of liquid water at 25 °C with 12 decades in frequency, generated from the complex permittivity using Eq. (3.405). The absorption coefficient increases steadily through the microwave region to a maximum value of 1.2×10^6 m^{-1} around 100 THz in the infrared region.

In the near infrared there is a rich structure due to vibrational excitations. In the visible region, on the other hand, there is basically no absorption, which explains why water is transparent. In the ultraviolet region, with the onset of electronic excitations, absorption picks up again and reaches an absolute maximum around the plasma frequency. In the X-ray region, the absorption coefficients displayed in Figure 3.27 are compiled from atomic cross-sections

$$\alpha = \mathcal{N}(2\sigma_{\rm H} + \sigma_{\rm O}). \tag{3.460}$$

The absorption decreases steadily in the X-ray region, but a distinct edge is observed around the ionization energy of the oxygen 1*s* orbital (131 PHz).

From Beer–Lamberts law in Eq. (3.411), we see that the inverse of the absorption coefficient, called the *penetration depth*, corresponds to the distance inside the medium where the intensity of the incoming radiation has been reduced to $1/e$ or 36.8% of its original value. It may be observed that the penetration length at the ε_2 peak at 20 GHz, representing maximum

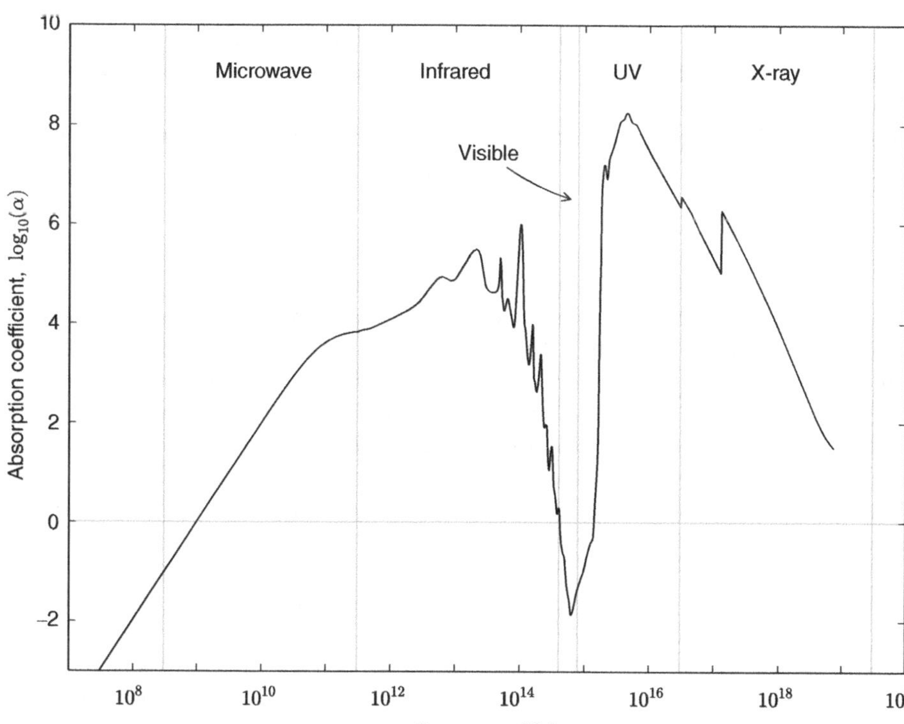

Figure 3.27 Log–log plot of the absorption coefficient of liquid water at 25 °C as a function of frequency.

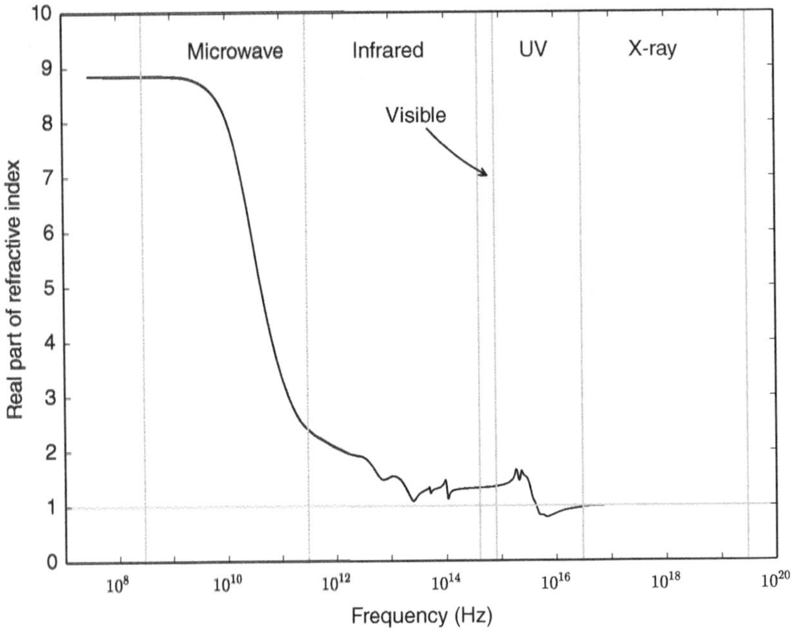

Figure 3.28 Real part of the refractive index of liquid water at 25 °C as a function of frequency.

energy loss in the microwave region, is only 0.5 mm. Commercial microwave ovens operate at 2.45 GHz, where the penetration depth is around 2 cm, thus allowing to heat the food more evenly throughout its volume.

In Figure 3.28, we show the real part of the refractive index of liquid water at 25 °C by using Eq. (3.395) and by assuming the relative permeability to be equal to one. Starting off from a static value of 8.85, the refractive index is seen to decrease to 1.33 in the visible region and finally become equal to unity in the high-frequency limit. In the optical region, dynamical polarizabilities $\alpha(\omega)$ are often extracted from measurements of the refractive index using the Lorentz–Lorenz relation

$$\frac{\mathcal{N}\alpha(\omega)}{3\varepsilon_0} = \frac{n^2(\omega) - 1}{n^2(\omega) + 2}, \tag{3.461}$$

which was derived independently by Ludvig Lorenz (1869) and Hendrik Lorentz (1878). For a nonmagnetic material, we obtain this equation straightforwardly from the Clausius–Mossotti relation in Eq. (3.355) by using Eq. (3.395). However, one should be aware of the fact that the Clausius–Mossotti relation was derived for the static case and it therefore only applies to non-polar media. The Lorentz–Lorenz relation, on the other hand, concerns the dynamic case and can also be applied to polar media at optical frequencies where the permanent dipoles can no longer follow the applied dynamic field. We may connect to the Lorentz model by considering media, for instance gases, for which the refractive index is close to one. The denominator on the right-hand side of Eq. (3.461) then becomes effectively three and we can write

$$n^2(\omega) = \varepsilon_r \approx 1 + \frac{\mathcal{N}\alpha(\omega)}{\varepsilon_0}. \tag{3.462}$$

If we now insert the dipole polarizability of Eq. (3.442), we recover the expression for ε_r in Eq. (3.444). Further refinement, Eq. (3.448), is obtained by the inclusion of damping. An intriguing aspect of this model is that it suggests that the refractive index can be below one, corresponding to phase velocities beyond the speed of light, as we actually see in the UV domain in Figure 3.28. However, as already discussed, although no signal may travel faster than the speed of light, the magnitudes of the phase (and even group) velocities may exceed the speed of light, as long as the signal velocity stays within the bound imposed by relativity.

3.5.5 Rotational Averages

When connecting microscopic polarizability to macroscopic polarization, the former quantity was given as a scalar quantity α. However, for a general molecule the first-order electric dipole $\boldsymbol{\mu}$ induced by an electric field $\mathbf{E}$ depends on the orientation of the field and takes the form

$$\mu_i^{(1)} = \sum_j \alpha_{ij} E_j,$$

where the polarizability is a two-index quantity. More precisely, it is a second-rank Cartesian tensor. It is Cartesian because the indices refer to Cartesian components, and it is a tensor because it has well-defined transformation properties upon the change of reference frame. Of particular importance is the connection between a reference frame attached to individual molecules, the molecular frame, and the reference frame of the observer, namely the laboratory frame. When deriving the Clausius–Mossotti relation, Eq. (3.355), we are assuming that we have a collection of molecules that may rotate freely and where all orientations have equal probability. The scalar polarizability α appearing in the formula is thereby a rotational average.

Before we actually derive expressions for rotational averages of tensors of various ranks, let us first see how a tensor changes upon rotation of the reference frame. We can focus on an individual molecule and let the origin of the molecular and laboratory frame coincide somewhere inside the molecule. A vector can be expressed as the three components of a first-rank Cartesian tensor spanned by Cartesian unit vectors

$$\mathbf{T} = T_x \boldsymbol{e}_x + T_y \boldsymbol{e}_y + T_z \boldsymbol{e}_z = T_i \boldsymbol{e}_i, \tag{3.463}$$

where the final term is expressed using the Einstein summation convention that we will use further. The three Cartesian unit vectors of a given frame are orthogonal, and so we may extract a component of our vector by projection with the corresponding Cartesian unit vector, for instance

$$T_x = \boldsymbol{e}_x^T \cdot \mathbf{T}. \tag{3.464}$$

Components of the vector in terms of some other frame, indicated by primed indices, can be found in the same manner

$$T_{x'} = \boldsymbol{e}_{x'}^T \cdot \mathbf{T}. \tag{3.465}$$

We can connect the vector components in the primed frame with those of the unprimed frame by simply expanding the vector in unframed unit vectors

$$T_{i'} = \boldsymbol{e}_{i'}^T \cdot \mathbf{T} = \boldsymbol{e}_{i'}^T \cdot T_j \boldsymbol{e}_j = l_{i'j} T_j, \tag{3.466}$$

where appear *directional cosines*

$$l_{i'j} = \boldsymbol{e}_{i'}^T \cdot \boldsymbol{e}_j = \cos\theta_{i'j}. \tag{3.467}$$

To relate second-rank tensors such as the electric dipole polarizablity in different frames, we write

$$\mu_{i'} = \alpha_{i'j'} E_{j'} = l_{i'i}\mu_i = l_{i'i}\alpha_{ij}E_j = l_{i'i}\alpha_{ij} l_{jj'} E_{j'} \tag{3.468}$$

from which we extract

$$\alpha_{i'j'} = l_{i'i} l_{j'j} \alpha_{ij}. \tag{3.469}$$

Generally, we have

$$T_{i_1' i_2' \cdots i_n'} = l_{i_1' i_1} l_{i_2' i_2} \cdots l_{i_n' i_n} T_{i_1 i_2 \cdots i_n}. \tag{3.470}$$

Let primed indices refer to the laboratory frame and unprimed ones to the molecular frame. The rotational average of a tensor is then given by

$$\langle T_{i'_1 i'_2 \ldots i'_n} \rangle^{\text{iso}} = \langle l_{i'_1 i_1} l_{i'_2 i_2} \ldots l_{i'_n i_n} \rangle T_{i_1 i_2 \ldots i_n}, \tag{3.471}$$

where we average over all possible orientations of the molecular frame relative to the laboratory frame. For a first-rank tensor, the exercise is rather straightforward

$$\langle T_{i'} \rangle^{\text{iso}} = \langle l_{i'i} \rangle T_i = \langle \cos \theta_{i'i} \rangle T_i = 0. \tag{3.472}$$

For higher-rank tensors such calculations quickly become cumbersome, but are considerably simplified by the identification of *basic isotropic tensors* for given ranks.[17] These are rotational invariant tensors that serve as a basis in which any rotationally invariant tensor of the same rank may be expanded.

Any zeroth-rank tensor is rotational invariant since they are scalars; as basic isotropic tensors we may simply choose unity. On the other hand, as we have seen, there are no first-rank isotropic tensors. There is a single basic second-rank isotropic tensor, namely the unit matrix, whose elements are given by the Kronecker delta. Any isotropic second-rank tensor, including the rotational average of a general second-rank tensor, can therefore be expressed as

$$\langle T \rangle^{\text{iso}}_{ij} = a\delta_{ij}. \tag{3.473}$$

The proportionality constant a is a scalar. In order to determine a, we reduce the rank of the tensors in the above equation to zero by contraction with the basic second-rank isotropic tensor, giving

$$\langle T \rangle^{\text{iso}}_{ij} \delta_{ij} = \langle T_{ij} \delta_{ij} \rangle^{\text{iso}} = T_{ij}\delta_{ij} = a\delta_{ij}\delta_{ij} = 3a \quad \Rightarrow \quad a = \frac{1}{3} T_{ii}. \tag{3.474}$$

The second step above is crucial: Since the rotational average implies integration over angles for the two indices of the tensor, we can do the contraction under the integral signs. The resulting scalar is by its nature rotationally invariant and so we can remove brackets to give the third term above.

There is likewise a single third-rank basic isotropic tensor whose elements are given by the Levi–Civita symbol ϵ_{ijk}. Elements of the rotational average of a general third-rank tensor can therefore be written as

$$\langle T \rangle^{\text{iso}}_{ijk} = b\epsilon_{ijk}. \tag{3.475}$$

In order to determine the proportionality constant b, we proceed as before and contract with ϵ_{ijk}, giving

$$\langle T \rangle^{\text{iso}}_{ijk} \epsilon_{ijk} = T_{ijk}\epsilon_{ijk} = b\epsilon_{ijk}\epsilon_{ijk} = 6b \quad \Rightarrow \quad b = \frac{1}{6} T_{ijk}\epsilon_{ijk}. \tag{3.476}$$

There are three fourth-rank basic isotropic tensors, and so the elements of the rotational average of a general fourth-rank tensor are given as a linear combination of these

$$\langle T \rangle^{\text{iso}}_{ijkl} = a\delta_{ij}\delta_{kl} + b\delta_{ik}\delta_{jl} + c\delta_{il}\delta_{jk}. \tag{3.477}$$

In order to determine the expansion coefficients, we contract separately with the three basic isotropic tensors, which thereby leads to a set of linear equations

$$\begin{bmatrix} T_{ijkl}\delta_{ij}\delta_{kl} \\ T_{ijkl}\delta_{ik}\delta_{jl} \\ T_{ijkl}\delta_{il}\delta_{jk} \end{bmatrix} = \begin{bmatrix} 9 & 3 & 3 \\ 3 & 9 & 3 \\ 3 & 3 & 9 \end{bmatrix} \begin{bmatrix} a \\ b \\ c \end{bmatrix}, \tag{3.478}$$

17 See P. G. Hodge. *Am. Math. Mon.*, 68:793, 1961.

Figure 3.29 The chiral CHFClBr molecule (*R* isomer).

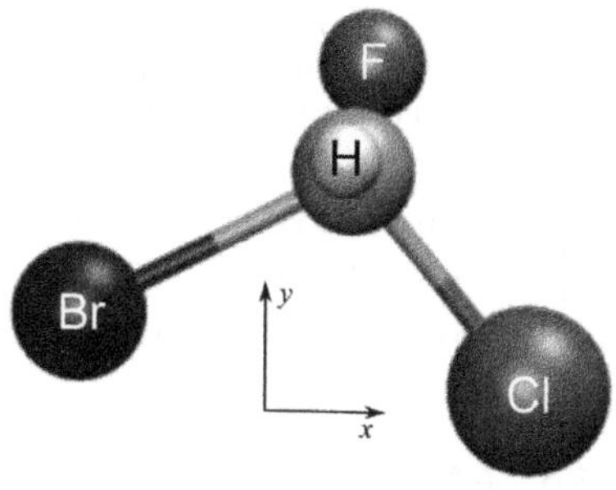

with solution

$$\begin{bmatrix} a \\ b \\ c \end{bmatrix} = \frac{1}{30}\begin{bmatrix} 4 & -1 & -1 \\ -1 & 4 & -1 \\ -1 & -1 & 4 \end{bmatrix}\begin{bmatrix} T_{ijkl}\delta_{ij}\delta_{kl} \\ T_{ijkl}\delta_{ik}\delta_{jl} \\ T_{ijkl}\delta_{il}\delta_{jk} \end{bmatrix}. \tag{3.479}$$

Higher-order basic isotropic tensors are given in terms of Kronecker deltas and Levi-Civita symbols, and their determination becomes an exercise in combinatorics. One can show that the number of basic isotropic tensors of rank $n = 0, 1, 2, \ldots$ is given by the Motzkin sum integer sequence $1, 0, 1, 1, 3, 6, 15, 36, 91, 232, \ldots$, with the recurrence formula

$$a_n = \frac{n-1}{n+1}(2a_{n-1} + 3a_{n-2}). \tag{3.480}$$

As an illustration of the effect of rotational averaging, let us consider a specific example. We have seen that macroscopic polarization is related to microscopic electric dipole moments. For the chiral molecule CHFClBr, we calculate a permanent electric dipole moment of 4.46 Debye in the direction $\mathbf{n} = (0.30, 0.95, 0.10)$ with respect to the orientation given in Figure 3.29 and therefore almost aligned with the C—F bond. However, if the CHFClBr molecules are in a medium (gas phase, solution, etc.,) where they can rotate freely, the electric dipole moment averages out to zero with respect to the laboratory frame. An electric dipole moment can also be induced by a (uniform) electric field **E** according to

$$\mu_i^E = \sum_j \alpha_{ij} E_j, \tag{3.481}$$

where appear components α_{ij} of the electric dipole–electric dipole polarizability. Calculated electronic contributions to α are shown in Table 3.1 and one may note that all components are nonzero such that polarization arising from this mechanism in an oriented medium will not be aligned with the electric field. However, upon rotational averaging, only the isotropic polarizability $\overline{\alpha} = 58.1083$ a.u. will be retained and the resulting polarization will be parallel to the electric field. An electric dipole moment can also be induced by a time-dependent magnetic field **B** according to

$$\mu_i^B = \frac{1}{\omega}\sum_j G'_{ij}\dot{B}_j, \tag{3.482}$$

where appear components G'_{ij} of the electric dipole–magnetic dipole polarizability. Calculated values for these tensor elements are found in Table 3.1 as well. Again, only the isotropic contribution $\overline{G'} = 0.0016$ a.u. will contribute in an isotropic medium. However, a fundamental difference with respect to μ_i^E is that the resulting polarization will be aligned with the magnetic and not the electric field. This is the origin of birefringence in isotropic media.

Table 3.1 Calculated (B3LYP/aug-cc-pVDZ) α and G' tensors, corresponding to the electric dipole and magnetic dipole polarizabilities, respectively, (in atomic units) of the *R*-CHFClBr molecule at $\lambda = 589$ nm, corresponding to the Fraunhofer D-line in sodium.

α	x	y	z
x	78.2270	−2.6433	0.9873
y	−2.6433	50.1867	3.3741
z	0.9873	3.3741	45.9111

G'	x	y	z
x	−0.1015	0.7162	−3.0470
y	−0.8895	0.1137	1.8328
z	2.9506	−1.7669	−0.0075

3.5.6 A Note About Dimensions, Units, and Magnitudes

Before we close this chapter, let us carry out a dimensional analysis in terms of mass M, electric charge Q, length L, and time T of some of the physical quantities introduced in this chapter. A good starting point is the expression for the Lorentz force given in Eq. (3.123). On the left-hand side we have force which is mass times acceleration and which therefore has dimension $M \cdot L \cdot T^{-2}$. This allows us to deduce that the dimension of electric and magnetic fields, **E** and **B**, is $M \cdot Q^{-1} \cdot L \cdot T^{-2}$ and $M \cdot Q^{-1} \cdot T^{-1}$, respectively. We now continue to the inhomogeneous pair of Maxwell's equations, Eqs. (3.5) and (3.6). On the right-hand side we have the sources, charge density ρ, and current density $\mathbf{j}$, with dimensions $Q \cdot L^{-3}$ and $Q \cdot L^{-2} \cdot T^{-1}$, respectively. This allows us to conclude that the dimensions of the electric constant ε_0 and the magnetic constant μ are $M^{-1} \cdot Q^2 \cdot L^{-3} \cdot T^2$ and $M \cdot Q^{-2} \cdot L$, respectively. As a check we note that the dimension of their product is $L^{-2} \cdot T^2$, corresponding to the dimension of inverse square speed, as we see in Eq. (3.7). Continuing with Eqs. (3.278) and (3.282), we see that the dimension of both polarization **P** and the electric displacement **D** is $Q \cdot L^{-2}$, whereas the dimension of magnetization **M** and the auxilliary field **H** is $Q \cdot L^{-1} \cdot T^{-1}$. Those are the dimensions of surface charges and currents, respectively.

Looking now at the expression for electric and magnetic fields in terms of potentials, Eqs. (3.49) and (3.50), we conclude that the dimensions of the scalar potentials ϕ and the vector potential A are $M \cdot Q^{-1} \cdot L^2 \cdot T^{-2}$ and $M \cdot Q^{-1} \cdot L \cdot T^{-1}$, respectively. Let us also have a look at Poynting's theorem, Eq. (3.131). The integrand of the second term of the left-hand side has the dimension $M \cdot L^{-1} \cdot T^{-2}$ and integrates to a quantity of dimension $M \cdot L^2 \cdot T^{-2}$, which is that of energy (recall that kinetic energy is $\frac{1}{2}mv^2$). The Poynting vector, and therefore also radiant intensity, has dimension $M \cdot T^{-3}$, which we can also express as $(M \cdot L^2 \cdot T^{-2}) \cdot L^{-2} \cdot T^{-1}$, which more clearly displays that this is energy flux. Finally, from Eqs. (3.187) and (3.190), we see that the dimensions of electric and magnetic multipoles of order n are $Q \cdot L^n$ and $Q \cdot L^{n+1} \cdot T^{-1}$, respectively.

In the International System of Units (SI), the units for mass M, electric charge Q, length L, and time T are the kilogram (kg), the coulomb (C), the meter (m) and the second (s), respectively. The strength of electric and magnetic fields are reported in volts per meter (V m^{-1}) and Teslas (T), respectively. In SI-based atomic units[18] the units for mass, electric charge, length and time are the electron mass m_e ($9.10938291 \times 10^{-31}$ kg), the elementary charge e ($1.602176565 \times 10^{-19}$ C), the Bohr radius a_0 ($5.2917721092 \times 10^{-11}$ m) and $\hbar/E_h$ ($2.418884326505 \times 10^{-17}$ s), respectively. The strength of electric and magnetic fields are given by $E_h/(ea_0)$ (5.1423×10^{11} V m^{-1}) and $\hbar/(ea_0)$ ($2.3505{\times}10^5$ T). These correspond to incredibly strong fields! The atomic unit of electric field strength is found at the Bohr radius in the hydrogen atom, but, as we already pointed out in the introduction, this should be compared to the electric field under power lines (up to 10^4 V m^{-1}) and the maximum field strength 3.0×10^6 V m^{-1} in air. The atomic unit of magnetic field strength may be compared to the magnetic field of the earth, which is around

18 D. H. Whiffen. *Pure Appl. Chem.*, 50:75, 1978.

60 μT, and NMR magnets, that are typically around 10 T. The strongest man-made magnetic fields, created in nondestructive manner, are around 100 T. To find one atomic unit of magnetic field strength, one has to go into space, to the so-called magnetic white dwarfs. Exceptionally strong magnetic fields, on the order of 10^8–10^{11} T are found in special neutron stars called magnetars.

Further Reading

1 G. B. Arfken, H. J. Weber, and F. E. Harris. *Mathematical Methods for Physicists*. Elsevier, 2013.
2 D. P. Craig and T. Thirunamachandran. *Molecular Quantum Electrodynamics*. Dover Publications, 1998.
3 D. J. Griffiths. *Introduction to Electrodynamics*. Prentice Hall, 1999.
4 J. D. Jackson. *Classical Electrodynamics*. Wiley, 1999.
5 H. M. Schey. *Div, Grad, Curl, and All That*. W. W. Norton & Company, 2005.

4

Symmetry

One day, while we were walking on the beach, he [Dirac] told me that he would teach me a saying: 'It is easy, if you remember the symmetry. Watch the symmetry.' He went on, 'When a man says yes, he means perhaps; when he says perhaps, he means no; when he says no, he is no diplomat. When a lady says no, she means perhaps; when she says perhaps, she means yes; when she says yes, she is no lady'. With a couple of repetitions, I learned it, and he was pleased.

S.A. Kursunoglu (1987)

4.1 Fundamentals

Symmetry is a powerful concept in natural sciences. In everyday life, symmetry operations are given in terms of the rotations, reflections, and inversions of a physical object that leaves the object unchanged—that is, the properties of the object are *invariant* under these transformations. Symmetry operations, however, can also be applied to more abstract objects, such as wave functions, operators, and equations, but we still require the invariance of observables. In a quantum-mechanical framework, this means that symmetry operations are those operations that leave expectation values and intensities unchanged. These quantities are quantum-mechanically represented by integrals, and symmetry can thus tell us that certain integrals are zero, providing us with selection rules for molecular excitations. In chemistry, the most straightforward use of symmetry lies in the study of spatial transformations, such as rotations and reflections, that move the nuclear framework into a position indistinguishable from the original one. At a more advanced level, it is also relevant to study how these symmetry operations act in spin space. For instance, in the relativistic domain, symmetry operations in spin and space are coupled through the spin-orbit interaction. Spin symmetry is therefore lost, but can to some extent be recovered using time-reversal symmetry, as is discussed in Section 4.2.

4.1.1 Symmetry Operations and Groups

When discussing the effect of symmetry operations, one has to carefully distinguish the active and passive points of view. The active point of view means that the symmetry operation moves the system, that is, transforms the function representing it, whereas the passive point of view means that the symmetry operation moves the adopted coordinate frame. We will here employ the active point of view in which the transformation of a function is given by

$$\hat{G}f(\mathbf{r}) = f(\hat{G}^{-1}\mathbf{r}), \tag{4.1}$$

Principles and Practices of Molecular Properties: Theory, Modeling and Simulations, First Edition.
Patrick Norman, Kenneth Ruud, and Trond Saue.

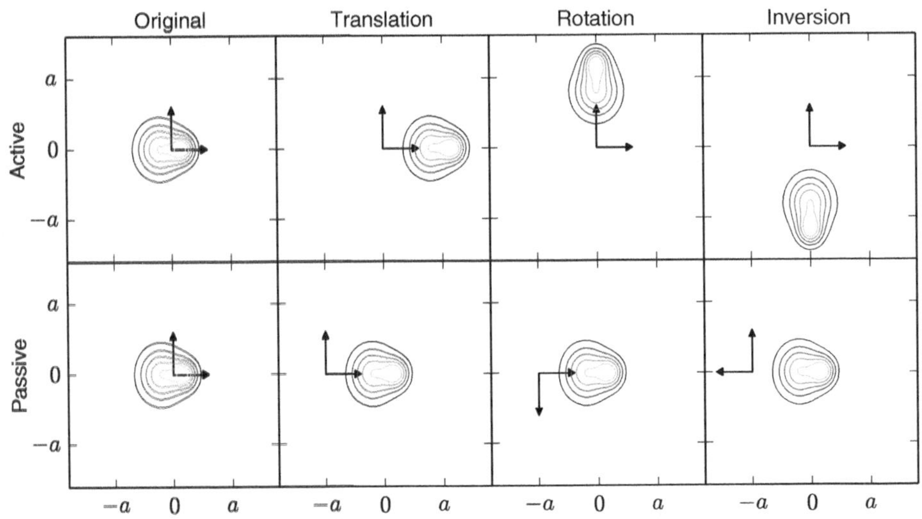

Figure 4.1 A sequence of symmetry operations transforms a function step by step. The original and resulting functions are illustrated by contour plots. The upper and lower sets of panels provide the *active* and *passive* points of view, respectively.

that is, we map the function values of f from one point in space $\mathbf{r}' = \hat{G}^{-1}\mathbf{r}$ to another point $\hat{G}\mathbf{r}' = \mathbf{r}$, thereby creating a new function $\hat{G}f$. This active view of a transformation is illustrated in the upper sequence of panels in Figure 4.1 and compared to the corresponding passive view in the lower set of panels.

Let us begin with a brief discussion of the use of symmetry to transform and analyze the spatial parts of wave functions. The mathematical theory of groups will in this case be helpful to take advantage of any spatial symmetry in the system. By limiting ourselves to spatial symmetry, *symmetry operations* will be restricted to geometric actions that leaves the nuclear molecular framework invariant. A *symmetry element* is a geometrical entity, for example, a point, line, or plane, about which an action takes place, and for every symmetry operation there is a corresponding symmetry element. There are five different kinds of symmetry operations:

1) The identity operation does nothing and is denoted as $\hat{I}$. An alternative and commonly adopted symbol for the identity operation is $\hat{E}$.
2) The rotation operation rotates the system by an angle $2\pi/n$ about an axis and is denoted as $\hat{C}_n$. We note that if $\hat{C}_n$ is a symmetry operation, so is $\hat{C}_n^k$ ($\hat{C}_n$ applied k times), and we also note that $\hat{C}_n^n = \hat{I}$. Within a point group, the axis with the largest n is called the *principal axis*. Later in this chapter we will consider more general rotations and will then use the symbol $\hat{\mathcal{R}}(\phi, \mathbf{n})$ to denote the operation of rotating an angle ϕ about the axis pointing along the unit vector $\mathbf{n}$.
3) The inversion, or parity, operation inverts the system through a point and is commonly denoted as $\hat{i}$. We note that $\hat{i}^2 = \hat{I}$.
4) The reflection operation mirrors the system in a plane and is Denoted as $\hat{\sigma}$. If the plane is perpendicular to the principal axis, the operation is denoted as $\hat{\sigma}_h$; if instead it contains the principal axis, the operation is denoted as $\hat{\sigma}_v$. If a symmetry plane bisects two $\hat{C}_2$ axes that are perpendicular to the principal axis, it is commonly referred to as a dihedral plane. The corresponding operation is denoted as $\hat{\sigma}_d$. We note that $\hat{\sigma}^2 = \hat{I}$.

5) The improper rotation, or rotoreflection, operation performs a combination of a rotation an angle $2\pi/n$ followed by a reflection in a plane perpendicular to the rotation axis and is denoted by $\hat{S}_n = \hat{\sigma}_h \hat{C}_n$.

The *symmetry group* $\mathcal{G}$ of a system such as a molecule, is defined as the group whose elements $\hat{G}$ commute with the Hamiltonian of the system

$$[\hat{G}, \hat{H}] = 0. \tag{4.2}$$

Alternatively, we can say that the symmetry operations $\hat{G}$ leave the Hamiltonian invariant, that is

$$\hat{G}\hat{H}\hat{G}^{-1} = \hat{H}. \tag{4.3}$$

These equations are equivalent provided that the inverse $\hat{G}^{-1}$ of the symmetry operation $\hat{G}$ exists, which in this case is guaranteed by the fact that symmetry operations form a group. The requirements needed to be fulfilled by the symmetry operations in order to form a group are:

1) If operations $\hat{P}$ and $\hat{Q}$ are members of the group, then their combination $\hat{R} = \hat{P}\hat{Q}$ is a member of the group as well. The rule for combining two symmetry operations is simply to perform operation $\hat{P}$ after operation $\hat{Q}$ has been completed.
2) The set of operations in the group must contain the identity $\hat{I}$.
3) The associate law applies, that is, $\hat{P}(\hat{Q}\hat{R}) = (\hat{P}\hat{Q})\hat{R}$ for all operations.
4) The inverse $\hat{R}^{-1}$ to all operations must belong to the set of operations in the group. The inverse operation is defined as the operation that satisfies the relation $\hat{R}^{-1}\hat{R} = \hat{I}$.

In the study of molecules, the set of symmetry operations form a *point group* since at least one point is left unchanged by all operations. This stands in contrast to crystal groups, where translational symmetry is present. A given molecule can be determined to belong to a specific point group by following the flowchart in Figure 4.2.

With a line of reasoning introduced by Wigner using intensities, it can be shown that symmetry operations are either unitary or antiunitary. This argument introduces the concept of linear and antilinear operators, from which the reader probably being most familiar with the former kind. For a linear operator, we have

$$\hat{\Omega}(c_1|\psi_1\rangle + c_2|\psi_2\rangle) = c_1\hat{\Omega}|\psi_1\rangle + c_2\hat{\Omega}|\psi_2\rangle, \tag{4.4}$$

whereas for an antilinear, or conjugate-linear, operator the corresponding relation is

$$\hat{\Omega}(c_1|\psi_1\rangle + c_2|\psi_2\rangle) = c_1^*\hat{\Omega}|\psi_1\rangle + c_2^*\hat{\Omega}|\psi_2\rangle. \tag{4.5}$$

The eigenstates of a Hermitian linear operator $\hat{\Omega}$ form a basis for the Hilbert space in which the action of the operator can be expressed as

$$\hat{\Omega}|\psi\rangle = \hat{\Omega}\sum_n c_n|\omega_n\rangle = \sum_n c_n\,\omega_n|\omega_n\rangle. \tag{4.6}$$

As can be seen by expanding a presumed eigenvector in a different basis, antilinear operators can have no associated eigenvalues and therefore cannot be associated with observables. However, the operation of time reversal is achieved by an antilinear operator.

Let us return to Wigner's argument. As will be discussed in Chapter 5, the intensity of a transition from an initial state $|\psi_i\rangle$ to a final state $|\psi_f\rangle$ is proportional to the squared norm of a transition moment

$$\sigma_{f\leftarrow i} \propto |\langle\psi_f|\hat{\Omega}|\psi_i\rangle|^2 = \langle\psi_f|\hat{\Omega}|\psi_i\rangle\langle\psi_f|\hat{\Omega}|\psi_i\rangle^*, \tag{4.7}$$

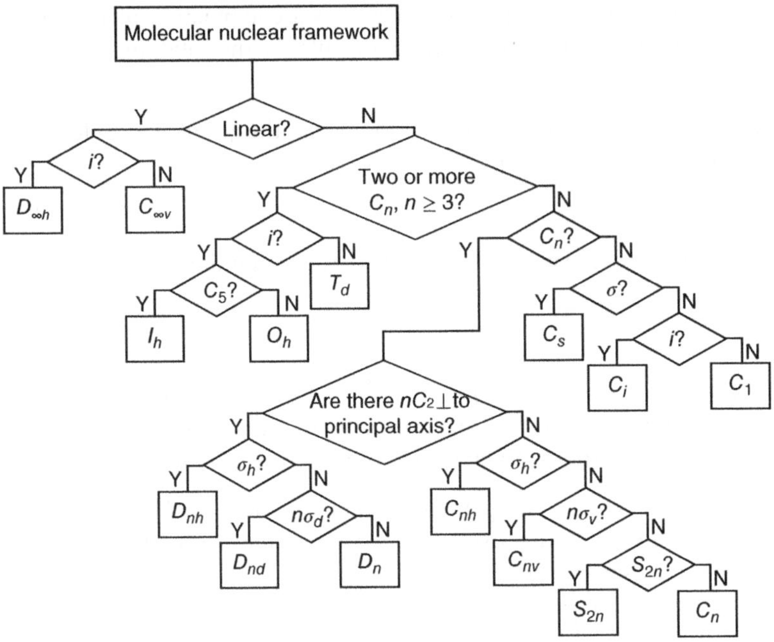

Figure 4.2 Flowchart for determining molecular point groups.

where the Hermitian operator $\hat{\Omega}$ represents the relevant coupling operator between the quantum-mechanical system and the external field that induces the transition. The action of the operator $\hat{\Omega}$ is to map the initial state vector into another state vector

$$\hat{\Omega}|\psi_i\rangle = |\psi_{\Omega i}\rangle, \tag{4.8}$$

whereas the bra vector $\langle\psi_f|$, representing the final state in dual space, maps the ket $|\psi_{\Omega i}\rangle$ into a general complex number, corresponding to the transition moment $\langle\psi_f|\hat{\Omega}|\psi_i\rangle$. Let us now consider the action of a symmetry operator $\hat{G}$ on this transition moment. In order to conserve the transition probability in Eq. (4.7), two possibilities exist namely

$$\hat{G}\langle\psi_f|\hat{\Omega}|\psi_i\rangle = \begin{cases} \langle\psi_f|\psi_{\Omega i}\rangle & \text{(i)} \\ \langle\psi_f|\psi_{\Omega i}\rangle^* & \text{(ii)} \end{cases}. \tag{4.9}$$

To investigate these further, we can act with the symmetry operator on the quantities under the integral sign. It can be noted that we can separate the action of the symmetry operator $\hat{G}$ (or any operator) on the general operator $\hat{\Omega}$ and the function ψ_i by inserting $\hat{G}^{-1}\hat{G}$ between them

$$\hat{G}|\psi_{\Omega i}\rangle = \hat{G}\hat{\Omega}\hat{G}^{-1}\hat{G}|\psi_i\rangle. \tag{4.10}$$

The expression $\hat{G}\hat{\Omega}\hat{G}^{-1}$ thus represents the action of $\hat{G}$ on the operator $\hat{\Omega}$. We can understand this expression by keeping in mind that an operator acts on anything to its right, but here $\hat{G}$ meets its inverse, which blocks further action to the right. This will be useful in Section 4.2, but in the present case we simply write

$$\hat{G}\langle\psi_f|\hat{\Omega}|\psi_i\rangle = \langle\hat{G}\psi_f|\hat{G}\psi_{\Omega i}\rangle. \tag{4.11}$$

In order to arrive at alternative (i) above, we recall the definition of the adjoint of a linear operator, Eq. (2.54). We then have

$$\hat{G}\langle\psi_f|\hat{\Omega}|\psi_i\rangle = \langle\hat{G}\psi_f|\hat{G}\psi_{\Omega i}\rangle = \langle\psi_f|\hat{G}^\dagger\hat{G}\psi_{\Omega i}\rangle = \langle\psi_f|\psi_{\Omega i}\rangle, \tag{4.12}$$

which implies $\hat{G}^\dagger \hat{G} = \hat{I}$, corresponding to a unitary operator. In the second case, we have to modify the definition of the adjoint operator to be

$$\langle \hat{G}^\dagger \psi_i | \psi_j \rangle = \langle \psi_i | \hat{G} \psi_j \rangle^*. \tag{4.13}$$

We then get

$$\hat{G}\langle \psi_f | \hat{\Omega} | \psi_i \rangle = \langle \hat{G}\psi_f | \hat{G}\psi_{\Omega i} \rangle = \langle \psi_f | \hat{G}^\dagger \hat{G} \psi_{\Omega i} \rangle^* = \langle \psi_f | \psi_{\Omega i} \rangle^*. \tag{4.14}$$

Again, the operator fulfills the relation $\hat{G}^\dagger \hat{G} = \hat{I}$, but the complex conjugation implies that the operator is antiunitary (and antilinear). Unitary transformations correspond to familiar spatial transformations such as rotations, reflections, and translations, whereas an example of an antiunitary symmetry operation is time reversal as we shall see in Section 4.2.

There is a subtle point concerning antilinear operators that should be noted. For a linear operator, when writing a matrix element $\langle \psi_f | \hat{\Omega} | \psi_i \rangle$, we can think of the operator acting either to the right on the ket vector or to the left on the bra vector, in the latter case becoming its adjoint according to Eq. (2.55). In both cases we obtain identical results, which we will indicate using brackets as

$$\langle \psi_f | [\hat{\Omega} | \psi_i \rangle] = [\langle \psi_f | \hat{\Omega}] | \psi_i \rangle. \tag{4.15}$$

However, for antilinear operators the corresponding equivalence is

$$\langle \psi_f | [\hat{\Omega} | \psi_i \rangle] = [\langle \psi_f | \hat{\Omega}] | \psi_i \rangle^*, \tag{4.16}$$

and so in this case it makes a difference whether the operator acts to the left or to the right.

4.1.2 Group Representation

A group table lists all combinations of elements in the symmetry group, as illustrated in Figure 4.3 for a group with four elements. A group for which all elements commute is known as an Abelian group, and as a consequence, the corresponding group table becomes symmetric and each symmetry operation is its own inverse.

When comparing two groups, they are said to be isomorphic if their respective group tables have the same structure and there is a one-to-one correspondence between the elements of each group, as shown in Figure 4.3. A group homomorphism is a function $h : \mathcal{G} \to \mathcal{H}$ that maps from the group $\mathcal{G}$ to the group $\mathcal{H}$. The one-to-one correspondence between group elements may be lost in such a group homomorphism, but the structure of the group is retained in the sense that for any two operations $\hat{A}$ and $\hat{B}$ in a group $\mathcal{G}$, it holds that the result of combining these two

	A	B	C	D
A	AA	AB	AC	AD
B	BA	BB	BC	BD
C	CA	CB	CC	CD
D	DA	DB	DC	DD

(a)

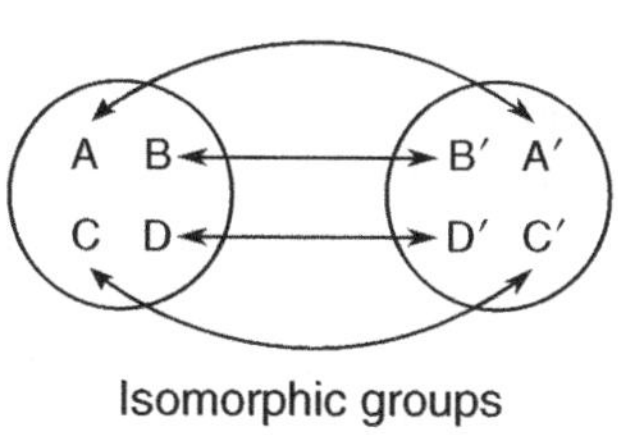

(b)

Figure 4.3 Group table (a) for a group containing four elements {A,B,C,D} and illustration (b) of two isomorphic groups.

operations matches the corresponding element obtained by combining the respective elements in the group $\mathcal{H}$

$$h(\hat{A} \times \hat{B}) = h(\hat{A}) \otimes h(\hat{B}), \tag{4.17}$$

where the operation $\times$ on the left-hand side is that of group $\mathcal{G}$ and that on the right-hand side $\otimes$ is that of group $\mathcal{H}$. The use of group theory in chemistry is largely concerned with finding homomorphic groups of matrices to our groups of symmetry operations. In other words, we seek matrix representations of our symmetry groups. If we for instance consider a basis consisting of the atomic 1*s* orbitals in the water molecule (O_{1s}, H^A_{1s}, H^B_{1s}), the representation matrices for the four symmetry operations of the C_{2v} point group of this molecule are

$$E = \begin{pmatrix} 1 & 0 & 0 \\ 0 & 1 & 0 \\ 0 & 0 & 1 \end{pmatrix} \qquad C_2 = \begin{pmatrix} 1 & 0 & 0 \\ 0 & 0 & 1 \\ 0 & 1 & 0 \end{pmatrix} \tag{4.18}$$
$$\sigma_v = \begin{pmatrix} 1 & 0 & 0 \\ 0 & 1 & 0 \\ 0 & 0 & 1 \end{pmatrix} \qquad \sigma'_v = \begin{pmatrix} 1 & 0 & 0 \\ 0 & 0 & 1 \\ 0 & 1 & 0 \end{pmatrix}.$$

We note that the oxygen 1*s* orbital is always transformed into itself, and it is therefore said to be a *symmetry-adapted basis function*. This is not the case for the hydrogen 1*s* orbitals. However, we can create the two symmetry-adapted orbitals $H^s_{1s} = H^A_{1s} + H^B_{1s}$ and $H^a_{1s} = H^A_{1s} - H^B_{1s}$. In this symmetry-adapted basis, the representation matrices are all diagonal

$$E = \begin{pmatrix} 1 & 0 & 0 \\ 0 & 1 & 0 \\ 0 & 0 & 1 \end{pmatrix} \qquad C_2 = \begin{pmatrix} 1 & 0 & 0 \\ 0 & 1 & 0 \\ 0 & 0 & -1 \end{pmatrix} \tag{4.19}$$
$$\sigma_v = \begin{pmatrix} 1 & 0 & 0 \\ 0 & 1 & 0 \\ 0 & 0 & 1 \end{pmatrix} \qquad \sigma'_v = \begin{pmatrix} 1 & 0 & 0 \\ 0 & 1 & 0 \\ 0 & 0 & -1 \end{pmatrix}.$$

In this case, the dimension of the space needed to represent the transformations of the symmetry-adapted orbitals under the different symmetry operations are all one-dimensional and cannot be reduced further. The smallest matrix representation that can be obtained is called an *irreducible representation*. In contrast, the original basis used in Eq. (4.18) is said to be a reducible representation. Abelian groups are particularly simple to handle, as they can be represented by one-dimensional matrices with elements that are the integers +1 and −1, as in the case above. The number of different irreducible representations of the group will be equal to the number of symmetry elements in the group.

For general point groups, we can also have multidimensional irreducible representations, but the largest dimension possible for molecular point groups is three (ignoring here the continuous groups that occur for atoms and linear molecules). Two-dimensional irreducible representations are denoted by the symbol *E*, and three-dimensional irreducible representations are denoted by the symbol *T*. In two- or three-dimensional irreducible representations, any linear combination of basis functions for this irreducible representation will also be a valid symmetry-adapted basis function. Multidimensional irreducible representations thus indicate degeneracies, and orbitals transforming as the *E* or *T* irreducible representations are double or triple degenerate, respectively.

A character table collects the traces of the matrices representing the elements in the different irreducible representations, which for Abelian groups will become equal to the numbers +1 or −1. Every row in the character table refers to an irreducible representation and every column

	A	B	C	D
Γ^1	+1	+1	+1	+1
Γ^2	+1	−1	+1	−1
Γ^3	+1	−1	−1	+1
Γ^4	+1	+1	−1	−1

(a)

	Γ^1	Γ^2	Γ^3	Γ^4
Γ^1	Γ^1	Γ^2	Γ^3	Γ^4
Γ^2	Γ^2	Γ^1	Γ^4	Γ^3
Γ^3	Γ^3	Γ^4	Γ^1	Γ^2
Γ^4	Γ^4	Γ^3	Γ^2	Γ^1

(b)

Figure 4.4 Character table (a) and direct product table (b) for an Abelian group containing four elements {A,B,C,D} and irreducible representations $\{\Gamma^1, \Gamma^2, \Gamma^3, \Gamma^4\}$.

refers to an element of the group, as illustrated in Figure 4.4. The character of the identity operation for a multidimensional irreducible representation is equal to the matrix dimension, thus $\Gamma^{\mathrm{E}}(\hat{E}) = 2$ and $\Gamma^{\mathrm{T}}(\hat{E}) = 3$.

Objects such as operators, symmetry-adapted basis functions, and canonical molecular orbitals belong to, or span, the irreducible representations of the symmetry group, as do products of such objects. If we for instance associate A, B, C, and D in the group table in Figure 4.4 with the operations $\hat{E}$, $\hat{C}_2$, $\hat{\sigma}_v$ and $\hat{\sigma}_v'$, we see that the basis functions O_{1s} and $\mathrm{H}^{\mathrm{s}}_{1s}$ both transform as Γ^1, whereas $\mathrm{H}^{\mathrm{a}}_{1s}$ transforms as Γ^2. The direct product table lists information to which irreducible representation products of objects belong, see Figure 4.4.

There are immediate and powerful practical benefits following from our study of symmetry groups of molecular systems. It turns out that a given matrix element, such as for example the transition matrix element in Eq. (4.7), is nonzero only if the direct product of the irreducible representations involved is equal to the all-symmetric representation

$$\Gamma(\psi_f) \otimes \Gamma(\hat{\Omega}) \otimes \Gamma(\psi_i) = \Gamma(\hat{E}). \tag{4.20}$$

Two immediate consequences from this result are that (i) the matrix representation of an operator in a symmetry-adapted basis set will be block diagonal, giving rise to substantial computational savings, and (ii) the derivation of selection rules in spectroscopy becomes straightforward and, in principle, trivial.

Let us view a simple and concrete example of the second point, namely how symmetry can be used to determine a selection rule. In Table 4.1, we have collected the character table for the C_{2v} point group to which the water molecule belongs. In water, the closed-shell electronic ground state spans the totally symmetric irreducible representation A_1. In order to recover the totally symmetric irreducible representation in Eq. (4.20), the only excited states that are accessible by symmetry in the electric-dipole approximation are those that transform as the irreducible representations of the different components of the electric-dipole operators, which transform

Table 4.1 Character table for the C_{2v} point group.

	$\hat{E}$	$\hat{C}_2(z)$	$\hat{\sigma}(xz)$	$\hat{\sigma}(yz)$	
A_1	1	1	1	1	z, x^2, y^2, z^2
A_2	1	1	−1	−1	xy, R_z
B_1	1	−1	1	−1	x, xz, R_y
B_2	1	−1	−1	1	y, yz, R_x

Table 4.2 Character table for the D_{2h} point group.

	$\hat{E}$	$\hat{C}_2(z)$	$\hat{C}_2(y)$	$\hat{C}_2(x)$	$\hat{i}$	$\hat{\sigma}(xy)$	$\hat{\sigma}(xz)$	$\hat{\sigma}(yz)$	
A_g	1	1	1	1	1	1	1	1	x^2, y^2, z^2
B_{1g}	1	1	−1	−1	1	1	−1	−1	R_z, xy
B_{2g}	1	−1	1	−1	1	−1	1	−1	R_y, xz
B_{3g}	1	−1	−1	1	1	−1	−1	1	R_x, yz
A_u	1	1	1	1	−1	−1	−1	−1	
B_{1u}	1	1	−1	−1	−1	−1	1	1	z
B_{2u}	1	−1	1	−1	−1	1	−1	1	y
B_{3u}	1	−1	−1	1	−1	1	1	−1	x

as the coordinates. From Table 4.1, we see that these states are of either A_1, B_1, or B_2 symmetries, whereas excited states of A_2 symmetry will not be reached.

Let us note that larger groups can be constructed from smaller groups by creating *direct product groups*. Given two groups $\mathcal{G}$ and $\mathcal{H}$, we can define a direct product group $\mathcal{G} \otimes \mathcal{H}$ if the following two conditions are satisfied

1) The only common operation in the two groups is the identity operation.
2) All symmetry operations of the group $\mathcal{G}$ commute with all symmetry operations of the group $\mathcal{H}$.

With these conditions satisfied, then the products of symmetry operations $\hat{A}_\mathcal{G}\hat{B}_\mathcal{H}$ with elements from the two groups satisfy all the conditions for a group. If the number of elements in the group $\mathcal{G}$ is g and the number of elements in $\mathcal{H}$ is h, then the number of elements in the direct product group is gh. The groups $\mathcal{G}$ and $\mathcal{H}$ can be denoted as *subgroups* of the group $\mathcal{G} \otimes \mathcal{H}$, where a subgroup is a set of symmetry operations in a group that is smaller than the total set of symmetry operations, and which by itself satisfies the conditions for a group.

To illustrate the concept of a direct product group, let us consider the direct product group of the C_{2v} point group in Table 4.1 with the group C_s containing only the identity operation $\hat{E}$ and the inversion operation $\hat{i}$. The direct product of the identity operator and the elements of C_{2v} obviously gives us back C_{2v}, whereas the direct product of the inversion operator with the elements of C_{2v} gives us four additional symmetry elements, that is, $\{\hat{i}\} \otimes \{\hat{E}, \hat{C}_2(z), \hat{\sigma}(xz), \hat{\sigma}(yz)\} = \{\hat{i}, \hat{\sigma}(xy), \hat{C}_2(y), \hat{C}_2(x)\}$. The resulting direct product group is denoted as D_{2h} and has consequently eight symmetry operations. The character table for a direct product group is given by the direct product of the character tables of the two groups. In the case of the D_{2h} point group, the character table is collected in Table 4.2. Direct product group is a convenient way of building increasingly complex groups and their corresponding character tables. From this exercise, we also note that C_{2v} and C_s are subgroups of the D_{2h} point group.

4.2 Time Symmetries

One of the important facts of physics is that for every symmetry in a system, that is, in the Hamiltonian, there is an associated conservation law. To illustrate this, we first recall that according to the Ehrenfest theorem, Eq. (2.82), the time derivative of the expectation value of an operator is

$$\frac{d}{dt}\langle\hat{\Omega}\rangle = \frac{1}{i\hbar}\langle[\hat{\Omega},\hat{H}]\rangle + \left\langle\frac{\partial\hat{\Omega}}{\partial t}\right\rangle, \tag{4.21}$$

where we note that the last term vanishes for time-independent operators.

There can only exist two possible symmetries with respect to time, namely invariance with respect to translation and/or inversion. Time-translational symmetry means that the system Hamiltonian is time independent, and from the Ehrenfest theorem with $\hat{\Omega} = \hat{H}$, we see that the expectation value of the Hamiltonian is then constant in time. Note that this is a general result, valid not only for stationary states. To the extent that the Hamiltonian can be identified with energy, as discussed in Chapter 2, this result accordingly implies conservation of energy. For most parts in this book, however, the Hamiltonian is time dependent due to the presence of external time-oscillating electromagnetic fields, and there will occur an exchange of photons, and thus energy, between the radiation field and the molecular system. Of greater significance here is therefore a discussion of time-inversion symmetry.

Consider the observation of a physical process that takes place during the time interval $-T < t < T$. An illustration in terms of the motion of a particle is given in Figure 4.5 where, at time t, the particle is observed at coordinate $\mathbf{r}(t)$ with velocity $\mathbf{v}(t)$ (a). The time-reversed process is illustrated in the panel (b) with particle coordinate and velocity given by $\mathbf{r}_T(t)$, $\mathbf{v}_T(t)$, respectively, and it is clear that the following relations hold

$$\mathbf{r}_T(t) = \mathbf{r}(-t), \tag{4.22}$$

$$\mathbf{v}_T(t) = -\mathbf{v}(-t). \tag{4.23}$$

More caution is required when considering the total momentum $\mathbf{p} = \boldsymbol{\pi} + q\mathbf{A}$. According to Eq. (3.123), the Lorentz force is the total time derivative of the mechanical momentum

$$\mathbf{F}_T(t) = \frac{d\boldsymbol{\pi}_T(t)}{dt} = q\mathbf{E}_T(t) + q[\mathbf{v}_T(t) \times \mathbf{B}_T(t)], \tag{4.24}$$

and, since both the mechanical momentum $\boldsymbol{\pi}_T$, which is proportional to $\mathbf{v}_T$, and time itself change sign under the operation of time reversal, the Lorentz force must be time-reversal symmetric, that is

$$\mathbf{F}_T(t) = \mathbf{F}(-t). \tag{4.25}$$

This leads to the conclusion that the electric and magnetic fields must fulfill

$$\mathbf{E}_T(t) = \mathbf{E}(-t); \quad \mathbf{B}_T(t) = -\mathbf{B}(-t), \tag{4.26}$$

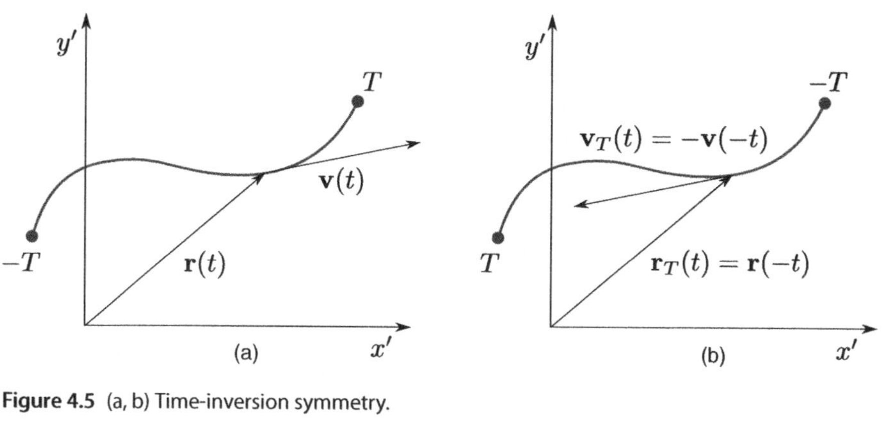

Figure 4.5 (a, b) Time-inversion symmetry.

or, equivalently, in terms of the potentials

$$\phi_T(t) = \phi(-t); \quad \mathbf{A}_T(t) = -\mathbf{A}(-t). \tag{4.27}$$

The momentum, and thus also the angular momentum, should therefore change sign as well, that is

$$\mathbf{p}_T(t) = -\mathbf{p}(-t), \tag{4.28}$$

$$\mathbf{l}_T(t) = -\mathbf{l}(-t). \tag{4.29}$$

However, there is a catch: If the potentials are introduced as *external* perturbations, they will be unaffected by the time-reversal operation since their sources are *outside* the system, that is, the vector potential of an external homogeneous magnetic field does not change sign under time reversal. This means that we can guarantee Eqs. (4.28) and (4.29) to hold true only in the special case where there is no external vector potential, that is, when $\mathbf{A} = \mathbf{0}$.

In the quantum-mechanical case, the situation is somewhat more complicated since the system is described by a wave function $\psi(t)$ and the particle trajectory in Figure 4.5(a) is replaced by the distribution $\psi(t)^\dagger \mathbf{r}\psi(t)$. We can, however, define operators $\hat{\mathcal{T}}$ and $\hat{\mathcal{K}}_T$, which connect $\psi(t)$ to the wave function of the time-reversed system $\psi_T(t)$ according to the relation

$$\psi_T(t) = \hat{\mathcal{T}}\psi(t) = \hat{\mathcal{K}}_T\psi(-t). \tag{4.30}$$

The total operation of time inversion of the wave function thus involves a change of sign of the time argument as well as the operation of $\hat{\mathcal{K}}_T$. We will now seek an explicit representation of this operator.

An elegant bridge between classical and quantum mechanics is provided by the Ehrenfest theorem, Eq. (4.21), and it seems therefore natural that the classical relations in Eqs. (4.22) and (4.28) should be replaced by a set of relations for expectation values in the quantum-mechanical case

$$\langle\psi_T(t)|\ \hat{\mathbf{r}}\ |\psi_T(t)\rangle = \langle\psi(-t)|\ \hat{\mathbf{r}}\ |\psi(-t)\rangle, \tag{4.31}$$

$$\langle\psi_T(t)|\ \hat{\mathbf{p}}\ |\psi_T(t)\rangle = -\langle\psi(-t)|\ \hat{\mathbf{p}}\ |\psi(-t)\rangle. \tag{4.32}$$

With the use of Eq. (4.30), we obtain a corresponding set of operator relations

$$\hat{\mathcal{K}}_T^\dagger \hat{\mathbf{r}} \hat{\mathcal{K}}_T = \hat{\mathbf{r}}, \tag{4.33}$$

$$\hat{\mathcal{K}}_T^\dagger \hat{\mathbf{p}} \hat{\mathcal{K}}_T = -\hat{\mathbf{p}}. \tag{4.34}$$

The above operator relations, which can only determine the operator $\hat{\mathcal{K}}_T$ to an arbitrary complex phase, hold in the sense that every possible expectation value of one of the operators to the left will equal the expectation value of the corresponding operator to the right. These relations suggest the complex conjugation operator $\hat{\mathcal{K}}_0$, which is an antiunitary operator, as the time-reversal operator. This hypothesis can be tested by direct application of the operator to the time-dependent Schrödinger equation

$$\hat{\mathcal{K}}_T \left[\hat{H} - i\frac{\partial}{\partial t}\right] \hat{\mathcal{K}}_T^{-1} \hat{\mathcal{K}}_T \psi(\mathbf{r}, t) = 0, \tag{4.35}$$

which, in the case of the one-electron Schrödinger equation, gives

$$\left[\frac{\hat{p}^2}{2m} - \frac{e}{m}(\hat{\mathbf{p}} \cdot \mathbf{A} + \mathbf{A} \cdot \hat{\mathbf{p}}) + \frac{e^2A^2}{2m} - e\phi + i\hbar\frac{\partial}{\partial t}\right] \hat{\mathcal{K}}_0 \psi_{\rm NR}(\mathbf{r}, t) = 0. \tag{4.36}$$

This equation does not have the form of the time-dependent Schrödinger equation because of the wrong sign in the time derivative as well as the paramagnetic contribution from the external vector potential $\mathbf{A}$ have the wrong sign. However, we recover the correct form of the wave equation by setting $\mathbf{A} = \mathbf{0}$, just as in the classical case, and by reversing the sign of the time variable. We can therefore conclude that the complex conjugation operator $\hat{\mathcal{K}}_0$ at this level of theory is indeed the time-reversal operator.

Moving to the Dirac equation, we find that the application of the complex conjugation operator gives

$$\left[\beta mc^2 - c(\boldsymbol{\alpha}^* \cdot \hat{\mathbf{p}}) + ec(\boldsymbol{\alpha}^* \cdot \mathbf{A}) - e\phi + i\hbar\frac{\partial}{\partial t}\right]\hat{\mathcal{K}}_0\psi_{\mathrm{R}}(\mathbf{r}, t) = 0, \tag{4.37}$$

where $\boldsymbol{\alpha}^* = (\alpha_x, -\alpha_y, \alpha_z)$. We conclude that the complex conjugation operator cannot represent time reversal at this level of theory since reversing the sign of the time variable and setting the vector potential to zero is insufficient to recover the original form of the Dirac equation. We can, however, write the desired time-reversal operator as

$$\hat{\mathcal{K}}_T = \hat{U}_T\hat{\mathcal{K}}_0, \tag{4.38}$$

where $\hat{U}_T$ is a unitary matrix. Application of this form in Eq. (4.35) leads to conditions satisfied by the choice

$$U_T = -i\,\Sigma_y, \tag{4.39}$$

where the phase factor has been chosen for convenience, as will be shown in the following. This form seems to single out the y-component of the four-component Pauli spin matrix, but the difference is canceled by the effect of complex conjugation, so that the total time-reversal operator treats the three components in an identical fashion. We can show this explicitly by the relation

$$\hat{\mathcal{K}}_T^\dagger \boldsymbol{\Sigma} \hat{\mathcal{K}}_T = -\boldsymbol{\Sigma}, \tag{4.40}$$

which demonstrates that the three components of the spin operator transform as the orbital angular momentum operator.

After having established the correct form of the time-reversal operator, it is of interest to explore the applicability of this operator. Consider first the effect of the time-reversal operator on a single orbital. For a purely spatial nonrelativistic orbital, the time-reversal operator is simply complex conjugation, but there is usually no effect from this operation since such orbitals can be chosen real in the absence of magnetic fields. For a two-component spin orbital, the four-component Pauli spin matrix of the time-reversal operator $\hat{\mathcal{K}}_T$ should be replaced by its two-component counterpart. We then obtain

$$\hat{\mathcal{K}}_T\psi = \overline{\psi} = \begin{bmatrix} 0 & -1 \\ 1 & 0 \end{bmatrix}\hat{\mathcal{K}}_0\begin{bmatrix} \psi^\alpha \\ \psi^\beta \end{bmatrix} = \begin{bmatrix} -\psi^{\beta*} \\ \psi^{\alpha*} \end{bmatrix}; \quad \hat{\mathcal{T}}^2\psi(t) = \hat{\mathcal{K}}_T^2\psi(t) = -\psi(t), \tag{4.41}$$

which is straightforwardly generalized to the four-component case. The second identity in Eq. (4.41) is interesting. Intuitively, one would expect that applying the time-reversal operator twice would bring us back to the original state, but the above result tells us that intuition must be used with some caution here. For an N-electron system, the explicit form of the time-reversal operator is

$$\hat{\mathcal{T}} = \prod_{i=1}^{N}\hat{\mathcal{T}}(i), \tag{4.42}$$

and it is therefore clear that $\hat{\mathcal{T}}^2$ will be -1 and 1 for systems with odd and even number of electrons, respectively.

The orbitals ψ and $\overline{\psi} = \hat{\mathcal{K}}_T\psi$ are denoted Kramers partners. The overbar on the latter orbital suggests that there exists a connection between Kramers partners and spin partners, and they are indeed identical in the case of nonrelativistic, spin-pure, orbitals with real spatial parts such as the atomic $p_x\alpha$ and $p_x\beta$ orbitals. However, this is no longer true for complex orbitals, and as an example, we note that the spin partner of the atomic $p_{+1}\beta$ orbital is the $p_{+1}\alpha$ orbital whereas its Kramers partner is the $p_{-1}\alpha$ orbital.

Starting from some orbital basis $\{\phi_i\}$, we may apply the time-reversal operator $\hat{\mathcal{K}}_T$ to obtain a *Kramers basis* $\{\phi_i\} \cup \{\overline{\phi}_i\}$. Let us now consider the matrix representation of an operator $\hat{\Omega}$ in such a basis. We will assume that the operator has a well-defined time-reversal symmetry, that is $\hat{\mathcal{K}}_T\hat{\Omega}\hat{\mathcal{K}}_T^{-1} = \theta_t\hat{\Omega}$ where the phases $\theta_t = +1$ and -1 indicate a time-reversal symmetric and antisymmetric operator, respectively. The general form of this matrix representation is given by

$$\begin{bmatrix} \mathbf{A} & \mathbf{B} \\ \mathbf{C} & \mathbf{D} \end{bmatrix}; \quad \begin{matrix} A_{pq} = \langle \phi_p|\hat{\Omega}|\phi_q\rangle & B_{pq} = \langle \phi_p|\hat{\Omega}|\overline{\phi}_q\rangle \\ C_{pq} = \langle \overline{\phi}_p|\hat{\Omega}|\phi_q\rangle & D_{pq} = \langle \overline{\phi}_p|\hat{\Omega}|\overline{\phi}_q\rangle \end{matrix}. \tag{4.43}$$

However, we can easily show that time-reversal symmetry adds structure to this matrix. Consider for example the action of the time-reversal operator on the matrix element A_{pq}. If we consider the matrix element as a (complex) number, the appropriate time-reversal operator is complex conjugation as expressed by

$$\hat{\mathcal{K}}_T A_{pq} = A^*_{pq}. \tag{4.44}$$

On the other hand, if we let the time-reversal operator $\hat{\mathcal{K}}_T$ act on the individual parts of the integrand [according to Eq. (4.10)], we get

$$\hat{\mathcal{K}}_T A_{pq} = \langle \hat{\mathcal{K}}_T\phi_p|\hat{\mathcal{K}}_T\hat{\Omega}\hat{\mathcal{K}}_T^{-1}|\hat{\mathcal{K}}_T\phi_q\rangle = \langle \overline{\phi}_p|\theta_t\hat{\Omega}|\overline{\phi}_q\rangle = \theta_t\, D_{pq}, \tag{4.45}$$

and we thus conclude that $D_{pq} = \theta_t A^*_{pq}$. In a similar manner we can connect the matrix blocks **B** and **C**, and so the overall matrix structure imposed by time-reversal symmetry becomes

$$\begin{bmatrix} \mathbf{A} & \mathbf{B} \\ -\theta_t\mathbf{B}^* & \theta_t\mathbf{A}^* \end{bmatrix}. \tag{4.46}$$

For a Hermitian operator $\hat{\Omega}$, it is then easy to show that eigenvectors come in pairs

$$\mathbf{c} = \begin{bmatrix} \mathbf{c}^\alpha \\ \mathbf{c}^\beta \end{bmatrix}; \quad \overline{\mathbf{c}} = \begin{bmatrix} -\mathbf{c}^{\beta*} \\ \mathbf{c}^{\alpha*} \end{bmatrix}, \tag{4.47}$$

and notably, if **c** is an eigenvector with eigenvalue ε, then $\overline{\mathbf{c}}$ is an eigenvector with eigenvalue $\theta_t\varepsilon$. A time-symmetric Hermitian operator such as the Hamiltonian, describing a system in the absence of external magnetic fields, is thus doubly degenerate. From the above discussion of N-electron systems, we see that a system with an odd number of electrons, for which $\hat{\mathcal{K}}_T^2\psi(t) = -\psi(t)$, has doubly degenerate solutions—this is referred to as Kramers degeneracy. We note that, in the nonrelativistic case with spin-pure orbitals, the matrix block **B** vanishes because of spin integration, and the operator matrix is therefore manifestly block diagonal. In the general relativistic case, block diagonalization can also be accomplished for a time-symmetric operator, but in this case at the cost of introducing quaternion algebra.[1]

1 T. Saue and H. J. A. Jensen. *J. Chem. Phys.*, 111:6211–6222, 1999.

4.3 Spatial Symmetries

A first encounter of spatial symmetries in studies of quantum mechanics is provided by the free particle, for which the Hamiltonian is manifestly translational invariant. It is often shown for this system that translational symmetry in the Hamiltonian leads to the conservation of linear momentum. In atomic and molecular physics, where translational symmetry is absent, spatial inversion and rotation are of greater interest.

4.3.1 Spatial Inversion

Let us study spatial inversion, or parity, in a similar manner as we did with time inversion. Consider a physical process that occurs in the time interval $-T < t < T$ in accordance with Figure 4.6(a). The same process viewed after spatial inversion has taken place is illustrated in panel (b). At a given point t in time, the positions and velocities of the particles in the left and right line drawings are related by

$$\begin{aligned} \mathbf{r}_i(t) &= -\mathbf{r}(t), \\ \mathbf{v}_i(t) &= -\mathbf{v}(t), \end{aligned} \tag{4.48}$$

and thus both observables are antisymmetric with respect to spatial inversion.

As was the case of time-reversal symmetry, the action of the spatial symmetry operation on electric and magnetic fields is deduced from a consideration of the Lorentz force. Since velocity, and thereby mechanical momentum, but not time, changes sign under inversion, the force itself must be antisymmetric with respect to parity, and as far as fields are concerned, we can therefore conclude that

$$\mathbf{E}_i(t) = -\mathbf{E}(t); \quad \mathbf{B}_i(t) = \mathbf{B}(t), \tag{4.49}$$

or

$$\phi_i(t) = \phi(t); \quad \mathbf{A}_i(t) = -\mathbf{A}(t). \tag{4.50}$$

The opposite parity symmetry of a field and its potential is because the former is a position derivative of the latter, and position is parity antisymmetric. From the above relations, we conclude that momentum, but not angular momentum, changes sign under inversion, that is

$$\mathbf{p}_i(t) = -\mathbf{p}(t), \tag{4.51}$$

$$\mathbf{l}_i(t) = \mathbf{l}(t). \tag{4.52}$$

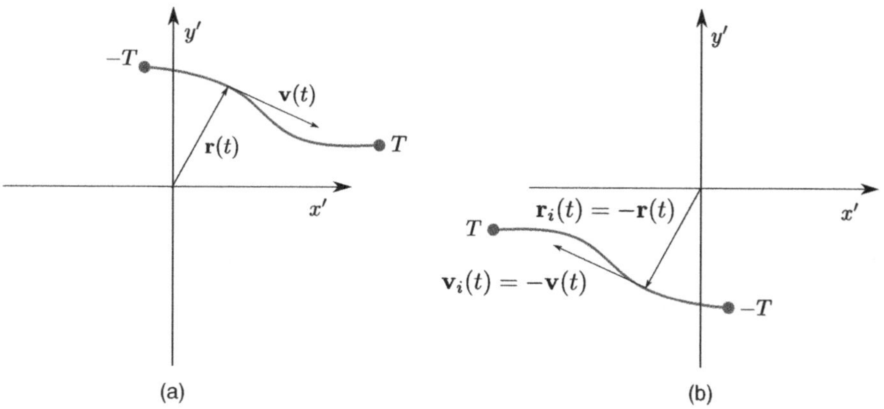

Figure 4.6 (a, b) Space inversion, or parity, symmetry.

A vector changing sign under inversion is called a *polar* vector, whereas a vector that does not change sign is called an *axial vector* or pseudovector.

In the study of a quantum system, we ask ourselves: if a system is described by a wave function $\psi(\mathbf{r}, t)$, how is it related to the wave function $\psi_i(\mathbf{r}, t)$ of the corresponding space-inverted system. As for the case of time-inversion symmetry, we (implicitly) define operators that connect the two wave functions as

$$\psi_i(\mathbf{r}, t) = \hat{\imath}\psi(\mathbf{r}, t) = \hat{\mathcal{K}}_i\psi(-\mathbf{r}, t), \tag{4.53}$$

where $\hat{\mathcal{K}}_i$, which is responsible for the appropriate change of the spin-angular momentum, is to be determined. In accordance with the classical relation between positions and momenta, we use as a starting point the corresponding quantum-mechanical expectation values

$$\langle\psi_i(t)|\ \hat{\mathbf{r}}\ |\psi_i(t)\rangle = -\langle\psi(t)|\ \hat{\mathbf{r}}\ |\psi(t)\rangle, \tag{4.54}$$

$$\langle\psi_i(t)|\ \hat{\mathbf{p}}\ |\psi_i(t)\rangle = -\langle\psi(t)|\ \hat{\mathbf{p}}\ |\psi(t)\rangle, \tag{4.55}$$

$$\langle\psi_i(t)|\ \boldsymbol{\Sigma}\ |\psi_i(t)\rangle = \langle\psi(t)|\ \boldsymbol{\Sigma}\ |\psi(t)\rangle, \tag{4.56}$$

$$\langle\psi_i(t)|\ \hat{H}\ |\psi_i(t)\rangle = \langle\psi(t)|\ \hat{H}\ |\psi(t)\rangle. \tag{4.57}$$

In the quantum-mechanical case, we choose to study the expectation value of the spin rather than orbital angular momentum, and we have here also added the requirement that the energy of a free particle with relativistic and nonrelativistic forms of the Hamiltonian given in Eqs. (3.151) and (3.153), respectively, should remain unaltered by the inversion operation. We note that this holds true for the electromagnetic forces governing chemistry, but not for the weak force that gives rise to the minute energy difference between enantiomers of chiral molecules. After inserting Eq. (4.53) into Eqs. (4.54)–(4.57), we arrive at a set of operator relations

$$\hat{\imath}^\dagger\hat{\mathbf{r}}\hat{\imath} = -\hat{\mathbf{r}}, \tag{4.58}$$

$$\hat{\imath}^\dagger\hat{\mathbf{p}}\hat{\imath} = -\hat{\mathbf{p}}, \tag{4.59}$$

$$\hat{\imath}^\dagger\boldsymbol{\Sigma}\hat{\imath} = \boldsymbol{\Sigma}, \tag{4.60}$$

$$\hat{\imath}^\dagger\hat{H}\hat{\imath} = \hat{H}. \tag{4.61}$$

The relations for the position and linear momentum operators affect the part of $\hat{\imath}$, which inverts signs of coordinates, see Eq. (4.53), and therefore provide no information on the appropriate form of $\hat{\mathcal{K}}_i$. The relations for the spin-angular momentum and the Hamiltonian, on the other hand, involve the operator $\hat{\mathcal{K}}_i$. From Eqs. (4.60) and (4.61), we conclude that the following relations must be satisfied

$$\hat{\imath}^\dagger\sigma_i\hat{\imath} = \hat{\mathcal{K}}_i^\dagger\sigma_k\hat{\mathcal{K}}_i = \sigma_i, \quad k = \{x, y, z\}, \tag{4.62}$$

$$\hat{\imath}^\dagger\hat{\alpha}_k\hat{p}_k\hat{\imath} = -\hat{\mathcal{K}}_i^\dagger\hat{\alpha}_k\hat{\mathcal{K}}_i\hat{p}_k = \hat{\alpha}_i\hat{p}_i, \quad k = \{x, y, z\}, \tag{4.63}$$

$$\hat{\imath}^\dagger\beta\hat{\imath} = \hat{\mathcal{K}}_i^\dagger\beta\hat{\mathcal{K}}_i = \beta, \tag{4.64}$$

and from these relations we deduce that the spin part of the parity operator is

$$\hat{\mathcal{K}}_i = \beta, \tag{4.65}$$

and therefore that the inverse parity operator is

$$\hat{\imath}^{-1} = \hat{\imath}. \tag{4.66}$$

In cases where the Hamiltonian is invariant with respect to spatial inversion, that is, when

$$\hat{\imath}\hat{H}\hat{\imath}^{-1} = \hat{H}, \tag{4.67}$$

we are guaranteed that it is in principle possible to find a complete set of eigenstates of $\hat{H}$ with definite parity.

4.3.2 Rotations

In the study of planetary motion we learn that, because the Hamiltonian is invariant with respect to rotations, orbital angular momentum is a conserved quantity. As a consequence, the linear momentum (and therefore the speed) of the particle is large close to the center of the potential and small at large distances away from it, and the area swept over by the planet per unit time remains constant. As we will see, an analogous argument can be made in the case of quantum mechanics.

Let us consider the infinitesimal rotation $d\phi$ about a unit axis $\mathbf{n}$. To linear order, we have

$$\hat{\mathcal{R}}(d\phi, \mathbf{n})f(\mathbf{r}) = f(\mathbf{r} - d\mathbf{r}) = f(\mathbf{r}) - d\mathbf{r} \cdot \nabla f(\mathbf{r}). \tag{4.68}$$

By simple geometric construction as illustrated in Figure 4.7, we can see that

$$d\mathbf{r} = d\phi(\mathbf{n} \times \mathbf{r}), \tag{4.69}$$

which then gives

$$\hat{\mathcal{R}}(d\phi, \mathbf{n})f(\mathbf{r}) = [1 - d\phi\, \mathbf{n} \cdot (\mathbf{r} \times \nabla)]f(\mathbf{r}) = \left[1 - i\frac{d\phi}{\hbar}(\mathbf{n} \cdot \hat{\mathbf{l}})\right] f(\mathbf{r}), \tag{4.70}$$

where $\hat{\mathbf{l}}$ is the orbital angular momentum operator. A finite rotation by an angle ϕ about an axis represented by the unit vector $\mathbf{n}$ is then given by

$$\hat{\mathcal{R}}(\phi, \mathbf{n}) = \lim_{k\to\infty} \hat{\mathcal{R}}\left(\frac{\phi}{k\hbar}, \mathbf{n}\right) = \lim_{k\to\infty}\left[1 - i\frac{\phi}{k\hbar}(\mathbf{n} \cdot \hat{\mathbf{l}})\right]^k = \exp\left[-i\frac{\phi}{\hbar}(\mathbf{n} \cdot \hat{\mathbf{l}})\right], \tag{4.71}$$

and we see that the rotation operator is an explicitly unitary exponential operator with an anti-Hermitian exponent containing the rotation angle and the component of the orbital angular momentum along the rotation axis.

Our derivation of an explicit form of the rotation operator is here made in terms of the rotation of a scalar function $f(\mathbf{r})$, whereas for the electronic wave functions that we are primarily concerned with, there is also a vectorial component associated with the spin degree of freedom of the electron, see Eq. (2.51). As an imperfect analogy, we may consider the infinitesimal rotation by an angle $d\phi$ of a two-component vector field $(f_x, f_y)^T$ about the z-axis as shown in Figure 4.8. The translation of the vector is a mere transformation of the scalar functions f_x and f_y, each in perfect analogy with our previous consideration that was achieved using the orbital angular momentum operator. The actual rotation of the vector, on the other hand, involves a mixing of its components, achieved through a multiplication by, in this case, a 2×2 matrix.

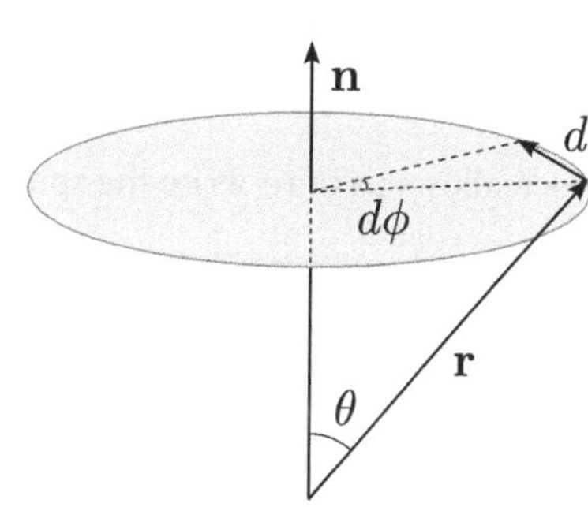

Figure 4.7 An infinitesimal rotation by an angle $d\phi$ about an arbitrary laboratory axis $\mathbf{n}$. The length of the displacement $d\mathbf{r}$ is $r \sin\theta d\phi$ and it is directed along $\mathbf{n} \times \mathbf{r}$.

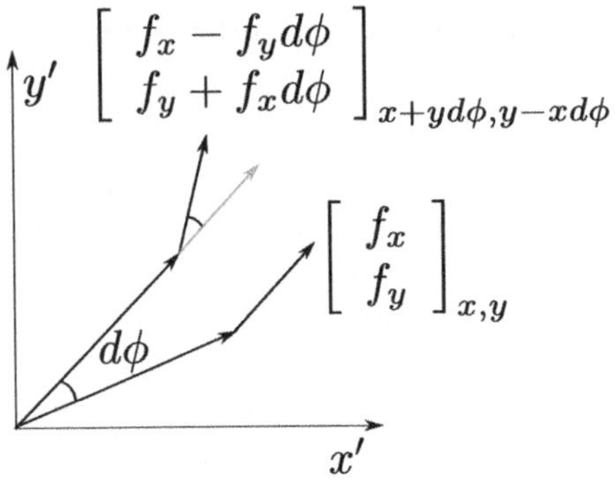

Figure 4.8 Infinitesimal rotation by an angle $d\phi$ of a two-dimensional vector field.

The analogy here is imperfect, albeit illustrative, because the wave function is not an observable that can be illustrated in our laboratory space; instead, the rotation takes place in the abstract Hilbert space and it is then achieved by the operators of the spin-angular momentum. The generator of rotations of an electronic wave function in Hilbert space is therefore the total angular momentum operator

$$\hat{j}_\alpha = \hat{l}_\alpha \otimes \hat{I}^{\rm s} \otimes \hat{I}^{\rm c} + \hat{I}^{\rm o} \otimes \frac{\hbar}{2}\sigma_\alpha \otimes \hat{I}^{\rm c}; \quad \alpha = \{x, y, z\}, \tag{4.72}$$

and we note that for an N-electron system, the rotation is carried out by the product operator

$$\hat{\mathcal{R}}(\phi, \mathbf{n}) = \prod_{i=1}^{N} e^{-i\phi\ \mathbf{n}\cdot\hat{\mathbf{j}}(i)/\hbar} = e^{-i\phi\ \mathbf{n}\cdot\hat{\mathbf{J}}/\hbar}; \quad \hat{\mathbf{J}} = \sum_{i=1}^{N} \hat{\mathbf{j}}(i), \tag{4.73}$$

where we have made use of the fact that the angular momenta operators of separate electrons commute.

The set of rotation operators $\hat{\mathcal{R}}(\phi, \mathbf{n})$ forms a continuous group known as the full rotation group, and irreducible representations of the full rotation group may be labeled by the quantum number j and are $(2j+1)$-fold degenerate. Matrix representations of the operators $\hat{\mathcal{R}}(\phi, \mathbf{n})$ may be constructed in a basis of eigenfunctions of the total angular momentum operator $\{\psi_{jm}; m = -j, \dots, j\}$ for a given j.

As an illustration, let us consider the representation matrices of $\hat{\mathcal{R}}(\phi, \mathbf{n})$ based on the eigenfunctions with $j = 1/2$, that is, zero orbital angular momentum and spin angular momentum of one half. The eigenfunctions are the well-known spin-up and spin-down functions that are commonly referred to as the α- and β-spinors. The representation matrix on this basis becomes

$$\hat{\mathcal{R}}^{\frac{1}{2}}(\phi, \mathbf{n}) = e^{-i\phi\ \mathbf{n}\cdot\hat{\sigma}/2} = \sum_{m}^{\infty} (-i)^m \frac{(\phi/2)^m}{m!} (\mathbf{n}\cdot\boldsymbol{\sigma})^m. \tag{4.74}$$

A considerable simplification is provided by noting that

$$(\mathbf{n}\cdot\boldsymbol{\sigma})^2 = I_2, \tag{4.75}$$

which allows us to re-write the spin-1/2 rotation operator as

$$\hat{\mathcal{R}}^{\frac{1}{2}}(\phi, \mathbf{n}) = \cos(\phi/2)\, I_2 - i\sin(\phi/2)\, \mathbf{n}\cdot\boldsymbol{\sigma}. \tag{4.76}$$

An example is given by the rotation of a spinor by an angle 2π about an arbitrary axis, an operation which is accomplished by the matrix

$$\hat{\mathcal{R}}^{\frac{1}{2}}(2\pi, \mathbf{n}) = I_2 \cos\pi - i(\mathbf{n}\cdot\boldsymbol{\sigma})\sin\pi = -I_2. \tag{4.77}$$

This result has been verified by experiment and implies the nonintuitive fact that in spin space a rotation 2π about an axis is not equivalent to the identity operation. Standard group character tables therefore need to be modified when considering irreducible representations spanned by wave functions of half-integer spin systems, also known as fermions.

Another example is provided by two subsequent rotations for which the rotation matrix is given by

$$\begin{aligned}&\hat{\mathcal{R}}^{\frac{1}{2}}(\phi_2,\mathbf{n}_2)\hat{\mathcal{R}}^{\frac{1}{2}}(\phi_1,\mathbf{n}_1)\\&\quad=[\cos(\phi_2/2)\cos(\phi_1/2)-\mathbf{n}_2\cdot\mathbf{n}_1\ \sin(\phi_2/2)\sin(\phi_1/2)]I_2\\&\qquad-i\boldsymbol{\sigma}\cdot[\mathbf{n}_1\cos(\phi_2/2)\sin(\phi_1/2)+\mathbf{n}_2\sin(\phi_2/2)\cos(\phi_1/2)\\&\qquad-i(\mathbf{n}_2\times\mathbf{n}_1)\sin(\phi_2/2)\sin(\phi_1/2)].\end{aligned}$$

The factor $(\mathbf{n}_2\times\mathbf{n}_1)=-(\mathbf{n}_1\times\mathbf{n}_2)$ shows that the two rotation operations do not commute unless the rotation axes are collinear.

Let us consider the rotation matrices for $j=1$. In this case, we may by choice employ a basis consisting of the eigenfunctions to $l=1$ and zero spin-angular momentum. These functions are the corresponding spherical harmonic functions

$$Y_{1,-1}=\sqrt{\frac{3}{8\pi}}\,\frac{x-iy}{r};\quad Y_{1,1}=-\sqrt{\frac{3}{8\pi}}\,\frac{x+iy}{r};\quad Y_{1,0}=\sqrt{\frac{3}{4\pi}}\,\frac{z}{r}. \tag{4.78}$$

We shall, however, for reasons that will be clear later, use a basis of real spherical harmonics corresponding to p orbitals directed along the Cartesian coordinate axes and with explicit expressions given by

$$\begin{aligned}p_x&=\frac{1}{\sqrt{2}}(Y_{1,-1}-Y_{1,1}),\\p_y&=\frac{i}{\sqrt{2}}(Y_{1,-1}+Y_{1,1}),\\p_z&=Y_{1,0}.\end{aligned} \tag{4.79}$$

In this basis, the matrix representation of the components of the orbital angular momentum operator takes the form

$$(\tau_i)_{jk}=\frac{1}{\hbar}\,\langle p_j|\,\hat{l}_i\,|p_k\rangle=-i\epsilon_{ijk}, \tag{4.80}$$

where the Levi–Civita symbol ϵ_{ijk} appears. Further manipulation shows that

$$(\mathbf{n}\cdot\boldsymbol{\tau})^{2n}=(\mathbf{n}\cdot\boldsymbol{\tau})^2=I_3-\mathbf{n}\mathbf{n}^T;\quad(\mathbf{n}\cdot\boldsymbol{\tau})^{2n+1}=(\mathbf{n}\cdot\boldsymbol{\tau}), \tag{4.81}$$

which, from Eq. (4.73), leads to an expression for the rotation matrix that is

$$\hat{\mathcal{R}}^1(\phi,\mathbf{n})=I_3-i(\mathbf{n}\cdot\boldsymbol{\tau})\sin\phi+(\mathbf{n}\cdot\boldsymbol{\tau})^2(\cos\phi-1). \tag{4.82}$$

As an example, we can consider the rotation about the laboratory z-axis, for which we have

$$\mathbf{n}=\begin{pmatrix}0\\0\\1\end{pmatrix};\quad\tau_z=\begin{pmatrix}0&-i&0\\i&0&0\\0&0&0\end{pmatrix}, \tag{4.83}$$

and, consequently, a rotation matrix that is

$$\hat{\mathcal{R}}^1(\phi,\mathbf{e}_z)=\begin{pmatrix}\cos\phi&-\sin\phi&0\\\sin\phi&\cos\phi&0\\0&0&1\end{pmatrix}. \tag{4.84}$$

The group of all rotation matrices, Eq. (4.82), in coordinate space is called SO(3).

By choosing the real spherical harmonics as basis functions, we are able to construct rotation matrices for regular coordinate transformations in laboratory, or coordinate, space. Rotations in spin space, Eq. (4.76), form the group SU(2). Interestingly, we can use rotations in spin space to generate the rotation matrices in coordinate space.[2] A general rotation, transforming the coordinate $\mathbf{r}$ into the coordinate $\mathbf{r}'$, can be expressed by

$$\boldsymbol{\sigma}\cdot\mathbf{r}' = \hat{\mathcal{R}}^{\frac{1}{2}}(\phi,\mathbf{n})(\boldsymbol{\sigma}\cdot\mathbf{r})\hat{\mathcal{R}}^{\frac{1}{2}\dagger}(\phi,\mathbf{n}) = \hat{\mathcal{R}}^{\frac{1}{2}}(\phi,\mathbf{n})(\boldsymbol{\sigma}\cdot\mathbf{r})\hat{\mathcal{R}}^{\frac{1}{2}}(-\phi,\mathbf{n}). \tag{4.85}$$

We split the coordinate $\mathbf{r}$ into its parallel $\mathbf{r}_\parallel$ and perpendicular $\mathbf{r}_\perp$ components with respect to the rotation axis $\mathbf{n}$, and by using the Dirac identity [Eq. (F.2)] and its corollaries [Eqs. (F.5) and (F.6)], we get

$$\begin{aligned}\boldsymbol{\sigma}\cdot\mathbf{r}' &= \hat{\mathcal{R}}^{\frac{1}{2}}(\phi,\mathbf{n})[(\boldsymbol{\sigma}\cdot\mathbf{r}_\parallel)+(\boldsymbol{\sigma}\cdot\mathbf{r}_\perp)]\hat{\mathcal{R}}^{\frac{1}{2}}(-\phi,\mathbf{n}) \\ &= (\boldsymbol{\sigma}\cdot\mathbf{r}_\parallel)+\hat{\mathcal{R}}^{\frac{1}{2}}(2\phi,\mathbf{n})(\boldsymbol{\sigma}\cdot\mathbf{r}_\perp) \\ &= (\boldsymbol{\sigma}\cdot\mathbf{r}_\parallel)+(\boldsymbol{\sigma}\cdot\mathbf{r}_\perp)\cos\phi+\boldsymbol{\sigma}\cdot(\mathbf{n}\times\mathbf{r}_\perp)\sin\phi \\ &= (\boldsymbol{\sigma}\cdot\mathbf{r})\cos\phi+(\boldsymbol{\sigma}\cdot\mathbf{r}_\parallel)(1-\cos\phi)+\boldsymbol{\sigma}\cdot(\mathbf{n}\times\mathbf{r}_\perp)\sin\phi \\ &= (\boldsymbol{\sigma}\cdot\mathbf{r})\cos\phi+(\boldsymbol{\sigma}\cdot\mathbf{n})(\mathbf{n}\cdot\mathbf{r})(1-\cos\phi)+\boldsymbol{\sigma}\cdot(\mathbf{n}\times\mathbf{r})\sin\phi \\ &= \boldsymbol{\sigma}\cdot[I_3\cos\phi+\mathbf{n}\mathbf{n}^T(1-\cos\phi)+\sin\phi\;\mathbf{n}\times]\mathbf{r}.\end{aligned} \tag{4.86}$$

Because we also have that

$$\mathbf{r}' = \hat{\mathcal{R}}^1(\phi,\mathbf{n})\mathbf{r}, \tag{4.87}$$

we can extract the rotation matrix in coordinate space from Eq. (4.86), and we note that it matches that given in Eq. (4.82).

As our final example of constructing explicit rotation matrices, we consider a system of two coupled spin one-half particles with zero orbital angular momentum. It is clear that such a system constitutes a bosonic system with total angular momentum equal to either zero or one, and we therefore immediately expect that for a rotation about an arbitrary laboratory axis, an angle 2π corresponds to the identity operation. For simplicity, each of the two particles (electrons) is here assumed to be represented by a two-component wave function, but the generalization to the four-component case is straightforward. The rotation of this system in our laboratory space is accomplished by the direct product matrix

$$\begin{aligned}\hat{\mathcal{R}}(\phi,\mathbf{n}) &= \hat{\mathcal{R}}^{\frac{1}{2}}(\phi,\mathbf{n})\otimes\hat{\mathcal{R}}^{\frac{1}{2}}(\phi,\mathbf{n}) \\ &= I_2\otimes I_2\cos^2(\phi/2)-(\mathbf{n}\cdot\boldsymbol{\sigma})\otimes(\mathbf{n}\cdot\boldsymbol{\sigma})\sin^2(\phi/2) \\ &\quad -\frac{i}{2}[(\mathbf{n}\cdot\boldsymbol{\sigma})\otimes I_2+I_2\otimes(\mathbf{n}\cdot\boldsymbol{\sigma})]\;\sin\phi.\end{aligned} \tag{4.88}$$

This 4×4 matrix acts on ket vectors of the form

$$\begin{bmatrix}\alpha_1\\ \beta_1\end{bmatrix}\otimes\begin{bmatrix}\alpha_2\\ \beta_2\end{bmatrix}=\begin{bmatrix}\alpha_1\alpha_2\\ \alpha_1\beta_2\\ \beta_1\alpha_2\\ \beta_1\beta_2\end{bmatrix}. \tag{4.89}$$

However, we know that the coupling of two spins will lead to eigenfunctions of singlet- and triplet-spin symmetry. Let us therefore introduce a unitary transformation $\hat{U}$ that transforms

2 Technically, SU(2) is a covering group of SO(3) as described in S. L. Altmann. *Rotations, Quaternions and Double Groups*. Oxford University Press, New York, 1986.

the original functions in Eq. (4.89) into these spin eigenfunctions

$$\begin{bmatrix} T_1 \\ T_0 \\ T_{-1} \\ S_0 \end{bmatrix} = \begin{bmatrix} 1 & 0 & 0 & 0 \\ 0 & 1/\sqrt{2} & 1/\sqrt{2} & 0 \\ 0 & 0 & 0 & 1 \\ 0 & 1/\sqrt{2} & -1/\sqrt{2} & 0 \end{bmatrix} \begin{bmatrix} \alpha_1\alpha_2 \\ \alpha_1\beta_2 \\ \beta_1\alpha_2 \\ \beta_1\beta_2 \end{bmatrix}. \tag{4.90}$$

On the basis of the spin eigenfunctions, the rotation matrix in Eq. (4.88) becomes block diagonal

$$\hat{U}\,\hat{\mathcal{R}}(\phi, \mathbf{n})\,\hat{U}^\dagger = \begin{bmatrix} \hat{\mathcal{R}}^{\mathrm{T}}(\phi, \mathbf{n}) & 0 \\ 0 & \hat{\mathcal{R}}^{\mathrm{S}}(\phi, \mathbf{n}) \end{bmatrix}. \tag{4.91}$$

Since the rotation matrix is block diagonal with $\hat{\mathcal{R}}^{\mathrm{T}}$ and $\hat{\mathcal{R}}^{\mathrm{S}}$ having dimensions three and one, respectively, there is no mixing of components of different spin symmetry upon rotation. The scalar singlet rotation operator $\hat{\mathcal{R}}^{\mathrm{S}}$ is equal to 1 whereas the triplet rotation operator $\hat{\mathcal{R}}^{\mathrm{T}}$ is given by

$$\hat{\mathcal{R}}^{\mathrm{T}}(\phi, \mathbf{n}) = \begin{bmatrix} \cos^2(\phi/2) - n_z^2\sin^2(\phi/2) - in_z\sin\phi & -\sqrt{2}n_z n_-\sin^2(\phi/2) - \frac{i}{\sqrt{2}}n_-\sin\phi & -n_-^2\sin^2(\phi/2) \\ -\sqrt{2}n_z n_+\sin^2(\phi/2) - \frac{i}{\sqrt{2}}n_+\sin\phi & \cos^2(\phi/2) + (2n_z^2 - 1)\sin^2(\phi/2) & \sqrt{2}n_z n_-\sin^2(\phi/2) - \frac{i}{\sqrt{2}}n_-\sin\phi \\ -n_+^2\sin^2(\phi/2) & \sqrt{2}n_z n_+\sin^2(\phi/2) - \frac{i}{\sqrt{2}}n_+\sin\phi & \cos^2(\phi/2) - n_z^2\sin^2(\phi/2) + in_z\sin\phi \end{bmatrix}, \tag{4.92}$$

where

$$n_\pm = n_x \pm in_y. \tag{4.93}$$

The derivation of rotation matrices may appear complicated and there is a risk that the physical reality may become obscured by matrix algebra. For this reason, it can be illuminating to employ the rotation matrix in Eq. (4.92) in a concrete example. Assume that a two-electron system is found in a state represented by the spin ket

$$T_1 = |1,\ 1\rangle = \begin{bmatrix} \alpha_1 \\ 0 \end{bmatrix} \otimes \begin{bmatrix} \alpha_2 \\ 0 \end{bmatrix}. \tag{4.94}$$

The spin-magnetic moment of the system is given by the expectation value of the spin operator and, in this case, it is clear that it will be directed along the laboratory z-axis as illustrated in Figure 4.9. The components of the spin operator takes the form

$$\hat{S}_\alpha = \hat{S}_\alpha(1) \otimes I_2(2) + I_2(1) \otimes \hat{S}_\alpha(2), \tag{4.95}$$

and we get expectation values for the collection of Cartesian components that are given by

$$\langle\hat{\mathbf{S}}\rangle = \langle 1,\ 1|\ \hat{\mathbf{S}}\ |1,\ 1\rangle = \hbar\mathbf{e}_z. \tag{4.96}$$

Let us now perform a rotation of coordinate space by an angle π about the x axis. The mapping of the T_1 ket is given by the first column in the rotation matrix in Eq. (4.92) and, with $\mathbf{n} = (1, 0, 0)^T$, we get a transformed ket that is

$$\hat{\mathcal{R}}(\pi, \mathbf{e}_x)\ T_1 = -T_{-1}. \tag{4.97}$$

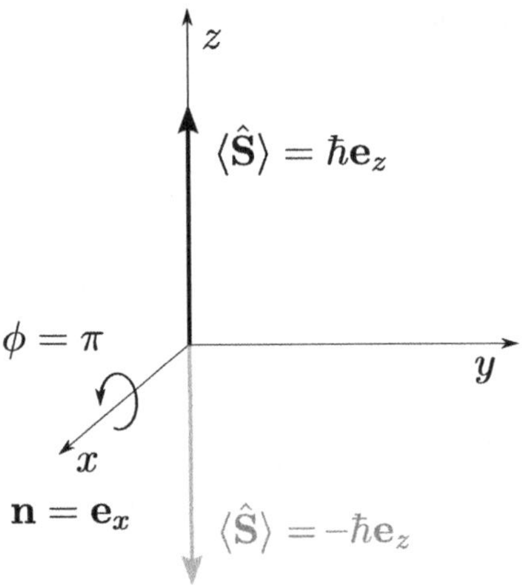

Figure 4.9 Rotation of a system in a triplet T_1 state by an angle π about the laboratory x-axis.

After rotation, the spin-magnetic moment becomes

$$\langle\hat{\mathbf{S}}\rangle = \langle 1,\ 1|\hat{\mathcal{R}}^{\dagger}(\pi,\mathbf{e}_x)\ \hat{\mathbf{S}}\ \hat{\mathcal{R}}(\pi,\mathbf{e}_x)\ |1,\ 1\rangle = -\hbar\mathbf{e}_z, \tag{4.98}$$

which is in agreement with our expectations. We also note that the rotation by an angle 2π about an arbitrary axis will always leave the boson system invariant, that is, $\hat{\mathcal{R}}(2\pi,\mathbf{n}) = \hat{I}$, and so the standard group character tables do apply to the study of systems with an even number of electrons.

The above example of a rotation by an angle π also gives an important signal on the appropriate representation of triplet states. In the nonrelativistic realm, where spin and spatial degrees of freedom separate, it is customary to choose the simultaneous eigenfunctions to the $\hat{S}^2$ and $\hat{S}_z$ operators, that is, the T_0 and $T_{\pm 1}$ states. However, our analysis shows that in a situation where the Hamiltonian commutes with, for example, a C_2 operation, this choice fails to produce wave functions that are simultaneous eigenfunctions to both $\hat{H}$ and the C_2 rotation operator, see Eq. (4.97). The class of C_2 operations are elements of Abelian point groups and of paramount importance in quantum chemistry. We will therefore take the time to discuss an alternative representation of the triplet states that provides us with eigenfunctions to the Hamiltonian as well as to this class of rotation operators.

In analogy with the construction of real spherical harmonics, as exemplified by the p orbitals in Eq. (4.79), it is possible to construct linear combinations of the triplet states that transform as the Cartesian coordinates upon rotation. The appropriate connection between coordinates and angular momenta is

$$x \sim \frac{1}{\sqrt{2}}(|1,-1\rangle - |1,\ 1\rangle); \quad y \sim \frac{i}{\sqrt{2}}(|1,-1\rangle + |1,\ 1\rangle); \quad z \sim |1,\ 0\rangle, \tag{4.99}$$

which suggests a transformation of spin states as

$$\begin{bmatrix} T_x \\ T_y \\ T_z \\ S_0 \end{bmatrix} = \frac{1}{\sqrt{2}} \begin{bmatrix} -1 & 0 & 0 & 1 \\ i & 0 & 0 & i \\ 0 & 1 & 1 & 0 \\ 0 & 1 & -1 & 0 \end{bmatrix} \begin{bmatrix} \alpha_1\alpha_2 \\ \alpha_1\beta_2 \\ \beta_1\alpha_2 \\ \beta_1\beta_2 \end{bmatrix}. \tag{4.100}$$

As in Eq. (4.90), the above equation defines a unitary transformation matrix $\hat{U}$, and in this case we get a rotation matrix that is

$$\hat{U}\,\hat{\mathcal{R}}(\phi,\mathbf{n})\,\hat{U}^{\dagger} = \begin{bmatrix} \hat{\mathcal{R}}^{\mathrm{T}}(\phi,\mathbf{n}) & 0 \\ 0 & \hat{\mathcal{R}}^{\mathrm{S}}(\phi,\mathbf{n}) \end{bmatrix}. \tag{4.101}$$

with $\hat{\mathcal{R}}^{\mathrm{S}} = 1$ and

$$\hat{\mathcal{R}}^{\mathrm{T}}(\phi,\mathbf{n}) \tag{4.102}$$
$$= \begin{bmatrix} \cos\phi + 2n_x^2\sin^2(\phi/2) & 2n_xn_y\sin^2(\phi/2) - n_z\sin\phi & 2n_xn_z\sin^2(\phi/2) + n_y\sin\phi \\ 2n_xn_y\sin^2(\phi/2) + n_z\sin\phi & \cos\phi + 2n_y^2\sin^2(\phi/2) & 2n_yn_z\sin^2(\phi/2) - n_x\sin\phi \\ 2n_zn_x\sin^2(\phi/2) - n_y\sin\phi & 2n_zn_y\sin^2(\phi/2) + n_x\sin\phi & \cos\phi + 2n_z^2\sin^2(\phi/2) \end{bmatrix}.$$

We emphasize that the rotation matrix for the Cartesian triplets is identical to the result seen in Eq. (4.82) for the rotation of coordinates. For C_2 rotations about the Cartesian axes, the corresponding rotation matrices are thus diagonal with elements equal to ± 1, and we conclude that while singlet states are symmetric, triplet states transform as rotations. The reader working in the nonrelativistic realm may regard this result to be of minor practical concern, but the discussion is relevant in any application where spin-orbit interactions need to be considered. A most fundamental example is provided by the molecular zero-field splitting that refers to the energy splitting between the Cartesian triplet states in the absence of external fields.

As a simple example, let us consider the ethylene molecule depicted in Figure 4.10 in its planar D_{2h} conformation. For this system, the intensive dipole-allowed electronic $\pi\pi^*$-transition corresponds to an excitation from the closed-shell ground state of A_g symmetry to the lowest excited state of B_{1u} symmetry. In a nonrelativistic framework, these states are both of singlet spin symmetry and the labeling of the excited state becomes ${}^1B_{1u}$. In the same framework, the corresponding triplet state would be denoted as ${}^3B_{1u}$ and refer to the spin kets given in Eq. (4.90) that are *not* eigenfunctions of the C_2 rotation operators. From a symmetry point of view, the appropriate wave functions should in this case be constructed from the set of Cartesian triplet states given in Eq. (4.100) and which span the irreducible representations B_{1g}, B_{2g}, and B_{3g}, respectively. The corresponding *total* wave functions are the combinations of the spatial and spin parts and they will consequently span the following irreducible representations:

$$B_{1u} \otimes \{B_{1g}, B_{2g}, B_{3g}\} = \{A_u, B_{3u}, B_{2u}\}.$$

This example illustrates that the conventional nonrelativistic picture provides an unsuitable symmetry representation of triplet states, which can be misleading in the formation of selection rules.

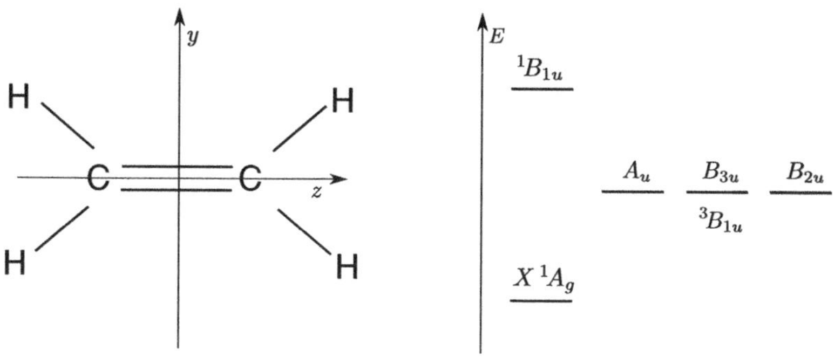

Figure 4.10 Ethylene molecule in D_{2h} point group.

Further Reading

1 S. L. Altmann. *Rotations, Quaternions and Double Groups.* Oxford University Press, New York, 1986.
2 D. M. Bishop. *Group Theory and Chemistry.* Dover Publications, 1993.
3 T. Saue and H. J. A. Jensen. Quaternion symmetry in relativistic molecular calculations: the Dirac–Hartree–Fock method. *J. Chem. Phys.*, 111:6211–6222, 1999.

5

Exact-State Response Theory

Whether you can observe a thing or not depends on the theory which you use. It is the theory which decides what can be observed.

A. Einstein (1926)

5.1 Responses in Two-Level System

In a simplistic and generalized view of spectroscopy, the observable is the number of detected particles (e.g., photons or electrons) per unit time in a narrow energy interval and into a small solid angle. The observable is recorded under certain experimental conditions defined by parameters such as temperature, pressure, and concentration, and there exists a model (or connection) from which one can deduce molecular properties from the set of measured data. Under normal circumstances, these connections can be viewed as measures of changes in the molecular properties due to the presence of electromagnetic fields with origins attributed to external sources used in the different spectroscopic techniques. In practice, molecular properties can here, for example, be the molecular energy or the molecular polarization, and external sources the beams of radiation or internal nuclear magnetic moments. The latter source is to be considered external only after the Born–Oppenheimer approximation has been applied and the nuclear and electronic degrees of freedom have been separated, in which case the nuclear magnetic moments are external sources of magnetic fields that affect the electronic motions (an effect that depends parametrically on the nuclear positions).

The view of a molecular property as a measure of change of another property may appear to be a circular definition: are we to view a polarization as a second-order change in the energy or as a first-order change in a moment? To some extent, this is a matter of preference and can be left as a choice of the individual, but there is one point that has an impact on the very core of the theoretical formulation: in the presence of a time-dependent electromagnetic field, the system is nonconservative and the energy is not well defined due to the exchange of energy between the molecular system and the radiation field. In such a situation, for physical reasons it appears more appropriate to use the polarization or magnetization as the point of reference and to study fluctuations about the field-free values of these properties. However, an alternative exists, as one can define a time-dependent *quasi-energy* that refers to the expectation value of the difference between the Hamiltonian and the energy (time differentiation) operator. However, regardless of the point of view, it is customary to view the energy (of a conservative system) as a zeroth-order property, electric and magnetic moments as first-order properties, polarization and magnetization as second-order properties, and so on.

Principles and Practices of Molecular Properties: Theory, Modeling and Simulations, First Edition.
Patrick Norman, Kenneth Ruud, and Trond Saue.

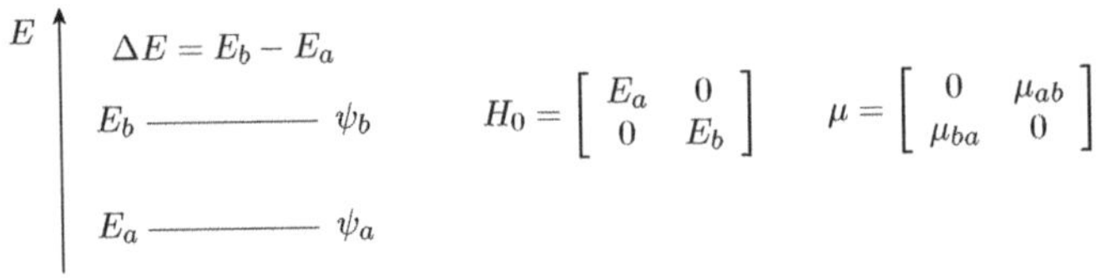

Figure 5.1 Energy levels and corresponding wave functions of a two-level atom together with matrix representations of the Hamiltonian and electric dipole moment operators.

The framework that we are about to set up is quite general and is based on a separation of the Hamiltonian as one part that describes the system in the absence of external fields and one part that describes the interactions between the system and the said fields. We have described the form of the interaction Hamiltonian in Chapter 3 and seen that interactions between a quantum-mechanical system and external electromagnetic fields are to be introduced by means of the scalar and vector potentials with an accompanying gauge freedom. In this chapter, we will adopt a few specific setups for the external fields and treat these in the electric dipole approximation where magnetic field and higher-order electric field interactions are ignored. We do so in order to produce explicit illustrations of the theory, but it is to be understood that this choice does not impose any limitations on the theory *per se*.

To illustrate these ideas, let us consider a two-level atom subjected to an external electric field. Such a system is depicted in Figure 5.1. In the electric dipole approximation, the Hamiltonian of the system is

$$\hat{H} = \hat{H}_0 - \hat{\mu}F(t), \tag{5.1}$$

where $\hat{H}_0$ is the Hamiltonian of the isolated atom, $F(t)$ is the amplitude of the external electric field, and $\hat{\mu}$ is the electric dipole operator along the direction of the electric field. With respect to our notation, we use a hat to denote quantum-mechanical operators such as $\hat{H}_0$ but omit the hat when we refer to any one specific matrix representation of an operator. The symbol H_0 seen in Figure 5.1 thus refers to the 2×2 matrix representing the Hamiltonian in the basis of the eigenstates ψ_a and ψ_b. We further note that in Chapter 3 the electric field was associated with the letter "E" as this is the conventional choice in electrodynamics. We will, however, allow ourselves to use "F" for the electric field in order to avoid confusion with the molecular energy E. Let us assume the energy separation of the electronic states to be $\Delta E = E_b - E_a$ and the transition dipole moment μ_{ab} (with $\mu_{ba} = \mu_{ab}$) between the two levels to be 0.5 and 1.0 a.u., respectively. It is also a common standard to introduce the transition angular frequency, or transition frequency for short, defined as $\omega_{ba} = \Delta E/\hbar$. The permanent dipole moment of the ground as well as the excited state is zero for the atom, due to symmetry, but in the presence of an external electric field, an *induced* polarization will arise.

Compared to atomic fields, the electric fields of conventional lasers are relatively weak. A laser delivering pulses of 10 ns duration and 1 mJ in energy and with a spot size (or circle radius) of 100 μm produces an intensity of about 0.3 GW cm^{-2}. This intensity corresponds to an electric field amplitude of some $F^{\omega} = 5 \times 10^{-5}$ a.u., which is several orders of magnitude smaller than the internal electric fields that bind electrons in atomic and molecular systems. It is thus reasonable to treat the external field by means of perturbation theory. Before entering the details of perturbation theory we will, however, first determine the responses for this simple system by more direct means.

Let us first consider the situation where the external electric field is nonoscillating (also referred to as static), so that $F(t) = F^0$. In this case, the system is time independent and the energies of the ground and excited states are well defined and given as the eigenvalues of the

Hamiltonian—that is, as solutions to the eigenvalue equation

$$\det(H - \hbar\omega I_2) = 0. \tag{5.2}$$

The two eigenvalues of our two-level system are

$$\hbar\omega = \frac{E_a + E_b}{2} \pm \sqrt{\frac{(E_b - E_a)^2}{4} + (\mu_{ab}F^0)^2}, \tag{5.3}$$

from which the electric field-dependent energies for small fields are found by a Taylor expansion of the square root

$$E_a'(F^0) = E_a - \frac{(\mu_{ab}F^0)^2}{\Delta E} + \frac{(\mu_{ab}F^0)^4}{(\Delta E)^3} + \cdots, \tag{5.4}$$

$$E_b'(F^0) = E_b + \frac{(\mu_{ab}F^0)^2}{\Delta E} - \frac{(\mu_{ab}F^0)^4}{(\Delta E)^3} + \cdots. \tag{5.5}$$

From these energy expansions, we determine the electric polarization properties by taking the field derivatives of the energy in the limit of zero field strength. For the ground state, we get

$$\mu_a = -\left.\frac{\partial E_a'}{\partial F^0}\right|_{F^0=0} = 0, \tag{5.6}$$

$$\alpha_a = -\left.\frac{\partial^2 E_a'}{\partial (F^0)^2}\right|_{F^0=0} = 2\frac{(\mu_{ab})^2}{\Delta E}, \tag{5.7}$$

$$\beta_a = -\left.\frac{\partial^3 E_a'}{\partial (F^0)^3}\right|_{F^0=0} = 0, \tag{5.8}$$

$$\gamma_a = -\left.\frac{\partial^4 E_a'}{\partial (F^0)^4}\right|_{F^0=0} = -24\frac{(\mu_{ab})^4}{(\Delta E)^3}, \tag{5.9}$$

where we have introduced the electric dipole moment (μ), the polarizability (α), the first-order hyperpolarizability (β), and the second-order hyperpolarizability (γ). We note that the corresponding properties for the system in the excited state are equal in magnitude but of opposite signs. Although we have here chosen to study the change of the total energy with respect to the external static field, we could equally well have chosen to study changes in any of the other properties listed. We would in this case have obtained the following expansions

$$\mu_a'(F^0) = \mu_a + \alpha_a F^0 + \frac{1}{2}\beta_a (F^0)^2 + \frac{1}{6}\gamma_a (F^0)^3 + \cdots, \tag{5.10}$$

$$\alpha_a'(F^0) = \alpha_a + \beta_a F^0 + \frac{1}{2}\gamma_a (F^0)^2 + \cdots, \tag{5.11}$$

$$\beta_a'(F^0) = \beta_a + \gamma_a F^0 + \cdots, \tag{5.12}$$

$$\gamma_a'(F^0) = \gamma_a + \cdots. \tag{5.13}$$

For instance, Eq. (5.10) describes the electric field-dependent dipole moment of the system. It is clear that the permanent dipole moment of the atom is zero ($\mu_a = \mu_b = 0$), but in the presence of the external field there will appear an *induced* polarization. In our example, the static ground-state polarizability α_a is 4.0 a.u., and, with $F^0 = 5 \times 10^{-5}$ a.u., the induced polarization amounts to $\mu_a' = 2.0 \times 10^{-4}$ a.u. (ignoring nonlinear terms). A static electric field with the strength of a conventional laser field is thus capable of causing a displacement of an elementary charge in this system by a distance of 2.0×10^{-4} a.u. Needless to say, such a displacement is small compared to the atomic size.

The previous example considered an atomic system in a static electric field. In most spectroscopies, however, experiments are conducted with the use of time-oscillating electromagnetic fields and as mentioned above, energies become undefined, prohibiting the use of Eq. (5.2) as a starting point. Instead, we can turn our attention to the polarization of the system, that is, the time-dependent equivalent of Eq. (5.10), which we obtain by substituting F^0 with $F(t)$. With a time-dependent perturbation, the wave function of the system will not be stationary but will rather reflect the induced motions of the electrons and, for a general molecular system, also the nuclei. It is again illustrative to consider our simple two-level atom for which $\psi(t)$ will denote the time-dependent wave function that is a solution to the time-dependent Schrödinger equation described in postulate D in Section 2.1

$$i\hbar\frac{\partial}{\partial t}\psi(t) = \hat{H}\psi(t). \tag{5.14}$$

We will assume that for $t < 0$, the external perturbation is zero, whereas for $t > 0$ it will consist of a nonresonant ($\omega = 0.1$ a.u.) monochromatic, time-oscillating electric field

$$F(t) = F^{\omega} \sin \omega t \times \text{erf}(at), \tag{5.15}$$

which is slowly switched on using the error function ($a = 1/100$ a.u.). The electric-field amplitude F^{ω} is that of the conventional laser described earlier. The initial condition for our system is that it resides in the ground state prior to exposure to the perturbation and with a phase that is zero at $t = 0$

$$\psi(t) = \psi_a\, e^{-iE_a t/\hbar} \quad \text{for } t \leq 0, \tag{5.16}$$

where ψ_a is the time-independent eigenstate of $\hat{H}_0$ with energy eigenvalue E_a. With a time step Δt that is small compared to the time variations in the Hamiltonian, we can approximate the Hamiltonian to be constant from time t to $t + \Delta t$. It is then possible to perform a time integration of the Schrödinger equation using

$$\psi(t_0 + \Delta t) = \hat{U}(t_0 + \Delta t, t_0)\psi(t_0), \tag{5.17}$$

where the time-evolution propagator is defined as in Eq. (2.75) and we insert the Hamiltonian at time t_0. The explicit form of the propagator then becomes

$$\hat{U}(t_0 + \Delta t, t_0) = e^{-i\hat{H}(t_0)\Delta t/\hbar}. \tag{5.18}$$

As discussed in Section 2.2, repeated applications of $\hat{U}$ allow us to determine the wave function in the region $t > 0$ and, given the time-dependent wave function, the polarization is obtained as the expectation value of the electric dipole operator

$$\mu(t) = \langle \psi(t)|\hat{\mu}|\psi(t)\rangle. \tag{5.19}$$

As a measure of the effect of the perturbation on the system, we may study the population of the ground and excited states, which we will denote by $\rho_a(t)$ and $\rho_b(t)$, respectively. The populations are given by the projections of the wave function on the eigenfunctions of $\hat{H}_0$

$$\rho_a(t) = |\langle \psi_a|\psi(t)\rangle|^2, \quad \rho_b(t) = |\langle \psi_b|\psi(t)\rangle|^2, \tag{5.20}$$

and it is clear that they fulfill the normalization condition

$$\rho_a(t) + \rho_b(t) = 1. \tag{5.21}$$

The electric field in Eq. (5.15) is depicted in Figure 5.2(c), and the associated population of the excited state is shown in panel (b). In Appendix I, using the perturbation theory that we

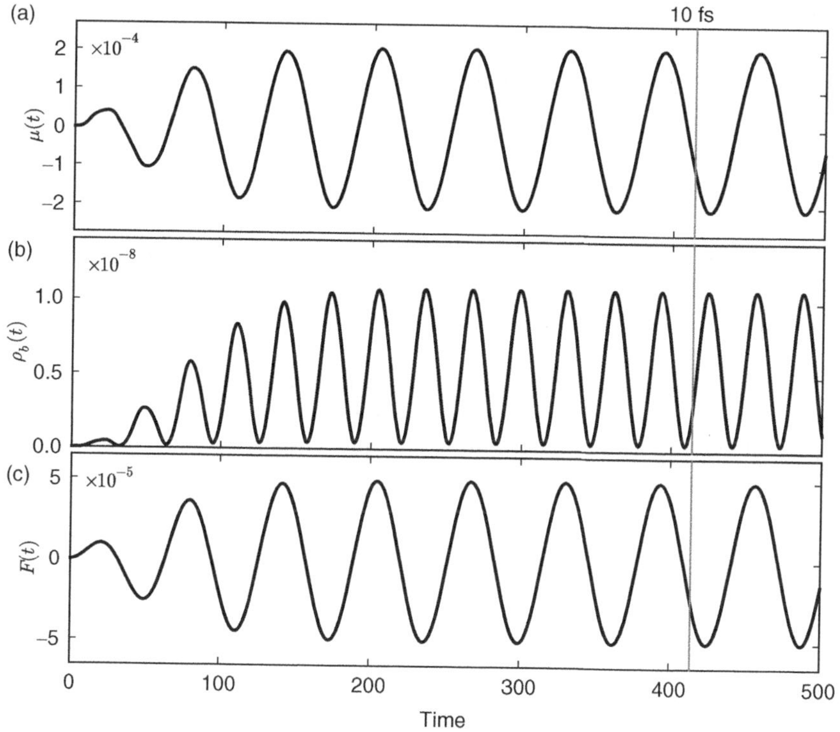

Figure 5.2 Induced electric dipole moment in a two-level atom. The external field $F(t)$, induced dipole moment $\mu(t)$, and the time are given in a.u., and $\rho_b(t)$ denotes the population of excited state ψ_b. The vertical bar indicates the time of 10 fs. A time-integration step length of $\Delta t = 0.1$ a.u. is used.

develop in Section 5.4.1, we show that, to second order in the field, the excited-state population is given by

$$\rho_b(t) = \left(\frac{\mu_{ab}F^{\omega}}{\hbar}\right)^2 \left[\frac{\omega^2 + (\omega_{ba}^2 - \omega^2)\sin^2\omega t}{(\omega_{ba}^2 - \omega^2)^2}\right] \mathrm{erf}^2(at). \tag{5.22}$$

The oscillations of $\rho_b(t)$ follow those of $F(t)$, but the frequency is twice that of the external field since the population refers to the squared absolute value of the projection coefficients. We note that a steady-state solution to the Schrödinger equation is reached around $t = 200$ a.u. The amplitude in the oscillations of $\rho_b(t)$ is small (on the order of 10^{-8}).

In Figure 5.2(a), the induced dipole moment of the system is shown. To first order in the electric field, the induced motions of the electrons will follow the oscillations of the external electric field, a fact that is clearly seen in Figure 5.2. Electric dipole molecular properties are defined by an expansion of the induced dipole moment in orders of the time-dependent electric field. The coupling parameters between polarization and fields will be dependent on the frequencies of the fields. In our case, the external perturbation is, to a good approximation, monochromatic (although not truly so due to the switch-on function), and the relevant expansion will therefore be

$$\mu(t) = \alpha F(t) + \frac{1}{6}\gamma F^3(t) + \cdots . \tag{5.23}$$

We have here omitted terms that are of even orders in the field since these vanish because of symmetry in the case of an atomic system. Owing to the cubic dependence on the small external field, the second term in the expansion in Eq. (5.23) is not visible in the plot. The numerical value of the linear polarizability, on the other hand, can be read off directly as the ratio of the amplitudes of the polarization and the electric field. With the present data we obtain a value of α that is[1]

$$\alpha = \frac{\max[\mu(t)]}{F^{\omega}} \approx 4.17 \text{ a.u.} \tag{5.24}$$

We recall that the value of α in the limit of zero frequency (static field) is given by Eq. (5.7), with a numerical value of 4.0 a.u. It is thus clear that the amplitude of the oscillations of the charges increases when the frequency becomes larger. The dependence of the molecular property (in this case the polarizability) on the frequency of the perturbation is referred to as the *dispersion* of the property.

We have so far considered two quite different ways of formulating and determining molecular properties. The first technique, employing expansions of the total energy, seemed appropriate for time-independent systems while the second technique, employing expansions of the polarization, was applicable also to time-dependent systems. We will now turn to a third alternative, which is also applicable to time-dependent systems and which employs expansions of a time-dependent phase function $\phi(t)$, or rather of its time derivative.

Let us write the wave function $\psi(t)$ as the product of two time-dependent functions

$$\psi(t) = e^{-i\phi(t)}\overline{\psi}(t). \tag{5.25}$$

Without further specifications, this division is of course arbitrary, but it is made unique by requiring that $\phi(t)$ is a real function (a phase) and that the phase of the projection of $\overline{\psi}(t)$ onto the unperturbed function ψ_a is zero. While the first requirement seems natural, the second one appears less intuitive. However, remember that in the absence of the external field, the wave function would become

$$\psi(t) = \psi_a e^{-iE_a t/\hbar}, \tag{5.26}$$

and our two requirements on $\phi(t)$ and $\overline{\psi}(t)$ would in this case lead to the division

$$\overline{\psi}(t) = \psi_a, \tag{5.27}$$

$$\phi(t) = E_a t/\hbar. \tag{5.28}$$

This is the reason why $\overline{\psi}(t)$ sometimes is referred to as the phase-isolated wave function. In the case of a nonzero perturbation, our requirements will guarantee that at least the zeroth-order contribution to $\overline{\psi}(t)$ will be time independent and equal to the eigenfunction ψ_a. As sensible as the division of the wave function in Eq. (5.25) may be, it remains to be seen if convenient equations of motion for the phase-isolated wave function and the phase function can be identified. We stress that while $\psi(t)$ fulfills the Schrödinger equation, $\overline{\psi}(t)$ will not.

In fact, if we substitute Eq. (5.25) into the Schrödinger equation, we get

$$\left(\hat{H} - i\hbar\frac{\partial}{\partial t}\right)\overline{\psi}(t) = Q(t)\,\overline{\psi}(t), \tag{5.29}$$

where the time-dependent quasi-energy $Q(t)$ has been introduced for the time derivative of the phase function

$$Q(t) = \hbar\dot{\phi}(t). \tag{5.30}$$

1 We shall later find in Eq. (5.169) an analytical expression for the linear polarizability from which we get $\alpha = 2\mu_{ab}^2\Delta E/[(\Delta E)^2 - (\hbar\omega)^2] = 1/(0.5^2 - 0.1^2) = 4.166\cdots$.

Given $Q(t)$, the phase function $\phi(t)$ can be obtained by time integration

$$\phi(t) = \frac{1}{\hbar} \int^{t} Q(t')\, dt'. \tag{5.31}$$

Note that, in the absence of an external field, the quasi-energy will be equal to the time-independent ground-state energy, $Q(t) = E_a$. We are now left in a situation where we have one equation of motion, Eq. (5.29), from which we are to determine not only the wave function $\overline{\psi}(t)$ but also $Q(t)$. This stands in some contrast to our previous propagator approach where we employed the Schrödinger equation for the determination of the single unknown function $\psi(t)$. So are we asking for more information than is available from Eq. (5.29)? The answer to this question is, of course, no. We should remember that although we did not explicitly state this, there were also two unknown time-dependent functions in the determination of $\psi(t)$ and these were associated with the matrix representation of the wave function. In the time propagation of $\psi(t)$, we made use of an expansion of the type

$$\psi(t) = \frac{a(t)\,\psi_a + b(t)\,\psi_b}{\sqrt{|a|^2 + |b|^2}}, \tag{5.32}$$

whereas, in the present case and due to the requirement of a zero phase of $\langle \psi_a | \overline{\psi}(t) \rangle$, we are left with an expansion of the type

$$\overline{\psi}(t) = \frac{\psi_a + b(t)\,\psi_b}{\sqrt{1 + |b|^2}}. \tag{5.33}$$

We can construct separate differential equations by projections of Eq. (5.29) onto state vectors of our choice. A first and obvious choice is to project the equation onto $\overline{\psi}$ itself, which isolates $Q(t)$ as

$$Q(t) = \left\langle \overline{\psi} \right| \left(\hat{H} - i\hbar \frac{\partial}{\partial t} \right) \left| \overline{\psi} \right\rangle. \tag{5.34}$$

For our continued analysis, a useful relation is

$$\left\langle \overline{\psi}_1 \right| \left(\hat{H} - i\hbar \frac{\partial}{\partial t} \right) \left| \overline{\psi}_2 \right\rangle = \left\langle \overline{\psi}_2 \right| \left(\hat{H} - i\hbar \frac{\partial}{\partial t} \right) \left| \overline{\psi}_1 \right\rangle^* - i\hbar \frac{\partial}{\partial t} \langle \overline{\psi}_1 | \overline{\psi}_2 \rangle. \tag{5.35}$$

From this relation, we can deduce that the quasi-energy must be real as follows:

$$Q(t) = \left\langle \overline{\psi} \right| \left(\hat{H} - i\hbar \frac{\partial}{\partial t} \right) \left| \overline{\psi} \right\rangle = \left\langle \overline{\psi} \right| \left(\hat{H} - i\hbar \frac{\partial}{\partial t} \right) \left| \overline{\psi} \right\rangle^* - i\hbar \frac{\partial}{\partial t} \langle \overline{\psi} | \overline{\psi} \rangle = Q^*(t), \tag{5.36}$$

where, in the last step, we have made use of the fact that the phase-isolated wave function remains normalized at all times [this follows from Eq. (2.84)].

A second desirable projection of Eq. (5.29) would be onto the orthogonal complement of $\overline{\psi}$ since such a projection would remove $Q(t)$

$$\left\langle \overline{\psi}_\perp \right| \left(\hat{H} - i\hbar \frac{\partial}{\partial t} \right) \left| \overline{\psi} \right\rangle = 0. \tag{5.37}$$

As an orthogonal complement vector, we may, in the case of our two-level atom, choose

$$\overline{\psi}_\perp(t) = \psi_b - b^*(t)\,\psi_a. \tag{5.38}$$

We thus arrive at

$$\hbar \dot{b}(t) = -i\Delta E\, b(t) + i\mu_{ab} F(t)[1 - b^2(t)]. \tag{5.39}$$

By using the initial condition $b(0) = 0$, it is straightforward to time propagate Eq. (5.39) and to retrieve $b(t)$. The solution to this exercise is already known to us, since $|b(t)|^2/(1+|b(t)|^2)$ is equal to what we denoted as $\rho_b(t)$ in Figure 5.2; however, it should be emphasized that Eq. (5.39) provides the time evolution of the phase-isolated wave function $\overline{\psi}(t)$ *without* knowledge of $\psi(t)$. From Eq. (5.39) it is also clear that there will appear nonlinear contributions in the responses to the external field, although, as already mentioned, these contributions are not visible in the plots due to the small amplitude of the perturbation.

Having determined $b(t)$ from Eq. (5.39) and thereby $\overline{\psi}(t)$ via Eq. (5.33), it is time to turn our attention to the quasi-energy obtained as an expectation value in Eq. (5.34). The quasi-energy is composed of two terms, one associated with the Hamiltonian and one associated with minus the energy operator. We denote these two contributions to $Q(t)$ by $Q_1(t)$ and $Q_2(t)$, respectively. In Appendix I, again using perturbation theory that we develop in Section 5.4.2, we show that, to second order in the field, these contributions are given by

$$Q_1(t) = \langle\overline{\psi}|\hat{H}|\overline{\psi}\rangle = \frac{\langle\psi_a + b(t)\psi_b|\ \hat{H}\ |\psi_a + b(t)\psi_b\rangle}{1+|b|^2} = E_a - \hbar\omega_{ba}\left(\frac{\mu_{ab}F^{\omega}}{\hbar}\right)^2\left[\frac{\omega^2 - (\omega_{ba}^2 - \omega^2)\sin^2\omega t}{(\omega_{ba}^2 - \omega^2)^2}\right]\mathrm{erf}^2(at) \tag{5.40}$$

and

$$Q_2(t) = -i\hbar\langle\overline{\psi}|\frac{\partial}{\partial t}|\overline{\psi}\rangle = -\hbar\omega_{ba}\left(\frac{\mu_{ab}F^{\omega}}{\hbar}\right)^2\frac{\omega^2}{(\omega_{ba}^2 - \omega^2)^2}\,\mathrm{erf}^2(at). \tag{5.41}$$

They combine to the simpler expression

$$Q(t) = E_a - \hbar\omega_{ba}\left(\frac{\mu_{ab}F^{\omega}}{\hbar}\right)^2\frac{\sin^2\omega t}{(\omega_{ba}^2 - \omega^2)}\,\mathrm{erf}^2(at). \tag{5.42}$$

We plot the two contributions to the quasi-energy separately in Figure 5.3. For small perturbations, it is clear from Eq. (5.40) that the expectation value of $\hat{H}$ will be close to the ground-state energy E_a. In fact, we may even say that it should be slightly smaller due to the coupling of the induced dipole moment and the external field as expressed in Eq. (5.1), and when the amplitude of the external field is zero we expect that Q_1 reaches its maximum close to the value of E_a. Since $F(t)$ crosses zero twice per full period of oscillation, the main frequency of Q_1 is twice that of $F(t)$. The second contribution to the quasi-energy is zero in the limit of a zero perturbation, and consequently expected to be small in the present case. From Figure 5.3, it is seen that $Q_2(t)$ becomes periodic in the steady-state region with a mean value that is about an order of magnitude smaller than the mean value of $Q_1(t) - E_a$. In the uppermost panel, we present the quasi-energy less the ground-state energy. The question now is how this quantity should be related to the polarization properties of the system.

Let us by $\delta\overline{\psi}$ denote a variation in the phase-isolated wave function. The variation must fulfill the normalization condition

$$\delta\langle\overline{\psi}|\overline{\psi}\rangle = \langle\delta\overline{\psi}|\overline{\psi}\rangle + \langle\overline{\psi}|\delta\overline{\psi}\rangle = 0. \tag{5.43}$$

This means that an allowed variation can be written as

$$\delta\overline{\psi}(t) = \delta\overline{\psi}_{\perp} + \delta\overline{\psi}_{\parallel}, \tag{5.44}$$

where

$$\delta\overline{\psi}_{\parallel} = i\epsilon(t)\overline{\psi}; \quad \epsilon(t) \in \mathcal{R}. \tag{5.45}$$

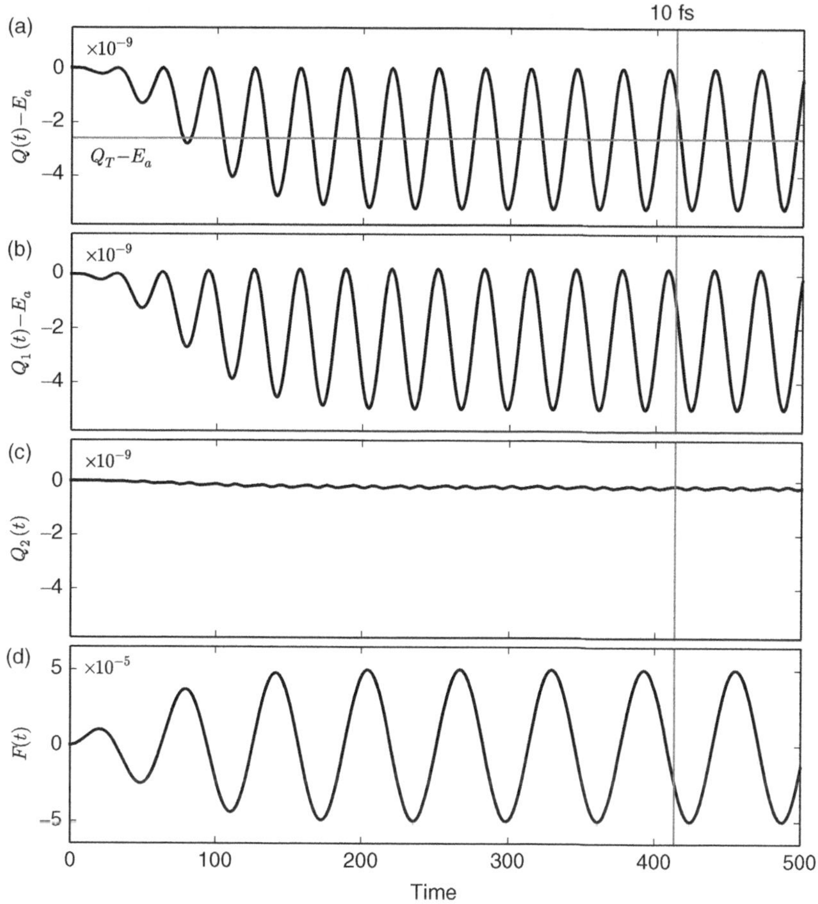

Figure 5.3 Quasi-energy in a two-level atom. The external field $F(t)$ and quasi-energy $Q(t)$ are given in a.u. The vertical bar indicates the time of 10 fs. The horizontal bar in the upper panel indicates the time-averaged quasi-energy minus the ground-state energy, that is, $Q_T - E_a$, in the steady-state region. A time-integration step length of $\Delta t - 0.1$ a.u. is used.

Allowed variations can thus be decomposed into terms that are orthogonal and parallel to $\overline{\psi}$, respectively, but with the requirement on ϵ to be real. For such variations, we have

$$\begin{aligned}\delta Q &= \langle \delta\overline{\psi}| \left(\hat{H} - i\hbar\frac{\partial}{\partial t}\right)|\overline{\psi}\rangle + \langle\overline{\psi}|\left(\hat{H} - i\hbar\frac{\partial}{\partial t}\right)|\delta\overline{\psi}\rangle \\ &= \langle \delta\overline{\psi}| \left(\hat{H} - i\hbar\frac{\partial}{\partial t}\right)|\overline{\psi}\rangle + \langle\delta\overline{\psi}|\left(\hat{H} - i\hbar\frac{\partial}{\partial t}\right)|\overline{\psi}\rangle^* - i\hbar\frac{\partial}{\partial t}\langle\overline{\psi}|\delta\overline{\psi}\rangle \\ &= Q(t)[\langle\delta\overline{\psi}|\overline{\psi}\rangle + \langle\delta\overline{\psi}|\overline{\psi}\rangle^*] - i\hbar\frac{\partial}{\partial t}\langle\overline{\psi}|\delta\overline{\psi}\rangle \\ &= -i\hbar\frac{\partial}{\partial t}\langle\overline{\psi}|\delta\overline{\psi}\rangle = \hbar\frac{\partial\epsilon(t)}{\partial t}.\end{aligned}$$

The hereby obtained relation

$$\delta Q(t) + i\hbar\frac{\partial}{\partial t}\langle\overline{\psi}|\delta\overline{\psi}\rangle = 0, \tag{5.46}$$

may be used as the starting point for the formulation of a time-dependent variational principle since, with a variation of the form

$$|\delta\overline{\psi}\rangle = \left|\frac{d\overline{\psi}}{dF^{\omega}}\right\rangle, \tag{5.47}$$

we get

$$\frac{dQ(t)}{dF^{\omega}} = \langle\overline{\psi}|\frac{\partial\hat{H}}{\partial F^{\omega}}|\overline{\psi}\rangle + \delta Q(t) = \langle\overline{\psi}|\frac{\partial\hat{H}}{\partial F^{\omega}}|\overline{\psi}\rangle - i\hbar\frac{\partial}{\partial t}\langle\overline{\psi}|\frac{d\overline{\psi}}{dF^{\omega}}\rangle, \tag{5.48}$$

which is known as the time-dependent Hellmann–Feynman theorem. The notable difference between Eq. (5.48) and its time-independent counterpart in Eq. (2.107) is the last term, representing the time derivative of the inner product between the wave function and its response. We note that this inner product will be a time-periodic function, which suggests that we can remove this term by a simple trick. Since

$$\frac{1}{T}\int_0^T \dot{f}(t)dt = 0, \tag{5.49}$$

for any periodic function $f(t)$ with period time T (i.e., $f(t+T) = f(t)$), the last term in Eq. (5.48) will vanish after we perform an operation of time averaging. For the time average of the field derivative of the quasi-energy, we have

$$\frac{1}{T}\int_t^{t+T}\frac{dQ(t')}{dF^{\omega}}dt' = \frac{d}{dF^{\omega}}\left[\frac{1}{T}\int_t^{t+T} Q(t')dt'\right] = \frac{dQ_T}{dF^{\omega}}, \tag{5.50}$$

where we have introduced the time-averaged quasi-energy Q_T

$$Q_T = \frac{1}{T}\int_t^{t+T} Q(t')\,dt', \tag{5.51}$$

and the time-dependent variational principle, Eq. (5.46), will for Q_T reduce to

$$\delta Q_T = 0. \tag{5.52}$$

The time averaging in Eq. (5.51) leads to a function Q_T that in principle is time dependent, but if we are concerned with observation times in the steady-state region, where only the fundamental frequency (the frequency of the external field) and the harmonics remain, any dependence on time will have vanished. We can therefore view Q_T as a function of the field amplitude F^{ω}, and in this time region the time-dependent Hellmann–Feynman theorem, Eq. (5.48), assumes the form

$$\frac{dQ_T}{dF^{\omega}} = \frac{1}{T}\int_t^{t+T}\langle\overline{\psi}(t')|\frac{\partial\hat{H}}{\partial F^{\omega}}|\overline{\psi}(t')\rangle\,dt'. \tag{5.53}$$

In Figure 5.4, the derivative of the quasi-energy with respect to the external field amplitude is shown for our two-level system and the time average of this quantity is depicted by a horizontal line. From Eq. (5.53), it is clear that, to linear order in the external electric field, we have

$$\begin{aligned}\frac{dQ_T}{dF^{\omega}} &= \frac{1}{T}\int_t^{t+T} -\langle\overline{\psi}|\hat{\mu}|\overline{\psi}\rangle \sin\omega t' \times \mathrm{erf}(at')\,dt' \\ &= -\alpha\,F^{\omega}\frac{1}{T}\int_t^{t+T}\sin^2\omega t'\,dt' = -\frac{1}{2}\alpha\,F^{\omega},\end{aligned} \tag{5.54}$$

where, in the intermediate step, it is recognized that the error function is equal to unity at sufficiently large times t, and thus can be removed from the integrand. The expectation value

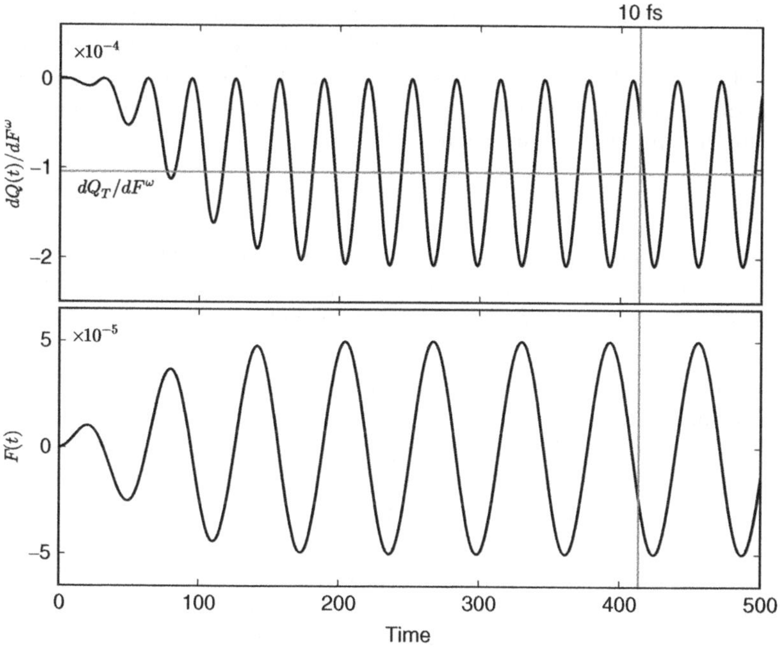

Figure 5.4 Field derivative of the quasi-energy in a two-level atom. The external field $F(t)$ and the derivative of the quasi-energy $dQ(t)/dF^{\omega}$ are given in a.u. The vertical bar indicates the time of 10 fs. The horizontal bar indicates the time-averaged quasi-energy derivative in the steady-state region. A time-integration step length of $\Delta t = 0.1$ a.u. is used and the field derivative is determined by finite differences using $\Delta F^{\omega} = 10^{-5}$ a.u.

of the dipole moment operator is identified as the polarizability times the time-dependent field strength. It is thus straightforward to determine the numerical value of the polarizability as

$$\alpha = -2\frac{dQ_T/dF^{\omega}}{F^{\omega}} \approx 4.17 \text{ a.u.,} \tag{5.55}$$

which of course is the same result as we obtained from the analysis of the polarization, see Eq. (5.24).

Our example has illustrated two commonly adopted theoretical formulations that are employed in calculations of time-dependent molecular properties. Which of the two methods to favor is to some extent a matter of preference, but the choice is also dependent on the electronic-structure method at hand.

5.2 Molecular Electric Properties

Since light can be considered either as a classical electromagnetic wave or as a stream of photons, we may describe the interaction between light and matter in two apparently different ways. In this section, we will consider how linear and nonlinear phenomena can be described in these two frameworks.

In nonresonant spectroscopies, light–matter interactions can be regarded as the scattering of photons by the electrons of the molecule. Each photon carries a linear momentum $p = E/c$, which is partially transferred to the molecule since the electrons remain bound, but the great mass of the molecule effectively prohibits energy transfer so that the scattered photons will have, for all practical purposes, identical frequency as the incoming ones, a process known as elastic scattering. At the instant of interaction, the photon can be regarded as absorbed and the molecule as being in a virtual excited state, intermediate in energy to the stationary states of the system. However, the time scale τ for this interaction, or the lifetime of the virtual state, is short enough not to violate the time-energy uncertainty relation

$$\tau\,\Delta E \sim \frac{\hbar}{2}, \tag{5.56}$$

where ΔE denotes the energy difference between the nearest eigenstate of the Hamiltonian, in this case $|1\rangle$, and the virtual state. The de-excitation of the system from the virtual state back to the ground state $|0\rangle$ is associated with the emission of the scattered photon, a process referred to as being linear, see Figure 5.5.

A high intensity of the incident radiation enhances the probability for simultaneous multiphoton interactions with a single molecule, that is, two or more photons are annihilated and absorbed by the molecule in a single quantum-mechanical process. In such cases, the frequency of the scattered photon does not have to be equal to that of the absorbed photons, for example, two quanta with frequency ω may be annihilated, creating a third photon with frequency 2ω. As indicated in Figure 5.6, the system returns to its ground state $|0\rangle$ after the interaction has occurred, and the intermediate virtual state is separated from the first excited state by an energy ΔE. This is an example of a nonlinear optical process known as second-harmonic generation, which can be used to accomplish frequency conversion of light. From these basic principles, a large number of multiphoton interactions can be envisaged when considering incident photons of different frequencies. We also note that when the laser detuning decreases, that is, when the difference ΔE between the laser frequency and the resonance frequency becomes smaller, the process can no longer be considered as an instantaneous scattering since the molecular state $|1\rangle$ will become absorbing.

In the complementary view of light–matter interaction, the laser field is described as an electro magnetic plane wave in which the molecular system resides, and the coupling between the two is to a first approximation the classical electric-dipole coupling. In the presence of the external electric field, all charged particles in the molecule, electrons as well as atomic nuclei, will

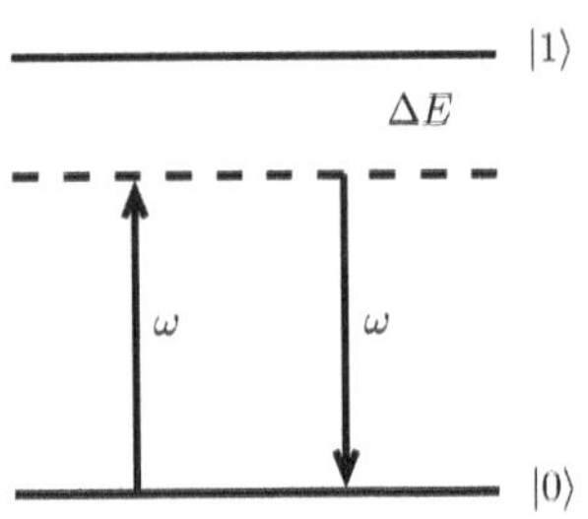

Figure 5.5 Elastic scattering of incident photons of frequency ω.

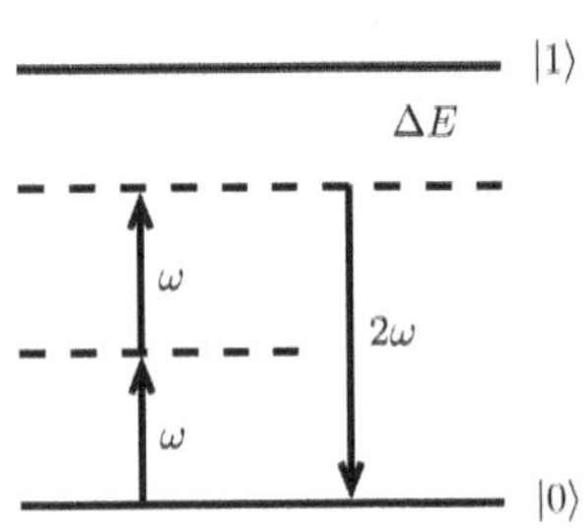

Figure 5.6 Second-harmonic generation involving two incident photons of frequency ω and a sum-frequency-generated photon of frequency 2ω.

experience a force that perturbs their motion. In a classical sense, one would expect the charges to follow the time oscillations of the electric field and thereby act as small antennas from which electro magnetic radiation would be transmitted; the frequency of the transmitted wave would of course be the same as that of the external field. We can compare this classical picture to the elastic scattering process described in Figure 5.5. The oscillating charges in the molecule give rise to an induced dipole moment, and the degree to which the external field $F(t)$ manages to set the charges in motion to first order is expressed in terms of the linear electric polarizability α. In general, the response of the charges depends on the frequency of the electric field, and the polarizability is therefore frequency dependent, as we have already seen in our simple example of a two-level system. The time-dependent polarization is described by an expression such as

$$\mu(t) = \mu^0 + \alpha F(t), \tag{5.57}$$

where μ^0 is the permanent electric dipole moment of the molecule. As discussed in Section 3.3.2, we are often concerned with optical fields, and in such cases the wavelength of $F(t)$ is at least in the order of a few hundred nanometers, whereas the size of the molecule is no greater than a few nanometers. The electric field is therefore often regarded as uniform over the spatial extent of the molecule.

There is one immediate and important question to be addressed: if the particle and wave pictures of light are two versions of the same thing, how can we understand nonlinear light–matter interactions when the electric field is represented by a plane wave? The answer lies in a generalization of Eq. (5.57) which lets the dipole moment be expressed as a Taylor series in the electric field strength

$$\mu(t) = \mu^0 + \alpha F(t) + \frac{1}{2}\beta F^2(t) + \frac{1}{6}\gamma F^3(t) + \cdots . \tag{5.58}$$

This equation introduces the first-order (nonlinear) hyperpolarizability β, the second-order hyperpolarizability γ, and so forth. Just as the linear polarizability, the nonlinear coupling constants depend on the frequency of the applied field. It is clear that the time-dependent polarization can have frequency components different from those of the external field due to its power dependence on the electric field strength, and the molecule can thus emit sum-frequency-generated radiation in correspondence with for example Figure 5.6. In optics, Eq. (5.58) provides the fundamental origin of nonlinearities, and at the microscopic level it is the expansion coefficients in this equation, the hyperpolarizabilities, that govern the nonlinear optical performance of the material. Hence, theoretical modeling of nonlinear optical properties is concerned with the determination of these quantities given the structure of the system.

Let us now examine Eq. (5.58) in some detail and, henceforth, we will consider coupling parameters between the external fields and the dipole moment as defined in the frequency domain. The electric field is vectorial and can in general considered to be composed of a static component and one or more time-oscillating components according to

$$F_\alpha(t) = \sum_\omega F_\alpha^\omega e^{-i\omega t}, \tag{5.59}$$

where F_α^ω are the Fourier amplitudes of the electric field along axis α in the molecular frame. The summation includes both positive and negative frequencies, and since the external field is real, we have $F^\omega = [F^{-\omega}]^*$. Furthermore, a vectorial electric field implies that the linear polarizability is a second-rank tensor, the first-order hyperpolarizability a third-rank tensor, and so forth. It is clear that the notation for the polarizabilities that was introduced in Eq. (5.58) needs to accommodate this tensorial nature as well as being defined in the frequency domain. In most cases, we will choose to define property tensors based on a Taylor expansion of the dipole moment and by means of a complex exponential Fourier decomposition. In a few exceptional cases, we will make use of real cosine expansions to make the argument more transparent, but the general starting point for the definition of frequency-dependent molecular properties will be an expression of the form

$$\begin{aligned}\mu_\alpha(t) = \mu_\alpha^0 &+ \sum_\omega \alpha_{\alpha\beta}(-\omega;\omega)F_\beta^\omega e^{-i\omega t} \\ &+\frac{1}{2}\sum_{\omega_1,\omega_2} \beta_{\alpha\beta\gamma}(-\omega_\sigma;\omega_1,\omega_2)F_\beta^{\omega_1}F_\gamma^{\omega_2}e^{-i\omega_\sigma t} \\ &+\frac{1}{6}\sum_{\omega_1,\omega_2,\omega_3} \gamma_{\alpha\beta\gamma\delta}(-\omega_\sigma;\omega_1,\omega_2,\omega_3)F_\beta^{\omega_1}F_\gamma^{\omega_2}F_\delta^{\omega_3}e^{-i\omega_\sigma t}+\cdots,\end{aligned} \tag{5.60}$$

where ω_σ denotes the sum of optical frequencies. For terms involving β, $\omega_\sigma = (\omega_1 + \omega_2)$ and for terms involving γ, $\omega_\sigma = (\omega_1 + \omega_2 + \omega_3)$, and so on. The frequency before a semicolon is introduced with a minus sign so that the sum of this frequency together with others becomes equal to zero as to reflect conservation of energy in photon scattering processes. The absolute value of this frequency is the frequency of the induced dipole that will be the source of the scattered radiation. We recall that we use the Einstein summation convention for repeated indices. We note that any pairwise interchange of indices and frequencies $\{\beta,\omega_1\}$, $\{\gamma,\omega_2\}$, and $\{\delta,\omega_3\}$ can be made without altering the physically observable polarization $\mu(t)$. It is therefore customary, but not necessary, to demand that the individual tensor elements are intrinsically symmetric

$$\beta_{\alpha\beta\gamma}(-\omega_\sigma;\omega_1,\omega_2) = \beta_{\alpha\gamma\beta}(-\omega_\sigma;\omega_2,\omega_1), \tag{5.61}$$

$$\begin{aligned}\gamma_{\alpha\beta\gamma\delta}(-\omega_\sigma;\omega_1,\omega_2,\omega_3) &= \gamma_{\alpha\beta\delta\gamma}(-\omega_\sigma;\omega_1,\omega_3,\omega_2) \\ &= \gamma_{\alpha\gamma\beta\delta}(-\omega_\sigma;\omega_2,\omega_1,\omega_3) = \gamma_{\alpha\gamma\delta\beta}(-\omega_\sigma;\omega_2,\omega_3,\omega_1) \\ &= \gamma_{\alpha\delta\gamma\beta}(-\omega_\sigma;\omega_3,\omega_2,\omega_1) = \gamma_{\alpha\delta\beta\gamma}(-\omega_\sigma;\omega_3,\omega_1,\omega_2).\end{aligned} \tag{5.62}$$

Furthermore, since the molecular polarization $\mu(t)$ as well as the electric field $F(t)$ is real, we have

$$\alpha(\omega;-\omega) = [\alpha(-\omega;\omega)]^*, \tag{5.63}$$

$$\beta(\omega_\sigma;-\omega_1,-\omega_2) = [\beta(-\omega_\sigma;\omega_1,\omega_2)]^*, \tag{5.64}$$

$$\gamma(\omega_\sigma;-\omega_1,-\omega_2,-\omega_3) = [\gamma(-\omega_\sigma;\omega_1,\omega_2,\omega_3)]^*. \tag{5.65}$$

The frequency ω_σ is that of the generated molecular polarization, and because the summations in Eq. (5.60) run over both positive and negative frequency components, the nonlinear hyperpolarizabilities will create both sum-frequency-generated as well as difference-frequency-generated polarization.

It is illustrative to view a few specific examples. First, let us consider two lasers A and B that operate at frequencies ω_A and ω_B, respectively. The external electric field experienced by the molecular system will in this case become

$$F_\alpha(t) = F_\alpha^{\omega_A}\cos(\omega_A t) + F_\alpha^{\omega_B}\cos(\omega_B t), \tag{5.66}$$

with a frequency decomposition that appears as in Figure 5.7. According to Eq. (5.60) and including terms up to second-order in the field, the time-dependent polarization will be

$$\begin{aligned}\mu_\alpha(t) = \mu_\alpha^0 &+ \alpha_{\alpha\beta}[F_\beta^{\omega_A}\cos(\omega_A t) + F_\beta^{\omega_B}\cos(\omega_B t)] \\ &+\frac{1}{2}\beta_{\alpha\beta\gamma}[F_\beta^{\omega_A}\cos(\omega_A t) + F_\beta^{\omega_B}\cos(\omega_B t)][F_\gamma^{\omega_A}\cos(\omega_A t) + F_\gamma^{\omega_B}\cos(\omega_B t)].\end{aligned} \tag{5.67}$$

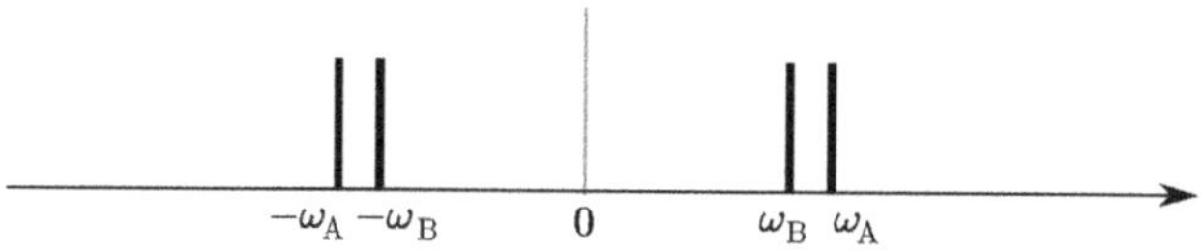

Figure 5.7 Frequency decomposition of two superimposed laser fields.

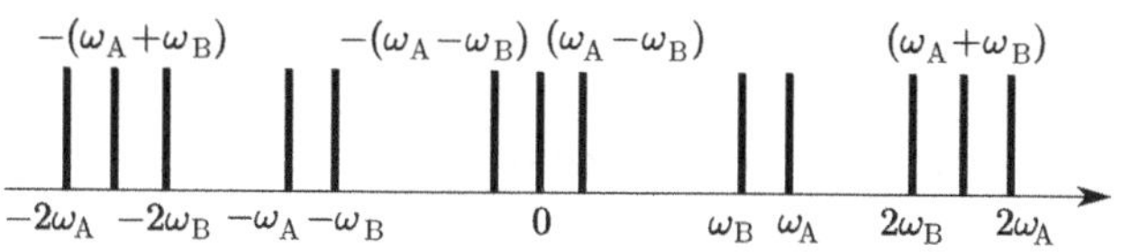

Figure 5.8 Frequency decomposition of the molecular polarization.

With the use of the trigonometric identity

$$\cos u \cos v = \frac{1}{2}[\cos(u+v) + \cos(u-v)], \tag{5.68}$$

we can rewrite the polarization as

$$\begin{aligned}\mu_\alpha(t) = \mu_\alpha^0 + \sum_{\omega=\{\omega_A,\omega_B\}} \Big[& \alpha_{\alpha\beta}(-\omega;\omega)F_\beta^\omega \cos(\omega t) \\ & + \frac{1}{2}\beta_{\alpha\beta\gamma}(0;\omega,-\omega)F_\beta^\omega F_\gamma^\omega + \frac{1}{2}\beta_{\alpha\beta\gamma}(-2\omega;\omega,\omega)F_\beta^\omega F_\gamma^\omega \cos(2\omega t)\Big] \\ & + \beta_{\alpha\beta\gamma}(-(\omega_A+\omega_B);\omega_A,\omega_B)F_\beta^{\omega_A}F_\gamma^{\omega_B}\cos([\omega_A+\omega_B]t) \\ & + \beta_{\alpha\beta\gamma}(-(\omega_A-\omega_B);\omega_A,-\omega_B)F_\beta^{\omega_A}F_\gamma^{\omega_B}\cos([\omega_A-\omega_B]t).\end{aligned} \tag{5.69}$$

The frequency decomposition of the polarization is illustrated in Figure 5.8, and in the figure we recognize the linear polarization at the frequencies of the external field and the second-harmonic generation frequencies as a result of two-photon absorption (see also Figure 5.6). In addition to these frequencies, we see that the induced molecular polarization will contain the frequencies $(\omega_A + \omega_B)$ (sum-frequency generation) and $(\omega_A - \omega_B)$ (difference-frequency generation) as well as a static component $\omega = 0$. Sum-frequency generation is similar to second-harmonic generation in that it involves the simultaneous absorption of two photons, although in this case one photon from each light source. Difference-frequency generation, on the other hand, has a fundamentally different microscopic origin. The corpuscular light–matter interaction in this nonlinear optical process is described in Figure 5.9. Since the energy is conserved in this process as well as in all others, the light quanta of frequency ω_A is annihilated and those of frequencies ω_B and $(\omega_A - \omega_B)$ are both created at this instant (two-photon emission). The creation of the photon with frequency ω_B is a result of stimulated emission induced by laser B, and the electric field amplitude F^{ω_B} is thereby enhanced by this process. It is thus in principle possible to amplify a weak light signal B with a pump laser A.

Let us analyze another example, namely when the external electric field is composed of a single laser field with frequency ω and amplitude F^ω in addition to a static field with amplitude F^0:

$$F_\alpha(t) = F_\alpha^0 + F_\alpha^\omega \cos(\omega t). \tag{5.70}$$

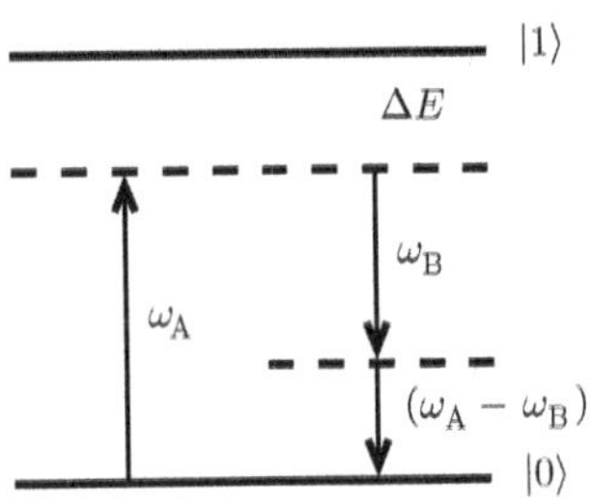

Figure 5.9 Difference-frequency generation involving photons of frequencies ω_A and ω_B.

Including terms up to third order in the electric field, the polarization in this case becomes

$$\begin{aligned}\mu_\alpha(t) = \mu_\alpha^0 &+ \alpha_{\alpha\beta}[F_\beta^0 + F_\beta^\omega \cos(\omega t)] \\ &+ \frac{1}{2}\beta_{\alpha\beta\gamma}[F_\beta^0 + F_\beta^\omega \cos(\omega t)][F_\gamma^0 + F_\gamma^\omega \cos(\omega t)] \\ &+ \frac{1}{6}\gamma_{\alpha\beta\gamma\delta}[F_\beta^0 + F_\beta^\omega \cos(\omega t)][F_\gamma^0 + F_\gamma^\omega \cos(\omega t)][F_\delta^0 + F_\delta^\omega \cos(\omega t)].\end{aligned} \tag{5.71}$$

We make repeated use of the trigonometric identity in Eq. (5.68) and rewrite the polarization as

$$\mu_\alpha(t) = \tilde{\mu}_\alpha^0 + \tilde{\mu}_\alpha^\omega \cos(\omega t) + \tilde{\mu}_\alpha^{2\omega} \cos(2\omega t) + \tilde{\mu}_\alpha^{3\omega} \cos(3\omega t), \tag{5.72}$$

where the Fourier amplitudes of the polarization become

$$\begin{aligned}\tilde{\mu}_\alpha^0 = \mu_\alpha^0 &+ \alpha_{\alpha\beta}(0;0)F_\beta^0 + \frac{1}{2}\beta_{\alpha\beta\gamma}(0;0,0)F_\beta^0 F_\gamma^0 \\ &+ \frac{1}{6}\gamma_{\alpha\beta\gamma\delta}(0;0,0,0)F_\beta^0 F_\gamma^0 F_\delta^0 \\ &+ \frac{1}{4}\beta_{\alpha\beta\gamma}(0;-\omega,\omega)F_\beta^\omega F_\gamma^\omega + \frac{1}{4}\gamma_{\alpha\beta\gamma\delta}(0;-\omega,\omega,0)F_\beta^0 F_\gamma^\omega F_\delta^\omega,\end{aligned} \tag{5.73}$$

$$\begin{aligned}\tilde{\mu}_\alpha^\omega = \alpha_{\alpha\beta}(-\omega;\omega)F_\beta^\omega &+ \beta_{\alpha\beta\gamma}(-\omega;\omega,0)F_\beta^\omega F_\gamma^0 \\ &+ \frac{1}{2}\gamma_{\alpha\beta\gamma\delta}(-\omega;\omega,0,0)F_\beta^\omega F_\gamma^0 F_\delta^0 \\ &+ \frac{1}{8}\gamma_{\alpha\beta\gamma\delta}(-\omega;\omega,-\omega,\omega)F_\beta^\omega F_\gamma^\omega F_\delta^\omega,\end{aligned} \tag{5.74}$$

$$\begin{aligned}\tilde{\mu}_\alpha^{2\omega} = &\frac{1}{4}\beta_{\alpha\beta\gamma}(-2\omega;\omega,\omega)F_\beta^\omega F_\gamma^\omega \\ &+ \frac{1}{4}\gamma_{\alpha\beta\gamma\delta}(-2\omega;\omega,\omega,0)F_\beta^\omega F_\gamma^\omega F_\delta^0,\end{aligned} \tag{5.75}$$

$$\tilde{\mu}_\alpha^{3\omega} = \frac{1}{24}\gamma_{\alpha\beta\gamma\delta}(-3\omega;\omega,\omega,\omega)F_\beta^\omega F_\gamma^\omega F_\delta^\omega. \tag{5.76}$$

An alternative summary of this result is sometimes expressed as an expansion of the polarization amplitudes in terms of the electric field amplitudes

$$\begin{aligned}\tilde{\mu}_\alpha^{\omega_\sigma} = \alpha(-\omega_\sigma;\omega_1)F^{\omega_1} &+ \frac{1}{2}K^{(2)}\beta(-\omega_\sigma;\omega_1,\omega_2)F^{\omega_1}F^{\omega_2} \\ &+ \frac{1}{6}K^{(3)}\gamma(-\omega_\sigma;\omega_1,\omega_2,\omega_3)F^{\omega_1}F^{\omega_2}F^{\omega_3} + \cdots,\end{aligned} \tag{5.77}$$

where the factors $K^{(n)}$ are required for the polarization related to the molecular response of order n to have the same static limit. By a direct comparison with Eqs. (5.72)–(5.76), we are able to identify these factors for some common nonlinear optical processes, see Table 5.1.

Table 5.1 Common nonlinear optical processes.

Process	Frequencies	Factor
	Second-order processes	$K^{(2)}$
Static	$0; 0, 0$	1
OR[a)]	$0; \omega, -\omega$	1/2
EOPE[b)]	$-\omega; \omega, 0$	2
SHG[c)]	$-2\omega; \omega, \omega$	1/2
	Third-order processes	$K^{(3)}$
Static	$0; 0, 0, 0$	1
dc-OR[d)]	$0; \omega, -\omega, 0$	3/2
EOKE[e)]	$-\omega; \omega, 0, 0$	3
IDRI[f)]	$-\omega; \omega, -\omega, \omega$	3/4
ESHG[g)]	$-2\omega; \omega, \omega, 0$	3/2
THG[h)]	$-3\omega; \omega, \omega, \omega$	1/4

a) Optical rectification.
b) Electro-optical Pockels effect.
c) Second-harmonic generation.
d) dc Optical rectification.
e) Electro-optical Kerr effect.
f) Intensity-dependent refractive index.
g) Electric field-induced second-harmonic generation.
h) Third-harmonic generation.

We have seen how the molecular electric properties are defined by the expansion of the molecular polarization in orders of the external electric field, Eq. (5.60). Beyond the linear polarization, this definition introduces the hyperpolarizabilities as coupling coefficients between the two quantities. The same equation also expresses an expansion in terms of the number of photons involved in simultaneous quantum-mechanical processes: α, β, γ, and so on involve emission or absorption of two, three, four, and so on photons. The cross-section for multiphoton absorption or emission, which takes place in nonlinear optical processes, is in typical cases relatively small and a high density of photons is required for these to occur.

In the introductory section of this chapter, we gave examples of how one can determine molecular properties from time-domain simulations of the polarization. Such a technique has certain pedagogical values and offers a general approach to time-resolved spectroscopies, but at the same time it also suffers from several limitations—for example, nonlinear responses are difficult to extract due to their small magnitudes. Instead, what is needed at this point is to determine quantum-mechanical expressions relating the molecular properties to other parameters such as excitation energies and transition moments. In order to achieve this, however, it is required to work out algebraic, albeit perturbational, solutions of the Schrödinger equation, and this will be the topic of the sections to follow.

5.3 Reference-State Parameterizations

When solving the Schrödinger equation, Eq. (5.14), we first need to choose a parametrization of the wave function $\psi(t)$. By parameterization, we refer to the expansion of $\psi(t)$ in a set of

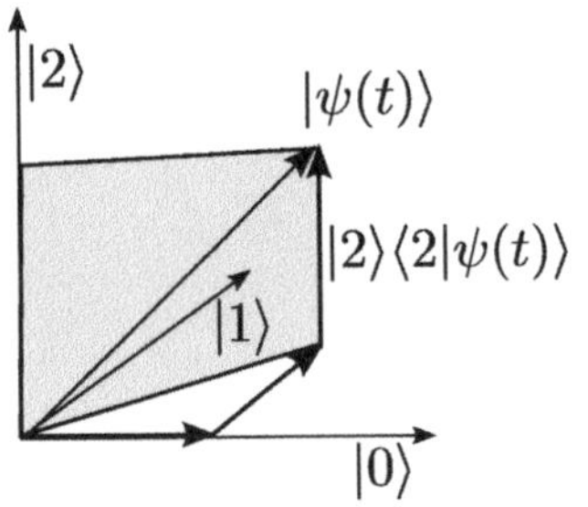

Figure 5.10 Parametrization of the wave function by projections.

known functions that under typical circumstances are time independent. The choice of parameterization is of course governed by the level of approximation one has chosen, that is, the electronic structure method at hand, but even when disregarding this fact, there are different paths to follow that impact the subsequent algebraic work that needs to be done. Perhaps, the most commonly adopted parameterization is given by

$$\psi(t) = \sum_n c_n(t)\psi_n, \tag{5.78}$$

where the summation runs over a set of time-independent functions ψ_n (the corresponding state vectors are denoted by $|n\rangle$ in Figure 5.10). The time-dependent parameters are here to be interpreted as projection coefficients according to

$$c_n(t) = \langle n|\psi(t)\rangle. \tag{5.79}$$

The set of time-independent functions does not have to be complete, in which case the solution of the Schrödinger equation is found in a linear subspace of the N-electron Hilbert space—as for example in configuration interaction based methods—and we refer in this case to an *approximate-state theory*. However, the set may be complete and include, for example, the true eigenstates of the unperturbed Hamiltonian, and we will refer to this situation as *exact-state theory*.

It is clear that not all values of the expansion coefficients in Eq. (5.78) are permissible since the normalization condition requires

$$\sum_n |c_n(t)|^2 = 1. \tag{5.80}$$

In exact-state theory, a conditional parameterization such as that in Eq. (5.78) can be of minor concern since the Schrödinger equation is norm conserving and guarantees that Eq. (5.80) is fulfilled. Approximate-state theories, on the other hand, involve some sort of projection of the true wave function onto a subspace of the full Hilbert space and the normalization condition then becomes an issue. It therefore seems worthwhile to consider an alternative parameterization where the normalization condition is fulfilled by construction by employing an explicitly unitary parameterization. Let us consider the following parameterization

$$|\overline{\psi}(t)\rangle = e^{-i\hat{P}(t)}|0\rangle, \tag{5.81}$$

where the Hermitian operator $\hat{P}(t)$ is given by

$$\hat{P}(t) = \sum_{n>0} [P_n(t)|n\rangle\langle 0| + P_n^*(t)|0\rangle\langle n|]. \tag{5.82}$$

Since the exponential of any anti-Hermitian operator (in this case $-i\hat{P}$) is guaranteed to be unitary, we note that this parameterization is unconstrained with respect to the time-dependent, complex-valued, parameters $P_n(t)$. We stress that the set of parameters $P_n(t)$, which are referred

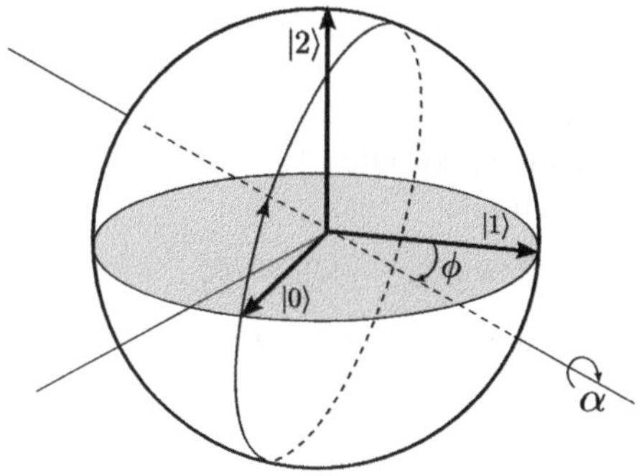

Figure 5.11 Parametrization of the wave function by rotations.

to as state-transfer amplitudes, are unknown functions that are to be determined from the Schrödinger equation (as was the case for the coefficients $c_n(t)$ in the previous example). This stands in contrast to the propagator in Eq. (5.18), where the operator in the exponent is known at any given time t. Common to the exponential parametrization and the propagator, however, is our view of these as rotations of the state vector, as illustrated in Figure 5.11.

How does the parametrization by rotations in Eq. (5.81) compare with that by projections in Eq. (5.78)? In order to best address this question, we evaluate the exponential operator. The linear and quadratic terms in a Taylor expansion of the exponential function are, respectively,

$$\hat{P}(t)|0\rangle = \sum_{n>0} P_n |n\rangle, \tag{5.83}$$

$$\hat{P}(t)^2|0\rangle = |0\rangle \sum_{n>0} |P_n|^2, \tag{5.84}$$

and from induction it is clear that a closed-form expression for the action of the exponential operator on the ground state will be

$$e^{-i\hat{P}(t)}|0\rangle = |0\rangle \cos\alpha - i \sum_{n>0} P_n |n\rangle \frac{\sin\alpha}{\alpha}, \tag{5.85}$$

$$\alpha = \sqrt{\sum_{n>0} |P_n|^2}. \tag{5.86}$$

A comparison of Eq. (5.85) with Eq. (5.78) shows that the two approaches are in some, but not all, respects equivalent. Among the differences, we first note that the exponential parametrization represents a many-to-one mapping of state-transfer amplitudes to state vectors. Second, the exponential parametrization as given here should be supplemented with a phase factor if we are to allow for an alteration of the phase of the ground state. A complete parametrization of $|0\rangle$ can thus instead be written as

$$|\psi(t)\rangle = e^{-i\hat{P}(t)}|0\rangle e^{-i\phi(t)}, \tag{5.87}$$

where $\phi(t)$ is a real function. If we, however, are concerned with quantum-mechanical averages, the determination of the overall phase factor is not needed (although it is of course possible). Further, if we are concerned with the parameterization of the phase-isolated function that was introduced in Eq. (5.25), the exponential parameterization provides sufficient flexibility so that

$$|\overline{\psi}(t)\rangle = e^{-i\hat{P}(t)}|0\rangle. \tag{5.88}$$

From Eq. (5.85), we note that the projection of the phase-isolated wave function $|\overline{\psi}(t)\rangle$ onto the reference state $|0\rangle$ is equal to a real number, $\cos\alpha$, or in other words, the phase of this projection is zero. We note that this is in accordance with our stated requirement below Eq. (5.25).

Even if the set of ground-to-excited state-transfer operators is incomplete in the expression for $\hat{P}$, Eq. (5.82), the images of the unitary transformation will span a linear vector space. An example of such a situation is given in Figure 5.11, where the excited-state manifold is limited to include two states, $|1\rangle$ and $|2\rangle$. Rotations in $\mathcal{R}^3$ are provided by letting P_1 and P_2 be purely imaginary

$$P_1 = iP_1^{\mathrm{I}}; \qquad P_2 = iP_2^{\mathrm{I}}, \tag{5.89}$$

so that a matrix representation of $-i\hat{P}$ reads as

$$-iP = \begin{pmatrix} 0 & -P_1^{\mathrm{I}} & -P_2^{\mathrm{I}} \\ P_1^{\mathrm{I}} & 0 & 0 \\ P_2^{\mathrm{I}} & 0 & 0 \end{pmatrix}. \tag{5.90}$$

In this case, we have

$$|\psi\rangle = e^{-i\hat{P}}|0\rangle = \begin{pmatrix} \cos\alpha \\ P_1^{\mathrm{I}} \sin\alpha/\alpha \\ P_2^{\mathrm{I}} \sin\alpha/\alpha \end{pmatrix}, \tag{5.91}$$

which is equivalent to a clockwise rotation of $|0\rangle$ by an angle α about an axis in the plane spanned by the vectors $|1\rangle$ and $|2\rangle$. The rotation axis makes an angle ϕ with vector $|1\rangle$, where ϕ is given by

$$\phi = \arctan(P_1^{\mathrm{I}}/P_2^{\mathrm{I}}), \tag{5.92}$$

as illustrated in Figure 5.11. Let us assume that the matrix representation of the Hamiltonian of the three-state system is

$$H = \begin{pmatrix} E_0 & 0 & 0 \\ 0 & E_1 & 0 \\ 0 & 0 & E_2 \end{pmatrix}, \tag{5.93}$$

with numerical values of E_0, E_1, and E_2 set to 0, 1.5, and 2.0 (in a unit system of choice). The energy as a function of the transition-state amplitudes is then

$$E(P_1, P_2) = \langle 0|e^{i\hat{P}}\hat{H}e^{-i\hat{P}}|0\rangle. \tag{5.94}$$

For variations in the amplitudes between $-\pi$ and π, this energy is depicted in Figure 5.12, where panel (b) provides the contour plot of the surface plot to panel (a). In the contour plot; at some points in the parameter space we have also given the corresponding wave functions.

Using the Baker–Campbell–Hausdorff (BCH) expansion [see Eq. (E.6)], Eq. (5.94) can be expressed as

$$E(P_1, P_2) = \langle 0|\hat{H}|0\rangle + i\langle 0|[\hat{P}, \hat{H}]|0\rangle - \frac{1}{2}\langle 0|[\hat{P}, [\hat{P}, \hat{H}]]|0\rangle + \cdots, \tag{5.95}$$

which gives an expression for the second-order derivatives of the energy (or Hessian) with respect to the transition state amplitudes as

$$\begin{aligned} \left.\frac{\partial^2 E}{\partial P_n \partial P_m}\right|_{\mathbf{P}=0} &= -\frac{1}{2}(\langle 0|[|n\rangle\langle 0|, [|m\rangle\langle 0|, \hat{H}]]|0\rangle \\ &\quad + \langle 0|[|m\rangle\langle 0|, [|n\rangle\langle 0|, \hat{H}]]|0\rangle) \\ &= 0, \end{aligned} \tag{5.96}$$

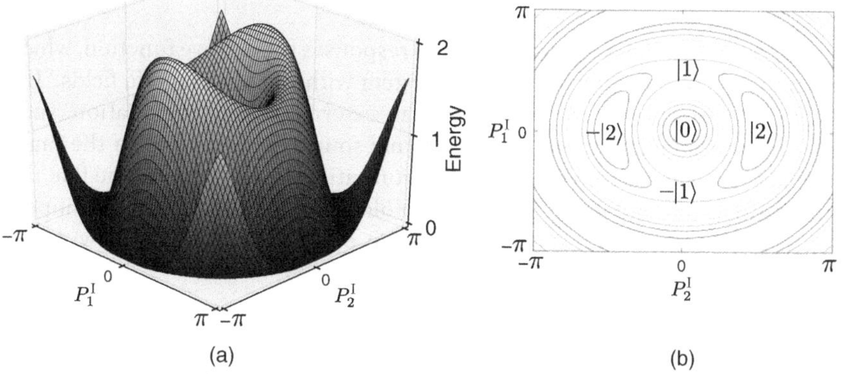

Figure 5.12 Energy as a function of rotation parameters.

$$\left.\frac{\partial^2 E}{\partial P_n \partial P_m^*}\right|_{\mathbf{P}=0} = -\frac{1}{2}(\langle 0|[|n\rangle\langle 0|,[|0\rangle\langle m|,\hat{H}]]|0\rangle + \langle 0|[|0\rangle\langle m|,[|n\rangle\langle 0|,\hat{H}]]|0\rangle) = (E_n - E_0)\,\delta_{nm}, \tag{5.97}$$

$$\left.\frac{\partial^2 E}{\partial P_n^* \partial P_m^*}\right|_{\mathbf{P}=0} = -\frac{1}{2}(\langle 0|[|0\rangle\langle n|,[|0\rangle\langle m|,\hat{H}]]|0\rangle + \langle 0|[|0\rangle\langle m|,[|0\rangle\langle n|,\hat{H}]]|0\rangle) = 0. \tag{5.98}$$

We have here collected the entire set of transition-state amplitudes in the vector $\mathbf{P}$. The Hessian is seen to contain the set of excitation energies of the system. We have regarded P_n and P_n^* as separate variables, although it may be more intuitive to instead consider the real and imaginary parts of P_n as separate variables since these may be varied independently. In Figure 5.12 for instance, the curvatures at the bottom of the well are given by the second-order derivatives of the energy with respect to the imaginary parts of P_1 and P_2, which are denoted by P_1^{I} and P_2^{I}, respectively. It is, however, straightforward to obtain the Hessian in this representation given the results in Eqs. (5.96)–(5.98). The transformation between the two representations is obtained by chain-rule differentiation, and the resulting formulas are

$$\frac{\partial^2 E}{\partial P_n^{\mathrm{R}} \partial P_m^{\mathrm{R}}} = \frac{\partial^2 E}{\partial P_n \partial P_m} + \frac{\partial^2 E}{\partial P_n^* \partial P_m} + \frac{\partial^2 E}{\partial P_n \partial P_m^*} + \frac{\partial^2 E}{\partial P_n^* \partial P_m^*}, \tag{5.99}$$

$$\frac{\partial^2 E}{\partial P_n^{\mathrm{R}} \partial P_m^{\mathrm{I}}} = i\left[\frac{\partial^2 E}{\partial P_n \partial P_m} + \frac{\partial^2 E}{\partial P_n^* \partial P_m} - \frac{\partial^2 E}{\partial P_n \partial P_m^*} - \frac{\partial^2 E}{\partial P_n^* \partial P_m^*}\right], \tag{5.100}$$

$$\frac{\partial^2 E}{\partial P_n^{\mathrm{I}} \partial P_m^{\mathrm{I}}} = -\frac{\partial^2 E}{\partial P_n \partial P_m} + \frac{\partial^2 E}{\partial P_n^* \partial P_m} + \frac{\partial^2 E}{\partial P_n \partial P_m^*} - \frac{\partial^2 E}{\partial P_n^* \partial P_m^*}. \tag{5.101}$$

From Eq. (5.101) we see that the curvature in Figure 5.12 at $\mathbf{P} = 0$ (the bottom of the well) is given by $2(E_n - E_0)$ in the direction of state $|n\rangle$, that is, when $m = n$.

To summarize, we have discussed two types of parameterizations that are based on projections and rotations, respectively. We will now shift our focus to the determination of the parameters as functions of time in these parameterizations.

5.4 Equations of Motion

In this section, we shall derive explicit expressions for the responses in the wave function, which describe the interactions of the quantum-mechanical system with electromagnetic fields. The fields are under typical circumstances external due to an external source of radiation, such as a laser, but they can in principle also be due to internal sources such as due to the magnetic moments of nuclei. We will use time-dependent perturbation theory and assume that the Hamiltonian can be divided into a part $\hat{H}_0$ that is the molecular Hamiltonian of the unperturbed system, and a part $\hat{V}(t)$ that is a small perturbation due to the interactions with the electromagnetic fields

$$\hat{H} = \hat{H}_0 + \hat{V}(t). \tag{5.102}$$

The explicit form of $\hat{V}(t)$ can be found by means of multipole expansions of the external fields as done in Section 3.3.2, where to lowest order we adopt the electric dipole approximation. We further assume that the solutions to the eigenvalue problem of $\hat{H}_0$ are known

$$\hat{H}_0|n\rangle = E_n|n\rangle, \tag{5.103}$$

where $|n\rangle$ are the exact rovibronic eigenstates and E_n the respective energies. Before being exposed to the perturbation, we assume the molecule to be in a reference state $|0\rangle$—in most cases the molecular ground state. If the perturbing field is truly periodic as described by Eq. (5.59), there does not exist a time before the perturbation was switched on and as a consequence the system has never resided in a stationary reference state. In our propagation of the wave function of the two-level system in Section 5.1, we addressed this issue by using the error function to slowly switch on the perturbation at $t = 0$. This choice is not the most convenient from a mathematical point of view, as the aim is to work out algebraic expressions for the wave function response, and for this reason we here adopt a different approach: we let the perturbation be adiabatically switched on in the long gone past. Enough time has since passed for the system to have lost any memory of the process of the perturbation being switched on, that is, we have reached a situation where there remains only fundamental frequencies in the oscillations of the electron density. The mathematical form of the perturbation is chosen as

$$F_\alpha(t) = \sum_\omega F_\alpha^\omega e^{-i\omega t} e^{\epsilon t}, \tag{5.104}$$

where ϵ is a positive infinitesimal that ensures a slow switch-on of the field as $t \to -\infty$. It is clear that we are interested in observation times for which t remains finite, so at the time of observation, $e^{\epsilon t}$ is close to unity. The classical field will couple to the quantum-mechanical system via a coupling operator. In the example we used in Section 5.2, this coupling operator was equal to minus the electric-dipole operator, but in a more general context it need not be. In fact, there may be different fields and operators associated with different frequencies (and spatial directions), and to accommodate such possibilities we write the perturbation operator as

$$\hat{V}(t) = \sum_\omega \hat{V}_\alpha^\omega F_\alpha^\omega e^{-i\omega t} e^{\epsilon t}. \tag{5.105}$$

We remind the reader that we use the convention of implied summation over the repeated index α referring to the Cartesian axes in the molecular frame, and the requirement of a real-valued perturbation corresponds not only to $F_\alpha^{-\omega} = [F_\alpha^\omega]^*$, but also to $\hat{V}_\alpha^{-\omega} = [\hat{V}_\alpha^\omega]^\dagger = \hat{V}_\alpha^\omega$.

We now wish to determine the wave function at times of observations, in other words, the probability amplitudes for the molecule to be in a state other than the reference state. These amplitudes are quite different in appearance depending on how we choose to parameterize our wave function, and the algebraic techniques for obtaining the amplitudes differ accordingly.

5.4.1 Time Evolution of Projection Amplitudes

Let us first consider the parameterization using projection amplitudes, Eq. (5.78). In order to determine the time evolution of the time-dependent amplitude functions $c_n(t)$, we substitute this expression for $\psi(t)$ into the Schrödinger equation. However, we first note that in the absence of a perturbation, $c_n(t) = c_n(0)\exp(-iE_nt/\hbar)$, and it is therefore appropriate to rewrite the parametrization in Eq. (5.78) as

$$|\psi(t)\rangle = \sum_n d_n(t)e^{-iE_nt/\hbar}|n\rangle, \tag{5.106}$$

Because, in this way, we isolate the changes of d_n in time so that the changes will be caused only by the presumably small perturbation. In terms of these coefficients, the initial condition of starting in the reference state $|0\rangle$ translates into $d_n(-\infty) = \delta_{n0}$ (Kronecker delta function). By inserting Eq. (5.106) into the Schrödinger equation and after performing a projection of the equation onto the mth state vector (i.e. by multiplying the left- and right-hand sides of the equation by the bra-vector $\langle m|$), we obtain

$$i\hbar\frac{\partial}{\partial t}d_m(t) = \sum_n V_{mn}(t)d_n(t)e^{i(E_m-E_n)t/\hbar}, \tag{5.107}$$

where V_{mn} is the matrix element of the perturbation operator $\hat{V}(t)$. We solve this equation by means of perturbation theory and by considering the expansions of the coefficients in a power series of the perturbation

$$d_n(t) = d_n^{(0)} + d_n^{(1)}(t) + d_n^{(2)}(t) + \cdots, \tag{5.108}$$

and given the solution to order $(N-1)$, the Nth order solution is after a straightforward time integration given by

$$d_m^{(N)}(t) = \frac{1}{i\hbar}\int_{-\infty}^{t}\sum_n V_{mn}(t')d_n^{(N-1)}(t')e^{i(E_m-E_n)t'/\hbar}dt'. \tag{5.109}$$

As is customary, the superscript indicates the order of the dependence on $\hat{V}$, and the corresponding order-by-order partitioning of the wave function is

$$|\psi(t)\rangle = |\psi^{(0)}(t)\rangle + |\psi^{(1)}(t)\rangle + |\psi^{(2)}(t)\rangle + \cdots, \tag{5.110}$$

where

$$|\psi^{(N)}(t)\rangle = \sum_n d_n^{(N)}e^{-iE_nt/\hbar}|n\rangle. \tag{5.111}$$

The zeroth-order solution for the parameters corresponds to the initial condition, that is, $d_n^{(0)} = \delta_{n0}$, which after insertion into Eq. (5.109) gives the first-order solution

$$\begin{aligned} d_m^{(1)}(t) &= \frac{1}{i\hbar}\int_{-\infty}^{t}\sum_n\langle m|\sum_{\omega_1}\hat{V}_\alpha^{\omega_1}F_\alpha^{\omega_1}e^{-i\omega_1t'}e^{\epsilon t'}|n\rangle e^{i\omega_{mn}t'}\delta_{n0}\,dt' \\ &= -\frac{1}{\hbar}\sum_{\omega_1}\frac{\langle m|\hat{V}_\alpha^{\omega_1}|0\rangle F_\alpha^{\omega_1}}{\omega_{m0}-\omega_1-i\epsilon}e^{i(\omega_{m0}-\omega_1)t}e^{\epsilon t}, \end{aligned} \tag{5.112}$$

where ω_{mn} has been introduced as the transition angular frequency $(E_m - E_n)/\hbar$. It is here recognized that the contribution from the lower integration limit vanishes due to ϵ, and because the coefficients $d_m^{(1)}(-\infty)$ are zero, the result is consistent with the molecule being in the reference state at $t = -\infty$.

The second-order response is obtained from Eq. (5.109) by inserting the result for the first-order response

$$\begin{aligned}
d_m^{(2)}(t) &= -\frac{1}{i\hbar}\int_{-\infty}^{t} \sum_n \langle m|\sum_{\omega_1} \hat{V}_\alpha^{\omega_1} F_\alpha^{\omega_1} e^{-i\omega_1 t'} e^{\epsilon t'}|n\rangle e^{i\omega_{mn}t'} \\
&\quad \times \frac{1}{\hbar}\sum_{\omega_2} \frac{\langle n|\hat{V}_\beta^{\omega_2}|0\rangle F_\beta^{\omega_2}}{\omega_{n0}-\omega_2-i\epsilon} e^{i(\omega_{n0}-\omega_2)t'} e^{\epsilon t'}\, dt' \\
&= \frac{1}{\hbar^2}\sum_{\omega_1\omega_2}\sum_n \frac{\langle m|\hat{V}_\alpha^{\omega_1}|n\rangle\langle n|\hat{V}_\beta^{\omega_2}|0\rangle F_\alpha^{\omega_1} F_\beta^{\omega_2}}{(\omega_{m0}-\omega_1-\omega_2-i2\epsilon)(\omega_{n0}-\omega_2-i\epsilon)} \\
&\quad \times e^{i(\omega_{m0}-\omega_1-\omega_2)t} e^{2\epsilon t}.
\end{aligned} \tag{5.113}$$

Repeating the procedure once more, we obtain the third-order response as

$$\begin{aligned}
d_m^{(3)}(t) &= \frac{1}{i\hbar}\int_{-\infty}^{t} \sum_n \langle m|\sum_{\omega_1} \hat{V}_\alpha^{\omega_1} F_\alpha^{\omega_1} e^{-i\omega_1 t'} e^{\epsilon t'}|n\rangle e^{i\omega_{mn}t'} \\
&\quad \times \frac{1}{\hbar^2}\sum_{\omega_2\omega_3}\sum_p \frac{\langle n|\hat{V}_\beta^{\omega_2}|p\rangle\langle p|\hat{V}_\gamma^{\omega_3}|0\rangle F_\beta^{\omega_2} F_\gamma^{\omega_3}}{(\omega_{n0}-\omega_2-\omega_3-i2\epsilon)(\omega_{p0}-\omega_3-i\epsilon)} e^{i(\omega_{n0}-\omega_2-\omega_3)t'} e^{2\epsilon t'}\, dt' \\
&= -\frac{1}{\hbar^3}\sum_{\omega_1\omega_2\omega_3}\sum_{np} \frac{\langle m|\hat{V}_\alpha^{\omega_1}|n\rangle\langle n|\hat{V}_\beta^{\omega_2}|p\rangle\langle p|\hat{V}_\gamma^{\omega_3}|0\rangle F_\alpha^{\omega_1} F_\beta^{\omega_2} F_\gamma^{\omega_3}}{(\omega_{m0}-\omega_1-\omega_2-\omega_3-i3\epsilon)(\omega_{n0}-\omega_2-\omega_3-i2\epsilon)(\omega_{p0}-\omega_3-i\epsilon)} \\
&\quad \times e^{i(\omega_{m0}-\omega_1-\omega_2-\omega_3)t} e^{3\epsilon t}.
\end{aligned} \tag{5.114}$$

By now, the pattern is clear as how to proceed to obtain higher-order responses.

Before we close this section, we take a closer look at the first-order-corrected wave function. By combining Eqs. (5.111) and (5.112) and adding to it the zeroth-order wave function, we get a wave function that to first order reads

$$\begin{aligned}
|\psi(t)\rangle &= |\psi^{(0)}(t)\rangle + |\psi^{(1)}(t)\rangle \\
&= |0\rangle e^{-iE_0 t/\hbar} - \frac{1}{\hbar}\sum_n\sum_\omega \frac{\langle n|\hat{V}_\alpha^{\omega}|0\rangle F_\alpha^{\omega}}{\omega_{n0}-\omega-i\epsilon} e^{-i\omega t} e^{\epsilon t}|n\rangle e^{-iE_0 t/\hbar}.
\end{aligned} \tag{5.115}$$

What is slightly disconcerting about this equation is the fact that since the summation is to include *all* states (also $n = 0$), it appears as if the wave function breaks the condition of normalization already to first order in the perturbation. To first order in $\hat{V}$, the squared norm of the wave function becomes

$$\langle\psi(t)|\psi(t)\rangle = 1 + \frac{1}{\hbar}\sum_\omega \left[\frac{F_\alpha^{\omega} e^{-i\omega t}}{-\omega-i\epsilon} + \frac{[F_\alpha^{\omega}]^* e^{i\omega t}}{-\omega+i\epsilon}\right] \langle 0|\hat{V}_\alpha^{\omega}|0\rangle e^{\epsilon t}. \tag{5.116}$$

We recall that the summation over the frequency of the perturbation runs over both positive and negative frequencies so that we are free to substitute ω for $-\omega$ in the second term of the summation. This operation, together with the use of the properties of the perturbation discussed in relation to Eq. (5.105), leads us to conclude that the summation vanishes altogether and the wave function is indeed normalized to first order in $\hat{V}$. Another way to see this is to note that $|\psi^{(1)}\rangle$ contains one term that is parallel to $|\psi^{(0)}\rangle$ (namely the $n = 0$ term) whereas all other terms are orthogonal to $|\psi^{(0)}\rangle$. However, the parallel term can at any given time be written as a purely imaginary number times $|\psi^{(0)}\rangle$, which guarantees that normalization is maintained—as expressed in Eq. (5.45) in the discussion of the time-dependent variational principle. It is also clear that the parallel term in $|\psi^{(1)}\rangle$ does not contribute to the expectation values of Hermitian operators as long as these do not contain operations of time differentiation. Examples will be given later in terms of the determination of the induced polarization and magnetization.

5.4.2 Time Evolution of Rotation Amplitudes

We will now determine the expansion coefficients for the perturbed wave function, using a parameterization based on rotations as given by Eq. (5.87). A straightforward substitution of Eq. (5.87) into the Schrödinger equation is in this case not necessarily the best way to proceed since it will be difficult to isolate the overall phase function $\phi(t)$ from the rotation amplitudes $P_n(t)$ due to the operation of time differentiation. Further, if we are only interested in quantum-mechanical averages, there is no need to determine $\phi(t)$. This last statement is actually a key point for finding a suitable equation of motion for the rotation amplitudes. We recall that the Ehrenfest theorem, Eq. (2.82), provides us with an equation of motion, fully equivalent to the Schrödinger equation, but written on the form of expectation values. The time evolution of the expectation value of the operator $\hat{\Omega}$ is governed by

$$\frac{\partial}{\partial t}\langle\psi(t)|\hat{\Omega}|\psi(t)\rangle = \frac{1}{i\hbar}\langle\psi(t)|[\hat{\Omega},\hat{H}]|\psi(t)\rangle + \langle\psi(t)|\frac{\partial\hat{\Omega}}{\partial t}|\psi(t)\rangle. \tag{5.117}$$

We immediately notice two things: First, if the operator $\hat{\Omega}$ is time independent, the last term vanishes, and, second, if $\hat{\Omega}$ at least does not involve the operation of time differentiation, then we may substitute the wave function with its phase-isolated counterpart in Eq. (5.81). If both criteria are fulfilled, we have

$$\frac{\partial}{\partial t}\langle\overline{\psi}(t)|\hat{\Omega}|\overline{\psi}(t)\rangle = \frac{1}{i\hbar}\langle\overline{\psi}(t)|[\hat{\Omega},\hat{H}]|\overline{\psi}(t)\rangle. \tag{5.118}$$

Finding an equation of motion for the phase-isolated wave function itself is highly beneficial since we may then completely disregard the overall phase function. We previously achieved this by a projection onto the orthogonal complement of $\overline{\psi}$ in Eq. (5.37), but the utilization of Eq. (5.118) is easier in the general case since we do not need to form $\overline{\psi}_\perp$. The question however remains on how we can best employ Eq. (5.118) to determine the amplitudes $P_n(t)$, or in other words, what operator $\hat{\Omega}$ should be employed. Before we provide an answer, let us consider the exponential operator in some detail.

Inserting Eq. (5.81) into Eq. (5.118), we obtain

$$\frac{\partial}{\partial t}\langle 0|e^{i\hat{P}(t)}\hat{\Omega}e^{-i\hat{P}(t)}|0\rangle = \frac{1}{i\hbar}\langle 0|e^{i\hat{P}(t)}[\hat{\Omega},\ \hat{H}_0 + \hat{V}(t)]e^{-i\hat{P}(t)}|0\rangle. \tag{5.119}$$

A BCH expansion [see Eq. (E.6)] of the operator in the left-hand-side gives

$$e^{i\hat{P}(t)}\hat{\Omega}e^{-i\hat{P}(t)} = \hat{\Omega} + i[\hat{P},\hat{\Omega}] - \frac{1}{2}[\hat{P},[\hat{P},\hat{\Omega}]] - \frac{i}{6}[\hat{P},[\hat{P},[\hat{P},\hat{\Omega}]]] + \cdots. \tag{5.120}$$

We will therefore be concerned with the expectation values

$$\langle 0|[\hat{P},\hat{\Omega}]|0\rangle = \sum_{m>0} P_m^*\langle m|\hat{\Omega}|0\rangle - P_m\langle 0|\hat{\Omega}|m\rangle, \tag{5.121}$$

$$\langle 0|[\hat{P},[\hat{P},\hat{\Omega}]]|0\rangle = -2\sum_{m,p>0} P_m^*P_p\langle m|\overline{\hat{\Omega}}|p\rangle, \tag{5.122}$$

$$\langle 0|[\hat{P},[\hat{P},[\hat{P},\hat{\Omega}]]]|0\rangle = 4\left[\sum_{m>0}|P_m|^2\right]\left[\sum_{m>0} P_m^*\langle m|\hat{\Omega}|0\rangle - P_m\langle 0|\hat{\Omega}|m\rangle\right], \tag{5.123}$$

where the fluctuation operator $\overline{\hat{\Omega}} = \hat{\Omega} - \langle 0|\hat{\Omega}|0\rangle$ has been introduced. We can thus single out any one amplitude P_n from Eq. (5.121) by choosing $\hat{\Omega}$ to be the state transition operator $|0\rangle\langle n|$. For the particular choice of $\hat{\Omega} = |0\rangle\langle n|$, we get

$$\langle 0|[\hat{P}, \hat{\Omega}]|0\rangle = -P_n, \tag{5.124}$$

$$\langle 0|[\hat{P}, [\hat{P}, \hat{\Omega}]]|0\rangle = 0, \tag{5.125}$$

$$\langle 0|[\hat{P}, [\hat{P}, [\hat{P}, \hat{\Omega}]]]|0\rangle = -4P_n \sum_{m>0} |P_m|^2, \tag{5.126}$$

as well as

$$[\hat{\Omega}, \hat{H}_0] = \hbar\omega_{n0}\hat{\Omega}. \tag{5.127}$$

We will solve Eq. (5.119) using perturbation theory, and we therefore introduce a perturbation expansion of the amplitudes in terms of $\hat{V}$

$$P_n(t) = P_n^{(1)} + P_n^{(2)} + P_n^{(3)} + \cdots, \tag{5.128}$$

which up to third order corresponds to an expansion of the wave function as

$$|\overline{\psi}(t)\rangle = |\overline{\psi}^{(0)}\rangle + |\overline{\psi}^{(1)}\rangle + |\overline{\psi}^{(2)}\rangle + |\overline{\psi}^{(3)}\rangle + \cdots, \tag{5.129}$$

$$|\overline{\psi}^{(0)}\rangle = |0\rangle, \tag{5.130}$$

$$|\overline{\psi}^{(1)}\rangle = -i \sum_{n>0} P_n^{(1)}|n\rangle, \tag{5.131}$$

$$|\overline{\psi}^{(2)}\rangle = -i \sum_{n>0} P_n^{(2)}|n\rangle - \frac{1}{2}\sum_{n>0} |P_n^{(1)}|^2|0\rangle, \tag{5.132}$$

$$\begin{aligned} |\overline{\psi}^{(3)}\rangle = {} & -i \sum_{n>0} P_n^{(3)}|n\rangle - \frac{1}{2}\sum_{n>0}(P_n^{(1)}P_n^{*(2)} + P_n^{*(1)}P_n^{(2)})|0\rangle \\ & + \frac{i}{6}\left[\sum_{n>0}|P_n^{(1)}|^2\right]\sum_{n>0} P_n^{(1)}|n\rangle. \end{aligned} \tag{5.133}$$

Our equation of motion, Eq. (5.119), must be fulfilled to each order in the perturbation, and to zeroth-order we get a not very surprising result

$$\frac{\partial}{\partial t}\langle 0|\hat{\Omega}|0\rangle = 0, \tag{5.134}$$

which reflects our use of a time-independent operator $\hat{\Omega}$. The first-order terms in the equation of motion are

$$\frac{\partial}{\partial t}\langle 0|[\hat{P}^{(1)}, \hat{\Omega}]|0\rangle = \frac{1}{i\hbar}\langle 0|[\hat{P}^{(1)}, [\hat{\Omega}, \hat{H}_0]]|0\rangle - \frac{1}{\hbar}\langle 0|[\hat{\Omega}, \hat{V}(t)]|0\rangle. \tag{5.135}$$

Inserting the explicit form of the state-transition operator $\hat{\Omega} = |0\rangle\langle n|$ in Eq. (5.135) and using Eqs. (5.124) and (5.127) result in

$$\frac{\partial}{\partial t}P_n^{(1)} = -i\omega_{n0}P_n^{(1)} + \frac{1}{\hbar}\langle n|\hat{V}(t)|0\rangle, \tag{5.136}$$

which by direct time integration yields

$$\begin{aligned} P_n^{(1)} &= e^{-i\omega_{n0}t}\int_{-\infty}^{t}\frac{1}{\hbar}\langle n|\hat{V}(t')|0\rangle e^{i\omega_{n0}t'}dt' \\ &= \frac{1}{i\hbar}\sum_{\omega}\frac{\langle n|\hat{V}_\alpha^{\omega}|0\rangle F_\alpha^{\omega}e^{-i\omega t}e^{\epsilon t}}{\omega_{n0} - \omega - i\epsilon}. \end{aligned} \tag{5.137}$$

Inserting Eq. (5.137) into Eq. (5.131), we obtain the first-order correction to the phase-isolated wave function

$$|\overline{\psi}^{(1)}\rangle = -\frac{1}{\hbar}\sum_{n>0}\sum_{\omega}\frac{\langle n|\hat{V}_{\alpha}^{\omega}|0\rangle F_{\alpha}^{\omega}e^{-i\omega t}e^{\epsilon t}}{\omega_{n0}-\omega-i\epsilon}|n\rangle. \tag{5.138}$$

We have here made a point of the fact that we do not need to determine the phase function $\phi(t)$ since we will eventually only be concerned with the calculation of expectation values, but evidently the zeroth-order phase function is

$$\phi^{(0)}(t) = E_0 t/\hbar, \tag{5.139}$$

in line with Eq. (5.28) in our simple example of the two-level system. Multiplying $|\overline{\psi}^{(1)}\rangle$ by $\exp(-i\phi^{(0)})$ confirms that, apart from the term parallel to $|\psi^{(0)}\rangle$ in Eq. (5.115) that does not contribute to the expectation values, the first-order correction to the wave function $|\psi(t)\rangle$ as determined with rotation amplitudes is identical to that determined with projection amplitudes.

Let us now continue and address the second-order terms in the equation of motion for the phase-isolated wave function. The collection of these terms gives

$$\begin{aligned}&\frac{\partial}{\partial t}\langle 0|(i[\hat{P}^{(2)},\hat{\Omega}] - \frac{1}{2}[\hat{P}^{(1)},[\hat{P}^{(1)},\hat{\Omega}]])|0\rangle\\ &= \frac{1}{i\hbar}\left[\langle 0|(i[\hat{P}^{(2)},[\hat{\Omega},\hat{H}_0]] - \frac{1}{2}[\hat{P}^{(1)},[\hat{P}^{(1)},[\hat{\Omega},\hat{H}_0]]])|0\rangle + \langle 0|i[\hat{P}^{(1)},[\hat{\Omega},\hat{V}(t)]]|0\rangle\right],\end{aligned} \tag{5.140}$$

or, equivalently,

$$\begin{aligned}&\frac{\partial}{\partial t}[\langle 0|\hat{\Omega}|\overline{\psi}^{(2)}\rangle + \langle\overline{\psi}^{(2)}|\hat{\Omega}|0\rangle + \langle\overline{\psi}^{(1)}|\hat{\Omega}|\overline{\psi}^{(1)}\rangle]\\ &= \frac{1}{i\hbar}[\langle 0|[\hat{\Omega},\hat{H}_0]|\overline{\psi}^{(2)}\rangle + \langle\overline{\psi}^{(2)}|[\hat{\Omega},\hat{H}_0]|0\rangle + \langle\overline{\psi}^{(1)}|[\hat{\Omega},\hat{H}_0]|\overline{\psi}^{(1)}\rangle\\ &\quad + \langle 0|[\hat{\Omega},\hat{V}(t)]|\overline{\psi}^{(1)}\rangle + \langle\overline{\psi}^{(1)}|[\hat{\Omega},\hat{V}(t)]|0\rangle].\end{aligned} \tag{5.141}$$

Inserting $\hat{\Omega} = |0\rangle\langle n|$ into Eq. (5.140) and using Eqs. (5.124), (5.125), and (5.127), we obtain

$$\frac{\partial}{\partial t}P_n^{(2)} = -i\omega_{n0}P_n^{(2)} + \frac{1}{i\hbar}\sum_{m>0}P_m^{(1)}\langle n|\overline{\hat{V}(t)}|m\rangle, \tag{5.142}$$

where the fluctuation operator $\overline{\hat{V}} = \hat{V} - \langle 0|\hat{V}|0\rangle$ has been introduced. As we did for the first-order amplitudes, we solve the differential equation for second-order amplitudes that vanish at $t = -\infty$

$$\begin{aligned}P_n^{(2)} &= e^{-i\omega_{n0}t}\int_{-\infty}^{t}\frac{1}{i\hbar}\sum_{m>0}P_m^{(1)}\langle n|\overline{\hat{V}(t')}|m\rangle e^{i\omega_{n0}t'}dt'\\ &= -\frac{1}{i\hbar^2}\sum_{\omega_1\omega_2}\sum_{m>0}\frac{\langle n|\overline{\hat{V}_{\alpha}^{\omega_1}}|m\rangle\langle m|\hat{V}_{\beta}^{\omega_2}|0\rangle F_{\alpha}^{\omega_1}F_{\beta}^{\omega_2}}{(\omega_{n0}-(\omega_1+\omega_2)-i2\epsilon)(\omega_{m0}-\omega_2-i\epsilon)}e^{-i(\omega_1+\omega_2)t}e^{2\epsilon t}.\end{aligned} \tag{5.143}$$

The second-order correction to the wave function is found by combining Eqs. (5.132), (5.137), and (5.143). We note that the order of the optical frequencies may be interchanged. It is common to choose a symmetric combination so that the polarization is symmetric in the optical frequencies, that is, the response functions possess intrinsic but not overall symmetry.

The third-order terms are

$$\frac{\partial}{\partial t}\langle 0|(i[\hat{P}^{(3)},\hat{\Omega}] - \frac{1}{2}\{[\hat{P}^{(2)},[\hat{P}^{(1)},\hat{\Omega}]] + [\hat{P}^{(1)},[\hat{P}^{(2)},\hat{\Omega}]]\} - \frac{i}{6}[\hat{P}^{(1)},[\hat{P}^{(1)},[\hat{P}^{(1)},\hat{\Omega}]]])|0\rangle$$
$$= \frac{1}{i\hbar}\Bigg[\langle 0|i[\hat{P}^{(3)},\ [\hat{\Omega},\hat{H}_0]]|0\rangle - \frac{1}{2}\langle 0|([\hat{P}^{(2)},[\hat{P}^{(1)},[\hat{\Omega},\hat{H}_0]]] + [\hat{P}^{(1)},[\hat{P}^{(2)},[\hat{\Omega},\hat{H}_0]]])|0\rangle$$
$$- \frac{i}{6}\langle 0|[\hat{P}^{(1)},[\hat{P}^{(1)},[\hat{P}^{(1)},[\hat{\Omega},\hat{H}_0]]]]|0\rangle + \langle 0|i[\hat{P}^{(2)},[\hat{\Omega},\hat{V}(t)]]|0\rangle +$$
$$- \frac{1}{2}\langle 0|[\hat{P}^{(1)},[\hat{P}^{(1)},[\hat{\Omega},\hat{V}(t)]]]|0\rangle\Bigg]. \tag{5.144}$$

or, equivalently,

$$\frac{\partial}{\partial t}[\langle 0|\hat{\Omega}|\overline{\psi}^{(3)}\rangle + \langle\overline{\psi}^{(3)}|\hat{\Omega}|0\rangle + \langle\overline{\psi}^{(2)}|\hat{\Omega}|\overline{\psi}^{(1)}\rangle + \langle\overline{\psi}^{(1)}|\hat{\Omega}|\overline{\psi}^{(2)}\rangle]$$
$$= \frac{1}{i\hbar}[\langle 0|[\hat{\Omega},\hat{H}_0]|\overline{\psi}^{(3)}\rangle + \langle\overline{\psi}^{(3)}|[\hat{\Omega},\hat{H}_0]|0\rangle + \langle\overline{\psi}^{(1)}|[\hat{\Omega},\hat{H}_0]|\overline{\psi}^{(2)}\rangle + \langle\overline{\psi}^{(2)}|[\hat{\Omega},\hat{H}_0]|\overline{\psi}^{(1)}\rangle$$
$$+\langle 0|[\hat{\Omega},\hat{V}(t)]|\overline{\psi}^{(2)}\rangle + \langle\overline{\psi}^{(2)}|[\hat{\Omega},\hat{V}(t)]|0\rangle + \langle\overline{\psi}^{(1)}|[\hat{\Omega},\hat{V}(t)]|\overline{\psi}^{(1)}\rangle]. \tag{5.145}$$

Inserting $\hat{\Omega} = |0\rangle\langle n|$ into Eq. (5.144), we obtain

$$\frac{\partial}{\partial t}D_n^{(3)} = -i\omega_{n0}D_n^{(3)} + \frac{1}{i\hbar}\sum_{m>0} P_m^{(2)}\langle n|\overline{\hat{V}}(t)|m\rangle$$
$$-\frac{1}{\hbar}\left[P_n^{(1)}\sum_{m>0} P_m^{*(1)}\langle m|\hat{V}(t)|0\rangle + \langle n|\hat{V}(t)|0\rangle\sum_{m>0}|P_m^{(1)}|^2\right], \tag{5.146}$$

where we have collected the terms inside the time differentiation in the auxiliary amplitudes

$$D_n^{(3)} = P_n^{(3)} - \frac{2}{3}P_n^{(1)}\sum_{m>0}|P_m^{(1)}|^2. \tag{5.147}$$

In order to arrive at Eq. (5.146), it is convenient to make use of Eqs. (5.124)–(5.127) for all terms that do not involve the perturbation operator, and Eqs. (5.121) and (5.122) with $\hat{\Omega} = [|0\rangle\langle n|, \hat{V}]$ for the two terms that involve the perturbation operator. The solution of Eq. (5.146) is given by

$$D_n^{(3)} = e^{-i\omega_{n0}t}\int_{-\infty}^{t}\Bigg[\frac{1}{i\hbar}\sum_{m>0} P_m^{(2)}\langle n|\overline{\hat{V}}(t')|m\rangle$$
$$+ \frac{1}{\hbar}P_n^{(1)}\sum_{m>0} P_m^{*(1)}\langle m|\hat{V}(t')|0\rangle + \frac{1}{\hbar}\langle n|\hat{V}(t')|0\rangle\sum_{m>0}|P_m^{(1)}|^2\Bigg] e^{i\omega_{n0}t'}dt'$$
$$= \frac{1}{i\hbar^3}\sum_{\omega_1\omega_2\omega_3} e^{-i(\omega_1+\omega_2+\omega_3)t}e^{3\epsilon t}\Bigg[$$
$$\times \sum_{m,p>0}\frac{\langle n|\overline{\hat{V}_\alpha^{\omega_1}}|m\rangle\langle m|\overline{\hat{V}_\beta^{\omega_2}}|p\rangle\langle p|\hat{V}_\gamma^{\omega_3}|0\rangle F_\alpha^{\omega_1}F_\beta^{\omega_2}F_\gamma^{\omega_3}}{(\omega_{n0}-(\omega_1+\omega_2+\omega_3)-i3\epsilon)(\omega_{m0}-(\omega_2+\omega_3)-i2\epsilon)(\omega_{p0}-\omega_3-i\epsilon)}$$
$$- \frac{\langle n|\hat{V}_\alpha^{\omega_1}|0\rangle F_\alpha^{\omega_1}}{(\omega_{n0}-(\omega_1+\omega_2+\omega_3)-i3\epsilon)}\sum_{m>0}\frac{\langle 0|\hat{V}_\beta^{\omega_2}|m\rangle\langle m|\hat{V}_\gamma^{\omega_3}|0\rangle F_\beta^{\omega_2}F_\gamma^{\omega_3}}{(\omega_{m0}+\omega_2+i\epsilon)(\omega_{n0}-\omega_1-i\epsilon)}$$
$$- \frac{\langle n|\hat{V}_\alpha^{\omega_1}|0\rangle F_\alpha^{\omega_1}}{(\omega_{n0}-(\omega_1+\omega_2+\omega_3)-i3\epsilon)}\sum_{m>0}\frac{\langle 0|\hat{V}_\beta^{\omega_2}|m\rangle\langle m|\hat{V}_\gamma^{\omega_3}|0\rangle F_\beta^{\omega_2}F_\gamma^{\omega_3}}{(\omega_{m0}+\omega_2+i\epsilon)(\omega_{m0}-\omega_3-i\epsilon)}\Bigg]. \tag{5.148}$$

Given $D_n^{(3)}$, the third-order correction to the wave function is found by forming the rotation amplitudes $P_n^{(3)}$ in Eq. (5.147) and inserting them together with lower-order amplitudes into Eq. (5.133).

5.5 Response Functions

As discussed in Section 5.2, a straightforward definition of time-dependent molecular properties is given by the time-dependent expectation value of an observable, such as the electric or magnetic dipole moment. If $\hat{\Omega}$ denotes the operator corresponding to the observable of interest, then we can consider the expansion

$$\begin{aligned}\langle\psi(t)|\hat{\Omega}|\psi(t)\rangle &= \langle\psi^{(0)}|\hat{\Omega}|\psi^{(0)}\rangle \\ &+\langle\psi^{(1)}|\hat{\Omega}|\psi^{(0)}\rangle + \langle\psi^{(0)}|\hat{\Omega}|\psi^{(1)}\rangle \\ &+\langle\psi^{(2)}|\hat{\Omega}|\psi^{(0)}\rangle + \langle\psi^{(1)}|\hat{\Omega}|\psi^{(1)}\rangle + \langle\psi^{(0)}|\hat{\Omega}|\psi^{(2)}\rangle \\ &+\langle\psi^{(3)}|\hat{\Omega}|\psi^{(0)}\rangle + \langle\psi^{(2)}|\hat{\Omega}|\psi^{(1)}\rangle + \langle\psi^{(1)}|\hat{\Omega}|\psi^{(2)}\rangle + \langle\psi^{(0)}|\hat{\Omega}|\psi^{(3)}\rangle \\ &+\cdots . \end{aligned} \tag{5.149}$$

As already mentioned, we can just as well consider the expectation value with respect to the phase-isolated wave function, and we will illustrate both alternatives in the subsequent sections. By a comparison of this expansion with an expansion that defines molecular properties, as in Eq. (5.60), we are able to obtain explicit quantum-mechanical expressions for the properties after the responses are inserted in the wave function. While Eq. (5.60) is concerned with the specific example of the electric polarization in the presence of an external electric field, a more general expansion, known as the Kubo relation, can be written

$$\begin{aligned}\langle\psi(t)|\hat{\Omega}|\psi(t)\rangle &= \langle 0|\hat{\Omega}|0\rangle \\ &+\sum_{\omega_1}\langle\langle\hat{\Omega};\hat{V}_\beta^{\omega_1}\rangle\rangle F_\beta^{\omega_1} e^{-i\omega_1 t}e^{\epsilon t} \\ &+\frac{1}{2}\sum_{\omega_1\omega_2}\langle\langle\hat{\Omega};\hat{V}_\beta^{\omega_1},\hat{V}_\gamma^{\omega_2}\rangle\rangle F_\beta^{\omega_1}F_\gamma^{\omega_2} e^{-i(\omega_1+\omega_2)t}e^{2\epsilon t} \\ &+\frac{1}{6}\sum_{\omega_1\omega_2\omega_3}\langle\langle\hat{\Omega};\hat{V}_\beta^{\omega_1},\hat{V}_\gamma^{\omega_2},\hat{V}_\delta^{\omega_3}\rangle\rangle F_\beta^{\omega_1}F_\gamma^{\omega_2}F_\delta^{\omega_3} e^{-i(\omega_1+\omega_2+\omega_3)t}e^{3\epsilon t} \\ &+\cdots , \end{aligned} \tag{5.150}$$

where $\langle\langle\hat{\Omega};\hat{V}^{\omega}\rangle\rangle$ is referred to as the linear response function and collects all terms to the expectation value of $\hat{\Omega}$ that are linear in the perturbation, $\langle\langle\hat{\Omega};\hat{V}^{\omega_1},\hat{V}^{\omega_2}\rangle\rangle$ is the first-order nonlinear response function, and so on.

We have seen that an alternative approach to identify molecular properties is provided by the time-dependent Hellmann–Feynman theorem, Eq. (5.48), and the quasi-energy. We see the time-averaged quasi-energy as a function of the amplitudes of the perturbation $Q_T(F^{\omega})$—the symbol F^{ω} collectively denotes amplitudes at all the different perturbation frequencies and along the three different Cartesian molecular axes. In the process of performing the time average in Eq. (5.51), the only component in $Q(t)$ that survives is the nonoscillating one, and Q_T may thus also be thought of as the zero-frequency Fourier amplitude of the quasi-energy. The function Q_T is expanded in the amplitudes according to

$$
\begin{aligned}
Q_T(F^\omega) = {} & E_0 \\
& + \sum_{\omega_1} \left. \frac{dQ_T}{dF_\alpha^{\omega_1}} \right|_{F^\omega=0} F_\alpha^{\omega_1} \\
& + \frac{1}{2} \sum_{\omega_1\omega_2} \left. \frac{d^2Q_T}{dF_\alpha^{\omega_1} dF_\beta^{\omega_2}} \right|_{F^\omega=0} F_\alpha^{\omega_1} F_\beta^{\omega_2} \\
& + \frac{1}{6} \sum_{\omega_1\omega_2\omega_3} \left. \frac{d^3Q_T}{dF_\alpha^{\omega_1} dF_\beta^{\omega_2} dF_\gamma^{\omega_3}} \right|_{F^\omega=0} F_\alpha^{\omega_1} F_\beta^{\omega_2} F_\gamma^{\omega_3} \\
& + \frac{1}{24} \sum_{\omega_1\omega_2\omega_3\omega_4} \left. \frac{d^4Q_T}{dF_\alpha^{\omega_1} dF_\beta^{\omega_2} dF_\gamma^{\omega_3} dF_\delta^{\omega_4}} \right|_{F^\omega=0} F_\alpha^{\omega_1} F_\beta^{\omega_2} F_\gamma^{\omega_3} F_\delta^{\omega_4} \\
& + \cdots ,
\end{aligned}
\tag{5.151}
$$

where E_0 is the energy of the reference state $|0\rangle$. The expansion coefficients in this expansion of Q_T can be identified as corresponding to molecular properties through the time-dependent Hellmann–Feynman theorem, Eq. (5.53),

$$
\frac{dQ_T}{dF_\alpha^\omega} = \frac{1}{T} \int_t^{t+T} \langle \overline{\psi}(t') | \frac{\partial \hat{H}}{\partial F_\alpha^\omega} | \overline{\psi}(t') \rangle \, dt' = \frac{1}{T} \int_t^{t+T} \langle \overline{\psi}(t') | \hat{V}_\alpha^\omega | \overline{\psi}(t') \rangle \, e^{-i\omega t'} e^{\epsilon t'} \, dt'. \tag{5.152}
$$

By choosing our perturbation operator $\hat{V}_\alpha^\omega$ to correspond to the property of interest, that is, to let it represent the operator $\hat{\Omega}$ in Eq. (5.149), we obtain the time-averaged equivalent of Eq. (5.149) in Eq. (5.152). In the time-integration step we make use of the fact that, at the observation times, the function $\exp(\epsilon t)$ is close to unity and therefore carries no time dependence, and reference to this function is therefore omitted in the quasi-energy derivations provided in the following text. The result for the first-order derivative at zero field strength is obtained directly from Eq. (5.152) by making the replacement $|\overline{\psi}\rangle = |0\rangle$, giving

$$
\left. \frac{dQ_T}{dF_\alpha^{\omega_1}} \right|_{F^\omega=0} = \langle 0 | \hat{V}_\alpha^{\omega_1} | 0 \rangle \, \delta_{\omega_1}, \tag{5.153}
$$

where δ_{ω_1} represents the generalization of the Kronecker delta function for the continuous variable ω_1. We note that to first order in the perturbation, Q_T will differ from E_0 only for static perturbations, that is, when $\omega_1 = 0$.

We will view Q_T as a function of external field strengths and wave function parameters (in this case P_n and P_n^*), or to be more precise, the Fourier amplitudes of the wave function parameters that are defined by the expansions

$$
P_n(t) = \sum_{\omega_1} P_n^{(1)}(\omega_1) e^{-i\omega_1 t} + \sum_{\omega_1,\omega_2} P_n^{(2)}(\omega_1, \omega_2) e^{-i(\omega_1+\omega_2)t} + \cdots , \tag{5.154}
$$

$$
P_n^*(t) = \sum_{\omega_1} [P_n^{(1)}(-\omega_1)]^* e^{-i\omega_1 t} + \sum_{\omega_1,\omega_2} [P_n^{(2)}(-\omega_1, -\omega_2)]^* e^{-i(\omega_1+\omega_2)t} + \cdots , \tag{5.155}
$$

that is, we understand Q_T as a function according to

$$
Q_T(P_n^{(1)}(\omega_1), [P_n^{(1)}(-\omega_1)]^*, P_n^{(2)}(\omega_1, \omega_2), [P_n^{(2)}(-\omega_1, -\omega_2)]^*, \ldots, F_\alpha^{\omega_1}, F_\beta^{\omega_2}, \ldots).
$$

Owing to the variational principle for Q_T in Eq. (5.52), we have *for all values of the field strengths* that all first-order derivatives with respect to the Fourier amplitudes vanish at all frequencies:

$$
\left. \frac{\partial Q_T}{\partial P_n^{(1)}} \right|_{F^\omega} = 0; \quad \left. \frac{\partial Q_T}{\partial [P_n^{(1)}]^*} \right|_{F^\omega} = 0; \quad \left. \frac{\partial Q_T}{\partial P_n^{(2)}} \right|_{F^\omega} = 0; \quad \left. \frac{\partial Q_T}{\partial [P_n^{(2)}]^*} \right|_{F^\omega} = 0; \quad \cdots . \tag{5.156}
$$

This set of equations means that the wave function parameters are functions of the field strengths and our task at hand is to determine the responses of the parameters with respect to the external fields. Starting from Eq. (5.156), we accomplish this by the derivation of a corresponding set of *response equations*. The key operation in this derivation is to perform a differentiation of the partial derivatives in Eq. (5.156) with respect to field strengths, which in practice is accomplished by means of chain rule differentiation. We obtain

$$\begin{aligned}
&\frac{d}{dF_\alpha^\omega}\frac{\partial Q_T}{\partial P_n}\bigg|_{F^\omega=0} \\
&= \frac{\partial^2 Q_T}{\partial F_\alpha^\omega \partial P_n}\bigg|_{F^\omega=0} + \sum_m \left[\frac{\partial^2 Q_T}{\partial P_n \partial P_m}\frac{\partial P_m}{\partial F_\alpha^\omega} + \frac{\partial^2 Q_T}{\partial P_n \partial P_m^*}\frac{\partial P_m^*}{\partial F_\alpha^\omega}\right]\bigg|_{F^\omega=0} = 0, \\
&\frac{d}{dF_\alpha^\omega}\frac{\partial Q_T}{\partial P_n^*}\bigg|_{F^\omega=0} \\
&= \frac{\partial^2 Q_T}{\partial F_\alpha^\omega \partial P_n^*}\bigg|_{F^\omega=0} + \sum_m \left[\frac{\partial^2 Q_T}{\partial P_n^* \partial P_m}\frac{\partial P_m}{\partial F_\alpha^\omega} + \frac{\partial^2 Q_T}{\partial P_n^* \partial P_m^*}\frac{\partial P_m^*}{\partial F_\alpha^\omega}\right]\bigg|_{F^\omega=0} = 0,
\end{aligned} \tag{5.157}$$

from which we can determine the responses in the wave function parameters to the perturbations $\partial P_m/F_\alpha^\omega$ and $\partial P_m^*/F_\alpha^\omega$ (we have here suppressed the superscripts of the Fourier amplitudes). It is clear that we can perform additional differentiations of Eq. (5.157) and in this way determine higher-order derivatives of the parameters with respect to the fields. However, in order to not clutter our notation, we introduce an efficient meta-notation for the relevant derivatives, writing them compactly as

$$Q_T^{N;\alpha\beta} = \frac{\partial^{N+2} Q_T}{\partial P^N \partial F_\alpha^{\omega_1} \partial F_\beta^{\omega_2}}\bigg|_{F^\omega=0}, \tag{5.158}$$

$$P^{\alpha\beta\gamma} = \frac{\partial^3 P^{(3)}}{\partial F_\alpha^{\omega_1} \partial F_\beta^{\omega_2} \partial F_\gamma^{\omega_3}}\bigg|_{F^\omega=0}, \tag{5.159}$$

where P is understood as a vector collecting all wave function Fourier parameters (both P_n and P_n^*). We have suppressed the superscript of P in the expression for $Q_T^{N;\alpha\beta}$, since the order will be clear from the parameter vectors that contract the matrix.

Using this short-hand notation, the zeroth-order response equations, Eq. (5.156), may be written as

$$Q_T^{1;0} = 0, \tag{5.160}$$

and the first-order response equations, Eq. (5.157), as

$$Q_T^{2;0} P^\alpha + Q_T^{1;\alpha} = 0, \tag{5.161}$$

where the organization of the parameters is made as

$$Q_T^{1;\alpha} = \begin{pmatrix} \dfrac{\partial^2 Q_T}{\partial F_\alpha^{\omega_1} \partial P_n^*}\bigg|_{F^\omega=0} \\ \dfrac{\partial^2 Q_T}{\partial F_\alpha^{\omega_1} \partial P_n}\bigg|_{F^\omega=0} \end{pmatrix}; \qquad P^\alpha = \begin{pmatrix} \dfrac{\partial P_m}{\partial F_\alpha^{\omega_1}}\bigg|_{F^\omega=0} \\ \dfrac{\partial P_m^*}{\partial F_\alpha^{\omega_1}}\bigg|_{F^\omega=0} \end{pmatrix}, \tag{5.162}$$

and

$$Q_T^{2;0} = \begin{pmatrix} \left.\dfrac{\partial^2 Q_T}{\partial P_n^* \partial P_m}\right|_{F^\omega=0} & \left.\dfrac{\partial^2 Q_T}{\partial P_n^* \partial P_m^*}\right|_{F^\omega=0} \\ \left.\dfrac{\partial^2 Q_T}{\partial P_n \partial P_m}\right|_{F^\omega=0} & \left.\dfrac{\partial^2 Q_T}{\partial P_n \partial P_m^*}\right|_{F^\omega=0} \end{pmatrix}. \tag{5.163}$$

The elements of the $Q_T^{2;0}$ matrix have been organized such that the matrix becomes diagonal dominant, as can be inferred from Eqs. (5.96)–(5.98). We stress that this ordering is a matter of preference but our choice is motivated by the fact that numerical methods for solving matrix equations are often based on a diagonal dominant character of matrices. Our adopted vector and matrix element ordering means that if a parameter vector is to multiply a matrix, for example, $Q_T^{2;0}$, from the left, its elements need to be re-organized as

$$[P^a]_-^\dagger = \left(\left.\frac{\partial P_n^*}{\partial F_\alpha^{\omega_1}}\right|_{F^\omega=0} \quad \left.\frac{\partial P_n}{\partial F_\alpha^{\omega_1}}\right|_{F^\omega=0} \right). \tag{5.164}$$

The dagger superscript is used here in the conventional manner to turn the column vector into a row vector with elements that are complex conjugated. In addition, the "−"-subscript denotes that optical frequencies are sign shifted.

The response functions are identified from a comparison of Eq. (5.151) to the expansion

$$\begin{aligned} Q_T(F^\omega) = E_0 &+ \sum_{\omega_1} \langle 0|\hat{V}_\alpha^{\omega_1}|0\rangle\, \delta_{\omega_1} F_\alpha^{\omega_1} \\ &+ \frac{1}{2} \sum_{\omega_1\omega_2} \langle\langle \hat{V}_\alpha^{\omega_1}; \hat{V}_\beta^{\omega_2} \rangle\rangle\, \delta_{\omega_1+\omega_2} F_\alpha^{\omega_1} F_\beta^{\omega_2} \\ &+ \frac{1}{6} \sum_{\omega_1\omega_2\omega_3} \langle\langle \hat{V}_\alpha^{\omega_1}; \hat{V}_\beta^{\omega_2}, \hat{V}_\gamma^{\omega_3} \rangle\rangle\, \delta_{\omega_1+\omega_2+\omega_3} F_\alpha^{\omega_1} F_\beta^{\omega_2} F_\gamma^{\omega_3} \\ &+ \frac{1}{24} \sum_{\omega_1\omega_2\omega_3\omega_4} \langle\langle \hat{V}_\alpha^{\omega_1}; \hat{V}_\beta^{\omega_2}, \hat{V}_\gamma^{\omega_3}, \hat{V}_\delta^{\omega_4} \rangle\rangle\, \delta_{\omega_1+\omega_2+\omega_3+\omega_4} F_\alpha^{\omega_1} F_\beta^{\omega_2} F_\gamma^{\omega_3} F_\delta^{\omega_4} \\ &+ \cdots, \end{aligned} \tag{5.165}$$

where the result for the first-order derivative given in Eq. (5.153) has been directly inserted into this equation. The machinery is now set up to derive explicit expressions for response functions to various orders, and we will pursue this exercise up to third order (describing fourth-order properties) using the Ehrenfest as well as the quasi-energy formalism.

5.5.1 First-Order Properties

From Eq. (5.149) as well as Eq. (5.165), we see that a first-order property is given by the expectation value with respect to the reference state of the operator corresponding to the observable of interest. In other words, the contribution to the observable that is of zeroth-order in the perturbation is identified as

$$\langle\hat{\Omega}\rangle^{(0)} = \langle 0|\hat{\Omega}|0\rangle, \tag{5.166}$$

where we introduce the notation $\langle\hat{\Omega}\rangle^{(n)}$ to collect terms of nth-order in the perturbation.

5.5.2 Second-Order Properties

The first-order correction to the expectation value of $\hat{\Omega}$ is obtained from Eq. (5.149) by inserting the zeroth-order as well as the first-order correction to the wave function, where

$|\psi^{(0)}\rangle = \exp(-iE_0t/\hbar)|0\rangle$ and $|\psi^{(1)}\rangle$ is given by a combination of Eqs. (5.111) and (5.112)

$$\begin{aligned}
\langle\hat{\Omega}\rangle^{(1)} &= \langle\psi^{(0)}|\hat{\Omega}|\psi^{(1)}\rangle + \langle\psi^{(1)}|\hat{\Omega}|\psi^{(0)}\rangle \\
&= -\langle 0|e^{iE_0t/\hbar}\hat{\Omega}\sum_n \frac{1}{\hbar}\sum_{\omega_1} \frac{\langle n|\hat{V}_\beta^{\omega_1}|0\rangle F_\beta^{\omega_1}}{\omega_{n0}-\omega_1-i\epsilon} e^{i(\omega_{n0}-\omega_1)t}e^{\epsilon t}e^{-iE_nt/\hbar}|n\rangle \\
&\quad -\sum_n \frac{1}{\hbar}\sum_{\omega_1} \frac{\langle 0|\hat{V}_\beta^{\omega_1}|n\rangle [F_\beta^{\omega_1}]^*}{\omega_{n0}-\omega_1+i\epsilon} e^{-i(\omega_{n0}-\omega_1)t}e^{\epsilon t}e^{iE_nt/\hbar}\langle n|\hat{\Omega}e^{-iE_0t/\hbar}|0\rangle \\
&= -\sum_{\omega_1}\frac{1}{\hbar}\sum_n \frac{\langle 0|\hat{\Omega}|n\rangle\langle n|\hat{V}_\beta^{\omega_1}|0\rangle}{\omega_{n0}-\omega_1-i\epsilon} F_\beta^{\omega_1}e^{-i\omega_1t}e^{\epsilon t} \\
&\quad -\sum_{\omega_1}\frac{1}{\hbar}\sum_n \frac{\langle 0|\hat{V}_\beta^{\omega_1}|n\rangle\langle n|\hat{\Omega}|0\rangle}{\omega_{n0}-\omega_1+i\epsilon} [F_\beta^{\omega_1}]^* e^{i\omega_1t}e^{\epsilon t} \\
&= -\sum_{\omega_1}\frac{1}{\hbar}\sum_n \left[\frac{\langle 0|\hat{\Omega}|n\rangle\langle n|\hat{V}_\beta^{\omega_1}|0\rangle}{\omega_{n0}-\omega_1-i\epsilon} + \frac{\langle 0|\hat{V}_\beta^{\omega_1}|n\rangle\langle n|\hat{\Omega}|0\rangle}{\omega_{n0}+\omega_1+i\epsilon}\right] F_\beta^{\omega_1}e^{-i\omega_1t}e^{\epsilon t}.
\end{aligned} \tag{5.167}$$

In the last step, we have made use of the fact that $[F^{\omega_1}]^* = F^{-\omega_1}$ and that ω_1 runs over both positive and negative frequencies. By a direct comparison to property-defining expansions such as that in Eq. (5.60), or a general expansion as in Eq. (5.150), we can identify the quantum-mechanical formulas for second-order properties, or linear response functions. The sum-over-states expression for the linear response function is

$$\langle\langle\hat{\Omega};\hat{V}_\beta^{\omega}\rangle\rangle = -\frac{1}{\hbar}\sum_n \left[\frac{\langle 0|\hat{\Omega}|n\rangle\langle n|\hat{V}_\beta^{\omega}|0\rangle}{\omega_{n0}-\omega-i\epsilon} + \frac{\langle 0|\hat{V}_\beta^{\omega}|n\rangle\langle n|\hat{\Omega}|0\rangle}{\omega_{n0}+\omega+i\epsilon}\right], \tag{5.168}$$

and that of a component of the electric-dipole polarizability tensor $\alpha_{\alpha\beta}(-\omega;\omega)$ is obtained by the substitutions $\hat{\Omega} = \hat{\mu}_\alpha$ and $\hat{V}_\beta^{\omega} = -\hat{\mu}_\beta$. It is clear that this formula can be used directly for practical computations of the linear response function once the excitation energies and transition moments of the system are known. More importantly, it elucidates the dependence of the response function on quantum-mechanical properties of the molecule.

For the two terms in the sum-over-states expression in Eq. (5.168) that involve the ground state $n = 0$, the transition frequency ω_{n0} is zero. The two terms are thus of opposite sign and will therefore cancel, and it is common practice to exclude the ground state from the summation and to use a primed summation symbol $\sum_n'$ for the sum over excited states.

We have made a point of carrying along the positive infinitesimal ϵ in the perturbation not only to avoid singularities or divergences in the time-integration step of the expansion coefficients $d_n^{(1)}$ in Eq. (5.112), but also to avoid divergences in the linear response function itself. With a reasonable laser detuning, however, the imaginary term in the denominator of Eq. (5.168) can safely be neglected in the calculation of the linear response function. The working formula for $\alpha(-\omega;\omega)$ will in this case take the form

$$\alpha_{\alpha\beta}(-\omega;\omega) = \frac{1}{\hbar}\sum_n{}' \left[\frac{\langle 0|\hat{\mu}_\alpha|n\rangle\langle n|\hat{\mu}_\beta|0\rangle}{\omega_{n0}-\omega} + \frac{\langle 0|\hat{\mu}_\beta|n\rangle\langle n|\hat{\mu}_\alpha|0\rangle}{\omega_{n0}+\omega}\right]. \tag{5.169}$$

In nonresonant regions of the spectrum, diagonal elements of the polarizability tensor are accordingly real and with the use of real wave functions (as is customary in the nonrelativstic realm) this applies also to off-diagonal elements.

We note that after having excluded the ground state from the summation, the polarizability is convergent in the limit of nonoscillating fields also without the imaginary term in the denominator. The singularities that occur for static fields with $\epsilon = 0$ and inclusion of the ground state in the summation are called secular divergences, since they can be removed by mathematical manipulations of the sum-over-states expression. We will later see that secular divergences appear for the hyperpolarizabilities as well. However, with $\epsilon = 0$, the linear polarizability, or the linear polarization propagator as it is sometimes called, will also be divergent for perturbation frequencies matching the transition frequencies of the system. These singularities are true divergences and are known as resonances of the system, and we will discuss them in detail in Section 5.9.

Since the secular divergences are associated with the inclusion of the ground state in the summation over states, we may avoid this issue altogether if we instead consider the phase-isolated wave function. The first-order response in the expectation value is then obtained from the expression

$$\begin{aligned}\langle\hat{\Omega}\rangle^{(1)} &= \langle 0|\hat{\Omega}|\overline{\psi}^{(1)}\rangle + \langle\overline{\psi}^{(1)}|\hat{\Omega}|0\rangle \\ &= -\sum_{\omega_1}\frac{1}{\hbar}\sum_{n>0}\left[\frac{\langle 0|\hat{\Omega}|n\rangle\langle n|\hat{V}^{\omega_1}_{\beta}|0\rangle}{\omega_{n0}-\omega_1-i\epsilon}+\frac{\langle 0|\hat{V}^{\omega_1}_{\beta}|n\rangle\langle n|\hat{\Omega}|0\rangle}{\omega_{n0}+\omega_1+i\epsilon}\right]F^{\omega_1}_{\beta}e^{-i\omega_1 t}e^{\epsilon t},\end{aligned} \tag{5.170}$$

where the first-order response of the phase-isolated wave function is taken from Eq. (5.138). We can here identify a linear response function that is in perfect agreement with that in Eq. (5.169).

In the quasi-energy formalism, the linear response function is identified as being equal to the second-order derivative of Q_T with respect to the external fields, and with our short-hand notation for the derivatives, we get

$$\begin{aligned}\left.\frac{d^2Q_T}{dF^{\omega_1}_{\alpha}dF^{\omega_2}_{\beta}}\right|_{F^{\omega}=0} &= Q_T^{0;\alpha\beta} + [Q_T^{1;\alpha}]^{\dagger}_{-}\,P^{\beta} + [Q_T^{1;\beta}]^{\dagger}_{-}\,P^{\alpha} + [P^{\alpha}]^{\dagger}_{-}\,Q_T^{2;0}P^{\beta} + [Q_T^{1;0}]^{\dagger}_{-}\,P^{\alpha\beta} \\ &= [Q_T^{1;\alpha}]^{\dagger}_{-}\,P^{\beta},\end{aligned} \tag{5.171}$$

where, in the last step, we have made use of the fact that there is no second-order field dependence on the Hamiltonian (which leads to $Q_T^{0;\alpha\beta} = 0$) as well as the zeroth- and first-order response equations, Eqs. (5.160) and (5.161). The result is a manifestation of Wigner's $(2n + 1)$-rule for energy derivatives in variational wave-function theory, which states that knowledge of the n-order response of the wave function (here given by P^{α} for $n = 1$) is sufficient to determine up to $(2n + 1)$-order properties. This means that by using the first-order response in the wave function parameters, we will also be able to determine the first-order nonlinear response function in the next section.

By inserting the first-order response parameters in Eq. (5.161) into Eq. (5.171), we get an expression for the linear response function that reads

$$\langle\langle\hat{V}^{\omega_1}_{\alpha};\hat{V}^{\omega_2}_{\beta}\rangle\rangle = -[Q_T^{1;\alpha}]^{\dagger}_{-}\,[Q_T^{2;0}]^{-1}Q_T^{1;\beta}. \tag{5.172}$$

What remains is to evaluate explicit expressions for the elements of the two vectors and the matrix in this equation. Using the BCH expansion we obtain

$$\begin{aligned}Q(t) &= \langle\overline{\psi}|\hat{H}_0 + \hat{V}(t) - i\hbar\frac{\partial}{\partial t}|\overline{\psi}\rangle \\ &= \langle 0|\hat{Q}|0\rangle + i\langle 0|[\hat{P}(t),\hat{Q}]|0\rangle - \frac{1}{2}\langle 0|[\hat{P}(t),[\hat{P}(t),\hat{Q}]]|0\rangle + \cdots,\end{aligned} \tag{5.173}$$

where $\hat{Q}$ has been introduced as the quasi-energy operator $\hat{H}_0 + \hat{V}(t) - i\hbar\partial/\partial t$ and $\hat{P}(t)$ is given by Eq. (5.82). For the vector, we thus get

$$Q_T^{1;\beta} = i\begin{bmatrix} \langle n|\hat{V}_\beta^{\omega_2}|0\rangle \\ -\langle 0|\hat{V}_\beta^{\omega_2}|n\rangle \end{bmatrix}\delta_{\omega_1+\omega_2}, \tag{5.174}$$

and, in analogy with the derivations made in Eqs. (5.96)–(5.98), the matrix becomes

$$Q_T^{2;0} = (E^{[2]} - \hbar\omega_2 S^{[2]})\delta_{\omega_1+\omega_2}, \tag{5.175}$$

where

$$E^{[2]} = \begin{pmatrix} \hbar\omega_{n0}\delta_{nm} & 0 \\ 0 & \hbar\omega_{n0}\delta_{nm} \end{pmatrix}\delta_{\omega_1+\omega_2}; \qquad S^{[2]} = \begin{pmatrix} \delta_{nm} & 0 \\ 0 & -\delta_{nm} \end{pmatrix}\delta_{\omega_1+\omega_2}. \tag{5.176}$$

A division of terms is here made such that the $E^{[2]}$ (Hessian) and $S^{[2]}$ matrices stem from the operators $\hat{H}_0$ and $-i\hbar\partial/\partial t$, respectively, and there is of course no contribution from the perturbation to any matrix of the form $Q_T^{N;0}$. The field derivative of the first-order response in the wave function parameters thus takes the explicit form

$$P^\beta = \begin{pmatrix} \left.\dfrac{\partial P_n^{(1)}(\omega_2)}{\partial F_\beta^{\omega_2}}\right|_{F^\omega=0} \\ \left.\dfrac{\partial [P_n^{(1)}(-\omega_2)]^*}{\partial F_\beta^{\omega_2}}\right|_{F^\omega=0} \end{pmatrix} = \frac{1}{\hbar}\begin{pmatrix} -i\dfrac{\langle n|\hat{V}_\beta^{\omega_2}|0\rangle}{\omega_{n0}-\omega_2} \\ i\dfrac{\langle 0|\hat{V}_\beta^{\omega_2}|n\rangle}{\omega_{n0}+\omega_2} \end{pmatrix}. \tag{5.177}$$

We note that the upper part in the resulting vector is related to the lower part by the combined operations of complex conjugation and sign shift in the optical frequency. In addition, we see from a comparison with Eq. (5.137) that our response parameters obtained in the quasi-energy formalism agrees with those obtained previously.

By inserting these results into Eq. (5.172), we get an expression for the linear response equation that, after performing matrix multiplications, is

$$\langle\langle \hat{V}_\alpha^{\omega_1}; \hat{V}_\beta^{\omega_2}\rangle\rangle = -\frac{1}{\hbar}\sum_{n>0}\left[\frac{\langle 0|\hat{V}_\alpha^{\omega_1}|n\rangle\langle n|\hat{V}_\beta^{\omega_2}|0\rangle}{\omega_{n0}-\omega_2} + \frac{\langle 0|\hat{V}_\beta^{\omega_2}|n\rangle\langle n|\hat{V}_\alpha^{\omega_1}|0\rangle}{\omega_{n0}-\omega_1}\right]\delta_{\omega_1+\omega_2}. \tag{5.178}$$

We emphasize that this response function is to be evaluated for $\omega_1 = -\omega_2$, and given this constraint, the expression agrees with that of Eq. (5.170) in the limit of zero ϵ.

5.5.3 Third-Order Properties

Just as an explicit formula for the linear response function is identified from the linear polarization, we are able to retrieve the corresponding formula for the first-order nonlinear response function from the second-order polarization. We obtain the second-order polarization from the third-row terms of Eq. (5.149) by inserting the first- and second-order corrections to the wave function, where $\psi^{(1)}$ and $\psi^{(2)}$ are given by Eqs. (5.112) and (5.113), respectively, in combination with Eq. (5.111). Let us first consider these three terms individually:

$$\langle\psi^{(0)}|\hat{\Omega}|\psi^{(2)}\rangle = \langle 0|e^{iE_0t/\hbar}\hat{\Omega}$$
$$\times\sum_n \frac{1}{\hbar^2}\sum_{\omega_1\omega_2}\sum_p \frac{\langle n|\hat{V}_\beta^{\omega_1}|p\rangle\langle p|\hat{V}_\gamma^{\omega_2}|0\rangle F_\beta^{\omega_1}F_\gamma^{\omega_2}}{(\omega_{n0}-\omega_1-\omega_2-i2\epsilon)(\omega_{p0}-\omega_2-i\epsilon)}$$
$$\times e^{i(\omega_{n0}-\omega_1-\omega_2)t}e^{2\epsilon t}e^{-iE_nt/\hbar}|n\rangle$$
$$= \sum_{\omega_1\omega_2}\frac{1}{\hbar^2}\sum_{np}\frac{\langle 0|\hat{\Omega}|n\rangle\langle n|\hat{V}_\beta^{\omega_1}|p\rangle\langle p|\hat{V}_\gamma^{\omega_2}|0\rangle F_\beta^{\omega_1}F_\gamma^{\omega_2}}{(\omega_{n0}-\omega_1-\omega_2-i2\epsilon)(\omega_{p0}-\omega_2-i\epsilon)}e^{-i(\omega_1+\omega_2)t}e^{2\epsilon t}, \tag{5.179}$$

$$\langle\psi^{(1)}|\hat{\Omega}|\psi^{(1)}\rangle =$$
$$\sum_n\frac{1}{\hbar}\sum_{\omega_1}\frac{\langle 0|\hat{V}_\beta^{\omega_1}|n\rangle[F_\beta^{\omega_1}]^*}{\omega_{n0}-\omega_1+i\epsilon}e^{-i(\omega_{n0}-\omega_1)t}e^{\epsilon t}e^{iE_nt/\hbar}\langle n|\hat{\Omega}$$
$$\times\sum_p\frac{1}{\hbar}\sum_{\omega_2}\frac{\langle p|\hat{V}_\gamma^{\omega_2}|0\rangle F_\gamma^{\omega_2}}{\omega_{p0}-\omega_2-i\epsilon}e^{i(\omega_{p0}-\omega_2)t}e^{\epsilon t}e^{-iE_pt/\hbar}|p\rangle$$
$$= \sum_{\omega_1\omega_2}\frac{1}{\hbar^2}\sum_{np}\frac{\langle 0|\hat{V}_\beta^{\omega_1}|n\rangle\langle n|\hat{\Omega}|p\rangle\langle p|\hat{V}_\gamma^{\omega_2}|0\rangle F_\beta^{\omega_1}F_\gamma^{\omega_2}}{(\omega_{n0}+\omega_1+i\epsilon)(\omega_{p0}-\omega_2-i\epsilon)}e^{-i(\omega_1+\omega_2)t}e^{2\epsilon t}, \tag{5.180}$$

$$\langle\psi^{(2)}|\hat{\Omega}|\psi^{(0)}\rangle = \sum_n\frac{1}{\hbar^2}\sum_{\omega_1\omega_2}\sum_p\frac{\langle p|\hat{V}_\beta^{\omega_1}|n\rangle\langle 0|\hat{V}_\gamma^{\omega_2}|p\rangle[F_\beta^{\omega_1}]^*[F_\gamma^{\omega_2}]^*}{(\omega_{n0}-\omega_1-\omega_2+i2\epsilon)(\omega_{p0}-\omega_2+i\epsilon)}$$
$$\times e^{-i(\omega_{n0}-\omega_1-\omega_2)t}e^{2\epsilon t}e^{iE_nt/\hbar}\langle n|\hat{\Omega}e^{-iE_0t/\hbar}|0\rangle$$
$$= \sum_{\omega_1\omega_2}\frac{1}{\hbar^2}\sum_{np}\frac{\langle 0|\hat{V}_\gamma^{\omega_2}|p\rangle\langle p|\hat{V}_\beta^{\omega_1}|n\rangle\langle n|\hat{\Omega}|0\rangle F_\beta^{\omega_1}F_\gamma^{\omega_2}}{(\omega_{n0}+\omega_1+\omega_2+i2\epsilon)(\omega_{p0}+\omega_2+i\epsilon)}e^{-i(\omega_1+\omega_2)t}e^{2\epsilon t}, \tag{5.181}$$

where we again have used that $[F^\omega]^* = F^{-\omega}$ and the fact that ω_1 and ω_2 are dummy summation indices that run over all positive and negative frequencies. We note that none of the three equations above are symmetric in the indices β and γ, but we have already noted in connection with Eq. (5.60) that we would like the hyperpolarizability tensors to be intrinsically symmetric. However, remembering that also β and γ are summation indices due to the use of the Einstein summation convention, it is clear that we can force Eqs. (5.179)–(5.181) to be intrinsically symmetric without altering the physical polarization; we do so by operating with the operator $1/2\sum\mathcal{P}_{1,2}$ which performs the summation of terms obtained by permuting the pairs (β,ω_1) and (γ,ω_2). The factor of one half obviously causes the polarization to maintain its original value. The final expression for the second-order polarization then becomes

$$\langle\hat{\Omega}\rangle^{(2)} = \langle\psi^{(0)}|\hat{\Omega}|\psi^{(2)}\rangle + \langle\psi^{(1)}|\hat{\Omega}|\psi^{(1)}\rangle + \langle\psi^{(2)}|\hat{\Omega}|\psi^{(0)}\rangle$$
$$= \frac{1}{2}\sum_{\omega_1\omega_2}\frac{1}{\hbar^2}\sum\mathcal{P}_{1,2}\sum_{np}\left[\frac{\langle 0|\hat{\Omega}|n\rangle\langle n|\hat{V}_\beta^{\omega_1}|p\rangle\langle p|\hat{V}_\gamma^{\omega_2}|0\rangle}{(\omega_{n0}-\omega_1-\omega_2-i2\epsilon)(\omega_{p0}-\omega_2-i\epsilon)}\right.$$
$$+\frac{\langle 0|\hat{V}_\beta^{\omega_1}|n\rangle\langle n|\hat{\Omega}|p\rangle\langle p|\hat{V}_\gamma^{\omega_2}|0\rangle}{(\omega_{n0}+\omega_1+i\epsilon)(\omega_{p0}-\omega_2-i\epsilon)}$$
$$\left.+\frac{\langle 0|\hat{V}_\gamma^{\omega_2}|p\rangle\langle p|\hat{V}_\beta^{\omega_1}|n\rangle\langle n|\hat{\Omega}|0\rangle}{(\omega_{n0}+\omega_1+\omega_2+i2\epsilon)(\omega_{p0}+\omega_2+i\epsilon)}\right]F_\beta^{\omega_1}F_\gamma^{\omega_2}e^{-i(\omega_1+\omega_2)t}e^{2\epsilon t}, \tag{5.182}$$

and by comparing to Eq. (5.150), we identify the expression for the first-order nonlinear response function as

$$\langle\langle\hat{\Omega};\hat{V}_\beta^{\omega_1},\hat{V}_\gamma^{\omega_2}\rangle\rangle=\frac{1}{\hbar^2}\sum\mathcal{P}_{1,2}\sum_{np}\left[\frac{\langle 0|\hat{\Omega}|n\rangle\langle n|\hat{V}_\beta^{\omega_1}|p\rangle\langle p|\hat{V}_\gamma^{\omega_2}|0\rangle}{(\omega_{n0}-\omega_\sigma-i2\epsilon)(\omega_{p0}-\omega_2-i\epsilon)}\right.$$
$$+\frac{\langle 0|\hat{V}_\beta^{\omega_1}|n\rangle\langle n|\hat{\Omega}|p\rangle\langle p|\hat{V}_\gamma^{\omega_2}|0\rangle}{(\omega_{n0}+\omega_1+i\epsilon)(\omega_{p0}-\omega_2-i\epsilon)}$$
$$\left.+\frac{\langle 0|\hat{V}_\gamma^{\omega_2}|p\rangle\langle p|\hat{V}_\beta^{\omega_1}|n\rangle\langle n|\hat{\Omega}|0\rangle}{(\omega_{n0}+\omega_\sigma+i2\epsilon)(\omega_{p0}+\omega_2+i\epsilon)}\right]. \tag{5.183}$$

An element of the first-order hyperpolarizability tensor $\beta_{\alpha\beta\gamma}(-\omega_\sigma;\omega_1,\omega_2)$ is obtained from the general response function in Eq. (5.183) by substitution of $\hat{\Omega}=\hat{\mu}_\alpha$, $\hat{V}_\beta^\omega=-\hat{\mu}_\beta$, and $\hat{V}_\gamma^\omega=-\hat{\mu}_\gamma$. The use of Eq. (5.183) is valid only when the frequencies of the perturbing fields are one- and two-photon nonresonant. Under these conditions, we may at this point in the calculation let ϵ tend to zero in Eq. (5.183), and the expression for the first-order hyperpolarizability can then be written on a more compact form as

$$\beta_{\alpha\beta\gamma}(-\omega_\sigma;\omega_1,\omega_2)=\frac{1}{\hbar^2}\sum\mathcal{P}_{-\sigma,1,2}\sum_{np}\frac{\langle 0|\hat{\mu}_\alpha|n\rangle\langle n|\hat{\mu}_\beta|p\rangle\langle p|\hat{\mu}_\gamma|0\rangle}{(\omega_{n0}-\omega_\sigma)(\omega_{p0}-\omega_2)}, \tag{5.184}$$

where $\sum\mathcal{P}_{-\sigma,1,2}$ denotes the sum of the six terms obtained by permuting the pairs $(\alpha,-\omega_\sigma)$, (β,ω_1) and (γ,ω_2). The verification of Eq. (5.184) is done in a straightforward manner by a direct comparison of the six terms in Eq. (5.183) with $\epsilon=0$. The first-order hyperpolarizability tensor is said to possess full permutation symmetry in the nonresonant region.

Before reaching a final formula for the first-order hyperpolarizability, the apparent divergences of Eq. (5.184) in the limit of nonoscillating perturbing fields needs to be removed. Just as for the case of the linear response function, it is our intention to derive an equivalent expression for $\beta(-\omega_\sigma;\omega_1,\omega_2)$ that excludes terms involving the ground state in the summation. Let us begin by splitting Eq. (5.184) as

$$\beta_{\alpha\beta\gamma}(-\omega_\sigma;\omega_1,\omega_2)=\frac{1}{\hbar^2}\sum\mathcal{P}_{-\sigma,1,2}\left[\sum_{np}{}'\frac{\langle 0|\hat{\mu}_\alpha|n\rangle\langle n|\hat{\mu}_\beta|p\rangle\langle p|\hat{\mu}_\gamma|0\rangle}{(\omega_{n0}-\omega_\sigma)(\omega_{p0}-\omega_2)}\right.$$
$$+\sum_p{}'\frac{\langle 0|\hat{\mu}_\alpha|0\rangle\langle 0|\hat{\mu}_\beta|p\rangle\langle p|\hat{\mu}_\gamma|0\rangle}{-\omega_\sigma(\omega_{p0}-\omega_2)}+\sum_n{}'\frac{\langle 0|\hat{\mu}_\alpha|n\rangle\langle n|\hat{\mu}_\beta|0\rangle\langle 0|\hat{\mu}_\gamma|0\rangle}{-(\omega_{n0}-\omega_\sigma)\omega_2}$$
$$\left.+\frac{\langle 0|\hat{\mu}_\alpha|0\rangle\langle 0|\hat{\mu}_\beta|0\rangle\langle 0|\hat{\mu}_\gamma|0\rangle}{\omega_\sigma\omega_2}\right]. \tag{5.185}$$

We can write

$$\frac{1}{\omega_\sigma\omega_2}=\frac{1}{\omega_1\omega_2}-\frac{1}{\omega_\sigma\omega_1}, \tag{5.186}$$

and thus the six permutations generated from the last term in Eq. (5.185) will cancel each other. Furthermore, with the full permutation operator on the outside of the parenthesis, we are free to interchange any two pairs of indices in the respective terms in Eq. (5.185): in the second term we interchange $(\alpha,-\omega_\sigma)$ and (β,ω_1), and in the third term we interchange (γ,ω_2) and (β,ω_1).

The expression for the first-order hyperpolarizability is therefore

$$\beta_{\alpha\beta\gamma}(-\omega_\sigma;\omega_1,\omega_2) = \frac{1}{\hbar^2}\sum \mathcal{P}_{-\sigma,1,2}\left[\sum_{np}{}' \frac{\langle 0|\hat{\mu}_\alpha|n\rangle\langle n|\hat{\mu}_\beta|p\rangle\langle p|\hat{\mu}_\gamma|0\rangle}{(\omega_{n0}-\omega_\sigma)(\omega_{p0}-\omega_2)}\right.$$
$$\left. + \sum_n{}' \langle 0|\hat{\mu}_\alpha|n\rangle\langle n|\hat{\mu}_\gamma|0\rangle\langle 0|\hat{\mu}_\beta|0\rangle\left(\frac{1}{\omega_1(\omega_{n0}-\omega_2)} - \frac{1}{(\omega_{n0}-\omega_\sigma)\omega_1}\right)\right]$$
$$= \frac{1}{\hbar^2}\sum \mathcal{P}_{-\sigma,1,2}\left[\sum_{np}{}' \frac{\langle 0|\hat{\mu}_\alpha|n\rangle\langle n|\hat{\mu}_\beta|p\rangle\langle p|\hat{\mu}_\gamma|0\rangle}{(\omega_{n0}-\omega_\sigma)(\omega_{p0}-\omega_2)}\right.$$
$$\left. - \sum_n{}' \frac{\langle 0|\hat{\mu}_\alpha|n\rangle\langle 0|\hat{\mu}_\beta|0\rangle\langle n|\hat{\mu}_\gamma|0\rangle}{(\omega_{n0}-\omega_\sigma)(\omega_{n0}-\omega_2)}\right]$$
$$= \frac{1}{\hbar^2}\sum \mathcal{P}_{-\sigma,1,2}\sum_{np}{}' \frac{\langle 0|\hat{\mu}_\alpha|n\rangle\langle n|\overline{\hat{\mu}_\beta}|p\rangle\langle p|\hat{\mu}_\gamma|0\rangle}{(\omega_{n0}-\omega_\sigma)(\omega_{p0}-\omega_2)}, \tag{5.187}$$

where $\overline{\hat{\mu}}$ is the fluctuation dipole moment operator $\hat{\mu} - \langle 0|\hat{\mu}|0\rangle$. This equation represents an appropriate form of the quantum-mechanical formula for the first-order hyperpolarizability in the nonresonant region and includes nonoscillating external fields. The molecular parameters that enter this expression are the excitation energies, the ground- to excited-state transition moments, the excited-to-excited state transition moments, and the permanent dipole moment of the ground as well as the excited states. All this information is in principle contained in $\beta(-\omega_\sigma;\omega_1,\omega_2)$, and as we shall see in Section 5.9, it can be extracted from the second-order response function by a residue analysis.

Let us now turn to the derivation of the first-order nonlinear response function from the phase-isolated wave function. In this case, we expect to avoid the issue of secular divergences as noted in the derivation of the linear response function. For the second-order response, we have

$$\langle\overline{\psi}^{(0)}|\hat{\Omega}|\overline{\psi}^{(2)}\rangle + \langle\overline{\psi}^{(1)}|\hat{\Omega}|\overline{\psi}^{(1)}\rangle + \langle\overline{\psi}^{(2)}|\hat{\Omega}|\overline{\psi}^{(0)}\rangle$$
$$= i\sum_{n>0}[\langle 0|\hat{\Omega}|n\rangle P_n^{(2)} - P_n^{(2)*}\langle n|\hat{\Omega}|0\rangle] + \sum_{n,k>0} P_n^{(1)*}\langle n|\overline{\hat{\Omega}}|k\rangle P_k^{(1)}, \tag{5.188}$$

and after inserting the amplitudes, we identify the second-order response function as

$$\langle\langle\hat{\Omega};\hat{V}_\beta^{\omega_1},\hat{V}_\gamma^{\omega_2}\rangle\rangle = \frac{1}{\hbar^2}\sum \mathcal{P}_{1,2}\sum_{n,k>0}$$
$$\times\left[\frac{\langle 0|\hat{\Omega}|n\rangle\langle n|\overline{\hat{V}_\beta^{\omega_1}}|k\rangle\langle k|\hat{V}_\gamma^{\omega_2}|0\rangle}{[\omega_{n0}-(\omega_1+\omega_2)-2i\epsilon][\omega_{k0}-\omega_2-i\epsilon]}\right.$$
$$+\frac{\langle 0|\hat{V}_\gamma^{\omega_2}|n\rangle\langle n|\overline{\hat{V}_\beta^{\omega_1}}|k\rangle\langle k|\hat{\Omega}|0\rangle}{[\omega_{n0}+\omega_2+i\epsilon][\omega_{k0}+(\omega_1+\omega_2)+2i\epsilon]}$$
$$\left.+\frac{\langle 0|\hat{V}_\beta^{\omega_1}|n\rangle\langle n|\overline{\hat{\Omega}}|k\rangle\langle k|\hat{V}_\gamma^{\omega_2}|0\rangle}{[\omega_{n0}+\omega_1+i\epsilon][\omega_{k0}-\omega_2-i\epsilon]}\right]. \tag{5.189}$$

As expected, this equation is in agreement with the expression in Eq. (5.183), but without the need to explicitly remove secular divergences. Since the handling of these divergences turns out to be a rather lengthy algebraic exercise for the nonlinear response functions, it appears beneficial to work within the framework of phase-isolated wave functions. However, we recall that the the determination of the time evolution of the rotation amplitudes is crucial. While the arbitrary-order response in the projection amplitudes can easily be written down, it is considerably more difficult to determine explicit expressions for the responses in the rotation amplitudes.

Yet another approach for obtaining an expression for the second-order response function comes from the third-order derivative of Q_T with respect to the external fields. This derivative becomes

$$
\begin{aligned}
\left.\frac{d^3 Q_T}{dF_\alpha^{\omega_1} dF_\beta^{\omega_2} dF_\gamma^{\omega_3}}\right|_{F^\omega=0} &= Q_T^{0;\alpha\beta\gamma} \\
&+[Q_T^{1;\alpha\beta}]_-^\dagger\, P^\gamma + [Q_T^{1;\beta\gamma}]_-^\dagger\, P^\alpha + [Q_T^{1;\gamma\alpha}]_-^\dagger\, P^\beta \\
&+[Q_T^{1;\alpha}]_-^\dagger\, P^{\beta\gamma} + [Q_T^{1;\beta}]_-^\dagger\, P^{\gamma\alpha} + [Q_T^{1;\gamma}]_-^\dagger\, P^{\alpha\beta} \\
&+[Q_T^{1;0}]_-^\dagger\, P^{\alpha\beta\gamma} \\
&+[P^\alpha]_-^\dagger\, Q_T^{2;\beta} P^\gamma + [P^\beta]_-^\dagger\, Q_T^{2;\gamma} P^\alpha + [P^\gamma]_-^\dagger\, Q_T^{2;\alpha} P^\beta \\
&+[P^\alpha]_-^\dagger\, Q_T^{2;0} P^{\beta\gamma} + [P^\beta]_-^\dagger\, Q_T^{2;0} P^{\gamma\alpha} + [P^\gamma]_-^\dagger\, Q_T^{2;0} P^{\alpha\beta} \\
&+[P^\alpha]_-^\dagger\, Q_T^{3;0} P^\beta P^\gamma \\
&= [P^\alpha]_-^\dagger\, Q_T^{2;\beta} P^\gamma + [P^\beta]_-^\dagger\, Q_T^{2;\gamma} P^\alpha + [P^\gamma]_-^\dagger\, Q_T^{2;\alpha} P^\beta, \qquad (5.190)
\end{aligned}
$$

where, in the intermediate step, we have made use of the fact that (i) the terms in the two first lines are zero due to the linear dependence of $\hat{V}(t)$ in the field amplitudes, (ii) pairwise sums of terms in lines three and six are zero for the same reason as discussed after Eq. (5.171), (iii) the term in line four is zero due to the variational condition, and (iv) the term in line seven is zero in exact-state theory due to a vanishing expectation value of a commutator involving any set of three state-transfer operators. We note that the final expression for third-order properties will therefore involve only first-order responses in the wave function parameters to the external field, in agreement with Wigner's $(2n+1)$-rule for energy derivatives in variational wave-function theory.

The determination of an explicit expression for the matrix $Q_T^{2;\beta}$ closely follows the derivation of the contribution from $\hat{H}_0$ to $Q_T^{2;0}$, replacing $\hat{H}_0$ by $\partial\hat{Q}/\partial F_\beta^{\omega_2} = \hat{V}_\beta^{\omega_2}\exp(-i\omega_2 t)$. Organizing the matrix elements in accordance with Eq. (5.163), we get

$$
Q_T^{2;\beta} = \begin{pmatrix} \langle n|\hat{V}_\beta^{\omega_2}|m\rangle & 0 \\ 0 & \langle m|\overline{\hat{V}_\beta^{\omega_2}}|n\rangle \end{pmatrix} \delta_{\omega_1+\omega_2+\omega_3}. \qquad (5.191)
$$

This block-diagonal matrix is the equivalent of the Hessian for the property operator $\hat{V}_\beta^{\omega_2}$. For the first term in the final formula for the quadratic response function in Eq. (5.190), we get an explicit expression for the second-order response function that is

$$[P^\alpha]_-^\dagger \, Q_T^{2;\beta} P^\gamma = \frac{1}{\hbar^2}\delta_{\omega_1+\omega_2+\omega_3}$$
$$\times\begin{pmatrix} i\dfrac{\langle 0|\hat{V}_\alpha^{\omega_1}|n\rangle}{\omega_{n0}+\omega_1} & -i\dfrac{\langle n|\hat{V}_\alpha^{\omega_1}|0\rangle}{\omega_{n0}-\omega_1}\end{pmatrix}\begin{pmatrix}\langle n|\overline{\hat{V}_\beta^{\omega_2}}|m\rangle & 0 \\ 0 & \langle m|\overline{\hat{V}_\beta^{\omega_2}}|n\rangle\end{pmatrix}\begin{pmatrix} -i\dfrac{\langle m|\hat{V}_\gamma^{\omega_3}|0\rangle}{\omega_{m0}-\omega_3} \\ i\dfrac{\langle 0|\hat{V}_\gamma^{\omega_3}|m\rangle}{\omega_{m0}+\omega_3}\end{pmatrix}$$
$$= \frac{1}{\hbar^2}\sum_{n,m>0}\left[\frac{\langle 0|\hat{V}_\alpha^{\omega_1}|n\rangle\langle n|\overline{\hat{V}_\beta^{\omega_2}}|m\rangle\langle m|\hat{V}_\gamma^{\omega_3}|0\rangle}{(\omega_{n0}+\omega_1)(\omega_{m0}-\omega_3)} + \frac{\langle 0|\hat{V}_\alpha^{\omega_3}|m\rangle\langle m|\overline{\hat{V}_\beta^{\omega_2}}|n\rangle\langle n|\hat{V}_\gamma^{\omega_1}|0\rangle}{(\omega_{m0}+\omega_3)(\omega_{n0}-\omega_1)}\right]$$
$$\times\,\delta_{\omega_1+\omega_2+\omega_3}. \tag{5.192}$$

The two remaining terms in Eq. (5.190) are given by permutations of this first one, and by collecting terms it is clear that we obtain an overall symmetric expression for the quadratic response function in the quasi-energy formalism

$$\langle\langle \hat{V}_\alpha^{\omega_1}; \hat{V}_\beta^{\omega_2}, \hat{V}_\gamma^{\omega_3}\rangle\rangle = \frac{1}{\hbar^2}\sum P_{1,2,3}\sum_{n,m>0}\left[\frac{\langle 0|\hat{V}_\alpha^{\omega_1}|n\rangle\langle n|\overline{\hat{V}_\beta^{\omega_2}}|m\rangle\langle m|\hat{V}_\gamma^{\omega_3}|0\rangle}{(\omega_{n0}+\omega_1)(\omega_{m0}-\omega_3)}\right]\delta_{\omega_1+\omega_2+\omega_3}. \tag{5.193}$$

We can compare this result to that in Eq. (5.189). In the limit $\epsilon \to 0$, the two expressions are seen to be identical after accounting for the change made in the notation of operators and optical frequencies.

5.5.4 Fourth-Order Properties

We once more return to the perturbation expansion of the molecular polarization in Eq. (5.149) and consider the third-order responses from which we will identify a formula for the second-order nonlinear response function in analogy with what we have achieved for the linear and the first-order nonlinear response functions. The first-, second-, and third-order corrections to the wave function are given by Eqs. (5.112), (5.113), and (5.114), respectively, in combination with Eq. (5.111). The four terms that contribute to the third-order polarization are as follows:

$$\langle\psi^{(0)}|\hat{\Omega}|\psi^{(3)}\rangle = -\langle 0|e^{iE_0t/\hbar}\hat{\Omega}\sum_n \frac{1}{\hbar^3}\sum_{\omega_1\omega_2\omega_3}\sum_{mp}$$
$$\times\frac{\langle n|\hat{V}_\beta^{\omega_1}|m\rangle\langle m|\hat{V}_\gamma^{\omega_2}|p\rangle\langle p|\hat{V}_\delta^{\omega_3}|0\rangle F_\beta^{\omega_1}F_\gamma^{\omega_2}F_\delta^{\omega_3}}{(\omega_{n0}-\omega_1-\omega_2-\omega_3-i3\epsilon)(\omega_{m0}-\omega_2-\omega_3-i2\epsilon)(\omega_{p0}-\omega_3-i\epsilon)}$$
$$\times\, e^{i(\omega_{n0}-\omega_1-\omega_2-\omega_3)t}e^{3\epsilon t}e^{-iE_nt/\hbar}|n\rangle$$
$$= -\sum_{\omega_1\omega_2\omega_3}\frac{1}{\hbar^3}\sum_{nmp}$$
$$\times\frac{\langle 0|\hat{\Omega}|n\rangle\langle n|\hat{V}_\beta^{\omega_1}|m\rangle\langle m|\hat{V}_\gamma^{\omega_2}|p\rangle\langle p|\hat{V}_\delta^{\omega_3}|0\rangle F_\beta^{\omega_1}F_\gamma^{\omega_2}F_\delta^{\omega_3}}{(\omega_{n0}-\omega_1-\omega_2-\omega_3-i3\epsilon)(\omega_{m0}-\omega_2-\omega_3-i2\epsilon)(\omega_{p0}-\omega_3-i\epsilon)}$$
$$\times\, e^{-i(\omega_1+\omega_2+\omega_3)t}e^{3\epsilon t}, \tag{5.194}$$

$$\langle\psi^{(1)}|\hat{\Omega}|\psi^{(2)}\rangle = -\sum_n \frac{1}{\hbar}\sum_{\omega_1}\frac{\langle 0|\hat{V}_\beta^{\omega_1}|n\rangle[F_\beta^{\omega_1}]^*}{\omega_{n0}-\omega_1+i\epsilon}e^{-i(\omega_{n0}-\omega_1)t}e^{\epsilon t}e^{iE_nt/\hbar}\langle n|\hat{\Omega}$$

$$\times\sum_m \frac{1}{\hbar^2}\sum_{\omega_2\omega_3}\sum_p \frac{\langle m|\hat{V}_\gamma^{\omega_2}|p\rangle\langle p|\hat{V}_\delta^{\omega_3}|0\rangle F_\gamma^{\omega_2}F_\delta^{\omega_3}}{(\omega_{m0}-\omega_2-\omega_3-i2\epsilon)(\omega_{p0}-\omega_3-i\epsilon)}$$

$$\times e^{i(\omega_{m0}-\omega_2-\omega_3)t}e^{2\epsilon t}e^{-iE_mt/\hbar}|m\rangle$$

$$= -\sum_{\omega_1\omega_2\omega_3}\frac{1}{\hbar^3}\sum_{nmp}\frac{\langle 0|\hat{V}_\beta^{\omega_1}|n\rangle\langle n|\hat{\Omega}|m\rangle\langle m|\hat{V}_\gamma^{\omega_2}|p\rangle\langle p|\hat{V}_\delta^{\omega_3}|0\rangle F_\beta^{\omega_1}F_\gamma^{\omega_2}F_\delta^{\omega_3}}{(\omega_{n0}+\omega_1+i\epsilon)(\omega_{m0}-\omega_2-\omega_3-i2\epsilon)(\omega_{p0}-\omega_3-i\epsilon)}$$

$$\times e^{-i(\omega_1+\omega_2+\omega_3)t}e^{3\epsilon t}, \tag{5.195}$$

$$\langle\psi^{(2)}|\hat{\Omega}|\psi^{(1)}\rangle = -\sum_n \frac{1}{\hbar^2}\sum_{\omega_1\omega_2}\sum_m \frac{\langle m|\hat{V}_\beta^{\omega_1}|n\rangle\langle 0|\hat{V}_\gamma^{\omega_2}|m\rangle[F_\beta^{\omega_1}]^*[F_\gamma^{\omega_2}]^*}{(\omega_{n0}-\omega_1-\omega_2+i2\epsilon)(\omega_{m0}-\omega_2+i\epsilon)}$$

$$\times e^{-i(\omega_{n0}-\omega_1-\omega_2)t}e^{2\epsilon t}e^{iE_nt/\hbar}\langle n|\hat{\Omega}$$

$$\times\sum_p \frac{1}{\hbar}\sum_{\omega_3}\frac{\langle p|\hat{V}_\delta^{\omega_3}|0\rangle F_\delta^{\omega_3}}{\omega_{p0}-\omega_3-i\epsilon}e^{i(\omega_{p0}-\omega_3)t}e^{\epsilon t}e^{-iE_pt/\hbar}|p\rangle$$

$$= -\sum_{\omega_1\omega_2\omega_3}\frac{1}{\hbar^3}\sum_{nmp}\frac{\langle 0|\hat{V}_\gamma^{\omega_2}|m\rangle\langle m|\hat{V}_\beta^{\omega_1}|n\rangle\langle n|\hat{\Omega}|p\rangle\langle p|\hat{V}_\delta^{\omega_3}|0\rangle F_\beta^{\omega_1}F_\gamma^{\omega_2}F_\delta^{\omega_3}}{(\omega_{n0}+\omega_1+\omega_2+i2\epsilon)(\omega_{m0}+\omega_2+i\epsilon)(\omega_{p0}-\omega_3-i\epsilon)}$$

$$\times e^{-i(\omega_1+\omega_2+\omega_3)t}e^{3\epsilon t}, \tag{5.196}$$

$$\langle\psi^{(3)}|\hat{\Omega}|\psi^{(0)}\rangle = -\sum_n \frac{1}{\hbar^3}\sum_{\omega_1\omega_2\omega_3}\sum_{mp}$$

$$\times\frac{\langle m|\hat{V}_\beta^{\omega_1}|n\rangle\langle p|\hat{V}_\gamma^{\omega_2}|m\rangle\langle 0|\hat{V}_\delta^{\omega_3}|p\rangle[F_\beta^{\omega_1}]^*[F_\gamma^{\omega_2}]^*[F_\delta^{\omega_3}]^*}{(\omega_{n0}-\omega_1-\omega_2-\omega_3+i3\epsilon)(\omega_{m0}-\omega_2-\omega_3+i2\epsilon)(\omega_{p0}-\omega_3+i\epsilon)}$$

$$\times e^{-i(\omega_{n0}-\omega_1-\omega_2-\omega_3)t}e^{3\epsilon t}e^{iE_nt/\hbar}\langle n|\hat{\Omega}e^{-iE_0t/\hbar}|0\rangle$$

$$= -\sum_{\omega_1\omega_2\omega_3}\frac{1}{\hbar^3}\sum_{nmp}$$

$$\times\frac{\langle 0|\hat{V}_\delta^{\omega_3}|p\rangle\langle p|\hat{V}_\gamma^{\omega_2}|m\rangle\langle m|\hat{V}_\beta^{\omega_1}|n\rangle\langle n|\hat{\Omega}|0\rangle F_\beta^{\omega_1}F_\gamma^{\omega_2}F_\delta^{\omega_3}}{(\omega_{n0}+\omega_1+\omega_2+\omega_3+i3\epsilon)(\omega_{m0}+\omega_2+\omega_3+i2\epsilon)(\omega_{p0}+\omega_3+i\epsilon)}$$

$$\times e^{-i(\omega_1+\omega_2+\omega_3)t}e^{3\epsilon t}. \tag{5.197}$$

We have made use of the fact that $[F^{\omega}]^* = F^{-\omega}$ and that ω_1, ω_2, and ω_3 are dummy summation indices that run over both positive and negative frequencies. None of the Eqs. (5.195)–(5.197) are symmetric in the tensor indices β, γ, and δ. As pointed out in connection with Eq. (5.62), we normally choose our hyperpolarizability tensors to possess intrinsic symmetry, and it is clear that we can accomplish this without changing the polarization of the molecule by taking the average of the six terms generated by permuting pairs of the dummy indices (β, ω_1), (γ, ω_2), and (δ, ω_3); we denote this operation with the symbol $1/6\sum\mathcal{P}_{1,2,3}$, where the factor of one sixth is required to maintain the same value of the polarization. The third-order polarization in Eq. (5.149) can then be written as

$$
\begin{aligned}
\langle\hat{\Omega}\rangle^{(3)} &= \langle\psi^{(0)}|\hat{\Omega}|\psi^{(3)}\rangle + \langle\psi^{(1)}|\hat{\Omega}|\psi^{(2)}\rangle + \langle\psi^{(2)}|\hat{\Omega}|\psi^{(1)}\rangle + \langle\psi^{(3)}|\hat{\Omega}|\psi^{(0)}\rangle \\
&= \frac{1}{6}\sum_{\omega_1\omega_2\omega_3}\frac{1}{\hbar^3}\sum\mathcal{P}_{1,2,3}\sum_{nmp} \\
&\times\left[\frac{\langle 0|\hat{\Omega}|n\rangle\langle n|\hat{V}_\beta^{\omega_1}|m\rangle\langle m|\hat{V}_\gamma^{\omega_2}|p\rangle\langle p|\hat{V}_\delta^{\omega_3}|0\rangle}{(\omega_{n0}-\omega_1-\omega_2-\omega_3-i3\epsilon)(\omega_{m0}-\omega_2-\omega_3-i2\epsilon)(\omega_{p0}-\omega_3-i\epsilon)}\right. \\
&+\frac{\langle 0|\hat{V}_\beta^{\omega_1}|n\rangle\langle n|\hat{\Omega}|m\rangle\langle m|\hat{V}_\gamma^{\omega_2}|p\rangle\langle p|\hat{V}_\delta^{\omega_3}|0\rangle}{(\omega_{n0}+\omega_1+i\epsilon)(\omega_{m0}-\omega_2-\omega_3-i2\epsilon)(\omega_{p0}-\omega_3-i\epsilon)} \\
&+\frac{\langle 0|\hat{V}_\gamma^{\omega_2}|m\rangle\langle m|\hat{V}_\beta^{\omega_1}|n\rangle\langle n|\hat{\Omega}|p\rangle\langle p|\hat{V}_\delta^{\omega_3}|0\rangle}{(\omega_{n0}+\omega_1+\omega_2+i2\epsilon)(\omega_{m0}+\omega_2+i\epsilon)(\omega_{p0}-\omega_3-i\epsilon)} \\
&\left.+\frac{\langle 0|\hat{V}_\delta^{\omega_3}|p\rangle\langle p|\hat{V}_\gamma^{\omega_2}|m\rangle\langle m|\hat{V}_\beta^{\omega_1}|n\rangle\langle n|\hat{\Omega}|0\rangle}{(\omega_{n0}+\omega_1+\omega_2+\omega_3+i3\epsilon)(\omega_{m0}+\omega_2+\omega_3+i2\epsilon)(\omega_{p0}+\omega_3+i\epsilon)}\right] \\
&\times F_\beta^{\omega_1}F_\gamma^{\omega_2}F_\delta^{\omega_3}e^{-i(\omega_1+\omega_2+\omega_3)t}e^{3\epsilon t},
\end{aligned}
\tag{5.198}
$$

and by comparing to Eq. (5.150), we identify the expression for the third-order response function as

$$
\begin{aligned}
\langle\langle\hat{\Omega};\hat{V}_\beta^{\omega_1},\hat{V}_\gamma^{\omega_2},\hat{V}_\delta^{\omega_3}\rangle\rangle &= -\frac{1}{\hbar^3}\sum\mathcal{P}_{1,2,3} \\
&\times\sum_{nmp}\left[\frac{\langle 0|\hat{\Omega}|n\rangle\langle n|\hat{V}_\beta^{\omega_1}|m\rangle\langle m|\hat{V}_\gamma^{\omega_2}|p\rangle\langle p|\hat{V}_\delta^{\omega_3}|0\rangle}{(\omega_{n0}-\omega_\sigma-i3\epsilon)(\omega_{m0}-\omega_2-\omega_3-i2\epsilon)(\omega_{p0}-\omega_3-i\epsilon)}\right. \\
&+\frac{\langle 0|\hat{V}_\beta^{\omega_1}|n\rangle\langle n|\hat{\Omega}|m\rangle\langle m|\hat{V}_\gamma^{\omega_2}|p\rangle\langle p|\hat{V}_\delta^{\omega_3}|0\rangle}{(\omega_{n0}+\omega_1+i\epsilon)(\omega_{m0}-\omega_2-\omega_3-i2\epsilon)(\omega_{p0}-\omega_3-i\epsilon)} \\
&+\frac{\langle 0|\hat{V}_\gamma^{\omega_2}|m\rangle\langle m|\hat{V}_\beta^{\omega_1}|n\rangle\langle n|\hat{\Omega}|p\rangle\langle p|\hat{V}_\delta^{\omega_3}|0\rangle}{(\omega_{n0}+\omega_1+\omega_2+i2\epsilon)(\omega_{m0}+\omega_2+i\epsilon)(\omega_{p0}-\omega_3-i\epsilon)} \\
&\left.+\frac{\langle 0|\hat{V}_\delta^{\omega_3}|p\rangle\langle p|\hat{V}_\gamma^{\omega_2}|m\rangle\langle m|\hat{V}_\beta^{\omega_1}|n\rangle\langle n|\hat{\Omega}|0\rangle}{(\omega_{n0}+\omega_\sigma+i3\epsilon)(\omega_{m0}+\omega_2+\omega_3+i2\epsilon)(\omega_{p0}+\omega_3+i\epsilon)}\right].
\end{aligned}
\tag{5.199}
$$

An element of the second-order hyperpolarizability tensor $\gamma_{\alpha\beta\gamma\delta}(-\omega_\sigma;\omega_1,\omega_2,\omega_3)$ is obtained from the general response function in Eq. (5.199) by the substitutions $\hat{\Omega}=\hat{\mu}_\alpha$, $\hat{V}_\beta^\omega=-\hat{\mu}_\beta$, $\hat{V}_\gamma^\omega=-\hat{\mu}_\gamma$, and $\hat{V}_\delta^\omega=-\hat{\mu}_\delta$. As already discussed for the lower-order response functions, the use of an expression such as Eq. (5.199) is valid only under nonresonant conditions—in this case, we have the potential for one-, two-, and three-photon resonances in the expression, and only if none of these occur will the expression remain valid. If we let ϵ tend to be zero in Eq. (5.199), the expression for the second-order hyperpolarizability can be written more compactly as

$$
\begin{aligned}
\gamma_{\alpha\beta\gamma\delta}(-\omega_\sigma;\omega_1,\omega_2,\omega_3) &= \frac{1}{\hbar^3}\sum\mathcal{P}_{-\sigma,1,2,3} \\
&\times\sum_{nmp}\frac{\langle 0|\hat{\mu}_\alpha|n\rangle\langle n|\hat{\mu}_\beta|m\rangle\langle m|\hat{\mu}_\gamma|p\rangle\langle p|\hat{\mu}_\delta|0\rangle}{(\omega_{n0}-\omega_\sigma)(\omega_{m0}-\omega_2-\omega_3)(\omega_{p0}-\omega_3)},
\end{aligned}
\tag{5.200}
$$

where $\sum\mathcal{P}_{-\sigma,1,2,3}$ denotes the sum of the 24 terms one gets by permuting pairs of $(\alpha,-\omega_\sigma)$, (β,ω_1), (γ,ω_2), and (δ,ω_3). A comparison of Eq. (5.199) with $\epsilon=0$ and Eq. (5.200) verifies this result. The second-order hyperpolarizability tensor possesses full permutation symmetry in the nonresonant region.

The secular divergences in Eq. (5.200) can be removed in a similar manner as we did for the lower-order properties leading to the expression

$$\gamma_{\alpha\beta\gamma\delta}(-\omega_\sigma;\omega_1,\omega_2,\omega_3)=\frac{1}{\hbar^3}\sum\mathcal{P}_{-\sigma,1,2,3}$$
$$\times\left[\sum_{nmp}{}'\frac{\langle 0|\hat{\mu}_\alpha|n\rangle\langle n|\overline{\hat{\mu}_\beta}|m\rangle\langle m|\overline{\hat{\mu}_\gamma}|p\rangle\langle p|\hat{\mu}_\delta|0\rangle}{(\omega_{n0}-\omega_\sigma)(\omega_{m0}-\omega_2-\omega_3)(\omega_{p0}-\omega_3)}\right.$$
$$\left.-\sum_{nm}{}'\frac{\langle 0|\hat{\mu}_\alpha|n\rangle\langle n|\hat{\mu}_\beta|0\rangle\langle 0|\hat{\mu}_\gamma|m\rangle\langle m|\hat{\mu}_\delta|0\rangle}{(\omega_{n0}-\omega_\sigma)(\omega_{m0}-\omega_3)(\omega_{m0}+\omega_2)}\right]. \tag{5.201}$$

This quantum-mechanical expression is applicable when all one-, two-, and three-photon combinations of the fields are nonresonant, as well as to the case of static fields.

Let us also for the third-order response function compare the above derivation to that made using the phase-isolated wave function and rotation amplitudes. The third-order corrections to an expectation value are

$$\begin{aligned}\langle\hat{\Omega}\rangle^{(3)}&=\langle\overline{\psi}^{(0)}|\hat{\Omega}|\overline{\psi}^{(3)}\rangle+\langle\overline{\psi}^{(1)}|\hat{\Omega}|\overline{\psi}^{(2)}\rangle+\langle\overline{\psi}^{(2)}|\hat{\Omega}|\overline{\psi}^{(1)}\rangle+\langle\overline{\psi}^{(3)}|\hat{\Omega}|\overline{\psi}^{(0)}\rangle\\&=\langle 0|(i[\hat{P}^{(3)},\hat{\Omega}]-\frac{1}{2}\{[\hat{P}^{(2)},[\hat{P}^{(1)},\hat{\Omega}]]+[\hat{P}^{(1)},[\hat{P}^{(2)},\hat{\Omega}]]\}-\frac{i}{6}[\hat{P}^{(1)},[\hat{P}^{(1)},[\hat{P}^{(1)},\hat{\Omega}]]])|0\rangle\\&=i\sum_{n>0}(D_n^{*(3)}\langle n|\hat{\Omega}|0\rangle-D_n^{(3)}\langle 0|\hat{\Omega}|n\rangle)+\sum_{n,m>0}(P_n^{*(1)}P_m^{(2)}+P_n^{*(2)}P_m^{(1)})\langle n|\overline{\hat{\Omega}}|m\rangle.\end{aligned} \tag{5.202}$$

If we insert the amplitudes given in Eqs. (5.137), (5.143), and (5.148), we can identify the cubic response function

$$\langle\langle\hat{\Omega};\hat{V}_\beta^{\omega_1},\hat{V}_\gamma^{\omega_2},\hat{V}_\delta^{\omega_3}\rangle\rangle=-\frac{1}{\hbar^3}\sum\mathcal{P}_{1,2,3}\Bigg\{$$
$$\sum_{n,m,p>0}\left[\frac{\langle 0|\hat{\Omega}|n\rangle\langle n|\overline{\hat{V}_\beta^{\omega_1}}|m\rangle\langle m|\overline{\hat{V}_\gamma^{\omega_2}}|p\rangle\langle p|\hat{V}_\delta^{\omega_3}|0\rangle}{[\omega_{n0}-\omega_\sigma-i3\epsilon][\omega_{m0}-(\omega_2+\omega_3)-2i\epsilon][\omega_{p0}-\omega_3-i\epsilon]}\right.$$
$$+\frac{\langle 0|\hat{V}_\beta^{\omega_1}|n\rangle\langle n|\overline{\hat{\Omega}}|m\rangle\langle m|\overline{\hat{V}_\gamma^{\omega_2}}|p\rangle\langle p|\hat{V}_\delta^{\omega_3}|0\rangle}{[\omega_{n0}+\omega_1+i\epsilon][\omega_{m0}-(\omega_2+\omega_3)-2i\epsilon][\omega_{p0}-\omega_3-i\epsilon]}$$
$$+\frac{\langle 0|\hat{V}_\delta^{\omega_3}|n\rangle\langle n|\overline{\hat{V}_\gamma^{\omega_2}}|m\rangle\langle m|\overline{\hat{\Omega}}|p\rangle\langle p|\hat{V}_\beta^{\omega_1}|0\rangle}{[\omega_{n0}+\omega_3+i\epsilon][\omega_{m0}+(\omega_2+\omega_3)+2i\epsilon][\omega_{p0}-\omega_1-i\epsilon]}$$
$$\left.+\frac{\langle 0|\hat{V}_\delta^{\omega_3}|n\rangle\langle n|\overline{\hat{V}_\gamma^{\omega_2}}|m\rangle\langle m|\overline{\hat{V}_\beta^{\omega_1}}|p\rangle\langle p|\hat{\Omega}|0\rangle}{[\omega_{n0}+\omega_3+i\epsilon][\omega_{m0}+(\omega_2+\omega_3)+2i\epsilon][\omega_{p0}+\omega_\sigma+i3\epsilon]}\right]$$
$$-\sum_{n,m>0}\left[\frac{\langle 0|\hat{\Omega}|n\rangle\langle n|\hat{V}_\beta^{\omega_1}|0\rangle\langle 0|\hat{V}_\gamma^{\omega_2}|m\rangle\langle m|\hat{V}_\delta^{\omega_3}|0\rangle}{[\omega_{n0}-\omega_\sigma-i3\epsilon][\omega_{m0}+\omega_2+i\epsilon][\omega_{m0}-\omega_3-i\epsilon]}\right.$$
$$+\frac{\langle 0|\hat{V}_\beta^{\omega_1}|n\rangle\langle n|\hat{\Omega}|0\rangle\langle 0|\hat{V}_\delta^{\omega_3}|m\rangle\langle m|\hat{V}_\gamma^{\omega_2}|0\rangle}{[\omega_{n0}+\omega_\sigma+i3\epsilon][\omega_{m0}+\omega_3+i\epsilon][\omega_{m0}-\omega_2-i\epsilon]}$$
$$+\frac{\langle 0|\hat{V}_\gamma^{\omega_2}|n\rangle\langle n|\hat{V}_\delta^{\omega_3}|0\rangle\langle 0|\hat{\Omega}|m\rangle\langle m|\hat{V}_\beta^{\omega_1}|0\rangle}{[\omega_{n0}+\omega_2+i\epsilon][\omega_{m0}-\omega_\sigma-i3\epsilon][\omega_{m0}-\omega_1-i\epsilon]}$$
$$\left.\left.+\frac{\langle 0|\hat{V}_\delta^{\omega_3}|n\rangle\langle n|\hat{V}_\gamma^{\omega_2}|0\rangle\langle 0|\hat{V}_\beta^{\omega_1}|m\rangle\langle m|\hat{\Omega}|0\rangle}{[\omega_{n0}-\omega_2-i\epsilon][\omega_{m0}+\omega_1+i\epsilon][\omega_{m0}+\omega_\sigma+i3\epsilon]}\right]\right\}. \tag{5.203}$$

Table 5.2 Permutations of terms A1 and B1 in Eq. (5.203) needed to obtain terms A2–A4, B2′, B3, and B4′.

Term	Permutations		
A2	$-\sigma \leftrightarrow 1$		
A3	$-\sigma \leftrightarrow 2$	$1 \leftrightarrow 3$	$2 \leftrightarrow 3$
A4	$-\sigma \leftrightarrow 3$	$1 \leftrightarrow 2$	
B2′	$-\sigma \leftrightarrow 1$	$2 \leftrightarrow 3$	
B3	$-\sigma \leftrightarrow 2$	$1 \leftrightarrow 3$	
B4′	$-\sigma \leftrightarrow 3$	$1 \leftrightarrow 2$	

Permutations are to be performed sequentially in the order listed from left to right.

By using the phase-isolated wave function, we again arrive directly at an expression that is free of secular divergences and which in the limit of $\epsilon \to 0$ can be compared to the overall symmetric expression in Eq. (5.201). However, it is not obvious that the cubic response function as written in Eq. (5.203) is identical to the overall symmetric expression in Eq. (5.201). Let us denote the terms in the first and second summation of Eq. (5.203) by A1–A4 and B1–B4, respectively. Since we have ordered terms so that A1 and B1 refer to the two terms in Eq. (5.201), we need to show that the remaining six terms in Eq. (5.203) are given by permutations of the first ones. For terms A2–A4 and B3, it is straightforward to identify the permutations needed, and we also list them in Table 5.2.

Terms B2 and B4 are considerably more complicated, since it is for instance clear that B2 is not simply given by a $-\sigma \leftrightarrow 1$ permutation of B1. Instead, we are forced to consider a few algebraic manipulations of the sum of terms B2 and B4. We first note that

$$\frac{1}{(\omega_{n0}+\omega_3)(\omega_{n0}-\omega_2)} = \frac{1}{\omega_2+\omega_3}\left[\frac{1}{\omega_{n0}-\omega_2} - \frac{1}{\omega_{n0}+\omega_3}\right],$$

$$\frac{1}{(\omega_{n0}+\omega_1)(\omega_{n0}+\omega_\sigma)} = \frac{1}{\omega_2+\omega_3}\left[\frac{1}{\omega_{n0}+\omega_1} - \frac{1}{\omega_{n0}+\omega_\sigma}\right],$$

and we can thus, with an interchange of the dummy summation indices n and m in B4, write the sum of B2 and B4 as

$$\begin{aligned}
\text{B2}+\text{B4} &= \langle 0|\hat{V}_\beta^{\omega_1}|n\rangle\langle n|\hat{\Omega}|0\rangle\langle 0|\hat{V}_\delta^{\omega_3}|m\rangle\langle m|\hat{V}_\gamma^{\omega_2}|0\rangle \\
&\quad\times\frac{1}{\omega_2+\omega_3}\left[\frac{1}{\omega_{n0}+\omega_\sigma}\left(\frac{1}{\omega_{m0}-\omega_2}-\frac{1}{\omega_{m0}+\omega_3}\right)\right. \\
&\quad\left.+\frac{1}{\omega_{m0}-\omega_2}\left(\frac{1}{\omega_{n0}+\omega_1}-\frac{1}{\omega_{n0}+\omega_\sigma}\right)\right] \\
&= \langle 0|\hat{V}_\beta^{\omega_1}|n\rangle\langle n|\hat{\Omega}|0\rangle\langle 0|\hat{V}_\delta^{\omega_3}|m\rangle\langle m|\hat{V}_\gamma^{\omega_2}|0\rangle \\
&\quad\times\frac{1}{\omega_2+\omega_3}\left[\frac{1}{\omega_{n0}+\omega_1}\left(\frac{1}{\omega_{m0}-\omega_2}-\frac{1}{\omega_{m0}+\omega_3}\right)\right. \\
&\quad\left.+\frac{1}{\omega_{m0}+\omega_3}\left(\frac{1}{\omega_{n0}+\omega_1}-\frac{1}{\omega_{n0}+\omega_\sigma}\right)\right] \\
&= \frac{\langle 0|\hat{V}_\beta^{\omega_1}|n\rangle\langle n|\hat{\Omega}|0\rangle\langle 0|\hat{V}_\delta^{\omega_3}|m\rangle\langle m|\hat{V}_\gamma^{\omega_2}|0\rangle}{[\omega_{n0}+\omega_1][\omega_{m0}+\omega_3][\omega_{m0}-\omega_2]}
\end{aligned}$$

$$+\frac{\langle 0|\hat{V}_{\delta}^{\omega_3}|m\rangle\langle m|\hat{V}_{\gamma}^{\omega_2}|0\rangle\langle 0|\hat{V}_{\beta}^{\omega_1}|n\rangle\langle n|\hat{\Omega}|0\rangle}{[\omega_{m0}+\omega_3][\omega_{n0}+\omega_1][\omega_{n0}+\omega_\sigma]}$$
$$= \mathrm{B2}' + \mathrm{B4}'. \tag{5.204}$$

After this re-write, it is straightforward to identify the permutations of B1 needed to obtain B2′ as well as B4′ (see Table 5.2), and which thereby also completes the proof that in the limit of $\epsilon \to 0$, the cubic response function in Eq. (5.203) is overall symmetric and equal to that in Eq. (5.201)

5.5.5 Higher-Order Properties

It is clear that the route followed hitherto can be extended to arbitrary order in the perturbation, and thus provide the description of general multiphoton interactions. However, since very high light intensities are needed to observe these nonlinear responses, the predominant interest in optics are focused on processes incorporated in the linear polarizability and the first- and second-order hyperpolarizability. The expression for the general-order nonresonant response $X^n_{\alpha\beta\ldots}(-\omega_\sigma;\omega_1,\omega_2,\ldots,\omega_n)$ can be written as

$$X^n_{\alpha\beta\ldots}(-\omega_\sigma;\omega_1,\omega_2,\ldots,\omega_n) = \hbar^{-n}\sum \mathcal{P}_{\alpha,\beta,\ldots}\sum_{a_1}\sum_{a_2}\cdots\sum_{a_n}\langle 0|\hat{\mu}_\alpha|a_1\rangle\langle a_1|\hat{\mu}_\beta|a_2\rangle\cdots$$
$$\times[(\omega_{a_1}-\omega_\sigma)(\omega_{a_2}-\omega_\sigma+\omega_1)\cdots(\omega_{a_n}-\omega_n)]^{-1}. \tag{5.205}$$

In this very general formula, we are using a notation in which $X^1_{\alpha\beta} = \alpha_{\alpha\beta}$ (linear polarizability), $X^2_{\alpha\beta\gamma} = \beta_{\alpha\beta\gamma}$ (first-order hyperpolarizability), and $X^3_{\alpha\beta\gamma\delta} = \gamma_{\alpha\beta\gamma\delta}$ (second-order hyperpolarizability). It thus expresses the higher-order responses to the electric polarization in the presence of external electric fields, and can be generalized to arbitrary molecular properties.

5.6 Dispersion

The derived response functions that were discussed in the previous section provide the detailed dependence of the molecular properties on the excitation energies and transition moments of the system. They also reveal the dependence of the induced polarization on the frequencies of the external electromagnetic fields. We recall that the frequency dependence of a molecular property, or response function, is known as the *dispersion*, and in this section we will focus on the dispersions associated with different optical processes. The discussion will be limited to the nonresonant region where optical frequencies are lower than the first transition frequency of the system. If one restricts the calculation to electronic responses, the frequency limit is thus set by the first electronic excitation energy of the system, which in a typical molecular system lies somewhere in the visible or ultra-violet region of the electromagnetic spectrum.

The dispersion of the linear electric-dipole polarizability is given by the sum-over-states expression in Eq. (5.169). In the nonrelativistic limit, the unperturbed eigenfunctions of the molecular Hamiltonian can be chosen as real and the numerators of the two terms in Eq. (5.169) are then equal and can be separated. In the general case, the two numerators are connected via the operation of complex conjugation, but the diagonal components (with indices $\alpha = \beta$) are real and thus equal. In cases where the numerators of the two terms are equal, we can re-write the expression for the polarizability as

$$\alpha_{\alpha\beta}(-\omega;\omega) = \frac{1}{\hbar}\sum_n{}' \frac{2}{\omega_{n0}}\frac{\langle 0|\hat{\mu}_\alpha|n\rangle\langle n|\hat{\mu}_\beta|0\rangle}{1-(\omega/\omega_{n0})^2}. \tag{5.206}$$

Using the Taylor expansion

$$\frac{1}{1-x} = \sum_{k=0}^{\infty} x^k \quad \text{convergent for } |x| < 1, \tag{5.207}$$

we get

$$\alpha_{\alpha\beta}(-\omega;\omega) = \hbar^{-1} \sum_{k=0}^{\infty} \omega^{2k} S(-2k-2) \quad \text{for } |\omega/\omega_{on}| < 1, \tag{5.208}$$

where the Cauchy moments $S(-k)$ have been introduced

$$S(-k) = \sum_n{}' \frac{f_{n0}^{\alpha\beta}}{\omega_{n0}^k}; \quad f_{n0}^{\alpha\beta} = 2\omega_{n0}\langle 0|\mu_\alpha|n\rangle\langle n|\mu_\beta|0\rangle. \tag{5.209}$$

The Cauchy moments are independent of the optical frequency and thus constitute a set of frequency-independent molecular properties. From Eq. (5.208), we see that the expansion of the polarizability in the optical frequency contains only terms of even power, and since the Cauchy moments with $\alpha = \beta$ are positive, the dispersion of the diagonal components of the polarizability has to be positive.

The dispersions of the nonlinear polarizabilities are considerably more complicated due to the larger number of terms and multiphoton resonances as well as the larger number of components in the property tensors. The dispersions of the first- and second-order electric-dipole hyperpolarizabilities are given by the expressions in Eqs. (5.187) and (5.201), respectively. If we first consider a diagonal component of the β-tensor, it is clear that the frequency dependence rests in

$$\sum P_{-\sigma,1,2} \left[\frac{1}{(\omega_{n0}-\omega_\sigma)(\omega_{p0}-\omega_2)} \right] = \frac{1}{\omega_{n0}\omega_{p0}} \sum P_{-\sigma,1,2} \left[\sum_{k,l=0}^{\infty} \left(\frac{\omega_\sigma}{\omega_{n0}}\right)^k \left(\frac{\omega_2}{\omega_{p0}}\right)^l \right],$$

and for higher-order polarizabilities the corresponding expression is found by straightforward generalization. In the general case, ω_σ is the sum of n optical frequencies and we introduce the notation $\omega_{-\sigma} = -\omega_\sigma$ as well as $\sum P_{-\sigma,1,2,\ldots,n} = \sum_P$. If we intend to study the frequency dependence up to second order in the optical frequencies, we benefit from the relations

$$\sum_P 1 = (n+1)!, \tag{5.210}$$

$$\sum_P \omega_1 = n! \sum_{k=-\sigma}^{n} \omega_k = 0, \tag{5.211}$$

$$\sum_P \omega_1^2 = n! \sum_{k=-\sigma}^{n} \omega_k^2 = n!\ \omega_L^2, \tag{5.212}$$

$$\sum_P \omega_1\omega_2 = (n-1)! \sum_{k=-\sigma}^{n} \sum_{l\neq k} \omega_k\omega_l = -(n-1)!\ \omega_L^2, \tag{5.213}$$

where the sums over k and l run over $\omega_{-\sigma}$ in addition to the n optical frequencies, and Eq. (5.212) defines the auxiliary frequency variable ω_L^2. Given these results, we can, up to second order in the optical frequencies, write a diagonal component of the β-tensor as

$$\beta_{zzz}(-\omega_\sigma;\omega_1,\omega_2) = \beta_{zzz}(0;0,0) + A\,\omega_L^2, \tag{5.214}$$

where the coefficient A is defined as

$$A = \sum_{n,p}' \frac{\overline{T}_{np}^{zzz}}{\omega_{n0}\omega_{p0}} \left(\frac{2}{\omega_{n0}^2} + \frac{2}{\omega_{p0}^2} + \frac{1}{\omega_{n0}\omega_{p0}} \right), \tag{5.215}$$

with

$$\overline{T}_{np}^{\alpha\beta\gamma} = \langle 0|\hat{\mu}_\alpha|n\rangle\langle n|\overline{\hat{\mu}_\beta}|p\rangle\langle p|\hat{\mu}_\gamma|0\rangle. \tag{5.216}$$

The coefficient A of the hyperpolarizability is the counterpart of the Cauchy moment $S(-4)$ for the polarizability, and because the coefficient itself is frequency independent, one will find a universal dependence of β with respect to ω_L^2 *regardless* of the optical process at hand. This means that if one has available data for $\beta(-\omega;\omega,0)$, a fit can be made to Eq. (5.214) with $\omega_L^2 = 2\omega^2$ to obtain the coefficient A. One can then obtain the results for $\beta(-2\omega;\omega,\omega)$ using the same value for A but with $\omega_L^2 = 6\omega^2$. We illustrate this feature in Figure 5.13 by time-dependent Hartree–Fock calculations of β_{zzz} for hydrogen fluoride. We have not yet discussed the details of such approximate-state response theory calculations, but we may understand these calculations to represent the derived sum-over-states expressions in approximate-state theory with inclusion made only of electronic states (nuclei are kept frozen in the calculations). Figure 5.13(b) depicts $\beta_{zzz}(-2\omega;\omega,\omega)$ and $\beta_{zzz}(-\omega;\omega,0)$ plotted versus the optical frequency ω. The dispersion in the former optical process is stronger than that in the latter due to closer proximity to two-photon resonances in Eq. (5.187). When plotted against the auxiliary frequency variable

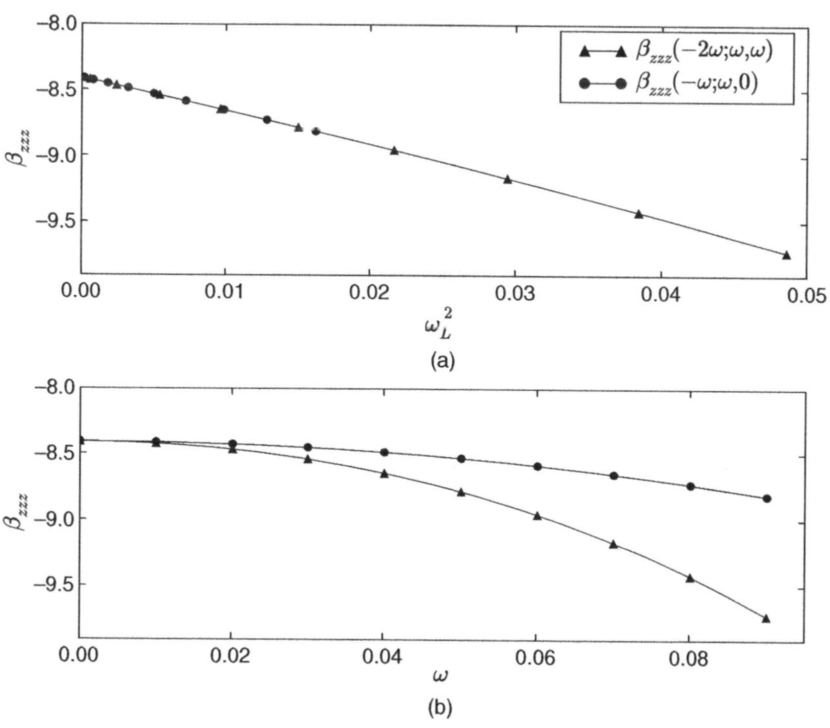

Figure 5.13 Electronic first-order hyperpolarizability β_{zzz} of hydrogen fluoride for the SHG and EOPE optical processes. Data refer to time-dependent Hartree–Fock calculations using Dunning's t-aug-cc-pVTZ basis set. The dipole moment is directed along the positive z-axis and the experimental bond length of 1.733 a.u. is employed. All quantities are given in a.u.

ω_L^2 in Figure 5.13(a), it is clear that the results are in agreement with the universal dispersion formula in Eq. (5.214)—the coefficient A is given by the slope of the line.

For nondiagonal components of the β-tensor, one cannot extract a common numerator from all six terms in Eq. (5.187) to isolate the frequency dependence in the way done for the diagonal components. However, in many cases, the interest is not in individual components of the property tensors but rather in averaged quantities. A relevant quantity for the first-order hyperpolarizability is

$$\overline{\beta}(-\omega_\sigma;\omega_1,\omega_2) = \frac{1}{5}\sum_{\alpha=x,y,z}(\beta_{z\alpha\alpha} + \beta_{\alpha z\alpha} + \beta_{\alpha\alpha z}), \tag{5.217}$$

where the z-axis is assumed to be aligned with the molecular dipole moment axis, and where the suppressed frequency arguments on the right-hand side of the equation are implicitly understood to be the same as those expressed in the left-hand side. For $\overline{\beta}$ one can, due to the averaging over tensor components, carry out the algebra in the same way as for the diagonal components and arrive at a universal dispersion expression

$$\overline{\beta}(-\omega_\sigma;\omega_1,\omega_2) = \overline{\beta}(0;0,0) + A\omega_L^2, \tag{5.218}$$

where the frequency-independent coefficient A in this case becomes

$$A = \frac{1}{5}\sum_{n,p}{}' \frac{\overline{T}_{np}^{z\alpha\alpha} + \overline{T}_{np}^{\alpha z\alpha} + \overline{T}_{np}^{\alpha\alpha z}}{\omega_{n0}\omega_{p0}}\left(\frac{2}{\omega_{n0}^2} + \frac{2}{\omega_{p0}^2} + \frac{1}{\omega_{n0}\omega_{p0}}\right). \tag{5.219}$$

We recall that there is an implied summation of α over the Cartesian indices x, y, and z in the above equation.

Let us now turn to the dispersion of the second-order hyperpolarizability. The expression for $\gamma(-\omega_\sigma;\omega_1,\omega_2,\omega_3)$ in Eq. (5.201) contains two terms. For the diagonal components of the tensor, the frequency dependencies of the two terms result in

$$\begin{aligned}&\sum \mathcal{P}_{-\sigma,1,2,3}\left[\frac{1}{(\omega_{n0}-\omega_\sigma)(\omega_{m0}-\omega_2-\omega_3)(\omega_{p0}-\omega_3)}\right]\\ &= \frac{1}{\omega_{n0}\omega_{m0}\omega_{p0}}\sum \mathcal{P}_{-\sigma,1,2,3}\left[\sum_{j,k,l=0}^{\infty}\left(\frac{\omega_\sigma}{\omega_{n0}}\right)^j\left(\frac{\omega_2+\omega_3}{\omega_{m0}}\right)^k\left(\frac{\omega_3}{\omega_{p0}}\right)^l\right],\end{aligned}$$

and

$$\begin{aligned}&\sum \mathcal{P}_{-\sigma,1,2,3}\left[\frac{1}{(\omega_{n0}-\omega_\sigma)(\omega_{m0}-\omega_3)(\omega_{m0}+\omega_2)}\right]\\ &= \frac{1}{\omega_{n0}\omega_{m0}\omega_{m0}}\sum \mathcal{P}_{-\sigma,1,2,3}\left[\sum_{j,k,l=0}^{\infty}\left(\frac{\omega_\sigma}{\omega_{n0}}\right)^j\left(\frac{\omega_3}{\omega_{m0}}\right)^k\left(\frac{-\omega_2}{\omega_{m0}}\right)^l\right],\end{aligned}$$

respectively. Using Eqs. (5.210)–(5.213), we get an expression for the dispersion of the second hyperpolarizability to second order in the optical frequencies

$$\gamma_{zzzz}(-\omega_\sigma;\omega_1,\omega_2,\omega_3) = \gamma_{zzzz}(0;0,0,0) + A\omega_L^2, \tag{5.220}$$

where the frequency-independent coefficient A is

$$A = \sum_{n,m,p}{}' \frac{\overline{T}_{nmp}^{zzzz}}{\omega_{n0}\omega_{m0}\omega_{p0}} \left(\frac{6}{\omega_{n0}^2} + \frac{8}{\omega_{m0}^2} + \frac{6}{\omega_{p0}^2} + \frac{4}{\omega_{n0}\omega_{m0}} + \frac{2}{\omega_{n0}\omega_{p0}} + \frac{4}{\omega_{m0}\omega_{p0}} \right)$$
$$- \sum_{n,m}{}' \frac{T_{nm}^{zzzz}}{\omega_{n0}\omega_{m0}^2} \left(\frac{6}{\omega_{n0}^2} + \frac{14}{\omega_{m0}^2} \right), \quad (5.221)$$

with

$$\overline{T}_{nmp}^{\alpha\beta\gamma\delta} = \langle 0|\hat{\mu}_\alpha|n\rangle\langle n|\overline{\hat{\mu}_\beta}|m\rangle\langle m|\overline{\hat{\mu}_\gamma}|p\rangle\langle p|\hat{\mu}_\delta|0\rangle, \quad (5.222)$$

$$T_{nm}^{\alpha\beta\gamma\delta} = \langle 0|\hat{\mu}_\alpha|n\rangle\langle n|\hat{\mu}_\beta|0\rangle\langle 0|\hat{\mu}_\gamma|m\rangle\langle m|\hat{\mu}_\delta|0\rangle. \quad (5.223)$$

For the second-order hyperpolarizability, an often considered averaged quantity is

$$\overline{\gamma}(-\omega_\sigma, \omega_1, \omega_2, \omega_3) = \frac{1}{15} \sum_{\alpha,\beta=x,y,z} (\gamma_{\alpha\alpha\beta\beta} + \gamma_{\alpha\beta\alpha\beta} + \gamma_{\alpha\beta\beta\alpha}), \quad (5.224)$$

where the suppressed frequency arguments on the right-hand side of the above equation are equal to those on the left-hand side. The universal dispersion formula for this averaged quantity is

$$\overline{\gamma}(-\omega_\sigma; \omega_1, \omega_2, \omega_3) = \overline{\gamma}(0; 0, 0, 0) + A\,\omega_L^2, \quad (5.225)$$

where A is given by Eq. (5.221) together with a substitution of the transition matrix elements according to

$$\overline{T}_{nmp}^{zzzz} \rightarrow \frac{1}{15} \sum_{\alpha,\beta=x,y,z} (\overline{T}_{nmp}^{\alpha\alpha\beta\beta} + \overline{T}_{nmp}^{\alpha\beta\alpha\beta} + \overline{T}_{nmp}^{\alpha\beta\beta\alpha}), \quad (5.226)$$

and likewise for T_{nm}^{zzzz}.

It should be emphasized that because the coefficient A is independent of the optical frequencies, the relations in Eqs. (5.220) and (5.225) remain unaltered for different optical processes. Given data for one optical process, we can thus predict the results for any other process. Let us illustrate this fact by calculations of a selection of second-order hyperpolarizabilities for hydrogen fluoride, employing the same computational parameters as used to generate Figure 5.13. In Figure 5.14(b), the results for γ_{zzzz} are plotted against the optical frequency ω, and the strongest dispersion is observed for the THG optical process due to the three-photon resonances in the sum-over-states expression for γ, Eq. (5.201). In Figure 5.14(a), we plot the same data but with respect to the auxiliary frequency parameter ω_L^2. For the THG, ESHG, IDRI, and EOKE optical processes (see Table 5.1), the frequency ω_L^2 is $12\omega^2$, $6\omega^2$, $4\omega^2$, and $2\omega^2$, respectively. It is clear that Eqs. (5.220) and (5.225) are exact only in the limit of small frequencies since higher-order terms are ignored, and in Figure 5.14(a) we see that a deviation from a linear relation becomes significant at about $\omega_L^2 = 0.04$ a.u.

5.7 Oscillator Strength and Sum Rules

Let us go back to the expression Eq. (5.169) for the elements of the polarizability tensor in nonresonant regions. The diagonal elements can be written in the more compact form

$$\alpha_{\alpha\alpha}(-\omega; \omega) = \frac{1}{\hbar} \sum_n{}' |\langle n|\hat{\mu}_\alpha|0\rangle|^2 \frac{2\omega_{n0}}{(\omega_{n0}^2 - \omega^2)}. \quad (5.227)$$

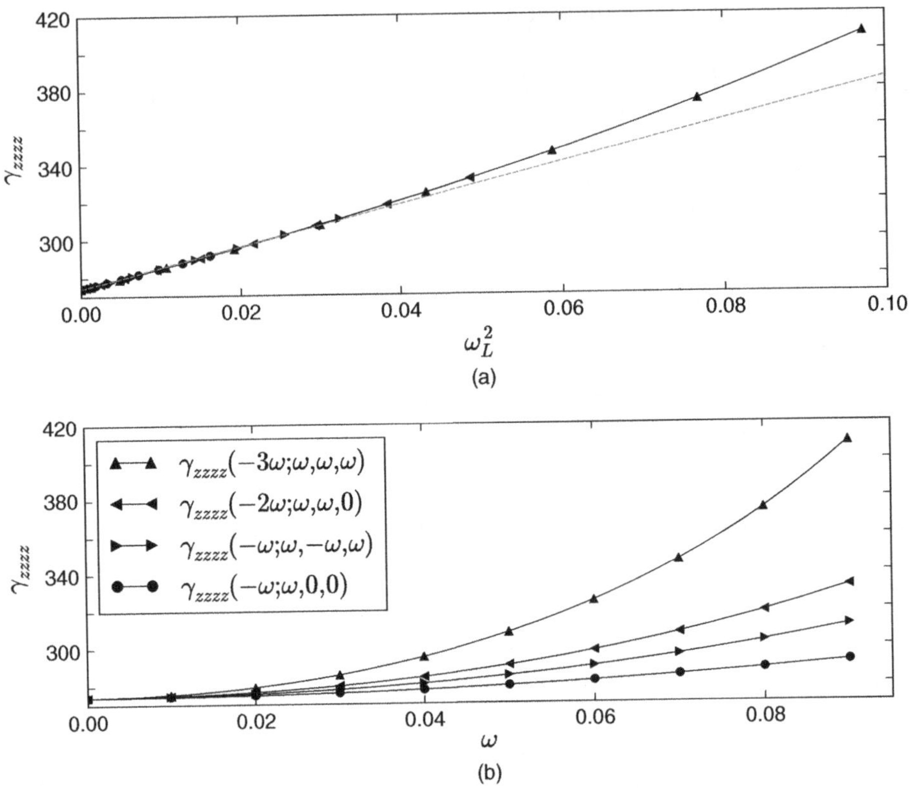

Figure 5.14 Electronic second-order hyperpolarizability γ_{zzzz} of hydrogen fluoride for the THG, ESHG, IDRI, and EOKE optical processes (see Table 5.1). Data refer to time-dependent Hartree–Fock calculations using Dunning's t-aug-cc-pVTZ basis set. The dipole moment is directed along the positive z-axis and the experimental bond length of 1.733 a.u. is employed. All quantities are given in a.u.

Comparing with the expression in Eq. (3.443) for the dipole polarizability in the Lorentz model, we can extract for each excitation $0 \rightarrow n$ a dimensionless quantity known as the oscillator strength

$$f_{n0}^{\alpha} = \frac{2m_e\omega_{n0}}{\hbar e^2}|\langle 0|\hat{\mu}_{\alpha}|n\rangle|^2. \tag{5.228}$$

In the case of freely rotating molecules, an experiment provides access only to the isotropic average of the polarizability

$$\overline{\alpha}(-\omega;\omega) = \frac{1}{3}(\alpha_{xx} + \alpha_{yy} + \alpha_{zz}), \tag{5.229}$$

from which we may extract the isotropic oscillator strength as

$$f_{n0} = \frac{2m_e\omega_{n0}}{3\hbar e^2}\sum_{\alpha=x,y,z}|\langle 0|\hat{\mu}_{\alpha}|n\rangle|^2 = \frac{2m_e\omega_{n0}}{3\hbar e^2}\boldsymbol{\mu}_{0n}\cdot\boldsymbol{\mu}_{n0}. \tag{5.230}$$

In the Lorentz model, the corresponding quantity f_i refers to the number of oscillators associated with a given frequency ω_i. We thus expect that the total number of oscillators, which in this case is the total number of electrons N_e, is obtained by summing over all resonance frequencies.

We proceed to show that this is indeed true, starting from the relation

$$\langle 0|[\hat{\mu}_\alpha, \hat{H}]|n\rangle = \hbar\omega_{n0}\langle 0|\hat{\mu}_\alpha|n\rangle = -\frac{i\hbar e}{m_e}\langle 0|\hat{p}_\alpha|n\rangle. \tag{5.231}$$

In the last step, we have employed the Heisenberg equation of motion, Eq. (2.88), which gives us the nonrelativistic velocity operator, and the above relation allows for a conversion between the dipole length and dipole velocity representations. It is also important to keep in mind that the one-electron operators that appear contain summations over all electrons in the system, that is

$$\hat{\mu}_\alpha = \sum_{i=1}^{N_e} \hat{\mu}_\alpha(i); \quad \hat{p}_\alpha = \sum_{i=1}^{N_e} \hat{p}_\alpha(i). \tag{5.232}$$

In the dipole velocity representation, the isotropic oscillator strength takes the form

$$f_{n0} = \frac{2}{3m_e\hbar\omega_{n0}}\mathbf{p}_{0n}\cdot\mathbf{p}_{n0}, \tag{5.233}$$

or, equivalently, in the mixed representation

$$f_{n0} = \frac{i}{3e\hbar}[\boldsymbol{\mu}_{0n}\cdot\mathbf{p}_{n0} - \mathbf{p}_{0n}\cdot\boldsymbol{\mu}_{n0}]. \tag{5.234}$$

By summing over the excited states, we obtain the Thomas–Reiche–Kuhn sum rule

$$\sum_n{}' f_{n0} = \frac{i}{3e\hbar}\sum_n[\boldsymbol{\mu}_{0n}\cdot\mathbf{p}_{n0} - \mathbf{p}_{0n}\cdot\boldsymbol{\mu}_{n0}] = \frac{i}{3e\hbar}\langle 0|\hat{\boldsymbol{\mu}}\cdot\hat{\mathbf{p}} - \hat{\mathbf{p}}\cdot\hat{\boldsymbol{\mu}}|0\rangle = N_e, \tag{5.235}$$

where, in the intermediate step, we have extended the summation to include $n = 0$, as is allowed since this contribution is zero. After this operation, the resolution of the identity can be identified in order to reach the final result, using the commutator relation

$$[\hat{\mu}_\alpha, \hat{p}_\beta] = -ei\hbar\delta_{\alpha\beta} \tag{5.236}$$

that follows from the postulates in Table 2.1.

5.8 Absorption

So far we have determined time-dependent corrections to wave functions of the unperturbed systems, see Eq. (5.106):

$$|\psi^{(0)}(t)\rangle = |0\rangle e^{-iE_0 t/\hbar}. \tag{5.237}$$

We have applied external fields under nonresonant conditions and considered field amplitudes that allow us to use perturbation theory. If, for small field amplitudes, we relax the former constraint and turn to frequencies in the fields that are near-resonant or resonant, the situation will be different. The populations of the excited states will not necessarily remain small due to absorption of photons by the molecular system.

We can illustrate this new situation by the two-level system that was introduced in Section 5.1. For the same amplitude of the perturbation that was considered previously, we change the angular frequency to match the transition frequency by setting $\omega = 0.5$ a.u. The response of the system to this field is given in Figure 5.15. The most notable effect, as compared to the nonresonant case in Figure 5.2 is that the population of the excited state $\rho_b(t)$ appears to grow with time. There are also ripples in $\rho_b(t)$, which follow the oscillations of the field and which give rise to the oscillations in the induced polarization, but it is the growth of the *time-averaged* excited-state population that signifies absorption of photons by the molecular system.

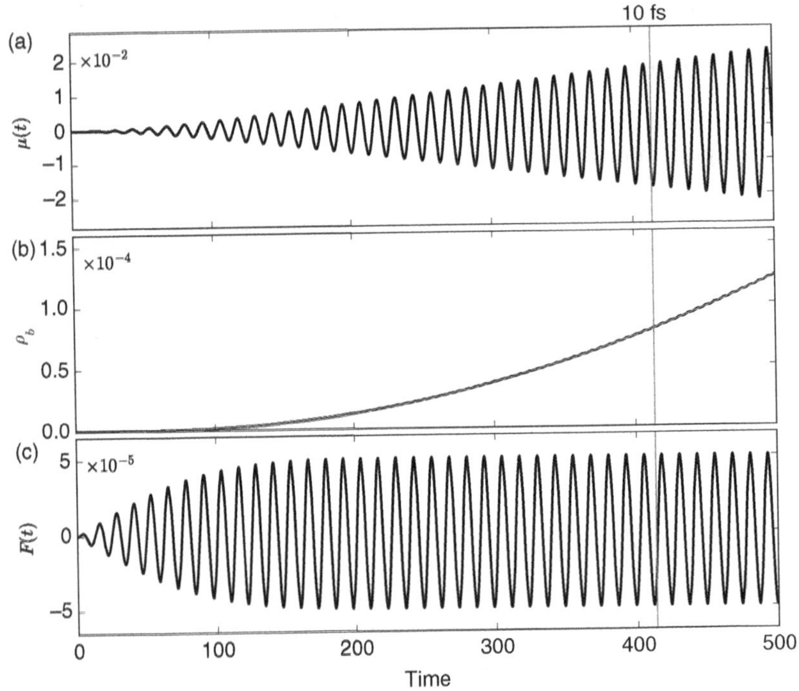

Figure 5.15 Induced electric dipole moment in a two-level atom for a frequency of the perturbation that is in resonance with the electronic transition. The external field $F(t)$, induced dipole moment $\mu(t)$, and the time are given in a.u., and $\rho_b(t)$ denotes the population of excited state ψ_b. The gray curve in panel (b) indicates the excited-state population obtained by using perturbation theory, that is, Eq. (5.239). The vertical bar indicates a time of 10 fs after the start. A time-integration step length of $\Delta t = 0.1$ a.u. is used.

In view of Figure 5.15, it is natural to ask what will happen with the populations and the polarization on a longer timescale. One way to provide an answer to this question, is to speed up the process by increasing the field amplitude, keeping the other characteristics in the responses the same, albeit at the price of moving out of the perturbative regime. With a field amplitude that is three orders of magnitude larger than in the previous case, we obtain a response in the system as seen in Figure 5.16. From this result, we see that the time-averaged population of the excited state keeps growing as a consequence of absorption until it reaches unity, after which it again returns to zero as a consequence of stimulated emission. This cycle is known as the Rabi cycle and a classical analogue is the cyclic energy transfer between two weakly coupled pendulums. In Appendix I, we provide a nonperturbative solution for our two-level atom in the vicinity of resonance. At resonance, the excited-state population is given by

$$\rho_b(t) = \sin^2(\Omega t/2) = \frac{1}{2}[1 - \cos \Omega t]; \quad \Omega = \frac{\mu_{ab} F^{\omega}}{\hbar}, \tag{5.238}$$

where appears the Rabi frequency Ω. In the present strong-field case, the excited-state population oscillates at an angular frequency of $\Omega = 5 \cdot 10^{-2}$ a.u., corresponding to a period time of $T = 2\pi/\Omega \approx 125.7$ a.u., in agreement with what we see in Figure 5.16.[2] In the weak-field case,

2 The alert reader notes that Eq. (5.238) does not feature the ripples in the excited-state population seen in Figure 5.16. This is because it was derived using the rotating wave approximation explained in Appendix I and which we shall meet again in Section 8.8. Also, when the field is sufficiently strong, the resonance condition will be modified, as we can infer from Eqs. (5.4) and (5.5). This is known as the Bloch–Siegert shift.

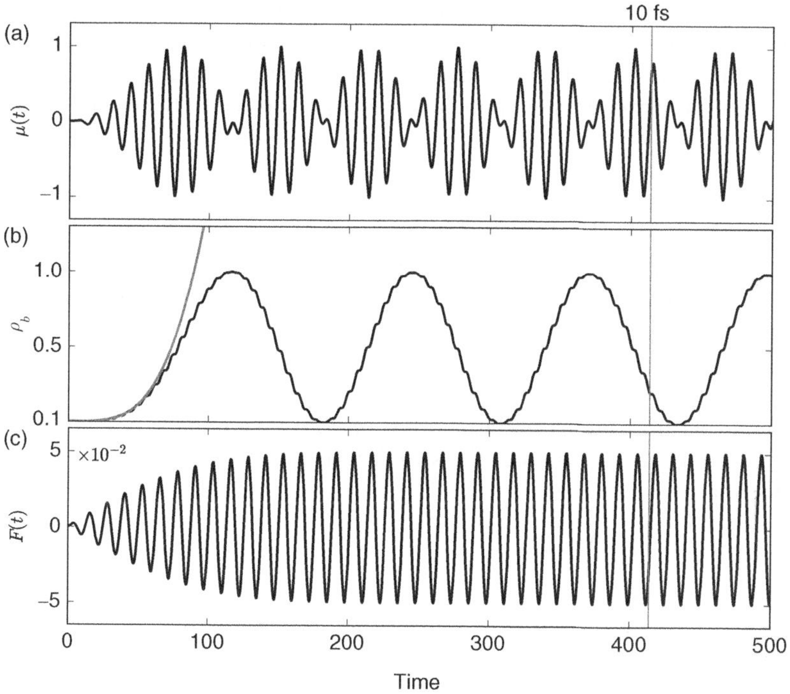

Figure 5.16 Induced electric dipole moment in a two-level atom for a frequency of the perturbation that is in resonance with the electronic transition. The external field $F(t)$, induced dipole moment $\mu(t)$, and the time are given in a.u., and $\rho_b(t)$ denotes the population of the excited state ψ_b. The gray curve in panel (b) indicates the excited-state population obtained by using perturbation theory, that is, Eq. (5.239). The vertical bar indicates a time of 10 fs after the start. A time-integration step length of $\Delta t = 0.1$ a.u. is used.

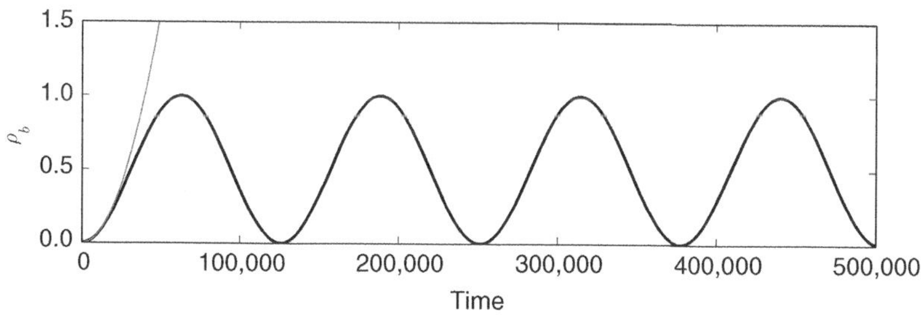

Figure 5.17 Excited-state population in a two-level atom for a frequency of the perturbation that is in resonance with the electronic transition, that is, $\omega = 0.5$ a.u. The amplitude of the electric field is equal to 5.0×10^{-5} a.u. The time are given in a.u. and $\rho_b(t)$ denotes the population of the excited state ψ_b. The gray curve indicates the result obtained by using perturbation theory, that is, Eq. (5.239). A time-integration step length of $\Delta t = 0.1$ a.u. is used.

corresponding to Figure 5.15, the corresponding period is $T \approx 125{,}663$ a.u., so it is no wonder that the Rabi oscillations are not apparent in Figure 5.15. In Figure 5.17, we show the result of a much longer simulation, indeed confirming the presence of Rabi oscillations in the weak-field case as well.

How do we then treat the situation of resonant fields and absorption in the general case where we cannot afford computationally to propagate the wave function in time as done for the two-level system? The limits of perturbation theory can be illustrated by our two-level system. In Appendix I, we derive the excited-state population to second order in the electric field and obtain

$$\rho_b(t) = \frac{1}{4}\left(\frac{\mu_{ab}F^{\omega}}{\hbar}\right)^2\left(t\ \mathrm{erf}(at) + \frac{e^{-(at)^2}-1}{a\sqrt{\pi}}\right)^2. \tag{5.239}$$

In Figure 5.15, we trace the excited-state population given by Eq. (5.239). It follows the simulated population very well, although ripples are absent since we only considered the resonant part of the first-order projection amplitude (see Appendix I for details). However, if we increase the strength of the electric field (Figure 5.16) or extend the time of the simulation (Figure 5.17), the limitations of perturbation theory becomes evident. In the latter case, we note that with increasing time the second term in the parenthesis of Eq. (5.239) rapidly becomes constant (~ -56 a.u. in the present simulation), whereas the switch (error) function in the first term goes to one. We therefore have

$$\rho_b(t) \sim \frac{1}{4}\left(\frac{\mu_{ab}F^{\omega}}{\hbar}\right)^2\left(t - \frac{1}{a\sqrt{\pi}}\right)^2. \tag{5.240}$$

The excited-state population is accordingly seen to grow quadratically with time, and perturbation theory can certainly not be trusted beyond times on the order of the inverse Rabi frequency, which in the weak-field case is around 500 a.u.

Let us now generalize somewhat and consider excitation between two discrete states. We limit the perturbation operator of Eq. (5.105) to a single frequency

$$\hat{V}(t) = [\hat{V}^{\omega}e^{-i\omega t} + \hat{V}^{-\omega}e^{+i\omega t}]e^{\epsilon t}, \tag{5.241}$$

where we as before have introduced an adiabatic switch function in terms of a positive infinitesimal ϵ that will be set to zero at the end of the calculation. We assume that the system is in some initial state $|0\rangle$ at $t = -\infty$. The first-order final-state amplitudes of Eq. (5.112) can then be expressed as

$$d_f^{(1)}(t) = -\frac{1}{\hbar}\left[V_{f0}(\omega)\frac{e^{i(\omega_{f0}-\omega)t}}{\omega_{f0}-\omega-i\epsilon} + V_{0f}^{*}(\omega)\frac{e^{i(\omega_{f0}+\omega)t}}{\omega_{f0}+\omega-i\epsilon}\right]e^{\epsilon t}, \tag{5.242}$$

where $V_{f0}(\omega) = \langle f|\hat{V}^{\omega}|0\rangle$. We observe that when $\epsilon \to 0$, singularities will occur when $\omega = \pm\omega_{f0}$, corresponding to absorption and emission, respectively. We shall assume that the final state is higher in energy than the initial state and that ω is in the vicinity of the resonance, and therefore only retain the first term in Eq. (5.242), that is, the absorptive part. The probability of finding the system in the final state $|f\rangle$ at time t is then

$$P_{f\leftarrow 0}(t) = |d_f^{(1)}(t)|^2 = \frac{1}{\hbar^2}|V_{f0}(\omega)|^2\frac{e^{2\epsilon t}}{(\omega_{f0}-\omega)^2+\epsilon^2}. \tag{5.243}$$

The corresponding transition rate is

$$w_{f\leftarrow 0}^{\epsilon} = \frac{dP_{f\leftarrow 0}}{dt} = \frac{1}{\hbar^2}|V_{f0}(\omega)|^2\frac{2\epsilon}{(\omega_{f0}-\omega)^2+\epsilon^2}e^{2\epsilon t} = \frac{2\pi}{\hbar^2}|V_{f0}(\omega)|^2 f(\omega;\omega_{f0},\epsilon)e^{2\epsilon t}, \tag{5.244}$$

where appears the Cauchy distribution of Eq. (3.455). To conclude, we consider the limit $\epsilon \to 0$. As seen in Eq. (3.459), the Cauchy distribution then goes into a Dirac delta function

$$w_{f\leftarrow 0} = \lim_{\epsilon\to 0} w_{f\leftarrow 0}^{\epsilon} = \frac{2\pi}{\hbar^2}|V_{f0}(\omega)|^2\delta(\omega_{f0}-\omega) = \frac{2\pi}{\hbar}|V_{f0}(\omega)|^2\delta(E_f - E_i - \hbar\omega), \tag{5.245}$$

where, in the final step, we used a scaling relation for the Dirac delta function.

The above expression is problematic because the Dirac delta function is not a proper function. It is often thought of as a function taking on an infinite value at the origin and being zero elsewhere, but at the same time it is known that changing the value of a function in one point does not change its integral over some interval. Therefore, such a function integrates to zero, whereas the Dirac delta function integrates to one. Instead, the Dirac delta function is a generalized function (or distribution) and it has meaning only under an integral sign. One way to circumvent this issue is to consider transitions to a continuum F_c of states with a corresponding density of states, resulting in

$$\begin{aligned} w_{F_c\leftarrow 0} &= \sum_{f\in F_c} w_{f\leftarrow 0} = \frac{2\pi}{\hbar}|V_{f0}(\omega)|^2 \int_{F_c} \delta(E - E_i - \hbar\omega)\rho(E)dE \\ &= \frac{2\pi}{\hbar}|V_{f0}(\omega)|^2 \rho(E_i + \hbar\omega), \end{aligned} \tag{5.246}$$

where it is noted that the transition rate is proportional to the density of states $\rho(E)$ evaluated at the final energy $E_f = E_i + \hbar\omega$ as to reflect the conservation of energy in the absorption process. This is the celebrated *Fermi golden rule.*

It should be noted that Fermi's golden rule refers explicitly to excitations from a discrete state to a continuum of states. Yet, it is often applied to transitions between discrete states, and it becomes natural to ask the question how can this be justified. Another note to be made is that our simulations of a two-level atom shows that under resonant conditions our system exhibits Rabi oscillations rather than irreversible absorption, so there is clearly something missing in our description of resonance fields and the process of state transitions. We will defer a discussion about these points until Section 5.10.

Up until now, we have considered transitions from the initial $|0\rangle$ (often, but not necessarily, equal to the ground state) to a final state $|f\rangle$ as due to the absorption of a single light quanta. Such a transition, however, can also occur by the absorption of several photons in a single quantum-mechanical process referred to as multiphoton absorption. While one-photon absorption is governed by the first-order amplitudes $d^{(1)}$ in Eq. (5.112), two-photon absorption is described by the second-order amplitudes $d^{(2)}$ in Eq. (5.113), and so forth. If we are only concerned with the overall dependence of absorption strengths on the quantum-mechanical transition moments, we may arrive at results in a straightforward manner. Let us consider absorption in the electric-dipole approximation with a first-order correction to the amplitude that is

$$d_f^{(1)}(t) = \frac{1}{\hbar}\sum_{\omega_1}\langle f|\hat{\mu}_\beta|0\rangle F_\beta^{\omega_1} F(t, \omega_{f0} - \omega_1), \tag{5.247}$$

where $F(t, \omega_{f0} - \omega_1)$ is a function that depends on the time t and the separation of the frequency of the perturbing field from the transition frequency of the system ω_{f0}, c.f. Eq. (5.112). For large times t compared to $(2\pi/\omega_{f0})$, the function F is sharply peaked at $\omega_{f0} = \omega_1$, which is equivalent to saying that one-photon absorption occurs only when the photon energy matches the excitation energy, in accordance with Fermi's golden rule. The probability for one-photon transitions per unit time will be proportional to the square of the one-photon matrix element

$$M_{0f}^{(1)} = \langle 0|\hat{\mu}_\beta|f\rangle, \tag{5.248}$$

in agreement with our more detailed analysis made earlier. We note that the probability for one-photon absorption, since it depends on the square of the amplitude and therefore also the square of the amplitude of the electric field, is linearly dependent on the intensity of the light. One-photon absorption is for that reason synonymous with linear absorption.

The second-order amplitude as defined by Eq. (5.113) is

$$d_f^{(2)}(t) = \frac{1}{\hbar^2} \sum_{\omega_1\omega_2} \sum_n \frac{\langle f|\hat{\mu}_\beta|n\rangle\langle n|\hat{\mu}_\gamma|0\rangle}{\omega_{n0} - \omega_2} F_\beta^{\omega_1} F_\gamma^{\omega_2} F(t, \omega_{f0} - \omega_1 - \omega_2), \tag{5.249}$$

where the function $F(t, \omega_{f0} - \omega_1 - \omega_2)$ depends on the time t and the separation of the sum of the perturbing frequencies and the transition frequency of the system ω_{f0}. In analogy with the case of linear absorption, for large times t compared to $(2\pi/\omega_{f0})$, the function F is sharply peaked at $\omega_{f0} = \omega_1 + \omega_2$, which means that two-photon absorption occurs only when the total energy of the two photons matches the excitation energy. We deduce that the two-photon transition matrix element can be written as

$$M_{0f}^{(2)} = \frac{1}{\hbar} \sum_n \left[\frac{\langle 0|\hat{\mu}_\beta|n\rangle\langle n|\hat{\mu}_\gamma|f\rangle}{\omega_{n0} - \omega_1} + \frac{\langle 0|\hat{\mu}_\gamma|n\rangle\langle n|\hat{\mu}_\beta|f\rangle}{\omega_{n0} - \omega_2} \right], \tag{5.250}$$

where we have symmetrized the summations over dummy indices. This symmetrization obviously does not affect the value of the amplitude $d^{(2)}$ as long as we multiply with a factor of one half. The complex conjugation associated with reversing the order of the states in the numerator is irrelevant since transition probabilities depend on the absolute square of the matrix elements. One of the interesting characteristics of two-photon absorption is that since the square of the second-order amplitude, that is, the probability of two-photon absorption, depends on the perturbing electric field to the fourth power, it will exhibit a square dependence on the light intensity. Two-photon absorption is therefore nonlinear to first order with respect to the intensity of the incident laser beam.

Carried to yet a higher order in the perturbation, we will be able to obtain an expression for the three-photon absorption matrix element. The third-order amplitude as defined by Eq. (5.114) is

$$d_f^{(3)}(t) = \frac{1}{\hbar^3} \sum_{\omega_1\omega_2\omega_3} \sum_{mn} \frac{\langle f|\hat{\mu}_\beta|n\rangle\langle n|\hat{\mu}_\gamma|m\rangle\langle m|\hat{\mu}_\delta|0\rangle}{(\omega_{n0} - \omega_2 - \omega_3)(\omega_{m0} - \omega_3)} F_\beta^{\omega_1} F_\gamma^{\omega_2} F_\delta^{\omega_3}$$
$$\times F(t, \omega_{f0} - \omega_1 - \omega_2 - \omega_3). \tag{5.251}$$

The three-photon absorption matrix element, symmetrized in the indices, can thus be written as

$$M_{0f}^{(3)} = \frac{1}{\hbar^2} \sum \mathcal{P}_{1,2,3} \sum_{mn} \frac{\langle 0|\hat{\mu}_\beta|m\rangle\langle m|\hat{\mu}_\gamma|n\rangle\langle n|\hat{\mu}_\delta|f\rangle}{(\omega_{n0} - \omega_1 - \omega_2)(\omega_{m0} - \omega_1)}. \tag{5.252}$$

We see that the probability of three-photon absorption depends on the light intensity to the third power.

5.9 Residue Analysis

If

$$f(z) = \frac{g(z)}{(z-a)^n}; \quad g(a) \neq 0, \tag{5.253}$$

then a is a pole of order n to the function f. The residue of such a pole is given by

$$\text{Res}(f, a) = \frac{1}{(n-1)!} \lim_{z\to a} \frac{d^{n-1}}{dz^{n-1}} [(z-a)^n f(z)]. \tag{5.254}$$

Figure 5.18 A resonance of the linear response function.

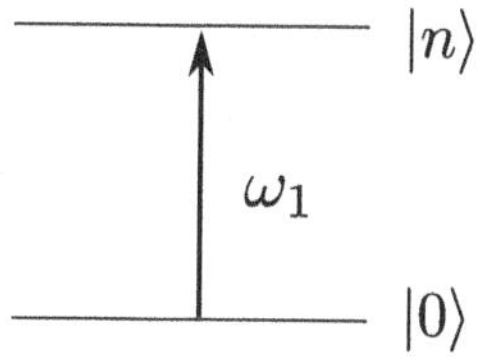

Figure 5.19 Resonances of the first-order nonlinear response function.

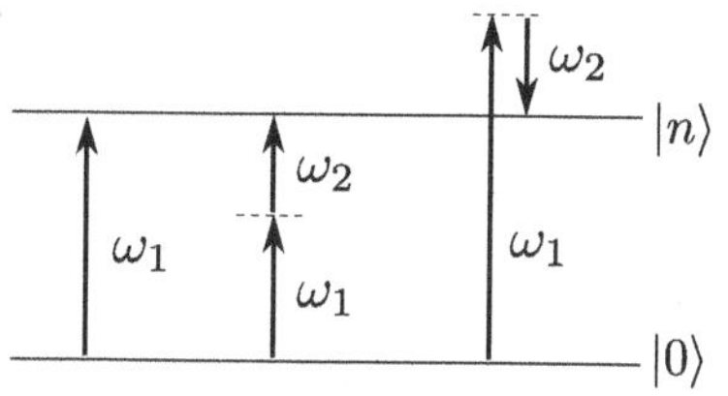

So, for poles of order one, the residue becomes equal to $g(a)$. In this section, we will explore the idea of poles and residues to see how it can help us to extract transition and excited-state properties from ground-state response functions.

We noted already in our study of the ground-state electronic Hessian, which was illustrated in Figure 5.12, that a ground-state property may contain information about the excited states—the diagonal elements of the Hessian contained the complete set of excitation energies. Analogously, it has become clear during the development of the linear and nonlinear response functions that these reference-state properties hold information about all other excited states as well. In this section, we will provide some detailed examples of this fact. For the sake of clarity, we will restrict our discussion to electric-dipole properties, but it is straightforward to generalize the analysis to the case of general response functions.

The poles of the linear polarizability are seen from Eq. (5.169) to match the excitation energies of the system (as illustrated in Figure 5.18), whereas the corresponding residues become

$$\lim_{\omega\to\omega_{f0}} (\omega_{f0} - \omega)\, \alpha_{\alpha\beta}(-\omega;\omega) = \langle 0|\hat{\mu}_\alpha|f\rangle\langle f|\hat{\mu}_\beta|0\rangle, \tag{5.255}$$

where $|f\rangle$ denotes a selected excited state. In the calculation of the polarizability, we can thus search for the divergences in the response function to determine the excitation energies of the system, and from the residues we obtain the transition moments that govern the intensities of one-photon absorption, see Eq. (5.248).

The first-order hyperpolarizability has both single and double residues, see Figure 5.19 for an illustration of the resonance frequencies. From Eq. (5.184), we see that one of the first-order residues becomes

$$\begin{aligned}
&\lim_{\omega_2\to\omega_{f0}} (\omega_{f0} - \omega_2)\beta(-\omega_\sigma;\omega_1,\omega_2) \\
&= \frac{1}{\hbar^2}\sum_n \left[\frac{\langle 0|\hat{\mu}_\alpha|n\rangle\langle n|\hat{\mu}_\beta|f\rangle\langle f|\hat{\mu}_\gamma|0\rangle}{(\omega_{n0}-\omega_1-\omega_{f0})} + \frac{\langle 0|\hat{\mu}_\beta|n\rangle\langle n|\hat{\mu}_\alpha|f\rangle\langle f|\hat{\mu}_\gamma|0\rangle}{(\omega_{n0}+\omega_1)}\right] \\
&= \frac{1}{\hbar^2}\sum_n \left[\frac{\langle 0|\hat{\mu}_\alpha|n\rangle\langle n|\hat{\mu}_\beta|f\rangle}{(\omega_{nf}-\omega_1)} + \frac{\langle 0|\hat{\mu}_\beta|n\rangle\langle n|\hat{\mu}_\alpha|f\rangle}{(\omega_{n0}+\omega_1)}\right]\langle f|\hat{\mu}_\gamma|0\rangle,
\end{aligned} \tag{5.256}$$

where it includes two terms due to the permutation of pairs of indices $(\alpha, -\omega_\sigma)$ and (β, ω_1). This residue, when evaluated for $\omega_1 = -\omega_{f0}/2$, is closely connected to the two-photon matrix element describing absorption of two monochromatic light quanta in the transition from the ground $|0\rangle$ to the excited state $|f\rangle$, see Eq. (5.250), which is analogous to the correspondence between the one-photon matrix element and the residue of the linear polarizability.

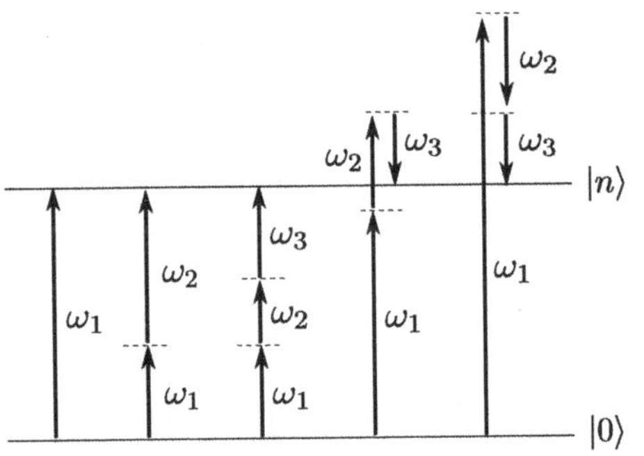

Figure 5.20 Resonances of the second-order nonlinear response function.

Turning to the double residues, we focus on the case when $\omega_1 = -\omega_{f0}$ and $\omega_2 = \omega_{g0}$. Since we are also interested in the situation when $|f\rangle = |g\rangle$, we choose the expression in Eq. (5.187) for the hyperpolarizability and where the secular divergences have been removed. This particular double residue then becomes

$$\lim_{\omega_1 \to -\omega_{f0}} (\omega_{f0} + \omega_1)[\lim_{\omega_2 \to \omega_{g0}} (\omega_{g0} - \omega_2)\, \beta(-\omega_\sigma; \omega_1, \omega_2)]$$
$$= \langle 0|\hat{\mu}_\beta|f\rangle\langle f|\overline{\hat{\mu}_\alpha}|g\rangle\langle g|\hat{\mu}_\gamma|0\rangle. \tag{5.257}$$

When $|f\rangle \neq |g\rangle$, the matrix element in the middle will match the one-photon transition moment between the two excited states $|f\rangle$ and $|g\rangle$. On the other hand, when $|f\rangle = |g\rangle$ the same matrix element will equal the difference between the dipole moment of the excited state $|f\rangle$ and the ground state $|0\rangle$. This provides a first example of the possibility of extracting lower-order properties of molecular excited states from higher-order ground-state properties.

The second-order hyperpolarizability has single, double, and triple residues, see Figure 5.20 for an illustration of the resonance frequencies. However, the triple residue contains no additional information to that already discussed in connection with lower-order response functions. We will consider a few of the residues that correspond to certain ground- and excited-state properties. From Eq. (5.200), we see that one of the first-order residues is

$$\lim_{\omega_3 \to \omega_{f0}} (\omega_{f0} - \omega_3)\gamma(-\omega_\sigma; \omega_1, \omega_2, \omega_3)$$
$$= \frac{1}{\hbar^3} \sum \mathcal{P}_{-\sigma,1,2} \sum_{nm} \frac{\langle 0|\hat{\mu}_\alpha|n\rangle\langle n|\hat{\mu}_\beta|m\rangle\langle m|\hat{\mu}_\gamma|f\rangle\langle f|\hat{\mu}_\delta|0\rangle}{(\omega_{n0} - \omega_\sigma)(\omega_{m0} - \omega_2 - \omega_{f0})}$$
$$= \frac{1}{\hbar^3} \sum \mathcal{P}_{-\sigma,1,2} \sum_{nm} \frac{\langle 0|\hat{\mu}_\alpha|n\rangle\langle n|\hat{\mu}_\beta|m\rangle\langle m|\hat{\mu}_\gamma|f\rangle}{(\omega_{nf} - \omega_1 - \omega_2)(\omega_{mf} - \omega_2)} \langle f|\hat{\mu}_\delta|0\rangle. \tag{5.258}$$

This expression is closely related to the three-photon matrix element describing an excitation from the molecular ground state to the excited state $|f\rangle$ by simultaneous absorption of three photons, see Eq. (5.252). When this residue is evaluated at $\omega_1 = \omega_2 = -\omega_{f0}/3$, it will provide the matrix element corresponding to the absorption of three monochromatic photons. We have just seen that, from the first residue of the first-order hyperpolarizability in Eq. (5.256), it is possible to identify the matrix element describing two-photon absorption. If we now consider one higher order of response as well as one higher order of the residue, it is possible to identify the expression for the matrix element describing two-photon absorption between two excited states, say $|g\rangle$ and $|f\rangle$. The following double residue of the second-order hyperpolarizability needs to be considered in this case

$$
\begin{aligned}
&\lim_{\omega_2\to-\omega_{g0}}(\omega_{g0}+\omega_2)[\lim_{\omega_3\to\omega_{f0}}(\omega_{f0}-\omega_3)\gamma(-\omega_\sigma;\omega_1,\omega_2,\omega_3)]\\
&=\frac{1}{\hbar^3}\sum\mathcal{P}_{-\sigma,1}\sum_m\frac{\langle 0|\hat\mu_\gamma|g\rangle\langle g|\hat\mu_\beta|m\rangle\langle m|\hat\mu_\alpha|f\rangle\langle f|\hat\mu_\delta|0\rangle}{(\omega_{m0}+\omega_\sigma-\omega_{f0})}\\
&=\frac{1}{\hbar^3}\sum\mathcal{P}_{-\sigma,1}\sum_m\frac{\langle g|\hat\mu_\beta|m\rangle\langle m|\hat\mu_\alpha|f\rangle}{(\omega_{mg}+\omega_1)}\langle 0|\hat\mu_\gamma|g\rangle\langle f|\hat\mu_\delta|0\rangle\\
&=\frac{1}{\hbar^3}\sum_m\left[\frac{\langle g|\hat\mu_\alpha|m\rangle\langle m|\hat\mu_\beta|f\rangle}{(\omega_{mf}-\omega_1)}+\frac{\langle g|\hat\mu_\beta|m\rangle\langle m|\hat\mu_\alpha|f\rangle}{(\omega_{mg}+\omega_1)}\right]\langle 0|\hat\mu_\gamma|g\rangle\langle f|\hat\mu_\delta|0\rangle. \qquad (5.259)
\end{aligned}
$$

Of the 24 permutations in Eq. (5.200) for $\gamma(-\omega_\sigma;\omega_1,\omega_2,\omega_3)$, only those obtained by permuting the pairs $(\alpha,-\omega_\sigma)$ and (γ,ω_2) contribute to this double residue. The residue is to be evaluated at $\omega_1=-\omega_{fg}/2$ in order to be related to monochromatic two-photon absorption in the excited state $|g\rangle$.

It is also possible to have $|g\rangle=|f\rangle$ in the double residue. However, this situation corresponds to a secular singularity in the second-order hyperpolarizability and it is therefore appropriate to use Eq. (5.201) as a starting point:

$$
\begin{aligned}
&\lim_{\omega_2\to-\omega_{f0}}(\omega_{f0}+\omega_2)[\lim_{\omega_3\to\omega_{f0}}(\omega_{f0}-\omega_3)\gamma(-\omega_\sigma;\omega_1,\omega_2,\omega_3)]\\
&=\frac{1}{\hbar^3}\sum\mathcal{P}_{-\sigma,1}\left[\sum_m{}'\frac{\langle 0|\hat\mu_\gamma|f\rangle\langle f|\overline{\hat\mu_\beta}|m\rangle\langle m|\overline{\hat\mu_\alpha}|f\rangle\langle f|\hat\mu_\delta|0\rangle}{(\omega_{m0}+\omega_\sigma-\omega_{f0})}\right.\\
&\quad-\sum_n{}'\frac{\langle 0|\hat\mu_\alpha|n\rangle\langle n|\overline{\hat\mu_\beta}|0\rangle\langle 0|\overline{\hat\mu_\gamma}|f\rangle\langle f|\hat\mu_\delta|0\rangle}{(\omega_{n0}-\omega_\sigma)}\\
&\quad\left.-\frac{\langle 0|\hat\mu_\gamma|f\rangle\langle f|\overline{\hat\mu_\beta}|0\rangle\langle 0|\overline{\hat\mu_\alpha}|f\rangle\langle f|\hat\mu_\delta|0\rangle}{(\omega_{f0}-\omega_\sigma)}\right]. \qquad (5.260)
\end{aligned}
$$

The relevance of this double residue to molecular properties is not obvious, but let us nevertheless first mention how the surviving terms in the residue arose. By considering the permutation of the pairs $(\alpha,-\omega_\sigma)$ and (γ,ω_2) in the first summation in Eq. (5.201), we see that it gives rise to the first term of the double residue. The second and third terms in the residue both come from the second summation in Eq. (5.201). We recall that the primed summations exclude the ground state $|0\rangle$. However, the third term in the residue is identical to the contribution from the omitted ground state in the first summation, and we can therefore write the double residue as

$$
\begin{aligned}
&\lim_{\omega_2\to-\omega_{f0}}(\omega_{f0}+\omega_2)[\lim_{\omega_3\to\omega_{f0}}(\omega_{f0}-\omega_3)\gamma(-\omega_\sigma;\omega_1,\omega_2,\omega_3)]\\
&=\frac{1}{\hbar^3}\sum\mathcal{P}_{-\sigma,1}\left[\sum_m\frac{\langle 0|\hat\mu_\gamma|f\rangle\langle f|\overline{\hat\mu_\beta}|m\rangle\langle m|\overline{\hat\mu_\alpha}|f\rangle\langle f|\hat\mu_\delta|0\rangle}{(\omega_{mf}+\omega_1)}\right.\\
&\quad\left.-\sum_n{}'\frac{\langle 0|\hat\mu_\alpha|n\rangle\langle n|\hat\mu_\beta|0\rangle\langle 0|\hat\mu_\gamma|f\rangle\langle f|\hat\mu_\delta|0\rangle}{(\omega_{n0}-\omega_1)}\right], \qquad (5.261)
\end{aligned}
$$

where we have used that $\omega_1=\omega_\sigma$ and $\langle 0|\overline{\hat\mu}|f\rangle=\langle 0|\hat\mu|f\rangle$. The first summation in Eq. (5.261) has a secular singularity at $\omega_1=0$ due to the term for which $|m\rangle=|f\rangle$. However, this term will cancel with that generated by the permutation of the pairs $(\alpha,-\omega_\sigma)$ and (β,ω_1), and a nondivergent

Table 5.3 A selection of electric molecular properties described by the first-, second-, and third-order response functions.

Response	Residue	Molecular property
$\alpha(-\omega;\omega)$	—	Linear electric dipole polarizability
	$\omega=\omega_{f0}$	One-photon transition matrix elements between the ground state $\|0\rangle$ and the excited state $\|f\rangle$
$\beta(-\omega_\sigma;\omega_1,\omega_2)$	—	First-order nonlinear electric dipole hyperpolarizability
	$\omega_2=\omega_{f0}$	Two-photon transition matrix element between the ground state $\|0\rangle$ and the excited state $\|f\rangle$
	$\omega_1=-\omega_{f0}$ $\omega_2=\omega_{g0}$	One-photon transition matrix element between the excited states $\|f\rangle$ and $\|g\rangle$
	$\omega_1=-\omega_{f0}$ $\omega_2=\omega_{f0}$	Permanent electric dipole moment of the excited state $\|f\rangle$
$\gamma(-\omega_\sigma;\omega_1,\omega_2,\omega_3)$	—	Second-order nonlinear electric dipole hyperpolarizability
	$\omega_3=\omega_{f0}$	Three-photon transition matrix element between the ground state $\|0\rangle$ and the excited state $\|f\rangle$
	$\omega_2=-\omega_{f0}$ $\omega_3=\omega_{g0}$	Two-photon transition matrix element between the excited states $\|f\rangle$ and $\|g\rangle$
	$\omega_2=-\omega_{f0}$ $\omega_3=\omega_{f0}$	Linear electric dipole polarizability of the excited state $\|f\rangle$

formula would therefore exclude the excited state $|f\rangle$ in the summation. We thus finally obtain

$$\lim_{\omega_2\to-\omega_{f0}}(\omega_{f0}+\omega_2)[\lim_{\omega_3\to\omega_{f0}}(\omega_{f0}-\omega_3)\gamma(-\omega_\sigma;\omega_1,\omega_2,\omega_3)]$$
$$=\frac{1}{\hbar^3}\sum\mathcal{P}_{-\sigma,1}\left[\sum_{m\neq f}\frac{\langle f|\hat{\mu}_\alpha|m\rangle\langle m|\hat{\mu}_\beta|f\rangle}{(\omega_{mf}-\omega_1)}-\sum_{n\neq 0}\frac{\langle 0|\hat{\mu}_\alpha|n\rangle\langle n|\hat{\mu}_\beta|0\rangle}{(\omega_{n0}-\omega_1)}\right]$$
$$\times\langle 0|\hat{\mu}_\gamma|f\rangle\langle f|\hat{\mu}_\delta|0\rangle, \tag{5.262}$$

where we permuted the pairs $(\alpha,-\omega_\sigma)$ and (β,ω_1) in the first term. If we compare Eq. (5.262) with Eq. (5.169), we see that the double residue of the second-order hyperpolarizability matches the excited-to-ground state difference in the linear polarizability except for some factors.

We complete this section by providing in Table 5.3 a list of common ground- and excited-state electric properties that can be extracted from the linear, quadratic, and cubic response functions.

5.10 Relaxation

In this section, the aim is to allow for weak external perturbing fields to be in near resonance or even in resonance with the transition frequencies of the system and study the induced polarization by means of perturbation theory. We have seen in the previous sections that on a longer time scale, perturbation theory is questionable because the excited-state populations may no longer remain small, which was illustrated by the Rabi oscillations depicted in Figure 5.16. Under vacuum conditions, that is, in the absence of external fields and intermolecular interactions, the system will remain in any of its eigenstates (ground or excited states) over time as the eigenstates are stationary. In the presence of external fields but in the absence of intermolecular

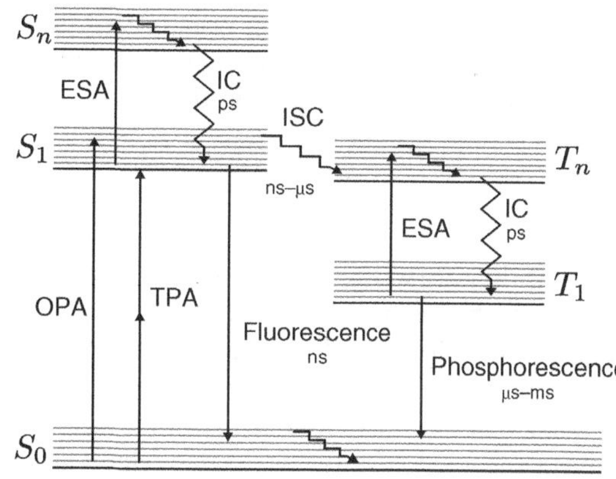

Figure 5.21 A selection of absorption and relaxation channels in molecular systems: one-photon absorption (OPA), two-photon absorption (TPA), excited-state absorption (ESA), internal conversion (IC), and inter-system crossing (ISC). Typical interaction times for the relaxation processes are indicated. Singlet (denoted by S_n) and triplet (denoted by T_n) electronic states are included in the diagram, which is known as a Jablonski diagram. For each electronic state, several ro-vibrational states are indicated. Wiggly and straight lines refer to nonradiative thermal and photon interactions, respectively.

interactions, the populations will change due to *absorption* and *stimulated emission*, which are the two phenomena that give rise to the Rabi oscillations. But even in the absence of external fields, the excited states have finite lifetimes due to various mechanisms. One such mechanism is *spontaneous emission* that was predicted by Einstein in 1916, before the development of quantum mechanics. The background for this early proposal of a phenomenon that is atomic and inherently quantum mechanical in nature is that it appears naturally by taking into account the quantized nature of electromagnetic fields. This consideration falls outside the scope of the present book in which a semi-classical approach is adopted. With spontaneous relaxation, there are associated natural linewidths of all transitions in atoms and molecules. Furthermore, in the presence of intermolecular interactions such as collisions, molecular systems may also undergo fast relaxation via both radiative and nonradiative relaxation channels as illustrated in Figure 5.21 and which further contributes to the observed spectral line broadenings. The total decay rates may greatly exceed the stimulated transition rates, so that excited-state populations remain small and a perturbational approach is still applicable under resonance conditions. The various underlying physical processes responsible for the decays give rise to spectral line broadening, and in the simulation of spectroscopies carried out under near-resonant or resonant conditions, it becomes important to incorporate relaxation into the theory.

5.10.1 Density Operator

A formulation of quantum mechanics that is well suited for the treatment of relaxation and finite lifetimes of the excited states is the density matrix formalism. The density operator is defined as

$$\hat{\rho} = \sum_s p(s)\, |\psi_s(t)\rangle\langle\psi_s(t)|, \tag{5.263}$$

where the summation runs over a statistical ensemble of system configurations and $p(s)$ denotes the respective probabilities. With wave function expansions in the form of projections onto the eigenstates

$$|\psi_s(t)\rangle = \sum_n c_n^s(t)|n\rangle,$$

the matrix elements of the density operator become

$$\rho_{mn} = \sum_s p(s)\, c_m^s c_n^{s*}. \tag{5.264}$$

It is clear that the diagonal elements of the density matrix are to be associated with state populations, but the interpretation of the off-diagonal elements is somewhat less intuitive. If we are considering weakly interacting systems that undergo thermally induced collisions that can only alter overall phases of wave functions but not induce excitations, then a diagonal element of the density matrix can be written on the form

$$\rho_{mn} = |c_m||c_n| \sum_s p(s)\, e^{i(\phi^s_m - \phi^s_n)}, \tag{5.265}$$

where ϕ^s_m and ϕ^s_n are the arguments (or phases) of the complex coefficients c^s_m and c^s_n, respectively, as they are written on polar form. We have made use of the fact that the absolute values of the two coefficients are identical in between different systems (molecules) in the weakly interacting ensemble. For a large number of systems with a stochastic distribution of phases, the summation in Eq. (5.265) will quickly go to zero.

An incoherent ensemble of two-level atoms (see Figure 5.1) with equal populations of the ground and excited states is thus described by the density matrix

$$\rho = \frac{1}{2} \times \begin{pmatrix} 1 & 0 \\ 0 & 1 \end{pmatrix}. \tag{5.266}$$

An ensemble that can be described by a single wave function, on the other hand, is fully coherent and is said to be in a *pure* state. A two-level system in a pure state with equal population of the ground and excited state is described by a wave function

$$|\psi\rangle = \frac{1}{\sqrt{2}}(e^{i\phi_a}|\psi_a\rangle + e^{i\phi_b}|\psi_b\rangle), \tag{5.267}$$

and by a density matrix

$$\rho = \frac{1}{2} \times \begin{pmatrix} 1 & e^{i(\phi_a - \phi_b)} \\ e^{i(\phi_b - \phi_a)} & 1 \end{pmatrix}. \tag{5.268}$$

Since we are here concerned with the development of wave function methods in approximate-state theory, our focus will be on situations where the system being studied can be described by pure states. We will, however, investigate how such a restriction affects the resulting formulas for the response functions.

5.10.2 Liouville Equation

The Schrödinger equation for the state vector in the bra form is

$$\frac{\partial}{\partial t}\langle\psi(t)| = -\frac{1}{i\hbar}\langle\psi(t)|\hat{H}. \tag{5.269}$$

We note the change of signs on the right-hand side with respect to the corresponding equation in the ket-form, which means that the equation-of-motion for the density operator can be compactly written by means of a commutator

$$\frac{\partial}{\partial t}\hat{\rho} = \frac{1}{i\hbar}[\hat{H}, \hat{\rho}]. \tag{5.270}$$

This equation is known as the Liouville equation, and it is equivalent to the Schrödinger equation. Response functions derived for pure states based on Eq. (5.270) will therefore be identical to those derived by us in previous sections based on wave-function theory. It appears to be of little value to carry out this exercise, and we may even argue that the density-matrix formalism seems somewhat redundant since, for a system with N degrees of freedom, there are $2N$ parameters to be determined in the wave-function approach (in terms of the N complex expansion coefficients $c_n(t)$) but no less than N^2 parameters in the density matrix approach (in terms of the nonredundant complex off-diagonal elements and real-diagonal elements of the Hermitian density matrix). The reason why we are nevertheless interested in the use of the density matrix formalism is the possibility it affords to modify Eq. (5.270) to include interactions that are not readily included in the Hamiltonian. This will provide us with a phenomenological way to include the effects of relaxation in the molecular system.

For a specific matrix element of the density operator, the Liouville equation is given as

$$\frac{\partial}{\partial t}\rho_{mn} = \frac{1}{i\hbar}[\hat{H}, \hat{\rho}]_{mn} - \gamma_{mn}(\rho_{mn} - \rho_{mn}^{\text{eq}}), \tag{5.271}$$

where the damping term γ_{mn} corresponds to the rate at which ρ_{mn} relaxes to its equilibrium value ρ_{mn}^{eq}. Since our main concern is the formulation of response theories in electronic structure theory, we will assume the equilibrium to be that of the ground state (i.e., $\rho_{mn}^{\text{eq}} = \delta_{n0}\delta_{m0}$). This assumption is reasonable for a molecular system where thermal electronic excitations are unlikely events.

We note that the diagonal elements of the damping parameter matrix will determine the population decays, and the average lifetime, or simply lifetime, of excited state ψ_n is given by the inverse of the damping parameter, that is

$$\tau_n = 1/\gamma_{nn} \quad \text{for } n > 0. \tag{5.272}$$

The inverse lifetime of excited states (or total decay rate) is often also denoted by $\Gamma_n = \gamma_{nn}$. The ground state needs a separate treatment, because, since there is no decay out of the ground state ($n = 0$), we have $\Gamma_0 = 0$ but $\gamma_{00} \neq 0$. This result is a consequence of the conservation of the summed population that will inflict a dependence of γ_{00} on other parameters γ_{nn}, $n > 0$.

An excited-state level n that has a population $|c_n(0)|^2$ at time zero will experience an exponential population decay according to

$$|c_n(t)|^2 = |c_n(0)|^2 e^{-\Gamma_n t}. \tag{5.273}$$

Since the decay rates of states n and m are Γ_n and Γ_m, respectively, we can thus for a pure state draw the conclusion that the off-diagonal elements of the density matrix will depend on time according to

$$|\rho_{mn}(t)| = |c_m(t)c_n^*(t)| = |c_m(0)c_n^*(0)|e^{-(\Gamma_m+\Gamma_n)t/2}, \tag{5.274}$$

and we must therefore have

$$\gamma_{mn} = (\Gamma_m + \Gamma_n)/2. \tag{5.275}$$

As discussed, intermolecular interactions such as weak collisions that do not alter the state populations but just the overall phases of wave functions will lead to dephasing of the system, and in such situations an additional term (sometimes denoted as γ_{mn}^{col}) will be present in Eq. (5.275). Dephasing will, however, lead to incoherence, and we will not be concerned with such a generalization. A practical way to check the coherence of a given density operator (and thus also the density matrix) is to make use of the fact that for a pure state (but not others), the density operator is idempotent

$$\hat{\rho}^2(t) = \hat{\rho}(t). \tag{5.276}$$

The property of idempotency follows directly from the definition in Eq. (5.263).

In order to illustrate the introduction of damping terms, let us again consider the two-level system depicted in Figure 5.1 with an inverse lifetime of the excited state that is $\Gamma = 0.01$ a.u. If we, at $t = 0$, start in the excited state ψ_b, our initial density matrix and damping matrix are, respectively,

$$\rho(0) = \begin{pmatrix} 0 & 0 \\ 0 & 1 \end{pmatrix}; \qquad \gamma = \Gamma \times \begin{pmatrix} 1 & 1/2 \\ 1/2 & 1 \end{pmatrix}, \tag{5.277}$$

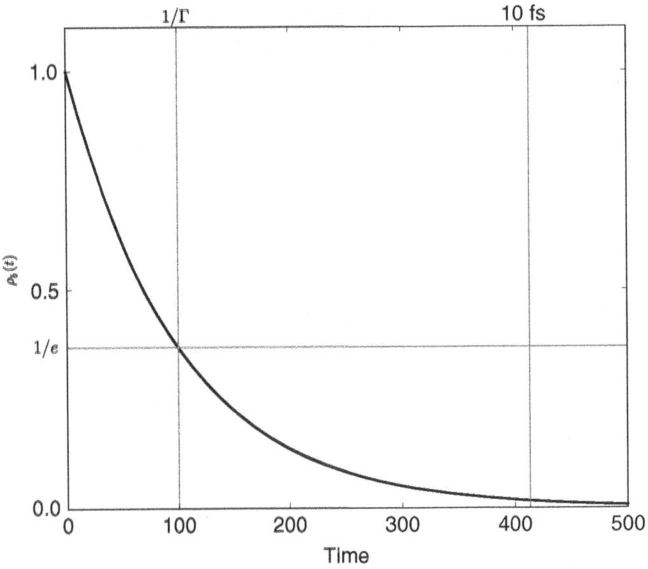

Figure 5.22 Relaxation in a two-level atom where $\rho_b(t)$ denotes the population of the excited state ψ_b. The vertical bars indicate times of $1/\Gamma$ and 10 fs, respectively, and the horizontal bar indicates a population of $1/e$. A time-integration step length of $\Delta t = 0.1$ a.u. is used.

then, using Eq. (5.271) in the absence of external fields, we get an exponentially decaying population of the excited state as illustrated in Figure 5.22. In the figure, we have indicated the time of $1/\Gamma$ after which the population of the excited state has decreased due to spontaneous decay to a value of $1/e$ in accordance with the finite lifetime of the excited state.

Let us now turn to a situation of greater interest to us, namely the resonant perturbation discussed in connection with the illustration of absorption in Figure 5.15. We now have the tool to investigate this situation with the addition of a finite lifetime of the excited state, and we expect the absorption to be saturated at a level where it is counterbalanced by the decay from the excited to the ground state. We will study this new situation of the two-level system in the same manner as before, by direct numerical propagation of the relevant equation of motion, Eq. (5.271), and we do not base our analysis on perturbation theory. Once the time-dependent density matrix has been obtained, we determine the polarization from the expectation value of the dipole operator, and we will assume it to match that used previously and which is shown in Figure 5.1. In the density matrix formalism, an expectation value is given as

$$\langle \hat{\Omega} \rangle = \mathrm{Tr}(\hat{\rho}\hat{\Omega}) = \mathrm{Tr}(\hat{\Omega}\hat{\rho}), \tag{5.278}$$

where "Tr" denotes the operation of trace, and refers to the sum of the diagonal elements of the matrix representing the product operator.

With

$$\rho(0) = \begin{pmatrix} 1 & 0 \\ 0 & 0 \end{pmatrix}; \qquad \gamma = \Gamma \times \begin{pmatrix} 1 & 1/2 \\ 1/2 & 1 \end{pmatrix}, \tag{5.279}$$

and a resonant perturbation as illustrated in the lower panel of Figure 5.23, the excited-state population is seen in the mid panel to reach a steady state with small oscillations about a value of about 4×10^{-6}. This tells us that with the introduction of relaxation in the theory, perturbation theory is applicable also in regions where the weak external electromagnetic fields are in resonance with transition frequencies of the molecular system. The time-dependent dipole moment

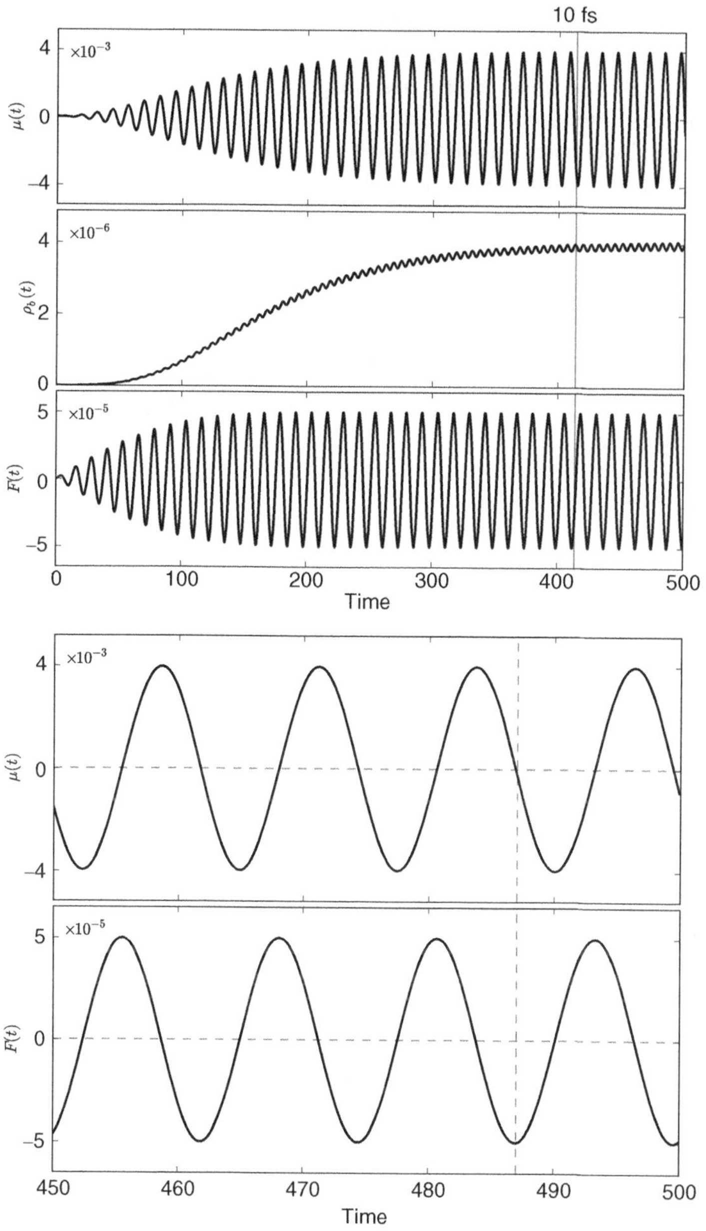

Figure 5.23 Induced electric dipole moment in a two-level atom for a frequency of the perturbation that is in resonance with the electronic transition. Relaxation from the upper to the lower level is described by the parameter $\Gamma = 0.025$ a.u. The external field $F(t)$, induced dipole moment $\mu(t)$, and the time are given in a.u., and $\rho_b(t)$ denotes the population of the excited state ψ_b. The vertical bar indicates the time of 10 fs. A time-integration step length of $\Delta t = 0.1$ a.u. is used. The lower illustration shows the same results in the zoomed-in time interval from 450 to 500 a.u.

is depicted in the upper panel of the figure, and it is seen that the amplitude of the induced polarization is much greater in the resonance region than in the nonresonant region treated in Figure 5.2. In the present case, the amplitude becomes 4.0×10^{-3} a.u. once the steady-state region is reached. More interestingly, we observe at closer inspection that there is a phase delay in the induced polarization as compared to the external electric field. This phase delay amounts to an angle of $\pi/2$ which, by considering the linear term in Eq. (5.60), indicates that the polarizability is purely imaginary. From the data in Figure 5.23, we get a numerical value of the polarizability that is[3]

$$\alpha = i \frac{\max[\mu(t)]}{F^{\omega}} \approx 80i \text{ a.u.} \tag{5.280}$$

5.10.3 Density Matrix from Perturbation Theory

We will now determine the solution to Eq. (5.271) be means of perturbation theory, and the order expansion of the density matrix is

$$\rho_{mn}(t) = \rho_{mn}^{(0)} + \rho_{mn}^{(1)} + \rho_{mn}^{(2)} + \dots, \tag{5.281}$$

with

$$\rho_{mn}^{(0)} = \delta_{m0}\delta_{n0}. \tag{5.282}$$

Since

$$[\hat{H}_0, \hat{\rho}]_{mn} = \hbar\omega_{mn}\rho_{mn}, \tag{5.283}$$

the solution to Eq. (5.271) is obtained by straightforward time integration, and is

$$\rho_{mn}^{(N)} = e^{-(i\omega_{mn}+\gamma_{mn})t} \int_{-\infty}^{t} \frac{1}{i\hbar}[\hat{V}, \hat{\rho}^{(N-1)}]_{mn}\, e^{(i\omega_{mn}+\gamma_{mn})t'} dt'. \tag{5.284}$$

After inserting the perturbation in Eq. (5.105), we get a first-order response in the density matrix

$$\begin{aligned} \rho_{mn}^{(1)} &= \frac{1}{i\hbar} \sum_{\omega} \frac{[\hat{V}_{\beta}^{\omega}, \hat{\rho}^{(0)}]_{mn} F_{\beta}^{\omega}}{i\omega_{mn} - i\omega + \gamma_{mn} + \epsilon} e^{-i\omega t} e^{\epsilon t} \\ &= \frac{1}{i\hbar} \sum_{\omega} \left[\frac{\langle m|\hat{V}_{\beta}^{\omega}|0\rangle\delta_{n0}}{i\omega_{m0} - i\omega + \gamma_{m0} + \epsilon} - \frac{\langle 0|\hat{V}_{\beta}^{\omega}|n\rangle\delta_{m0}}{-i\omega_{n0} - i\omega + \gamma_{n0} + \epsilon} \right] F_{\beta}^{\omega} e^{-i\omega t} e^{\epsilon t}. \end{aligned} \tag{5.285}$$

The structures of the lower-order density matrices are shown in Figure 5.24, and we note that there is no change in the populations to first order in the perturbation.

$$\rho^{(0)} = \begin{pmatrix} \blacksquare & 0 & \cdots & 0 \\ 0 & 0 & \cdots & 0 \\ \vdots & \vdots & \ddots & \vdots \\ 0 & 0 & \cdots & 0 \end{pmatrix}; \quad \rho^{(1)} = \begin{pmatrix} 0 & \blacksquare & \cdots & \blacksquare \\ \blacksquare & 0 & \cdots & 0 \\ \vdots & \vdots & \ddots & \vdots \\ \blacksquare & 0 & \cdots & 0 \end{pmatrix}; \quad \rho^{(2)} = \begin{pmatrix} \blacksquare & \blacksquare & \cdots & \blacksquare \\ \blacksquare & \blacksquare & \cdots & \blacksquare \\ \vdots & \vdots & \ddots & \vdots \\ \blacksquare & \blacksquare & \cdots & \blacksquare \end{pmatrix}$$

Figure 5.24 Structures of zeroth-, first-, and second-order density matrices. The filled squares denote nonzero matrix elements.

---

3 We shall later find in Eq. (5.288) an analytical expression for the linear polarizability also valid under resonance conditions, from which we get $\alpha \approx \mu_{ab}^2/(-i\hbar\Gamma/2)$.

Repeated use of Eq. (5.284) gives a second-order response that is

$$\rho_{mn}^{(2)} = -\frac{1}{\hbar^2}\sum_{\omega_1\omega_2}\left[\frac{\langle m|\hat{V}_\beta^{\omega_1}|0\rangle\langle 0|\hat{V}_\gamma^{\omega_2}|n\rangle}{i\omega_{n0}+i\omega_2-\gamma_{n0}-\epsilon} - \frac{\langle m|\hat{V}_\gamma^{\omega_2}|0\rangle\langle 0|\hat{V}_\beta^{\omega_1}|n\rangle}{i\omega_{m0}-i\omega_2+\gamma_{m0}+\epsilon}\right.$$
$$\left.+\sum_p \frac{\langle m|\hat{V}_\beta^{\omega_1}|p\rangle\langle p|\hat{V}_\gamma^{\omega_2}|0\rangle\delta_{n0}}{i\omega_{p0}-i\omega_2+\gamma_{p0}+\epsilon} - \frac{\langle 0|\hat{V}_\gamma^{\omega_2}|p\rangle\langle p|\hat{V}_\beta^{\omega_1}|n\rangle\delta_{m0}}{i\omega_{p0}+i\omega_2-\gamma_{p0}-\epsilon}\right]$$
$$\times\frac{1}{i\omega_{mn}-i\omega_\sigma+\gamma_{mn}+2\epsilon}F_\beta^{\omega_1}F_\gamma^{\omega_2}e^{-i(\omega_1+\omega_2)t}e^{2\epsilon t}, \tag{5.286}$$

which introduces changes in the off-diagonal as well as the diagonal elements of the density matrix, in agreement with the illustration in Figure 5.24.

5.10.4 Linear Response Functions from the Density Matrix

The linear response function will as usual be identified from the first-order correction to the expectation value of an operator. In the density matrix formalism, this correction will be

$$\langle\hat{\Omega}\rangle^{(1)} = \mathrm{Tr}(\hat{\rho}^{(1)}\hat{\Omega}) = \sum_{mn}\rho_{mn}^{(1)}\Omega_{nm}$$
$$= -\frac{1}{\hbar}\sum_\omega\sum_n\left[\frac{\langle 0|\hat{\Omega}|n\rangle\langle n|\hat{V}_\beta^\omega|0\rangle}{\omega_{n0}-\omega-i\gamma_{n0}-i\epsilon}+\frac{\langle 0|\hat{V}_\beta^\omega|n\rangle\langle n|\hat{\Omega}|0\rangle}{\omega_{n0}+\omega+i\gamma_{n0}+i\epsilon}\right]F_\beta^\omega e^{-i\omega t}e^{\epsilon t}, \tag{5.287}$$

and after comparing with Eq. (5.150), we identify the linear response function in the limit $\epsilon \to 0$ as being

$$\langle\langle\hat{\Omega};\hat{V}_\beta^\omega\rangle\rangle = -\frac{1}{\hbar}\sum_n\left[\frac{\langle 0|\hat{\Omega}|n\rangle\langle n|\hat{V}_\beta^\omega|0\rangle}{\omega_{n0}-\omega-i\gamma_{n0}}+\frac{\langle 0|\hat{V}_\beta^\omega|n\rangle\langle n|\hat{\Omega}|0\rangle}{\omega_{n0}+\omega+i\gamma_{n0}}\right]. \tag{5.288}$$

The striking difference between this response function and that derived previously, see Eq. (5.168) comes from the imaginary damping terms in the denominators, which correspond to the inverse lifetimes of the excited states ($\gamma_{n0} = \Gamma_n/2$). From a mathematical point of view, the effect of the damping terms is to make the response functions well behaved for all values of the optical frequencies, at the cost of making them complex. From a physical point of view, as we shall see, the real and imaginary parts of such complex response functions will have quite different interpretations and correspond to dispersive and absorptive spectroscopies interrelated by the Kramers–Kronig relations in Eq. (3.426).

If we assume that a common damping term is used for all excited states, that is, $\gamma_{n0} = \gamma$, then we can choose to identify the damping term not together with the transition angular frequency but instead with the optical frequency and to define a response function for a complex frequency argument $z = \omega + i\gamma$:

$$\langle\langle\hat{A};\hat{B}\rangle\rangle_z = -\frac{1}{\hbar}\sum_n\left[\frac{\langle 0|\hat{A}|n\rangle\langle n|\hat{B}|0\rangle}{\omega_{n0}-z}+\frac{\langle 0|\hat{B}|n\rangle\langle n|\hat{A}|0\rangle}{\omega_{n0}+z}\right]. \tag{5.289}$$

Since it is not clear from the general operator labels, we choose to introduce the frequency as a subscript of the response function. We will use this convention elsewhere also for real frequencies when the frequency is not given by the context.

We recall that the polarizability $\alpha_{\alpha\beta}(-\omega;\omega)$ is obtained from Eq. (5.288) by letting $\hat{\Omega} = \hat{\mu}_\alpha$ and $\hat{V}_\beta^\omega = -\hat{\mu}_\beta$. If we assume a set of common damping parameters ($\gamma_{n0} = \gamma$), the formula for the polarizability will take the form

$$\alpha_{\alpha\beta}(-\omega;\omega) = \frac{1}{\hbar}\sum_n \left[\frac{\langle 0|\hat{\mu}_\alpha|n\rangle\langle n|\hat{\mu}_\beta|0\rangle}{\omega_{n0}-\omega-i\gamma} + \frac{\langle 0|\hat{\mu}_\beta|n\rangle\langle n|\hat{\mu}_\alpha|0\rangle}{\omega_{n0}+\omega+i\gamma}\right], \tag{5.290}$$

or, equivalently, for real wave functions

$$\alpha_{\alpha\beta}(-\omega;\omega) = \alpha^R_{\alpha\beta}(-\omega;\omega) + i\,\alpha^I_{\alpha\beta}(-\omega;\omega), \tag{5.291}$$

with

$$\alpha^R_{\alpha\beta} = \frac{1}{\hbar}\sum_n \left[\frac{\langle 0|\hat{\mu}_\alpha|n\rangle\langle n|\hat{\mu}_\beta|0\rangle(\omega_{n0}-\omega)}{(\omega_{n0}-\omega)^2+\gamma^2} + \frac{\langle 0|\hat{\mu}_\beta|n\rangle\langle n|\hat{\mu}_\alpha|0\rangle(\omega_{n0}+\omega)}{(\omega_{n0}+\omega)^2+\gamma^2}\right], \tag{5.292}$$

$$\alpha^I_{\alpha\beta} = \frac{\gamma}{\hbar}\sum_n \left[\frac{\langle 0|\hat{\mu}_\alpha|n\rangle\langle n|\hat{\mu}_\beta|0\rangle}{(\omega_{n0}-\omega)^2+\gamma^2} - \frac{\langle 0|\hat{\mu}_\beta|n\rangle\langle n|\hat{\mu}_\alpha|0\rangle}{(\omega_{n0}+\omega)^2+\gamma^2}\right]. \tag{5.293}$$

For optical frequencies in the proximity of a transition frequency, the dispersion of the real and imaginary parts of the polarizability are dictated by the first terms in Eqs. (5.292) and (5.293), respectively. At least the diagonal components of the α^R-tensor are positive below the first resonance. At resonance, with neglect made of the second term in Eq. (5.292), the real part of the polarizability crosses zero and changes sign. The imaginary part of the polarizability, on the other hand, is zero in the limit of static fields, and it has maxima at the resonance frequencies. We also note that in the limit of small frequencies, α^I has a linear dependence on the damping parameter γ. With respect to sign inversion of the optical frequency, it can be seen that

$$\alpha^R_{\alpha\beta}(\omega;-\omega) = \alpha^R_{\alpha\beta}(-\omega;\omega), \tag{5.294}$$

$$\alpha^I_{\alpha\beta}(\omega;-\omega) = -\alpha^I_{\alpha\beta}(-\omega;\omega). \tag{5.295}$$

This result was previously obtained in Eq. (5.63).

The resonance term of $\alpha^I_{\alpha\beta}$ is identified as Af with

$$f(\omega;\omega_{n0},\gamma) = \frac{1}{\pi}\frac{\gamma}{(\omega_{n0}-\omega)^2+\gamma^2}; \quad A = \frac{\langle 0|\hat{\mu}_\alpha|n\rangle\langle n|\hat{\mu}_\beta|0\rangle}{\hbar/\pi}, \tag{5.296}$$

where we recognize f as the Cauchy distribution introduced in Eq. (3.455) that provides a Lorentzian profile with amplitude $1/(\gamma\pi)$ and a half width at half maximum HWHM $= \gamma$. In Chapter 3 we also learned that the Cauchy distribution provides a representation of the Dirac delta function [cf. Eq. (3.459)]. This mathematical property of the resonance term allows us to relate the isotropic average of the α^I-tensor to the oscillator strength, Eq. (5.230), according to

$$\frac{2m_e\omega}{\pi e^2}\lim_{\gamma\to 0}\overline{\alpha}^I(-\omega;\omega) = \sum_n f_{n0}\delta(\omega_{n0}-\omega). \tag{5.297}$$

Due to the connection between absorption strengths and oscillator strengths, there is reason to expect the imaginary part of the polarizability to be associated with linear absorption.

For the two-level system described in Figure 5.1, we obtain a polarizability as illustrated in Figure 5.25. In this case, we have chosen an off-diagonal damping parameter $\gamma_{ab} = \Gamma/2 = 0.02$. The properties of the complex polarizability are nicely illustrated in this simple example by the upper plot of the real and imaginary parts of the polarizability. However, when considering the induced polarization, it will be beneficial to view the polarizability on polar form

$$\alpha(-\omega;\omega) = |\alpha(-\omega;\omega)|e^{i\phi}. \tag{5.298}$$

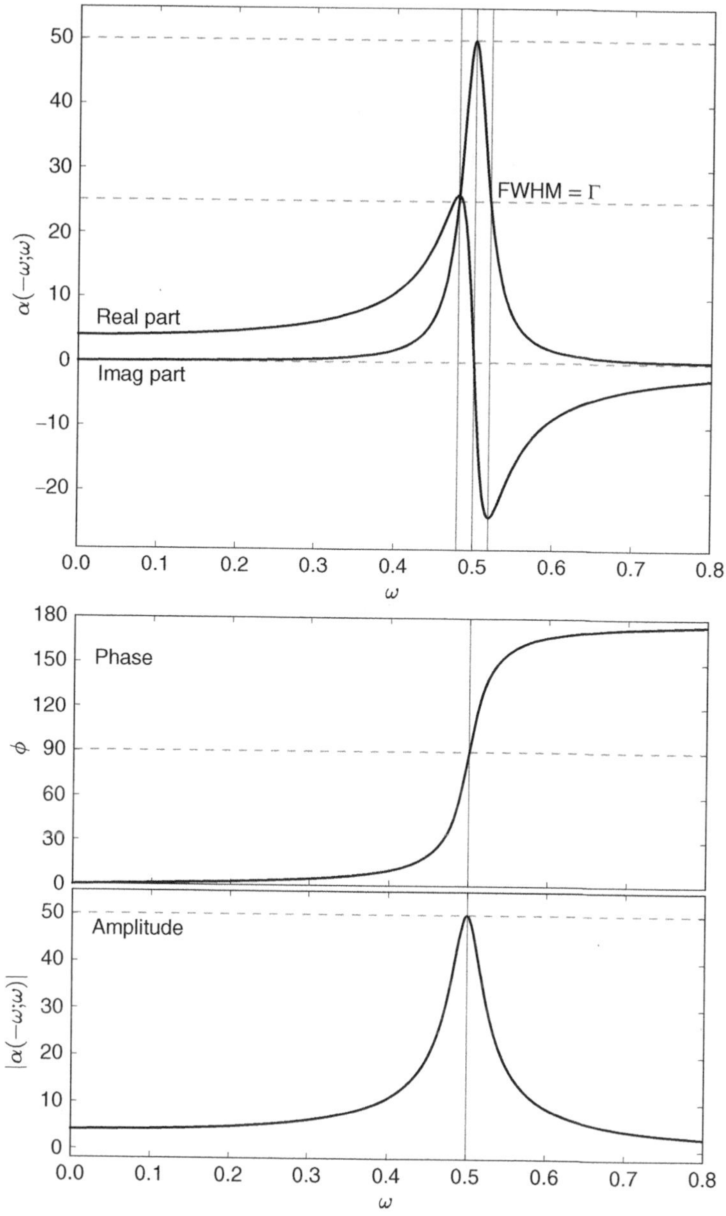

Figure 5.25 The upper illustration shows the real and imaginary parts of the linear polarizability for the two-level system. The full width at half maximum of the imaginary part is equal to Γ. The numerical value of Γ is here chosen as 0.04. The lower illustration shows the same polarizability but plotted on a polar form.

If we consider a monochromatic perturbation with a real field amplitude, we get an induced linear polarization that is

$$\mu^{\text{ind}}(t) = 2|\alpha(-\omega;\omega)|F^{\omega}\cos(\omega t - \phi), \tag{5.299}$$

which better illustrates the fact that $\mu^{\text{ind}}(t)$ is proportional to the absolute value of the polarizability and that in resonance regions where absorption is significant, there is a phase delay in the polarization as compared to the external field. This was previously discussed in relation to the numerical propagation of the Liouville equation, as shown in Figure 5.23.

5.10.5 Nonlinear Response Functions from the Density Matrix

The first-order nonlinear response function is identified from the second-order correction to the expectation value. In the limit of $\epsilon \to 0$, it becomes

$$
\begin{aligned}
\langle \hat{\Omega} \rangle^{(2)} &= \mathrm{Tr}(\hat{\rho}^{(2)} \hat{\Omega}) = \sum_{mn} \rho_{mn}^{(2)} \Omega_{nm} \\
&= \frac{1}{\hbar^2} \sum_{\omega_1 \omega_2} \frac{1}{2} \sum \mathcal{P}_{1,2} \sum_{mn} \\
&\times \left[\frac{\langle 0|\hat{V}_\beta^{\omega_1}|m\rangle\langle m|\hat{\Omega}|n\rangle\langle n|\hat{V}_\gamma^{\omega_2}|0\rangle}{(\omega_{m0} + \omega_1 + i\gamma_{m0})(\omega_{nm} - \omega_\sigma - i\gamma_{nm})} + \frac{\langle 0|\hat{V}_\beta^{\omega_1}|m\rangle\langle m|\hat{\Omega}|n\rangle\langle n|\hat{V}_\gamma^{\omega_2}|0\rangle}{(\omega_{n0} - \omega_2 - i\gamma_{n0})(\omega_{mn} + \omega_\sigma + i\gamma_{mn})} \right. \\
&\left. + \frac{\langle 0|\hat{\Omega}|m\rangle\langle m|\hat{V}_\beta^{\omega_1}|n\rangle\langle n|\hat{V}_\gamma^{\omega_2}|0\rangle}{(\omega_{n0} - \omega_2 - i\gamma_{n0})(\omega_{m0} - \omega_\sigma - i\gamma_{m0})} + \frac{\langle 0|\hat{V}_\beta^{\omega_1}|m\rangle\langle m|\hat{V}_\gamma^{\omega_2}|n\rangle\langle n|\hat{\Omega}|0\rangle}{(\omega_{m0} + \omega_1 + i\gamma_{m0})(\omega_{n0} + \omega_\sigma + i\gamma_{n0})} \right] \\
&\times F_\beta^{\omega_1} F_\gamma^{\omega_2} e^{-i(\omega_1+\omega_2)t},
\end{aligned}
\tag{5.300}
$$

where we have enforced intrinsic symmetry in the external fields by application of the operator $\frac{1}{2}\sum \mathcal{P}_{1,2}$ and in some terms interchanged the dummy summation indices n and m. After comparison with Eq. (5.150), we identify the first-order nonlinear response function as

$$
\begin{aligned}
\langle\langle \hat{\Omega}; \hat{V}_\beta^{\omega_1}, \hat{V}_\gamma^{\omega_2} \rangle\rangle &= \frac{1}{\hbar^2} \sum \mathcal{P}_{1,2} \sum_{mn} \\
&\times \left[\frac{\langle 0|\hat{V}_\beta^{\omega_1}|m\rangle\langle m|\hat{\Omega}|n\rangle\langle n|\hat{V}_\gamma^{\omega_2}|0\rangle}{(\omega_{m0} + \omega_1 + i\gamma_{m0})(\omega_{nm} - \omega_\sigma - i\gamma_{nm})} + \frac{\langle 0|\hat{V}_\beta^{\omega_1}|m\rangle\langle m|\hat{\Omega}|n\rangle\langle n|\hat{V}_\gamma^{\omega_2}|0\rangle}{(\omega_{n0} - \omega_2 - i\gamma_{n0})(\omega_{mn} + \omega_\sigma + i\gamma_{mn})} \right. \\
&\left. + \frac{\langle 0|\hat{\Omega}|m\rangle\langle m|\hat{V}_\beta^{\omega_1}|n\rangle\langle n|\hat{V}_\gamma^{\omega_2}|0\rangle}{(\omega_{n0} - \omega_2 - i\gamma_{n0})(\omega_{m0} - \omega_\sigma - i\gamma_{m0})} + \frac{\langle 0|\hat{V}_\beta^{\omega_1}|m\rangle\langle m|\hat{V}_\gamma^{\omega_2}|n\rangle\langle n|\hat{\Omega}|0\rangle}{(\omega_{m0} + \omega_1 + i\gamma_{m0})(\omega_{n0} + \omega_\sigma + i\gamma_{n0})} \right].
\end{aligned}
\tag{5.301}
$$

In comparison to our previously derived expression for this response function Eq. (5.183), there is, in addition to the damping terms that appear, also a striking difference in the number of terms in the response function. In the nonresonant case, the first-order nonlinear response function contained six terms whereas there appear eight terms in the above expression. Terms number one and two in Eq. (5.301) together replace term number two in Eq. (5.183), and the two formulas must of course be identical in the limit of zero relaxation rates. We will postpone the comparison of the eighth-term and six-term formulas until we have derived the equivalent expression of Eq. (5.301) in wave function theory. We also note that the introduction of relaxation parameters not only makes the quadratic response function complex but also makes it lose the property of being overall symmetric; that is, in contrast to the nonresonant case we cannot re-write Eq. (5.301) by means of an operator $\sum \mathcal{P}_{\sigma,1,2}$. However, the intrinsic symmetry with respect to the perturbing fields remains, since as in the nonresonant case we are free to impose this symmetry on the response function without altering the observable.

5.10.6 Relaxation in Wave Function Theory

In wave function theory, we are concerned with systems in pure states for which the time evolution is governed by the Hamiltonian. However, while interactions with internal and external electromagnetic fields are readily incorporated into the Hamiltonian, there is no well-defined and rigorous way to include relaxation into the operator. Nevertheless, in exact-state theory, where eigenstates and eigenvalues of the Hamiltonian are available, relaxation can be introduced by modifying the matrix representation of the Hamiltonian of the isolated system as follows:

$$H_0 = \begin{pmatrix} E_0 & 0 & \cdots & 0 \\ 0 & E_1 + i\Gamma_1/2 & \cdots & 0 \\ \vdots & \vdots & \ddots & \vdots \\ 0 & 0 & \cdots & E_n + i\Gamma_n/2 \end{pmatrix}, \tag{5.302}$$

where Γ_n is the previously defined inverse lifetime (or decay rate) of state $|n\rangle$. There is (at least) one severe issue of concern with this approach, namely that by substituting the energies by their complex counterparts as carried out in Eq. (5.302), the Hamiltonian will no longer be Hermitian and, as one of many consequences, the norm of the wave function will not be conserved. In order to resolve this issue, we return to our formulation of wave function dynamics based on explicitly unitary parameterizations and the Ehrenfest theorem.

Let us apply the Ehrenfest theorem Eq. (5.117) to a state-transfer operator $\hat{\Omega}_{nm} = |n\rangle\langle m|$. After expanding the commutator with the Hamiltonian and some re-arrangements, we get

$$\frac{\partial}{\partial t}\langle m|\psi(t)\rangle\langle\psi(t)|n\rangle = \frac{1}{i\hbar}[\langle m|\hat{H}|\psi(t)\rangle\langle\psi(t)|n\rangle - \langle m|\psi(t)\rangle\langle\psi(t)|\hat{H}|n\rangle]. \tag{5.303}$$

In this equation, we may identify the density operator $\hat{\rho}(t)$ as $|\psi(t)\rangle\langle\psi(t)|$, and the above equation is thus a reformulation of the Liouville equation Eq. (5.270) for density matrix element ρ_{mn}. This equation therefore serves as a suitable starting point to introduce relaxation into a wave function-based theory, and the resulting equation-of-motion becomes

$$\frac{\partial}{\partial t}\langle\psi|\hat{\Omega}_{nm}|\psi\rangle = \frac{1}{i\hbar}\langle\psi|[\hat{\Omega}_{nm}, \hat{H}]|\psi\rangle - \gamma_{mn}[\langle\psi|\hat{\Omega}_{nm}|\psi\rangle - \langle\psi^{\text{eq}}|\hat{\Omega}_{nm}|\psi^{\text{eq}}\rangle], \tag{5.304}$$

to be compared with Eq. (5.271). As pointed out for the case of a pure state, there is a redundancy connected to the density operator description and this is reflected by the coherent off-diagonal elements of the density matrix. This is the reason why we only need to consider state-transfer operators $\hat{\Omega}_{n0}$ and $\hat{\Omega}_{0n}$ with a set of corresponding relaxation parameters $\gamma_{n0} = \gamma_{0n}$ when using a state-vector description of the system. Situations that require a statistical mixture of states and involve incoherent density matrix elements are thus not included in a wave-function approach based on Eq. (5.304).

The solution of Eq. (5.304) is found by means of perturbation theory, and as an algebraic exercise it differs very little from the work carried out in Section 5.4.2. We will therefore not go through the algebraic details once more, but rather provide some intermediate and the final results. The time dependencies of the rotation amplitudes with inclusion of relaxation are given by

$$P_n^{(1)} = \frac{1}{i\hbar}\sum_{\omega}\frac{\langle n|\hat{V}_\alpha^\omega|0\rangle F_\alpha^\omega e^{-i\omega t}e^{\epsilon t}}{\omega_{n0} - \omega - i\gamma_{n0} - i\epsilon}, \tag{5.305}$$

$$P_n^{(2)} = -\frac{1}{i\hbar^2}\sum_{\omega_1,\omega_2}\sum_{m>0}\frac{\langle n|\overline{\hat{V}_\alpha^{\omega_1}}|m\rangle\langle m|\hat{V}_\beta^{\omega_2}|0\rangle F_\alpha^{\omega_1}F_\beta^{\omega_2}e^{-i(\omega_1+\omega_2)t}e^{2\epsilon t}}{(\omega_{n0} - (\omega_1+\omega_2) - i\gamma_{n0} - i2\epsilon)(\omega_{m0} - \omega_2 - i\gamma_{m0} - i\epsilon)}, \tag{5.306}$$

and the first-order correction to the wave function is

$$|\overline{\psi}^{(1)}\rangle = -\frac{1}{\hbar}\sum_{n>0}\sum_{\omega}\frac{\langle n|\hat{V}_\alpha^\omega|0\rangle F_\alpha^\omega e^{-i\omega t}e^{\epsilon t}}{\omega_{n0} - \omega - i\gamma_{n0} - i\epsilon}|n\rangle. \tag{5.307}$$

The expression for the second-order correction to the wave function is omitted here but can be found by combining Eqs. (5.132), (5.305), and (5.306). With the first- and second-order corrections of the wave function at hand, we can collect the corrections to the expectation value of an operator $\hat{\Omega}$ and identify the linear and first-order nonlinear response functions. The expressions for these response functions are

$$\langle\langle \hat{\Omega}; \hat{V}_\beta^\omega \rangle\rangle = -\frac{1}{\hbar}\sum_{n>0}\left[\frac{\langle 0|\hat{\Omega}|n\rangle\langle n|\hat{V}_\beta^\omega|0\rangle}{\omega_{n0}-\omega-i\gamma_{n0}-i\epsilon} + \frac{\langle 0|\hat{V}_\beta^\omega|n\rangle\langle n|\hat{\Omega}|0\rangle}{\omega_{n0}+\omega+i\gamma_{n0}+i\epsilon}\right], \tag{5.308}$$

$$\begin{aligned}\langle\langle \hat{\Omega}; \hat{V}_\beta^{\omega_1}, \hat{V}_\gamma^{\omega_2} \rangle\rangle = \frac{1}{\hbar^2}\sum \mathcal{P}_{1,2}\sum_{m,n>0} &\times\left[\frac{\langle 0|\hat{\Omega}|m\rangle\langle m|\overline{\hat{V}_\beta^{\omega_1}}|n\rangle\langle n|\hat{V}_\gamma^{\omega_2}|0\rangle}{(\omega_{m0}-\omega_\sigma-i\gamma_{m0}-2i\epsilon)(\omega_{n0}-\omega_2-i\gamma_{n0}-i\epsilon)}\right.\\ &+\frac{\langle 0|\hat{V}_\beta^{\omega_1}|m\rangle\langle m|\overline{\hat{\Omega}}|n\rangle\langle n|\hat{V}_\gamma^{\omega_2}|0\rangle}{(\omega_{m0}+\omega_1+i\gamma_{m0}+i\epsilon)(\omega_{n0}-\omega_2-i\gamma_{n0}-i\epsilon)}\\ &\left.+\frac{\langle 0|\hat{V}_\beta^{\omega_1}|m\rangle\langle m|\overline{\hat{V}_\gamma^{\omega_2}}|n\rangle\langle n|\hat{\Omega}|0\rangle}{(\omega_{m0}+\omega_1+i\gamma_{m0}+i\epsilon)(\omega_{n0}+\omega_\sigma+i\gamma_{n0}+2i\epsilon)}\right].\end{aligned} \tag{5.309}$$

The linear response function derived here is seen to be identical to that obtained using the density matrix formalism, Eq. (5.288), but a discrepancy is noted when the same comparison is made for the nonlinear response function, Eq. (5.301). We have here presented the expressions based on summations that exclude the ground state, whereas those obtained from the density matrix formalism include the ground state. However, for the linear response function, the $n = 0$ term vanishes so that the different summations may be interchanged freely for oscillating fields, and for the first-order nonlinear response function, they may be interchanged if one also removes the overbars denoting fluctuation operators. Apart from the issue of including the ground state or not in the summations, there is still a difference in the number of terms in the nonlinear response function. In wave function theory we obtain six terms but in density matrix theory we obtain eight. Let us make a detailed comparison of the two formulas to identify the reason for this difference.

The first-order hyperpolarizability as determined from Eq. (5.309) will take the form

$$\begin{aligned}\beta_{\alpha\beta\gamma}(-\omega_\sigma;\omega_1,\omega_2) = \frac{1}{\hbar^2}\sum \mathcal{P}_{1,2}{\sum_{m,n}}' &\left[\frac{\langle 0|\hat{\mu}_\alpha|m\rangle\langle m|\overline{\hat{\mu}_\beta}|n\rangle\langle n|\hat{\mu}_\gamma|0\rangle}{(\omega_{m0}-\omega_\sigma-i\gamma_{m0})(\omega_{n0}-\omega_2-i\gamma_{n0})}\right.\\ &+\frac{\langle 0|\hat{\mu}_\beta|m\rangle\langle m|\overline{\hat{\mu}_\alpha}|n\rangle\langle n|\hat{\mu}_\gamma|0\rangle}{(\omega_{m0}+\omega_1+i\gamma_{m0})(\omega_{n0}-\omega_2-i\gamma_{n0})}\\ &\left.+\frac{\langle 0|\hat{\mu}_\beta|m\rangle\langle m|\overline{\hat{\mu}_\gamma}|n\rangle\langle n|\hat{\mu}_\alpha|0\rangle}{(\omega_{m0}+\omega_1+i\gamma_{m0})(\omega_{n0}+\omega_\sigma+i\gamma_{n0})}\right],\end{aligned} \tag{5.310}$$

whereas the expression for the same property based on Eq. (5.301) will take the form

$$\begin{aligned}\beta_{\alpha\beta\gamma}(-\omega_\sigma;\omega_1,\omega_2) = \frac{1}{\hbar^2}\sum \mathcal{P}_{1,2}\sum_{m,n} &\left[\frac{\langle 0|\hat{\mu}_\alpha|m\rangle\langle m|\hat{\mu}_\beta|n\rangle\langle n|\hat{\mu}_\gamma|0\rangle}{(\omega_{m0}-\omega_\sigma-i\gamma_{m0})(\omega_{n0}-\omega_2-i\gamma_{n0})}\right.\\ &+\frac{\langle 0|\hat{\mu}_\beta|m\rangle\langle m|\hat{\mu}_\alpha|n\rangle\langle n|\hat{\mu}_\gamma|0\rangle}{(\omega_{nm}-\omega_\sigma-i\gamma_{nm})(\omega_{m0}+\omega_1+i\gamma_{m0})}\end{aligned}$$

$$+\frac{\langle 0|\hat{\mu}_\beta|m\rangle\langle m|\hat{\mu}_\alpha|n\rangle\langle n|\hat{\mu}_\gamma|0\rangle}{(\omega_{mn}+\omega_\sigma+i\gamma_{mn})(\omega_{n0}-\omega_2-i\gamma_{n0})}$$
$$\left.+\frac{\langle 0|\hat{\mu}_\beta|m\rangle\langle m|\hat{\mu}_\gamma|n\rangle\langle n|\hat{\mu}_\alpha|0\rangle}{(\omega_{m0}+\omega_1+i\gamma_{m0})(\omega_{n0}+\omega_\sigma+i\gamma_{n0})}\right]. \tag{5.311}$$

The second term in Eq. (5.310) is replaced by terms number two and three in Eq. (5.311), and we rewrite the second and third terms of the latter equation in order to identify what separates the two formulas. The numerators are identical and can therefore be left aside, and for the denominators, we get

$$\begin{aligned}
&\frac{1}{(\omega_{nm}-\omega_\sigma-i\gamma_{nm})(\omega_{m0}+\omega_1+i\gamma_{m0})}+\frac{1}{(\omega_{mn}+\omega_\sigma+i\gamma_{mn})(\omega_{n0}-\omega_2-i\gamma_{n0})}\\
&=\frac{1}{(\omega_{nm}-\omega_\sigma-i\gamma_{nm})}\left[\frac{1}{(\omega_{m0}+\omega_1+i\gamma_{m0})}-\frac{1}{(\omega_{n0}-\omega_2-i\gamma_{n0})}\right]\\
&=\frac{1}{(\omega_{nm}-\omega_\sigma-i\gamma_{nm})}\left[\frac{\omega_{nm}-\omega_\sigma-i(\gamma_{n0}+\gamma_{m0})}{(\omega_{m0}+\omega_1+i\gamma_{m0})(\omega_{n0}-\omega_2-i\gamma_{n0})}\right]\\
&=\frac{1}{(\omega_{m0}+\omega_1+i\gamma_{m0})(\omega_{n0}-\omega_2-i\gamma_{n0})}\left[1+\frac{i(\gamma_{nm}-\gamma_{n0}-\gamma_{m0})}{(\omega_{nm}-\omega_\sigma-i\gamma_{nm})}\right].
\end{aligned} \tag{5.312}$$

The first term is identified as the second term in Eq. (5.310), and the second term thereby constitutes the difference between the formula based on wave function theory and the formula based on density matrix theory. The difference term will thus be

$$\Delta=\frac{1}{\hbar^2}\sum\mathcal{P}_{1,2}\sum_{m,n}\frac{i(\gamma_{nm}-\gamma_{n0}-\gamma_{m0})\langle 0|\hat{\mu}_\beta|m\rangle\langle m|\hat{\mu}_\alpha|n\rangle\langle n|\hat{\mu}_\gamma|0\rangle}{(\omega_{nm}-\omega_\sigma-i\gamma_{nm})(\omega_{m0}+\omega_1+i\gamma_{m0})(\omega_{n0}-\omega_2-i\gamma_{n0})}. \tag{5.313}$$

This term displays two-photon resonances at frequencies that correspond to differences in transition frequencies between two excited states, whereas the "ordinary" resonances always occur at excited-to-ground state transition frequencies. However, the strengths of these additional resonances are proportional to the factor

$$(\gamma_{nm}-\gamma_{n0}-\gamma_{m0}),$$

which, if there are no mechanisms of dephasing in the system, is zero. Terms in Eq. (5.311) with either n or m equal to zero are readily shown to correspond to the contributions from the $\langle 0|\bar{\hat{\mu}}|0\rangle$ component in the fluctuation operator in Eq. (5.310). Therefore, our conclusion is that for systems in coherent pure states, the introduction of relaxation parameters in the Ehrenfest theorem is a sound approach to take in wave function theory, yielding response functions in agreement with those obtained under the same conditions in a density matrix theory.

5.10.7 Absorption Cross Section

In Section 5.8 we derived Fermi's golden rule, but also pointed out that it strictly only applies to transitions from a discrete state to a continuum of states, yet it is often applied to transitions between discrete states. This may be justified by taking into consideration that the excited states have finite lifetimes, leading to spectral broadenings that can be viewed as an "energy smearing." We have seen how this situation can be handled by the introduction of damping parameters γ_{f0} and, in particular, the first-order wave function amplitudes in this framework will be given by

$$d_f^{(1)}(t)=-\frac{1}{\hbar}\left[V_{f0}(\omega)\frac{e^{i(\omega_{f0}-\omega)t}}{\omega_{f0}-i\gamma_{f0}-\omega-i\epsilon}+V_{0f}^*(\omega)\frac{e^{i(\omega_{f0}+\omega)t}}{\omega_{f0}-i\gamma_{f0}+\omega-i\epsilon}\right]e^{\epsilon t}. \tag{5.314}$$

We note that the corresponding undamped wave function amplitude is found in Eq. (5.242). When taking the limit $\epsilon \to 0$ for transition rates and keeping only the first resonant term, we obtain

$$w_{f\leftarrow 0}(\omega) = \lim_{\epsilon\to 0} w^{\epsilon}_{f\leftarrow 0} = \frac{2\pi}{\hbar^2}|V_{f0}(\omega)|^2 f(\omega;\omega_{f0},\gamma_{f0}), \tag{5.315}$$

which provides the Lorentzian lineshape that, in a resonant-divergent formalism, would have to be added by hand.

Let us consider a monochromatic light source in the electric dipole approximation. The interaction operator is then given by

$$\hat{V}(t) = -\hat{\mu}_\alpha F^{\omega}_{\alpha}[e^{-i\omega t} + e^{+i\omega t}], \tag{5.316}$$

and the transition rate in Eq. (5.315) becomes

$$w_{f\leftarrow 0} = \frac{2\pi}{\hbar^2}|\langle f|\hat{\mu}_\alpha F^{\omega}_{\alpha}|0\rangle|^2 f(\omega;\omega_{f0},\gamma_{f0}). \tag{5.317}$$

We will consider the situation of freely rotating, noninteracting molecules and therefore carry out a rotational average. Using our results from Section 3.5.5, we find that the isotropic transition rate is

$$w^{\mathrm{iso}}_{f\leftarrow 0}(\omega) = \frac{2\pi}{3\hbar^2}\sum_{\alpha=x,y,z}|\langle f|\hat{\mu}_\alpha|0\rangle|^2 (F^{\omega})^2 f(\omega;\omega_{f0},\gamma_{f0}), \tag{5.318}$$

where F^{ω} denotes the norm of the electric field strength vector. Comparing with the expression for the intensity of an electromagnetic plane wave given in Eq. (3.134), we see that the isotropic transition rate can be written as

$$w^{\mathrm{iso}}_{f\leftarrow 0}(\omega) = \frac{\pi}{3\epsilon_0 c\hbar^2}\sum_{\alpha=x,y,z}|\langle f|\hat{\mu}_\alpha|0\rangle|^2 I(\omega) f(\omega;\omega_{f0},\gamma_{f0}). \tag{5.319}$$

We remind the reader that we are here working with an electric field expanded in terms of the complex exponential function, whereas the expression in Eq. (3.134) is based on the use of a cosine expansion. This difference gives rise to an additional factor of 4. The rate of energy transfer from the electromagnetic radiation to the molecule is

$$\hbar\omega w^{\mathrm{iso}}_{f\leftarrow 0} = I(\omega)\sigma(\omega), \tag{5.320}$$

showing that the absorption cross-section can be expressed as

$$\sigma(\omega) = \frac{\pi\omega}{3\epsilon_0 c\hbar}\sum_{\alpha=x,y,z}|\langle f|\hat{\mu}_\alpha|0\rangle|^2 f(\omega;\omega_{f0},\gamma_{f0}), \tag{5.321}$$

or, in terms of the dimensionless oscillator strength f_{f0} defined in Eq. (5.228), we get

$$\sigma(\omega) = \frac{\pi e^2}{2\epsilon_0 c m_e} f_{f0}\, f(\omega;\omega_{f0},\gamma_{f0}). \tag{5.322}$$

In the last step, we have made use of the fact that the lineshape function is sharply peaked around the transition frequency, allowing us to replace ω with ω_{f0} before substituting for the oscillator strength. We again see that by accounting for relaxation in response theory, we automatically get the line broadening that otherwise would have had to be added by hand.

We can alternatively introduce absorption of photons by molecular systems in terms of the mechanical work performed on the particles by the electromagnetic radiation. The rate of this work is given in Eq. (3.126) and, based on the argument of energy conservation, we can conclude that the rate of absorbed electromagnetic radiation energy must equal this rate of work. To remove the oscillatory behavior of the rate of absorption in Eq. (3.126), it is time averaged over one period of the oscillation, and by expanding the current density, $\mathbf{j} = \mathbf{j}^{(0)} + \mathbf{j}^{(1)} + \cdots$, in orders of the electric field strength, the rate of absorbed energy can be expressed as a power series in the light intensity

$$\left\langle \frac{dW}{dt} \right\rangle_T = \mathcal{N}\sigma(\omega)I(\omega) + \ldots, \tag{5.323}$$

where $\mathcal{N}$ is the number density of the sample. Note that terms of odd order in the perturbation have only periodic contributions, which vanish upon time averaging and thus do not contribute to light absorption.

The current density for a sample of nonmagnetic, closed-shell molecules is given by

$$\mathbf{j}(\mathbf{r}, t) = \mathcal{N}\langle \psi(t)|\hat{\mathbf{j}}(\mathbf{r}, t)|\psi(t)\rangle, \tag{5.324}$$

where the current density operator is

$$\hat{\mathbf{j}}(\mathbf{r}, t) = -\frac{e}{2m_e}\sum_{i=1}^{N_e}[\hat{\mathbf{p}}_i + e\mathbf{A}(\mathbf{r}_i, t), \delta(\mathbf{r}_i - \mathbf{r})]_+. \tag{5.325}$$

In linear absorption, the second term in the commutator involving the vector potential $\mathbf{A}(\mathbf{r}, t)$ can be disregarded and we are left with

$$\mathbf{j}(\mathbf{r}, t) = -\mathcal{N}\sum_{i=1}^{N_e}\left[\frac{e}{2m_e}\langle \psi(t)|\hat{\mathbf{p}}_i\delta(\mathbf{r}_i - \mathbf{r}) + \delta(\mathbf{r}_i - \mathbf{r})\hat{\mathbf{p}}_i|\psi(t)\rangle\right]. \tag{5.326}$$

The first-order current density $\mathbf{j}^{(1)}$ is obtained by means of the first-order correction to the phase-isolated wave function $|\overline{\psi}^{(1)}\rangle$ in Eq. (5.131) and the accompanying amplitudes $P_n^{(1)}$ in Eq. (5.137). Using the phase-isolated wave function for the evaluation of the above expectation value is legitimate since the operator does not involve an operation of time differentiation, in fact it is, in this case, time independent. All things taken into account, we arrive at the following first-order expression for the absorbed energy per unit time

$$\begin{aligned}
\frac{dW}{dt} &= \int \mathbf{E}\cdot\mathbf{j}^{(1)}\, d^3\mathbf{r} \\
&= \mathcal{N}\sum_{\omega_1}\omega_1\sum_{n>0}\left\{P_n^{(1)}(t)\int\langle 0|\mathbf{A}^{\omega_1}\cdot\hat{\mathbf{j}}(\mathbf{r})|n\rangle\, d\mathbf{r} - P_n^{(1)*}(t)\int\langle n|\mathbf{A}^{\omega_1}\cdot\hat{\mathbf{j}}(\mathbf{r})|0\rangle\, d\mathbf{r}\right\}e^{-i\omega_1 t} \\
&= i\mathcal{N}\sum_{\omega_1,\omega_2}\omega_1\frac{e}{2m_e\hbar}\sum_{n>0}\{(\langle 0|\hat{\mathbf{p}}\cdot\hat{\mathbf{A}}^{\omega_1}|n\rangle + \langle 0|\hat{\mathbf{A}}^{\omega_1}\cdot\hat{\mathbf{p}}|n\rangle)\left[\frac{\langle n|\hat{V}^{\omega_2}|0\rangle}{\omega_{n0} - \omega_2 - i\gamma_{n0}}\right] \\
&\quad + \left[\frac{\langle 0|\hat{V}^{\omega_2}|n\rangle}{\omega_{n0} + \omega_2 + i\gamma_{n0}}\right](\langle n|\hat{\mathbf{p}}\cdot\hat{\mathbf{A}}^{\omega_1}|0\rangle + \langle n|\hat{\mathbf{A}}^{\omega_1}\cdot\hat{\mathbf{p}}|0\rangle)\}e^{-i(\omega_1+\omega_2)t} \\
&= -i\mathcal{N}\sum_{\omega_1,\omega_2}\omega_1\left\{-\frac{1}{\hbar}\sum_{n>0}\left[\frac{\langle 0|\hat{V}^{\omega_1}|n\rangle\langle n|\hat{V}^{\omega_2}|0\rangle}{\omega_{n0} - \omega_2 - i\gamma_{n0}} + \frac{\langle 0|\hat{V}^{\omega_2}|n\rangle\langle n|\hat{V}^{\omega_1}|0\rangle}{\omega_{n0} + \omega_2 + i\gamma_{n0}}\right]\right\}e^{-i(\omega_1+\omega_2)t} \\
&= -i\mathcal{N}\sum_{\omega_1,\omega_2}\omega_1\langle\langle \hat{V}^{\omega_1}; \hat{V}^{\omega_2}\rangle\rangle e^{-i(\omega_1+\omega_2)t}.
\end{aligned} \tag{5.327}$$

When we perform the operation of time averaging over one period of oscillation in the mono-chromatic external field, only two terms will survive, both with $\omega_1 = -\omega_2$. We get

$$\left\langle \frac{dW}{dt} \right\rangle_T = i\mathcal{N}\omega[\langle\langle \hat{V}^{-\omega}; \hat{V}^{\omega}\rangle\rangle - \langle\langle \hat{V}^{-\omega}; \hat{V}^{\omega}\rangle\rangle] = -2\mathcal{N}\omega\,\mathrm{Im}\{\langle\langle \hat{V}^{-\omega}; \hat{V}^{\omega}\rangle\rangle\}. \tag{5.328}$$

In the above derivation, we have made use of the expression for the Hamiltonian in the Coulomb gauge in Eq. (3.154). The term that is linear in the external field is in our case given by

$$\hat{V}^{\omega} = \frac{e}{m_e}\hat{\mathbf{A}}^{\omega} \cdot \hat{\mathbf{p}} = \frac{e}{im_e\omega}\hat{p}_\alpha F_\alpha^{\omega}. \tag{5.329}$$

In Section 7.2, we will derive dipole equivalence relations for response functions based on the commutator relation between position and momentum operators. For the complex response function with a common damping term $\gamma_{n0} = \gamma$, the dipole equivalence relation corresponding to Eq. (7.66) takes the form

$$\langle\langle \hat{p}_\alpha; \hat{p}_\beta \rangle\rangle_\omega = m_e[m_e(\omega + i\gamma)^2 \langle\langle \hat{r}_\alpha; \hat{r}_\beta \rangle\rangle_\omega - N_e\delta_{\alpha\beta}]. \tag{5.330}$$

In the limit when $\gamma \ll \omega$, we have

$$\mathrm{Im}\{\langle\langle \hat{p}_\alpha; \hat{p}_\beta \rangle\rangle_\omega\} = m_e^2\omega^2\, \mathrm{Im}\{\langle\langle \hat{r}_\alpha; \hat{r}_\beta \rangle\rangle_\omega\}, \tag{5.331}$$

so that we arrive at

$$\left\langle \frac{dW}{dt} \right\rangle_T = -2\mathcal{N} e^2\omega\, \mathrm{Im}\{\langle\langle \hat{r}_\alpha; \hat{r}_\beta \rangle\rangle_\omega\}(F_\alpha^{\omega})^* F_\beta^{\omega}. \tag{5.332}$$

In this expression, we can identify the electric dipole polarizability

$$\alpha_{\alpha\beta}(-\omega; \omega) = -e^2\langle\langle \hat{r}_\alpha; \hat{r}_\beta \rangle\rangle_\omega, \tag{5.333}$$

which, for an isotropic sample of freely rotating molecules, leaves us at

$$\left\langle \frac{dW}{dt} \right\rangle_T = \mathcal{N}\frac{\omega}{\epsilon_0 c}\mathrm{Im}\{\overline{\alpha}(-\omega; \omega)\}I(\omega) \tag{5.334}$$

or, by comparison to Eq. (5.323),

$$\sigma(\omega) = \frac{\omega}{\epsilon_0 c}\mathrm{Im}\{\overline{\alpha}(-\omega; \omega)\}. \tag{5.335}$$

By means of Eqs. (5.296) and (5.297), it is seen that the above expression for the linear absorption cross-section in terms of the imaginary part of the polarizability is in perfect agreement with that in Eq. (5.322). The technique used here can be carried out to higher orders in the external field and one thereby obtains the corresponding relations for multiphoton absorption cross-sections. Specifically, the two-photon absorption cross-section will become proportional to the imaginary part of the second-order hyperpolarizability $\gamma(-\omega; \omega, -\omega, \omega)$.

5.10.8 Einstein Coefficients

We now consider a collection of molecules in a radiation field causing transitions between levels i and f

$$i + \gamma \rightleftharpoons f. \tag{5.336}$$

We assume that the system is in thermal equilibrium. Following Einstein, we write the transition rate of absorption as

$$w_{f\leftarrow i} = B_{f\leftarrow i}I(\omega_{fi}). \tag{5.337}$$

Lineshapes are ignored and, in the electric dipole approximation, we can immediately from Eq. (5.319) read off the Einstein coefficient of absorption as

$$B_{f\leftarrow i} = \frac{\pi}{3\varepsilon_0 c\hbar^2}\sum_{\alpha=x,y,z} |\langle f|\hat{\mu}_\alpha|i\rangle|^2. \tag{5.338}$$

The transition rate of the corresponding emission rate is expressed as

$$w_{f\to i} = B_{f\to i} I(\omega_{fi}), \tag{5.339}$$

and it is found by switching initial and final states and by using the emission part of Eq. (5.314). We find that the Einstein coefficient of stimulated emission is equal to that of absorption, that is

$$B_{f\to i} = B_{f\leftarrow i}. \tag{5.340}$$

A problem occurs if no further processes are taken into account, since, at equilibrium, we would have

$$w_{f\leftarrow i} N_i = w_{f\to i} N_f, \tag{5.341}$$

where N_i and N_f are the number of molecules in each state. However, since the two transition rates are the same, the populations of the initial and final states would be identical as well, and this situation would be in conflict with the Boltzmann distribution

$$\frac{N_f}{N_i} = e^{-\hbar\omega_{fi}/kT}. \tag{5.342}$$

Such considerations led Einstein to introduce a third process, namely spontaneous emission, which does not depend on any external field. Upon equilibrium, we now have

$$B_{f\leftarrow i} I(\omega_{fi}) N_i = (A_{f\to i} + B_{f\to i} I(\omega_{fi})) N_f. \tag{5.343}$$

Rearrangement gives

$$A_{f\to i} = (B_{f\leftarrow i} e^{E_{fi}/kT} - B_{f\to i}) I(\omega_{fi}), \tag{5.344}$$

or, equivalently

$$I(\omega_{fi}) = \frac{A_{f\to i}/B_{f\to i}}{(B_{f\leftarrow i}/B_{f\to i}) e^{\hbar\omega_{fi}/kT} - 1}. \tag{5.345}$$

Under condition of thermal equilibrium, the intensity is given by Planck's radiation law

$$I(\omega) = \frac{\hbar\omega^3/\pi^2 c^2}{e^{\hbar\omega/kT} - 1}. \tag{5.346}$$

Note that, at this point, we go beyond the semi-classical approach employed in this book, since the radiation law has been derived by imposing quantization of electromagnetic fields. A comparison of the equations above gives

$$A_{f\to i} = \frac{1}{3\pi\varepsilon_0 \hbar}\left(\frac{\omega}{c}\right)^3 \sum_{\alpha=x,y,z} |\langle f|\hat{\mu}_\alpha|i\rangle|^2, \tag{5.347}$$

which shows that the rate of spontaneous emission has a very strong cubic dependence on the radiation frequency.

Further Reading

1 R. W. Boyd. *Nonlinear Optics.* Academic Press, 2003.

2 O. Christiansen, P. Jørgensen, and C. Hättig. Response functions from Fourier component variational perturbation theory applied to a time-averaged quasienergy. *Int. J. Quantum Chem.*, 68:1–52, 1998.

3 T. Helgaker, S. Coriani, P. Jørgensen, K. Kristensen, J. Olsen, and K. Ruud. Recent advances in wave function-based methods of molecular-property calculations. *Chem. Rev.*, 112:543–631, 2012.
4 N. H. List, J. Kauczor, T. Saue, H. J. A. Jensen, and P. Norman. Beyond the electric-dipole approximation: a formulation and implementation of molecular response theory for the description of absorption of electromagnetic field radiation. *J. Chem. Phys.*, 142:244111, 2015.
5 S. Mukamel. *Principles of Nonlinear Optical Spectroscopy*. Oxford University Press, 1995.
6 P. Norman. A perspective on nonresonant and resonant electronic response theory for time-dependent molecular properties. *Phys. Chem. Chem. Phys.*, 13:20519–20535, 2011.
7 P. Norman, D. M. Bishop, H. J. A. Jensen, and J. Oddershede. Nonlinear response theory with relaxation: the first-order hyperpolarizability. *J. Chem. Phys.*, 123:194103, 2005.
8 J. Olsen and P. Jørgensen. Linear and nonlinear response functions for an exact state and for an MCSCF state. *J. Chem. Phys.*, 82:3235–3264, 1985.
9 T. Saue. Post Dirac–Hartree–Fock methods—properties. In P. Schwerdtfeger, editor, *Relativistic Electronic Structure Theory. Part 1. Fundamentals*, p. 332. Elsevier, Amsterdam, 2002.

6

Electronic and Nuclear Contributions to Molecular Properties

Nature isn't classical, dammit, and if you want to make a simulation of nature, you'd better make it quantum mechanical, and by golly it's a wonderful problem, because it doesn't look so easy.

R. Feynman (1982)

We are in this book concerned with motions of molecular charges and their interactions with internal and external sources of electromagnetic fields. In its most general form, this is a problem of insurmountable complexity and the introduction of approximations is required. It would be erroneous to believe that we somehow could disregard the motions of nuclei since the nuclei in many cases can play as important a role as the electrons, giving rise to prominent electromagnetic field interactions in quite separate frequency ranges. In this chapter, we will devote our attention to the nuclear motions and their interactions with external electromagnetic fields. We will consider both the scattering and absorption of electromagnetic radiation when interacting with nuclei, as well as the nuclear contributions to linear and nonlinear optical properties.

6.1 Born–Oppenheimer Approximation

In electronic structure theory, the presence of nuclei merely gives rise to a static external potential experienced by the electrons through the electron–nuclear attraction, as well as a constant nuclear repulsion energy contribution that is independent of the electronic charge density. As the nuclei in general will be moving rather slowly compared to the electrons, we can consider the nuclei to be nonrelativistic and assume the nuclear wave functions to be determined by the Schrödinger equation, even in cases where the motions of electrons are described by the Dirac equation due to their relativistic nature.

For a molecular system, the Hamiltonian of the total system will, in addition to the potential energy contributions and the kinetic energy of the electrons, also involve the kinetic energy of all nuclei, and the generalization of the electronic Hamiltonian in Eq. (3.136) to the description of the motion of both the nuclei and the electrons can be written as

$$\hat{H}^{\text{tot}} = \hat{T}_{\text{N}} + \hat{H}^{\text{e}}, \tag{6.1}$$

where the nuclear kinetic energy operator $\hat{T}_{\text{N}}$ is given by

$$\hat{T}_{\text{N}} = -\sum_{N} \frac{\hbar^2}{2m_N} \nabla_N^2, \tag{6.2}$$

Principles and Practices of Molecular Properties: Theory, Modeling and Simulations, First Edition.
Patrick Norman, Kenneth Ruud, and Trond Saue.

and the electronic Hamiltonian $\hat{H}^e$ contains the electronic kinetic energy, electron–nuclear attraction, and electron–electron repulsion terms, as well as the nuclear–nuclear repulsion interactions. In solving the electronic problem, the latter serves as a mere constant that can be added at the end of the calculation.

Since the forces acting on the nuclei and electrons in the system are electromagnetic in origin and are thus determined by the charges and spins of the electrons and nuclei, the forces acting on the individual particles in the molecular system are approximately of the same order of magnitude. However, the nuclei are at least 1800 times[1] heavier than the electrons. As a consequence, the nuclear velocities are much smaller than those of the electrons, and it would appear plausible that the electrons can adjust almost instantaneously to the motion of the nuclei. It should therefore be possible to decouple the wave function of the electrons from that of the nuclei, the electrons moving in the "static" potential generated by the "stationary" nuclei (as seen by the electrons). In a similar manner, the motion of the nuclei can be considered as governed by an *effective electronic potential.*

If the nuclei can be considered as being fixed from the point of view of the electrons, we may consider the nuclei as static sources of potentials in the electronic Schrödinger (or Dirac) equation given in Eqs. (3.136) and (3.141). For any given *fixed* nuclear configuration **R**, the exact solution to the electronic eigenvalue problem

$$\hat{H}^e \psi_K(\mathbf{r};\mathbf{R}) = V_K(\mathbf{R})\psi_K(\mathbf{r};\mathbf{R}), \tag{6.3}$$

gives us access to a basis in which we can expand the total wave function as[2]

$$\Psi(\mathbf{r},\mathbf{R}) = \sum_K \Phi^K(\mathbf{R})\psi_K(\mathbf{r};\mathbf{R}). \tag{6.4}$$

In the electronic wave function ψ_K, the nuclear coordinates are given after the semicolon to indicate implicit dependence. In contrast, the expansion coefficient Φ^K carries a *direct* dependence on **R**. In the adiabatic approximation, the summation in Eq. (6.4) is assumed to contain a single term, which corresponds to a physical situation of slow nuclear motions not giving rise to any electronic transitions. This is an accurate approximation as long as the states are well separated but less so in cases of near-degeneracies.

Before we insert the total wavefunction, Eq. (6.4), into the Schrödinger equation for the total system, let us first consider the effect of the nuclear kinetic energy operator on the total wave function

$$\begin{aligned} \hat{T}_N\Psi(\mathbf{r},\mathbf{R}) &= -\hbar^2 \sum_N \frac{1}{2m_N}\nabla_N^2 \sum_K \Phi^K(\mathbf{R})\psi_K(\mathbf{r};\mathbf{R}) \\ &= \sum_K \Bigg[[\hat{T}_N\Phi^K(\mathbf{R})]\psi_K(\mathbf{r};\mathbf{R}) + \Phi^K(\mathbf{R})[\hat{T}_N\psi_K(\mathbf{r};\mathbf{R})] \\ &\quad -\hbar^2 \sum_N \frac{1}{m_N}[\nabla_N\Phi^K(\mathbf{R})]\cdot[\nabla_N\psi_K(\mathbf{r};\mathbf{R})] \Bigg]. \end{aligned} \tag{6.5}$$

In order to proceed with this equation, it is customary to introduce an approximation following the original proposal of Born and Oppenheimer. In view of the difference in velocities of

1 The ratio of the proton mass to the electron mass is $m_p/m_e = 1836.15267247(80)$.

2 This expansion is possibly problematic because the molecular Schrödinger equation as its stands has no bound solutions due to the contribution from translation. The Born–Oppenheimer is therefore often derived by first separating out translation. We may then define n electron coordinates relative to the center of mass of the system. This avoids the mixing of electronic and nuclear degrees of freedom in the kinetic energy operator, but leads to the appearance of a mass polarization term.

the electrons and the nuclei, we assume that the electronic wave functions are slowly varying functions with respect to nuclear displacements, that is, we assume that we can ignore the first- and second-order variations of the electronic wave functions appearing in the expansion with respect to the nuclear coordinates

$$\nabla_N \psi_K(\mathbf{R}) \approx 0; \quad \hat{T}_N \psi_K(\mathbf{R}) \approx 0. \tag{6.6}$$

Ignoring these contributions, the action of the nuclear kinetic energy operator on the total wave function is given by

$$\hat{T}_N \Psi(\mathbf{r}, \mathbf{R}) = \sum_K \psi_K(\mathbf{r}; \mathbf{R})[\hat{T}_N \Phi^K(\mathbf{R})]. \tag{6.7}$$

We are now ready to return to the Schrödinger equation for the total system, which in the adiabatic Born–Oppenheimer approximation will reduce to

$$[\hat{T}_N + V_K(\mathbf{R})]\Phi_k^K(\mathbf{R}) = E_k^K \Phi_k^K(\mathbf{R}), \tag{6.8}$$

where we have explicitly indicated that the total molecular energy depends not only on the particular nuclear state k but also on the specific electronic state K for which this nuclear state has been determined.

This is the conventional Born–Oppenheimer expression for the motion of the nuclei in the presence of an effective potential due to the electrons in the molecule.[3] An important consequence of this equation is the concept of *potential energy surfaces*. We note that Eq. (6.8) defines a complete set of nuclear wave functions for each electronic state ψ_K. Having selected a specific electronic state of the molecule, the potential in which the nuclei move is well defined and the motion of the nuclei will not be able to induce changes in the electronic state of the molecule. While this approximation in most cases is very accurate, we know from experiment that it sometimes will not suffice. Consider, for instance, a molecule in an electronic excited state, which cannot directly relax to the electronic ground state by means of emission of light, as this transition may be forbidden by spin and/or spatial symmetries. The lifetime of such an excited state is in many cases nevertheless found to be short for polyatomic molecules, due to interactions between this excited state and a second excited state that can couple to the ground state through an allowed transition. The interaction between the two electronic states is induced by the nuclear wave function through nonadiabatic effects and conical intersections, the latter being illustrated in Figure 6.1.

3 An alternative approximation (Born 1951) is obtained in the following manner: We insert the expansion Eq. (6.4) into the molecular Schrödinger equation without approximation and next project with an electronic solution from the left. We then obtain the equation

$$[\hat{T}_N + V_K(\mathbf{R}) - E_k^K]\Phi_k^K(\mathbf{R}) = \sum_L C_{KL}(\mathbf{R}), \tag{6.9}$$

where the coupling coefficients

$$C_{KL}(\mathbf{R}) = \langle \psi_K | \hat{T}_N | \psi_L \rangle - \hbar^2 \sum_N \frac{1}{2m_N} \langle \psi_K | \nabla_N | \psi_L \rangle \cdot \nabla_N \Phi^K(\mathbf{R}), \tag{6.10}$$

appear. The electronic states decouple if we set all off-diagonal ($K \neq L$) coupling coefficients to zero. This is sometimes known as the adiabatic approximation, whereas setting diagonal elements C_{KK} to zero as well leads to Eq. (6.8), sometimes called the clamped-nucleus approximation. The difference between the two approximations is that in the former, the nuclei move in an effective potential that depends on their mass and is therefore different for each isotopologue.

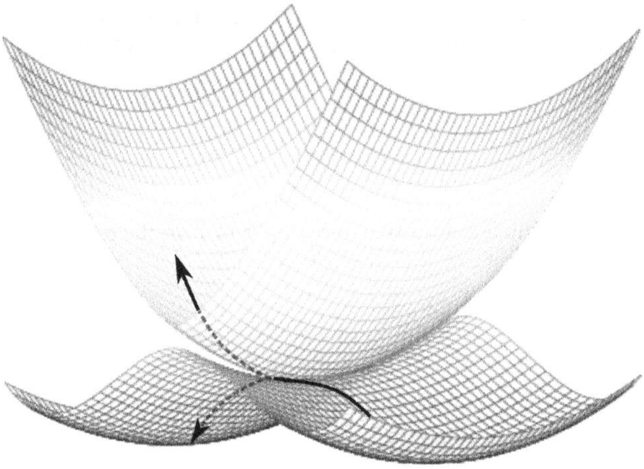

Figure 6.1 Illustration of a seam intersection occurring when two potential energy surfaces cross, allowing electrons to pass from one electronic state to another at nuclear geometries giving rise to very similar (or identical) energies for two different electronic states.

6.2 Separation of Response Functions

We will start by considering the exact-state theory for polarizabilities when discussing the separation of the polarizabilities into electronic and nuclear contributions. In Section 6.4, a perturbation theory approach for calculating vibrational wave functions will be introduced, which we will subsequently develop into working perturbation theory expressions for the so-called zero-point and pure vibrational contributions to molecular properties. We will limit ourselves to considering (hyper)polarizabilities arising from the application of external electromagnetic fields in the visible and ultraviolet regions. This restriction is particularly important when making certain approximations for the pure vibrational contributions to the hyperpolarizabilities. However, the general theory is also applicable to longer wavelengths, and many vibrational spectroscopies can, for instance, be understood from an analysis of the pole and residue structure of the vibrational (hyper)polarizabilities for electromagnetic fields in the infrared region.

The Born–Oppenheimer approximation allows us to treat separately the electronic and nuclear degrees of freedom, keeping in mind that the nuclei move in the effective potential of the electrons. In the following, we will largely focus on the motion of the nuclei relative to each other, that is, the molecular vibrations, and we will in general not consider the overall translation and rotational motion of the molecule. Furthermore, we will focus our attention on incoming light with frequencies in the UV/vis region, thus avoiding resonances with the excitations in the vibrational degrees of freedom.

The Born–Oppenheimer approximation allows us to characterize the total vibrational and electronic state of a molecule by the state of the electronic and nuclear wave functions, where the nuclear wave function depends on the electronic state of the molecule. Furthermore, we may write the total molecular wave function as a product of an electronic and a nuclear wave function, see Eq. (6.4). In the conventional bracket notation, this situation refers to

$$|K,k\rangle = |K\rangle_{\rm e}|k\rangle_K, \tag{6.11}$$

which for convenience will be adopted here. In cases where it is clear from the context, the electronic state index on the nuclear state vector $|k\rangle$ will be omitted, and also the subscript e used to emphasize that $|K\rangle_{\rm e}$ corresponds to the electronic excited state K.

The linear polarizability for a molecule in the ground electronic and nuclear (vibronic) state is given by

$$\alpha_{\alpha\beta}(-\omega_\sigma;\omega_1) = \frac{1}{\hbar}\sum \mathcal{P}_{-\sigma,1}\sum_{K,k}{}' \frac{\langle 0,0|\hat{\mu}_\alpha|K,k\rangle\langle k,K|\hat{\mu}_\beta|0,0\rangle}{\omega_{Kk,00}-\omega_1}, \tag{6.12}$$

where the prime on the summation here will be understood as excluding the vibronic ground state only, that is $K,k \neq 0,0$. We have here, as well as in the following, assumed the infinite lifetime approximation, and thus do not include imaginary damping terms in the expression for the polarizability. It should also be emphasized that the dipole moment operators here refer to electrons as well as nuclei, that is,

$$\hat{\mu}_\alpha = \sum_{i=1}^{N_e}\hat{\mu}_\alpha(i) + \sum_{N=1}^{N_n}\hat{\mu}_\alpha(N). \tag{6.13}$$

We start by splitting the expression for the polarizability into two terms: one term that only involves the electronic ground state and another containing the remaining contributions

$$\alpha_{\alpha\beta}(-\omega_\sigma;\omega_1) = \frac{1}{\hbar}\sum \mathcal{P}_{-\sigma,1}\left[\sum_{k\neq 0}\frac{\langle 0,0|\hat{\mu}_\alpha|0,k\rangle\langle k,0|\hat{\mu}_\beta|0,0\rangle}{\omega_{0k,00}-\omega_1} + \sum_{K\neq 0}\sum_{k}\frac{\langle 0,0|\hat{\mu}_\alpha|K,k\rangle\langle k,K|\hat{\mu}_\beta|0,0\rangle}{\omega_{Kk,00}-\omega_1}\right]. \tag{6.14}$$

In the first term, we can integrate over the electronic coordinates in each of the vibrational transition moments, so that the vibronic transition moments in the numerator reduce to

$$\langle 0,0|\hat{\mu}_\alpha|0,k\rangle = \langle 0|\mu_\alpha(\mathbf{R})|k\rangle, \tag{6.15}$$

that is, to a vibrational transition moment whose strength is governed by the geometry dependence of the molecular dipole moment. Clearly, if the dipole moment $\mu_\alpha(\mathbf{R}) = \int \psi_0(\mathbf{r};\mathbf{R})\hat{\mu}_\alpha\psi_0(\mathbf{r};\mathbf{R})d^3\mathbf{r}$ is independent of the molecular geometry, the integral over the nuclear coordinates will vanish for $k \neq 0$.

Because the summation in the first term of Eq. (6.14) only runs over the vibrational manifold of the molecule and only involves quantities related to the electronic ground state, this contribution is referred to as the *pure vibrational contribution* to the molecular polarizability and is given by

$$\alpha^{\mathrm{v}}_{\alpha\beta}(-\omega_\sigma;\omega_1) = \frac{1}{\hbar}\sum \mathcal{P}_{-\sigma,1}\sum_{k\neq 0}\frac{\langle 0|\mu_\alpha(\mathbf{R})|k\rangle\langle k|\mu_\beta(\mathbf{R})|0\rangle}{\omega_{k0}-\omega_1} = [\mu^2], \tag{6.16}$$

where ω_{k0} denotes the difference in the energy for the kth vibrationally excited state and the vibrational ground state, both defined for the electronic ground state. We see that the pure vibrational contribution to the polarizability involves two vibrational transition moments induced (as we will see in Section 6.6) by the geometry variation of the dipole moment, and it is customary to denote this expression by $[\mu^2]$.

For the second contribution in Eq. (6.14), we will assume that the electronic states are energetically well separated compared to the vibrational levels, and that we can invoke the approximation $\omega_{Kk,00} \approx \omega_{K0}$, where the latter denotes the *vertical* electronic transition angular frequency. We also note that the summation runs over all vibrational states, and we can thus replace this summation by the resolution of the identity, arriving at an expression for this contribution that is identified as the *zero-point vibrationally averaged polarizability*

$$\alpha_{\alpha\beta}^{\text{zpva}}(-\omega_\sigma;\omega_1) = \frac{1}{\hbar}\sum \mathcal{P}_{-\sigma,1}\sum_{K\neq 0}\frac{\langle 0,0|\hat{\mu}_\alpha|K\rangle\langle K|\hat{\mu}_\beta|0,0\rangle}{\omega_{K0}-\omega_1} = \langle 0|\alpha_{\alpha\beta}^{\text{e}}(\mathbf{R})|0\rangle, \quad (6.17)$$

where $\alpha_{\alpha\beta}^{\text{e}}(\mathbf{R})$ denotes the conventional frequency-dependent electronic polarizability at a nuclear configuration given by the set of coordinates $\mathbf{R}$, and where the integration in the last expression in Eq. (6.17) is performed over the vibrational ground state of the electronic ground state. We note that the nuclear part of the dipole moment operator, that is, the second term in Eq. (6.13), does not contribute, as it amounts to a mere constant in the integration over electronic coordinates and because the states $|0\rangle$ and $|K\rangle$ are orthogonal.

Within the Born–Oppenheimer approximation, the polarizability of a nuclear–electronic wave function thus contains a contribution that arises from excitations within the nuclear part of the wave function only, and a contribution arising largely from excitations within the electronic states of the molecule, averaged over the ground-state vibrational wave function. It is often customary to split the latter contribution into a purely electronic contribution that is evaluated for a fixed nuclear geometry, in most cases the equilibrium geometry of the molecule, and the correction term due to the zero-point vibrational motion of the nuclei. These different contributions are easily separated if a perturbation theory approach is used for evaluating these corrections, as we will return to in Section 6.5. With this partitioning scheme, we write

$$\alpha_{\alpha\beta}(-\omega;\omega) = \alpha_{\alpha\beta}^{\text{e}}(-\omega;\omega) + \Delta\alpha_{\alpha\beta}^{\text{zpva}}(-\omega;\omega) + \alpha_{\alpha\beta}^{\text{v}}(-\omega;\omega). \quad (6.18)$$

The expressions for the pure vibrational contributions to the first- and second-order hyperpolarizabilities can be derived in a similar manner. As a starting point for the first-order hyperpolarizability, we choose the sum-over-states expression that is free from secular divergences and is given by

$$\beta_{\alpha\beta\gamma}(-\omega_\sigma;\omega_1,\omega_2) = \frac{1}{\hbar^2}\sum \mathcal{P}_{-\sigma,1,2}\sum_{K,k}{}'\sum_{L,l}{}'\frac{\langle 0,0|\hat{\mu}_\alpha|K,k\rangle\langle k,K|\overline{\hat{\mu}_\beta}|L,l\rangle\langle l,L|\hat{\mu}_\gamma|0,0\rangle}{(\omega_{Kk,00}-\omega_\sigma)(\omega_{Ll,00}-\omega_2)}, \quad (6.19)$$

where $\overline{\hat{\mu}}$ is the fluctuation dipole moment operator $\hat{\mu} - \langle 0,0|\hat{\mu}|0,0\rangle$ and the primes indicate omission of the vibronic ground state in the summations. We identify the pure vibrational contribution to the first-order hyperpolarizability from terms in Eq. (6.19) with $(K\neq 0, L=0)$, $(K=0, L\neq 0)$, or $(K=0, L=0)$, that is, those involving the electronic ground state. As a result, we get

$$\begin{aligned}\beta_{\alpha\beta\gamma}^{\text{v}}(-\omega_\sigma;\omega_1,\omega_2) &= \frac{1}{\hbar^2}\sum \mathcal{P}_{-\sigma,1,2}\\ &\left[\sum_K{}'\sum_k\sum_l{}'\frac{\langle 0|\mu_\alpha^{0K}|k\rangle\langle k|\mu_\beta^{K0}|l\rangle\langle l|\mu_\gamma(\mathbf{R})|0\rangle}{(\omega_{Kk,00}-\omega_\sigma)(\omega_{l0}-\omega_2)}\right.\\ &+\sum_L{}'\sum_k{}'\sum_l\frac{\langle 0|\mu_\alpha(\mathbf{R})|k\rangle\langle k|\mu_\beta^{0L}|l\rangle\langle l|\mu_\gamma^{L0}|0\rangle}{(\omega_{k0}-\omega_\sigma)(\omega_{Ll,00}-\omega_2)}\\ &\left.+\sum_k{}'\sum_l{}'\frac{\langle 0|\mu_\alpha(\mathbf{R})|k\rangle\langle k|\overline{\mu_\beta(\mathbf{R})}|l\rangle\langle l|\mu_\gamma(\mathbf{R})|0\rangle}{(\omega_{k0}-\omega_\sigma)(\omega_{l0}-\omega_2)}\right],\end{aligned} \quad (6.20)$$

where μ^{0K} denotes the nuclear coordinate-dependent electronic transition moment between the electronic ground and Kth excited state. Now introducing the approximation $\omega_{Kk,00}\approx\omega_{K0}$

and applying the closure relation over vibrational states $|k\rangle$, we get

$$\beta^{\mathrm{v}}_{\alpha\beta\gamma}(-\omega_\sigma;\omega_1,\omega_2) = \frac{1}{\hbar^2}\sum \mathcal{P}_{-\sigma,1,2}$$
$$\left[\sum_K{}'\sum_l{}'\frac{\langle 0|\mu_\alpha^{0K}\mu_\beta^{K0}|l\rangle\langle l|\mu_\gamma(\mathbf{R})|0\rangle}{(\omega_{K0}-\omega_\sigma)(\omega_{l0}-\omega_2)}\right.$$
$$+\sum_L{}'\sum_k{}'\frac{\langle 0|\mu_\alpha(\mathbf{R})|k\rangle\langle k|\mu_\beta^{0L}\mu_\gamma^{L0}|0\rangle}{(\omega_{k0}-\omega_\sigma)(\omega_{L0}-\omega_2)}$$
$$\left.+\sum_k{}'\sum_l{}'\frac{\langle 0|\mu_\alpha(\mathbf{R})|k\rangle\langle k|\overline{\mu_\beta(\mathbf{R})}|l\rangle\langle l|\mu_\gamma(\mathbf{R})|0\rangle}{(\omega_{k0}-\omega_\sigma)(\omega_{l0}-\omega_2)}\right]. \tag{6.21}$$

By comparing with Eq. (5.169), we identify in the first two expressions the frequency-dependent polarizability. We can now in principle insert the frequency-dependent polarizability in these expressions, but often the optical frequencies ω_σ and ω_2 are ignored, so that we in the first two terms introduce the *static* electronic polarizability α^{e}, arriving at the expression

$$\beta^{\mathrm{v}}_{\alpha\beta\gamma}(-\omega_\sigma;\omega_1,\omega_2) = [\mu\alpha] + [\mu^3], \tag{6.22}$$

where

$$[\mu\alpha] = \frac{1}{2\hbar}\sum \mathcal{P}_{-\sigma,1,2}\sum_k{}'\left[\frac{\langle 0|\alpha^{\mathrm{e}}_{\alpha\beta}(\mathbf{R})|k\rangle\langle k|\mu_\gamma(\mathbf{R})|0\rangle}{(\omega_{k0}-\omega_2)} + \frac{\langle 0|\mu_\alpha(\mathbf{R})|k\rangle\langle k|\alpha^{\mathrm{e}}_{\beta\gamma}(\mathbf{R})|0\rangle}{(\omega_{k0}-\omega_\sigma)}\right], \tag{6.23}$$

$$[\mu^3] = \frac{1}{\hbar^2}\sum \mathcal{P}_{-\sigma,1,2}\sum_k{}'\sum_l{}'\frac{\langle 0|\mu_\alpha(\mathbf{R})|k\rangle\langle k|\overline{\mu_\beta(\mathbf{R})}|l\rangle\langle l|\mu_\gamma(\mathbf{R})|0\rangle}{(\omega_{k0}-\omega_\sigma)(\omega_{l0}-\omega_2)}. \tag{6.24}$$

To ignore optical frequencies, or to set them to zero, may appear as a drastic approximation. However, we should remember that the optical frequencies come with negative as well as positive signs after the operation with the permutation operator, and the approximation we make is rather to be viewed as to replace the polarizability at different dynamic frequencies with a common static value. With optical frequencies well separated from electronic resonances, this is a fair approximation since the dispersion is then rather modest. This static approximation is more important when addressing pure vibrational contributions to the second hyperpolarizability, where this approximation is needed in order to identify the contribution from the first hyperpolarizability.

All transition matrix elements in Eq. (6.23) are assumed to be real, so after an operation of permutation with $\mathcal{P}_{-\sigma,2}$ on the first term it can be combined with the second term and we get

$$[\mu\alpha] = \frac{1}{\hbar}\sum \mathcal{P}_{-\sigma,1,2}\sum_k{}'\frac{\langle 0|\mu_\alpha(\mathbf{R})|k\rangle\langle k|\alpha^{\mathrm{e}}_{\beta\gamma}(\mathbf{R})|0\rangle}{(\omega_{k0}^2-\omega_\sigma^2)}\omega_{k0}. \tag{6.25}$$

In the same manner, a compact expression for the pure vibrational contribution to the second-order hyperpolarizability can be derived and written as

$$\gamma^{\mathrm{v}}_{\alpha\beta\gamma\delta}(-\omega_\sigma;\omega_1,\omega_2,\omega_3) = [\alpha^2] + [\mu\beta] + [\mu^2\alpha] + [\mu^4], \tag{6.26}$$

where we refer to Table 6.1 for a detailed account of the respective terms.

Before proceeding to consider approximate ways of determining the pure vibrational and zero-point vibrational corrections to the polarizabilities, we make a few general observations from the expressions given in Table 6.1.

Table 6.1 Expressions for the different contributions to the pure vibrational contributions to α, β, and γ for exact nuclear–electronic wave functions within the Born–Oppenheimer approximation.

$[\mu^2]$	$\frac{1}{\hbar}\sum P_{-\sigma,1}\sum_{k\neq0}(\mu_\alpha)_{0k}(\mu_\beta)_{k0}(\omega_{k0}-\omega_\sigma)^{-1}$
$[\mu\alpha]$	$\frac{1}{\hbar}\sum P_{-\sigma,1,2}\sum_{k\neq0}(\mu_\alpha)_{0k}(\alpha_{\beta\gamma})_{k0}\omega_{k0}(\omega_{k0}^2-\omega_\sigma^2)^{-1}$
$[\mu^3]$	$\frac{1}{\hbar^2}\sum P_{-\sigma,1,2}\sum_{k,l\neq0}(\mu_\alpha)_{0k}(\overline{\mu}_\beta)_{kl}(\mu_\gamma)_{l0}(\omega_{k0}-\omega_\sigma)^{-1}(\omega_{l0}-\omega_2)^{-1}$
$[\mu\beta]$	$\frac{1}{3\hbar}\sum P_{-\sigma,1,2,3}\sum_{k\neq0}(\mu_\alpha)_{0k}(\beta_{\beta\gamma\delta})_{k0}\omega_{k0}(\omega_{k0}^2-\omega_\sigma^2)^{-1}$
$[\alpha^2]$	$\frac{1}{4\hbar}\sum P_{-\sigma,1,2,3}\sum_{k\neq0}(\alpha_{\alpha\beta})_{0k}(\alpha_{\gamma\delta})_{k0}(\omega_{k0}-\omega_2-\omega_3)^{-1}$
$[\mu^2\alpha]$	$\frac{1}{\hbar^2}\sum P_{-\sigma,1,2,3}\sum_{k,l\neq0}\left[\frac{1}{2}(\mu_\alpha)_{0k}(\overline{\alpha}_{\beta\gamma})_{kl}(\mu_\delta)_{l0}(\omega_{k0}-\omega_\sigma)^{-1}(\omega_{l0}-\omega_3)^{-1}\right.$ $\left.+(\mu_\alpha)_{0k}(\overline{\mu}_\beta)_{kl}(\alpha_{\gamma\delta})_{l0}(\omega_{k0}\omega_{l0}+\omega_\sigma[\omega_2+\omega_3])(\omega_{k0}^2-\omega_\sigma^2)^{-1}(\omega_{l0}^2-[\omega_2+\omega_3]^2)^{-1}\right]$
$[\mu^4]$	$\frac{1}{\hbar^3}\sum P_{-\sigma,1,2,3}\left[\sum_{k,l,m\neq0}(\mu_\alpha)_{0k}(\overline{\mu}_\beta)_{kl}(\overline{\mu})_{lm}(\mu_\delta)_{m0}\times(\omega_{k0}-\omega_\sigma)^{-1}(\omega_{l0}-\omega_2-\omega_3)^{-1}(\omega_m-\omega_3)^{-1}\right.$ $\left.-\sum_{k,l\neq0}(\mu_\alpha)_{0k}(\mu_\beta)_{k0}(\mu_\gamma)_{0l}(\mu_\delta)_{l0}(\omega_{k0}-\omega_\sigma)^{-1}(\omega_{l0}-\omega_3)^{-1}(\omega_{l0}+\omega_2)^{-1}\right]$

For an electronic property P, we have used the notation $(P)_{kl}=\langle k|P(\mathbf{R})|l\rangle$ with an overbar to indicate fluctuation quantities.

A first observation to be made is that the pure vibrational contributions arise from the lower order responses of the molecule. In cases where these lower order responses vanish, as is, for instance, the case for magnetic moments of closed-shell states, there will be no pure vibrational contributions. As a consequence, one expects pure vibrational corrections to be more important for electric response properties than for (mixed) magnetic field-dependent properties.

A second observation to be made regarding the pure vibrational contributions is that, when considering optical (or UV/vis) spectroscopies, external field frequencies are in general more than an order of magnitude larger than the transition frequencies separating the nuclear states. There are therefore no poles appearing in the frequency-dependent pure vibrational (hyper)polarizabilities for the photon wavelengths used in the experiment. In this situation, the optical frequency of the incoming light will dampen the pure vibrational effects because the optical frequencies are so much larger than the vibrational transition energies, making the pure vibrational corrections small. Pure vibrational contributions are therefore most important for static fields, where it may even dominate over the electronic contribution. For selected optical processes, such as the electro-optical Kerr effect or the electric-field-induced second-harmonic generation (see Table 5.1), the pure vibrational contributions may be significant even though a frequency-dependent electromagnetic field is involved due to the presence of at least one static field, either as part of the perturbing fields, or arising as a result of the mixing of the different incoming frequency-dependent electromagnetic fields.

The expressions in Table 6.1 apply to the general case where we can determine all vibrational excited states for a given electronic potential energy surface. In practice, this is very difficult to determine for anything but diatomic or few-atomic molecules, and a more general approach for determining the vibrational contributions will in general be needed in order to treat polyatomic molecules. There have been several approaches presented in the literature for approximate evaluation of pure vibrational contributions. We will here use the perturbation theoretical approach. Although such an approach can become cumbersome and lead to lengthy algebraic expressions, it has the advantage of retaining a formal similarity to the sum-over-states formalisms in use for the evaluation of the electronic response functions. It also allows for a more physical interpretation in terms of harmonic and anharmonic contributions, and allows for an analysis in terms of dominating vibrational modes. In addition, we can also treat pure vibrational and zero-point vibrational contributions on an equal footing.

However, before discussing the perturbation-theoretical approach for solving vibrational matrix elements, we first need to discuss the reference system for molecular vibrations, namely, the harmonic oscillator.

6.3 Molecular Vibrations and Normal Coordinates

In Eq. (6.8), we gave the form of the Schrödinger equation for the nuclear part of the vibronic wave function within the Born–Oppenheimer approximation, where the electrons create an effective potential in which the nuclei move. The dependence of this potential on the nuclear coordinates can be quite complicated, but a natural starting point for a given nuclear configuration is to expand the potential in a Taylor series around an adopted expansion point

$$V(\mathbf{R}) = V^{(0)} + \sum_i V_i^{(1)}\Delta_i + \frac{1}{2}\sum_{i,j} V_{ij}^{(2)}\Delta_i\Delta_j + \frac{1}{6}\sum_{ijk} V_{ijk}^{(3)}\Delta_i\Delta_j\Delta_k + \cdots, \tag{6.27}$$

where Δ_i denotes a displacement along a Cartesian axis of a given nucleus. Each nucleus can be displaced along three Cartesian directions, and the summations in Eq. (6.27) thus run over $3N$ coordinates with N being the number of atoms in the molecule. The expansion coefficients $V_{ij..}^{(n)}$ are defined as the nth order derivative of the electronic potential in terms of these Cartesian nuclear displacements, evaluated at the expansion point with all $\Delta_i = 0$. The first term in Eq. (6.27) is independent of the nuclear coordinates and, without loss of generality, we can set this term to zero. We also note that a common choice made for the expansion point is the *molecular equilibrium geometry*, which corresponds to the nuclear configuration in which the electronic energy is stationary with respect to all nuclear displacements, that is, $V_i^{(1)} = 0$.

In the vicinity of the equilibrium geometry, at least for the electronic ground state and stable (nondissociative) electronically excited states, the electronic potential is fairly quadratic and we may then as an approximation truncate the expansion in Eq. (6.27) after the quadratic term. Inserting the resulting quadratic potential into the nuclear Schrödinger equation in Eq. (6.8), we obtain the *harmonic approximation* to the vibrational Schrödinger equation

$$\left[\hat{T}_{\mathrm{N}} + \frac{1}{2}\sum_{i,j} V_{ij}^{(2)}\Delta_i\Delta_j\right]\Phi_k(\mathbf{R}) = E_k\Phi_k(\mathbf{R}), \tag{6.28}$$

where it is understood that this equation is to be solved for a particular potential energy surface although the electronic index K has been omitted. We note that we here include the full kinetic operator, including also translational and rotational degrees of freedom. Solving Eq. (6.28) is difficult, as all displacements are coupled through the harmonic potential term. In order to simplify the problem, we first introduce a set of *mass-weighted coordinates* q_i by the transformation

$$q_i = \sqrt{m_i}\Delta_i, \tag{6.29}$$

where m_i is the mass of the nucleus whose Cartesian displacement Δ_i is being considered. Introducing the mass-weighted coordinates in Eq. (6.28), we get

$$\left[-\frac{\hbar^2}{2}\sum_i \frac{\partial^2}{\partial q_i^2} + \frac{1}{2}\sum_{i,j} \frac{\partial^2 V^{(2)}}{\partial q_i \partial q_j} q_i q_j\right] \Phi_k(\mathbf{q}) = E_k \Phi_k(\mathbf{q}). \tag{6.30}$$

If the Hamiltonian had been diagonal, this equation would be separable, with each term in the form of the well-known Schrödinger equation for a harmonic oscillator. Let us therefore introduce a set of *mass-weighted normal coordinates* defined such that they simultaneously diagonalize the kinetic energy operator and the operator for the potential. These mass-weighted normal coordinates Q are written as linear combinations of the mass-weighted coordinates

$$Q_i = \sum_{j=1}^{3N} L_{ij} q_j. \tag{6.31}$$

The matrix L contains the eigenvectors to the force constant matrix in mass-weighted coordinates, which thus needs to be diagonalized in this step of the calculation. A certain number of the corresponding eigenvalues will be zero, or at least close to zero, as they correspond to rotation and overall translation of the molecule. For a molecule with N nuclei, six (or five for linear molecules) of the eigenvalues will be zero—three associated with translation and three (or two) associated with rotation. While the translation of a molecule in general is not quantized, the rotational motion will in general be quantized, and although not shown here, the rotational motion of the molecule is described by a Hamiltonian that is similar in structure to that of a particle on a sphere. For linear molecules there can be no overall rotation about the molecular axis and thus there are only two rotational coordinates in this case. Instead, for polyatomic linear molecules, one of the bending vibrational modes will be twofold degenerate as illustrated in Figure 6.2.

In the following, we will only be concerned with nuclear motion associated with vibration and this will be described by the $3N - 6$ mass-weighted normal coordinates with nonzero eigenvalues. In these coordinates, the Hamiltonian for the vibrational motion in the harmonic approximation has a particularly simple form, namely,

$$\hat{H}^{(0)} = -\frac{\hbar^2}{2}\sum_a \frac{\partial^2}{\partial Q_a^2} + \frac{1}{2}\sum_a V_{aa}^{(2)} Q_a^2, \tag{6.32}$$

where it has been implied that the summation running over a takes on values from 1 to $3N - 6$ and the superscript on the Hamiltonian indicates that it is chosen to correspond to our zeroth-order description of nuclear vibrations. This equation is valid as an equation for the vibrational motions assuming that the displacements of the nuclei are small—that is, we assume that the true motion of the nuclei can be well described by a linearized form for their displacements from the equilibrium geometry. In many cases this is a fairly good approximation, and

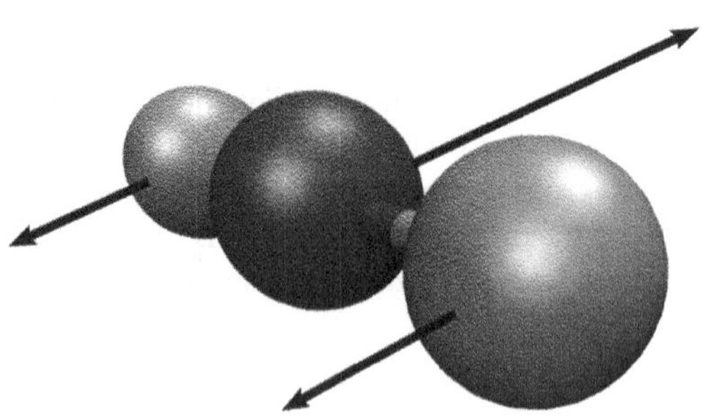

Figure 6.2 One of the twofold degenerate bending modes in CO_2.

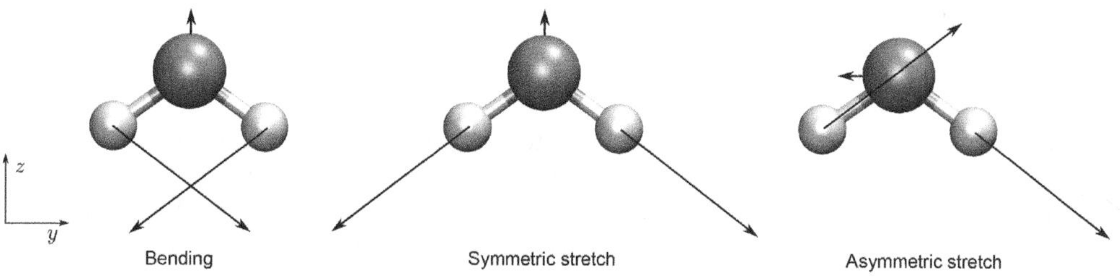

Figure 6.3 The three normal modes in water. Note that it is largely the light nuclei (hydrogen) that move in all of the three normal modes as a consequence of the fact that the center of mass is spatially fixed.

in Figure 6.3 we have illustrated the normal modes for the water molecule. We note, however, that for low-frequency modes, for which the amplitude of the nuclear displacements may be large, the linear approximation may break down, and we also note that for the bending mode in the water molecule in Figure 6.3, the linear approximation may be less accurate and that the bending motion would be more accurately described by curvilinear coordinates as shown in Figure 6.4.

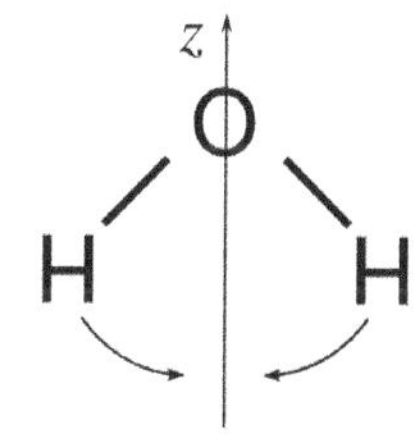

Figure 6.4 The curvilinear motion associated with the bending mode in water.

As intended with the transformation to normal coordinates, we see that the Hamiltonian for nuclear motion in Eq. (6.32) is separable, and the $3N$-dimensional vibrational problem has thus been converted into $3N - 6$ identical one-dimensional eigenvalue equations for the harmonic oscillator in addition to equations for the three translations and three rotational degrees of freedom. The total vibrational wave function becomes the product of the solutions for each normal coordinate and the total vibrational energy will equal the sum of the corresponding harmonic oscillator energies. These are given by

$$\varepsilon_{n_a} = \left(n_a + \frac{1}{2}\right)\hbar\omega_a \qquad n_a = 0, 1, 2, \ldots, \tag{6.33}$$

where $\omega_a = \sqrt{V_{aa}^{(2)}}$ and the eigenfunctions are given as

$$\psi_{n_a}(\xi_a) = N_{n_a} H_{n_a}(\xi_a) \exp(-\xi_a^2/2), \tag{6.34}$$

where $\xi_a = \sqrt{\omega_a/\hbar}Q_a$ is the so-called reduced dimensionless normal coordinate, N_{n_a} is a normalization constant, and $H_{n_a}(\xi_a)$ is the Hermite polynomial of degree n_a. The total nuclear wave function is equal to the product function

$$\Phi^{(0)}(Q) = \prod_a \psi_{n_a}(\xi_a), \tag{6.35}$$

and the associated total vibrational energy is

$$E^{(0)} = \sum_a \left(n_a + \frac{1}{2}\right)\hbar\omega_a, \tag{6.36}$$

where the superscript on the total wave function and energy indicates that the harmonic oscillator approximation is to be considered as the zeroth-order solution to the vibrational problem.

In the following we shall make extensive use of the recursion relation

$$\langle\psi_m|\xi^k|\psi_n\rangle = \sqrt{\frac{n}{2}}\langle\psi_m|\xi^{k-1}|\psi_{n-1}\rangle + \sqrt{\frac{n+1}{2}}\langle\psi_m|\xi^{k-1}|\psi_{n+1}\rangle, \tag{6.37}$$

as well as the orthonormality of the harmonic oscillator wave functions

$$\langle \psi_m | \psi_n \rangle = \delta_{mn}. \tag{6.38}$$

The above two simple relationships make it easy to evaluate arbitrary matrix elements in the basis of the reduced normal coordinates $\{\xi_a\}$, which is precisely what is needed when the vibrational Hamiltonian is expressed in this set of coordinates—as will be done in the subsequent sections where we will consider anharmonic corrections. If, on the other hand, one prefers to work in the basis of the regular normal coordinates $\{Q_a\}$, as in the above discussion, then there is a need to convert matrix elements from one basis to the other. This conversion is achieved with the relation

$$\langle \psi_m(Q_a) | Q_a^k | \psi_n(Q_a) \rangle = \left(\sqrt{\frac{\hbar}{\omega_a}} \right)^k \langle \psi_m(\xi_a) | \xi_a^k | \psi_n(\xi_a) \rangle. \tag{6.39}$$

As an example of the theory developed for treating vibrational motions in molecules, let us consider a diatomic molecule composed of atoms A and B with masses m_A and m_B, respectively, and positioned along the Cartesian z-axis. The Cartesian force constant matrix $V_\Delta^{(2)}$ has nonzero elements only for nuclear displacements along the z-direction. The symmetry of this simple system allows us to take advantage of the fact that $\Delta_{z_\mathrm{A}} = -\Delta_{z_\mathrm{B}}$, and it therefore suffices to study the two-by-two submatrix:

$$V_\Delta^{(2)} = V_{\Delta_{z_\mathrm{A}}\Delta_{z_\mathrm{A}}}^{(2)} \begin{pmatrix} 1 & -1 \\ -1 & 1 \end{pmatrix}, \tag{6.40}$$

The corresponding force constant matrix in mass-weighted coordinates $V_q^{(2)}$ will be

$$V_q^{(2)} = V_{\Delta_{z_\mathrm{A}}\Delta_{z_\mathrm{A}}}^{(2)} \begin{pmatrix} 1/m_\mathrm{A} & -1/\sqrt{m_\mathrm{A} m_\mathrm{B}} \\ -1/\sqrt{m_\mathrm{A} m_\mathrm{B}} & 1/m_\mathrm{B} \end{pmatrix}. \tag{6.41}$$

This matrix has two eigenvalues, one of which is associated with translation and thus is zero, so the force constant matrix in normal coordinates $V_q^{(2)}$ becomes

$$V_Q^{(2)} = V_{\Delta_{z_\mathrm{A}}\Delta_{z_\mathrm{A}}}^{(2)} \begin{pmatrix} 0 & 0 \\ 0 & \dfrac{m_\mathrm{A} + m_\mathrm{B}}{m_\mathrm{A} m_\mathrm{B}} \end{pmatrix}, \tag{6.42}$$

where we can identify the vibrational frequency of the sole vibrational mode for a diatomic as the square root of the nonzero eigenvalue:

$$\omega_a = \sqrt{V_{z_\mathrm{A} z_\mathrm{A}}^{(2)} \frac{m_\mathrm{A} + m_\mathrm{B}}{m_\mathrm{A} m_\mathrm{B}}} = \sqrt{\frac{k}{\mu}}, \tag{6.43}$$

where $\mu = m_\mathrm{A} m_\mathrm{B}/(m_\mathrm{A} + m_\mathrm{B})$ is the reduced mass and k the force constant. The corresponding eigenvector of unit norm is

$$Q_a = \frac{1}{\sqrt{m_\mathrm{A} + m_\mathrm{B}}} \begin{pmatrix} \sqrt{m_\mathrm{B}} \\ -\sqrt{m_\mathrm{A}} \end{pmatrix}, \tag{6.44}$$

as expressed in the set of mass-weighted coordinates q_a and q_b. As done in Figures 6.2 and 6.3, it is a common practice to illustrate the normal modes in the set of Cartesian displacements Δ_i, which requires a mere scaling of the eigenvector with the inverse square root of atomic masses:

$$Q_a = \frac{1}{\sqrt{m_A + m_B}} \begin{pmatrix} \sqrt{\dfrac{m_B}{m_A}} \\ -\sqrt{\dfrac{m_A}{m_B}} \end{pmatrix}. \tag{6.45}$$

We note that after this scaling is performed, the vector norm is no longer equal to unity.

The harmonic oscillator approximation for the vibrational motion of the nuclei works surprisingly well considering the complicated form of the potential energy surfaces. The calculated vibrational frequencies (assuming high-level *ab initio* calculations) are typically found to be within 5–10% of the true values. However, it also suffers from serious deficiencies. In particular, excitations are only allowed between neighboring vibrational levels, as follows from Eq. (6.37), excluding overtones as being forbidden. Furthermore, effects such as Fermi resonances, in which a "dark" state[4] gains intensity through interactions with other vibrational states or their overtones, cannot be recovered within the regular harmonic approximation for the electronic potential. In order to treat such experimentally observable effects, we need to consider anharmonicities. The most common way of going beyond the harmonic approximation is to use vibrational perturbation theory. We will give a brief introduction to the principles of vibrational perturbation theory in the next section.

6.4 Perturbation Theory for Vibrational Wave Functions

In order to calculate corrections to the harmonic oscillator approximation for the nuclear motion, we define the contributions in Eq. (6.27) that do not enter the harmonic oscillator Hamiltonian in Eq. (6.32) as perturbations to the harmonic oscillator reference model system. We will therefore define

$$\hat{H}^{(1)} = \frac{1}{6} \sum_{a,b,c} V^{(3)}_{abc} \xi_a \xi_b \xi_c, \tag{6.46}$$

$$\hat{H}^{(2)} = \frac{1}{24} \sum_{a,b,c,d} V^{(4)}_{abcd} \xi_a \xi_b \xi_c \xi_d, \tag{6.47}$$

as the even higher order corrections are typically too small to be of practical relevance. We have here defined the potential in terms of an expansion in the reduced dimensionless normal coordinates.

We do a perturbation expansion of the energy expression using conventional Rayleigh–Schrödinger perturbation theory to arrive at the expression[5]

$$\begin{aligned} \langle \Phi | \hat{H} - E | \Phi \rangle = \langle \Phi^{(0)} + \lambda \Phi^{(1)} + \lambda^2 \Phi^{(2)} + \cdots | (\hat{H}^{(0)} - E^{(0)}) + \lambda (\hat{H}^{(1)} - E^{(1)}) \\ + \lambda^2 (\hat{H}^{(2)} - E^{(2)}) + \cdots | \Phi^{(0)} + \lambda \Phi^{(1)} + \lambda^2 \Phi^{(2)} + \cdots \rangle = 0. \end{aligned} \tag{6.48}$$

We collect contributions of the same order in the perturbation expansion and solve the resulting equations for each order in the order parameter λ. The zeroth-order equation is that of

4 By "dark" state, we here refer to a vibrationally excited state whose vibrational transition moment from the ground state is forbidden by symmetry.

5 This expression implicitly assumes that both the zeroth-order and the perturbed wave functions are normalized. For our purposes, however, this is not of any concern, since it can be shown that corrections due to normalization of the perturbed wave function only appear to higher than second order in the perturbation expansion, and we will not consider corrections beyond second order in this book.

the harmonic oscillator in Eq. (6.32). The first-order energy correction has two contributions, one involving the first-order perturbed Hamiltonian $\hat{H}^{(1)}$ and one involving the first-order perturbed wave function, and both vanish

$$E^{(1)} = \langle\Phi^{(0)}|\hat{H}^{(1)}|\Phi^{(0)}\rangle + 2\langle\Phi^{(1)}|\hat{H}^{(0)} - E^{(0)}|\Phi^{(0)}\rangle = 0. \tag{6.49}$$

The first term vanishes because $\hat{H}^{(1)}$ is odd with respect to at least one geometrical displacement, and the second part of this equation vanishes because $\Phi^{(0)}$ is an eigenfunction of $\hat{H}^{(0)}$, and because $\Phi^{(1)}$ is orthogonal to $\Phi^{(0)}$.

The lowest order nonvanishing correction to the vibrational energy is thus given by

$$E^{(2)} = \langle\Phi^{(0)}|\hat{H}^{(2)}|\Phi^{(0)}\rangle + 2\langle\Phi^{(0)}|\hat{H}^{(1)}|\Phi^{(1)}\rangle + \langle\Phi^{(1)}|\hat{H}^{(0)} - E^{(0)}|\Phi^{(1)}\rangle, \tag{6.50}$$

where we have assumed the vibrational wave functions to be real so that, for Hermitian operators, ket and bra vectors may be interchanged. In order to determine the second-order energy correction, we need to find the first-order corrections to the vibrational wave functions. We expand $\Phi^{(1)}$ in terms of the harmonic oscillator wave functions

$$\Phi^{(1)} = \sum_{a}\sum_{n=1}^{\infty} c_a^n \Phi_a^n + \sum_{a\neq b}\sum_{n,m=1}^{\infty} c_{ab}^{nm}\Phi_{ab}^{nm} + \sum_{a\neq b\neq c}\sum_{n,m,p=1}^{\infty} c_{abc}^{nmp}\Phi_{abc}^{nmp}, \tag{6.51}$$

where $\Phi_{ab\ldots}^{nm\ldots}$ denotes a vibrational wave function with modes a and b in the nth and mth eigenstate of the harmonic oscillator, respectively, and other modes in the ground states. The first-order wave function cannot contain higher than three-mode excitations due to the form of the Hamiltonian in Eq. (6.46). We obtain the expansion coefficients by evaluating the matrix elements

$$c_{ab\ldots}^{nm\ldots} = -\frac{\langle\Phi_{ab\ldots}^{nm\ldots}|H^{(1)}|\Phi^{(0)}\rangle}{E_{ab\ldots}^{nm\ldots} - E^{(0)}}. \tag{6.52}$$

Using Eq. (6.37), the only nonzero expansion coefficients in Eq. (6.51) are found to be

$$\begin{aligned} c_a^1 &= -\frac{1}{2\hbar\omega_a}\sum_b V_{abb}^{(3)}\langle 0|\xi_b^2|0\rangle\langle 0|\xi_a|1_a\rangle \\ &= -\frac{1}{4\sqrt{2}\hbar\omega_a}\sum_b V_{abb}^{(3)}, \end{aligned} \tag{6.53}$$

$$\begin{aligned} c_a^3 &= -\frac{V_{aaa}^{(3)}}{18\hbar\omega_a}\langle 0|\xi_a^3|3_a\rangle \\ &= -\frac{\sqrt{3}V_{aaa}^{(3)}}{36\hbar\omega_a}, \end{aligned} \tag{6.54}$$

$$\begin{aligned} c_{ab}^{21} &= -\frac{V_{aab}^{(3)}}{2\hbar(2\omega_a + \omega_b)}\langle 0|\xi_a^2|2_a\rangle\langle 0|\xi_b|1_b\rangle \\ &= -\frac{V_{aab}^{(3)}}{4\hbar(2\omega_a + \omega_b)}, \end{aligned} \tag{6.55}$$

$$\begin{aligned} c_{abc}^{111} &= -\frac{V_{abc}^{(3)}}{6\hbar(\omega_a + \omega_b + \omega_c)}\langle 0|\xi_a|1_a\rangle\langle 0|\xi_b|1_b\rangle\langle 0|\xi_c|1_c\rangle \\ &= -\frac{\sqrt{2}V_{abc}^{(3)}}{24\hbar(\omega_a + \omega_b + \omega_c)}, \end{aligned} \tag{6.56}$$

where we use the notation $|2_b\rangle$ to denote the second excited state of normal mode b. These expressions for the perturbation coefficients are only valid for the vibrational ground state and corresponding expressions have to be derived for the vibrationally excited states in order to get anharmonic corrections of transition energies between vibrationally excited states.

With these expansion coefficients in place, we have a representation of the anharmonic vibrational wave function that will suffice for most practical calculations of molecular vibrational properties, and we will in the next two sections apply these perturbed vibrational wave functions to the calculation of zero-point and pure vibrational corrections to molecular properties. Before closing this section, however, we note that there exist other approaches for calculating vibrational frequencies and perturbed vibrational wave functions. Conceptually, these approaches are based on the same principles for describing vibrational wave functions as is used in electronic structure theory. Schemes based on the self-consistent field approximation, as well as more accurate schemes based on perturbation theory, configuration interaction, and coupled-cluster approaches have been introduced. All of these methods can be applied to the study of vibrational spectroscopies, but we will limit ourselves to the use of the perturbation approach as it is the predominant technique used in calculations, in addition to lending itself better to a qualitative analysis.

6.5 Zero-Point Vibrational Contributions to Properties

6.5.1 First-Order Anharmonic Contributions

Let us now consider the calculation of contributions arising from the zero-point vibrational averaging (ZPVA) to a given molecular property P using the zeroth- and first-order perturbed wave functions presented in the previous section. We will base this analysis on an expansion of the electronic property in terms of the mass weighted normal coordinates, which up to third order takes the form

$$P(Q) = P^{(0)} + \sum_a P_a^{(1)} Q_a + \frac{1}{2} \sum_{a,b} P_{ab}^{(2)} Q_a Q_b + \frac{1}{6} \sum_{a,b,c} P_{abc}^{(3)} Q_a Q_b Q_c. \tag{6.57}$$

Before we make use of this property expansion, let us reflect a bit on the symmetry properties of $P(Q)$, which is either an electronic transition moment or an expectation value at a molecular geometry given by the set of normal coordinates Q. We will limit this symmetry analysis to Abelian groups, noting that the following analysis is not valid for general point groups.

For every property, there is an associated electronic operator $\hat{P}$ that belongs to an irreducible representation $\Gamma(\hat{P})$. For a first-order property, such as the electric dipole moment, $\hat{P}$ equals the operator corresponding to the observable itself, for example, $\hat{\mu}_\alpha$ for the electric dipole moment along the axis α in the molecular frame. For a second-order property, such as the electric dipole polarizability, the situation becomes a bit more complicated and the operator $\hat{P}$ is to be considered as the sum-over-states expression that is identified inside the expectation value over the electronic ground state. For the dynamic polarizability, we, for instance, have

$$\hat{\alpha}^{\mathrm{e}}_{\alpha\beta}(-\omega_\sigma;\omega_1) = \frac{1}{\hbar} \sum \mathcal{P}_{-\sigma,1} \sum_{K\neq 0} \frac{\hat{\mu}_\alpha |K\rangle\langle K| \hat{\mu}_\beta}{\omega_{K0} - \omega_1}. \tag{6.58}$$

In this case the irreducible representation associated with the property operator is

$$\Gamma(\hat{\alpha}_{\alpha\beta}) = \Gamma(\hat{\mu}_\alpha) \otimes \Gamma(\hat{\mu}_\beta). \tag{6.59}$$

The symmetry properties of P with respect to nuclear displacements along the symmetry-adapted normal coordinates are summarized in Table 6.2. We distinguish between two cases:

Table 6.2 Dependence of electronic property $P(Q_a)$ with respect to nuclear displacements along a normal coordinate Q_a belonging to the irreducible representation $\Gamma(Q_a)$.

	$P(0) = 0$	$P(0) \neq 0$
$\Gamma(\hat{P}) = \Gamma(Q_a)$	$P(Q_a) = -P(-Q_a)$	–
$\Gamma(\hat{P}) \neq \Gamma(Q_a)$	$P(Q_a) = 0$	$P(Q_a) = P(-Q_a)$

(i) $P(0) = 0$, denoting the situation when the property at the equilibrium geometry is zero for reasons of symmetry as, for example, α^{e}_{xy} in C_{2v} symmetry; and (ii) $P(0) \neq 0$, denoting the contrary situation as, for example, α^{e}_{xx} or a nonzero permanent dipole moment $\mu_\alpha = \langle 0|\hat{\mu}_\alpha|0\rangle$. From the table, we note that, in the former case, $P(Q_a)$ is an odd function when the symmetry of the operator equals that of the normal mode and $P(Q_a) = 0$ if symmetries do not match. In the latter case, we note that $P(Q_a)$ is an even function if symmetries do not match. These simple observations have far-reaching consequences in vibrational spectroscopies as they allow for the prediction of selection rules. They also reduce computational costs since in the property expansion in Eq. (6.57) there will only be linear contributions from modes with $\Gamma(\hat{P}) = \Gamma(Q_a)$.

The ZPVA contribution to the property P is obtained by integrating the property expansion in Eq. (6.57) over the zeroth- and first-order perturbed vibrational wave function according to

$$\Delta P^{\mathrm{zpva}} = \langle \Phi^{(0)} + \Phi^{(1)} | \sum_a P^{(1)}_a Q_a + \frac{1}{2} \sum_{a,b} P^{(2)}_{ab} Q_a Q_b | \Phi^{(0)} + \Phi^{(1)} \rangle. \tag{6.60}$$

We have here left out $P^{(0)}$ in the property expansion since it is independent of Q and results in a term corresponding to the electronic part of the property determined at the equilibrium geometry ($Q = 0$). The first-order contributions to the ZPVA contribution then becomes

$$[P]^{1,0} = \frac{1}{2} \langle \Phi^{(0)} | \sum_{a,b} P^{(2)}_{ab} Q_a Q_b | \Phi^{(0)} \rangle = \frac{\hbar}{4} \sum_a \frac{P^{(2)}_{aa}}{\omega_a}, \tag{6.61}$$

$$[P]^{0,1} = 2 \langle \Phi^{(0)} | \sum_a P^{(1)}_a Q_a | \Phi^{(1)} \rangle = -\frac{\hbar}{4} \sum_{a,b} \frac{P^{(1)}_a V^{(3)}_{abb}}{\omega_a^2 \omega_b}, \tag{6.62}$$

where $[P]^{n,m}$ denotes the contribution due to the nth- and mth-order property and mechanical anharmonicity, respectively (see Table 6.3 for further details and examples). We have here used Eq. (6.53), taking into account also the transformation from the reduced dimensionless normal coordinates to the normal coordinate basis, Eq. (6.39). The first term $[P]^{1,0}$ is commonly referred to as the (mechanical) harmonic contribution because it is determined by the zeroth-order harmonic oscillator wave function, whereas the second term $[P]^{0,1}$, that is, the correction due to mechanical anharmonicity, couples the zeroth- and first-order perturbed wave functions. The next contribution to the zero-point vibrational correction of a property is of third order and it is common to neglect this and higher order corrections when dealing with polyatomic molecules. In the following we adhere to this convention and set

$$\Delta P^{\mathrm{zpva}} = [P]^{1,0} + [P]^{0,1}. \tag{6.63}$$

We can generalize the zero-point vibrational energy $E^{\mathrm{zpv}}(Q)$ to be a function of nuclear coordinates, referring to one-half times the sum of the square root of eigenvalues of the

Table 6.3 Formulas for the pure vibrational contributions to dynamic vibrational polarizabilities and hyperpolarizabilities.[a)]

Term	Formula
$[\mu^2]^{0,0}$	$= \frac{1}{2}\sum \mathcal{P}_{\alpha,\beta}\sum_{a}\frac{\partial\mu_\alpha}{\partial Q_a}\frac{\partial\mu_\beta}{\partial Q_a}\lambda_a^{\pm\sigma}$
$[\mu^2]^{2,0}$	$= \frac{\hbar}{8}\sum \mathcal{P}_{\alpha,\beta}\sum_{a,b}\left[\frac{\partial^2\mu_\alpha}{\partial Q_a\partial Q_b}\frac{\partial^2\mu_\beta}{\partial Q_a\partial Q_b}\left(\frac{1}{\omega_a}+\frac{1}{\omega_b}\right)\lambda_{ab}^{\pm\sigma} + \frac{\partial^3\mu_\alpha}{\partial Q_a\partial Q_a\partial Q_b}\frac{\partial\mu_\beta}{\partial Q_b}\frac{\lambda_b^{\pm\sigma}}{\omega_a} + \frac{\partial\mu_\alpha}{\partial Q_b}\frac{\partial^3\mu_\beta}{\partial Q_a\partial Q_a\partial Q_b}\frac{\lambda_b^{\pm\sigma}}{\omega_a}\right]$
$[\mu^2]^{1,1}$	$= -\frac{\hbar}{4}\sum \mathcal{P}_{\alpha,\beta}\sum_{a,b,c}\left[V_{abc}^{(3)}\frac{\partial^2\mu_\alpha}{\partial Q_a\partial Q_b}\frac{\partial\mu_\beta}{\partial Q_c}\left(\frac{1}{\omega_a}+\frac{1}{\omega_b}\right)\lambda_{ab}^{\pm\sigma}\lambda_c^{\pm\sigma} + V_{bcc}^{(3)}\frac{\partial^2\mu_\alpha}{\partial Q_a\partial Q_b}\frac{\partial\mu_\beta}{\partial Q_a}\frac{\lambda_a^{\pm\sigma}}{\omega_b^2\omega_c}\right]$
$[\mu^2]^{0,2}$	$= -\frac{\hbar}{8}\sum \mathcal{P}_{\alpha,\beta}\sum_{a,b,c}\frac{1}{\omega_a}\left[V_{aabc}^{(4)}\frac{\partial\mu_\alpha}{\partial Q_b}\frac{\partial\mu_\beta}{\partial Q_c}\lambda_b^{\pm\sigma}\lambda_c^{\pm\sigma} - \sum_d\left\{V_{aab}^{(3)}V_{bcd}^{(3)}\frac{\partial\mu_\alpha}{\partial Q_c}\frac{\partial\mu_\beta}{\partial Q_d}\frac{\lambda_c^{\pm\sigma}\lambda_d^{\pm\sigma}}{\omega_b^2} + 2V_{abc}^{(3)}V_{abd}^{(3)}\frac{\partial\mu_\alpha}{\partial Q_c}\frac{\partial\mu_\beta}{\partial Q_d}\lambda_{ab}^{\pm\sigma}\lambda_c^{\pm\sigma}\lambda_d^{\pm\sigma}\right\}\right]$
$[\mu^3]^{1,0}$	$= \frac{1}{2}\sum \mathcal{P}_{\alpha,\beta,\gamma}\sum_{a,b}\frac{\partial\mu_\alpha}{\partial Q_a}\frac{\partial^2\mu_\beta}{\partial Q_a\partial Q_b}\frac{\partial\mu_\gamma}{\partial Q_b}\lambda_a^{\pm\sigma}\lambda_b^{\pm2}$
$[\mu^3]^{0,1}$	$= -\frac{1}{6}\sum \mathcal{P}_{\alpha,\beta,\gamma}\sum_{a,b,c}V_{abc}^{(3)}\frac{\partial\mu_\alpha}{\partial Q_a}\frac{\partial\mu_\beta}{\partial Q_b}\frac{\partial\mu_\gamma}{\partial Q_c}\lambda_a^{\pm\sigma}\lambda_b^{\pm1}\lambda_c^{\pm2}$
$[\mu^2\alpha]^{1,0}$	$= \frac{1}{4}\sum \mathcal{P}_{\alpha,\beta,\gamma,\delta}\sum_{a,b}\left[\frac{\partial\mu_\alpha}{\partial Q_a}\frac{\partial^2\alpha_{\beta\gamma}}{\partial Q_a\partial Q_b}\frac{\partial\mu_\delta}{\partial Q_b}\lambda_a^{\pm\sigma}\lambda_b^{\pm3} + 2\frac{\partial\mu_\alpha}{\partial Q_a}\frac{\partial^2\mu_\beta}{\partial Q_a\partial Q_b}\frac{\partial\alpha_{\gamma\delta}}{\partial Q_b}\lambda_a^{\pm\sigma}\lambda_b^{\pm23}\right]$
$[\mu^2\alpha]^{0,1}$	$= -\frac{1}{4}\sum \mathcal{P}_{\alpha,\beta,\gamma,\delta}\sum_{a,b,c}V_{abc}^{(3)}\frac{\partial\mu_\alpha}{\partial Q_a}\frac{\partial\mu_\beta}{\partial Q_b}\frac{\partial\alpha_{\gamma\delta}}{\partial Q_c}\lambda_a^{\pm\sigma}\lambda_b^{\pm1}\lambda_c^{\pm23}$
$[\mu^4]^{2,0}$	$= \frac{1}{6}\sum \mathcal{P}_{\alpha,\beta,\gamma,\delta}\sum_{a,b,c}\left[3\frac{\partial\mu_\alpha}{\partial Q_a}\frac{\partial^2\mu_\beta}{\partial Q_a\partial Q_b}\frac{\partial^2\mu_\gamma}{\partial Q_b\partial Q_c}\frac{\partial\mu_\delta}{\partial Q_c}\lambda_a^{\pm\sigma}\lambda_b^{\pm23}\lambda_c^{\pm3} + \frac{\partial^3\mu_\alpha}{\partial Q_a\partial Q_b\partial Q_c}\frac{\partial\mu_\beta}{\partial Q_a}\frac{\partial\mu_\gamma}{\partial Q_b}\frac{\partial\mu_\delta}{\partial Q_c}\lambda_a^{\pm1}\lambda_b^{\pm2}\lambda_c^{\pm3}\right]$
$[\mu^4]^{1,1}$	$= -\frac{1}{2}\sum \mathcal{P}_{\alpha,\beta,\gamma,\delta}\sum_{a,b,c,d}V_{abc}^{(3)}\frac{\partial\mu_\alpha}{\partial Q_a}\frac{\partial\mu_\beta}{\partial Q_b}\frac{\partial^2\mu_\gamma}{\partial Q_c\partial Q_d}\frac{\partial\mu_\delta}{\partial Q_d}\lambda_a^{\pm\sigma}\lambda_b^{\pm1}\lambda_c^{\pm23}\lambda_d^{\pm3}$
$[\mu^4]^{0,2}$	$= -\frac{1}{24}\sum \mathcal{P}_{\alpha,\beta,\gamma,\delta}\sum_{a,b,c,d}\left[V_{abcd}^{(4)}\frac{\partial\mu_\alpha}{\partial Q_a}\frac{\partial\mu_\beta}{\partial Q_b}\frac{\partial\mu_\gamma}{\partial Q_c}\frac{\partial\mu_\delta}{\partial Q_d}\lambda_a^{\pm\sigma}\lambda_b^{\pm1}\lambda_c^{\pm2}\lambda_d^{\pm3} - 3\sum_{e}V_{abc}^{(3)}V_{cde}^{(3)}\frac{\partial\mu_\alpha}{\partial Q_a}\frac{\partial\mu_\beta}{\partial Q_b}\frac{\partial\mu_\gamma}{\partial Q_d}\frac{\partial\mu_\delta}{\partial Q_e}\lambda_a^{\pm\sigma}\lambda_b^{\pm1}\lambda_c^{\pm23}\lambda_d^{\pm2}\lambda_e^{\pm3}\right]$

a) (i) For the $[\mu^2]$ terms, $\omega_\sigma = \omega_1$; for the $[\mu^3]$ and $[\mu^2\alpha]$ terms, $\omega_\sigma = \omega_1 + \omega_2$; for the $[\mu^4]$ terms, $\omega_\sigma = \omega_1 + \omega_2 + \omega_3$. (ii) $\sum \mathcal{P}_{\alpha,\beta}$ indicates the sum over terms generated by the permutation of the pairs of optical frequencies and indices: $(-\omega_\sigma, \alpha)$ and (ω_1, β); $\sum \mathcal{P}_{\alpha,\beta,\gamma}$ indicates the permutations of $(-\omega_\sigma, \alpha)$, (ω_1, β), and (ω_2, γ); $\sum \mathcal{P}_{\alpha,\beta,\gamma,\delta}$ indicates the permutations of $(-\omega_\sigma, \alpha)$, (ω_1, β), (ω_2, γ), and (ω_3, δ). (iii) $\lambda_x^{\pm i} = (\omega_x^2 - \omega_i^2)^{-1}$; $\lambda_{xy}^{\pm i} = [(\omega_x + \omega_y)^2 - \omega_i^2]^{-1}$; $\lambda_x^{\pm ij} = [\omega_x^2 - (\omega_i + \omega_j)^2]^{-1}$.

b) The notation $[P]^{n,m}$ indicates the order of the property (or products of such) P and mechanical anharmonicity, respectively. More precisely, each term in the right-hand expression has total order n in derivatives of property P beyond first derivatives and total order m beyond second derivatives (harmonic contribution) of electronic energy, both with respect to mass-weighted normal coordinates. Note that some contributions to the pure vibrational contributions are given in Table 6.4.

mass-weighted force constant matrix at a given geometry. At the equilibrium geometry ($Q = 0$), we of course have $E^{zpv}(0) = E^{(0)}$. We can then identify

$$\frac{\hbar}{4}\sum_b \frac{V^{(3)}_{abb}}{\omega_b} = \left.\frac{\partial E^{zpv}(Q)}{\partial Q_a}\right|_{Q=0}, \tag{6.64}$$

and write the mechanical anharmonic contribution to the zero-point vibrational correction of the property as

$$[P]^{0,1} = -\sum_a \frac{P^{(1)}_a}{\omega_a^2}\frac{\partial E^{zpv}}{\partial Q_a}. \tag{6.65}$$

This expression can be advantageous to use since the zero-point vibrational energy is often more easily accessible in the calculation as compared to the cubic force constants, enabling a simple numerical differentiation scheme.

To get a more qualitative understanding of these formulas, let us first consider the Morse oscillator, which is a simple, analytic model for a one-dimensional anharmonic oscillator. The potential of the Morse oscillator is given by

$$V(r) = D_e(1 - e^{-a(r-r_e)})^2, \tag{6.66}$$

where r is the distance between the atoms, r_e their equilibrium distance, and D_e the dissociation energy, as illustrated in Figure 6.5. It can easily be verified that the first derivative of this potential is zero at the equilibrium geometry, and by relating the force constant k to the second derivative of the Morse potential, we find that $a = \sqrt{k/2D_e}$. Evaluating the third derivative of the Morse potential at the equilibrium geometry, we find

$$\left.\frac{\partial^3 V(r)}{\partial r^3}\right|_{r=r_e} = -2a^3 D_e. \tag{6.67}$$

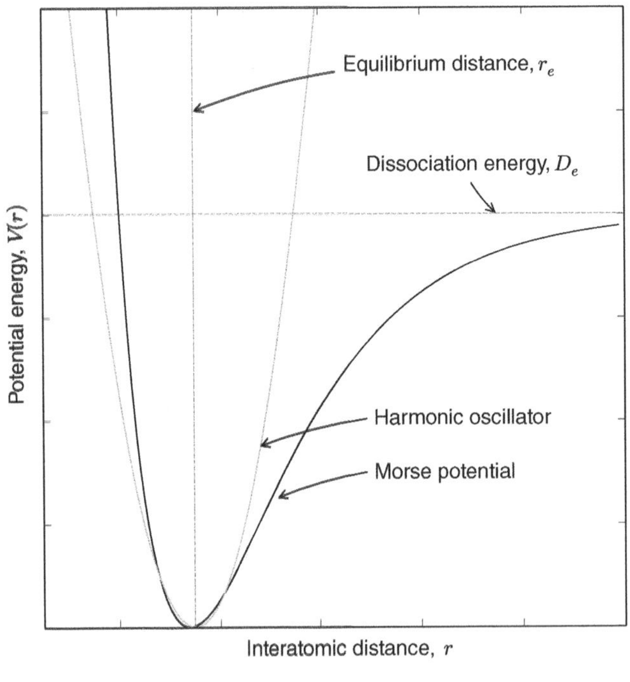

Figure 6.5 Morse and harmonic oscillator potentials.

The third derivative of an anharmonic potential is thus in general negative, as can also be seen from Figure 6.5.

Keeping this in mind, let us now return to the expression for the vibrationally averaged property as given by Eq. (6.60), and consider nuclear displacements along the normal coordinates as the property of interest. Only the first derivatives of the property will then be nonvanishing, and Eq. (6.60) reduces to

$$\Delta Q^{\text{zpva}} = -\frac{\hbar}{4}\sum_{a,b}\frac{V^{(3)}_{abb}}{\omega_a^2\omega_b}, \tag{6.68}$$

and we can thus in general expect that the vibrationally averaged geometry will lead to an elongation along the positive normal coordinates because $V^{(3)}$ is negative, and thus to longer bond lengths.

The perturbation theory expansion used here relies on the electronic potential to be fairly close to a harmonic function of nuclear displacements with respect to the equilibrium geometry. For low-frequency vibrational motions—arising from low-barrier internal rotations (e.g., methyl group rotations in organic molecules), torsional modes that break planar symmetry (e.g., the twisting of the CH_2-group in ethylene), or a vibrational motion connecting local minima through a low-energy barrier (e.g., inter- or intra-molecular hydrogen bonds)—the harmonic approximation will not suffice. In these cases, a nonlocal description of the potential must be used for an accurate account of nuclear motions, which often leads to a need to solve the vibrational problem numerically. This can sometimes be achieved by considering a full one-dimensional potential energy surface along the large-amplitude motion, while ignoring the coupling to the other harmonic modes in the molecule and treating these vibrational modes as independent harmonic oscillators. As an illustration, we consider the dihedral rotation about the O—O bond axis in hydrogen peroxide. The rotational barrier for this motion is very low for passage between the two degenerate minima on the potential energy surface, so the barrier is easily overcome at room temperature—the potential energy curve for this rotation is shown in Figure 6.6. In the trans-configuration, the nuclear framework displays inversion symmetry and parity selection rules apply. The second electronically excited state is parity forbidden at this point and the corresponding oscillator strength therefore vanishes in the electric dipole approximation. However, from the figure it is clear that this intramolecular rotational motion will induce a nonzero intensity for this absorption band.

6.5.2 Importance of Zero-Point Vibrational Corrections

The contributions to molecular properties from zero-point vibrational averaging are typically small, amounting to about 2–5% of the electronic contribution. However, it should be noted that this can vary quite strongly between different properties. It has, for example, been shown that zero-point vibrational contributions can be as large as 10–20% for indirect nuclear spin–spin coupling constants. More typically, however, ZPVA contributions to molecular properties are smaller than residual errors in quantum chemical calculations. This fact, combined with the rather large computational cost due to the need to determine geometrical derivatives of the molecular property of interest, means that ZPVA contributions often are neglected in computational work.

There are, however, a few experimental observables that can only be addressed if zero-point vibrational contributions are taken into account. One such example is given by isotope shifts for properties that do not depend explicitly on the nuclear structure, that is, properties that do not depend on the nuclear magnetic moments or nuclear multipole moments. Such isotope shifts can occur in a number of different spectroscopic techniques. While the effects in some cases,

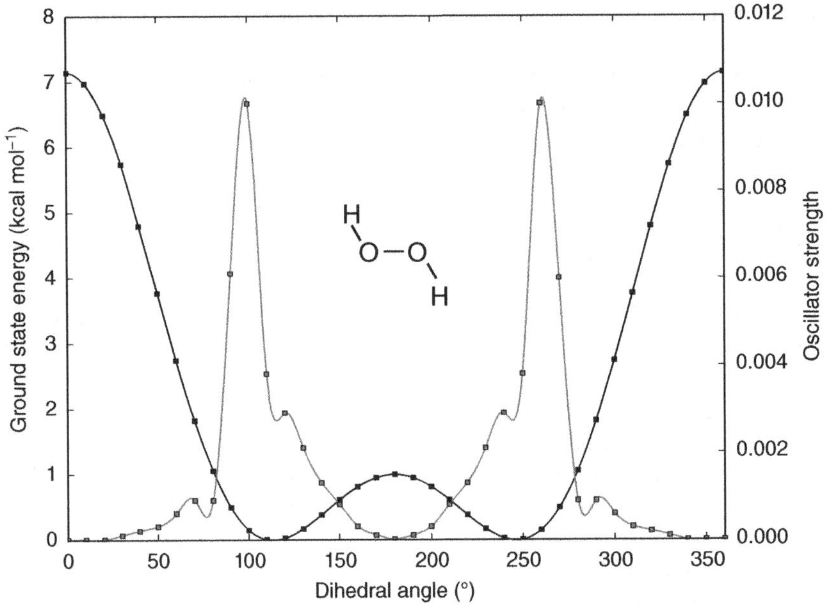

Figure 6.6 Property dependencies with respect to the low-frequency rotational mode in hydrogen peroxide. The electronic ground-state potential energy curve (in kilocalories per mole) and oscillator strength for the transition to the second electronically excited state are plotted in black and grey color, respectively. Results are obtained at the B3LYP/aug-cc-pVTZ level of theory.

such as rotational spectroscopies, are due to the direct change in the molecular mass, affecting the moment of inertia tensor of the molecule, the isotope shifts are, in other cases, of a more quantum nature. The most well-known isotope effect is the isotope shifts seen in vibrational frequencies. As the normal coordinates are defined using mass-weighted coordinates, the normal coordinates and eigenvalues are directly affected by a change in nuclear masses. This is perhaps most easily realized in the case of a diatomic molecule for which the eigenvalues are given by Eq. (6.43). It is seen that the vibrational frequencies of different isotopologues in this case are related by the ratio of the corresponding reduced masses. This also means that an isotopic substitution that leads to the largest change in the reduced mass, for instance, the substitution of hydrogen with deuterium, can be expected to lead to the largest change in vibrational frequencies. If we consider hydrogen fluoride as a specific case, the ratio between the $H^{19}F$ and $D^{19}F$ frequencies is expected to be

$$\frac{\omega_{\mathrm{HF}}}{\omega_{\mathrm{DF}}} = \sqrt{\frac{m_{\mathrm{D}}(m_{\mathrm{H}} + m_{\mathrm{F}})}{m_{\mathrm{H}}(m_{\mathrm{D}} + m_{\mathrm{F}})}} \approx 1.379, \tag{6.69}$$

which is indeed confirmed by the experimental result for the ratio of vibrational frequencies (in cm^{-1}) that equals $4138/2998 = 1.380$.

Another example of nuclear effects on the vibrational frequencies is the asymmetric substitution of a nucleus in a molecule with atoms that are related by symmetry, as for example H_2O and HDO. All electronic states of the two isotopologues are identical and belong to the same point group, namely, C_{2v}. The two molecules therefore also have the same equilibrium geometry if we ignore zero-point vibrational corrections to the nuclear geometry. However, the symmetry of the nuclear wave function of HDO is lower than that of H_2O, as it belongs to the C_s point group. The difference in the mass between the hydrogen and deuterium will lead to changes in the vibrational frequencies and normal modes. In particular, the normal modes can no longer be described as truly symmetric and antisymmetric bending or stretching vibrations.

If we also include zero-point vibrational corrections to the molecular structure, the averaged molecular geometry in HDO will reflect the lower C_s symmetry of this isotopologue, giving rise to different bond lengths for the O—H and O—D bonds, although the difference will be very small. Nevertheless, the relaxation of the nuclear framework into a C_s geometry will as a consequence also make the symmetry of the electronic wave function of the HDO isotopologue at the vibrationally averaged geometry to be C_s, and this will have consequences for the properties of this isotopologue.

Let us consider the ethylene molecule in which we substitute both hydrogen atoms attached to one of the carbons with deuteriums, that is, we consider 1,1-dideuteroethene. Performing a vibrational averaging of the molecular geometry, as in the case of the example with water above, will lower the symmetry of the molecule from D_{2h} to C_{2v} by making the C—H and C—D bonds slightly unequal. In the C_{2v} point group, inversion is not a symmetry operation, which means that the molecule can sustain a permanent dipole moment. This is an example of a vibrationally induced dipole moment in a molecule that is electronically nonpolar. The effect is, as can be expected, rather small, but the dipole moment of this molecule has been measured in highly accurate microwave rotational spectroscopy studies and found to be 0.0091 ± 0.0004 Debye.

In a similar manner, we can also induce chirality in a molecule by isotope substitutions. Chiral molecules are characterized by the absence of improper rotation axes, that is, the molecule cannot be superimposed onto its own mirror image. By repeated substitutions of different isotopologues, in principle even highly symmetric molecules can be made vibrationally chiral, although the effect can be expected to be most pronounced in the case of substitutions of isotopes with as large a ratio as possible between their isotopic masses. An example of a vibrationally induced chirality would be the molecule CFHDT, as illustrated in Figure 6.7.

As a final example, let us consider the so-called secondary isotope effects on nuclear magnetic shielding constants. While the observed resonances of NMR-active nuclei are determined primarily by the nuclear magnetic moment of the nucleus in question, which may vary significantly for different isotopes, small changes can also be observed on the chemical shift of a nucleus when an isotope substitution is carried out on a different nucleus in the molecule. Consider, for instance, the water molecule with focus on the ^{17}O shielding. When comparing H_2O to HDO, the absolute ^{17}O shielding constant differs by about 1.5 ppm as a result of the averaging of the shielding constant over the zero-point vibrational motion in the two different isotopic species.

In general, the calculation of zero-point vibrational corrections is a time-consuming task, as the leading-order anharmonic correction involves the semidiagonal part of the cubic force field and thus couples all the vibrational modes in the molecule, and a complete determination of

Figure 6.7 The two enantiomers of the CFHDT molecule.

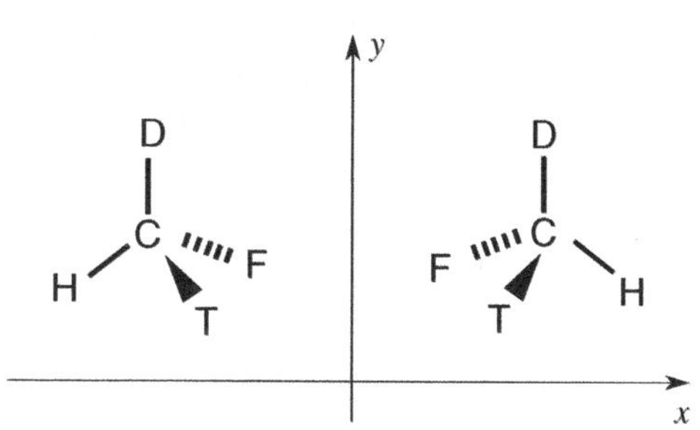

the vibrational modes in the molecule is therefore necessary. However, if we are considering local properties in contrast to extensive properties, such as the shielding constant of a nucleus or its electric field gradient at the nucleus, and this nucleus is a proton, then it may suffice to only consider the local vibrational motion in which this nucleus is involved. This observation can be rationalized as being due to the fact that the proton is a very light nucleus, and thus the vibrational motion of this nucleus will, to only a very limited extent, be able to perturb the motion of heavier nuclei in the molecule. The coupling of the vibrational motion of the hydrogen to other vibrational modes through the cubic force field will therefore in general be smaller, making this vibrational motion rather isolated and involving only a very small part of the molecular structure.

6.5.3 Temperature Effects

When considering the effects of temperature, we will restrict ourselves to generalize the expressions in Eqs. (6.61) and (6.62) to provide the ZPVA contribution to a molecular property in the case of a finite temperature. In doing so, there are two separate effects that we need to consider, namely, (i) the population of vibrational excited states and (ii) centrifugal distortions.

From statistical mechanics, we know that the partition function at temperature T is given by

$$Z(T) = \sum_{n=0}^{\infty} \exp(-E_n/kT), \tag{6.70}$$

and the probability of populating a given level n with energy E_n in the system is given by

$$p_n(T) = e^{-E_n/kT}/Z(T). \tag{6.71}$$

For a harmonic oscillator, we have $E_n = (n + 1/2)\hbar\omega$, resulting in a partition function that becomes

$$Z(T) = \sum_{n=0}^{\infty} \exp\left[-(n+1/2)\frac{\hbar\omega}{kT}\right] = \frac{e^{-\hbar\omega/(2kT)}}{1 - e^{-\hbar\omega/kT}}, \tag{6.72}$$

and the population probabilities become

$$p_n(T) = \frac{(1 - e^{-\hbar\omega/kT})\, e^{-n\hbar\omega/kT}}{e^{-\hbar\omega/(2kT)}}. \tag{6.73}$$

In Figure 6.8, we plot the population of the vibrational ground state at room temperature as a function of the vibrational frequency. It is seen that at frequencies of 500 and 1000 cm^{-1}, the populations of the vibrational ground state exceed 90% and 99%, respectively. As a consequence, temperature effects on vibrational properties as due to the population of higher vibrational states are very small and can typically be ignored.

A more important temperature effect on vibrational corrections is due to centrifugal distortions. As the temperature increases, higher rotational states become populated as these are temperature accessible even for rather low temperatures because the separation between rotational energy levels is of the order of kT. As a molecule rotates faster due to thermally induced rotational excitations, there will be a reactive centrifugal force that will lead to bond elongations. This will in turn change the effective geometry—or alternatively, the anharmonic contribution to the molecular property—leading to a temperature-averaged geometry that differs from that obtained by a consideration of the population of vibrational states alone.

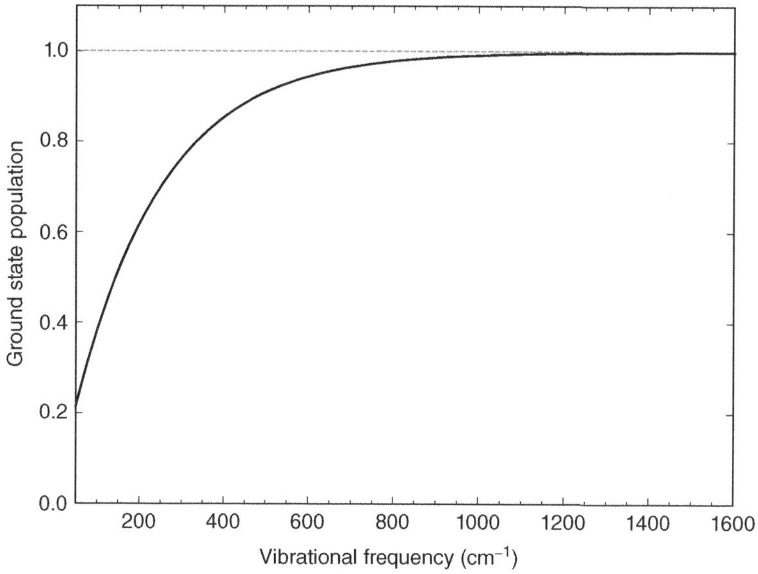

Figure 6.8 Population of the ground vibrational state at a temperature of 298.15 K as a function of the harmonic vibrational frequency.

6.6 Pure Vibrational Contributions to Properties

6.6.1 Perturbation Theory Approach

We will now return to the contribution to the polarizability given by the first term in Eq. (6.14), which is known as the pure vibrational contribution. Pure vibrational contributions are of particular concern for nonlinear optical properties such as hyperpolarizabilities. For nonlinear optical processes, in particular when one or more static electric fields are involved, the pure vibrational contributions to molecular properties can be of similar size or even larger than the corresponding electronic contributions.

The derivation of explicit expressions for pure vibrational contributions to response functions using perturbation theory is quite lengthy, and we will restrict ourselves to providing the final expressions in Tables 6.3 and 6.4. But it is instructive to demonstrate the general procedure by considering in some detail the derivation of the pure vibrational contributions to the polarizability, as well as the leading-order correction to the first hyperpolarizability.

One reason for taking a perturbation theory approach is that it allows for a more direct analysis of which modes are the most important for the pure vibrational contributions. In addition, it allows us to keep a rather close correspondence to the exact-state expressions. On the negative side, it is also clear that this approach offers very little in terms of physical insights and becomes quickly more of an algebraic exercise. As we will see later, a more intuitive understanding of the these vibrational contributions to molecular properties can instead come from an analysis of the geometry changes induced in a molecule when applying an external electric field.

The perturbation theory approach for calculating the pure vibrational contributions follows the same methodology as used in the previous section for the calculation of zero-point

Table 6.4 Conversion of formulas for $[\mu^2]^{n,m}$ contributions given in Table 6.3 to those for $[\mu\alpha]^{n,m}$, $[\alpha^2]^{n,m}$, and $[\mu\beta]^{n,m}$.

	Conversion			
Initial term	**Multiply by**	**Permutation sum change**	**Electrical property and frequency change**	**Final term**
$[\mu^2]^{0,0}$	1	$\mathcal{P}_{\alpha,\beta} \to \mathcal{P}_{\alpha,\beta,\gamma}$	$\mu_\beta \to \alpha_{\beta\gamma}$	$[\mu\alpha]^{0,0}$
$[\mu^2]^{0,0}$	$\frac{1}{4}$	$\mathcal{P}_{\alpha,\beta} \to \mathcal{P}_{\alpha,\beta,\gamma,\delta}$	$\mu_\alpha \to \alpha_{\alpha\beta}, \mu_\beta \to \alpha_{\gamma\delta}, \lambda_a^{\pm\sigma} \to \lambda_a^{\pm 23}$	$[\alpha^2]^{0,0}$
$[\mu^2]^{0,0}$	$\frac{1}{3}$	$\mathcal{P}_{\alpha,\beta} \to \mathcal{P}_{\alpha,\beta,\gamma,\delta}$	$\mu_\beta \to \beta_{\beta\gamma\delta}$	$[\mu\beta]^{0,0}$
$[\mu^2]^{2,0}$	1	$\mathcal{P}_{\alpha,\beta} \to \mathcal{P}_{\alpha,\beta,\gamma}$	$\mu_\beta \to \alpha_{\beta\gamma}$	$[\mu\alpha]^{2,0}$
$[\mu^2]^{2,0}$	$\frac{1}{4}$	$\mathcal{P}_{\alpha,\beta} \to \mathcal{P}_{\alpha,\beta,\gamma,\delta}$	$\mu_\alpha \to \alpha_{\alpha\beta}, \mu_\beta \to \alpha_{\gamma\delta}, \lambda_{ab}^{\pm\sigma} \to \lambda_{ab}^{\pm 23}$, $\lambda_b^{\pm\sigma} \to \lambda_b^{\pm 23}$	$[\alpha^2]^{2,0}$
$[\mu^2]^{2,0}$	$\frac{1}{3}$	$\mathcal{P}_{\alpha,\beta} \to \mathcal{P}_{\alpha,\beta,\gamma,\delta}$	$\mu_\beta \to \beta_{\beta\gamma\delta}$	$[\mu\beta]^{2,0}$
$[\mu^2]^{1,1}$	$\frac{1}{2}$	$\mathcal{P}_{\alpha,\beta} \to \mathcal{P}_{\alpha,\beta,\gamma}$	Each term is replaced by two: in the first $\mu_\beta \to \alpha_{\beta\gamma}$ and in the second $\mu_\alpha \to \alpha_{\beta\gamma}$ and $\mu_\beta \to \mu_\alpha$	$[\mu\alpha]^{1,1}$
$[\mu^2]^{1,1}$	$\frac{1}{4}$	$\mathcal{P}_{\alpha,\beta} \to \mathcal{P}_{\alpha,\beta,\gamma,\delta}$	$\mu_\alpha \to \alpha_{\alpha\beta}, \mu_\beta \to \alpha_{\gamma\delta}, \lambda_{ab}^{\pm\sigma} \to \lambda_{ab}^{\pm 23}$, $\lambda_a^{\pm\sigma} \to \lambda_a^{\pm 23}$, $\lambda_c^{\pm\sigma} \to \lambda_c^{\pm 23}$	$[\alpha^2]^{1,1}$
$[\mu^2]^{1,1}$	$\frac{1}{6}$	$\mathcal{P}_{\alpha,\beta} \to \mathcal{P}_{\alpha,\beta,\gamma,\delta}$	Each term is replaced by two: in the first $\mu_\beta \to \beta_{\beta\gamma\delta}$ and in the second $\mu_\alpha \to \beta_{\beta\gamma\delta}$ and $\mu_\beta \to \mu_\alpha$	$[\mu\beta]^{1,1}$
$[\mu^2]^{0,2}$	1	$\mathcal{P}_{\alpha,\beta} \to \mathcal{P}_{\alpha,\beta,\gamma}$	$\mu_\beta \to \alpha_{\beta\gamma}$	$[\mu\alpha]^{0,2}$
$[\mu^2]^{0,2}$	$\frac{1}{4}$	$\mathcal{P}_{\alpha,\beta} \to \mathcal{P}_{\alpha,\beta,\gamma,\delta}$	$\mu_\alpha \to \alpha_{\alpha\beta}, \mu_\beta \to \alpha_{\gamma\delta}, \lambda_{ab}^{\pm\sigma} \to \lambda_{ab}^{\pm 23}$, $\lambda_b^{\pm\sigma} \to \lambda_b^{\pm 23}$, $\lambda_c^{\pm\sigma} \to \lambda_c^{\pm 23}$, $\lambda_d^{\pm\sigma} \to \lambda_d^{\pm 23}$	$[\alpha^2]^{0,2}$
$[\mu^2]^{0,2}$	$\frac{1}{3}$	$\mathcal{P}_{\alpha,\beta} \to \mathcal{P}_{\alpha,\beta,\gamma,\delta}$	$\mu_\beta \to \beta_{\beta\gamma\delta}$	$[\mu\beta]^{0,2}$

vibrational corrections, but instead of considering an expectation value over a vibrational wave function, we need here to consider vibrational ground- to excited-state transition moments. We will refer to the linear term in the geometry dependence of the dipole moment [see Eq. (6.57)] as the harmonic electric contribution and the leading-order correction to the pure vibrational polarizability is thus the double-harmonic contribution $[\mu^2]^{0,0}$ involving only the zeroth-order harmonic oscillator wave functions of the ground and excited vibrational levels.

We have to determine transition moments of the form $\langle\Phi_0^{(0)}|\mu_\alpha(Q)|\Phi_k^{(0)}\rangle$, where k is used as a concise notation for a vibrationally excited state specified by the set of vibrational quantum numbers $k = \{n_1, n_2, \ldots, n_{3N-6}\}$; as discussed above, the number of vibrational quantum numbers changes to become $3N - 5$ for linear molecules but the derivation to follow is not affected by this. The unperturbed vibrational wave functions are given in Eq. (6.35) and we note that the linear contribution to the geometry dependence of the dipole moment operator can only couple the ground state to excited states with a single vibrational mode in the first excited level. For such matrix elements, we have

$$\langle\Phi_0^{(0)}|\mu_\alpha^{e}(Q)|\Phi_k^{(0)}\rangle = \sqrt{\frac{\hbar}{2\omega_a}}\frac{\partial\mu_\alpha}{\partial Q_a}; \qquad k = \{0_1, 0_2, \ldots, 1_a, \ldots, 0_{3N-6}\}. \tag{6.74}$$

Inserting this result into the expression for the pure vibrational contribution to the polarizability in Eq. (6.16) results in the following expression for $\alpha^{\mathrm{v}}_{\alpha\beta}(-\omega;\omega)$ in the double-harmonic approximation

$$[\mu^2]^{0,0} = \sum_a \frac{\partial \mu_\alpha}{\partial Q_a}\frac{\partial \mu_\beta}{\partial Q_a}\frac{1}{2\omega_a}\left[\frac{1}{(\omega_a-\omega)}+\frac{1}{(\omega_a+\omega)}\right] = \sum_a \frac{\partial \mu_\alpha}{\partial Q_a}\frac{\partial \mu_\beta}{\partial Q_a}\frac{1}{\omega_a^2-\omega^2}. \tag{6.75}$$

Let us now consider electrical anharmonicities to the pure vibrational polarizability. As it is necessary that the vibrationally excited states of the two multiplying transition dipole moments in the numerator of the expression for the polarizability are the same, there can be no contribution arising from a mix of a harmonic and a first-order anharmonic expansion term of the electronic dipole moment, that is, $[\mu^2]^{1,0}$ has to vanish.

The next-order contribution to the pure vibrational polarizability due to electrical anharmonicity is $[\mu^2]^{2,0}$ and it will contain two types of terms: One set of terms that are quadratic in the first-order electric anharmonicities and a second set of terms that mix the electric harmonic contribution with the second-order anharmonic correction in the expansion of the geometry dependence of the dipole moment; see Eq. (6.57). We therefore need to consider transition matrix elements of the form

$$\langle\Phi_0^{(0)}|Q_a|\Phi_k^{(0)}\rangle\langle\Phi_k^{(0)}|Q_bQ_cQ_d|\Phi_0^{(0)}\rangle = \frac{\hbar^2}{4\sqrt{\omega_a\omega_b\omega_c\omega_d}}(\delta_{ab}\delta_{cd}+\delta_{ac}\delta_{bd}+\delta_{ad}\delta_{bc}). \tag{6.76}$$

and

$$\langle\Phi_0^{(0)}|Q_aQ_b|\Phi_k^{(0)}\rangle\langle\Phi_k^{(0)}|Q_cQ_d|\Phi_0^{(0)}\rangle = \frac{\hbar^2}{4\omega_a\omega_b}(\delta_{ac}\delta_{bd}+\delta_{ad}\delta_{bd}), \tag{6.77}$$

We are now ready to carry out the sum over all unperturbed excited vibrational states $\Phi_k^{(0)}$ in Eq. (6.16) and we get

$$\begin{aligned}[\mu^2]^{2,0} = \frac{\hbar}{4}\sum_{a,b}\Bigg[&\frac{\partial^2\mu_\alpha}{\partial Q_a\partial Q_b}\frac{\partial^2\mu_\beta}{\partial Q_a\partial Q_b}\left(\frac{1}{\omega_a}+\frac{1}{\omega_b}\right)\frac{1}{(\omega_a+\omega_b)^2-\omega^2}\\ &+\left(\frac{\partial\mu_\alpha}{\partial Q_a}\frac{\partial^3\mu_\beta}{\partial Q_a\partial^2 Q_b}+\frac{\partial\mu_\beta}{\partial Q_a}\frac{\partial^3\mu_\alpha}{\partial Q_a\partial^2 Q_b}\right)\frac{1}{\omega_b}\frac{1}{(\omega_a^2-\omega^2)}\Bigg].\end{aligned} \tag{6.78}$$

The derivation of contributions arising from mechanical (as compared to electrical) anharmonicity is algebraically more tedious and involves perturbational corrections to the ground and excited vibrational states in accordance with the expansion made in Eq. (6.51). One thing that is easily seen, however, is that

$$\langle\Phi_0^{(0)}|Q_a|\Phi_k^{(0)}\rangle\langle\Phi_k^{(0)}|Q_b|\Phi_0^{(1)}\rangle = 0. \tag{6.79}$$

This result follows immediately from Eqs. (6.53)–(6.56), from which it is clear that both c_{ab}^{11} and c_a^2 are zero. Had we derived the first-order correction also to the excited vibrational states, we would also have seen that

$$\langle\Phi_0^{(0)}|Q_a|\Phi_k^{(0)}\rangle\langle\Phi_k^{(1)}|Q_b|\Phi_0^{(0)}\rangle = 0, \tag{6.80}$$

which together with Eq. (6.79) guarantees that $[\mu^2]^{0,1}$ vanishes. To second order, the remaining terms not yet discussed and contributing to the pure vibrational polarizability are $[\mu^2]^{1,1}$ and $[\mu^2]^{0,2}$. Explicit expressions for these two contributions are collected in Table 6.3. All leading-order contributions to pure vibrational hyperpolarizabilities are given by using the conversion formulas in Table 6.4.

For the hyperpolarizabilities, the derivations follow the steps indicated for the polarizability demonstrated above. We will demonstrate this procedure by considering the pure vibrational contribution to the first-order hyperpolarizability in the double-harmonic approximation. One merely needs to insert the linear expansions of the electronic dipole moment and static polarizability into Eqs. (6.24) and (6.25) together with the zeroth-other vibrational wave functions. We then obtain

$$[\mu\alpha]^{0,0} = \frac{1}{2}\sum P_{-\sigma,1,2} \sum_a \frac{\partial \mu_\alpha}{\partial Q_a} \frac{\partial \alpha_{\alpha\beta}}{\partial Q_a} \frac{1}{(\omega_a^2 - \omega_\sigma^2)}, \tag{6.81}$$

$$[\mu^3]^{0,0} = 0. \tag{6.82}$$

All terms to second order in electric and mechanical anharmonicities for the pure vibrational contributions to polarizabilities and first and second hyperpolarizabilities are collected in Tables 6.3 and 6.4.

6.6.2 Pure Vibrational Effects from an Analysis of the Electric-Field Dependence of the Molecular Geometry

The perturbation theory approach presented in the previous section for the study of pure vibrational contributions to molecular properties is in principle straightforward to apply to any order, and is derived using conventional perturbation theory. However, the approach lends itself poorly to a more intuitive understanding of the origins of the pure vibrational contributions to nonlinear properties.

We will therefore take a step back and consider an alternative approach to calculating and understanding the effects of molecular vibrations on the electric properties. We will do so by considering the changes in the electronic energy and molecular structure in the presence of an external static electric field. Our focus will be on achieving an intuitive understanding of the molecular origins of the pure vibrational effects, but the analysis also lends itself to computation through the use of a finite field approach.

In Figure 6.9, we have plotted the change in the molecular geometry of the hydrogen fluoride molecule as a function of the strength of an applied static electric field. As we turn on the electric field along the direction of the dipole, we will increase the electronic charge transfer from the hydrogen to the fluorine atom, and in this manner increase the dipole moment and the ionic character of the bond. Thinking of the molecule in terms of atomic charges, one might expect the molecule to reduce the bond length when the atomic charges increase. However, such a bond length reduction would also reduce the dipole moment (since the dipole moment is proportional to the charge separation) and a greater energy gain is obtained by increasing the bond distance with increasing electric field in the direction of the dipole moment. An electric field opposing the direction of the dipole moment conversely leads to a reduction in the bond length, but due to the stronger repulsive potential at shorter bond distances, the bond length reduction for negative fields is much weaker than the elongation for positive fields; see Figure 6.9.

These effects are also illustrated in Figure 6.9(b), where we plot both the total and the electronic dipole moment as a function of the electric field strength. The slope of the curves is proportional to the polarizability and we note that the slope of the dipole moment curve is larger when the nuclei are allowed to relax than when we consider only the electronic dipole moment. This is in accordance with expectation, as we would expect that a molecule that is allowed to fully relax and adapt to the applied electric field will have an increased energy gain in this process. From this simple picture we therefore see that the total polarizability (electronic and vibrational) will in general be larger than the electronic polarizability.

Before proceeding with this analysis, let us also recall that in addition to the electronic energy of the molecule, there is also an energy contribution arising from the zero-point vibrational

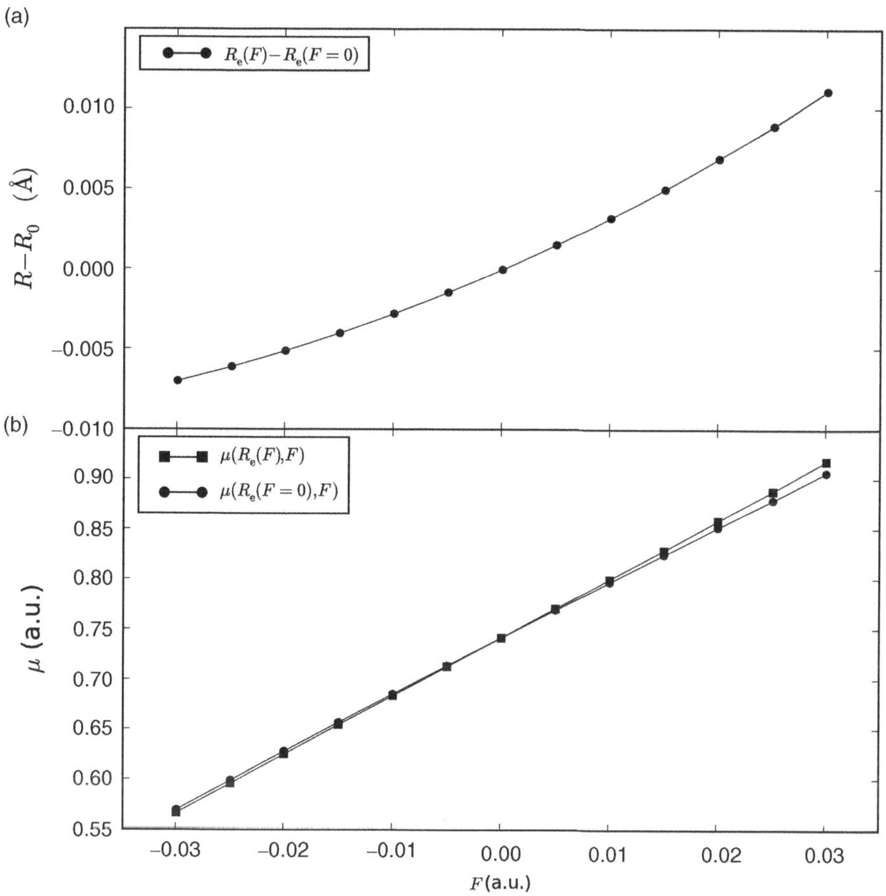

Figure 6.9 Dependence of the bond distance and dipole moment of the HF molecule as a function of the field strength of an external static electric field. The molecule is located along the *z* axis with the direction of the dipole moment being in the positive *z* direction. Calculations have been carried out using the taug-cc-pVTZ basis set at the Hartree–Fock level of theory.

motion of the nuclei, and when we consider the relaxation of the molecular geometry when an external electric field is applied, we also have to consider the changes in the zero-point vibrational corrections in the presence of the external electric field in order to treat the electric field effects on the nuclei in a consistent manner. In particular, our starting point should not be the electronic dipole moment at a field-relaxed or unrelaxed geometry, but rather the vibrationally averaged dipole moment at the electric-field relaxed or unrelaxed geometry. However, as the zero-point vibrational and electronic contributions are additive, we can treat the electric field dependence of the zero-point vibrationally averaged dipole moment separately in order to identify the different contributions to the polarizability.

If we now summarize this analysis of how the dipole moment changes due to the changes in the electronic structure of the molecule, relaxation of the nuclear geometry, and the changes in the zero-point vibrational corrections to the energy and the dipole moment, we can identify three distinct contributions to the polarizability

$$\alpha^{e} = -\left.\frac{d^2E^{e}(R_0,F)}{dF^2}\right|_{F=0}, \qquad (6.83)$$

$$\alpha^{\mathrm{nr}} = -\left.\frac{d^2[E^{\mathrm{e}}(R,F) - E^{\mathrm{e}}(R_0,F)]}{dF^2}\right|_{F=0}, \tag{6.84}$$

$$\alpha^{\mathrm{curv}} = -\left.\frac{d^2 E^{\mathrm{zpv}}(R,F)}{dF^2}\right|_{F=0}, \tag{6.85}$$

where α^{e} is the electronic polarizability, α^{nr} is the nuclear relaxation polarizability, and α^{curv} the contribution to the polarizability due to the curvature of the potential energy surface with respect to nuclear distortions. This analysis can of course be carried out to higher orders in the electric field, and we will obtain a similar separation of the total hyperpolarizabilities. We emphasize that even though we have discussed a separation of the polarizabilities into three distinct contributions when analyzing the relaxation of the nuclear structure in the presence of the electric field, these are distinctly different from the separation of the vibrational contributions to the polarizabilities in terms of pure and zero-point vibrational corrections due to the different manner in which the analysis is performed.

It is instructive to compare this approach based on the concept of relaxing the molecular electronic and nuclear structure in the presence of a static electric field to the approach based on the vibronic analysis and separation of the polarizabilities as presented earlier in this chapter. In order to do so, we shall use variational perturbation theory, as outlined in Appendix H. We write the ground-state electronic energy as

$$E^{\mathrm{e}} \equiv V(Q,F) = E_0^{\mathrm{e}}(Q,F) - \mu_\alpha(Q)F_\alpha, \tag{6.86}$$

where Q collects the set of normal coordinates $\{Q_a\}$ and F denotes the applied *static* electric field. We assume that we have optimized the molecular geometry at all field strengths

$$\frac{\partial V(Q,F)}{\partial Q_a} = 0, \tag{6.87}$$

so that the normal coordinates become functions of the external fields

$$Q_a \equiv Q_a(F). \tag{6.88}$$

We can use the full machinery of variational perturbation theory and, following Eq. (H.7), we find that the first derivative of the electronic energy with respect to a component of the electric field simplifies to

$$-\left.\frac{dV}{dF_\alpha}\right|_{F=0} = -\left.\frac{\partial V}{\partial F_\alpha}\right|_{F=0} = \mu_\alpha. \tag{6.89}$$

To second order, we obtain

$$\begin{aligned} -\left.\frac{d^2 V}{dF_\alpha dF_\beta}\right|_{F=0} &= -\left.\left[\frac{\partial^2 V}{\partial F_\alpha \partial F_\beta} + \sum_a \frac{\partial^2 V}{\partial F_\alpha \partial Q_a}\frac{\partial Q_a}{\partial F_\beta}\right]\right|_{F=0} \\ &= \alpha^{\mathrm{e}}_{\alpha\beta} + \sum_a \left.\frac{\partial \mu_\alpha}{\partial Q_a}\frac{\partial Q_a}{\partial F_\beta}\right|_{F=0}, \end{aligned} \tag{6.90}$$

which is to be compared with Eq. (H.13). To find the first-order change of the normal coordinates to the electric field, we use the first-order response equation in Eq. (H.10), which in the present case reads

$$\frac{d}{dF_\beta}\left(\frac{\partial V}{\partial Q_a}\right) = \frac{\partial^2 V}{\partial F_\beta \partial Q_a} + \sum_b \frac{\partial^2 V}{\partial Q_a \partial Q_b}\frac{\partial Q_b}{\partial F_\beta} = 0. \tag{6.91}$$

When the above response equation is evaluated at zero field strength, we get

$$-\frac{\partial \mu_\beta}{\partial Q_a} + \omega_a^2 \left.\frac{\partial Q_a}{\partial F_\beta}\right|_{F=0} = 0, \tag{6.92}$$

where we used the fact that the second-order derivative of the potential V with respect to normal coordinates is diagonal when evaluated at $Q = 0$ or, in other words, $V_{ab} = V_{aa}\delta_{ab}$. By rearranging terms, we obtain

$$\frac{\partial Q_a}{\partial F_\beta} = \frac{1}{\omega_a^2}\frac{\partial \mu_\beta}{\partial Q_a}, \tag{6.93}$$

which gives the final result

$$-\left.\frac{d^2E^e}{dF_\alpha dF_\beta}\right|_{F=0} = \alpha_{\alpha\beta}^e + \sum_a \frac{1}{\omega_a^2}\frac{\partial \mu_\alpha}{\partial Q_a}\frac{\partial \mu_\beta}{\partial Q_a} = \alpha_{\alpha\beta}^e + [\mu^2]_{\alpha\beta}^{0,0} = \alpha_{\alpha\beta}^e + \alpha_{\alpha\beta}^{nr}. \tag{6.94}$$

Going back to Eq. (6.83), we see that the electronic polarizability α^e is defined as the second derivative of the electronic energy with respect to the electric field, keeping the nuclei fixed at the field-free geometry. It was therefore correctly identified as the second *partial derivative* of the electronic energy with respect to the electric field in Eq. (6.90). The remainder of the second *total derivative* of the electronic energy hence corresponds to the nuclear relaxation contribution to the polarizability, α^{nr}, which is precisely what is expressed by Eq. (6.84). By taking the static limit of Eq. (6.75), we see that it is identical to the pure vibrational polarizability in the double-harmonic approximation.

Following Eq. (H.17), we find that the third-order derivative with respect to components of the electric field reads

$$\begin{aligned}
&-\left.\frac{d^3V}{dF_\alpha dF_\beta dF_\gamma}\right|_{F=0} \\
&= \beta_{\alpha\beta\gamma}^e + \sum_a \left[\frac{\partial \alpha_{\alpha\beta}^e}{\partial Q_a}\frac{\partial Q_a}{\partial F_\gamma} + \frac{\partial \alpha_{\alpha\gamma}^e}{\partial Q_a}\frac{\partial Q_a}{\partial F_\beta} + \frac{\partial \alpha_{\beta\gamma}^e}{\partial Q_a}\frac{\partial Q_a}{\partial F_\alpha}\right] \\
&+ \sum_{a,b}\left[\frac{\partial^2 \mu_\alpha}{\partial Q_a \partial Q_b}\frac{\partial Q_a}{\partial F_\beta}\frac{\partial Q_b}{\partial F_\gamma} + \frac{\partial^2 \mu_\beta}{\partial Q_a \partial Q_b}\frac{\partial Q_a}{\partial F_\gamma}\frac{\partial Q_b}{\partial F_\alpha} + \frac{\partial^2 \mu_\gamma}{\partial Q_a \partial Q_b}\frac{\partial Q_a}{\partial F_\beta}\frac{\partial Q_b}{\partial F_\alpha}\right] \\
&+ \sum_{a,b,c} V_{abc}^{(3)}\frac{\partial Q_a}{\partial F_\beta}\frac{\partial Q_b}{\partial F_\alpha}\frac{\partial Q_c}{\partial F_\gamma} \\
&= \beta_{\alpha\beta\gamma}^e + \sum_a \frac{1}{\omega_a^2}\left[\frac{\partial \alpha_{\alpha\beta}^e}{\partial Q_a}\frac{\partial \mu_\gamma}{\partial Q_a} + \frac{\partial \alpha_{\alpha\gamma}^e}{\partial Q_a}\frac{\partial \mu_\beta}{\partial Q_a} + \frac{\partial \alpha_{\beta\gamma}^e}{\partial Q_a}\frac{\partial \mu_\alpha}{\partial Q_a}\right] \\
&+ \sum_{a,b}\frac{1}{\omega_a^2\omega_b^2}\left[\frac{\partial^2 \mu_\alpha}{\partial Q_a \partial Q_b}\frac{\partial \mu_\beta}{\partial Q_a}\frac{\partial \mu_\gamma}{\partial Q_b} + \frac{\partial^2 \mu_\beta}{\partial Q_a \partial Q_b}\frac{\partial \mu_\gamma}{\partial Q_a}\frac{\partial \mu_\alpha}{\partial Q_b} + \frac{\partial^2 \mu_\gamma}{\partial Q_a \partial Q_b}\frac{\partial \mu_\beta}{\partial Q_a}\frac{\partial \mu_\alpha}{\partial Q_b}\right] \\
&+ \sum_{a,b,c} V_{abc}^{(3)}\frac{1}{\omega_a^2\omega_b^2\omega_c^2}\frac{\partial \mu_\beta}{\partial Q_a}\frac{\partial \mu_\alpha}{\partial Q_b}\frac{\partial \mu_\gamma}{\partial Q_c} \\
&= \beta^e + [\mu\alpha]^{0,0} + [\mu^3]^{1,0} + [\mu^3]^{0,1} = \beta^e + \beta^{nr}.
\end{aligned} \tag{6.95}$$

Looking at the expression after the second equality sign we first see three terms in square brackets that involve first derivatives of components of the electronic polarizability and the electric dipole moment, thus in compact notation corresponding to $[\mu\alpha]^{0,0}$. Then follow three terms in square brackets containing one second and two first derivatives of components of the electric dipole moment, corresponding to $[\mu^3]^{1,0}$. Finally there is a term containing the cubic force constant and three first derivatives of components of the electric dipole moment, corresponding to $[\mu^3]^{0,1}$. These identifications are confirmed by taking the static limit of the corresponding terms in Table 6.3. The nuclear relaxation contribution to the first-order hyperpolarizability includes both the leading-order electric and geometric anharmonic contributions ($[\mu^3]^{1,0}$ and $[\mu^3]^{0,1}$, respectively) in addition to the double-harmonic contribution $[\mu\alpha]^{0,0}$. The extension to higher order hyperpolarizabilities follows the approach above, and one finds that the nuclear relaxation contribution to the second hyperpolarizability is

$$\gamma^{\mathrm{nr}} = [\alpha^2]^{0,0} + [\mu\beta]^{0,0} + [\mu^2\alpha]^{0,1} + [\mu^2\alpha]^{1,0} + [\mu^4]^{2,0} + [\mu^4]^{0,2} + [\mu^4]^{1,1}, \tag{6.96}$$

recovering an even larger number of the anharmonic contributions to the pure vibrational contribution of the second hyperpolarizability.

Let us now turn our attention to the curvature contribution. The expression for the field-dependent zero-point vibrational energy in the harmonic approximation is

$$E^{\mathrm{zpv}}(Q,F) = \frac{\hbar}{2}\sum_a \omega_a; \quad \omega_a = \sqrt{V_{aa}^{(2)}} \quad \text{(no implied summation)}, \tag{6.97}$$

so the curvature contribution to the molecular dipole moment, which is equal to minus the derivative of this energy with respect to a component of the electric field, becomes

$$\begin{aligned}\mu_\alpha^{\mathrm{curv}} &= -\left.\frac{dE^{\mathrm{zpv}}}{dF_\alpha}\right|_{F=0} \\ &= -\sum_a \frac{\hbar}{4\omega_a}\left.\frac{dV_{aa}^{(2)}}{dF_\alpha}\right|_{F=0} \\ &= -\sum_a \frac{\hbar}{4\omega_a}\left.\left(\frac{\partial V_{aa}^{(2)}}{\partial F_\alpha} + \sum_b \frac{\partial V_{aa}^{(2)}}{\partial Q_b}\frac{\partial Q_b}{\partial F_\alpha}\right)\right|_{F=0}.\end{aligned} \tag{6.98}$$

To proceed, we will need the first-order response of the normal coordinates, which we take from Eq. (6.93). We therefore arrive at

$$\mu_\alpha^{\mathrm{curv}} = \sum_a \frac{\hbar}{4\omega_a}\left(\frac{\partial^2\mu_\alpha}{\partial Q_a\partial Q_a} - \sum_b V_{aab}^{(3)}\frac{1}{\omega_b^2}\frac{\partial\mu_\alpha}{\partial Q_b}\right) = \Delta\mu_\alpha^{\mathrm{zpva}}, \tag{6.99}$$

which we identify as the zero-point vibrational averaging contribution to the dipole moment from the expressions given in Eqs. (6.61) and (6.62). By differentiating the zero-point vibrational energy a second time with respect to the electric field, we obtain

$$\begin{aligned}\alpha_{\alpha\beta}^{\mathrm{curv}} &= -\left.\frac{d^2E^{\mathrm{zpv}}}{dF_\alpha dF_\beta}\right|_{F=0} \\ &= \frac{\hbar}{8}\sum_a \left.[V_{aa}^{(2)}]^{-3/2}\frac{dV_{aa}^{(2)}}{dF_\alpha}\frac{dV_{aa}^{(2)}}{dF_\beta}\right|_{F=0} \\ &\quad -\frac{\hbar}{4}\sum_a [V_{aa}^{(2)}]^{-1/2}\left(\frac{\partial^2 V_{aa}^{(2)}}{\partial F_\alpha\partial F_\beta} + \sum_b \frac{\partial^2 V_{aa}^{(2)}}{\partial F_\alpha\partial Q_b}\frac{\partial Q_b}{\partial F_\beta} + \sum_b \frac{\partial^2 V_{aa}^{(2)}}{\partial F_\beta\partial Q_b}\frac{\partial Q_b}{\partial F_\alpha}\right. \\ &\quad \left.\left. + \sum_{b,c}\frac{\partial^2 V_{aa}^{(2)}}{\partial Q_b\partial Q_c}\frac{\partial Q_b}{\partial F_\alpha}\frac{\partial Q_c}{\partial F_\beta} + \sum_b \frac{\partial V_{aa}^{(2)}}{\partial Q_b}\frac{\partial^2 Q_b}{\partial F_\alpha\partial F_\beta}\right)\right|_{F=0}.\end{aligned} \tag{6.100}$$

In the final term appears the second-order response of the normal coordinates. This can be compared to the second derivative of Eq. (6.90), where only first-order response enters, in accordance with the Wigner $(2n + 1)$ rule (see Appendix H). The difference is the variational condition, Eq. (6.87), imposed on the ground-state electronic energy, but not on the zero-point vibrational energy in Eq. (6.97). We here remark that whereas it is most common to optimize the molecular structure with respect to the electronic energy only, giving us the molecular equilibrium geometry, we could have chosen to optimize the geometry with respect to the molecular free energy, that is, the sum of the electronic energy and the zero-point vibrational energy. This would give rise to an *effective molecular geometry*, and which corresponds to the vibrationally averaged geometry discussed in Section 6.5. We refer to the further reading for more information about the effective geometry.

The second-order response of the nuclear responses to the external fields are obtained with use of the response equation in Eq. (H.11), which in the present context reads

$$\begin{aligned}\frac{d^2}{dF_\alpha dF_\beta}\left(\frac{\partial V}{\partial Q_b}\right) &= \frac{\partial^3 V}{\partial F_\alpha \partial F_\beta \partial Q_b} \\ &+ \sum_c \left(\frac{\partial^3 V}{\partial F_\alpha \partial Q_b \partial Q_c}\frac{1}{\omega_b^2}\frac{\partial^2 V}{\partial F_\beta \partial Q_c} + \frac{\partial^3 V}{\partial F_\beta \partial Q_b \partial Q_c}\frac{1}{\omega_b^2}\frac{\partial^2 V}{\partial F_\alpha \partial Q_c}\right) \\ &+ \sum_{c,d} \frac{\partial^3 V}{\partial Q_b \partial Q_c \partial Q_d}\frac{1}{\omega_c^2}\frac{\partial^2 V}{\partial F_\alpha \partial Q_c}\frac{1}{\omega_d^2}\frac{\partial^2 V}{\partial F_\beta \partial Q_d} \\ &+ \frac{\partial^2 V}{\partial Q_b \partial Q_b}\frac{\partial^2 Q_b}{\partial F_\alpha \partial F_\beta} = 0, \end{aligned} \tag{6.101}$$

where, in the last term, we have used the fact that the second-order derivative of the potential with respect to normal coordinates is diagonal. When evaluated at zero field strength, the response equation gives

$$\begin{aligned}\left.\frac{\partial^2 Q_b}{\partial F_\alpha \partial F_\beta}\right|_{F=0} &= \frac{1}{\omega_b^2}\frac{\partial \alpha_{\alpha\beta}}{\partial Q_b} \\ &+ \frac{1}{\omega_b^2}\sum_c \frac{1}{\omega_c^2}\left(\frac{\partial^2 \mu_\alpha}{\partial Q_b \partial Q_c}\frac{\partial \mu_\beta}{\partial Q_c} + \frac{\partial^2 \mu_\beta}{\partial Q_b \partial Q_c}\frac{\partial \mu_\alpha}{\partial Q_c}\right) \\ &- \frac{1}{\omega_b^2}\sum_{c,d} V_{bcd}\frac{1}{\omega_c^2}\frac{\partial \mu_\alpha}{\partial Q_c}\frac{1}{\omega_d^2}\frac{\partial \mu_\beta}{\partial Q_d}. \end{aligned} \tag{6.102}$$

Inserting Eq. (6.102) into Eq. (6.100) gives

$$\begin{aligned}\alpha_{\alpha\beta}^{\text{curv}} &= \frac{\hbar}{8}\sum_a \frac{1}{\omega_a^3}\left(\frac{\partial^2 \mu_\alpha}{\partial Q_a \partial Q_a} - \sum_b V_{aab}^{(3)}\frac{1}{\omega_b^2}\frac{\partial \mu_\alpha}{\partial Q_b}\right)\left(\frac{\partial^2 \mu_\beta}{\partial Q_a \partial Q_a} - \sum_c V_{aac}^{(3)}\frac{1}{\omega_c^2}\frac{\partial \mu_\beta}{\partial Q_c}\right) \\ &+ \frac{\hbar}{4}\sum_a \frac{1}{\omega_a}\left(\frac{\partial^2 \alpha_{\alpha\beta}}{\partial Q_a \partial Q_a} + \sum_b \frac{1}{\omega_b^2}\left[\frac{\partial^3 \mu_\alpha}{\partial Q_a \partial Q_a \partial Q_b}\frac{\partial \mu_\beta}{\partial Q_b}\right.\right. \\ &+ \left.\left.\frac{\partial^3 \mu_\beta}{\partial Q_a \partial Q_a \partial Q_b}\frac{\partial \mu_\alpha}{\partial Q_b} - \sum_c V_{aabc}^{(4)}\frac{\partial \mu_\alpha}{\partial Q_b}\frac{1}{\omega_c^2}\frac{\partial \mu_\beta}{\partial Q_c}\right]\right)\end{aligned}$$

$$-\frac{\hbar}{4}\sum_a \frac{1}{\omega_a}\left(\sum_b V^{(3)}_{aab}\frac{1}{\omega_b^2}\left[\frac{\partial \alpha_{\alpha\beta}}{\partial Q_b}+\sum_c \frac{\partial^2 \mu_\alpha}{\partial Q_b \partial Q_c}\frac{1}{\omega_c^2}\frac{\partial \mu_\beta}{\partial Q_c}\right.\right.$$
$$\left.\left.+\sum_c \frac{\partial^2 \mu_\beta}{\partial Q_b \partial Q_c}\frac{1}{\omega_c^2}\frac{\partial \mu_\alpha}{\partial Q_c}-\sum_{c,d} V^{(3)}_{bcd}\frac{1}{\omega_c^2}\frac{\partial \mu_\alpha}{\partial Q_c}\frac{1}{\omega_d^2}\frac{\partial \mu_\beta}{\partial Q_d}\right]\right). \tag{6.103}$$

Expressed in the zero frequency limit of the terms given in Table 6.3, the curvature contribution to the polarizability is seen to correspond to[6]

$$\alpha^{\text{curv}} = \Delta\alpha^{\text{zpva}} + [\mu^2]^{2,0} + [\mu^2]^{1,1} + [\mu^2]^{0,2}. \tag{6.104}$$

By differentiating the zero-point vibrational energy a third time with respect to the electric field, we obtain the following expression for the curvature contribution to the first-order hyperpolarizability

$$\beta^{\text{curv}} = \Delta\beta^{\text{zpva}} + [\mu\alpha]^{2,0} + [\mu\alpha]^{1,1} + [\mu\alpha]^{0,2}$$
$$+[\mu^3]^{3,0} + [\mu^3]^{2,1} + [\mu^3]^{1,2} + [\mu^3]^{0,3}, \tag{6.105}$$

but we omit the grisly details of the derivation. The curvature contribution thus captures not only the contribution to the polarizabilities and hyperpolarizabilities due to the zero-point vibrational motion, but also the anharmonic corrections to the pure vibrational part that are missing in the nuclear relaxation contribution.

The perturbation theory and nuclear relaxation approaches offer different views on how an externally applied electric field alters the electronic and molecular structure, and it is noted that the nuclear relaxation approach has more to offer in terms of a physically intuitive picture but is limited to static external fields. Through the use of variational perturbation theory, we have derived working equations for the nuclear relaxation and curvature contributions that are in a form that allows for a direct comparison with terms obtained in the perturbation theory approach, although computationally a finite-field approach would be used. In contrast, the perturbation theory approach is applicable also to frequency-dependent external fields and leads to expressions that lend themselves well to computations, but the derivations are cumbersome and the final expressions difficult in case anharmonic contributions need to be included.

6.7 Adiabatic Vibronic Theory for Electronic Excitation Processes

Vibrational corrections to electronic excitations often play a crucial role in determining the fine structure of an absorption spectrum as well as the location of the maximum intensity of the absorption peak, and these effects are often referred to as *vibronic* effects. Calculated absorption spectra resulting from a simultaneous consideration of electronic and vibrational contributions are referred to as *vibronic absorption spectra*, in contrast to a purely electronic absorption spectrum. The perhaps most striking examples of vibronic effects on absorption spectra are observed in the case of highly symmetric molecules where only a few excited states are accessible by irradiation with electromagnetic light due to the selection rules imposed on the electric dipole-allowed transitions. In this case, the coupling to vibrational modes transforming

---

6 The observant reader will note that some terms in Eq. (6.103) are not fully identical to the corresponding terms in Table 6.3. These differences stem from us starting the above differentiation from a diagonal force field. Partial derivatives of these force constants therefore only give rise to semidiagonal cubic (and higher) force constants, Off-diagonal cubic force constants can only arise from the responses of the normal coordinates with respect to the electric field, as seen in Eq. (6.102).

according to specific irreducible representations may make otherwise dipole-forbidden transitions vibronically allowed. Such vibronically allowed transitions are often characterized by having absorption bands that are much weaker than the dipole-allowed transitions. However, even for low-symmetry molecules, nuclear motions affect the absorption spectrum, in part by broadening the spectrum and giving rise to vibrational fine structure, but also by shifting the absorption maximum to shorter wavelengths (higher energy) than the adiabatic electronic excitation energy.

The analysis of vibronic transitions takes as a starting point the *Franck–Condon principle*, which states that because the nuclei are much heavier than the mass of the electron being excited, the electronic excitation can be assumed to occur in the presence of a stationary nuclear framework. We note the similarity to the Born–Oppenheimer approximation, but in the case of the Franck–Condon principle, the electronic excitations happen so fast that the nuclei do not move during the excitation process and the geometry remains fixed. However, whereas the nuclei prior to the absorption process will experience the electronic ground-state potential, they will, after the absorption process, experience the excited-state electronic potential, and thus there will in general be a structural reorganization after the excitation process. Furthermore, the excited-state electronic potential will also lead to different vibrational wave functions in the electronic excited state than those in the electronic ground state.

We recall from Chapter 3 that the interaction between a molecular charge distribution and an electromagnetic field can be described by a multipolar expansion, in which the electric dipole contribution is the leading-order contribution, and we will in the following assume the electric dipole approximation. Electronic absorption spectroscopy is then governed by the electric dipole transition moment between the initial (most often the ground state) and the final states. Assuming vibronic wave functions within the Born–Oppenheimer approximation, we can write the relevant transition dipole moment between the vibronic states as $\langle k, K|\hat{\mu}_\alpha|0,0\rangle$. In order to find a suitable approximation to this vibronic transition moment, we follow the vibrational perturbation theory approach used for the pure vibrational contributions in Section 6.6, integrating over the electronic coordinates and expanding the electronic transition dipole moment in terms of the mass-weighted normal coordinates of the electronic ground state. We thereby get

$$\langle k, K|\hat{\mu}_\alpha|0,0\rangle = \langle k|\mu_\alpha^{K0}(\mathbf{Q})|0\rangle = \mu_\alpha^{K0}(0)S(k,0) + \sum_a \left.\frac{\partial \mu_\alpha^{K0}}{\partial Q_a}\right|_{Q=0} \langle k|Q_a|0\rangle + \cdots, \quad (6.106)$$

where $S(k, 0)$ has been introduced to denote the overlap between vibrational state $|k\rangle$ of electronic state $|K\rangle$ and vibrational state $|0\rangle$ belonging to the electronic ground state $|0\rangle$, or more generally

$$S(k,n) = \langle k|n\rangle. \quad (6.107)$$

By truncating the expansion after the first term we invoke the so-called Frank–Condon approximation, whereas the second term is referred to as the Herzberg–Teller correction. We note that it is only the Herzberg–Teller term that would allow dipole-forbidden transitions to gain intensity through a coupling to selected vibrational modes.

Let us first consider in some detail the Frank–Condon approximation, returning to the consequences of the Herzberg–Teller correction in Section 6.7.4. The first thing to note is that a purely electronic transition in the Franck–Condon approximation is modulated by the overlap of the vibrational wave functions, so that the intensity for a particular vibronic transition largely depends on the degree of overlap of the initial and final vibrational wave functions. Calculating the vibronic spectrum of an electronic excitation in the Frank–Condon approximation thus reduces to evaluating this overlap integral. The most likely final vibrational state in the process will be that with the largest overlap, as illustrated in Figure 6.10.

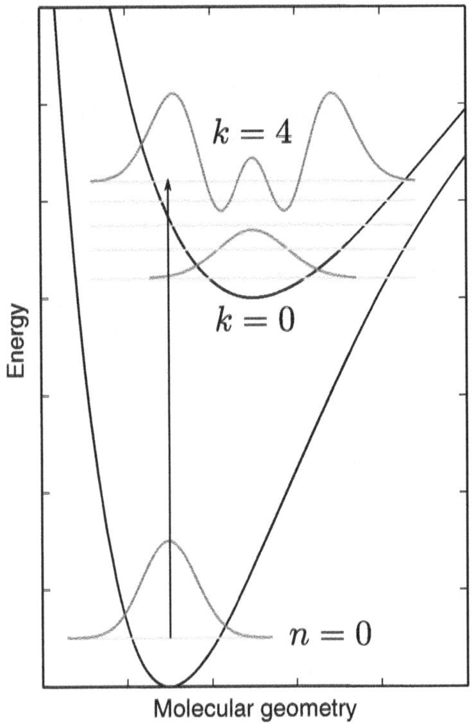

Figure 6.10 Illustration of the Franck–Condon principle with a fixed equilibrium ground-state geometry during the excitation process (vertical transition).

More quantitatively, the Franck–Condon factor $|S(k, 0)|^2$ represents the probability to end up in a specific vibrationally excited state $|k\rangle$, and thus for an ensemble of molecules, the resulting absorption spectrum will consist of a series of bands with intensities proportional to these factors and transition energies equal to the separation energy of vibronic states $|K, k\rangle$ and $|0, 0\rangle$.

In order to proceed with the evaluation of the Franck–Condon factors, we start by noting that because the normal coordinates are a set of linearized nuclear displacement coordinates, we can perform a linear transformation of one set of normal coordinates into another according to

$$\mathbf{Q}^0 = \mathbf{J}\mathbf{Q}^K + \mathbf{K}, \tag{6.108}$$

where $\mathbf{K}$ is a vector translating the equilibrium geometry of the electronic ground state to that of the electronic excited state, and $\mathbf{J}$ is a matrix that rotates the normal coordinates of the excited state into those of the ground state, as illustrated in Figure 6.11. For now, we assume that we have access to the equilibrium geometries and force constant matrices of the ground and excited states and we express the difference in equilibrium geometries in terms of the mass-weighted coordinates

$$\Delta q_i = \sqrt{m_i}(x^K_{\text{eq},i} - x^0_{\text{eq},i}). \tag{6.109}$$

It is then straightforward to determine the translation vector $\mathbf{K}$ with use of the eigenvectors of the mass-weighted force constant matrix of the electronic ground state in accordance with Eq. (6.31). We get

$$\mathbf{K} = \mathbf{L}^0 \Delta\mathbf{q}. \tag{6.110}$$

The role of the rotation matrix is to express the vector $\mathbf{Q}^K$ in Figure 6.11 in the normal coordinates of the ground state. This is readily achieved in two steps: first transform $\mathbf{Q}^K$ into the basis

Figure 6.11 Coordinate transformation between normal coordinates of electronic ground and excited states denoted by $\mathbf{Q}^0$ and $\mathbf{Q}^K$, respectively, and with origins at the respective equilibrium molecular geometries. Point P represents a particular nuclear configuration with Cartesian coordinates $\mathbf{x}$. Cartesian displacements of this point are given by $\boldsymbol{\Delta}^0 = \mathbf{x} - \mathbf{x}^0_{\text{eq}}$ and $\boldsymbol{\Delta}^K = \mathbf{x} - \mathbf{x}^K_{\text{eq}}$ for the ground and excited states, respectively.

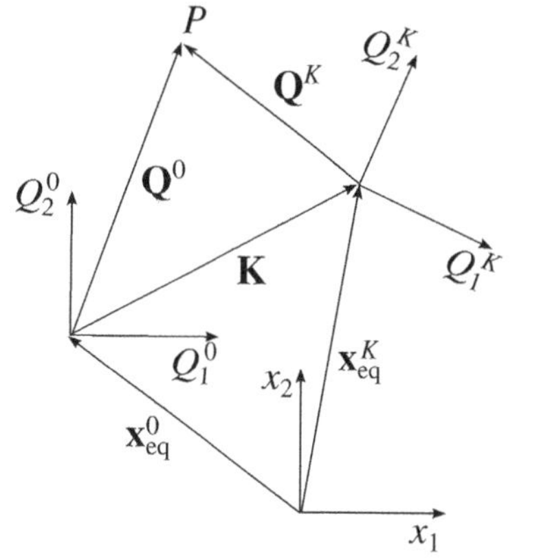

of mass-weighted coordinates and then, in a second step, transform the vector into the basis of the normal coordinates of the ground state. This procedure can be compactly expressed as

$$\mathbf{J} = \mathbf{L}^0(\mathbf{L}^K)^{-1}. \tag{6.111}$$

The rotation of the force field from one electronic state to the other is known as a Duschinsky rotation. It is worth noting that the transformations in Eqs. (6.110) and (6.111) eliminate any contribution from an overall translation of the molecule. In contrast, the same can in general not be achieved for overall molecular rotation, which inevitably will contaminate the transformed normal coordinates.

Although the transformation from one normal coordinate basis to another in Eq. (6.108) is straightforward to apply, it is not often used in practice. The main reasons for this rest in the challenges in determining the equilibrium structure and force field of the electronically excited state and, as we will return to momentarily, the cost of evaluating the Franck–Condon factors. For this reason, several approximations are often invoked, and we can create a hierarchy of different levels of approximations:

1) The adiabatic Franck–Condon approximation. In this approximation, we perform the full transformation in Eq. (6.108), including both Duschinsky rotation, shifts in vibrational frequencies, and the displacement of the equilibrium geometry.
2) As an intermediate approximation, we can ignore the rotation of the normal coordinates but retain the changes in curvature, that is, we make $\mathbf{J}$ diagonal with diagonal elements that scale the vibrational frequencies. There are few computational benefits to this approximation as the equilibrium structure and force field still need to be determined for the electronic excited state.
3) A further approximation would be to include the displacement $\mathbf{K}$ but keeping the ground-state vibrational force field, letting $\mathbf{L}^K = \mathbf{L}^0$. This approximation requires the determination of the equilibrium geometry, but not the force field, of the excited state.
4) In a final approximation known as the linear coupling model, or the vertical gradient approximation, we retain only the leading-order contribution to the Franck–Condon factor, assuming a common force field for the ground and excited states. In addition, we approximate the displacement of the equilibrium geometry by the excited-state gradient projected onto the ground-state force field. We will return to the linear coupling model in Section 6.7.3.

In order to better understand these approximations, we will consider in some detail the evaluation of Franck–Condon factors.

6.7.1 Franck–Condon Integrals

The evaluation of Franck–Condon integrals is far from straightforward in the general case. While the properties of the Hermite polynomials for a single potential energy function leave only a very limited number of nonvanishing integrals over vibrational wave functions, the displacement of the reference geometry of the ground and excited states as well as the necessary Duschinsky rotations required to transform the force field of the ground electronic state into that of the excited electronic state imply that in general there are no integrals that necessarily vanish, and it is only through a small overlap between specific vibrational states in the ground and excited electronic states that such integrals will become small enough to allow them to be discarded.

The Duschinsky rotations also imply that there will in general be a mixing of the normal modes of the ground and excited states, adding further to the difficulty of evaluating the Franck–Condon integrals. The complexity of the problem thus grows in a factorial manner. Indeed, even if we limit the number of quanta allowed in the different normal modes to five, for a molecule with as little as 12 atoms, the number of nonvanishing Franck–Condon integrals would become equal to 5^{30}, or 9×10^{20}, and it is obvious that clever schemes for selecting the most important Franck–Condon integrals are needed in such calculations. We will not discuss such selection schemes here, nor the summation rules that apply and that can be used to check the completeness of the mode selection. Instead we will outline a general strategy for evaluating Franck–Condon factors based on the generating function approach.

For simplicity, we will start by considering the evaluation of Franck–Condon factors for a diatomic molecule in the harmonic approximation. There will be no need to consider Duschinsky rotations, since, for a diatomic molecule, the normal coordinates are independent of the curvature of the potential energy surfaces, see Eq. (6.44). The only two considerations that need to be made for the two electronic states are the differences in vibrational frequencies and equilibrium geometries. For a one-dimensional system, Franck–Condon integrals can of course rather simply be evaluated by numerical integration, but in order to illustrate the principle, we will use the generating function approach.

We recall the harmonic oscillator wave functions in Eq. (6.34) and define these for the electronic ground $|0\rangle$ and excited $|K\rangle$ states. In reduced dimensionless normal coordinates, vibrational wave functions belonging to the respective electronic states thereby take the form, see Eq. (6.34)

$$|n\rangle_0 = \frac{1}{\pi^{\frac{1}{4}}\sqrt{2^n n!}} H_n(\xi_0) e^{-\frac{1}{2}\xi_0^2}, \tag{6.112}$$

$$|k\rangle_K = \frac{1}{\pi^{\frac{1}{4}}\sqrt{2^k k!}} H_k(\xi_K) e^{-\frac{1}{2}\xi_K^2}. \tag{6.113}$$

Before proceeding, we note that we can define a set of Hermite polynomials from their exponential generating function through

$$\sum_k \frac{T^k}{k!} H_k(x) = e^{2xT-T^2}, \tag{6.114}$$

where T is an arbitrary variable.

Let us now write a power series in a set of arbitrary variables T and U with expansion coefficients that contain the Franck–Condon integrals such that

$$P(T,U) = \sum_{k,n} T^k U^n \sqrt{\frac{2^k 2^n}{k!n!}} S(k,n)$$
$$= \frac{1}{\sqrt{\pi}} \int \left[e^{-\frac{1}{2}(\xi_K^2+\xi_0^2)} \sum_k \frac{T^k}{k!} H_k(\xi_K) \sum_n \frac{U^n}{n!} H_n(\xi_0) \right] d\xi_K, \tag{6.115}$$

where the summations run over all vibrational levels, from 0 to infinity. T and U are expansion coefficients for the vibrational wave functions in the electronic excited and electronic ground state, respectively. In all the following equations, we can at any time permute the T and U variables and the corresponding vibrational frequencies and normal coordinates. This would change the reference electronic state from the ground to the electronic excited state, and the resulting equations would then instead of absorption spectra correspond to emission spectra.

Inserting Eq. (6.114) into our power series, we obtain

$$P(T,U) = \frac{1}{\sqrt{\pi}} e^{-T^2-U^2} \int \left[e^{-\frac{1}{2}(\xi_K^2+\xi_0^2)+2\xi_K T+2\xi_0 U} \right] d\xi_K. \tag{6.116}$$

In order to solve this integral, we need to express the reduced dimensionless normal coordinates of the ground state in terms of the reduced dimensionless normal coordinates of the excited state. Because the reduced dimensionless normal coordinates depend on the frequencies of the modes, we can express the reduced dimensionless normal coordinates of the ground state in terms of those of the excited state as

$$\xi_0 = \sqrt{\frac{\omega_0}{\omega_K}} (\xi_K + \Delta_K), \tag{6.117}$$

where Δ_K denotes the shift in the reduced dimensionless mass-weighted normal coordinates *of the excited state* in order to bring the molecule from the equilibrium geometry of the excited state ($Q_K = 0$) to that of the equilibrium geometry of the ground state ($Q_0 = 0$).

Expressing ξ_0 in Eq. (6.116) in terms of ξ_K and Δ_K gives

$$P(T,U) = \frac{1}{\sqrt{\pi}} e^{-T^2-U^2-\frac{1}{2}\frac{\omega_0}{\omega_K}\Delta_K^2+2\sqrt{\frac{\omega_0}{\omega_K}}\Delta_K U} \int e^{-\frac{1}{2}\frac{(\omega_K+\omega_0)}{\omega_K}\xi_K^2+(2T+2\sqrt{\frac{\omega_0}{\omega_K}}U-\frac{\omega_0}{\omega_K}\Delta_K)\xi_K} d\xi_K$$
$$= \frac{1}{\sqrt{\pi}} e^{-T^2-U^2-\frac{1}{2}\frac{\omega_0}{\omega_K}\Delta_K^2+2\sqrt{\frac{\omega_0}{\omega_K}}\Delta_K U} \int e^{-\frac{1}{2}\Theta\xi_K^2+W\xi_K} dQ_K, \tag{6.118}$$

where for ease of notation we have introduced

$$W = 2T + 2\sqrt{\frac{\omega_0}{\omega_K}} U - \frac{\omega_0}{\omega_K}\Delta_K, \tag{6.119}$$

$$\Theta = \frac{\omega_K + \omega_0}{\omega_K}. \tag{6.120}$$

This last integral is a standard one, with solution

$$\int e^{-\frac{1}{2}\Theta Q_K^2+Q_K W} dQ_K = e^{\frac{1}{2}W^2\Theta} \int e^{-\frac{1}{2}\Theta[Q_K-W/\Theta]^2} dQ_K = e^{\frac{1}{2}W^2/\Theta} \sqrt{\frac{2\pi}{\Theta}}. \tag{6.121}$$

By inserting this result into Eq. (6.118), we get

$$P(T,U) = \sum_{k,n} T^k U^n \sqrt{\frac{2^k 2^n}{k!n!}} S(k,n) = I_0 e^{AT^2+BT+CU^2+DU+ETU}, \tag{6.122}$$

where we have introduced

$$
\begin{aligned}
I_0 &= \sqrt{\frac{2\omega_K}{\omega_0 + \omega_K}} e^{-\frac{1}{2}\frac{\omega_0}{\omega_K}\Delta_K^2[1-\omega_0/(\omega_K+\omega_0)]}, \\
A &= 2\omega_K(\omega_K + \omega_0)^{-1} - 1, \\
B &= -2\omega_0(\omega_K + \omega_0)^{-1}\Delta_K, \\
C &= 2\omega_0(\omega_K + \omega_0)^{-1} - 1, \\
D &= -2\sqrt{\frac{\omega_0}{\omega_K}}[(\omega_K + \omega_0)^{-1}\omega_0 - 1]\Delta_K, \\
E &= 4\sqrt{\frac{\omega_0}{\omega_K}}(\omega_K + \omega_0)^{-1}.
\end{aligned}
\tag{6.123}
$$

We note from Eq. (6.122) that we have succeeded in finding an expression that allows us to identify an arbitrary Franck–Condon integral. We can, for example, identify the integral for the 0–0 transition as being equal to I_0. In general, expressions for the Franck–Condon integrals can be obtained from Eq. (6.122), and we will give some examples of these for some special cases in the next section.

This procedure of determining Franck–Condon factors can be generalized to the case of a polyatomic molecule. The final equations will have much of the same structure as we have identified in the case of a diatomic system, and the interested reader is referred to original work.[7]

6.7.2 Vibronic Effects in a Diatomic System

It is clear that the vibronic features in an electronic absorption spectrum can be very rich due to the large number of vibronic transitions that can be active even if we only consider excitations from the vibronic ground state. In order to get a better understanding of the vibronic process, let us analyze in some detail the case of a diatomic molecule. To further reduce the complexity, we will assume that the potential energy surfaces of the ground and excited states have identical fundamental frequencies, that is, $\omega_K = \omega_0$. As a consequence, the expressions for the coefficients in Eq. (6.123) reduce to $A = C = 0$ together with

$$
\begin{aligned}
I_0 &= e^{-\frac{1}{4}\omega_0\Delta^2}, \\
B &= -\Delta_K, \\
D &= \Delta_K, \\
E &= 2\omega_0^{-1}.
\end{aligned}
\tag{6.124}
$$

We recall that by permuting T and U, we could obtain the contributions to the vibronic emission spectrum rather than the absorption spectrum. This would change the signs of the B and D coefficients. The vibronic emission spectrum in this approximation will therefore be the mirror image of the corresponding absorption spectrum, centered around the 0–0 transition as illustrated in Figure 6.12. In plotting these spectra, we have adopted a Lorentzian profile of individual vibronic transitions with a broadening that provides a high resolution. In situations where the broadening is larger due to, for instance, shorter excited-state lifetimes, the vibronic fine structure may partially disappear and it can be difficult to distinguish spectral peaks as being due to vibrations or close-lying electronic states. As a guiding rule of thumb, a homogeneous broadening suggests independent electronic states whereas an asymmetric and

7 T. E. Sharp and H. M. Rosenstock. *J. Chem. Phys.*, 41:3453, 1964.

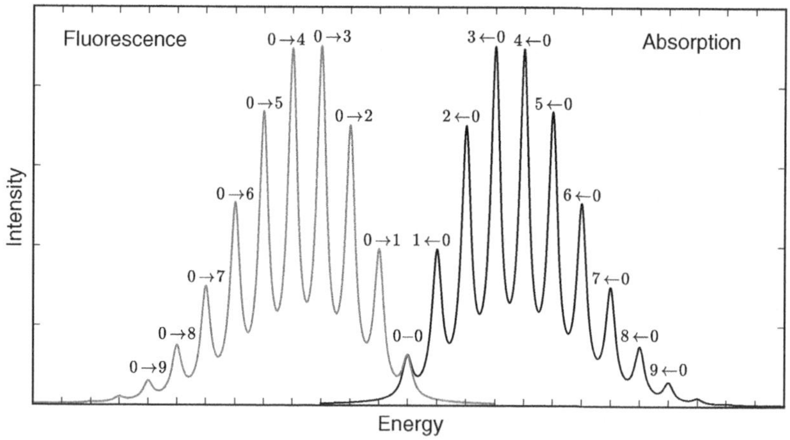

Figure 6.12 Absorption and fluorescence spectrum of a diatomic molecule.

nonhomogeneous broadening suggests vibronic fine structure. We will in Section 8.5.1 return to the origin and representation of different sources of spectral broadening.

Another and more obvious observation from Figure 6.12 is that the maximum of the absorption does not coincide with the 0–0 transition but instead occurs at a higher energy, or shorter wavelength. Adiabatic transition energies are thus in general lower than observed absorption maximum, normally about 0.2–0.3 eV for high-level electronic structure calculations. The calculations of a 0–0 transition energy is instead to be compared with the onset of a band in the experimental absorption spectrum. This 0–0 transition may be very weak due to the fact that the Franck–Condon factor decays very quickly as function of the shift in the equilibrium geometries according to

$$S(0,0) = I_0^2 = e^{-\omega_0 \Delta_K^2/2}. \tag{6.125}$$

In order to get a feeling for how Eq. (6.122) works in practice, it is instructive to consider explicitly the evaluation of the Franck–Condon factors in the case of a diatomic system with $\omega_K = \omega_0$. From the population analysis based on Figure 6.8, it is clear that it is reasonable to assume that only the ground vibronic state is populated at room temperature. In the context of Eq. (6.122), this means that $n = 0$ in the expression for generating the function and we arrive at

$$P(T, U) = \sum_k T^k \sqrt{\frac{2^k}{k!}} S(k, 0) = I_0\, e^{TB}. \tag{6.126}$$

By expanding both sides, we get

$$S(0,0) + T\sqrt{2}S(1,0) + T^2\sqrt{2}S(2,0) + T^3\frac{2}{\sqrt{3}}S(3,0) + \cdots$$
$$= I_0\left[1 + TB + \frac{1}{2!}T^2B^2 + \frac{1}{3!}T^3B^3 + \cdots\right], \tag{6.127}$$

which allows us determine the Franck–Condon integrals (without reference to an overall sign) as

$$S(k, 0) = I_0\sqrt{\frac{1}{2^k k!}}[\Delta_K]^k. \tag{6.128}$$

We note from Eq. (6.123) that even in the case of a diatomic molecule for which $\omega_K \neq \omega_0$, things already become more complicated as A and D enter into the equation. But as long as we only consider excitations from the vibronic ground state, the derivation of the Franck–Condon factors for a diatomic molecule remains tractable in this approach.

6.7.3 Linear Coupling Model

A commonly adopted approximation for the evaluation of Franck–Condon integrals is the linear coupling model. In this model, we invoke two approximations:

1) We assume that the force field of the electronically excited state is the same as the force field of the electronic ground state, that is, $\mathbf{L}^K = \mathbf{L}^0$ and $\mathbf{J} = \mathbf{I}$, the identity matrix. This approximation also implies that $\mathbf{\Gamma}^K = \mathbf{\Gamma}^0$, where $\mathbf{\Gamma}^0$ and $\mathbf{\Gamma}^K$ are diagonal matrices collecting the vibrational frequencies of the ground and excited electronic states, respectively.
2) We assume that the shift in geometry from the ground to the excited state can be approximated by the gradient of the excited state evaluated at the ground-state equilibrium geometry and projected onto its reduced normal coordinates, that is,

$$\mathbf{K} = \hbar^{-1/2}(\mathbf{\Gamma}^0)^{-3/2}\mathbf{L}^0\frac{\partial E^K}{\partial \xi}. \tag{6.129}$$

These approximations largely correspond to the approximations introduced in the case of diatomic molecules in Section 6.7.2 but with the additional assumption that the change in the equilibrium geometry of the ground and excited states can be approximated by the excited-state gradient evaluated at the ground-state geometry. The working equations of the linear coupling model thus correspond to the generalization of Eq. (6.124) to the case of polyatomic molecules, and are given by

$$\begin{aligned}
I_0 &= \frac{\sqrt[4]{|\mathbf{\Gamma}^0\mathbf{\Gamma}^0|^{4m}}}{\sqrt{|(2\mathbf{\Gamma}^0)|}}\exp\left[-\frac{1}{4}\mathbf{K}^\dagger\mathbf{\Gamma}^0\mathbf{K}\right],\\
\mathbf{B} &= \sqrt{\mathbf{\Gamma}^0}\mathbf{K},\\
\mathbf{D} &= -\sqrt{\mathbf{\Gamma}^0}\mathbf{K},\\
\mathbf{E} &= 2\mathbf{I},
\end{aligned} \tag{6.130}$$

with $\mathbf{A} = \mathbf{C} = \mathbf{0}$. As a consequence of the form of these equations, many of the observations made in the case of diatomic molecules carry over to the case of the linear coupling model with the simplification of not having to determine the equilibrium geometry of the excited state. We thus conclude that the full vibronic fine structure can be obtained using single-point calculations of the ground-state force field and the excited-state gradient, and the latter can be obtained from a residue analysis of the quadratic response function as discussed in Section 5.9.

As compared to the diatomic system, the polyatomic vibronic absorption spectrum becomes significantly more complicated as a progression of vibronic fine structure can arise from *all* vibrational modes in the molecule. However, Eq. (6.129) suggests that the most prominent vibronic bands are due to those with the largest excited-state gradients. An understanding of the nature of the excited state, and thus how the geometry can be expected to relax due to the promotion of an electron from an occupied to an unoccupied orbital, will provide a qualitative picture of the modes that can be anticipated to be most important for the vibronic fine structure.

6.7.4 Herzberg–Teller Corrections and Vibronically Induced Transitions

One important consequence of molecular vibrations is that electronic transitions forbidden by symmetry may become allowed, although typically weaker, as the full vibronic wave functions have the required symmetries. The most important contribution in this respect is the Herzberg–Teller correction, arising from the leading-order contribution beyond the Franck–Condon contribution in the expansion of the vibronic dipole transition moment in Eq. (6.106). The Herzberg–Teller correction to the vibronic transition moment is given by the second term in Eq. (6.106).

We need to evaluate vibrational matrix elements between different vibrational states and we will assume that we in the following are looking at an absorption process. The same need to determine integrals involving vibrational states belonging to different electronic states was considered in the evaluation of Franck–Condon integrals, but the approach presented in Section 6.7.1 has to be generalized slightly due to the presence of the normal coordinate inside the integral. Using the property of Hermite polynomials presented in Eq. (6.37), we can determine the action of Q_a on the vibrational ground state as to create a linear combination of Franck–Condon integrals. These can in turn be evaluated using one of the approaches described earlier—the full adiabatic Franck–Condon approximation or the simpler linear coupling model.

We first note that the strength of the Herzberg–Teller correction is proportional to the gradient of the electronic transition dipole moment at the equilibrium geometry of the electronic ground state. In other words, the larger the change in the transition dipole moment with nuclear distortions along a normal mode, the larger the Herzberg–Teller contribution will be. Second, all electronic states that are accessible from the symmetry of the transition dipole gradient along any of the normal modes can in principle be observed in the absorption spectrum.

As an illustration, let us consider metal $d \rightarrow d$ transitions in an octahedral transition metal complex. This system belongs to the O_h point group and all d-orbitals have *gerade* symmetry, belonging to irreducible representations E_g and T_{2g}. In contrast, all components of the dipole moment operator have *ungerade* symmetry and belong to T_{1u}. Within the Franck–Condon approximation, all $d \rightarrow d$ metal transitions are thus forbidden due to inversion symmetry, or parity, and this is known as Laporte's selection rule. With atomic ligands, the system has a total of 15 normal modes, spanning irreducible representations A_{1g}, E_g, T_{2g}, $2T_{1u}$, and T_{2u}. We focus on the three modes of *ungerade* symmetry that can help break the selection rule. The direct product of the irreducible representations of these normal modes with that of the dipole moment operator gives

$$T_{1u} \otimes T_{1u} = A_{1g} \oplus E_g \oplus T_{1g} \oplus T_{2g},$$
$$T_{2u} \otimes T_{1u} = A_{2g} \oplus E_g \oplus T_{1g} \oplus T_{2g},$$

which makes it clear that the Herzberg–Teller coupling will induce $d \rightarrow d$ transitions since the right-hand sides include the symmetries of the products of any two d-functions.

Further Reading

1 D. M. Bishop. Molecular vibration and nonlinear optics. *Adv. Chem. Phys.*, 104:1–40, 1998.

2 W. Domcke and G. Stock. Theory of ultrafast nonadiabatic excited-state processes and their spectroscopic detection in real time. *Adv. Chem. Phys.*, 100:1–169, 1997.

3 J. M. Luis, M. Duran, and J. L. Andrés. A systematic and feasible method for computing nuclear contributions to electrical properties of polyatomic molecules. *J. Chem. Phys.*, 107:1501–1512, 1997.

4 T. E. Sharp and H. M. Rosenstock. Franck–Condon factors for polyatomic molecules. *J. Chem. Phys.*, 41:3453–3463, 1964.

5 E. B. Wilson Jr., J. C. Decius, and P. C. Cross. *Molecular Vibrations: The Theory of Infrared and Raman Vibrational Spectroscopy*. Dover Publications, New York, 1980.

is the Herzberg–Teller correction, arising from the leading order contribution beyond the Franck–Condon contribution in the expansion of the vibronic dipole transition moment in Eq. (4.96). The Herzberg–Teller correction to the vibronic transition moment is given by the ... [illegible]

[illegible]

7

Approximate Electronic State Response Theory

Although this may seem a paradox, all exact science is dominated by the idea of approximation. When a man tells you that he knows the exact truth about anything, you are safe in inferring that he is an inexact man.

Bertrand Russell

7.1 Reference State Parameterizations

In Chapter 5 on exact-state response theory we explored three approaches for the generation of the equations of motion needed for the calculation of molecular properties: in addition to time-dependent perturbation theory, we also considered starting either from the time-averaged quasi-energy Q_{T}, Eq. (5.51), or the Ehrenfest theorem, Eq. (5.117), in both cases restricting attention to the phase-isolated part of the wave function, Eq. (5.25). In this chapter, we consider the more realistic situation where the eigenstates of the Hamiltonian for the unperturbed system are not available. Instead, the time-dependent reference state is represented by an approximate state, which is understood to be expressed in some way or another in terms of Slater determinants. The elements of the determinants are one-electron wave functions (orbitals), which in turn are often expressed in terms of a set of time-independent atomic orbitals with expansion parameters that are known as molecular orbital coefficients.

The task at hand is therefore to determine the time dependence of these approximate reference states, and one concern is the constraint of wave function normalization. We discussed this also in the case of exact-state response theory, and we designed an explicitly unitary parameterization of the wave function [see Eqs. (5.81) and (5.82)] that by construction preserved normalization. In the case of approximate-state response theories, we seek unconstrained parameterizations that preserve not only the normalization of the reference state vectors but also the orthonormality of orbitals.

7.1.1 Single Determinant

Let us begin by considering the parameterization of a single phase-isolated Slater determinant. This is relevant for the calculation of molecular properties at the Hartree–Fock and Kohn–Sham levels of theory. Based on second quantization, we employ an exponential parametrization

$$|\overline{\psi}(t)\rangle = e^{-i\hat{\kappa}(t)}|0\rangle; \quad \hat{\kappa}(t) = \sum_{p,q} \kappa_{pq}\hat{a}_p^{\dagger}\hat{a}_q; \quad \kappa_{qp}^{*} = \kappa_{pq}, \tag{7.1}$$

Principles and Practices of Molecular Properties: Theory, Modeling and Simulations, First Edition.
Patrick Norman, Kenneth Ruud, and Trond Saue.

referring to Appendix C for notation and a review of second quantization. The reference state determinant $|0\rangle$ is represented by an occupation number vector that is constructed with the use of creation operators $\{\hat{a}_p^\dagger\}$ defined with respect to an orthonormal orbital set $\{\varphi_p\}$ optimized with respect to the Hamiltonian of the unperturbed system

$$|0\rangle = \hat{a}_1^\dagger \hat{a}_2^\dagger \cdots \hat{a}_N^\dagger |\text{vac}\rangle. \tag{7.2}$$

The set of amplitudes $\{\kappa_{pq}\}$ form a Hermitian matrix such that the exponential operator is unitary. Let us first demonstrate that this implies that the exponential generates unitary transformations of the orbital set. We note that

$$e^{i\hat{\kappa}}|\text{vac}\rangle = \left(1 + i\hat{\kappa} - \frac{1}{2}\hat{\kappa}^2 + \cdots\right)|\text{vac}\rangle = |\text{vac}\rangle, \tag{7.3}$$

since all terms except the first one in the expansion of the exponential contains $\hat{\kappa}$ and thus an annihilation operator to the right (which gives zero when acting on the vacuum state). The phase-isolated Slater determinant can accordingly be expressed as

$$\begin{aligned} |\overline{\psi}(t)\rangle &= e^{-i\hat{\kappa}}\hat{a}_1^\dagger e^{i\hat{\kappa}} e^{-i\hat{\kappa}}\hat{a}_2^\dagger e^{i\hat{\kappa}} \cdots e^{-i\hat{\kappa}}\hat{a}_N^\dagger e^{i\hat{\kappa}}|\text{vac}\rangle \\ &= \hat{\tilde{a}}_1^\dagger \hat{\tilde{a}}_2^\dagger \cdots \hat{\tilde{a}}_N^\dagger |\text{vac}\rangle, \end{aligned} \tag{7.4}$$

where we have introduced time-transformed creation operators

$$\hat{\tilde{a}}_p^\dagger = e^{-i\hat{\kappa}}\hat{a}_p^\dagger e^{i\hat{\kappa}}. \tag{7.5}$$

Using the Baker–Campbell–Hausdorff (BCH) expansion, Eq. (E.6), we write these transformed creation operators as

$$\hat{\tilde{a}}_p^\dagger = \hat{a}_p^\dagger - i[\hat{\kappa}, \hat{a}_p^\dagger] - \frac{1}{2}[\hat{\kappa}, [\hat{\kappa}, \hat{a}_p^\dagger]] + \cdots. \tag{7.6}$$

From the algebra of creation and annihilation operators we find that

$$[\hat{\kappa}, \hat{a}_p^\dagger] = \hat{a}_r^\dagger \kappa_{rp}, \tag{7.7}$$

where we remind the reader that we, here and elsewhere, employ implicit summation over repeated indices. From induction, we can conclude that

$$\hat{\tilde{a}}_p^\dagger = \hat{a}_r^\dagger\left(\delta_{rp} - i\kappa_{rp} - \frac{1}{2}\kappa_{rp}^2 + \cdots\right) = \hat{a}_r^\dagger[e^{-i\kappa(t)}]_{rp}. \tag{7.8}$$

In Appendix C, we learn that creation and annihilation operators are obtained by projection of the wave operators on an orbital set, so, from Eq. (C.13), we obtain

$$\hat{\tilde{a}}_p^\dagger = \int \hat{\psi}^\dagger \tilde{\varphi}_p d1; \quad \tilde{\varphi}_p = \varphi_r[e^{-i\kappa(t)}]_{rp}, \tag{7.9}$$

where $d1$ denotes the infinitesimal volume associated with space and spin coordinates of electron 1. We have therefore demonstrated that the exponential operator in Eq. (7.1) generates rotations among the starting orbitals. The exponential parametrization therefore allows the variation of orbitals without having to impose constraints to maintain orthonormality of the orbitals.

Another advantage of the exponential parametrization is that it allows the straightforward identification and elimination of redundant parameters. A nonredundant set of parameters corresponds to the minimum number of parameters that still allows us to reach all states in the subspace at hand (in this case the single determinant subspace of the Hilbert space). Excluding

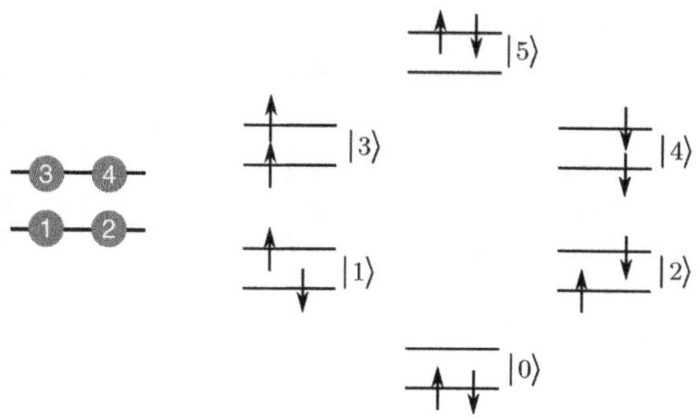

Figure 7.1 Complete set of Slater determinants for a system of two electrons in four orbitals.

rotations between occupied orbitals and likewise between unoccupied orbitals, we end up with a nonredundant time-dependent Hermitian operator $\hat{\kappa}(t)$ that takes the form

$$\hat{\kappa}(t) = \sum_{s}^{\text{unocc}} \sum_{i}^{\text{occ}} [\kappa_{si}(t)\hat{a}_s^\dagger \hat{a}_i + \kappa_{si}^*(t)\hat{a}_i^\dagger \hat{a}_s], \tag{7.10}$$

in terms of the pairs of creation and annihilation operators acting on unoccupied s and occupied i molecular orbitals. The unoccupied and occupied orbitals are often referred to as secondary and inactive orbitals, respectively, and, hence, the choice of indices to label them. There is a strong similarity in form between the operator $\hat{P}$, which was employed in the case of exact states, and the operator $\hat{\kappa}$ that is employed here. The former is expressed as a sum of state-transfer operators and the latter as a sum of one-electron transfer operators. But there are also differences to be noted. For instance, while there is an operator in $\hat{P}$ that couples the reference state and *all* excited states, there is in $\hat{\kappa}$ only a coupling between the reference state determinant and single-electron excited determinants. It is not quite so obvious how the operation of $\exp(-i\hat{\kappa})$ on $|0\rangle$ can result in multielectron excited determinants, and we will therefore explore a small, illustrative example consisting of two electrons in two orbitals. In this example, we will adopt a nonrelativistic notation with spin being a good quantum number for orbitals, but the principal arguments do not rely on this assumption and they can just as well be adopted in the relativistic case.

The complete set of six Slater determinants for a system of two electrons in four spin orbitals is illustrated in Figure 7.1. We assume the reference state to be closed shell with occupation of orbitals 1 and 2, whereas orbitals 3 and 4 are the virtual ones. In the basis of the six determinants, the matrix representation of the $\hat{\kappa}$-operator is

$$\hat{\kappa} = \begin{pmatrix} 0 & \kappa_{31}^* & \kappa_{42}^* & \kappa_{32}^* & \kappa_{41}^* & 0 \\ \kappa_{31} & 0 & 0 & 0 & 0 & \kappa_{42}^* \\ \kappa_{42} & 0 & 0 & 0 & 0 & \kappa_{31}^* \\ \kappa_{32} & 0 & 0 & 0 & 0 & \kappa_{41}^* \\ \kappa_{41} & 0 & 0 & 0 & 0 & \kappa_{32}^* \\ 0 & \kappa_{42} & \kappa_{31} & \kappa_{41} & \kappa_{32} & 0 \end{pmatrix}. \tag{7.11}$$

Our remark about the lack of coupling between the reference state and multielectron excited determinants by means of $\hat{\kappa}$ is reflected by the zeros in the lower left and upper right corners of the matrix. The coupling between the reference state and the double-electron excited determinant is instead found in $\hat{\kappa}^2$ as illustrated in Figure 7.2. Furthermore, there appear four nonzero elements in row (as well as column) six of $\hat{\kappa}$, due to the coupling between single- and double-electron excited determinants. In the exact-state case with the parametrization given

$$\hat{\kappa} = \begin{pmatrix} 0 & \blacksquare & \blacksquare & \blacksquare & \blacksquare & 0 \\ \blacksquare & 0 & 0 & 0 & 0 & \blacksquare \\ \blacksquare & 0 & 0 & 0 & 0 & \blacksquare \\ \blacksquare & 0 & 0 & 0 & 0 & \blacksquare \\ \blacksquare & 0 & 0 & 0 & 0 & \blacksquare \\ 0 & \blacksquare & \blacksquare & \blacksquare & \blacksquare & 0 \end{pmatrix}; \qquad \hat{\kappa}^2 = \begin{pmatrix} \blacksquare & 0 & 0 & 0 & 0 & \blacksquare \\ 0 & \blacksquare & \blacksquare & \blacksquare & \blacksquare & 0 \\ 0 & \blacksquare & \blacksquare & \blacksquare & \blacksquare & 0 \\ 0 & \blacksquare & \blacksquare & \blacksquare & \blacksquare & 0 \\ 0 & \blacksquare & \blacksquare & \blacksquare & \blacksquare & 0 \\ \blacksquare & 0 & 0 & 0 & 0 & \blacksquare \end{pmatrix}$$

Figure 7.2 Structures of the matrix representations of the operators $\hat{\kappa}$ and $\hat{\kappa}^2$. The filled squares denote nonzero matrix elements.

in Eq. (5.81), on the other hand, we have seen that there are only nonzero elements in the first row (and column).

Let us consider a parameterization involving only the spin-conserving electron-transfer operators, that is, setting the elements κ_{32} and κ_{41} to zero in the matrix in Eq. (7.11). We then get

$$\begin{aligned} |\overline{\psi}\rangle &= e^{-i\hat{\kappa}}|0\rangle \\ &= |0\rangle \cos|\kappa_{31}|\cos|\kappa_{42}| - |1\rangle\, i\frac{\kappa_{31}}{|\kappa_{31}|}\sin|\kappa_{31}|\cos|\kappa_{42}| \\ &\quad - |2\rangle\, i\frac{\kappa_{42}}{|\kappa_{42}|}\cos|\kappa_{31}|\sin|\kappa_{42}| - |5\rangle \frac{\kappa_{31}\kappa_{42}}{|\kappa_{31}\kappa_{42}|}\sin|\kappa_{31}|\sin|\kappa_{42}|. \end{aligned} \tag{7.12}$$

In accordance with the parameterization for exact states by rotations, as made in Eq. (5.81), Eq. (7.12) preserves normalization of the state vector $|\overline{\psi}\rangle$, it is a many-to-one mapping of amplitudes to state vectors, and it does not alter the overall phase of the unperturbed reference state $|0\rangle$. From Eq. (7.12) we also see that the first-order changes in the reference state with respect to the κ-parameters (the electronic gradient) are given by the single-electron (or singly) excited determinants $|1\rangle$ and $|2\rangle$, whereas the double-electron excited determinant $|5\rangle$ has a quadratic dependence on the κ-parameters. An illustration of the parameterization in Eq. (7.12) is made in Figure 7.3, where the absolute square of the respective expansion coefficients are plotted for κ_{31} and κ_{42} in the interval $[-\pi/2, \pi/2]$.

One worrying detail about the parameterization in Eq. (7.12) is the seemingly multideterminant nature of the resulting state vector, which stands in contrast to our desire to parameterize a single Slater determinant using an exponential operator. Let us therefore perform a closer inspection of $|\overline{\psi}\rangle$ in Eq. (7.12), using the standard notation for the Slater determinant [see also Eq. (2.60)]

$$|\chi_i, \chi_j\rangle = \frac{1}{\sqrt{2}}\begin{vmatrix} \chi_i(1) & \chi_j(1) \\ \chi_i(2) & \chi_j(2) \end{vmatrix}. \tag{7.13}$$

Owing to the property of linearity of determinants, we can in this notation rewrite Eq. (7.12) as

$$|\overline{\psi}\rangle = |\tilde{\chi}_1\tilde{\chi}_2\rangle = \left|\cos|\kappa_{31}|\,\chi_1 - i\frac{\kappa_{31}}{|\kappa_{31}|}\sin|\kappa_{31}|\,\chi_3,\quad \cos|\kappa_{42}|\,\chi_2 - i\frac{\kappa_{42}}{|\kappa_{42}|}\sin|\kappa_{42}|\,\chi_4\right\rangle, \tag{7.14}$$

in agreement with Eq. (7.9). The immediate conclusion to be drawn from this exercise is that $|\overline{\psi}\rangle$ is manifestly single determinant in nature. But we also note that orthonormality is maintained for the occupied orbitals, and the orbital variations (or rotations) are complete for a phase-isolated reference state where the overall phases of orbitals χ_1 and χ_2 are not to be altered.

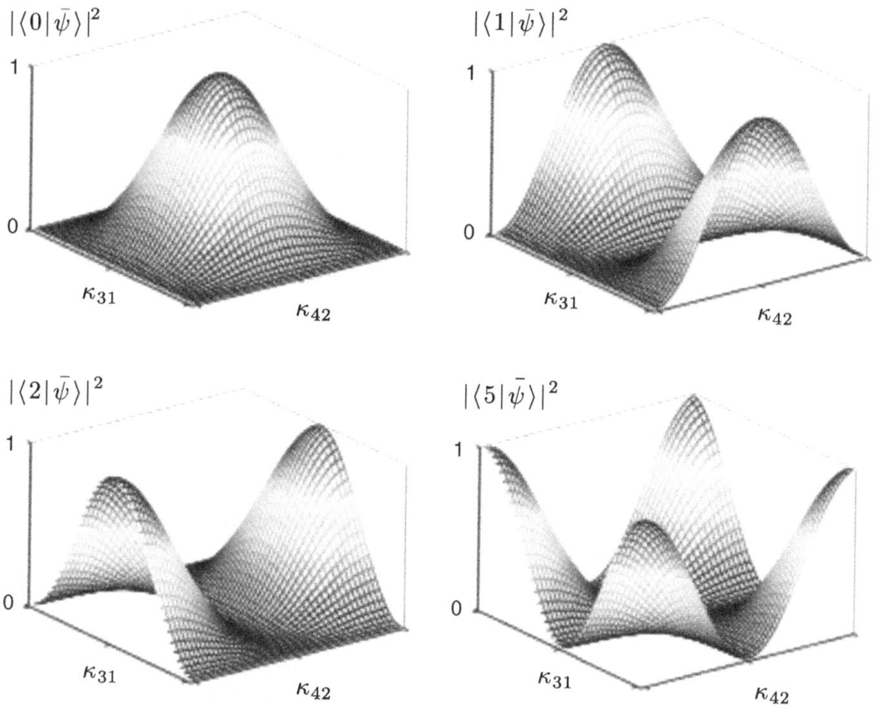

Figure 7.3 Parameterization of reference state determinant.

An alternative formulation of our findings is that when the exponential parameterization is viewed as a mapping of $|0\rangle$ onto $|\overline{\psi}\rangle$, it is clear that the range of this mapping includes all single determinants but only restricted linear combinations of them [as illustrated in Eq. (7.12)]. That is to say, the range of the mapping does not represent a linear vector space. This stands in contrast to the parameterization of the exact state where, even if the mapping includes only a limited number of excited states, the range of the mapping still represents a linear vector space, and this property of linearity of the range space is therefore also conveyed to approximate-state configuration interaction (CI) theories.

In our simple example of two electrons in four orbitals, we considered unconstrained variations of amplitudes for the spin-conserving electron-transfer operators. However, this does not mean that spin is preserved as a good quantum number for $|\overline{\psi}\rangle$ because the determinants $|1\rangle$ and $|2\rangle$ are not eigenfunctions of the spin operator $\hat{S}^2$. On the other hand, we know that eigenfunctions of singlet and triplet spin symmetry with $M_S = 0$ are readily constructed as the linear combinations

$$|S,0\rangle = \frac{1}{\sqrt{2}}(|\chi_3,\chi_2\rangle + |\chi_1,\chi_4\rangle) \tag{7.15}$$

and

$$|T,0\rangle = \frac{1}{\sqrt{2}}(|\chi_3,\chi_2\rangle - |\chi_1,\chi_4\rangle). \tag{7.16}$$

From a comparison with Eq. (7.12), we see that conservation of singlet spin symmetry in $|\overline{\psi}\rangle$ is achieved by requiring that $\kappa_{31} = \kappa_{42}$. In cases when the perturbation operator $\hat{V}(t)$ is of singlet spin symmetry and the reference state is closed shell in nature, this requirement is typically enforced to further reduce the number of parameters that need to be introduced, and the generator of orbital rotations then takes the form

$$\hat{\kappa}(t) = \sum_{s}^{\text{unocc}} \sum_{i}^{\text{occ}} [\kappa_{si}(t)\hat{E}_{si}^{\dagger} + \kappa_{si}^{*}(t)\hat{E}_{si}], \tag{7.17}$$

where the summations in this case run over *spatial* orbitals and

$$\hat{E}_{si}^{\dagger} = \hat{a}_{s\alpha}^{\dagger}\hat{a}_{i\alpha} + \hat{a}_{s\beta}^{\dagger}\hat{a}_{i\beta}. \tag{7.18}$$

The corresponding triplet excitation operators that conserve spin projection are given by

$$\hat{T}_{si}^{\dagger} = \hat{a}_{s\alpha}^{\dagger}\hat{a}_{i\alpha} - \hat{a}_{s\beta}^{\dagger}\hat{a}_{i\beta}. \tag{7.19}$$

We remind the reader that these simplifications are restricted to the nonrelativistic realm where spin is a good quantum number.

In exact-state theory, the excitation energies of the isolated system appear as the elements of the electronic Hessian in the diagonal representation, and from the perspective of molecular properties, it seems therefore relevant to determine the Hessian in approximate-state theories as well. The energy associated with state $|\overline{\psi}\rangle$ is given by the expectation value

$$\begin{aligned} E(\kappa) &= \langle\overline{\psi}|\hat{H}_0|\overline{\psi}\rangle = \langle 0|e^{i\hat{\kappa}}\hat{H}_0 e^{-i\hat{\kappa}}|0\rangle \\ &= \langle 0|\hat{H}_0|0\rangle + i\langle 0|[\hat{\kappa},\hat{H}_0]|0\rangle - \frac{1}{2}\langle 0|[\hat{\kappa},[\hat{\kappa},\hat{H}_0]]|0\rangle + \cdots, \end{aligned} \tag{7.20}$$

where the κ-symbol in the argument of the energy is to be understood as the entire set of parameters including κ_{si} as well as κ_{si}^{*}. With a general generator of orbital rotations as given in Eq. (7.10), we get an electronic gradient in the single-determinant approximation that is

$$E^{[1]} = \begin{pmatrix} g \\ g^{*} \end{pmatrix} = \begin{pmatrix} \left.\dfrac{\partial E(\kappa)}{\partial \kappa_{si}^{*}}\right|_{\kappa=0} \\ \left.\dfrac{\partial E(\kappa)}{\partial \kappa_{si}}\right|_{\kappa=0} \end{pmatrix} = i\begin{pmatrix} \langle 0|\ [\hat{a}_i^{\dagger}\hat{a}_s, \hat{H}_0]\ |0\rangle \\ \langle 0|\ [\hat{a}_s^{\dagger}\hat{a}_i, \hat{H}_0]\ |0\rangle \end{pmatrix} = i\begin{pmatrix} \langle 0_i^s|\hat{H}_0|0\rangle \\ -\langle 0|\hat{H}_0|0_i^s\rangle \end{pmatrix}, \tag{7.21}$$

where $|0_i^s\rangle$ denotes a single-electron excited determinant that differs from the reference state $|0\rangle$ by the presence of a hole in orbital i and an electron in orbital s. It is noted that the gradient in Eq. (7.21) can be divided into an upper (g^*) and a lower (g) component that are related by the operation of complex conjugation. In the presentation of the explicit expressions for the elements of g and g^*, we have given a single element and it is to be understood that the gradient components collect all single excitations. We note that the expression for the gradient is in agreement with the expansion made in Eq. (7.12) for the system of two electrons in four orbitals. Contributions to the gradient stem from transition matrix elements between the reference state $|0\rangle$ on the one hand and the single-electron excited states $|1\rangle$ and $|2\rangle$ on the other.

An expression for the electronic Hessian is obtained by further differentiation, which gives

$$\begin{aligned} E^{[2]} &= \begin{pmatrix} A & B \\ B^{*} & A^{*} \end{pmatrix} \\ &= \begin{pmatrix} \left.\dfrac{\partial^2 E(\kappa)}{\partial \kappa_{si}^{*}\partial \kappa_{tj}}\right|_{\kappa=0} & \left.\dfrac{\partial^2 E(\kappa)}{\partial \kappa_{si}^{*}\partial \kappa_{tj}^{*}}\right|_{\kappa=0} \\ \left.\dfrac{\partial^2 E(\kappa)}{\partial \kappa_{si}\partial \kappa_{tj}}\right|_{\kappa=0} & \left.\dfrac{\partial^2 E(\kappa)}{\partial \kappa_{si}\partial \kappa_{tj}^{*}}\right|_{\kappa=0} \end{pmatrix} \\ &= \begin{pmatrix} \langle 0_i^s|\hat{H}_0|0_j^t\rangle - \langle 0|\hat{H}_0|0\rangle\delta_{st}\delta_{ij} & -\langle 0_{ij}^{st}|\hat{H}_0|0\rangle(1-\delta_{ij})(1-\delta_{st}) \\ -\langle 0|\hat{H}_0|0_{ij}^{st}\rangle(1-\delta_{ij})(1-\delta_{st}) & \langle 0_j^t|\hat{H}_0|0_i^s\rangle - \langle 0|\hat{H}_0|0\rangle\delta_{st}\delta_{ij} \end{pmatrix}. \end{aligned} \tag{7.22}$$

In analogy with the gradient vector, it is noted that the Hessian matrix has an internal structure in terms of the matrix blocks A and B and their respective complex conjugates. We have also here provided single matrix elements for each of the sub-blocks and it is again understood that all electron excitations are to be considered. The Hessian is noted to be Hermitian, which also means that A is Hermitian and B is symmetric. From the above expression for the Hessian, we see that the diagonal elements correspond to differences in energy between the single-excited determinants $|0_i^s\rangle$ and the reference state $|0\rangle$. In fact, the upper left (and lower right) block of the Hessian matrix is equal to the matrix representation of the Hamiltonian in the configuration interaction singles (CIS) approximation, and diagonalization of this sub-block gives electronic excitation energies of the system in the CIS approximation. To ignore the matrix block B in Eq. (7.22) is also referred to as the Tamm–Dancoff approximation. In the full Hessian, however, there exist the off-diagonal coupling blocks B and B^* due to the nonzero transition matrix elements between the reference state and the double-electron excited determinants. Alternatively, the Hessian may be expressed in terms of the Fock matrix and two-electron integrals as

$$E^{[2]} = \begin{pmatrix} F_{st}\delta_{ij} - F_{ji}\delta_{st} - \langle sj||ti\rangle & -\langle st||ij\rangle \\ -\langle ij||st\rangle & F_{ts}\delta_{ij} - F_{ij}\delta_{st} - \langle ti||sj\rangle \end{pmatrix}, \tag{7.23}$$

where

$$F_{ab} = h_{ab} + \sum_i^{\text{occ}} \langle ai||bi\rangle, \tag{7.24}$$

and

$$\langle ab||cd\rangle = \int \chi_a^\dagger(1)\chi_b^\dagger(2)\, r_{12}^{-1}(1 - P_{12})\, \chi_c(1)\chi_d(2)\, d\tau. \tag{7.25}$$

In a basis of canonical self-consistent field (SCF) orbitals, we note that the diagonal elements of the Hessian depend on the difference in orbital energies

$$E^{[2]}_{si,si} = \varepsilon_s - \varepsilon_i - \langle si||si\rangle. \tag{7.26}$$

In exact-state theory, the excitation energies of the system are found along the diagonal of the Hessian (in the diagonal representation), whereas in the single-determinant approximation, the Hessian is a diagonal dominant matrix with diagonal elements that refer to differences in orbital energies (in the canonical orbital basis).

As a simple numerical example, let us consider the nonrelativistic Hartree–Fock wave function of helium in Dunning's aug-cc-pVDZ basis set of size [5s2p|3s2p], which is a basis set that provides a relatively accurate description of the atomic orbitals 1s, 2s, and 2p. We have included these and the remaining two orbitals in an orbital energy diagram in Figure 7.4, where we have chosen to order orbitals with respect to symmetry and energy. The s-type spatial orbitals are thus referred to as orbitals number 1, 2, and 3, and the corresponding part of the Hessian is obtained by considering a generator of orbital rotations according to

$$\hat{\kappa} = \sum_{s=2}^{3}\sum_{i=1}^{1}[\kappa_{si}\hat{E}_{si}^\dagger + \kappa_{si}^*\hat{E}_{si}]. \tag{7.27}$$

From Eq. (7.23), we get an explicit expression for sub-blocks A and B of the Hessian that equal

$$A = 2\times\begin{pmatrix} (F_{22} - F_{11}) - (22|11) + 2(21|12) & -(23|11) + 2(13|21) \\ -(32|11) + 2(31|12) & (F_{33} - F_{11}) - (33|11) + 2(31|13) \end{pmatrix}, \tag{7.28}$$

and

$$B = 2\times\begin{pmatrix} -(21|21) & -(21|31) \\ -(31|21) & -(31|31) \end{pmatrix}. \tag{7.29}$$

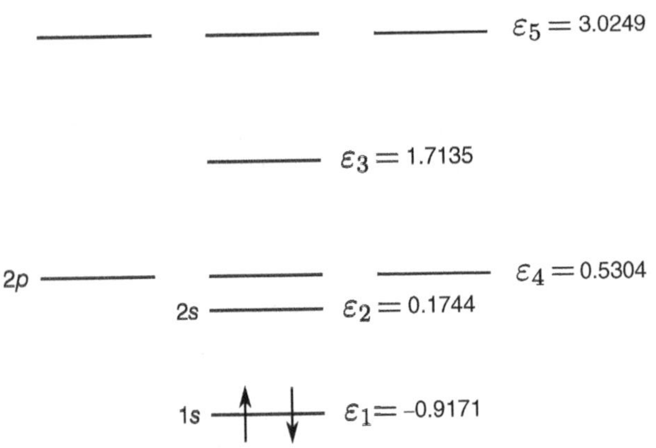

Figure 7.4 Hartree–Fock orbital energy diagram of helium using Dunning's aug-cc-pVDZ basis set. Energies are given in units of E_h.

Table 7.1 A selection of nonrelativistic two-electron integrals in the canonical Hartree–Fock molecular orbital basis for helium using Dunning's aug-cc-pVDZ basis set.

Type	Spatial integral	Value (a.u.)
Coulomb (2*s*, 1*s*)-interaction	(22|11)	0.3297
Exchange (2*s*, 1*s*)-interaction	(21|12)	0.0313
	(32|11)	0.1509
	(31|12)	0.0714

The factor of 2 that multiplies the matrices is due to the use of the non-normalized excitation and de-excitation operators in Eq. (7.27), so the eigenvalues of the A-matrix correspond in this case to twice the CIS excitation energies.

If we insert the numerical values for the orbital energies (equal to the diagonal elements of the Fock matrix) from Figure 7.4 and the two-electron integrals (some of which are listed in Table 7.1), we get

$$A = \begin{bmatrix} 1.6490 & -0.0161 \\ -0.0161 & 4.4119 \end{bmatrix}; \quad B = \begin{bmatrix} -0.0627 & -0.1428 \\ -0.1428 & -0.4182 \end{bmatrix}. \tag{7.30}$$

Among the integrals listed in Table 7.1, the largest value of 0.3297 E_h corresponds to the Coulomb interaction between the occupied 1s- and the virtual 2s-orbital, and since the corresponding exchange interaction is as small as 0.0313 E_h and therefore does not cancel the Coulomb term, the diagonal elements of the Hessian will both in this example and in general be strongly affected by the sum of the Coulomb and exchange interactions between virtual and occupied orbitals. In practice, this implies that the approximation of excitation energies by orbital energy differences is a poor one. The off-diagonal elements of the A-matrix are small due to effective integral cancelations. However, the degree of cancelation is somewhat special in the present case due to the very small coupling between the two single-electron excited determinants.

Half the smallest eigenvalue of the A-matrix in our helium example, which corresponds to the $1s \rightarrow 2s$ excitation energy, is equal to 0.8245 E_h (or, equivalently, 22.43 eV), whereas the corresponding difference in orbital energies amounts to 1.091 E_h (or 29.70 eV). In comparison with the result based on the difference between orbital energies, the CIS result is in far better agreement with the experimental result of 20.62 eV. Later in this chapter, we will explain in detail how to determine excitation energies in the time-dependent SCF approximation, which involves the solution of a generalized eigenvalue equation for the Hessian with a metric in terms of a generalized overlap matrix. Excitation energies obtained in the time-dependent Hartree–Fock (TDHF) approximation differ from the CIS ones because of the coupling matrix B in Eq. (7.22), and this coupling will, under normal circumstances, lower the estimates of valence excitation energies. In the numerical example considered here, the TDHF result for the excitation energy amounts to 22.37 eV—this result is later found in atomic units as the smaller of the two diagonal elements in the matrix in Eq. (7.134). With an improved basis set, the discrepancy between theory and experiment is further reduced, and using the t-aug-cc-pVTZ basis set, the TDHF result becomes 21.13 eV.

7.1.2 Configuration Interaction

In the CI approximations, the wave function is expressed as a linear combination of a set of time-independent determinants formed from the canonical Hartree–Fock orbitals, and the time dependence is therefore limited to the expansion coefficients. The various CI methods differ in the inclusion of determinants of different order of electron excitations; the CIS method includes determinants that are at the most single-electron excited, the configuration interaction singles and doubles (CISD) method in addition includes the double-electron excited determinants, and so on.

The matrix representation of the Hamiltonian is

$$H^{\mathrm{CI}} = \begin{pmatrix} E^{\mathrm{HF}} & 0 & \langle \mathrm{HF}|\hat{H}_0|\mathrm{HF}_{ij}^{st}\rangle & \cdots \\ 0 & \langle \mathrm{HF}_k^u|\hat{H}_0|\mathrm{HF}_i^s\rangle & \langle \mathrm{HF}_k^u|\hat{H}_0|\mathrm{HF}_{ij}^{st}\rangle & \cdots \\ \langle \mathrm{HF}_{kl}^{uv}|\hat{H}_0|\mathrm{HF}\rangle & \langle \mathrm{HF}_{kl}^{uv}|\hat{H}_0|\mathrm{HF}_i^s\rangle & \langle \mathrm{HF}_{kl}^{uv}|\hat{H}_0|\mathrm{HF}_{ij}^{st}\rangle & \cdots \\ \vdots & \vdots & \vdots & \ddots \end{pmatrix}, \tag{7.31}$$

where E^{HF} denotes the Hartree–Fock energy and the zeros are due to the fact that the Hartree–Fock state is variationally optimized with respect to the Hamiltonian, giving rise to a zero gradient in Eq. (7.21). The CI ground- and excited-state energies and corresponding state vectors are given by the eigenvalues and eigenvectors of the CI Hamiltonian matrix, and we note that there are contributions from the single-electron excited determinants to the ground state in CISD (and higher order CI methods) despite the gradient being zero. This is due to the couplings between single and higher-order electron excited determinants. If we denote the CI eigenvalues and eigenvectors by E_n and $|n\rangle$, respectively, the Hamiltonian can be written as

$$\hat{H}_0 = \sum_n E_n |n\rangle\langle n|, \tag{7.32}$$

which in form is identical to the exact-state case. As a consequence, we can directly apply the strategies from exact-state theory for the reference-state parameterization and response function derivation also in CI theories. The explicit form of an appropriate parameterization of the phase-isolated wave function is

$$|\overline{\psi}(t)\rangle = e^{-i\hat{P}(t)}|0\rangle, \tag{7.33}$$

where the Hermitian operator $\hat{P}(t)$ is given by

$$\hat{P}(t) = \sum_{n>0} [P_n(t)|n\rangle\langle 0| + P_n^*(t)|0\rangle\langle n|]. \tag{7.34}$$

We refer the reader to Chapter 5 for further details on the properties of this parameterization.

From a formal point of view, it is clear that the CI approximation is appealing in its analogy with exact-state theory. However, there is a catch, namely, that the dimensionality of the problem increases rapidly in practical calculations. For a system with N_e electrons in K orbitals, the number of m-electron excited determinants is

$$D = \binom{N_e}{m} \times \binom{K - N_e}{m},$$

so even for systems of modest size, the large number of determinants becomes a bottleneck in the calculation. One way to limit this number is to introduce the concept of an *active space*, which constitutes a set of occupied and virtual orbitals to be included in the formation of excited determinants. The most commonly used version of this idea is the complete active space (CAS), which includes all electron excitations within the given active space, but there exist also other variants of, in one way or another, restricted active spaces. The nomenclature used is that inactive orbitals are kept occupied in all determinants, active orbitals may be either occupied or unoccupied, and secondary orbitals are not occupied in any of the determinants. Setting up an active space that provides a balanced description of the electronic structure requires some experience and chemical intuition, but there are some standard choices to be made such as, for example, the full-valence and the $\pi\pi^*$ CAS where, respectively, either all valence orbitals or all π and π^* orbitals are included in the active spaces.

Our simple example of helium is not a system that illustrates the virtues of CAS-type wave functions since in this system there is largely dynamic electron correlation that needs to be accounted for, but we will nevertheless use it for a conceptual demonstration. Let us correlate the occupied 1s-orbital with the virtual 2s-orbital in the optimization of the ground electronic state. In this case, there are obviously no inactive orbitals, the 1s- and 2s-orbitals constitute the active orbitals, and the remaining ones are secondary; see Figure 7.5.

The nonrelativistic ground-state CI wave function based on the active space in Figure 7.5 involves three spin-adapted configuration state functions (CSFs), namely

$$|\Phi_0\rangle = |\psi_1^\alpha(1), \psi_1^\beta(2)\rangle,$$
$$|\Phi_1\rangle = \frac{1}{\sqrt{2}}[|\psi_1^\alpha(1), \psi_2^\beta(2)\rangle + |\psi_2^\alpha(1), \psi_1^\beta(2)\rangle],$$
$$|\Phi_2\rangle = |\psi_2^\alpha(1), \psi_2^\beta(2)\rangle,$$

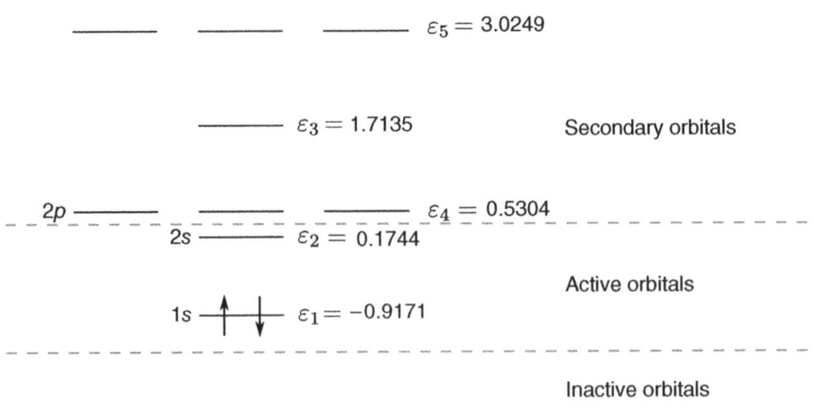

Figure 7.5 Active space for helium based on the Hartree–Fock orbitals using Dunning's aug-cc-pVDZ basis set. Energies are given in units of E_h.

where Φ_0 is recognized as the Hartree–Fock wave function. In this basis, the matrix representation of the Hamiltonian (in units of E_h) is

$$H^{\text{CI}} = \begin{bmatrix} -2.8557 & 0 & 0.0313 \\ 0 & -2.0312 & 0.1864 \\ 0.0313 & 0.1864 & -0.6571 \end{bmatrix}. \tag{7.35}$$

From the diagonalization of the CI Hamiltonian, we obtain a set of eigenvectors (CI expansion coefficients) and eigenvalues (CI energies). They are found to be

$$c = \begin{bmatrix} 0.9999 & -0.0052 & 0.0140 \\ 0.0033 & 0.9912 & 0.1320 \\ -0.0145 & -0.1320 & 0.9911 \end{bmatrix}; \quad E = [-2.8562, -2.0560, -0.6319]. \tag{7.36}$$

The above matrix with CI expansion coefficients is ordered so that the state vectors $|n\rangle$ corresponding to Eq. (7.32) are given by

$$|n\rangle = \sum_{i=0}^{2} c_{in}|\Phi_i\rangle, \quad n = \{0, 1, 2\}. \tag{7.37}$$

The CI estimate of the correlation energy of the ground state is given by the difference between the lowest eigenvalue and the upper left element of the H^{CI} matrix, and is thus

$$\Delta E_{\text{corr}} = -2.8562 - (-2.8557) = -0.5 \times 10^{-3}\ E_h = -0.01\ \text{eV},$$

whereas the transition energy to the first excited state is given by the difference in eigenvalues

$$\Delta E_{\text{transition}} = -2.0560 - (-2.8562) = 0.8001\ E_h = 21.77\ \text{eV}.$$

If we compare this estimate of the $1s \rightarrow 2s$ excitation energy with that obtained previously in the CIS approximation (which is to be considered as the uncorrelated result), we see that the correction from electron correlation to the excitation energy is

$$\Delta E_{\text{corr}} = 0.8001 - 0.8245 = -24.4 \times 10^{-3}\ E_h = -0.66\ \text{eV}.$$

The correlation treatment thus has a stronger impact on the excitation energy than on the ground-state energy, which can be understood by noting that the H^{CI} matrix has a larger coupling between the two excited-state CSFs than between the ground- and excited-state CSFs. Couplings between the single-electron excited determinants and the double-electron excited ones will, under quite general terms, lower the eigenvalues associated with the former states, and we therefore in general expect that the combined effect of electron correlation and excited-state orbital relaxation will be to lower the computed excitation energies of the lowest valence transitions.

Even with the introduction of active spaces in order to reduce the dimensionality of the CSF space, it often becomes impractical to form and diagonalize the CI Hamiltonian and one turns for that reason to *direct* CI techniques. However, with such a strategy the eigenvectors $|n\rangle$ become unavailable and the formation of the generator of rotations in Eq. (7.34) cannot be carried out. Instead, a generator of the form

$$\hat{R}(t) = \sum_{n} [R_n(t)|\Phi_n\rangle\langle 0| + R_n^*(t)|0\rangle\langle\Phi_n|], \tag{7.38}$$

can be employed, where the summation runs over all CSFs including the Hartree–Fock state ($n = 0$). It is clear that also this choice of parameterization is complete in the sense that the operator $\exp(-i\hat{R})$ can map $|0\rangle$ to any given point in the CI space. But the parameterization is redundant since it contains one more state transfer operator than what is used in Eq. (7.34), and this corresponds to the possibility of altering the overall phase of the reference state. But we are willing to accept this redundancy in order to avoid finding the complete set of eigenvectors of the matrix H^{CI}. With a parameterization based on Eq. (7.38), the Hessian takes the form of Eq. (7.22) with matrix elements that are

$$\begin{aligned} A_{nm} &= \left.\frac{\partial^2 E(R)}{\partial R_n^* \partial R_m}\right|_{R=0} \\ &= -\frac{1}{2}(\langle 0|[|\Phi_m\rangle\langle 0|,\ [|0\rangle\langle\Phi_n|,\ \hat{H}_0]]|0\rangle + \langle 0|[|0\rangle\langle\Phi_n|,\ [|\Phi_m\rangle\langle 0|, \hat{H}_0]]|0\rangle) \\ &= \begin{cases} \langle\Phi_m|\hat{H}_0|\Phi_m\rangle - E^{\mathrm{CI}}, & n = m \\ \langle\Phi_n|\hat{H}_0|\Phi_m\rangle, & n \neq m \end{cases}, \end{aligned} \tag{7.39}$$

and

$$B_{nm} = \left.\frac{\partial^2 E(R)}{\partial R_n^* \partial R_m^*}\right|_{R=0} = 0. \tag{7.40}$$

We note that element $n = m = 0$ of the A-matrix corresponds to minus the ground-state correlation energy. With the nonredundant parameterization using $\hat{P}(t)$, the Hessian becomes diagonal with elements that equal excitation energies [see Eqs. (5.96)–(5.98)], whereas with the redundant parameterization using $\hat{R}(t)$, the Hessian becomes block diagonal. In the latter case, one of the eigenvalues of the nonzero A-block will be zero and the corresponding eigenvector is equal to the CI expansion coefficients of the ground state. Other eigenvalues and eigenvectors refer to excitation energies and CI expansion coefficients of the excited states, respectively.

In our helium example, we get the result

$$A = \begin{bmatrix} 0.0005 & 0 & 0.0313 \\ 0 & 0.8250 & 0.1864 \\ 0.0313 & 0.1864 & 2.1990 \end{bmatrix}, \tag{7.41}$$

for which the three eigenvalues are 0, 0.8001, and 2.2243 E_h. These results are in perfect agreement with differences of the CI energies presented in Eq. (7.36).

7.1.3 Multiconfiguration Self-Consistent Field

Given an approximate wave function in a truncated CI model, there are two natural ways to improve on it that come directly to mind, namely, (i) an expansion of the determinant space by increasing the active space or by allowing for higher electron excitations, or (ii) a rotation of the molecular orbitals. The latter improvement leads us to the multiconfiguration self-consistent field (MCSCF) wave function, which is variationally optimized with respect to changes in CI expansion coefficients as well as molecular orbital (MO) coefficients. The parameterization of such an MCSCF state vector $|0\rangle$ is a seemingly straightforward affair given what we have learned about orbital rotations and CI parameterizations. The combination of the two techniques yields

$$|\overline{\psi}(t)\rangle = e^{-i\hat{\kappa}(t)}e^{-i\hat{R}(t)}|0\rangle, \tag{7.42}$$

where we have chosen to operate first in the configuration parameter space and subsequently in the orbital parameter space. In principle, it is possible to apply the exponential operators in reverse order, but this is not recommended due to the fact that it would introduce a dependence in $\hat{R}$ on the κ-parameters. This is seen by considering Eq. (7.38), in which, with a reversed-order application of the exponential operators in Eq. (7.42), the state transfer operators would no longer involve $|0\rangle$, but rather $\exp(-i\hat{\kappa})|0\rangle$.

In the case of a single determinant wave function, we needed to consider rotation parameters for all pairs of secondary–inactive (s, i) orbitals, but, in the case of a CAS wave function, pairs

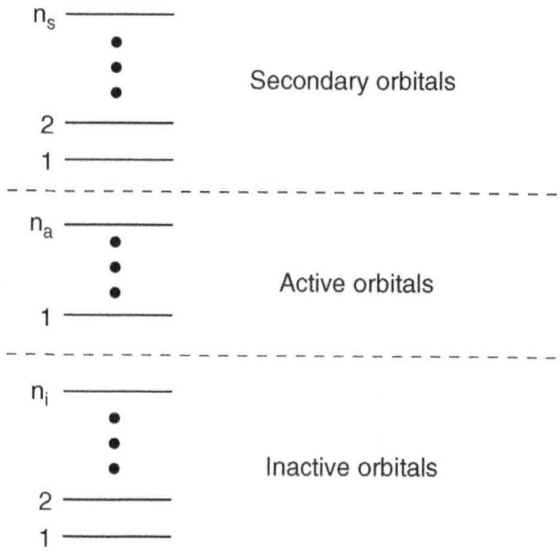

Figure 7.6 Complete active space.

of active–inactive (a, i) and secondary–active (s, a) orbitals must also be included in the parameterization. Orbital excitations of the type active–active (a, b), on the other hand, are included in the variations of configurations, which means that in the full configuration interaction (FCI) model there are no nonredundant orbital rotations left.

If we consider a CAS as shown in Figure 7.6, the total number of orbital rotation parameters will be

$$D_\mathrm{o} = n_\mathrm{i} \times n_\mathrm{s} + n_\mathrm{i} \times n_\mathrm{a} + n_\mathrm{a} \times n_\mathrm{s}, \tag{7.43}$$

and for a system of N_e electrons, the number of configuration parameters will be

$$D_\mathrm{c} = \binom{n_\mathrm{a}}{N_\mathrm{e} - n_\mathrm{i}}. \tag{7.44}$$

The Hessian will take the form given in Eq. (7.22) with matrix blocks A and B that contain derivatives with respect to orbital, configuration, and mixed orbital–configuration parameters. The structures of the blocks will be

$$A = \begin{pmatrix} \left.\dfrac{\partial^2 E(\kappa, R)}{\partial \kappa^* \partial \kappa}\right|_{\kappa,R=0} & \left.\dfrac{\partial^2 E(\kappa, R)}{\partial \kappa^* \partial R}\right|_{\kappa,R=0} \\ \left.\dfrac{\partial^2 E(\kappa, R)}{\partial R^* \partial \kappa}\right|_{\kappa,R=0} & \left.\dfrac{\partial^2 E(\kappa, R)}{\partial R^* \partial R}\right|_{\kappa,R=0} \end{pmatrix} \tag{7.45}$$

and

$$B = \begin{pmatrix} \left.\dfrac{\partial^2 E(\kappa, R)}{\partial \kappa^* \partial \kappa^*}\right|_{\kappa,R=0} & \left.\dfrac{\partial^2 E(\kappa, R)}{\partial \kappa^* \partial R^*}\right|_{\kappa,R=0} \\ \left.\dfrac{\partial^2 E(\kappa, R)}{\partial R^* \partial \kappa^*}\right|_{\kappa,R=0} & \left.\dfrac{\partial^2 E(\kappa, R)}{\partial R^* \partial R^*}\right|_{\kappa,R=0} \end{pmatrix}. \tag{7.46}$$

Explicit formulas for the energy derivatives are obtained by first forming the expectation value of the Hamiltonian with respect to the MCSCF state vector in Eq. (7.42) before carrying out the differentiation. From a book-keeping point of view, this exercise benefits from the introduction of a superoperator $\underline{\underline{\hat{\Omega}}}$ for which the action on an operator $\hat{\Lambda}$ is to form a commutator according to

$$\underline{\underline{\hat{\Omega}}}\,\hat{\Lambda} = [\hat{\Omega}, \hat{\Lambda}]. \tag{7.47}$$

Given this notation, the MCSCF energy can be written as

$$E(\kappa, R) = \langle \overline{\psi} | \hat{H}_0 | \overline{\psi} \rangle = \langle 0 | e^{i\hat{R}} e^{i\hat{\kappa}} \hat{H}_0 \, e^{-i\hat{\kappa}} e^{-i\hat{R}} | 0 \rangle$$
$$= \sum_{k=0}^{\infty} \sum_{l=0}^{\infty} \frac{i^{k+l}}{k! \, l!} \langle 0 | \underline{\underline{\hat{R}}}^{k} \underline{\underline{\hat{\kappa}}}^{l} \, \hat{H}_0 | 0 \rangle = \sum_{k=0}^{\infty} \sum_{l=0}^{k} \frac{i^{k}}{(k-l)! \, l!} \langle 0 | \underline{\underline{\hat{R}}}^{k-l} \underline{\underline{\hat{\kappa}}}^{l} \, \hat{H}_0 | 0 \rangle, \tag{7.48}$$

where the last step constitutes a mere reorganization of terms, in combined powers of $\hat{R}$ and $\hat{\kappa}$, with the result that contributions to the Hessian will come from the three terms with $k = 2$. Of these three terms, the $l = 0$ and $l = 2$ terms will contribute to derivatives with respect to configuration and orbital parameters, respectively, whereas the $l = 1$ term will contribute to the mixed orbital–configuration derivative. The differentiation of the former two terms follows closely the derivations made in the previous two sections concerned with the CI and the single-determinant wave functions, so let us here focus on the mixed derivative. We get

$$\left. \frac{\partial^2 E(\kappa)}{\partial \kappa_{si} \partial R_n^*} \right|_{\kappa, R=0} = -\langle 0 | [| 0 \rangle \langle \Phi_n |, \, [\hat{a}_s^\dagger \hat{a}_i, \hat{H}_0]] | 0 \rangle = \langle \Phi_n | \hat{H}_0 | 0_i^s \rangle, \tag{7.49}$$

where $|0_i^s\rangle$ denotes the single-electron excited MCSCF state vector. After the corresponding derivations are made for orbital parameters referring to inactive–active and active–secondary electron excitations, one can form the complete Hessian in the MCSCF approximation. However, in contrast to the Hartree–Fock and CI approximations, it is not clear how we associate the MCSCF Hessian with the excitation energies of the system. We will postpone a further discussion on this issue until we have derived the linear response function, and we will then adopt an approach for the determination of excitation energies in approximate-state theories that is based on a residue analysis of this response function.

7.1.4 Coupled Cluster

A method that shows the correct scaling of the energy with respect to the number of particles in the system is said to be *size extensive*. One obvious situation where the issue of size extensivity becomes crucial is in studies of fragmentation, and size extensivity is a necessary condition for describing the separated fragments correctly. However, it is not a sufficient condition for the method to be *size consistent*, that is, that the method correctly describes the dissociation of a molecule into its respective fragments. A spin-restricted method, such as Hartree–Fock will in general not be size consistent even though it is size extensive. Using the hydrogen molecule as a simple example, the spin-restricted Hartree–Fock method will not be able to dissociate this molecule into the correct hydrogen atom fragments. As response theory is normally not applied to situations where chemical reactions are induced, we will here primarily focus on the size-extensive nature of an electronic-structure method.

Let us consider a system that can be decomposed into subsystems A and B as shown in Figure 7.7. The separation distance R is sufficiently large for quantum-mechanical exchange

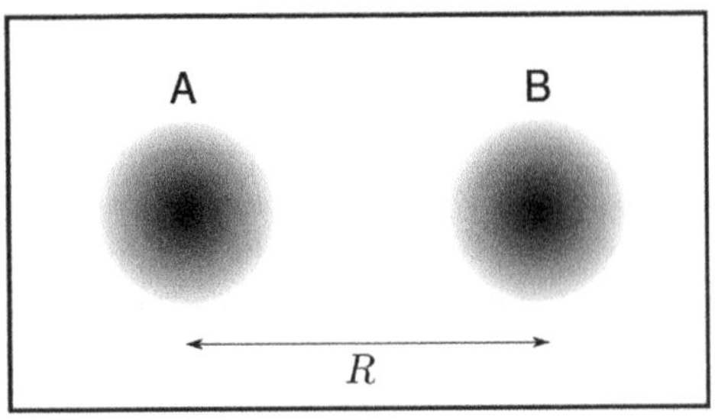

Figure 7.7 A system divided into subsystems A and B.

interactions to be negligible and the wave function of the combined system can thus be written as a product of wave functions for the individual systems A and B

$$|\psi\rangle = |\Phi\rangle_{\mathrm{A}} \otimes |\Phi\rangle_{\mathrm{B}}. \tag{7.50}$$

With a description of the subsystems in a truncated CI approach, for notational convenience we here consider that CIS, the total wave function, will be

$$|\psi\rangle = |\Phi_0\rangle_{\mathrm{A}} \otimes |\Phi_0\rangle_{\mathrm{B}} + \sum_{s,i} a_i^s\, |\Phi_i^s\rangle_{\mathrm{A}} \otimes |\Phi_0\rangle_{\mathrm{B}} + |\Phi_0\rangle_{\mathrm{A}} \otimes \sum_{t,j} b_j^t\, |\Phi_j^t\rangle_{\mathrm{B}} + \sum_{s,i}\sum_{t,j} a_i^s b_j^t\, |\Phi_i^s\rangle_{\mathrm{A}} \otimes |\Phi_j^t\rangle_{\mathrm{B}}. \tag{7.51}$$

We note that this (size-extensive) product wave function of the total system will contain terms that are double-electron excited, and it is clear that in the absence of these, our approach cannot be size extensive. In other words, the CIS wave function of the total system will *not* correspond to an energy that equals the sum of CIS energies of A and B. This line of argument can be pursued at any order of truncation in the CI expansion, and we have identified the single most important drawback of the CI approach, the lack of size extensivity.

The double-electron excited terms in Eq. (7.51) stem from products of single-electron excited configurations, and it is common to refer to such terms as *disconnected clusters* (as opposed to connected clusters). In CISD there will appear disconnected clusters that are triple and quadruple excited as well, and in the general case we need a more efficient way to generate the disconnected clusters that are required in order for size extensivity to be maintained. An elegant way to achieve this is to make use of an exponential *ansatz* for the coupled cluster wave function

$$|\overline{\psi}(t)\rangle = e^{\hat{T}(t)}|0\rangle, \tag{7.52}$$

where $|0\rangle$ here represents the Hartree–Fock state and $\hat{T}$ is an excitation operator of the form

$$\hat{T} = \hat{T}_1 + \hat{T}_2 + \cdots \tag{7.53}$$

with

$$\hat{T}_1 = \sum_{s,i} t_i^s\, \hat{a}_s^\dagger \hat{a}_i; \quad \hat{T}_2 = \sum_{s>t}\sum_{i>j} t_{ij}^{st}\, \hat{a}_t^\dagger \hat{a}_j \hat{a}_s^\dagger \hat{a}_i; \quad \text{and so on.} \tag{7.54}$$

The disconnected double excitations discussed above in relation to Eq. (7.51) will be generated by the $\hat{T}_1^2$ operator that appears after an expansion of the exponential operator is carried out. In passing, we note that if the excitation operator $\hat{T}$ in Eq. (7.53) is not truncated but includes up to N_e-electron excitations (with N_e being the number of electrons in the system), then the coupled cluster parameterization constitutes an alternative representation of the FCI state vector. It is also to be noted that we have here assumed that the coupled cluster expansion is made in terms of a TDHF state with fixed molecular orbitals. Such an approach is referred to as "unrelaxed." Orbital-relaxed coupled cluster theories would allow time-dependent rotations of the Hartree–Fock orbitals in line with the discussion above in connection to the parameterization of the SCF state. For simplicity, we will restrict ourselves to a discussion of unrelaxed coupled cluster theory.

The price to be paid for this elegant handling of the disconnected clusters is more complicated equations. We have seen that the exponential parametrization of the state vector in Hartree–Fock theory takes us from the reference determinant to a new, single determinant, expressed in terms of unitary transformed orbitals. We have also seen that the determinants of a CI expansion are limited to the chosen excitation level. The CC expansion, on the other hand, goes beyond the chosen excitation level and only truncates when all N_e electrons have been excited. We may consider the implications for the determination of the unperturbed reference state. We start from the time-independent Schrödinger equation

$$\hat{H}_0 e^{\hat{T}}|0\rangle = E e^{\hat{T}}|0\rangle. \tag{7.55}$$

If we now project from the left with the adjoint of the coupled cluster ket vector to set up an energy expression for the variational optimization of the coupled cluster parameters (amplitudes), we will get a metric on the right-hand side since the exponential excitation operator $\hat{T}$ is not unitary. Furthermore, an expansion of the exponentials inside the resulting matrix elements will not truncate until the N_e-electron limit since the right-hand excitation operators $\hat{T}$ and left-hand de-excitation operators $\hat{T}^\dagger$ do not commute. The energy is instead obtained by a projection of this equation onto the bra vector $\langle 0|\exp(-\hat{T})$ according to

$$E = \langle 0|e^{-\hat{T}}\hat{H}_0\, e^{\hat{T}}|0\rangle, \tag{7.56}$$

where appears the similarity-transformed Hamiltonian $e^{-\hat{T}}\hat{H}_0\, e^{\hat{T}}$. This implies, however, that the coupled cluster energy becomes nonvariational and that the state vector does not fulfill the Brillouin condition. We stress that the adopted bra vector is *not* the adjoint of the coupled cluster ket vector defined in Eq. (7.52). The similarity-transformed Hamiltonian is accordingly not Hermitian and the projection is therefore not equivalent to the formation of an expectation value. In fact, this expression can be further simplified and made more asymmetric by noting that

$$\langle 0|\hat{T} = 0, \tag{7.57}$$

which reduces the energy expression to

$$E = \langle 0|\hat{H}_0\, e^{\hat{T}}|0\rangle. \tag{7.58}$$

The annihilation of the Hartree–Fock bra vector by the $\hat{T}$-operator as expressed in Eq. (7.57) means that we could have directly performed a projection onto $\langle 0|$, rather than $\langle 0|\exp(-\hat{T})$, in order to isolate the energy from Eq. (7.55).

The evaluation of the coupled cluster energy in Eq. (7.58) requires knowledge of the amplitudes in order to form the $\hat{T}$-operator. These are, however, straightforwardly determined from a set of equations obtained by projections involving excited determinants

$$\langle 0_{ij\ldots}^{st\ldots}|e^{-\hat{T}}\hat{H}_0\, e^{\hat{T}}|0\rangle = 0. \tag{7.59}$$

This set of equations is decoupled from the coupled cluster energy equation, and they are to be solved before the evaluation of the energy is done.[1] For convenience, the time-independent CC reference state, which is obtained by inserting the CC amplitudes into Eq. (7.52), will be denoted by

$$|\mathrm{CC}\rangle = e^{\hat{T}}|0\rangle, \tag{7.60}$$

and the corresponding energy will be denoted by E^{CC}. In the presence of an external field, the coupled cluster amplitudes become time dependent, and the excitation operator employed in the formation of the reference state [Eq. (7.60)] then represents the zeroth-order solution $\hat{T}^{(0)}$.

1 An excellent introduction to coupled cluster theory is provided by D. T. Crawford and H. F. Schaefer. *Rev. Comput. Chem.*, 14:33, 2000.

7.2 Equations of Motion

In the previous section, we presented parameterizations of some common approximate wave functions. For a given wave function parameterization, one needs an equation of motion to determine the time-dependent parameters. We will here study the Ehrenfest and the quasi-energy approaches. Once the general principles have been established, the actual derivation of explicit formulas for the parameter responses and response functions is very much an algebraic exercise that carries limited physical insight beyond that gained in our exact-state theory treatment. For this reason, we will be selective in our presentation and keep the main focus on response theory to linear order.

7.2.1 Ehrenfest Theorem

The foundation principle for the use of the Ehrenfest theorem [Eq. (5.117)] to determine the time evolution of the phase-isolated wave function is that it is fulfilled through each order in the perturbation. For an operator $\hat{\Omega}$, we have

$$\frac{\partial}{\partial t}\langle\overline{\psi}(t)|\hat{\Omega}|\overline{\psi}(t)\rangle - \langle\overline{\psi}(t)|\frac{\partial\hat{\Omega}}{\partial t}|\overline{\psi}(t)\rangle$$
$$= \frac{1}{i\hbar}\left[\langle\overline{\psi}(t)|[\hat{\Omega},\hat{H}_0]|\overline{\psi}(t)\rangle + \sum_{\omega}\langle\overline{\psi}(t)|[\hat{\Omega},\hat{V}_\alpha^\omega]|\overline{\psi}(t)\rangle F_\alpha^\omega e^{-i\omega t}e^{\epsilon t}\right]. \tag{7.61}$$

Let us first consider the case when the operator $\hat{\Omega}$ is time independent, in which case the second term in the left-hand side of Eq. (7.61) vanishes. To linear order in the perturbation, an expansion of any time-dependent expectation value will be in general given as

$$\langle\psi(t)|\hat{\Lambda}|\psi(t)\rangle = \langle 0|\hat{\Lambda}|0\rangle + \sum_{\omega}\langle\langle\hat{\Lambda};\hat{V}_\alpha^\omega\rangle\rangle F_\alpha^\omega e^{-i\omega t}e^{\epsilon t}. \tag{7.62}$$

We will use this expansion individually for the terms in Eq. (7.61). For the first and remaining term on the left-hand side, we insert $\hat{\Lambda} = \hat{\Omega}$, and for the first and second terms on the right-hand side, we insert $\hat{\Lambda} = [\hat{\Omega},\hat{H}_0]$ and $\hat{\Lambda} = [\hat{\Omega},\hat{V}_\alpha^\omega]$, respectively. After collecting terms of zeroth order, we get

$$0 = \langle 0|[\hat{\Omega},\hat{H}_0]|0\rangle, \tag{7.63}$$

where the zero on the left-hand side is due to the fact that the time derivative of the expectation value vanishes. Continuing and collecting terms of first order in the perturbation, we get

$$(-i\omega+\epsilon)\,\langle\langle\hat{\Omega};\hat{V}_\alpha^\omega\rangle\rangle = \frac{1}{i\hbar}[\langle\langle[\hat{\Omega},\hat{H}_0];\hat{V}_\alpha^\omega\rangle\rangle + \langle 0|[\hat{\Omega},\hat{V}_\alpha^\omega]|0\rangle]. \tag{7.64}$$

The former of these two equations is fulfilled as long as the reference state $|0\rangle$ is variationally optimized against the Hamiltonian $\hat{H}_0$. The latter equation, however, is nontrivial and can be used to derive equivalence expressions for dipole properties based on the following operator equality

$$[\hat{r}_\alpha,\hat{H}_0] = \frac{i\hbar}{m_e}\,\hat{p}_\alpha. \tag{7.65}$$

A word of caution is required here when one is concerned with practical calculations, because the matrix representation of the commutator relation expressed in Eq. (7.65) holds true only in the limit of a complete basis set, as shown in Appendix C. With this in mind, let us employ

Table 7.2 Nonrelativistic time-dependent Hartree–Fock results for neon of the equivalent dipole expression in Eqs. (7.66) and (7.67).

Basis set	$-\langle\langle \hat{r}_\alpha;\hat{r}_\alpha\rangle\rangle_\omega$	$\frac{-i}{m_e\omega}\langle\langle \hat{p}_\alpha;\hat{r}_\alpha\rangle\rangle_\omega$	$\lvert\Delta\rvert$ (%)	$\lvert\langle 0\lvert\hat{r}_\alpha\rvert f\rangle\rvert$	$\frac{1}{m_e\omega_f}\lvert\langle 0\lvert\hat{p}_\alpha\rvert f\rangle\rvert$	$\lvert\Delta\rvert$ (%)
aug-cc-pVDZ	1.837	1.721	6.8	0.4715	0.4808	2.0
daug-cc-pVDZ	2.362	2.378	0.6	0.3442	0.3483	1.2
taug-cc-pVTZ	2.384	2.382	0.1	0.3489	0.3506	0.5
qaug-cc-pVQZ	2.389	2.386	0.1	0.3499	0.3505	0.2
qaug-cc-pV5Z	2.386	2.386	0.0	0.3504	0.3505	0.0

Response functions are evaluated at the angular frequency $\omega = 0.0656$ a.u. Transition matrix elements refer to the excited state $|f\rangle$ being the $1^1\ P_1^o$ state with electronic configuration $1s^2 2s^2 2p^5 3s$. Dunning's series of basis sets have been employed. All quantities are given in a.u.

this commutator relation inside Eq. (7.64) in the limit of $\epsilon \to 0$ and with $\hat{\Omega} = \hat{r}_\alpha$ and $\hat{V}_\alpha^\omega = \hat{r}_\beta$ to arrive at

$$\omega\langle\langle \hat{r}_\alpha;\hat{r}_\beta\rangle\rangle_\omega = \frac{i}{m_e}\langle\langle \hat{p}_\alpha;\hat{r}_\beta\rangle\rangle_\omega, \tag{7.66}$$

where the angular frequency of the field is indicated as a subscript of the response function. Given the sum-over-states expression for the linear response equation, Eq. (5.168), the residue of Eq. (7.66) at a transition frequency $\omega = \omega_f$ is seen to express the equivalence of transition moments in the dipole length and dipole velocity representations

$$\omega_f\ \langle 0|\hat{r}_\alpha|f\rangle = \frac{i}{m_e}\langle 0|\hat{p}_\alpha|f\rangle. \tag{7.67}$$

We illustrate these dipole length–dipole velocity equivalences in Table 7.2 by nonrelativistic TDHF calculations on neon. A slow convergence of the results with respect to improvements in the basis set is observed in the table, and when treated with caution, a comparison of dipole length and dipole velocity results can be used as a measure of basis set completeness in the calculation.

For future reference, we will also consider a second equivalence relation. For a static field perturbation, the time derivative of the expectation value is zero and the left-hand side of Eq. (7.64) vanishes. We thereby get

$$0 = \frac{1}{i\hbar}[\langle\langle [r_\alpha, \hat{H}_0], \hat{l}_\beta\rangle\rangle_0 + \langle [r_\alpha, \hat{l}_\beta]\rangle] = \frac{1}{i\hbar}\left[\frac{i\hbar}{m_e}\langle\langle \hat{p}_\alpha;\hat{l}_\beta\rangle\rangle_0 + \langle [r_\alpha, \hat{l}_\beta]\rangle\right], \tag{7.68}$$

or, equivalently,

$$\langle\langle \hat{p}_\alpha;\hat{l}_\beta\rangle\rangle_0 = \frac{im_e}{\hbar}\ \varepsilon_{\beta\gamma\delta}\langle [r_\alpha, r_\gamma\hat{p}_\delta]\rangle = -m_e\ \varepsilon_{\alpha\beta\gamma}\langle r_\gamma\rangle. \tag{7.69}$$

It may appear as if we are drifting away from our main focus, namely, to establish equations of motion in the Ehrenfest approach for approximate variational wave function models. The reason for this digression is that the equivalence relation expressed in Eq. (7.64) is a fundamental property of response functions that we wish to maintain also in approximate-state theories, and we therefore need to find the requirements that are needed to be imposed on the equations of motion in order to preserve this property. It is clear that we can only fulfill the dipole equivalence relations expressed in Eqs. (7.66) and (7.67) in the limit of a complete basis set, since the matrix representation of the operator relation in Eq. (7.65) is not valid in a finite basis set. But we are now concerned with requirements that go beyond the issue of basis set. To begin with, the reference state $|0\rangle$ must be variationally optimized in order to fulfill Eq. (7.63). Next, the time-dependent state $|\overline{\psi}\rangle$ will be such that the Ehrenfest theorem is fulfilled for an incomplete set of electron-transfer and/or state-transfer operators of our choice. The number of operators in this set will of course be equal to the number of parameters in the reference-state parameterization [see Eqs. (7.43) and (7.44)], and it will lead to a corresponding set of, in general, coupled, differential equations from which the time dependences of the parameters are determined. In approximate-state theory, however, this recipe does not necessarily ensure that the Ehrenfest theorem [Eq. (7.61)] is fulfilled for *all* operators.

If we consider the single-determinant SCF approximation as an example, then the set of operators in the parameterization ($\hat{a}_s^\dagger \hat{a}_i$ and $\hat{a}_i^\dagger \hat{a}_s$) transfer an electron between the inactive and secondary orbitals of the reference state, that is, the wave function at time zero. General one-electron operators, on the other hand, include electron-transfer operators not only of secondary–inactive but also of inactive–inactive and secondary–secondary types:

$$\hat{\Omega} = \sum_{p,q} \Omega_{pq}\, \hat{a}_p^\dagger \hat{a}_q. \tag{7.70}$$

If we choose to determine the time evolution of the κ-parameters by applying Eq. (7.61) to the incomplete set of one-electron operators used in the parameterization, that is, $\hat{a}_s^\dagger \hat{a}_i$ and $\hat{a}_i^\dagger \hat{a}_s$, then Eq. (7.61), and consequently the equivalence relation in Eq. (7.64), will in general not be fulfilled for $\hat{\Omega}$. We may ask whether there exists an alternative incomplete set of one-electron operators that we can employ while fulfilling Eq. (7.64) for a general $\hat{\Omega}$. In order to shed light on this question, let us return to our simple example of two electrons in four orbitals illustrated in Figure 7.1.

In this example, we focused on the parameterization of the ground state $|0\rangle$ and, in Eq. (7.14) we established explicit expressions for the rotations of the occupied orbitals

$$\chi_1(t) = \cos|\kappa_{31}|\; \chi_1(0) - i\frac{\kappa_{31}}{|\kappa_{31}|} \sin|\kappa_{31}|\; \chi_3(0), \tag{7.71}$$

$$\chi_2(t) = \cos|\kappa_{42}|\; \chi_2(0) - i\frac{\kappa_{42}}{|\kappa_{42}|} \sin|\kappa_{42}|\; \chi_4(0). \tag{7.72}$$

Here, $\chi_1(0)$ and $\chi_2(0)$ refer to the occupied orbitals of the reference state, that is, the wave function at time zero, and $\chi_3(0)$ and $\chi_4(0)$ are the associated virtual orbitals. Orbitals $\chi_1(t)$ and $\chi_2(t)$, on the other hand, are the occupied orbitals at a later time t for which the associated virtual orbitals can be obtained by considering the mappings of $|1\rangle$, $|2\rangle$, and $|5\rangle$. This gives the result

$$e^{-i\hat{\kappa}}|1\rangle = |\chi_3(t),\ \chi_2(t)\rangle; \quad e^{-i\hat{\kappa}}|2\rangle = |\chi_1(t),\ \chi_4(t)\rangle; \quad e^{-i\hat{\kappa}}|5\rangle = |\chi_3(t),\ \chi_4(t)\rangle,$$

where

$$\chi_3(t) = \cos|\kappa_{31}|\; \chi_3(0) - i\frac{\kappa_{31}^*}{|\kappa_{31}|} \sin|\kappa_{31}|\; \chi_1(0), \tag{7.73}$$

$$\chi_4(t) = \cos|\kappa_{42}|\; \chi_4(0) - i\frac{\kappa_{42}^*}{|\kappa_{42}|} \sin|\kappa_{42}|\; \chi_2(0). \tag{7.74}$$

It is clear that the set of orbitals $\{\chi_p(t)\}_{p=1}^K$ spans the same space as the original set $\{\chi(0)\}_{p=1}^K$ and either set can of course be used for expansions of operators and wave functions without loss or gain in accuracy. But it is with respect to the latter set that we have defined our creation and annihilation operators and in order to work in the former orbital basis, we need to perform a time transformation of the electron-transfer operators according to

$$\hat{t}_{pq}^\dagger = e^{-i\hat{\kappa}}\, \hat{q}_{pq}^\dagger\, e^{i\hat{\kappa}}, \tag{7.75}$$

where

$$\hat{q}^{\dagger}_{pq} = \hat{a}^{\dagger}_{p}\hat{a}_{q}. \tag{7.76}$$

As an example, we then get

$$\hat{t}^{\dagger}_{31}|\overline{\psi}(t)\rangle = \hat{t}^{\dagger}_{31}e^{-i\hat{\kappa}}|0\rangle = e^{-i\hat{\kappa}}\ \hat{q}^{\dagger}_{31}|0\rangle = e^{-i\hat{\kappa}}\ |1\rangle, \tag{7.77}$$

which shows how the operator $\hat{t}^{\dagger}_{31}$ transfers an electron from orbital $\chi_1(t)$ to orbital $\chi_3(t)$ in the time-transformed orbital basis.

In the time-transformed orbital basis, a general one-electron operator takes the form

$$\hat{\Omega} = \sum_{p,q} \Omega'_{pq}\ \hat{t}^{\dagger}_{pq}, \tag{7.78}$$

in analogy with Eq. (7.70). Let us form our equations of motion by requiring that the secondary–inactive set of operators $\hat{t}^{\dagger}_{si}$ and $\hat{t}_{si}$ appearing in Eq. (7.78) fulfill the Ehrenfest theorem, Eq. (7.61), and from the resulting equations we determine the time dependence of the κ parameters. All other operator blocks will automatically fulfill the Ehrenfest theorem; we will discuss those for which this is not self-evident. Let us first consider the secondary–secondary terms in Eq. (7.78). For an electron-transfer operator of this type, it is only the second term of Eq. (7.61) that requires special attention. Inserting $\hat{t}^{\dagger}_{st}$, we get

$$\begin{aligned}\langle\overline{\psi}(t)|\frac{\partial\hat{t}^{\dagger}_{st}}{\partial t}|\overline{\psi}(t)\rangle &= \langle 0|e^{i\hat{\kappa}}\frac{\partial}{\partial t}(e^{-i\hat{\kappa}}\hat{q}^{\dagger}_{st}e^{i\hat{\kappa}})e^{-i\hat{\kappa}}|0\rangle \\ &= \langle 0|\left[e^{i\hat{\kappa}}\frac{\partial}{\partial t}(e^{-i\hat{\kappa}})\hat{q}^{\dagger}_{st} + \hat{q}^{\dagger}_{st}\frac{\partial}{\partial t}(e^{i\hat{\kappa}})e^{-i\hat{\kappa}}\right]|0\rangle = 0,\end{aligned} \tag{7.79}$$

and we can thus conclude that operators of the type $\hat{t}^{\dagger}_{st}$ will fulfill the Ehrenfest theorem. Finally, we turn to the inactive–inactive terms in Eq. (7.78). For operators $\hat{t}^{\dagger}_{ij}$ with $i \neq j$, all terms in Eq. (7.61) can be shown to vanish—the proof of this follows closely that for $\hat{t}^{\dagger}_{st}$. For operators $\hat{t}^{\dagger}_{ii}$, it is again the second term in Eq. (7.61) that deserves a closer inspection, but for operators of this type we get

$$\begin{aligned}\langle\overline{\psi}(t)|\frac{\partial\hat{t}^{\dagger}_{ii}}{\partial t}|\overline{\psi}(t)\rangle &= \langle 0|\left[e^{i\hat{\kappa}}\frac{\partial}{\partial t}(e^{-i\hat{\kappa}})\ \hat{q}^{\dagger}_{ii} + \hat{q}^{\dagger}_{ii}\frac{\partial}{\partial t}(e^{i\hat{\kappa}})e^{-i\hat{\kappa}}\right]|0\rangle \\ &= \langle 0|\left[e^{i\hat{\kappa}}\frac{\partial}{\partial t}(e^{-i\hat{\kappa}}) + \frac{\partial}{\partial t}(e^{i\hat{\kappa}})e^{-i\hat{\kappa}}\right]|0\rangle = \langle 0|\left[\frac{\partial}{\partial t}(e^{i\hat{\kappa}}e^{-i\hat{\kappa}})\right]|0\rangle = 0.\end{aligned} \tag{7.80}$$

This completes the proof that all terms in the one-electron operator [Eq. (7.78)] fulfill the Ehrenfest theorem [Eq. (7.61)], and if our equations of motion are based on the set of secondary–inactive time-transformed electron-transfer operators ($\hat{t}^{\dagger}_{si}$ and $\hat{t}_{si}$), then the equivalence relations arising from Eq. (7.64) will be fulfilled also in approximate single-determinant SCF theory.

In MCSCF theory, the parameterization of the reference state involves in addition to electron-transfer operators also state-transfer operators

$$\hat{Q}^{\dagger}_{n} = |\Phi_n\rangle\langle 0|. \tag{7.81}$$

Organized as a row vector, the appropriate time-transformed operators in MCSCF response theory are

$$(\hat{t}) = \begin{bmatrix} e^{-i\hat{\kappa}}\hat{q}^{\dagger}e^{i\hat{\kappa}} \\ e^{-i\hat{\kappa}}e^{-i\hat{R}}\hat{Q}^{\dagger}e^{i\hat{R}}e^{i\hat{\kappa}} \\ e^{-i\hat{\kappa}}\hat{q}e^{i\hat{\kappa}} \\ e^{-i\hat{\kappa}}e^{-i\hat{R}}\hat{Q}e^{i\hat{R}}e^{i\hat{\kappa}} \end{bmatrix}, \tag{7.82}$$

and by *appropriate* we here mean that when the set of operators in Eq. (7.82) is used for the construction of the equations of motion for the MCSCF state, it is guaranteed that the general one-electron operator in Eq. (7.78) will fulfill the equivalence relation in Eq. (7.64).

7.2.2 Quasi-Energy Derivatives

In accordance with Eq. (5.34) from exact-state theory, the quasi-energy will in the SCF approximation be

$$Q(t) = \langle 0|e^{i\hat{\kappa}(t)} \left(\hat{H} - i\hbar\frac{\partial}{\partial t}\right) e^{-i\hat{\kappa}(t)}|0\rangle, \tag{7.83}$$

where the parameterization of the reference state has been adopted from Eq. (7.1). The equation that governs the responses of the wave function parameters with respect to the perturbing fields is the variational principle for the time average of $Q(t)$

$$\delta Q_{\mathrm{T}} = 0. \tag{7.84}$$

This variational principle will translate into equations of motion when considering that all partial derivatives of the Fourier amplitudes of the wave function parameters vanish. We will, for example, for the first-order amplitudes, have

$$\frac{\partial Q_{\mathrm{T}}}{\partial \kappa_n^{(1)}(\omega_1)} = 0; \qquad \frac{\partial Q_{\mathrm{T}}}{\partial [\kappa_n^{(1)}(-\omega_1)]^*} = 0, \tag{7.85}$$

which is in accordance with Eq. (5.156) in the case of exact-state theory.

We have thus seen that the transition from applying the quasi-energy approach in exact-state theory to applying it in variational approximate state theory is quite straightforward. The single most important point of the quasi-energy derivative technique, however, is that it allows for a transparent and clear extension to nonvariational electronic structure theory, such as the family of projected coupled cluster approaches. The counterpart of Eq. (5.29) in exact-state theory will, after insertion of the phase-isolated coupled cluster state given in Eq. (7.52), take the form

$$\left(\hat{H} - i\hbar\frac{\partial}{\partial t}\right) e^{\hat{T}(t)}|0\rangle = Q(t)\, e^{\hat{T}(t)}|0\rangle. \tag{7.86}$$

This equation is also the counterpart of Eq. (7.55), and the isolation of the quasi-energy is achieved in a way analogous to the isolation of the energy in the time-independent case, namely, by a projection onto the Hartree–Fock state. For the coupled cluster approaches we therefore obtain a quasi-energy that becomes

$$Q(t) = \langle 0| \left(\hat{H} - i\hbar\frac{\partial}{\partial t}\right) e^{\hat{T}(t)}|0\rangle = \langle 0|\hat{H}e^{\hat{T}(t)}|0\rangle, \tag{7.87}$$

where we in the intermediate step have made use of the fact that the time derivative of the phase-isolated coupled cluster state is orthogonal to the Hartree–Fock state. By isolation of $Q(t)$ through projection, we obtain an asymmetric expression and not an expectation value. As a consequence there is no guarantee that the coupled cluster quasi-energy (or energy for that matter) will be real, and care must be exercised in the subsequent definition of response functions. For example, if one is concerned with the nonresonant electric-dipole polarizability, it is permissible to extract the real part of the linear response function at the end of the calculation, or equivalently to define the response function from the real part of the quasi-energy.

The time-dependent coupled cluster amplitudes are defined by projections onto the set of electron excited determinants in accordance with Eq. (7.59); these equations take the form

$$\langle 0_{ij\ldots}^{st\ldots}|e^{-\hat{T}} \left(\hat{H} - i\hbar\frac{\partial}{\partial t}\right) e^{\hat{T}}|0\rangle = 0. \tag{7.88}$$

We may understand the time-dependent coupled cluster approach as defined by the quasi-energy in Eq. (7.87) under the constraints defined in Eq. (7.88). The incorporation of these constraints is conveniently achieved by means of the variational Lagrangian technique[2] with a coupled cluster Lagrangian defined as

$$L(t) = Q(t) + \sum_n \lambda_n(t) \langle n|e^{-\hat{T}} \left(\hat{H} - i\hbar\frac{\partial}{\partial t}\right) e^{\hat{T}}|0\rangle, \tag{7.89}$$

where λ_n is the time-dependent Lagrangian multiplier associated with the constraint for the electron-excited determinant $|n\rangle = |0_{ij\ldots}^{st\ldots}\rangle$.

The equations of motion in time-dependent coupled cluster theory are obtained from derivatives of the time-averaged Lagrangian with respect to cluster amplitudes and Lagrangian multipliers. These derivatives will vanish due to the variational condition for the Lagrangian

$$\delta L_{\mathrm{T}} = 0, \tag{7.90}$$

and in view of Eq. (7.84), it becomes clear how the quasi-energy formalism enables a common framework for the formulations of time-dependent variational and nonvariational response theories. The former case in comparison with the latter is characterized by the absence of constraints (and thereby Lagrangian multipliers), but apart from this fact, the equations of motion are derived from a common source, namely Eq. (7.90).

7.3 Response Functions

We have so far discussed the parameterization of the most commonly adopted electronic structure theories used to study time-dependent molecular properties. Our consideration has not included the second-order Møller–Plesset (MP2) approach and also not the orbital-relaxed coupled cluster (CC) approaches or CC approaches with perturbational treatment of selected levels of electron excitations, such as the CC with singly, doubly, and noniterative triply excited configuration (CCSD(T)) method, which includes a perturbative treatment of triple excitations. The main reason for this exclusion is that the pole structure of the corresponding response functions in these cases is not consistent with the one in exact-state theory. Instead the response functions will, in addition to correlated poles, also contain uncorrelated Hartree–Fock poles, which is cumbersome not only for calculations of excitation energies but also for frequency-dependent properties in general due to the effect of the pole structure on frequency dispersions. A pragmatic way to deal with this inconsistency in the pole structure is to combine the calculation of molecular properties in the limit of static fields together with dispersion corrections obtained with a separate electronic structure method, for example, by combining the determination of the static polarizability as an energy derivative at the MP2 level of theory with the evaluation of frequency dispersion at the Hartree–Fock level of theory in order to obtain an estimate for $\alpha(\omega)$.

7.3.1 Single Determinant Approaches

In the Ehrenfest approach, we determine the time dependence of the SCF wave function parameters from the equation

$$\frac{\partial}{\partial t}\langle\psi|\hat{t}_n|\psi\rangle - \langle\psi|\frac{\partial}{\partial t}\hat{t}_n|\psi\rangle = \frac{1}{i\hbar}(\langle\psi|[\hat{t}_n, \hat{H}_0]|\psi\rangle + \langle\psi|[\hat{t}_n, \hat{V}(t)]|\psi\rangle), \tag{7.91}$$

2 Further discussion is found in Appendix H.

where $\hat{t}_n$ is the time-transformed electron de-excitation operator corresponding to the excitation operator defined in Eq. (7.75), but here indexed with a single subscript n that refers to a pair of secondary–inactive orbital indices (s, i). Equation (7.91) is thus really a set of equations, one for each pair of indices (s, i), and there is in addition a complimentary set of equations arising from applying Eq. (7.91) to the adjoint of $\hat{t}_n$, that is, the excitation operator $\hat{t}_n^\dagger$. In order to facilitate a compact and convenient notation for the matrix multiplications that will come, it is beneficiary to define the integer index n to take on positive as well as negative values (but $n \neq 0$) and to let the sign of the index be associated with the adjoint operation according to

$$\hat{t}_{-n} = \hat{t}_n^\dagger; \quad \hat{q}_{-n} = \hat{q}_n^\dagger; \quad \kappa_{-n} = \kappa_n^*. \tag{7.92}$$

We will also follow the convention of implied summations over repeated indices, so that the generator of orbital rotations can be written as

$$\hat{\kappa}(t) = \kappa_n \hat{q}_n^\dagger = \kappa_n^* \hat{q}_n. \tag{7.93}$$

One at a time, we will now rewrite the terms in Eq. (7.91) by expanding the exponential operators and thereby cast them into the form of matrix multiplications that separates the wave function parameters that we intend to determine. Defined matrices should be time independent and possible to evaluate given the variationally optimized reference state $|0\rangle$. Let us start with the term involving the unperturbed Hamiltonian, for which we get

$$-i\langle \psi | [\hat{t}_n, \hat{H}_0] | \psi \rangle = -i\langle 0 | [\hat{q}_n, e^{i\hat{\kappa}} \hat{H}_0 e^{-i\hat{\kappa}}] | 0 \rangle = \sum_{k=1}^{\infty} \frac{i^{k+1}}{k!} E^{[k+1]}_{nl_1 \cdots l_k} \prod_{j=1}^{k} \kappa_{l_j}, \tag{7.94}$$

where the generalized Hessian matrices have been introduced as

$$E^{[k+1]}_{nl_1 \cdots l_k} = -\langle 0 | \left[\hat{q}_n, \prod_{j=1}^{k} \hat{q}_{l_j}^\dagger \hat{H}_0 \right] | 0 \rangle. \tag{7.95}$$

We note that the term $k = 0$ in the expansion of the exponential operators in Eq. (7.94) vanishes since the reference state is variationally optimized. Since implied summations run over positive as well as negative index values, there are other ways to order the elements in the defined matrices. The chosen order is made so that we understand $E^{[2]}$ as having the following structure

$$E^{[2]} = -\begin{pmatrix} \langle 0 | [\hat{q}_n, [\hat{q}_m^\dagger, \hat{H}_0]] | 0 \rangle & \langle 0 | [\hat{q}_n, [\hat{q}_m, \hat{H}_0]] | 0 \rangle \\ \langle 0 | [\hat{q}_n^\dagger, [\hat{q}_m^\dagger, \hat{H}_0]] | 0 \rangle & \langle 0 | [\hat{q}_n^\dagger, [\hat{q}_m, \hat{H}_0]] | 0 \rangle \end{pmatrix}, \tag{7.96}$$

which is in accordance with the Hessian matrix ordering in Eq. (7.22) as well as that adopted in exact-state theory in Eq. (5.163).

For the perturbation operator $\hat{V}(t)$, we proceed in a similar manner and obtain

$$-i\langle \psi | [\hat{t}_n, \hat{V}(t)] | \psi \rangle = -\sum_{k=0}^{\infty} \frac{i^{k+1}}{k!} \sum_{\omega} V^{\omega,[k+1]}_{nl_1 \cdots l_k} e^{-i\omega t} \prod_{j=1}^{k} \kappa_{l_j}, \tag{7.97}$$

where the product symbol should be understood to contribute only for $k > 0$ and

$$V^{\omega,[k+1]}_{nl_1 \cdots l_k} = \langle 0 | \left[\hat{q}_n, \prod_{j=1}^{k} \hat{q}_{l_j}^\dagger \hat{V}^\omega \right] | 0 \rangle. \tag{7.98}$$

For the perturbation, the lowest-order term $k = 0$ will of course not vanish, but rather result in a property vector that is given as

$$V^{\omega,[1]} = \begin{pmatrix} \langle 0 | [\hat{q}_n, \hat{V}^\omega] | 0 \rangle \\ \langle 0 | [\hat{q}_n^\dagger, \hat{V}^\omega] | 0 \rangle \end{pmatrix}. \tag{7.99}$$

Since the property operator is Hermitian, the structure of this property gradient is

$$V^{\omega,[1]} = \begin{pmatrix} g \\ -g^* \end{pmatrix}. \tag{7.100}$$

The two terms on the left-hand side of Eq. (7.91) differ from those on the right-hand side in that they contain the operation of time differentiation. The first term on the left-hand side vanishes since

$$\langle \psi | \hat{t}_n | \psi \rangle = \langle 0 | \hat{q}_n | 0 \rangle = 0. \tag{7.101}$$

In the evaluation of the second term on the left-hand side, we will need to consider the time derivative of an exponential operator. From Appendix E, we have

$$\frac{d}{dt}[e^{-i\hat{\kappa}}\hat{q}_n e^{i\hat{\kappa}}] = e^{-i\hat{\kappa}} \sum_{k=0}^{\infty} \frac{i^{k+1}}{(k+1)!} [\hat{q}_n, \underline{\underline{\hat{\kappa}}}^k \dot{\hat{\kappa}}] e^{i\hat{\kappa}}, \tag{7.102}$$

and so

$$-\langle \psi | \frac{\partial}{\partial t} \hat{t}_n | \psi \rangle = -\sum_{k=0}^{\infty} \frac{i^{k+1}}{(k+1)!} \langle 0 | [\hat{q}_n, \underline{\underline{\hat{\kappa}}}^k \dot{\hat{\kappa}}] | 0 \rangle = -\sum_{k=0}^{\infty} \frac{i^{k+1}}{(k+1)!} S^{[k+2]}_{nml_1\cdots l_k} \dot{\kappa}_m \prod_{j=1}^{k} \kappa_{l_j}, \tag{7.103}$$

with the introduction of a set of generalized overlap matrices defined as

$$S^{[k+2]}_{nml_1\cdots l_k} = \langle 0 | \left[\hat{q}_n, \underline{\underline{\prod_{j=1}^{k} \hat{q}^\dagger_{l_j}}} \hat{q}^\dagger_m \right] | 0 \rangle. \tag{7.104}$$

The lowest-order term ($k = 0$) becomes

$$S^{[2]} = \begin{pmatrix} \langle 0 | [\hat{q}_n, \hat{q}^\dagger_m] | 0 \rangle & \langle 0 | [\hat{q}_n, \hat{q}_m] | 0 \rangle \\ \langle 0 | [\hat{q}^\dagger_n, \hat{q}^\dagger_m] | 0 \rangle & \langle 0 | [\hat{q}^\dagger_n, \hat{q}_m] | 0 \rangle \end{pmatrix}. \tag{7.105}$$

This matrix is diagonal with plus and minus ones in the upper left and lower right blocks, respectively.

At this stage, perturbation expansions in the frequency domain are introduced for the time-dependent wave function parameters

$$\kappa_n(t) = \sum_{\omega_1} \kappa^{(1)}_n(\omega_1) e^{-i\omega_1 t} + \sum_{\omega_1,\omega_2} \kappa^{(2)}_n(\omega_1, \omega_2) e^{-i(\omega_1+\omega_2)t} + \cdots, \tag{7.106}$$

$$\kappa^*_n(t) = \sum_{\omega_1} [\kappa^{(1)}_n(-\omega_1)]^* e^{-i\omega_1 t} + \sum_{\omega_1,\omega_2} [\kappa^{(2)}_n(-\omega_1, -\omega_2)]^* e^{-i(\omega_1+\omega_2)t} + \cdots, \tag{7.107}$$

where it has been used that all zeroth-order parameters vanish in our parameterization, that is, $\kappa^{(0)}_n = 0$, and we note that the sum of optical frequencies $(\omega_1 + \omega_2)$ in the exponent is due to the squared field dependence in the second-order terms. If we collect terms to first order in the perturbation in Eq. (7.91), we get

$$\sum_{\omega} -i(-i\omega) S^{[2]} \kappa^{(1)}(\omega) e^{-i\omega t} = \sum_{\omega} \frac{1}{\hbar}(-E^{[2]}\kappa^{(1)}(\omega) - iV^{\omega,[1]}) e^{-i\omega t}, \tag{7.108}$$

where the first-order response parameters are collected in a row vector as

$$\kappa^{(1)}(\omega) = \begin{pmatrix} \kappa^{(1)}_n(\omega) \\ [\kappa^{(1)}_n(-\omega)]^* \end{pmatrix}. \tag{7.109}$$

Equation (7.108) can be solved by matrix inversion separately for each individual frequency, resulting in the expression

$$\kappa^{(1)}(\omega) = -i(E^{[2]} - \hbar\omega S^{[2]})^{-1} V^{\omega,[1]}. \tag{7.110}$$

This equation provides the first-order response in the wave function parameters in the single-determinant approximation, and it can be compared to Eq. (5.137) in exact-state theory. The difference amounts merely to the fact that the Hessian is not diagonal in the SCF approximation, but is instead diagonal dominant with diagonal elements dominated by differences in orbital energies as expressed by Eq. (7.26).

Having determined the first-order response in the wave function, we can form the first-order correction to an expectation value and thereby identify the linear response function. We get

$$\begin{aligned}\langle\psi|\hat{\Omega}|\psi\rangle &= \langle 0|e^{i\kappa(t)}\hat{\Omega}e^{-i\kappa(t)}|0\rangle \\ &= \langle 0|\hat{\Omega}|0\rangle + i\sum_{\omega}\langle 0|\ \underline{\underline{\hat{q}_n^{\dagger}\hat{\Omega}}}\ |0\rangle\ \kappa_n^{(1)}(\omega)\ e^{-i\omega t} + \cdots,\end{aligned} \tag{7.111}$$

which gives an expression for the linear response equation as

$$\langle\langle\hat{\Omega};\hat{V}^{\omega}\rangle\rangle = i\langle 0|\ \underline{\underline{\hat{q}_n^{\dagger}\hat{\Omega}}}\ |0\rangle\ \kappa_n^{(1)}(\omega) = -[\Omega^{[1]}]^{\dagger}(E^{[2]} - \hbar\omega S^{[2]})^{-1} V^{\omega,[1]}. \tag{7.112}$$

We have here introduced a property gradient for the Hermitian observable that parallels that of the perturbation, namely

$$\Omega^{[1]} = \begin{pmatrix}\langle 0|[\hat{q}_n,\hat{\Omega}]|0\rangle \\ \langle 0|[\hat{q}_n^{\dagger},\hat{\Omega}]|0\rangle\end{pmatrix};\quad [\Omega^{[1]}]^{\dagger} = -\left(\langle 0|[\hat{q}_n^{\dagger},\hat{\Omega}]|0\rangle,\quad \langle 0|[\hat{q}_n,\hat{\Omega}]|0\rangle\right). \tag{7.113}$$

Equation (7.112) is the analogue of Eq. (5.168) in exact-state theory, and it is commonly referred to as the random phase approximation (RPA) equation.

The derivation of the RPA equation in the quasi-energy formalism amounts to the determination of the vector and matrix elements in Eq. (5.172), where derivatives of Q_{T} are given with respect to electron-transfer rather than state-transfer amplitudes. We have

$$Q_{\mathrm{T}}^{1;\beta} = \begin{pmatrix}\left.\dfrac{\partial^2 Q_{\mathrm{T}}}{\partial F_{\beta}^{\omega_2}\partial\kappa_n^*}\right|_{F^{\omega}=0} \\ \left.\dfrac{\partial^2 Q_{\mathrm{T}}}{\partial F_{\beta}^{\omega_2}\partial\kappa_n}\right|_{F^{\omega}=0}\end{pmatrix} = i\begin{pmatrix}\langle 0|[\hat{q}_n,\hat{V}_{\beta}^{\omega_2}]|0\rangle \\ \langle 0|[\hat{q}_n^{\dagger},\hat{V}_{\beta}^{\omega_2}]|0\rangle\end{pmatrix}\delta_{\omega_1+\omega_2} \tag{7.114}$$

and

$$Q_{\mathrm{T}}^{2;0} = \begin{pmatrix}\left.\dfrac{\partial^2 Q_{\mathrm{T}}}{\partial\kappa_n^*\partial\kappa_m}\right|_{F^{\omega}=0} & \left.\dfrac{\partial^2 Q_{\mathrm{T}}}{\partial\kappa_n^*\partial\kappa_m^*}\right|_{F^{\omega}=0} \\ \left.\dfrac{\partial^2 Q_{\mathrm{T}}}{\partial\kappa_n\partial\kappa_m}\right|_{F^{\omega}=0} & \left.\dfrac{\partial^2 Q_{\mathrm{T}}}{\partial\kappa_n\partial\kappa_m^*}\right|_{F^{\omega}=0}\end{pmatrix} = (E^{[2]} - \hbar\omega_2 S^{[2]})\delta_{\omega_1+\omega_2}, \tag{7.115}$$

where $E^{[2]}$ is the Hessian in the SCF approximation [Eqs. (7.22) and (7.96)] and $S^{[2]}$ is the overlap matrix defined in Eq. (7.105).

The formulation of response functions in the SCF approximation as external field derivatives of the quasi-energy is particularly valuable in the case of time-dependent Kohn–Sham density functional theory (DFT), since it provides a straightforward way to derive computationally tractable expressions. The relevant time-averaged quasi-energy of the reference system is viewed as a functional of the time-dependent electron density $\rho(\mathbf{r},t)$ according to

$$Q_{\mathrm{T}}[\rho] = T_{\mathrm{s}}[\rho] + V[\rho] + J[\rho] + Q_{\mathrm{xc}}[\rho] - S_{\mathrm{s}}[\rho], \tag{7.116}$$

which assumes the conventional division into separate contributions from kinetic energy $T_{\mathrm{s}}[\rho]$, external and internal fields $V[\rho]$, classical Coulomb interaction $J[\rho]$, and exchange-correlation energy $Q_{\mathrm{xc}}[\rho]$, and $S_{\mathrm{s}}[\rho]$ (from the $i\hbar\partial/\partial t$ operator). In evaluating derivatives of this quasi-energy, we note that the sum of terms in Eq. (7.116) excluding Q_{xc} corresponds to Hartree–Fock theory without exchange interactions. Based on an implementation of the RPA equation [Eq. (7.112)], the extension to time-dependent Kohn–Sham theory involves merely turning off the exchange interaction (or partial turn-off for hybrid functionals) in the construction of the $E^{[2]}$ matrix together with the addition of the contribution from Q_{xc}. For this term, it is customary to adopt the adiabatic approximation[3] and employ the time-independent exchange-correlation functional as a substitute for Q_{xc}, so that

$$Q_{\mathrm{xc}}[\rho] \rightarrow E_{\mathrm{xc}}[\rho] = \int e_{\mathrm{xc}}(\rho,\ldots)d^3\mathbf{r}. \tag{7.117}$$

A time averaging is implied here as well. The additional matrix contributions needed for the construction of the Hessian in time-dependent DFT thus becomes

$$\begin{pmatrix} \left.\dfrac{\partial^2 E_{\mathrm{xc}}}{\partial\kappa_n^*\partial\kappa_m}\right|_{F^\omega=0} & \left.\dfrac{\partial^2 E_{\mathrm{xc}}}{\partial\kappa_n^*\partial\kappa_m^*}\right|_{F^\omega=0} \\ \left.\dfrac{\partial^2 E_{\mathrm{xc}}}{\partial\kappa_n\partial\kappa_m}\right|_{F^\omega=0} & \left.\dfrac{\partial^2 E_{\mathrm{xc}}}{\partial\kappa_n\partial\kappa_m^*}\right|_{F^\omega=0} \end{pmatrix}.$$

Let us for simplicity consider the local density approximation (LDA) in which the energy density e_{xc} is assumed to be a function solely of the density ρ, which in turn can be identified as the expectation value of the density operator

$$\hat{\rho} = \sum_{pq} \Omega_{pq}\hat{a}_p^\dagger\hat{a}_q; \quad \Omega_{pq} = \phi_p^\dagger(\mathbf{r})\phi_q(\mathbf{r}). \tag{7.118}$$

The electron density then takes the form

$$\begin{aligned} \rho(\mathbf{r}) &= \sum_{pq} \Omega_{pq}\langle 0|e^{i\hat{\kappa}}\,\hat{a}_p^\dagger\hat{a}_q\,e^{-i\hat{\kappa}}|0\rangle \\ &= \sum_{pq} \Omega_{pq}\left[\langle 0|\hat{a}_p^\dagger\hat{a}_q|0\rangle + i\langle 0|[\hat{\kappa},\hat{a}_p^\dagger\hat{a}_q]|0\rangle - \frac{1}{2}\langle 0|[\hat{\kappa},[\hat{\kappa},\hat{a}_p^\dagger\hat{a}_q]]|0\rangle + \cdots\right], \end{aligned} \tag{7.119}$$

which reveals the explicit dependence of the density on the κ-parameters. If we consider the A-block [see Eq. (7.22)] as an example, we obtain by chain rule differentiation of the integrand an explicit expression for the exchange-correlation contributions that equals

$$\begin{aligned} \left.\frac{\partial^2 E_{\mathrm{xc}}}{\partial\kappa_{si}^*\partial\kappa_{tj}}\right|_{F^\omega=0} &= \int \left.\left(\frac{\partial^2 e_{\mathrm{xc}}}{\partial\rho^2}\frac{\partial\rho}{\partial\kappa_{si}^*}\frac{\partial\rho}{\partial\kappa_{jt}} + \frac{\partial e_{\mathrm{xc}}}{\partial\rho}\frac{\partial^2\rho}{\partial\kappa_{si}^*\partial\kappa_{jt}}\right)\right|_{F^\omega=0} \mathrm{d3}\mathbf{r} \\ &= \int \left(-\frac{\partial^2 e_{\mathrm{xc}}}{\partial\rho^2}\Omega_{si}\Omega_{jt} + \frac{\partial e_{\mathrm{xc}}}{\partial\rho}[\Omega_{st}\delta_{ij} + \Omega_{ij}\delta_{st} - \rho^{(0)}(\mathbf{r})\delta_{st}\delta_{ij}]\right)\mathrm{d3}\mathbf{r}, \end{aligned} \tag{7.120}$$

where $\rho^{(0)}$ is the electron density of the unperturbed reference state $|0\rangle$. As usual, the time averaging implies that $\omega_1 + \omega_2 = 0$ in the above equation.

---

3 This is not to be confused with the adiabatic approximation introduced in Chapter 6. The adiabatic approximation in connection with DFT response theory implies that the exchange–correlation functional is local in time and thus that any time dependence in the functional is ignored.

7.3.2 Configuration Interaction

As mentioned when discussing the parameterization of CI states, the formulation of response theory in the basis of CI eigenvectors parallels the case of exact-state theory that was treated in Chapter 5. In practice, however, we will adopt the CI space parameterization given in Eq. (7.38), and we will here be concerned with the explicit form of the linear response function in this case. For simplicity we will ignore orbital relaxation, but a detailed discussion on how to handle this using a variational Lagrangian technique is found in Appendix H. The time dependence of the $R_n(t)$ CI parameters is found from the Ehrenfest theorem Eq. (7.91) applied to the set of operators $t_n = \exp(-i\hat{R})\hat{Q}_n \exp(i\hat{R})$ (and the corresponding adjoint operators $t_n^\dagger$)—with the state-transfer operator here given by

$$\hat{Q}_n^\dagger = |\Phi_n\rangle\langle 0|. \tag{7.121}$$

The algebraic work needed is analogous to that done for the SCF case, and the explicit expression for the linear response function will be identical to that found in Eq. (7.112) but with a difference in matrix and property definitions that amounts to mere interchanges of $\hat{q}$s by $\hat{Q}$s and κs by Rs. The $E^{[2]}$-matrix in the CI approximation will be equal to the CI Hessian with elements given in Eqs. (7.39) and (7.40), and the Hermitian $S^{[2]}$ matrix becomes block diagonal with elements given by

$$S^{[2]} = \begin{pmatrix} \langle 0|[\hat{Q}_n, \hat{Q}_m^\dagger]|0\rangle & \langle 0|[\hat{Q}_n, \hat{Q}_m]|0\rangle \\ \langle 0|[\hat{Q}_n^\dagger, \hat{Q}_m^\dagger]|0\rangle & \langle 0|[\hat{Q}_n^\dagger, \hat{Q}_m]|0\rangle \end{pmatrix} = \begin{pmatrix} \delta_{nm} - c_n c_m^* & 0 \\ 0 & c_n^* c_m - \delta_{nm} \end{pmatrix}, \tag{7.122}$$

where c_n are the CI expansion coefficients of the reference state $|0\rangle$.

7.3.3 Multiconfiguration Self-Consistent Field

In the MCSCF approximation, the time dependence of both orbital and configuration parameters must be determined, and the necessary equations of motion can be obtained by applying the Ehrenfest theorem to the set of $\hat{t}$ operators collected in Eq. (7.82). At the end of the calculation, the form of the linear response function in the MCSCF approximation turns out to be of the same form as in the RPA, Eq. (7.112), but where the explicit expressions for the matrix elements are modified. The $E^{[2]}$-matrix will be equal to the MCSCF Hessian given in Eqs. (7.45) and (7.46) and the Hermitian $S^{[2]}$-matrix will be

$$S^{[2]} = \begin{bmatrix} \Sigma & \Delta \\ -\Delta^* & -\Sigma^* \end{bmatrix}, \tag{7.123}$$

with

$$\Sigma = \begin{pmatrix} \langle 0|[\hat{q}_n, \hat{q}_m^\dagger]|0\rangle & \langle 0|[\hat{q}_n, \hat{Q}_m^\dagger]|0\rangle \\ \langle 0|[\hat{Q}_n, \hat{q}_m^\dagger]|0\rangle & \langle 0|[\hat{Q}_n, \hat{Q}_m^\dagger]|0\rangle \end{pmatrix}; \quad \Delta = \begin{pmatrix} \langle 0|[\hat{q}_n, \hat{q}_m]|0\rangle & \langle 0|[\hat{q}_n, \hat{Q}_m]|0\rangle \\ \langle 0|[\hat{Q}_n, \hat{q}_m]|0\rangle & \langle 0|[\hat{Q}_n, \hat{Q}_m]|0\rangle \end{pmatrix}, \tag{7.124}$$

where the submatrices Σ and Δ are Hermitian and antisymmetric, respectively.

7.3.4 Matrix Structure in the SCF, CI, and MCSCF Approximations

Owing to the fact that the states involved in the wave function parameterizations are not eigenstates of the unperturbed Hamiltonian, the resulting expressions for the response functions become matrix equations rather than sum-over-states expressions as in exact-state theory. The vector and matrix structures are common for the variational wave function approaches considered (SCF, CI, and MCSCF), and it is worthwhile to explore these structures in order to further compare approximate- and exact-state theories. To start with, we denote by X_e the eigenvectors of the generalized eigenvalue equation

$$E^{[2]}X_e = \lambda_e S^{[2]}X_e, \quad e = -n, \dots, -1, 1, \dots, n, \tag{7.125}$$

where the matrix dimension is $2n$. We find the set of eigenvalues and eigenvectors by diagonalizing the non-Hermitian matrix $(S^{[2]})^{-1}E^{[2]}$ according to

$$X^{-1}[(S^{[2]})^{-1}E^{[2]}]X = \begin{bmatrix} \lambda & 0 \\ 0 & -\lambda \end{bmatrix}, \tag{7.126}$$

where λ is a diagonal matrix of dimension n collecting the eigenvalues with positive index, and the columns of X store the eigenvectors (X is assumed to be nonsingular). This pairing of eigenvalues has its correspondence in the eigenvectors through

$$X_e = \begin{pmatrix} Z_e \\ Y_e^* \end{pmatrix} \text{ with eigenvalue } \lambda_e; \quad X_{-e} = \begin{pmatrix} Y_e \\ Z_e^* \end{pmatrix} \text{ with eigenvalue } -\lambda_e, \tag{7.127}$$

as can be seen by considering the matrix structures in Eqs. (7.22) and (7.123). The matrix X will therefore have the structure

$$X = \begin{bmatrix} Z & Y \\ Y^* & Z^* \end{bmatrix}. \tag{7.128}$$

The eigenvalues will be real as long as the matrix is sufficiently diagonal dominant, and we will, for the time being, assume that this is the case and return to a discussion of this point later on. Furthermore, the eigenvectors are unique if all eigenvalues are separate, whereas with sets of degenerate eigenvalues, we are free to choose a basis of eigenvectors within the respective subspaces. The important result for our discussion is that, with an appropriate scaling of the eigenvectors X_e, the nonunitary matrix X achieves a simultaneous diagonalization of the two noncommuting matrices $E^{[2]}$ and $S^{[2]}$ as

$$X^\dagger E^{[2]}X = \begin{bmatrix} \lambda & 0 \\ 0 & \lambda \end{bmatrix}; \quad X^\dagger S^{[2]}X = \begin{bmatrix} I & 0 \\ 0 & -I \end{bmatrix}. \tag{7.129}$$

For each pair of eigenvectors, the second equation involving the metric of the generalized eigenvalue equation defines which of the two should be indexed with a positive index. We also emphasize that $X^\dagger \neq X^{-1}$ and for this reason the X_es are eigenvectors neither of $E^{[2]}$ nor of $S^{[2]}$, just as the λ_es are not the eigenvalues of $E^{[2]}$. However, the number of negative eigenvalues of a Hermitian matrix is unchanged by a nonsingular transformation such as that made in Eq. (7.129), which implies that there are equally many negative elements in $X^\dagger E^{[2]}X$ as in the electronic Hessian $E^{[2]}$ itself.

In order to prove Eq. (7.129), we return to Eq. (7.126), which leads us to conclude that

$$X^\dagger E^{[2]}X = X^\dagger S^{[2]}X\Lambda, \tag{7.130}$$

where the diagonal matrix Λ collects all $2n$ eigenvalues. Since $E^{[2]}$ and $S^{[2]}$ are Hermitian and Λ is real, we must have

$$X_f^\dagger S^{[2]}X_e\lambda_e = X_f^\dagger E^{[2]}X_e = (X_e^\dagger E^{[2]}X_f)^* = (X_e^\dagger S^{[2]}X_f)^*\lambda_f = X_f^\dagger S^{[2]}X_e\lambda_f,$$

which for $\lambda_e \neq \lambda_f$ implies that $X_e^\dagger S^{[2]}X_f$ and thus $X_e^\dagger E^{[2]}X_f$ are zero. If, on the other hand, degenerate eigenvalues exist, then this proof does not hold. However, since $X^\dagger S^{[2]}X$ as well as $X^\dagger E^{[2]}X$ are Hermitian, we can diagonalize any given degenerate subspace by means of a unitary transformation common to the two matrices—which is obvious since, in the subspace, the two matrices differ only by the multiplication of the eigenvalue.

Let us return to the issue of the λ_es being real. In the SCF approximation, we have $(S^{[2]})^{-1} = S^{[2]}$ and consequently

$$(S^{[2]})^{-1}E^{[2]} = \begin{bmatrix} A & B \\ -B^* & -A^* \end{bmatrix}, \tag{7.131}$$

so the requirement that this matrix is diagonal dominant is equivalent to requiring that $E^{[2]}$ itself is diagonal dominant. In the SCF approximation, this issue was discussed in connection to Eq. (7.23), but it may also be illuminating to consider a 2×2 example with

$$E^{[2]} = \begin{bmatrix} a & b \\ b^* & a^* \end{bmatrix}; \quad S^{[2]} = \begin{bmatrix} 1 & 0 \\ 0 & -1 \end{bmatrix}.$$

Since $E^{[2]}$ is Hermitian, the diagonal element a is real, and we get

$$(S^{[2]})^{-1}E^{[2]} = \begin{bmatrix} a & b \\ -b^* & a \end{bmatrix},$$

with eigenvalues

$$\lambda_{\pm e} = \pm\sqrt{a^2 - |b|^2}.$$

We see that if $E^{[2]}$ is truly diagonal dominant, the eigenvalues are real and $E^{[2]}$ and $S^{[2]}$ can be diagonalized by a common transformation matrix X. But there exist pathological cases when this is not the case and which lead to the appearance of complex eigenvalues when solving the generalized eigenvalue equation.

Since $(\Omega\Lambda)^{-1} = \Lambda^{-1}\Omega^{-1}$, we can form the inverse of the RPA matrix in Eq. (7.112) as

$$(E^{[2]} - \omega S^{[2]})^{-1} = X[X^\dagger(E^{[2]} - \omega S^{[2]})X]^{-1}X^\dagger, \tag{7.132}$$

which by use of Eq. (7.129) yields an expression for the linear response function that reads

$$\langle\langle \hat{\Omega}; \hat{V}^\omega \rangle\rangle = -[\Omega^{[1]}]^\dagger \sum_{e=1}^{n} \left(\frac{X_e X_e^\dagger}{\lambda_e - \hbar\omega} + \frac{X_e X_e^\dagger}{\lambda_e + \hbar\omega} \right) V^{\omega,[1]}. \tag{7.133}$$

In this form, the linear response function for variational wave functions (SCF, CI, and MCSCF) displays a striking resemblance to that in exact-state theory [Eq. (5.168)], and the association of the λ_es (positive index) with approximations of excitation energies becomes self-evident.

For our simple numerical example in terms of helium in the Hartree–Fock approximation as illustrated in Figure 7.4, we obtain solution vectors to the generalized eigenvalue equation, Eq. (7.126), that are given by

$$Z = \begin{bmatrix} 0.7074 & -0.0063 \\ 0.0074 & 0.7081 \end{bmatrix}; \quad Y = \begin{bmatrix} 0.0139 & 0.0168 \\ 0.0172 & 0.0336 \end{bmatrix}.$$

We note that the eigenvectors are normalized in accordance with Eq. (7.129), with

$$\lambda = \begin{bmatrix} 0.8221 & 0 \\ 0 & 2.1944 \end{bmatrix}, \tag{7.134}$$

and we recall that non-normalized, spin-conserving, excitation and de-excitation operators, Eq. (7.18), have been employed in the construction of the response matrices. With this choice of operators, the diagonal elements of $S^{[2]}$ have a magnitude of 2 (instead of 1), and as a consequence, the dominant elements of the X-matrix become close to $1/\sqrt{2}$.

As in the great majority of practical calculations, our calculation on helium is based on a reference state that approximates the exact electronic ground state. In the general case, the adopted approximate reference state may be chosen to represent the Nth excited state, and if in this situation one finds more (or less) than N negative excitation energies, the calculation is said to be *unstable*. As already mentioned, the number of negative excitation energies equals half the number of negative eigenvalues of the electronic Hessian, so the focus on the stability of the calculation may be shifted toward the $E^{[2]}$ matrix rather than the RPA matrix. In fact, in the nonrelativistic realm, where the reference state is often real, instabilities can be detected from a study of the matrix

$$\begin{bmatrix} A+B & 0 \\ 0 & A-B \end{bmatrix} = \frac{1}{2}\begin{bmatrix} I & I \\ -I & I \end{bmatrix}\begin{bmatrix} A & B \\ B & A \end{bmatrix}\begin{bmatrix} I & -I \\ I & I \end{bmatrix}. \tag{7.135}$$

The reason for this is of course that the eigenvalues of $E^{[2]}$ are invariant with respect to unitary matrix transformations. The stability criterion can thus be formulated as a requirement that there are a total of $2N$ negative eigenvalues for the submatrices $A + B$ and $A - B$, and for the ground state this means that the calculation is stable only if both matrices are positive definite.

Particular caution is needed when carrying out time-dependent Hartree–Fock calculations involving excited states of a different spin symmetry than the reference state. We illustrate this aspect of the RPA by the calculation of the lowest singlet and triplet excitation energies in beryllium with results presented in Table 7.3. The orbital energy diagram of beryllium is depicted in Figure 7.8 with Hartree–Fock orbital energies reported as the leftmost values, and we are concerned with the singlet and triplet excitations from the 2s to the p-type orbitals. The energy gap between occupied and unoccupied orbitals is relatively large, so from this perspective, there is no reason to be suspicious, but a warning flag is raised when inspecting the Hessian.

The critical part of the Hessian is the section referring to triplet excitations Eq. (7.19) from the valence 2s orbital to the virtual p-type orbitals—using a valence double-ζ basis set, there

Table 7.3 Excitation energies (eV) for the lowest triplet and singlet states in beryllium.

	TDHF		TDDFT		
State	6-31G	t-aug-cc-pVTZ	6-31G	t-aug-cc-pVTZ	Expt
$2p\ ^3P^o$	—	—	2.05	2.08	2.32
$2p\ ^1P^o$	5.16	4.84	5.30	4.89	5.28

Theoretical results refer to the time-dependent Hartree–Fock and DFT/B3LYP approximations using a small (6-31G) and a large (t-aug-cc-pVTZ) basis set. The Hartree–Fock model shows a triplet instability, resulting in imaginary eigenvalues that are not presented.

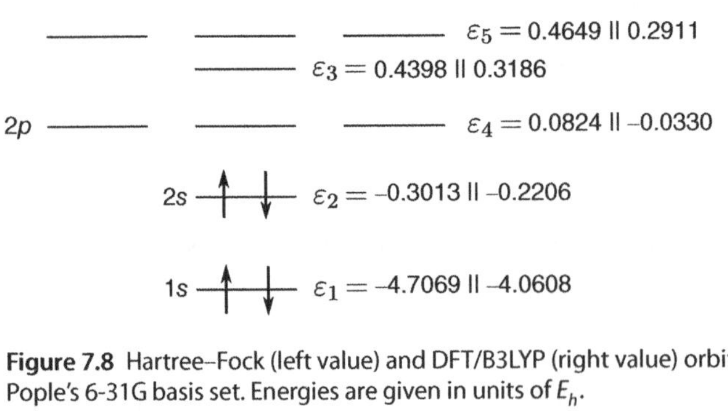

Figure 7.8 Hartree–Fock (left value) and DFT/B3LYP (right value) orbital energy diagram of beryllium using Pople's 6-31G basis set. Energies are given in units of E_h.

Table 7.4 Parts of the electronic Hessian for beryllium referring to valence $2s \rightarrow np$ spin-adapted excitations.

Hartree–Fock		DFT/B3LYP	
A	*B*	*A*	*B*
Singlet spin symmetry:			
$\begin{bmatrix} 0.4087 & 0.0440 \\ 0.0440 & 0.9606 \end{bmatrix}$	$\begin{bmatrix} -0.1261 & 0.0570 \\ 0.0570 & -0.0634 \end{bmatrix}$	$\begin{bmatrix} 0.4193 & -0.0249 \\ -0.0249 & 0.9099 \end{bmatrix}$	$\begin{bmatrix} -0.1481 & 0.0488 \\ 0.0488 & -0.0075 \end{bmatrix}$
Triplet spin symmetry:			
$\begin{bmatrix} 0.1565 & 0.1579 \\ 0.1579 & 0.8339 \end{bmatrix}$	$\begin{bmatrix} 0.1261 & -0.0570 \\ -0.0570 & 0.0634 \end{bmatrix}$	$\begin{bmatrix} 0.1842 & 0.0608 \\ 0.0608 & 0.8471 \end{bmatrix}$	$\begin{bmatrix} 0.0869 & -0.0369 \\ -0.0369 & 0.0553 \end{bmatrix}$

Results refer to the Hartree–Fock and DFT/B3LYP approximations employing a small (6-31G) basis set. Atomic units are used.

are two sets of p-orbitals for beryllium, as seen in the orbital diagram in Figure 7.8. The A and B blocks of this triplet part of the Hessian are presented in Table 7.4 together with a comparison of the corresponding singlet part. In the Hartree–Fock approximation, it is clear that the large off-diagonal elements in the A matrix will inflict one negative eigenvalue for the matrix $A - B$. The explicit expressions for the elements of the A and B matrices are given in Eq. (7.23) with a mere adaptation for the use of the spin-adapted excitation operators in Eqs. (7.18) and (7.19). Expressed in terms of spatial two-electron integrals, the resulting expressions become

$$A^{s}_{is,it} = -2(st|ii) + 4(si|it) = 0.0440\ E_h,$$
$$A^{t}_{is,it} = -2(st|ii) = 0.1579\ E_h,$$

where i and s, t refer to the inactive 2s and the two secondary p-type orbitals, respectively. In evaluating both expressions, we note that two of the four Coulomb integrals are zero due to spin integration, whereas in the evaluation of $A^{t}_{is,it}$, there is also a perfect cancelation of all four exchange integrals. The strong Coulomb interaction between the inactive and the two secondary orbitals thus leads to a negative eigenvalue in the electronic Hessian. In turn, this negative eigenvalue signals a triplet instability in the reference state, which means that the Hartree–Fock energy of the singlet ground state would be lowered by mixing in a small part of triplet character into the wave function. In the DFT approximation, where the exchange interaction is fully (or partly as in hybrid functionals) replaced by the exchange–correlation functional, the ground state of beryllium is stable, and the excitation energy of the lowest excited triplet state becomes well defined and is reported in Table 7.3 for the case of the B3LYP functional.

7.3.5 Coupled Cluster

For a nonvariational electronic structure method, such as coupled cluster theory, the first-order property (or, equivalently, the zeroth-order response function) cannot be identified with a simple expectation value. Instead, we turn to a definition of the zeroth-order response function as the first-order external field derivative of the variational time-averaged Lagrangian; see Eqs. (7.89) and (7.90). But before turning to this recipe, let us see what the variational condition itself tells us. The time-averaged coupled cluster Lagrangian is viewed as a function of amplitudes, multipliers, and external field amplitudes, according to

$$L_{\mathrm{T}}(t_n^{(0)}, t_n^{(1)}(\omega_1), t_n^{(2)}(\omega_1, \omega_2), \ldots, \lambda_n^{(0)}, \lambda_n^{(1)}(\omega_1), \lambda_n^{(2)}(\omega_1, \omega_2), \ldots, F_\alpha^{\omega_1}, F_\beta^{\omega_2}, \ldots).$$

From the variational condition applied to $\lambda_n^{(0)}$, we have

$$0 = \left.\frac{\partial L_{\mathrm{T}}}{\partial \lambda_n^{(0)}}\right|_{F^{\omega}=0} = \langle n|e^{-\hat{T}^{(0)}}\hat{H}_0\, e^{\hat{T}^{(0)}}|0\rangle, \tag{7.136}$$

which is equal to Eq. (7.59), that is, a set of equations (for the various excited determinants $|n\rangle = |0_{ij\ldots}^{st\ldots}\rangle$) from which the amplitudes for the coupled cluster reference state are determined. From the variational condition applied to $t_n^{(0)}$, on the other hand, we get

$$0 = \left.\frac{\partial L_{\mathrm{T}}}{\partial t_n^{(0)}}\right|_{F^{\omega}=0} = \langle 0|\hat{H}_0\hat{\tau}_n^{\dagger}|\mathrm{CC}\rangle + \sum_k \lambda_k^{(0)}\langle k|e^{-\hat{T}^{(0)}}[\hat{H}_0, \hat{\tau}_n^{\dagger}]\,|\mathrm{CC}\rangle, \tag{7.137}$$

where $\hat{\tau}_n^{\dagger}$ denotes the electron excitation operator associated with amplitude t_n, that is,

$$\hat{T} = \sum_n t_n\hat{\tau}_n^{\dagger}; \quad t_n(t) = t_n^{(0)} + \sum_{\omega_1} t_n^{(1)}e^{-i\omega_1 t} + \cdots. \tag{7.138}$$

In deriving Eq. (7.137), known as the Λ-equations, we have used the fact that the excitation operator $\hat{a}_s^{\dagger}\hat{a}_i$ (and therefore also $\hat{\tau}_n^{\dagger}$) commutes with the operator $\hat{T}$, which is a result of the more general commutator relation

$$[\hat{a}_a^{\dagger}\hat{a}_b, \hat{a}_s^{\dagger}\hat{a}_i] = \delta_{bs}\hat{a}_a^{\dagger}\hat{a}_i - \delta_{ai}\hat{a}_s^{\dagger}\hat{a}_b. \tag{7.139}$$

From this result, we can conclude that

$$\frac{\partial}{\partial t_n^{(1)}}e^{\hat{T}} = e^{-i\omega_1 t}\,\hat{\tau}_n^{\dagger}e^{\hat{T}} = e^{-i\omega_1 t}\, e^{\hat{T}}\hat{\tau}_n^{\dagger}. \tag{7.140}$$

We note that Eq. (7.137) allows us to determine the zeroth-order multipliers, which will be needed for the evaluation of first-order properties in the coupled cluster approximation. In matrix form, we may write this equation as

$$\lambda^{(0)} = -\eta^{[1]}[A^{[2]}]^{-1}; \quad A_{kn}^{[2]} = \langle k|e^{-\hat{T}^{(0)}}[\hat{H}_0, \hat{\tau}_n^{\dagger}]|\mathrm{CC}\rangle, \quad \eta_n^{[1]} = \langle 0|\hat{H}_0\hat{\tau}_n^{\dagger}|\mathrm{CC}\rangle. \tag{7.141}$$

The zeroth-order response function is then

$$\langle \hat{V}_\alpha^{\omega_1}\rangle^{(0)} = \left.\frac{dL_{\mathrm{T}}}{dF_\alpha^{\omega_1}}\right|_{F^{\omega}=0} = \left.\frac{\partial L_{\mathrm{T}}}{\partial F_\alpha^{\omega_1}}\right|_{F^{\omega}=0} + \sum_n \left.\left[\frac{\partial L_{\mathrm{T}}}{\partial t_n^{(1)}}\frac{\partial t_n^{(1)}}{\partial F_\alpha^{\omega_1}} + \frac{\partial L_{\mathrm{T}}}{\partial \lambda_n^{(1)}}\frac{\partial \lambda_n^{(1)}}{\partial F_\alpha^{\omega_1}}\right]\right|_{F^{\omega}=0}, \tag{7.142}$$

where we have used the fact that only the first-order responses in the parameters contribute, because the expression is evaluated at zero field strength. Since the time-averaged Lagrangian is variational, derivatives of it with respect to coupled cluster amplitudes t_n and multipliers λ_n vanish, and we are left with

$$\langle \hat{V}_\alpha^{\omega_1}\rangle^{(0)} = \left.\frac{\partial L_{\mathrm{T}}}{\partial F_\alpha^{\omega_1}}\right|_{F^{\omega}=0} = \left[\langle 0|\,\hat{V}_\alpha^{\omega_1}\, e^{\hat{T}^{(0)}}|0\rangle + \sum_n \lambda_n^{(0)}\,\langle n|e^{-\hat{T}^{(0)}}\hat{V}_\alpha^{\omega_1}\, e^{\hat{T}^{(0)}}|0\rangle\right]\delta_{\omega_1}, \tag{7.143}$$

into which we insert the zeroth-order amplitudes determined in Eq. (7.141). We note that a nonvanishing result is only obtained for $\omega_1 = 0$, and we remind the reader that, in the determination of an expectation value of an observable in projected coupled cluster approaches, one extracts the real part from the result in Eq. (7.143).

An expression for the linear response function in the coupled cluster approximation is obtained by further differentiation of Eq. (7.142) with respect to the external field. We enforce the expression to be symmetric with respect to the external field amplitudes by using a permutation operator ($\mathcal{P}_{\alpha,\beta}$ permutes field amplitudes $F_{\alpha}^{\omega_1}$ and $F_{\beta}^{\omega_2}$)

$$\begin{aligned}\langle\langle \hat{V}_{\alpha}^{\omega_1}; \hat{V}_{\beta}^{\omega_2}\rangle\rangle &= \left.\frac{d^2 L_{\mathrm{T}}}{dF_{\alpha}^{\omega_1} dF_{\beta}^{\omega_2}}\right|_{F^{\omega}=0} \\ &= \frac{1}{2}\sum \mathcal{P}_{\alpha,\beta}\left[2\left.\frac{\partial^2 L_{\mathrm{T}}}{\partial F_{\alpha}^{\omega_1}\partial t_n^{(1)}}\frac{\partial t_n^{(1)}}{\partial F_{\beta}^{\omega_2}}\right|_{F^{\omega}=0} + 2\left.\frac{\partial^2 L_{\mathrm{T}}}{\partial F_{\alpha}^{\omega_1}\partial \lambda_m^{(1)}}\frac{\partial \lambda_m^{(1)}}{\partial F_{\beta}^{\omega_2}}\right|_{F^{\omega}=0}\right. \\ &\left. + \left.\frac{\partial t_m^{(1)}}{\partial F_{\alpha}^{\omega_1}}\frac{\partial^2 L_{\mathrm{T}}}{\partial t_m^{(1)}\partial t_n^{(1)}}\frac{\partial t_n^{(1)}}{\partial F_{\beta}^{\omega_2}}\right|_{F^{\omega}=0} + 2\left.\frac{\partial \lambda_m^{(1)}}{\partial F_{\beta}^{\omega_2}}\frac{\partial^2 L_{\mathrm{T}}}{\partial \lambda_m^{(1)}\partial t_n^{(1)}}\frac{\partial t_n^{(1)}}{\partial F_{\alpha}^{\omega_1}}\right|_{F^{\omega}=0}\right],\end{aligned} \tag{7.144}$$

where implicit summation over repeated indices of parameters and multipliers is assumed. We have also used the fact that there is no explicit second-order field dependence in the Lagrangian nor in the variational conditions of the Lagrangian with respect to parameters and multipliers. This expression can be further simplified noting the cancelation of terms two and four. In order to see this, we make repeated use of the variational condition for the Lagrangian with respect to multipliers, and perform a differentiation with respect to the external field

$$0 = \frac{d}{dF_{\alpha}^{\omega_1}}\left[\frac{\partial L_{\mathrm{T}}}{\partial \lambda_m^{(1)}(\omega_2)}\right] = \frac{\partial^2 L_{\mathrm{T}}}{\partial F_{\alpha}^{\omega_1}\partial \lambda_m^{(1)}} + \frac{\partial^2 L_{\mathrm{T}}}{\partial \lambda_m^{(1)}\partial t_n^{(1)}}\frac{\partial t_n^{(1)}}{\partial F_{\alpha}^{\omega_1}}, \tag{7.145}$$

which, when substituted into Eq. (7.144), shows the cancelation of terms two and four. We thus arrive at

$$\begin{aligned}\langle\langle \hat{V}_{\alpha}^{\omega_1}; \hat{V}_{\beta}^{\omega_2}\rangle\rangle &= \left.\frac{d^2 L_{\mathrm{T}}}{dF_{\alpha}^{\omega_1} dF_{\beta}^{\omega_2}}\right|_{F^{\omega}=0} \\ &= \sum \mathcal{P}_{\alpha,\beta}\left[\left.\frac{\partial^2 L_{\mathrm{T}}}{\partial F_{\alpha}^{\omega_1}\partial t_n^{(1)}}\frac{\partial t_n^{(1)}}{\partial F_{\beta}^{\omega_2}}\right|_{F^{\omega}=0} + \frac{1}{2}\left.\frac{\partial t_m^{(1)}}{\partial F_{\alpha}^{\omega_1}}\frac{\partial^2 L_{\mathrm{T}}}{\partial t_m^{(1)}\partial t_n^{(1)}}\frac{\partial t_n^{(1)}}{\partial F_{\beta}^{\omega_2}}\right|_{F^{\omega}=0}\right].\end{aligned} \tag{7.146}$$

The first-order responses in the parameters are determined by evaluating Eq. (7.145) in the limit of zero field strengths

$$\left.\frac{\partial t^{(1)}(\omega_2)}{\partial F_{\beta}^{\omega_2}}\right|_{F^{\omega}=0} = -\left[\left.\frac{\partial^2 L_{\mathrm{T}}}{\partial \lambda^{(1)}\partial t^{(1)}}\right|_{F^{\omega}=0}\right]^{-1}\left.\frac{\partial^2 L_{\mathrm{T}}}{\partial F_{\beta}^{\omega_2}\partial \lambda^{(1)}}\right|_{F^{\omega}=0}, \tag{7.147}$$

where the suppression of indices on parameters and multipliers indicates that we have collected them into vector and matrix forms.

The explicit expressions for the vector and matrix elements that appear in the coupled cluster linear response equation become

$$\left.\frac{\partial^2 L_{\mathrm{T}}}{\partial \lambda_m^{(1)}(\omega_1)\partial t_n^{(1)}(\omega_2)}\right|_{F^{\omega}=0} = [A_{mn}^{[2]} - \hbar\omega_2\delta_{nm}]\delta_{\omega_1+\omega_2}, \tag{7.148}$$

$$\left.\frac{\partial^2 L_{\mathrm{T}}}{\partial F_{\beta}^{\omega_2}\partial \lambda_m^{(1)}(\omega_1)}\right|_{F^{\omega}=0} = \langle m|e^{-\hat{T}^{(0)}}\hat{V}_{\beta}^{\omega_2}|\mathrm{CC}\rangle\,\delta_{\omega_1+\omega_2}, \tag{7.149}$$

$$\left.\frac{\partial^2 L_T}{\partial F_\alpha^{\omega_1}\partial t_n^{(1)}(\omega_2)}\right|_{F^\omega=0} = \left[\langle 0|\hat{V}_\alpha^{\omega_1}\hat{\tau}_n^\dagger|\mathrm{CC}\rangle + \sum_k \lambda_k^{(0)}\langle k|e^{-\hat{T}^{(0)}}[\hat{V}_\alpha^{\omega_1},\hat{\tau}_n^\dagger]|\mathrm{CC}\rangle\right]\delta_{\omega_1+\omega_2}, \tag{7.150}$$

$$\left.\frac{\partial^2 L_T}{\partial t_m^{(1)}(\omega_1)\partial t_n^{(1)}(\omega_2)}\right|_{F^\omega=0} = \left[\langle 0|\hat{H}_0\hat{\tau}_m^\dagger\hat{\tau}_n^\dagger|\mathrm{CC}\rangle + \sum_k \lambda_k^{(0)}\langle k|e^{-\hat{T}^{(0)}}[[\hat{H}_0,\hat{\tau}_m^\dagger],\hat{\tau}_n^\dagger]|\mathrm{CC}\rangle\right]\delta_{\omega_1+\omega_2}. \tag{7.151}$$

The derivation of these expressions is rather straightforward using in particular Eq. (7.140), but it may be illustrative to focus on one detail in the derivation of Eq. (7.148), namely the handling of the time-differentiation operator in the Lagrangian, Eq. (7.89). For this specific term, we get

$$\left.\frac{\partial^2}{\partial\lambda_m^{(1)}(\omega_1)\partial t_n^{(1)}(\omega_2)}\left\{\sum_k \lambda_k(t)\langle k|e^{-\hat{T}}\left(-i\hbar\frac{\partial}{\partial t}\right)e^{\hat{T}}|0\rangle\right\}_T\right|_{F^\omega=0}$$

$$= \left.\frac{\partial}{\partial t_n^{(1)}(\omega_2)}\left\{e^{-i\omega_1 t}\langle m|e^{-\hat{T}}\left(-i\hbar\frac{\partial}{\partial t}\right)e^{\hat{T}}|0\rangle\right\}_T\right|_{F^\omega=0} = /\text{Lemma E.7 of Appendix E}/$$

$$= -i\hbar\left.\frac{\partial}{\partial t_n^{(1)}(\omega_2)}\left\{e^{-i\omega_1 t}\langle m|\sum_{k=0}^{\infty}\frac{(-1)^k}{(k+1)!}\underline{\hat{T}^k\dot{\hat{T}}}|0\rangle\right\}_T\right|_{F^\omega=0} = /[\hat{T},\dot{\hat{T}}]=0/$$

$$= -i\hbar\left.\frac{\partial}{\partial t_n^{(1)}(\omega_2)}\{e^{-i\omega_1 t}\dot{t}_m(t)\}_T\right|_{F^\omega=0} = -\hbar\omega_2\,\delta_{nm}\delta_{\omega_1+\omega_2}, \tag{7.152}$$

which is equal to the second term in Eq. (7.148).

One thing to note is that the frequency dependence of the linear response function in the coupled cluster approximation is contained in the linear response equation, Eq. (7.147), and it is clear that poles in the response functions correspond to singularities in the matrix $(A^{[2]} - \hbar\omega_2 I)$.

From the study of the diagonalization of this nonsymmetric coupled cluster Jacobian $A^{[2]}$ defined in Eq. (7.141), we can identify excitation energies and ground-to-excited-state transition matrix elements. We will omit the details of this analysis but note that because the Jacobian is nonsymmetric, it becomes an issue whether to solve for eigenvectors from the left or from the right.

7.4 Residue Analysis

In Section 5.9 we performed a detailed residue analysis of the linear and nonlinear response functions in exact-state theory, and in Table 5.3, we listed a selection of properties associated with this analysis. There is of course a one-to-one correspondence between these property expressions in exact-state theory and the corresponding ones in approximate-state theory—poles in sum-over-states expressions in the former case become replaced by matrix singularities in the latter case. In analogy with the previous section, we will limit our discussion of approximate-state theory to only include a residue analysis of the linear response function.

For the variational wave function theories considered (SCF, CI, and MCSCF), the key equation for a residue analysis of the linear response function is Eq. (7.133). Based on a comparison of this equation with that in exact-state theory, Eq. (5.168), we can make the following identifications

$$E_n - E_0 = \lambda_n, \tag{7.153}$$

$$\langle 0|\hat{\Omega}|n\rangle = [\Omega^{[1]}]^\dagger X_n, \tag{7.154}$$

$$\langle n|\hat{V}^\omega|0\rangle = X_n^\dagger V^{\omega,[1]}. \tag{7.155}$$

We note that with the exception of the CI approximation, a representation of the wave function of the excited state is not provided by the adopted residue analysis. We are instead restricted to a description of transition matrix elements and excitation energies. This holds true also when turning to a residue analysis of higher-order response functions as shown in Table 5.3, although this analysis allows for the identifications of not only transition-state properties but also excited-state properties such as excited-state multiple moments. Our response methodology thus enables us to identify excited-state electron densities but not wave functions, and as tempting as it may be to ascribe the eigenvector X_n to the excited state wave function, this is not admissible. Let us consider the RPA as an example. If we were to define an excitation operator from the eigenvector $X_n = (Z\ Y^*)^T$

$$\hat{O}_n^\dagger = \sum_{si} (Z_{si}\ \hat{a}_s^\dagger \hat{a}_i + Y_{si}^*\ \hat{a}_i^\dagger \hat{a}_s), \tag{7.156}$$

then the action of the corresponding de-excitation operator

$$\hat{O}_n = \sum_{si} (Z_{si}^*\ \hat{a}_i^\dagger \hat{a}_s + Y_{si}\ \hat{a}_s^\dagger \hat{a}_i), \tag{7.157}$$

on the Hartree–Fock reference state would be given by a sum of single-excited determinants according to

$$\hat{O}_n|0\rangle = \sum_{si} Y_{si}\ |0_i^s\rangle. \tag{7.158}$$

This stands in contrast to exact-state (or CI) theory, where $Y = 0$ and the corresponding action of the de-excitation operator on the reference state would lead to annihilation—the de-excitation operator is then said to fulfill the *killer condition*. In a situation where the killer condition is fulfilled, it is permissible to associate the excited state with the result of the excitation operator acting on the reference state.

As mentioned earlier, the dimension of the electronic Hessian is, in typical calculations, large ($\sim 10^4$–10^6) and the Hessian is, for this reason, never formed and stored in the computer memory. Instead, iterative algorithms capable of solving the generalized eigenvalue equation, Eq. (7.126), for a relatively small number of eigenvectors ($\sim$10–100) are used. The adopted procedure is of *bottom-up* type so that one retrieves eigenvectors with the smallest (positive) eigenvalues. On the other hand, if one performs a complete diagonalization of the Hessian and finds the complete set of eigenvectors X_e, then a complete spectral resolution is obtained. However, only a limited number of energy values correspond to transition energies observable in the experiment, whereas others fall in between the discrete spectral resonances, corresponding to bound or semibound excited states. When localized atomic orbitals are employed in the calculation, there will appear resonances in the continuum region of the spectrum that do not, at least directly, correspond to features in the experimental absorption spectrum.

As an illustration, we consider the electric-field-induced linear absorption spectrum of helium at the TDHF level of theory as depicted in Figure 7.9. We have here calculated oscillator strengths and shown them in the figure as vertical bars as well as with Lorentzian

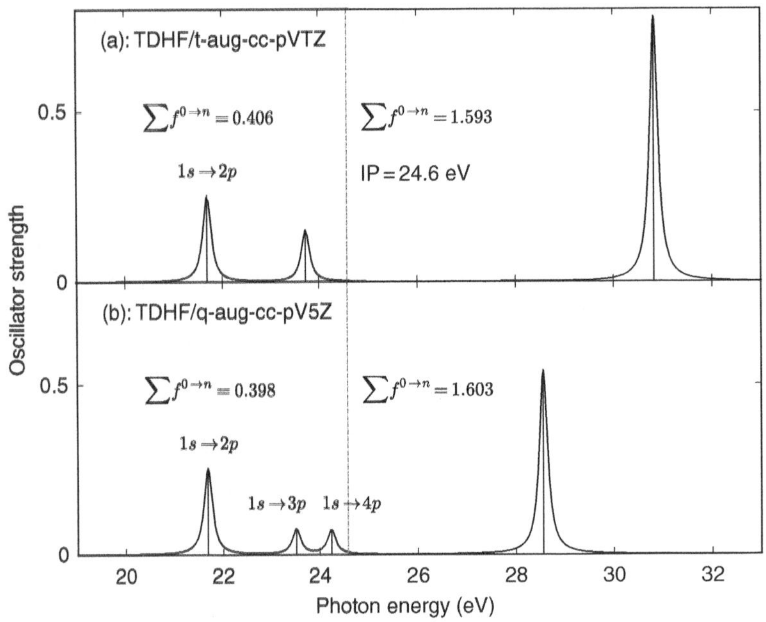

Figure 7.9 Electric dipole oscillator strength distribution for helium in the time-dependent Hartree–Fock approximation employing Dunning's (a) t-aug-cc-pVTZ and (b) q-aug-cc-pV5Z basis sets. Summed oscillator strengths are given for transitions below and above the experimental ionization potential, respectively. For visual clarity, the bars representing the oscillator strengths are broadened by Lorentzian profiles. The experimental $1s \to 2p$ transition energy is 21.22 eV.

line profiles that are added manually. As we shall see in the next section, this manual treatment of spectral profiles can be circumvented by the use of damped response theory, but, for the present discussion, it is not important which of the two approaches that has been employed.

The lowest electric dipole-allowed transition in helium is $1s \to 2p$, having an experimental excitation energy of 21.22 eV. Figure 7.9(a) and (b) refer to calculations using triple- and quintuple-ζ basis sets, respectively, augmented with diffuse functions. For the lowest transition, it can be seen that the theoretical results (obtained with the two different basis sets) are in close quantitative agreement with experiment for the excitation energy and oscillator strength. However, the corresponding results for the higher excited states are in qualitative disagreement with experiment, unable to separate the transitions $1s \to 3p$ and $1s \to 4p$ in the calculation employing the smaller basis set. On the other hand, it should be noted that the sum of oscillator strengths below (and also above) the experimental ionization potential (IP = 24.6 eV) is in close agreement between the two cases (a) and (b), so the single peak at 23.7 eV in (a) collects the intensities of the two peaks at 23.5 and 24.2 eV in (b).

In calculations on molecular systems, where the density of excited states can be high and where relatively small atomic orbital basis sets are used, it is likely that a similar poor resolution of individual excited states is obtained, but that band-integrated absorption cross sections are more accurate. Another example of similar character but different in nature is given by the comparison of absorption spectra that are calculated with and without inclusion of spin–orbit interactions. In the former case, the intensity acquired by the triplet states is zero whereas in the latter case, the triplet states will "borrow" intensity of near-lying singlet states of the same spinor symmetry while the total integrated intensity in a given energy region is more or less preserved.

This "conservation" of integrated cross sections is closely connected to the Thomas–Reiche–Kuhn sum rule, Eq. (5.235), which also holds in approximate state SCF, CI, and MCSCF calculations provided complete basis sets are used, and it is exemplified in our calculation on helium. The sums of oscillator strengths for transitions below and above the ionization potential amount to 1.999 and 2.001 in calculations (a) and (b), respectively.

7.5 Relaxation

Relaxation in approximate-state theory is introduced in an analogous manner as in exact-state theory, namely, via a damping term in the Ehrenfest equation of motion; see Eq. (5.304). If we consider the Hartree–Fock approximation as an example, the resulting master equation will become

$$\frac{\partial}{\partial t}\langle\psi|\hat{t}_n|\psi\rangle - \langle\psi|\frac{\partial}{\partial t}\hat{t}_n|\psi\rangle = \frac{1}{i\hbar}\langle\psi|[\hat{t}_n, \hat{H}]|\psi\rangle - \gamma_{n0}(\langle\psi|\hat{q}_n|\psi\rangle - \langle\psi^{\text{eq}}|\hat{q}_n|\psi^{\text{eq}}\rangle). \quad (7.159)$$

With the exception of the last term, this equation corresponds to Eq. (7.91) from which we derived the RPA equation. The last term within parentheses provides the relaxation of the system toward the state of thermal equilibrium and the rate of the relaxation is determined by the damping parameters γ_{n0}. With the state of thermal equilibrium chosen to be the electronic ground state $|0\rangle$—which is motivated by the fact that thermal excitations will typically not include molecular electronic transitions—the term $\langle\psi^{\text{eq}}|\hat{q}_n|\psi^{\text{eq}}\rangle$ will vanish, and we are thus left with the task of performing a perturbational expansion of the term $\langle\psi|\hat{q}_n|\psi\rangle$.

We obtain

$$\begin{aligned}\langle\psi|\hat{q}_n|\psi\rangle &= \langle 0|e^{i\hat{\kappa}}\hat{q}_n e^{-i\hat{\kappa}}|0\rangle \\ &= \sum_{k=0}^{\infty}\frac{i^k}{k!}\langle 0|\underline{\hat{\kappa}}^k\hat{q}_n|0\rangle = \sum_{k=0}^{\infty}\frac{i^k}{k!}\langle 0|\prod_{\underline{j=1}}^{k}\hat{q}^{\dagger}_{l_j}\,\hat{q}_n|0\rangle\prod_{j=1}^{k}\kappa_{l_j} = \sum_{k=1}^{\infty}\frac{i^k}{k!}R^{[k+1]}_{nl_1l_2\cdots l_k}\prod_{j=1}^{k}\kappa_{l_j},\end{aligned} \quad (7.160)$$

where we in the last step have recognized that the term $k = 0$ vanishes due to orthogonality of states, and where we have introduced the relaxation matrix

$$R^{[k+1]}_{nl_1l_2\cdots l_k} = \langle 0|\prod_{\underline{j=1}}^{k}\hat{q}^{\dagger}_{l_j}\,\hat{q}_n|0\rangle. \quad (7.161)$$

The relaxation matrix displays a strong resemblance to the overlap matrix $S^{[k+1]}$ but there is a shuffling of indices involved when comparing the two. We also mention that elements of $R^{[k+1]}$ with separate values of the primary index n are later to be multiplied by separate γ_n-values.

If we focus on the linear response function, it is clear that it will involve only the $R^{[2]}$-matrix. The explicit expression for this matrix is

$$R^{[2]} = \begin{pmatrix}\langle 0|[\hat{q}^{\dagger}_m, \hat{q}_n]|0\rangle & \langle 0|[\hat{q}_m, \hat{q}_n]|0\rangle \\ \langle 0|[\hat{q}^{\dagger}_m, \hat{q}^{\dagger}_n]|0\rangle & \langle 0|[\hat{q}_m, \hat{q}^{\dagger}_n]|0\rangle\end{pmatrix}, \quad (7.162)$$

where we again emphasize that n is the matrix row index and m is the column index so that, by a comparison to Eq. (7.105), it can be noted that $R^{[2]} = -S^{[2]}$. Directly from Eq. (7.159), we thereby get an equation for the determination of the first-order amplitudes, which reads

$$-\omega S^{[2]}\kappa^{(1)}(\omega) = \frac{1}{\hbar}(-E^{[2]}\kappa^{(1)}(\omega) - iV^{\omega,[1]}) - i\gamma R^{[2]}\kappa^{(1)}(\omega), \quad (7.163)$$

where γ is defined as a diagonal matrix collecting the damping parameters γ_{n0}. This equation for the first-order response parameters leads to an expression for the linear response function, which is

$$\langle\langle \hat{\Omega}; \hat{V}^{\omega} \rangle\rangle = i\langle 0| \underline{\underline{\hat{q}_n^{\dagger}\hat{\Omega}}} |0\rangle\, \kappa_n^{(1)}(\omega) = -[\Omega^{[1]}]^{\dagger}(E^{[2]} - \hbar(\omega + i\gamma)S^{[2]})^{-1}V^{\omega,[1]}, \tag{7.164}$$

where the relaxation matrix has been replaced by minus the overlap matrix. With use of the transformation matrix X from Eq. (7.128) that diagonalizes $E^{[2]}$ and $S^{[2]}$, we get an alternative form of the linear response equation

$$\langle\langle \hat{\Omega}; \hat{V}^{\omega} \rangle\rangle = -[\Omega^{[1]}]^{\dagger} \sum_{n>0} \left(\frac{X_n X_n^{\dagger}}{\varepsilon_n - \hbar\omega - i\hbar\gamma_{n0}} + \frac{X_n X_n^{\dagger}}{\varepsilon_n + \hbar\omega + i\hbar\gamma_{n0}} \right) V^{\omega,[1]}, \tag{7.165}$$

which is the RPA equivalent of Eq. (5.288) in exact-state theory.

As a numerical example, let us return to the example calculation on helium in the TDHF approximation with use of the t-aug-cc-pVTZ basis. We determine the linear response function [Eq. (7.164)] on a grid of frequencies with a common damping parameter $\gamma_{n0} = \gamma = 0.1$ eV, and the operators in the response functions are chosen as $\hat{\Omega} = \hat{\mu}_{\alpha}$ and $\hat{V}^{\omega} = -\hat{\mu}_{\beta}$ so that it corresponds to the $\alpha\beta$-component of the electric-dipole polarizability tensor. Figure 7.10(b) shows the result for the real part of the isotropic average of the polarizability defined in Eq. (5.229). The key features of the real and imaginary parts of the polarizability were shown in Figure 5.25 for the case of a two-level system, and the curve plotted here reflects this general principle but with one spectral feature for each of the resonances.

The imaginary part of the polarizability is closely connected with the phenomenon of linear absorption as illustrated by its relation, in the limit of zero damping parameters, to the oscillator strengths in Eq. (5.297) as well as its relation to the absorption cross section in Eq. (5.335). We plot the absorption cross section in Figure 7.10(a) and the resemblance to the corresponding

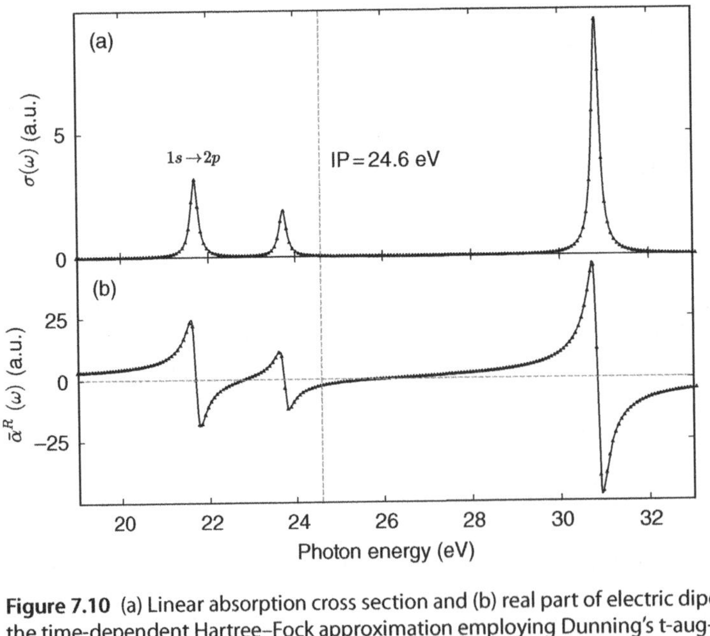

Figure 7.10 (a) Linear absorption cross section and (b) real part of electric dipole polarizability for helium in the time-dependent Hartree–Fock approximation employing Dunning's t-aug-cc-pVTZ basis set. A common γ-parameter equal to 0.1 eV is employed. The data points marked with triangles are interpolated with cubic splines.

plot of oscillator strengths in Figure 7.9(a) is striking. This example illustrates that the linear absorption spectrum may be determined from a consideration of oscillator strengths or the imaginary part of the polarizability tensor—the former alternative involves solving the generalized eigenvalue equation, Eq. (7.126), whereas the latter alternative involves the evaluation of the damped linear response function, Eq. (7.164). Which of the two approaches one adopts is largely a matter of taste, but it is clear that the latter technique can be advantageous in addressing frequency regions where the density of states is high since it circumvents the need to resolve any of the eigenvectors.

Further Reading

1 O. Christiansen, P. Jørgensen, and C. Hättig. Response functions from Fourier component variational perturbation theory applied to a time-averaged quasienergy. *Int. J. Quantum Chem.*, 68:1–52, 1998.
2 T. Helgaker, S. Coriani, P. Jørgensen, K. Kristensen, J. Olsen, and K. Ruud. Recent advances in wave function-based methods of molecular-property calculations. *Chem. Rev.*, 112:543–631, 2012.
3 T. Helgaker, P. Jørgensen, and J. Olsen. *Molecular Electronic-Structure Theory*. John Wiley & Sons, 2014.
4 P. Norman. A perspective on nonresonant and resonant electronic response theory for time-dependent molecular properties. *Phys. Chem. Chem. Phys.*, 13:20519–20535, 2011.
5 P. Norman, D. M. Bishop, H. J. A. Jensen, and J. Oddershede. Nonlinear response theory with relaxation: the first-order hyperpolarizability. *J. Chem. Phys.*, 123:194103, 2005.
6 J. Oddershede, P. Jørgensen, and D. L. Yeager. Polarization propagator methods in atomic and molecular calculations. *Comput. Phys. Rep.*, 2:33–92, 1984.
7 J. Olsen and P. Jørgensen. Linear and nonlinear response functions for an exact state and for an MCSCF state. *J. Chem. Phys.*, 82:3235–3264, 1985.

8

Response Functions and Spectroscopies

All the compliments that I have received from Arago, Laplace, and Biot never gave me so much pleasure as the discovery of a theoretical truth, or the confirmation of a calculation by experiment.

A.-J. Fresnel in a letter to Young (1824).

Many molecular properties are defined in terms of electric and magnetic multipoles, in particular induced ones. For instance, the first-order induced electric dipole moment due to an electric field defines the polarizability, see Eq. (5.60), which we will discuss in Section 8.3, and the first-order *magnetic* dipole induced by a magnetic field defines the magnetizability, which we will discuss in Section 8.4. But a time-dependent magnetic field can also induce an electric dipole moment and vice versa, as follows from Maxwell equations; see Eq. (3.3). These effects are central to the understanding of the phenomena of birefringence and dichroism that will be discussed in Section 8.6. We can thus generalize Eq. (5.60) to also include induced electric and magnetic dipole moments due to higher order corrections to the multipole expansion of the electromagnetic fields. Limiting ourselves here to first-order corrections to the induced dipole moment, we can write

$$\mu_\alpha(t) = \mu_\alpha^0 + \sum_\omega \alpha_{\alpha\beta}(-\omega;\omega)F_\beta^\omega e^{-i\omega t} - i\sum_\omega G'_{\alpha\beta}(-\omega;\omega)B_\beta^\omega e^{-i\omega t}$$
$$+\frac{1}{3}\sum_\omega A_{\alpha,\beta\gamma}(-\omega;\omega)F_{\beta\gamma}^\omega e^{-i\omega t} + \cdots, \tag{8.1}$$

where $G'_{\alpha\beta}(-\omega;\omega)$ and $A_{\alpha,\beta\gamma}(-\omega;\omega)$ are the mixed electric dipole–magnetic dipole polarizability and electric dipole–electric quadrupole polarizability, respectively, and $F_{\beta\gamma}^\omega$ denotes a component of the electric field gradient. We recall from Chapter 5 that response theory provides a framework for calculating induced molecular properties. In this chapter, we will relate these theoretical concepts more directly to observable quantities obtained from a variety of different experiments. The link between the two arises in many cases by relating the phenomenological expansion of the induced electric and magnetic dipole moments in Eq. (8.1) to the formal expansion of the expectation value of a property in orders of applied perturbations shown in Eq. (5.150). Finally, various spectroscopic techniques developed by physicists to probe electric and magnetic moments of nuclei have had an enormous impact on chemistry, due to the sensitivity of these spectroscopies to the chemical environment of the probed nucleus. A prime example is the nuclear magnetic resonance (NMR) experiment that was originally developed to determine nuclear magnetic dipole moments. Another example is the Mössbauer experiment, where nuclear electric quadrupole moments may come into play. In this chapter, we provide a taste of the rich world of molecular properties.

Principles and Practices of Molecular Properties: Theory, Modeling and Simulations, First Edition.
Patrick Norman, Kenneth Ruud, and Trond Saue.

8.1 Nuclear Interactions

The dominant electron–nucleus interaction is electrostatic and may be expressed as

$$E_{\mathrm{eN}} = \int \rho_{\mathrm{e}}(\mathbf{r}_{\mathrm{e}})\phi_{\mathrm{n}}(\mathbf{r}_{\mathrm{e}})d^3\mathbf{r}_{\mathrm{e}}, \tag{8.2}$$

where ϕ_{n} is the scalar potential associated with the nuclear charge distribution

$$\phi_{\mathrm{n}}(\mathbf{r}_{\mathrm{e}}) = \frac{1}{4\pi\varepsilon_0}\int \frac{\rho_{\mathrm{n}}(\mathbf{r}_{\mathrm{n}})}{|\mathbf{r}_{\mathrm{n}} - \mathbf{r}_{\mathrm{e}}|}d^3\mathbf{r}_{\mathrm{n}}. \tag{8.3}$$

Very often the nuclei are simply considered as point charges, which in practice means that the nuclear charge distribution is written in terms of Dirac delta functions for each nucleus K

$$\rho_{\mathrm{n}}(\mathbf{r}_{\mathrm{n}}) = \sum_K Z_K e\, \delta(\mathbf{r}_{\mathrm{n}} - \mathbf{R}_K), \tag{8.4}$$

such that the electron–nucleus interaction reduces to

$$E_{\mathrm{eN}} = \sum_K \frac{Z_K e}{4\pi\varepsilon_0}\int \frac{\rho_{\mathrm{e}}(\mathbf{r}_{\mathrm{e}})}{|\mathbf{r}_{\mathrm{e}} - \mathbf{R}_K|}d^3\mathbf{r}_{\mathrm{e}}. \tag{8.5}$$

In reality, the atomic nucleus is a composite particle built from Z protons and N neutrons, which in turn are built from quarks. Deviations from the simple point-charge model of the nucleus give the hyperfine structure of atomic and molecular spectra, and in the following subsections we shall have a closer look at what they are. Let us first introduce some terminology: Chemical elements are distinguished by the proton number Z and may have different *isotopes*, specified by the neutron number N. Isotopes, in turn, may have different nuclear energy states, denoted nuclear *isomers*. Protons and neutrons are both *nucleons* and their total number $A = N + Z$ is the nucleon, or mass, number.

8.1.1 Nuclear Charge Distribution

Information about the nuclear charge distribution has in part been obtained by electron scattering experiments. Such experiments do not give the charge distribution directly but rather the *form factor*

$$F(\mathbf{k}) = \frac{1}{Ze}\int \rho_{\mathrm{n}}(\mathbf{r}_{\mathrm{n}})e^{i\mathbf{k}\cdot\mathbf{r}_{\mathrm{n}}}d^3\mathbf{r}_{\mathrm{n}}, \tag{8.6}$$

which is seen to correspond to the Fourier transform of the charge density. In order to get $\rho_{\mathrm{n}}(\mathbf{r}_{\mathrm{n}})$ from the inverse transform of $F(\mathbf{k})$, one needs the full range of momentum transfer. Since this is not available, we can instead expand the form factor in orders of the momentum transfer. If we assume a spherically symmetric charge density, we find that

$$F(\mathbf{k}) = 1 - \frac{k^2}{6}\langle r_{\mathrm{n}}^2\rangle + O(k^4); \quad \langle r_{\mathrm{n}}^2\rangle = \frac{1}{Ze}\int_0^\infty r_{\mathrm{n}}^2\rho_{\mathrm{n}}(r_{\mathrm{n}})d^3\mathbf{r}_{\mathrm{n}}, \tag{8.7}$$

showing that the mean square radius of the nuclear charge distribution can be obtained in the low-energy limit. The square root of $\langle r_{\mathrm{n}}^2\rangle$ is defined as the nuclear radius and such radii are found to be on the order of a femtometer (1 fm $= 10^{-15}$ m), often called a *Fermi* in honor of Enrico Fermi, a pioneer of nuclear physics. Other scattering experiments, for instance with neutrons, reveal that the nuclear mass and charge radii agree to within about 0.1 fm. The nuclear mass density is therefore on the order of 2.3×10^{17} kg m^{-3}.

The simplest model for an extended spherical nuclear charge distribution is a homogeneously charged sphere

$$\rho_{\mathrm{n}}^{\mathrm{H}}(\mathbf{r}_{\mathrm{n}}) = \rho_0^{\mathrm{H}}\Theta(R_0 - r_{\mathrm{n}}); \quad \rho_0^{\mathrm{H}} = \frac{3Ze}{4\pi R_0^3}, \tag{8.8}$$

where the charge is uniformly distributed within a sphere of radius R_0, as expressed through the Heaviside step function Θ. The associated scalar potential is

$$\phi_{\mathrm{n}}^{\mathrm{H}}(\mathbf{r}_{\mathrm{n}}) = \begin{cases} \dfrac{Ze}{8\pi\varepsilon_0 R_0}\left(3 - \dfrac{r_{\mathrm{n}}^2}{R_0^2}\right); & r_{\mathrm{n}} \leq R_0, \\[2ex] \dfrac{Ze}{4\pi\varepsilon_0 r_{\mathrm{n}}}; & r_{\mathrm{n}} > R_0. \end{cases} \tag{8.9}$$

You may note that for an observer point outside the sphere, the potential is identical to that of a point charge Ze, a result that we could have deduced using a Gaussian sphere, as discussed in Chapter 3. The mean square nuclear charge radius of the homogeneously charged sphere is

$$\langle r_{\mathrm{n}}^2 \rangle^{\mathrm{H}} = \frac{3}{5}R_0^2. \tag{8.10}$$

If the nuclear mass is uniformly distributed within a sphere as well, then the nuclear volume would be proportional to the nucleon number A. If the nuclear mass and charge radii basically agree, as indicated by experiment, then both are proportional to $A^{1/3}$. In practice, we find that the nuclear charge radius is given by[1]

$$\langle r_{\mathrm{n}}^2 \rangle^{1/2} = (0.836A^{1/3} + 0.570)\ \mathrm{fm}. \tag{8.11}$$

In fact, experiment suggests that the nuclear charge distribution has a core region of constant density and then a skin region on the order of 2–3 fm where the density falls off to zero. Such a charge distribution can be modeled by a Fermi distribution

$$\rho^{\mathrm{F}}(\mathbf{r}_{\mathrm{n}}) = \rho_0^{\mathrm{F}}\left[1 + \exp\left(\frac{r_{\mathrm{n}} - c}{a}\right)\right]^{-1}; \quad \rho_0^{\mathrm{F}} = Ze\left\{\frac{4\pi c}{3}[(\pi a)^2 + c^2]\right\}^{-1}; \quad c \gg a. \tag{8.12}$$

The parameter c is denoted the half-charge radius since $\rho^{\mathrm{F}}(c) = \rho_0^{\mathrm{F}}/2$, whereas the parameter a controls the size of the skin region. It is reasonable to assume that a is very much smaller than c. If s is the interval around c where the charge distribution drops from $0.9\rho_0^{\mathrm{F}}$ to $0.1\rho_0^{\mathrm{F}}$, then simple math shows that $s = 4a\ln 3$. There is no analytic expression for the scalar potential of a Fermi distribution. The mean square nuclear charge radius of the Fermi distribution is

$$\langle r_{\mathrm{n}}^2 \rangle^{\mathrm{F}} = [7(\pi a)^2 + 3c^2]/5; \quad c \gg a, \tag{8.13}$$

and we see that c takes the place of the radius of the homogeneously charged sphere if a is set to zero.

In molecular applications, in particular using Gaussian basis functions, a Gaussian distribution

$$\rho^{\mathrm{G}}(\mathbf{r}_{\mathrm{n}}) = \rho_0^{\mathrm{G}}\exp\left(-\frac{r_{\mathrm{n}}^2}{R_G^2}\right); \quad \rho_0^{\mathrm{G}} = \frac{Ze}{(\pi^{1/2}R_G)^3}, \tag{8.14}$$

is preferred, as this leads to simpler integrals. Using the identity

$$\frac{1}{r} = \frac{2}{\sqrt{\pi}}\int_0^{+\infty} e^{-t^2 r^2}\,dt, \tag{8.15}$$

1 W. R. Johnson and G. Soff. *At. Data Nucl. Data Tables*, 33:405, 1985.

we find that the corresponding scalar potential is given in terms of the error function

$$\phi_{\mathrm{n}}^{\mathrm{G}}(\mathbf{r}_{\mathrm{n}}) = \frac{Ze}{4\pi\varepsilon_0}\mathrm{erf}(r_{\mathrm{n}}/R_G); \quad \mathrm{erf}(x) = \frac{2}{\sqrt{\pi}}\int_0^x e^{-t^2}dt. \tag{8.16}$$

In terms of the nuclear radius parameter R_G, the mean square nuclear charge radius is

$$\langle r_{\mathrm{n}}^2\rangle^{\mathrm{G}} = \frac{3}{2}R_G^2. \tag{8.17}$$

In Figure 8.1, three different charge distribution models are shown for ^{57}Fe and ^{197}Au. More precisely, we have plotted the nuclear charge distribution divided by nuclear charge. It is perhaps surprising to see the large difference in the shape of the Fermi and Gaussian distributions.

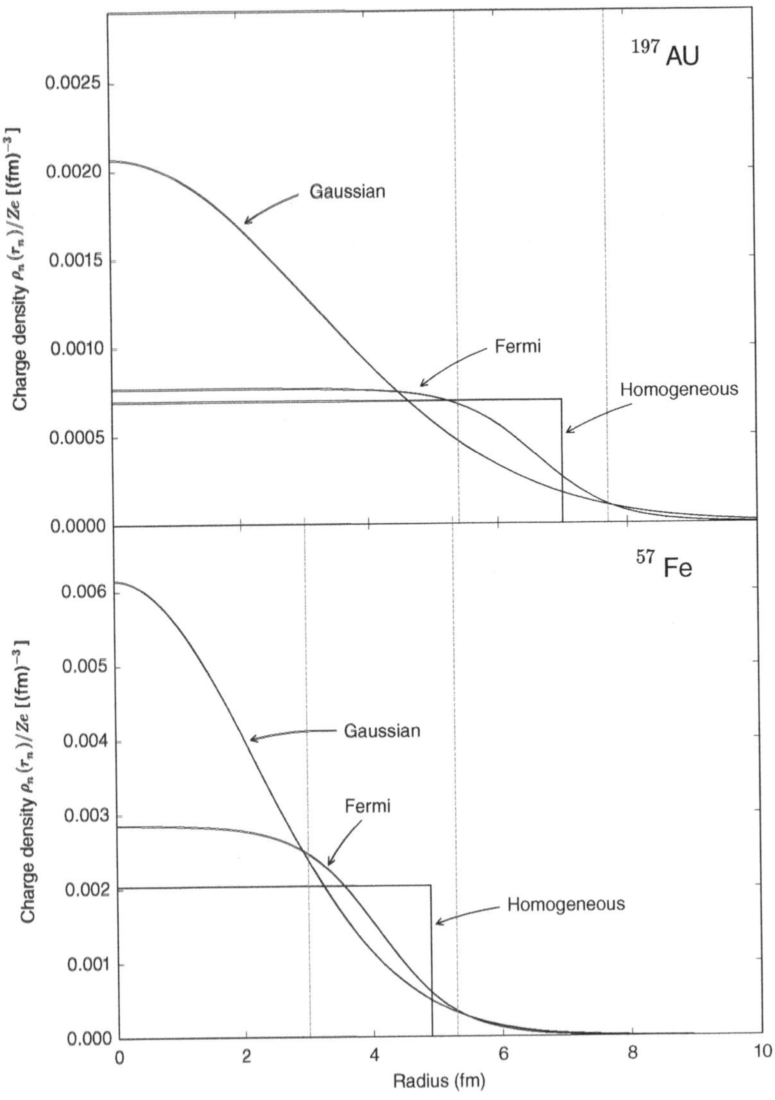

Figure 8.1 Illustration of different models for the nuclear charge distributions for ^{57}Fe (bottom) and ^{197}Au (top). The skin regions (s = 2.3 fm) of the Fermi distribution is between the dotted lines.

Table 8.1 Total ground-state Hartree–Fock energy (in Hartrees) using orbitals optimized for the ground-state configuration and different nuclear charge distribution models.

Model	^{57}Fe	^{197}Au
Point charge	−7865.781	−19039.844
Homogeneous sphere	−7865.540	−19035.554
Fermi distribution	−7865.541	−19035.562
Gaussian distribution	−7865.542	−19035.595

However, one should remember that when integrating up spherical charge distributions to total charge

$$\int \rho_{\mathrm{n}}(r_{\mathrm{n}}) d^3\mathbf{r}_{\mathrm{n}} = 4\pi \int \rho_{\mathrm{n}}(r_{\mathrm{n}}) r_{\mathrm{n}}^2 dr_{\mathrm{n}} = Ze, \tag{8.18}$$

the radial charge distribution $\rho_{\mathrm{n}}(r_{\mathrm{n}})$ is weighted by $4\pi r_{\mathrm{n}}^2$, such that large differences in the inner regions are compensated by small differences in the outer regions. Table 8.1 furthermore shows that the introduction of a finite nucleus model leads to a sizable change in total energy, but the energy difference between different nuclear models is quite small.

We can rework the expression for the electrostatic electron–nucleus interaction, Eq. (8.2), into a form suitable for assessing the effect of the finite nuclear size on energy and molecular properties. For simplicity we fix the origin of our coordinate system at the nucleus of interest. We then start from the observation that outside some radius, let us call it r_{c}, the nuclear charge distribution will be zero and the scalar potential identical to the scalar potential of a point charge. We therefore write the electron–nucleus interaction as

$$E_{\mathrm{eN}} = \int \rho_{\mathrm{e}}(\mathbf{r}_{\mathrm{e}}) \phi_{\mathrm{n}}^{\mathrm{point}}(\mathbf{r}_{\mathrm{n}}) d^3\mathbf{r}_{\mathrm{e}} + \Delta E_{\mathrm{eN}}, \tag{8.19}$$

where the second term is given as

$$\Delta E_{\mathrm{eN}} = \int_{r_{\mathrm{e}}<r_{\mathrm{c}}} \rho_{\mathrm{e}}(\mathbf{r}_{\mathrm{e}}) [\phi_{\mathrm{n}} - \phi_{\mathrm{n}}^{\mathrm{point}}](\mathbf{r}_{\mathrm{e}}) d^3\mathbf{r}_{\mathrm{e}}. \tag{8.20}$$

This may be thought of as a term correcting for the effect of replacing a point charge by a finite charge distribution inside the radius r_{c}, but this is not entirely correct since the electronic charge distribution ρ_{e} is the one obtained with a finite nucleus. On the other hand, the finite nucleus potential ϕ_{n} is always bounded by the point charge potential $\phi_{\mathrm{n}}^{\mathrm{point}}$ and so the expression inside the square brackets above is always negative and thus, more importantly, never changes sign. This allows us to invoke the first mean value theorem for definite integrals, which guarantees the existence of an r_{eff}, with $0 < r_{\mathrm{eff}} < r_{\mathrm{c}}$, such that the energy correction can be rewritten as

$$\Delta E_{\mathrm{eN}} = \overline{\rho}_{\mathrm{e}} \int_{r_{\mathrm{e}}<r_{\mathrm{c}}} [\phi_{\mathrm{n}} - \phi_{\mathrm{n}}^{\mathrm{point}}](\mathbf{r}_{\mathrm{e}}) d^3\mathbf{r}_{\mathrm{e}}, \tag{8.21}$$

where we have introduced the effective electronic charge density $\overline{\rho}_{\mathrm{e}} = \rho_{\mathrm{e}}(r_{\mathrm{eff}})$ as the value of the electron density at radius r_{eff} inside the nuclear volume.

Let us next consider the contribution involving the finite nucleus potential that reads

$$\Delta E_{\mathrm{eN,finite}} = \frac{\overline{\rho}_{\mathrm{e}}}{4\pi\varepsilon_0} \int_{r_{\mathrm{n}}<r_{\mathrm{c}}} \int_{r_{\mathrm{e}}<r_{\mathrm{c}}} \frac{\rho_{\mathrm{n}}(\mathbf{r}_{\mathrm{n}})}{|\mathbf{r}_{\mathrm{e}} - \mathbf{r}_{\mathrm{n}}|} d^3\mathbf{r}_{\mathrm{e}} d^3\mathbf{r}_{\mathrm{n}}. \tag{8.22}$$

It may be noted that the radial integration over both the electronic and nuclear coordinates can be restricted to be within the radius r_c since the nuclear charge distribution is zero outside. To proceed, we now invoke the addition theorem in Eq. (3.233). We will assume that the nuclear charge distribution is spherically symmetric (deviations will be addressed later). The expansion in angular momentum l therefore truncates at $l = 0$, such that

$$\Delta E_{\mathrm{eN,finite}} = \frac{\overline{\rho}_{\mathrm{e}}}{4\pi\varepsilon_0} \int_{r_{\mathrm{n}}<r_{\mathrm{c}}} \int_{r_{\mathrm{e}}<r_{\mathrm{c}}} \frac{\rho_{\mathrm{n}}(\mathbf{r}_{\mathrm{n}})}{r_>} d^3\mathbf{r}_{\mathrm{e}} d^3\mathbf{r}_{\mathrm{n}}, \tag{8.23}$$

where $r_> = \max(r_{\mathrm{n}}, r_{\mathrm{e}})$. When $r_{\mathrm{n}} > r_{\mathrm{e}}$, we get

$$\Delta E_{\mathrm{eN,finite},r_{\mathrm{n}}} = \frac{\overline{\rho}_{\mathrm{e}}}{4\pi\varepsilon_0} \int_{r_{\mathrm{n}}<r_{\mathrm{c}}} \int_{r_{\mathrm{e}}<r_{\mathrm{c}}} \frac{\rho_{\mathrm{n}}(\mathbf{r}_{\mathrm{n}})}{r_{\mathrm{n}}} d^3\mathbf{r}_{\mathrm{e}} d^3\mathbf{r}_{\mathrm{n}} = \frac{Ze\overline{\rho}_{\mathrm{e}}}{3\varepsilon_0}\langle r_{\mathrm{n}}^2\rangle, \tag{8.24}$$

and, when $r_{\mathrm{e}} > r_{\mathrm{n}}$, we get

$$\Delta E_{\mathrm{eN,finite},r_{\mathrm{e}}} = \frac{\overline{\rho}_{\mathrm{e}}}{4\pi\varepsilon_0} \int_{r_{\mathrm{n}}<r_{\mathrm{c}}} \int_{r_{\mathrm{e}}<r_{\mathrm{c}}} \frac{\rho_{\mathrm{n}}(\mathbf{r}_{\mathrm{n}})}{r_{\mathrm{e}}} d^3\mathbf{r}_{\mathrm{e}} d^3\mathbf{r}_{\mathrm{n}} = \frac{Ze\overline{\rho}_{\mathrm{e}}}{2\varepsilon_0} r_{\mathrm{c}}^2 - \frac{Ze\overline{\rho}_{\mathrm{e}}}{2\varepsilon_0}\langle r_{\mathrm{n}}^2\rangle. \tag{8.25}$$

The first term of $\Delta E_{\mathrm{eN,finite},r_{\mathrm{e}}}$ will be canceled by the contribution to ΔE_{eN} from the point nucleus potential, so when adding up all contributions we end up with the final expression

$$E_{\mathrm{eN}} = \int \rho_{\mathrm{e}}(\mathbf{r}_{\mathrm{e}})\phi_{\mathrm{n}}^{\mathrm{point}}(\mathbf{r}_{\mathrm{n}}) d^3\mathbf{r}_{\mathrm{e}} - \frac{Ze\overline{\rho}_{\mathrm{e}}}{6\varepsilon_0}\langle r_{\mathrm{n}}^2\rangle. \tag{8.26}$$

A nice feature of this expression is that it is independent of the specific choice of (spherical) finite nuclear model and it shows that the mean square radius of the nuclear charge distribution is the appropriate measure of nuclear size.

An example of a molecular property where the nuclear size comes into play is the Mössbauer isomer shift. Mössbauer spectroscopy probes transitions between nuclear energy levels, that is, between isomers. Such transitions are in the domain of gamma rays, and one of the initial difficulties of this spectroscopy was that the high-energy radiation would simply knock the sample out of position. Rudolf Mössbauer solved this problem in 1958 by using a solid sample, allowing the recoil energy to be absorbed by the surrounding lattice. Nowadays, the scope of Mössbauer spectroscopy has been extended by using frozen solutions or by trapping the sample in cavities (capillary Mössbauer spectroscopy), and it has become a prominent analysis tool in bioinorganic chemistry. The most widely studied Mössbauer-active nucleus is the ^{57}Fe isotope. The source of γ radiation is then the radioactive ^{57}Co isotope (half-life $\tau_{1/2} = 271.8$ days), which decays by electron capture to a metastable ($\tau_{1/2} = 98.3$ ns) excited nuclear state of ^{57}Fe, which is a source of γ-radiation of energy $E_\gamma = 14.4$ keV when decaying to the nuclear ground state. The "bare" transition energy E_γ is modified by the chemical environment of the source and the sample. To bring them into resonance, the Doppler effect is exploited by mounting the source on a vibrating shaft (see Figure 8.2). The Mössbauer *isomer shift*

$$\delta = \frac{c}{E_\gamma}(\Delta E_\gamma^a - \Delta E_\gamma^s) \tag{8.27}$$

is therefore reported in millimeters per second, where $\Delta E_\gamma^{\mathrm{a}}$ and $\Delta E_\gamma^{\mathrm{s}}$ represent the shift in γ-transition energy of sample (absorber) and source, respectively. Each shift is expressed as the change in the electron–nucleus interaction, Eq. (8.2), due to the change in nuclear size between the nuclear excited and ground states

$$\Delta E_\gamma = E_{\mathrm{eN}}(\langle r_{\mathrm{n}}^2\rangle^\star) - E_{\mathrm{eN}}(\langle r_{\mathrm{n}}^2\rangle_0). \tag{8.28}$$

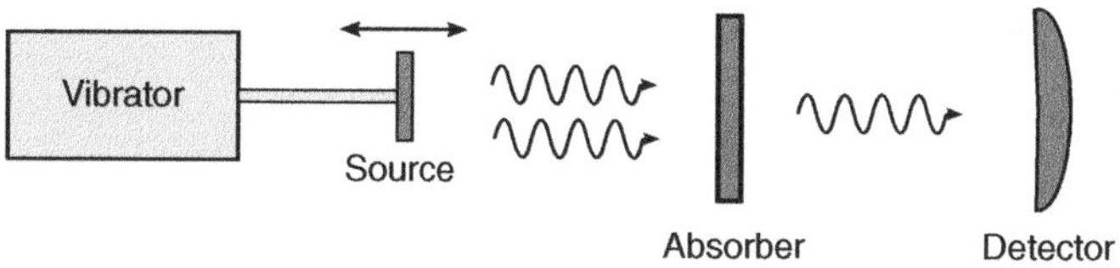

Figure 8.2 Schematic setup of a Mössbauer spectrometer.

Since this change is small compared to the ground-state mean square radius $\langle r_n^2 \rangle_0$, we can express this change by a first-order Taylor expansion. Using Eq. (8.26), we obtain

$$\Delta E_\gamma = \left.\frac{dE_{eN}}{d\langle r_n^2 \rangle}\right|_{\langle r_n^2 \rangle_0} \delta\langle r_n^2 \rangle = -\frac{Ze\bar{\rho}_e}{6\varepsilon_0}\delta\langle r_n^2 \rangle. \tag{8.29}$$

The Mössbauer isomer shift can therefore be expressed as

$$\delta = \alpha(\bar{\rho}_e^a - \bar{\rho}_e^s); \quad \alpha = -\frac{Zec}{6\varepsilon_0 E_\gamma}\delta\langle r_n^2 \rangle. \tag{8.30}$$

where all constants and nuclear information have been collected in the isomer shift calibration constant α. Since the effective density of the source is fixed and in fact usually eliminated by measuring the isomer shift of some well-defined reference, the isomer shift is seen to be a simple linear function of the effective density of the sample. We have also seen that the derivative of the nuclear potential with respect to nuclear radius is an extremely local quantity, so the effective density $\bar{\rho}_e$ is usually approximated by the *contact density* ρ_0, that is, the value of the electronic density at the nuclear position.[2] Only atomic *s* orbitals contribute to the contact density. One may wonder what chemical information can be extracted from a quantity at the very heart of an atom. Although it is true that the largest contribution to the absolute value of the contact density comes from core *s* orbitals, the change in the contact density between different samples is dictated by valence contributions. Combined with the extreme precision of the Mössbauer experiment this makes the isomer shift a sensitive probe, allowing, for instance, to distinguish oxidation states of the probe atom.[3]

8.1.2 Hyperfine Structure

8.1.2.1 Nuclear Magnetic Dipole Moment

Most atomic nuclei have spin. The term "spin" here refers to the total angular momentum of the nucleus, rather like the total angular momentum J of an atom, and it therefore has contributions from the intrinsic angular momenta of each constituent elementary particle as well as their orbital motion. The actual value of the spin for a given isotope thus depends on the nuclear structure and may vary from one nuclear state to another. Some rules of thumb are as follows:

- Nuclei with even (odd) mass number $A = Z + N$ have integer (half-integer) spin I
- Nuclei with Z and N both even have $I = 0$, for example, ^{12}C, ^{16}O, ^{56}Fe
- Nuclei with Z and N both odd have $I > 0$, for example, $^{2}H(I = 1)$, $^{10}B(I = 3)$, $^{14}N(I = 1)$, $^{40}K(I = 4)$.

---

2 This is a good approximation for light elements. For heavier nuclei, it may overestimate $\bar{\rho}_e$ by as much as 10%. One may also argue that the approximation $\bar{\rho}_e \approx \rho_0$ is not necessary since both quantities can be calculated at the same computational cost.

3 Nice examples are given in P. Gütlich and C. Schröder. *Bunsenmagazin*, 12:4, 2010.

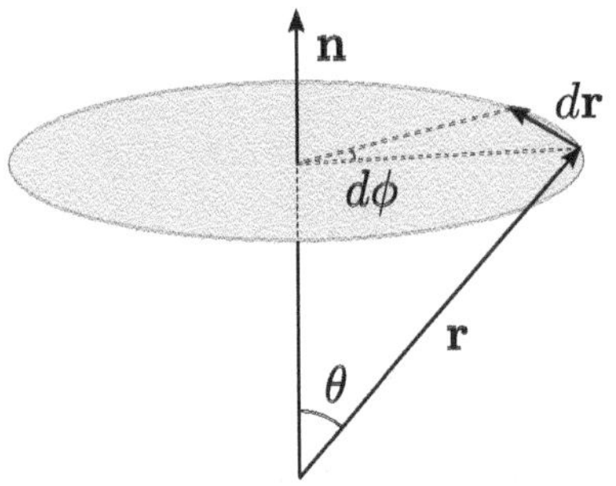

Figure 8.3 Classical model of a magnetic dipole.

Nuclei with spin also have a magnetic dipole moment **m**. The two quantities are related by the *gyromagnetic ratio* γ, that is,

$$\mathbf{m} = \gamma \mathbf{I}. \tag{8.31}$$

To get a feeling for the gyromagnetic ratio, let us consider the classical model shown in Figure 8.3 of a point charge q in circular motion. Starting from Eq. (3.191), we find that the magnetic dipole moment is

$$\mathbf{m}^{[1]} = \frac{1}{2}\mathbf{r} \times q\mathbf{v} = \frac{1}{2} rqv\mathbf{n}, \tag{8.32}$$

whereas the angular momentum is

$$\mathbf{L} = \mathbf{r} \times \mathbf{p} = rmv\mathbf{n}. \tag{8.33}$$

The gyromagnetic ratio is accordingly

$$\gamma = \frac{m^{[1]}}{L} = \frac{q}{2m}. \tag{8.34}$$

For an atomic nucleus, the gyromagnetic ratio is

$$\gamma = \frac{ge}{2m_{\mathrm{p}}} = \frac{g\mu_{\mathrm{N}}}{\hbar}, \tag{8.35}$$

where appears the *nuclear magneton*,

$$\mu_{\mathrm{N}} = \frac{e\hbar}{2m_{\mathrm{p}}} = 5.050783699(31) \times 10^{-27}\ \mathrm{J\ T^{-1}}, \tag{8.36}$$

as well as the nuclear g factor. In Table 8.2 we have given information about nuclear spin, abundance, and gyromagnetic ratio for selected isotopes. Of particular interest is perhaps to note that the gyromagnetic ratio of the proton (^{1}H) implies a g factor of 5.585694713(46), which indicates a very large deviation from the classical model outlined above. Even more striking is perhaps the neutron, which has no charge but has a gyromagnetic ratio of −18.3247172(43) 10^7 rad $\mathrm{s^{-1}\ T^{-1}}$. These observations can be explained by the composition of nucleons in terms of quarks (two up quarks and one down quark for the proton and vice versa for the neutron). When the proton and neutron combine to form deuterium (^{2}H), the resultant spin is one and the gyromagnetic ratio is significantly reduced. Upon adding a second neutron to form tritium (^{3}H), the spin is reduced to 1/2, but the gyromagnetic ratio surpasses that of the proton.

Let us now derive the vector potential associated with the nuclear magnetic dipole moment of a particular nucleus K. We write the current density of nucleus K as the curl of magnetization, as we see in Eq. (3.297)

$$\mathbf{j}_K(\mathbf{r}) = \boldsymbol{\nabla} \times \mathbf{m}_K(\mathbf{r}); \quad \mathbf{m}_K(\mathbf{r}) = \psi_K^*(\mathbf{r})\hat{\mathbf{m}}\psi_K(\mathbf{r}). \tag{8.37}$$

Table 8.2 Data for selected isotopes.

	Spin	Abundance (%)	γ (10^7 rad s^{-1} T^{-1})		Spin	Abundance (%)	γ (10^7 rad s^{-1} T^{-1})
^{1}H	1/2	~100	26.752219	^{23}Na	3/2	~100	7.0808493
^{2}H	1	0.015	4.10662791	^{27}Al	5/2	~100	6.9762715
^{3}H	1/2	0	28.5349779	^{29}Si	1/2	4.7	−5.3190
^{10}B	3	19.9	2.8746786	^{31}P	1/2	~100	10.8394
^{11}B	3/2	80.1	8.5847044	^{35}Cl	3/2	75.77	2.624198
^{12}C	0	98.9		^{37}Cl	3/2	24.23	2.184368
^{13}C	1/2	1.1	6.728284	^{63}Cu	3/2	69.17	7.1117890
^{14}N	1	99.6	1.9337792	^{65}Cu	3/2	30.83	7.60435
^{15}N	1/2	0.37	−2.71261804	^{107}Ag	1/2	51.84	−1.0889181
^{16}O	0	~100		^{109}Ag	1/2	48.16	−1.2518634
^{17}O	5/2	0.04	−3.62808	^{129}Xe	1/2	24.4	−7.452103
^{19}F	1/2	~100	25.18148	^{207}Pb	1/2	22.1	5.58046

Here ψ_K is the (unknown) nuclear wave function and $\hat{\mathbf{m}}$ the magnetic dipole operator. This is clearly a purely solenoidal quantity. We now insert this current density into the stationary vector potential of Eq. (3.198) and integrate by parts to obtain

$$\mathbf{A}_K(\mathbf{r}_1) = \frac{\mu_0}{4\pi}\int \frac{1}{r_{12}}(\nabla_2 \times \mathbf{m}_K(\mathbf{r}_2))d^3\mathbf{r}_2 = \frac{\mu_0}{4\pi}\int \frac{\mathbf{m}_K(\mathbf{r}_2)\times\mathbf{r}_{12}}{r_{12}^3}d^3\mathbf{r}_2. \tag{8.38}$$

If we assume a point-like nuclear magnetic dipole, we obtain the standard expression of Eq. (3.202)

$$\mathbf{m}_K(\mathbf{r}) = \mathbf{m}_K\delta^3(\mathbf{r}-\mathbf{R}_K) \quad \Rightarrow \quad \mathbf{A}_K(\mathbf{r}_1) = \frac{1}{4\pi\varepsilon_0 c^2}\frac{\mathbf{m}_K\times\mathbf{r}_{1K}}{r_{1K}^3}, \tag{8.39}$$

where we also replaced the magnetic constant by the electric constant and the squared speed of light using Eq. (3.7). One should be aware that this is, strictly speaking, an approximation; the effect of the finite size of the nuclear magnetization is known as the Bohr–Weisskopf effect. The magnetic field corresponding to the vector field in Eq. (8.39) is

$$\mathbf{B}_K(\mathbf{r}_1) = \frac{1}{4\pi\varepsilon_0 c^2}\left[\mathbf{m}_K\frac{8\pi}{3}\delta(\mathbf{r}_{1K}) - \frac{\mathbf{m}_K r_{1K}^2 - 3\mathbf{r}_{1K}(\mathbf{r}_{1K}\cdot\mathbf{m}_K)}{r_{1K}^5}\right], \tag{8.40}$$

corresponding to Eq. (3.204).

In the nonrelativistic case, substitution into the Hamiltonian of the vector potential associated with all nuclei in a molecule gives rise to the following interaction Hamiltonian:

$$\hat{H}^{\text{hfs}} = \sum_K \left[\hat{h}_K^{\text{pso}} + \hat{h}_K^{\text{fc}} + \hat{h}_K^{\text{sd}} + \sum_L \hat{h}_{KL}^{\text{dso}}\right], \tag{8.41}$$

where we have

$$\hat{h}_K^{\text{pso}} = \sum_i \frac{e}{4\pi\varepsilon_0 m_e c^2 r_{iK}^3} \mathbf{m}_K \cdot \hat{\mathbf{l}}_{iK}, \tag{8.42}$$

$$\hat{h}_K^{\text{fc}} = \sum_i \frac{g_e}{4\pi\varepsilon_0 c^2} \mu_B \hat{\mathbf{s}}_i \cdot \mathbf{m}_K \frac{8\pi}{3} \delta(\mathbf{r}_{iK}), \tag{8.43}$$

$$\hat{h}_K^{\text{sd}} = -\sum_i \frac{g_e}{4\pi\varepsilon_0 \hbar c^2} \mu_B \hat{\mathbf{s}}_i \cdot \left\{ \frac{\mathbf{m}_K r_{iK}^2 - 3\mathbf{r}_{iK}(\mathbf{r}_{iK} \cdot \mathbf{m}_K)}{r_{iK}^5} \right\}, \tag{8.44}$$

$$\hat{h}_{KL}^{\text{dso}} = \sum_i \left(\frac{e^2}{2m_e} \right) \left(\frac{1}{4\pi\varepsilon_0 \hbar c^2} \right)^2 \left[\frac{(\mathbf{m}_K \cdot \mathbf{m}_L)(\mathbf{r}_{iK} \cdot \mathbf{r}_{iL}) - (\mathbf{m}_K \cdot \mathbf{r}_{iL})(\mathbf{r}_{iK} \cdot \mathbf{m}_L)}{r_{iK}^3 r_{iL}^3} \right]. \tag{8.45}$$

We may compare this to the corresponding relativistic interaction Hamiltonian

$$\hat{H}_{\text{rel}}^{\text{hfs}} = \sum_K \hat{h}_{K,\text{rel}}^{\text{hfs}}(i), \tag{8.46}$$

where appears the operator

$$\hat{h}_{K,\text{rel}}^{\text{hfs}} = -\sum_i \mathbf{m}_K \cdot \hat{\mathbf{B}}_{K,\text{rel}}^{\text{el}}(i); \quad \hat{\mathbf{B}}_{K,\text{rel}}^{\text{el}}(i) = -\frac{1}{4\pi\varepsilon_0 c^2} \frac{\mathbf{r}_{iK} \times ec\boldsymbol{\alpha}}{r_{iK}^3}. \tag{8.47}$$

Comparing with the Biot–Savart expression, Eq. (3.34), keeping in mind that the relativistic current density operator is $\hat{\mathbf{j}}^{\text{R}} = -ec\boldsymbol{\alpha}$ (see Eq. (3.165)), we identify $\hat{\mathbf{B}}_{K,\text{rel}}^{\text{el}}$ as the operator giving the magnetic field due to electrons at the nuclear position, and so the operator $\hat{h}_{K,\text{rel}}^{\text{hfs}}$ describes the interaction of the nuclear magnetic dipole with this field.

The nonrelativistic interaction operator is analogous in that it describes the coupling of the nuclear magnetic dipole moment with the magnetic field associated with electrons at the nuclear position. It does take a bit more work, though, to extract the nonrelativistic operator for the magnetic field due to the electrons. Similarly to the nonrelativistic current density, Eq. (3.168), the operator splits into an orbital part and a spin part. The former describes the magnetic field generated by the motion of the electrons relative to the nucleus, and is furthermore split into a paramagnetic and a diamagnetic part, hence the names paramagnetic spin–orbit (pso), Eq. (8.42), and diamagnetic spin–orbit (dso), Eq. (8.45). The spin part distinguishes the contact term (Fermi contact (fc), Eq. (8.43)) and the general dipole part (spin-dipole (sd), Eq. (8.44)) of the magnetic field (compare with Eq. (3.204)).

To get some feeling for hyperfine interactions, let us consider an atom where we are helped by symmetry and can use angular momentum algebra. The name *hyperfine* correctly implies that the interaction of electrons with nuclear magnetic dipole moments leads to splittings in atomic and molecular spectra that are smaller than the *fine* structure due to spin–orbit splitting. In Appendix G, we look at the splitting of an atomic ^{2S+1}L term due to spin–orbit splitting. This leads to atomic states that are eigenfunctions of total electronic angular momentum and, if the spin–orbit interaction is not too strong, can be designated as $^{2S+1}L_J$.

Let us now look at the further splitting of atomic $|JM_J\rangle$ states due to hyperfine interaction. We can attack this problem by perturbation theory and diagonalize the hyperfine interaction Hamiltonian, Eq. (8.46) (or its nonrelativistic equivalent), in the space of degenerate or near-degenerate atomic states. If we limit our attention to the degenerate states $\{|JM_J\rangle\}$ arising for a particular J, then the microscopic hyperfine Hamiltonian, Eq. (8.46), can be replaced by an effective operator

$$\hat{H}_K = A_K(J)\hat{\mathbf{I}}_K \cdot \hat{\mathbf{J}}, \tag{8.48}$$

where we shall drop the subscript K in the following. We can see that the operator for the magnetic field of the electrons at the nuclear position has been replaced by the electronic total angular momentum $\hat{\mathbf{J}}$ times a hyperfine constant $A(J)$, which also contains the nuclear gyromagnetic ratio γ_K. This is possible because it can be shown that $\hat{\mathbf{B}}_K^{el}$ and $\hat{\mathbf{J}}$ transform in the same manner under rotations. When diagonalizing the effective hyperfine operator in the space of the unperturbed atomic states, we expect to obtain eigenfunctions of total angular momentum $\mathbf{F} = \mathbf{I} + \mathbf{J}$. Eigenstates of the resultant angular momentum can be expressed as

$$|I,J;F,M_F\rangle = \sum_{M_I}\sum_{M_J} |I,M_I,J,M_J\rangle\langle I,M_I,J,M_J|F,M_F\rangle, \tag{8.49}$$

with the restriction that $M_I + M_J = M_F$. The corresponding eigenvalues are

$$E_{\text{hfs}}(F) = \langle I,J;F,M_F|A(J)\hat{\mathbf{I}}\cdot\hat{\mathbf{J}}|I,J;F,M_F\rangle. \tag{8.50}$$

In order to evaluate the matrix elements, we proceed analogously to the treatment of spin–orbit interaction in Eq. (G.10) and write

$$\hat{F}^2 = (\hat{\mathbf{I}}+\hat{\mathbf{J}})^2 = \hat{I}^2 + \hat{J}^2 + 2\hat{\mathbf{I}}\cdot\hat{\mathbf{J}} \quad\Rightarrow\quad \hat{\mathbf{I}}\cdot\hat{\mathbf{J}} = \frac{1}{2}(\hat{F}^2 - \hat{I}^2 - \hat{J}^2). \tag{8.51}$$

We then obtain

$$E_{\text{hfs}}(F) = \frac{1}{2}\hbar^2 A(J)[F(F+1) - I(I+1) - J(J+1)]. \tag{8.52}$$

The spacing between energy levels follows the interval rule

$$E_{\text{hfs}}(F) - E_{\text{hfs}}(F-1) = \hbar^2 A(J)F. \tag{8.53}$$

As an example, we may consider the cesium atom. The only naturally occurring isotope ^{133}Cs has nuclear spin $I = 7/2$. The $^2S_{1/2}$ ground state is thereby split into components $F = 4$ and $F = 3$. The splitting was defined as being exactly 9.192631770 GHz and has since 1967 formed the basis for the definition of the second in the SI unit system. From the above interval rule, it follows that the hyperfine constant A is 2.298157942 GHz and in the microwave region.

In the more general case of molecules, the effective Hamiltonian, Eq. (8.48), is generalized to

$$\hat{H}_K = \hat{\mathbf{I}}_K \cdot A_K \cdot \hat{\mathbb{S}}. \tag{8.54}$$

Here $\hat{\mathbb{S}}$ is a *pseudo-spin operator* whose magnitude is chosen to give the same multiplicity as the number of states considered in the perturbative treatment, whereas A_K is a 3×3 matrix. This allows the introduction of anisotropy such that the effective magnetic field $A_K\cdot\hat{\mathbb{S}}$ need not be parallel to the pseudo-spin. There will be more about pseudo-spin in Section 8.2.

8.1.2.2 Nuclear Electric Quadrupole Moment

The first clear case of hyperfine structure due to a nuclear electric quadrupole moment was demonstrated by Schüler and Schmidt in 1935 when they could show deviations from the interval rule, Eq. (8.53), in the hyperfine splittings of the atomic spectra of the two naturally occurring isotopes of europium, ^{151}Eu and ^{153}Eu, both with nuclear spin $I = 5/2$. As we shall see, only nuclei of spin $I > 1/2$ may possess an electric quadrupole moment.

The energy of interaction of an electric quadrupole with electromagnetic fields is given in Eq. (3.228). We see that the quadrupole interacts with the gradient of the electric field at the expansion point, which in this case is the position of the quadrupole nucleus. Since in our notation $E^{[1]}_{\alpha;\beta} = -\phi^{[2]}_{\alpha\beta}$, we see that the general expression for the negative of the electric field gradient is given by Eq. (3.229). It has contributions both from the electrons and the other nuclei. The latter contribution simplifies to

$$\phi^{\mathrm{n}}_{\alpha\beta}(\mathbf{R}_K) = \sum_{A\neq K} \frac{Z_A e}{4\pi\varepsilon_0}\left[\frac{3r_{KA;\alpha}r_{KA;\beta} - \delta_{\alpha\beta}r^2_{KA}}{r^5_{KA}}\right], \tag{8.55}$$

whereas the former may be calculated as an expectation value using the electronic wave function.

For the nuclear electric quadrupole moments we shall employ the traceless forms, discussed in Section 3.3.5, that is,

$$\Theta_{\alpha\beta} = \int \rho_{\mathrm{n}}(\mathbf{r})\hat{\Theta}_{\alpha\beta}d^3\mathbf{r}, \quad \hat{\Theta}_{\alpha\beta} = \frac{1}{2}(3r_\alpha r_\beta - \delta_{\alpha\beta}r^2), \tag{8.56}$$

where appear *nuclear* electric quadrupole moment operators $\hat{\Theta}_{\alpha\beta}$. This rather simple-looking expression hides a serious problem: Nuclear structure theory is not as well advanced as the electronic one, and the nuclear charge distribution $\rho_{\mathrm{n}}(\mathbf{r})$ is not readily available. Fortunately, angular momentum algebra (see Appendix G) comes to our rescue. One can show that the nuclear coordinates transform under rotations as the nuclear spin operators, which leads to the proportionality

$$\Theta_{\alpha\beta} = C\langle I, m_I|\frac{1}{2}(3\hat{I}_\alpha\hat{I}_\beta - \delta_{\alpha\beta}\hat{I}^2)|I, m_I\rangle. \tag{8.57}$$

This connection to (spin) angular momentum explains why an electric quadrupole moment (multipole order $L = 2$) is only possible for nuclear spin $I > 1/2$, as explained in Section 3.3.5.

An expression for the constant of proportionality is found by defining

$$eQ = 2\Theta_{zz} = C\langle II|3\hat{I}_z^2 - I^2|II\rangle = \hbar^2 CI(2I-1); \quad C = \frac{eQ}{\hbar^2 I(2I-1)}, \tag{8.58}$$

where Q is called the electric quadrupole moment of the nucleus. The alert reader may note that this is twice the traceless form given in Eq. (3.227). This is simply a question of different conventions in different communities. In fact, in the nuclear physics community, the fundamental charge e is even extracted from the quadrupole moment, so that the field gradient q has units of area. It is usually reported in barns,[4] where 1 b $= 10^{-28}$ m^2, or 100 fm^2. A nonzero quadrupole moment indicates a nonspherical charge distribution of the nucleus. The nuclear charge distribution has axial symmetry about the nuclear spin axis and can be thought of as an ellipsoid. A positive sign indicates a *prolate* ("stretched") ellipsoid whereas a negative sign indicates an *oblate* ("squashed") ellipsoid, as illustrated in Figure 8.4.

Following Eq. (3.228), we can now write the interaction energy as

$$E_Q(I, m_I) = -\frac{eQ}{6\hbar^2 I(2I-1)}\langle I, m_I|3\hat{I}_\alpha\hat{I}_\beta - \delta_{\alpha\beta}\hat{I}^2|I, m_I\rangle E^{[1]}_{\alpha;\beta} - \frac{1}{6\varepsilon_0}\frac{eQ(I+1)}{(2I-1)}\rho_{\mathrm{e}}(0). \tag{8.59}$$

It should be noted that the second term is independent of m_I and therefore does not contribute to any splittings. We shall therefore omit it in the following, effectively assuming $E^{[1]}_{\alpha;\alpha} = 0$.

Further simplification is obtained by switching to a principal axis system in which the electric field gradient tensor is diagonal, that is, $E^{[1]}_{\alpha;\beta} = \delta_{\alpha\beta}E^{[1]}_{\alpha;\beta}$, to give

$$E_Q(I, m_I) = -\frac{eQ}{6\hbar^2 I(2I-1)}\langle I, m_I|(3\hat{I}_z^2 - \hat{I}^2)E^{[1]}_{z;z} + (3\hat{I}_y^2 - \hat{I}^2)E^{[1]}_{y;y} + (3\hat{I}_x^2 - \hat{I}^2)E^{[1]}_{x;x}|I, m_I\rangle.$$

4 This somewhat rural unit was introduced by H. G. Holloway and C. P. Baker in Chicago in 1942 as a unit for nuclear cross sections, where barn referred to something big.

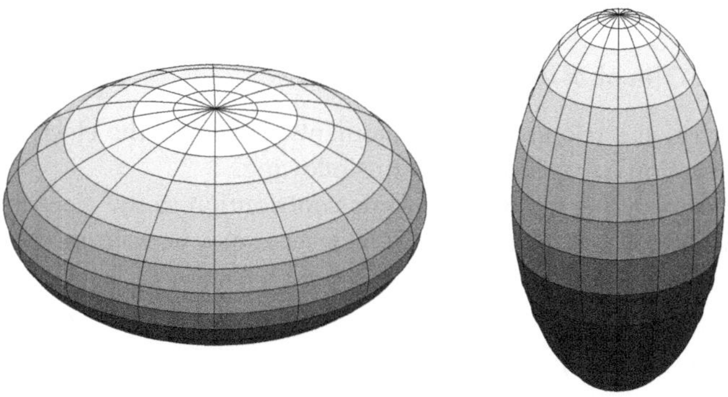

(a) (b)

Figure 8.4 Oblate (a) and prolate (b) ellipsoids.

(8.60)

We can also, without loss of generality, choose the axes according to

$$|E^{[1]}_{z;z}| \geq |E^{[1]}_{y;y}| \geq |E^{[1]}_{x;x}|. \tag{8.61}$$

Using the effective trace condition $E^{[1]}_{\alpha;\alpha} = 0$ and introducing the quantities

$$\eta = \frac{(E^{[1]}_{x;x} - E^{[1]}_{y;y})}{E^{[1]}_{z;z}}; \quad eq = \phi^{[2]}_{zz} = -E^{[1]}_{z;z}, \tag{8.62}$$

we arrive at the expression

$$E_Q(I, m_I) = \frac{e^2qQ}{4\hbar^2 I(2I-1)} \langle I, m_I|(3\hat{I}^2_z - \hat{I}^2) + (\hat{I}^2_x - \hat{I}^2_y)\eta|I, m_I\rangle, \tag{8.63}$$

or, in terms of the shift operators $I_\pm = \hat{I}_x \pm i\hat{I}_y$,

$$E_Q(I, m_I) = \frac{e^2qQ}{4\hbar^2 I(2I-1)} \langle I, m_I|(3\hat{I}^2_z - \hat{I}^2) + (\hat{I}^2_+ + \hat{I}^2_-)\eta/2|I, m_I\rangle. \tag{8.64}$$

The quantity q is what is usually reported as the field gradient, although it has the opposite sign and the fundamental charge has once again been extracted. The quantity η is called the *asymmetry parameter*. In atoms and diatomic molecules it vanishes due to spherical and axial symmetry, respectively, and the interaction energy simplifies to

$$E_Q(I, m_I) = \frac{e^2qQ}{4I(2I-1)}[3m_I^2 - I(I+1)]. \tag{8.65}$$

The hyperfine structure due to nuclear electric quadrupole moments is present in atomic and molecular spectra, but can also be probed directly by nuclear quadrupole resonance (NQR) spectroscopy. Transitions between nuclear sublevels $|I, m_I\rangle$ are then probed by applying a weak magnetic field. Transition moments are given by

$$\langle I', m'_I|-\gamma\hat{\mathbf{I}}\cdot\mathbf{B}|I, m_I\rangle = -\gamma\left\langle I', m'_I\left|\frac{1}{2}(I_-B_+ + I_+B_-) + I_zB_z\right|I, m_I\right\rangle, \tag{8.66}$$

showing that selection rules for such nuclear magnetic dipole transitions are

$$\Delta I = 0; \quad \Delta m_I = 0, \pm 1. \tag{8.67}$$

From experiment one may extract the *nuclear quadrupole coupling constant* (NQCC) e^2qQ/h in frequency units, whereas electronic structure calculations may provide the field gradient q.

The two quantities are related as

$$\text{NQCC [in MHz]} = 234.9647 \times Q\,[\text{in b}] \times q\,[\text{in atomic units } E_h/ea_0^2], \tag{8.68}$$

and allow the determination of nuclear electric quadrupole moments. However, the electric field gradient carries chemical information, as we shall see in the following.

We start by looking at the expectation value for the electronic contribution to the electric field gradient at some nucleus K. At the self-consistent field (SCF) level, it may be written as

$$\langle e\hat{q}_K^{e}\rangle = \sum_{i=1}^{\text{occ}} \left\langle \varphi_i \left| \frac{-e}{4\pi\varepsilon_0}\left(\frac{3z_{iK}^2 - r_{iK}^2}{r_{iK}^5}\right)\right| \varphi_i \right\rangle. \tag{8.69}$$

We place the target nucleus at the origin, and then note that we may write

$$\frac{3z^2 - r^2}{r^5} = \frac{3\cos^2\theta - 1}{r^3} = \sqrt{\frac{16\pi}{5}}\frac{Y_{20}}{r^3}. \tag{8.70}$$

We next introduce atomic basis functions $\varphi_{n\ell m}(\mathbf{r}) = R_{n\ell}(r)Y_{\ell m}(\theta, \phi)$. Atomic matrix elements then separate into radial and angular parts

$$\langle \varphi_{n'\ell'm'}| e\hat{q}_K^{e}|\varphi_{n\ell m}\rangle = \frac{-e}{4\pi\varepsilon_0}\sqrt{\frac{16\pi}{5}}\left\langle R_{n'\ell'}\left|\frac{1}{r^3}\right|R_{n\ell}\right\rangle_r \langle Y_{\ell'm'}|Y_{20}|Y_{\ell m}\rangle_{\theta,\phi}. \tag{8.71}$$

We see that the angular integral is a triple product of spherical harmonics. The general expression for such an integral may be given in terms of Clebsch–Gordan coefficients

$$\langle Y_{\ell_3 m_3}|Y_{\ell_2 m_2}|Y_{\ell_1 m_1}\rangle_{\theta,\phi} = \left[\frac{(2\ell_1+1)(2\ell_2+1)}{4\pi(2\ell_3+1)}\right]^{1/2}\langle \ell_1 m_1, \ell_2 m_2|\ell_3 m_3\rangle\langle \ell_1 0, \ell_2 0|\ell_3 0\rangle. \tag{8.72}$$

Their presence introduces some constraints

$$m_1 + m_2 = m_3 \quad \text{and} \quad \ell_3 \in [|\ell_1 - \ell_2|, \ell_1 + \ell_2]. \tag{8.73}$$

In our particular case, $\ell_2 = 2$ and $m_2 = 0$. This means that within an atomic shell ($\ell_1 = \ell_3$), only diagonal elements contribute ($m_1 = m_3$). There are no contributions from the spherically symmetric s shell ($\ell_1 = \ell_3 = 0$). For other shells, the nonzero matrix elements are

$$\langle \varphi_{n\ell m}|e\hat{q}_K^{e}|\varphi_{n\ell m}\rangle = \frac{e}{2\pi\varepsilon_0}\left\langle R_{n\ell}\left|\frac{1}{r^3}\right|R_{n\ell}\right\rangle_r \frac{3m^2 - \ell(\ell+1)}{(2\ell-1)(2\ell+3)}. \tag{8.74}$$

In particular, for the p shell we find

$$\langle p_m|e\hat{q}_K^{e}|p_m\rangle = \frac{-e}{4\pi\varepsilon_0}\left\langle R\left|\frac{1}{r^3}\right|R\right\rangle_r \times \begin{cases} \frac{4}{5}; & m = 0, \\ -\frac{2}{5}; & |m| = 1. \end{cases} \tag{8.75}$$

An important observation is that the sum over diagonal elements ($m = 0, \pm 1$) gives zero, and in fact this is a general feature of any shell. It is a consequence of *Unsöld's theorem*

$$\sum_{m=-\ell}^{+\ell} |Y_{\ell m}(\theta,\phi)|^2 = \frac{2\ell+1}{4\pi} \quad \Rightarrow \quad \sum_{m=-\ell}^{\ell}\langle Y_{\ell m}|Y_{20}|Y_{\ell m}\rangle_{\theta,\phi} = 0, \tag{8.76}$$

and means that any complete atomic shell becomes spherically symmetric and therefore does not contribute to the field gradient.

This sensitivity of the field gradient to deviations from spherical symmetry makes nonspherical nuclei probes of electronic structure. Consider for instance, as did Townes and Dailey,[5] the hydrogen halides. Consider first the ionic dissociation limit. The electric field gradient of the spherically symmetric anion of fluorine will be strictly zero, but as seen in Table 8.3, individual orbital contributions are not necessarily so. The contribution from the first two orbitals, $1s$ and $2s$, are strictly zero, whereas the contributions from the $2p$ shell follow Eq. (8.75) and allow us to deduce that the expectation value of the inverse cube distance from the atomic origin is 6.60 a_0^{-3}. As the proton approaches the halide, a bonding σ orbital and two nonbonding π orbitals are created. The latter, with four electrons, will contribute

$$4 \times \frac{e}{4\pi\varepsilon_0}\frac{2}{5}\left\langle R_{2p}\left|\frac{1}{r^3}\right|R_{2p}\right\rangle_r = \frac{e}{4\pi\varepsilon_0}\frac{8}{5}\left\langle R_{2p}\left|\frac{1}{r^3}\right|R_{2p}\right\rangle_r, \tag{8.77}$$

to the expectation value of the field gradient q at the fluorine nucleus. In fact, from Table 8.3 we can deduce that the expectation value $\langle r^{-3}\rangle$ is now 7.21 a_0^{-3}, indicating that the $2p$ orbital has shrunk. We may tentatively write the σ orbital as

$$\varphi_\sigma = N(c_1F2p_0 + c_2H1s); \quad N = \frac{1}{\sqrt{c_1^2 + c_2^2 + 2c_1c_2S}}, \tag{8.78}$$

where S is the overlap between the valence p_0 orbital of fluorine and the hydrogen $1s$ orbital. The contribution to the expectation value from this orbital will be

$$q^{e}_{F,\sigma} = N^2\{c_1^2\langle F2p_0|e\hat{q}^{e}_{F}|F2p_0\rangle + 2c_1c_2\langle F2p_0|e\hat{q}^{e}_{F}|H1s\rangle + c_2^2\langle H1s|e\hat{q}^{e}_{F}|H1s\rangle\}.$$

Since we have seen that the operator $e\hat{q}^{e}_{F}$ has an inverse cubic dependence with respect to the distance from the halide nucleus, we expect the second and third terms to be quite small. They will not be zero, though, but the third term will be partly canceled by the nuclear contribution, which has the opposite sign but is of approximately the same magnitude. It is therefore the first

Table 8.3 Orbital contributions, multiplied with orbital occupation numbers of two, to the field gradient eq_F (in E_h/ea_0^2) at the fluorine nucleus of fluoride and hydrogen fluoride.

	F^-	HF
1	0.000	−0.010
2	0.000	−0.403
3	−10.561	−8.680
4	5.281	5.768
5	5.281	5.768
q^{e}_{F}	0.000	2.444
q^{n}_{F}	0.000	0.385
q_F	0.000	2.828

Results are obtained at the restricted DFT/PBE0 level of theory.

5 C. H. Townes and B. P. Dailey. *J. Chem. Phys.*, 17:78296, 1949.

term that dominates, and we can approximate the total expectation value by

$$q_X = q_X^{\text{e}} + q_X^{\text{n}} \approx \frac{e}{4\pi\varepsilon_0}\frac{8}{5}(1 - N^2c_1^2)\left\langle R_{2p}\left|\frac{1}{r^3}\right|R_{2p}\right\rangle_r, \tag{8.79}$$

which clearly shows that the expectation value is sensitive to the degree of ionicity of the bond.

Accurate calculations of the electric field gradient are rather challenging. The operator probes the electron density in the vicinity of nuclei, which makes it sensitive to relativistic effects and requires the use of basis functions with flexibility in the core region. At the same time, the contribution from (closed) core shells is zero and so the operator probes the tails of valence orbitals, which makes it quite sensitive to electron correlation effects.[6]

8.2 Zeeman Interaction and Electron Paramagnetic Resonance

A colleague who met me strolling rather aimlessly in the beautiful streets of Copenhagen [1922] said to me in a friendly manner, "You look very unhappy"; whereupon I answered fiercely, "How can one look happy when he is thinking about the anomalous Zeeman effect?"

Wolfgang Pauli (1945)

In 1897, the Dutch physicist Pieter Zeeman reported a broadening of the sodium *D*-line upon the application of a magnetic field. His colleague Hendrik A. Lorentz, using his classical model of the electron that we encountered in Section 3.5.4 together with the Lorentz force, Eq. (3.123), attempted to provide an explanation of the effect: Upon application of a magnetic field, the single-frequency, isotropic oscillatory motion of the electron becomes a composite one. The frequency of oscillations along the axis of the field is unchanged, whereas the frequency of circular motion in the plane perpendicular to the field will increase in one direction and diminish in the other. Zeeman did not have sufficient resolution in his experiment to resolve the predicted triplet, but the polarization of the emitted radiation was in agreement with the predictions of Lorentz. Very soon, however, more complicated splittings were observed in atomic spectra when applying magnetic fields. This anomalous Zeeman effect could not be explained until the advent of quantum mechanics, and then in particular with the introduction of the electron spin.

Let us look at the quantum-mechanical description of the Zeeman effect. From our discussion of multipole expansions in Section 3.3.2, we may expect the interaction to be of the form

$$\langle \hat{H}_{\text{M1}} \rangle = -\mathbf{m}_{\text{e}}^{[1]} \cdot \mathbf{B}, \tag{8.80}$$

with the generic formula for the magnetic dipole moment $\mathbf{m}^{[1]}$ given in Eq. (3.191). The vector potential associated with a uniform magnetic field is given by Eq. (3.212) and when the vector potential is introduced into the nonrelativistic Hamiltonian, Eq. (3.158), three interaction terms appear:

$$\hat{H}^{\text{OZ}} = \sum_i \frac{e}{2m_{\text{e}}}\hat{\mathbf{l}}_{\text{G}}(i) \cdot \mathbf{B}, \tag{8.81}$$

$$\hat{H}^{\text{SZ}} = \sum_i \frac{e}{m_{\text{e}}}\hat{\mathbf{s}}(i) \cdot \mathbf{B}, \tag{8.82}$$

$$\hat{H}_{\text{BB}}^{\text{dia}} = \sum_i \frac{e^2}{8m_{\text{e}}}[B^2 r_{iG}^2 - (\mathbf{B} \cdot \mathbf{r}_{iG})(\mathbf{r}_{iG} \cdot \mathbf{B})]. \tag{8.83}$$

6 An illustrative example is given in J. N. P. van Stralen and L. Visscher. *J. Chem. Phys.*, 117:3103, 2002.

The first two terms, denoted the orbital and spin Zeeman terms, respectively, are linear in the magnetic field and allow us to extract the paramagnetic contributions to the magnetic dipole operator of the electron

$$\hat{\mathbf{m}}_{\mathrm{e}}^{[1]\mathrm{para}} = -\sum_i \frac{e}{2m_{\mathrm{e}}}[\hat{\mathbf{l}}_{\mathrm{G}}(i) + g_{\mathrm{e}}\hat{\mathbf{s}}(i)]. \tag{8.84}$$

The common prefactor is the gyromagnetic ratio as expected for a classical rotating body, see Eq. (8.34), but for the spin contribution it is corrected by the electronic g-factor labeled g_{e} and which has the value 2 at this level of theory. If quantum electrodynamical (QED) effects are included, however, the value is slightly modified and becomes

$$g_{\mathrm{e}} = 2.0023193043617(15). \tag{8.85}$$

The final, diamagnetic term of the interaction Hamiltonian arises from the diamagnetic contribution of the nonrelativistic current density, Eq. (3.168), and tells us that the orbital contribution of the electron magnetic dipole is proportional to the *kinetic angular momentum* $\mathbf{K} = \mathbf{r} \times m\mathbf{v}$ rather than the canonical angular momentum $\mathbf{l} = \mathbf{r} \times \mathbf{p}$. The two quantities are identical *only* in the absence of magnetic fields. This becomes clear by looking at the Zeeman interaction term arising from the Dirac Hamiltonian given in Eq. (3.151):

$$\hat{H}^{Z}_{rel} = \sum_i \frac{1}{2} ec\mathbf{B} \cdot (\mathbf{r}_{iG} \times \boldsymbol{\alpha}(i)) = -\hat{\mathbf{m}}_{\mathrm{e}}^{[1]} \cdot \mathbf{B}; \quad \hat{\mathbf{m}}_{\mathrm{e}}^{[1]} = -\sum_i \frac{e}{2m_{\mathrm{e}}}(\mathbf{r}_{iG} \times m_{\mathrm{e}}c\boldsymbol{\alpha}(i)), \tag{8.86}$$

Here we explicitly see that the relativistic velocity operator $c\boldsymbol{\alpha}$ (see Section 2.3) rather than the momentum operator $\hat{\mathbf{p}}$ appears in the expression for the magnetic dipole operator.

To get some feeling for the Zeeman interaction, we again turn to atoms where angular momentum algebra (see Appendix G) allows us to derive explicit expressions. In the weak-field case, we can ignore the diamagnetic term. We shall also place the gauge origin and the atom at the origin of our coordinate system and let the magnetic field be aligned along the z axis. The interaction Hamiltonian then becomes

$$\hat{H}_{\mathrm{M1}} = \frac{eB}{2m} \sum_i (\hat{l}_z(i) + g_{\mathrm{e}}\hat{s}_z(i)) = \frac{eB}{2m}(\hat{L}_z + g_{\mathrm{e}}\hat{S}_z). \tag{8.87}$$

We shall assume that the splittings due to the Zeeman interaction are smaller than the splitting of atomic ^{2S+1}L terms due to spin–orbit coupling. This is a very reasonable assumption as can be understood from the following argument: The spin–orbit interaction is the interaction of the electron spin with the magnetic field generated by charged particles, such as the atomic nucleus, in relative motion. We can therefore write the effective spin–orbit operator, see Eq. (G.9), as

$$\hat{H}^{\mathrm{eff}}_{\mathrm{SO}} = \zeta(^{2S+1}L)\hat{\mathbf{L}} \cdot \hat{\mathbf{S}} = \frac{e}{m_{\mathrm{e}}}\hat{\mathbf{S}} \cdot \mathbf{B}^{\mathrm{eff}}. \tag{8.88}$$

We then find that a spin–orbit constant ζ on the order of 100 cm^{-1} corresponds to an effective magnetic field B^{eff} of 100 T, which is far beyond typical laboratory field strengths.

We seek an expression for matrix elements of the interaction Hamiltonian $\hat{H}_{\mathrm{SO}}$ within the manifold $\{|L, S; J, M_J\rangle\}$ with fixed values of L, S, and J. To accomplish this, we make use of the projection theorem in Eq. (G.6). For the orbital and spin parts, the interaction matrix elements are given by

$$\langle L,S;J,M_J|\hat{L}_z|L,S;J,M'_J\rangle = \frac{\langle L,S;J,M_J|\hat{\mathbf{L}}\cdot\hat{\mathbf{J}}|L,S;J,M_J\rangle}{\hbar^2 J(J+1)}\langle L,S;J,M_J|\hat{J}_z|L,S;J,M'_J\rangle$$
$$= \frac{\langle L,S;J,M_J|\hat{\mathbf{L}}\cdot\hat{\mathbf{J}}|L,S;J,M_J\rangle}{\hbar^2 J(J+1)}\hbar M_J\delta_{M_J M'_J}, \tag{8.89}$$

$$\langle L,S;J,M_J|\hat{S}_z|L,S;J,M'_J\rangle = \frac{\langle L,S;J,M_J|\hat{\mathbf{S}}\cdot\hat{\mathbf{J}}|L,S;J,M_J\rangle}{\hbar^2 J(J+1)}\hbar M_J\delta_{M_J M'_J}, \tag{8.90}$$

which are clearly diagonal. To evaluate the matrix elements on the right-hand side, we use

$$S^2 = (\mathbf{J}-\mathbf{L})\cdot(\mathbf{J}-\mathbf{L}) = J^2 + L^2 - 2\mathbf{L}\cdot\mathbf{J}, \tag{8.91}$$
$$L^2 = (\mathbf{J}-\mathbf{S})\cdot(\mathbf{J}-\mathbf{S}) = J^2 + S^2 - 2\mathbf{S}\cdot\mathbf{J}, \tag{8.92}$$

such that the diagonal elements can be expressed as

$$\langle L,S;J,M_J|\hat{L}_z|L,S;J,M_J\rangle = \frac{J(J+1)+L(L+1)-S(S+1)}{2J(J+1)}M_J, \tag{8.93}$$
$$\langle L,S;J,M_J|\hat{S}_z|L,S;J,M_J\rangle = \frac{J(J+1)+S(S+1)-L(L+1)}{2J(J+1)}M_J. \tag{8.94}$$

Each atomic J-level is accordingly split on the azimuthal quantum number M_J with energy

$$E(J,M_J;L,S) = E(^{2S+1}L_J) + g_J\mu_B B M_J, \tag{8.95}$$

where appears the Landé g-factor defined as

$$g_J = \frac{J(J+1)+L(L+1)-S(S+1)}{2J(J+1)} + g_e\frac{J(J+1)+S(S+1)-L(L+1)}{2J(J+1)}, \tag{8.96}$$

and the Bohr magneton

$$\mu_B = \frac{e\hbar}{2m_e}. \tag{8.97}$$

Transition energies are then given by

$$h\nu = h\nu_0 + \mu_B B(g'_J M'_J - g_J M_J), \tag{8.98}$$

where ν_0 refers to the transition frequency in the absence of the magnetic field and primed quantities refer to the upper state. In passing, we note that at 1 T, the quantity $\mu_B B$ corresponds to about 14 GHz, so the Zeeman splittings are typically in the microwave region.

As a specific example, let us return to the sodium doublet.[7] The D_1 and D_2 lines refer to the transitions $3s(^2S_{1/2}) \rightarrow 3p(^2P_{1/2})$ (589.592 nm) and $3s(^2S_{1/2}) \rightarrow 3p(^2P_{3/2})$ (588.995 nm), respectively. Table 8.4 gives the values of g_J and $g_J M_J$ for each participating atomic level. In order to predict the spectrum, we need to know, in addition, the electric dipole selection rules for M_J

$$\Delta M_J = 0, \pm 1; \quad M_J = 0 \nleftrightarrow M'_J = 0 \text{ if } \Delta J = 0. \tag{8.99}$$

Using these, we find that in the presence of a magnetic field, the sodium D_1 and D_2 lines split into a quartet and a sextet, respectively,[8] as shown in Figure 8.5, and they are examples of the anomalous Zeeman effect.

---

7 This example was motivated by D. Weaire and S O'Connor. *Ann. Sci.*, 44:617, 1987.

8 In the case of linearly polarized light this strictly corresponds to transverse observation, where the electric polarization vector is perpendicular to the applied magnetic field. In longitudinal observation, the $\Delta M_J = 0$ transitions are absent.

Table 8.4 Landé g-factors for states in sodium related to the D-line transitions.

Level	J	L	S	g_J	$g_J M_J$			
$^2P_{3/2}$	3/2	1	1/2	4/3	−2	−2/3	2/3	2
$^2P_{1/2}$	1/2	1	1/2	2/3		−1/3	1/3	
$^2S_{1/2}$	1/2	0	1/2	2		−1	1	

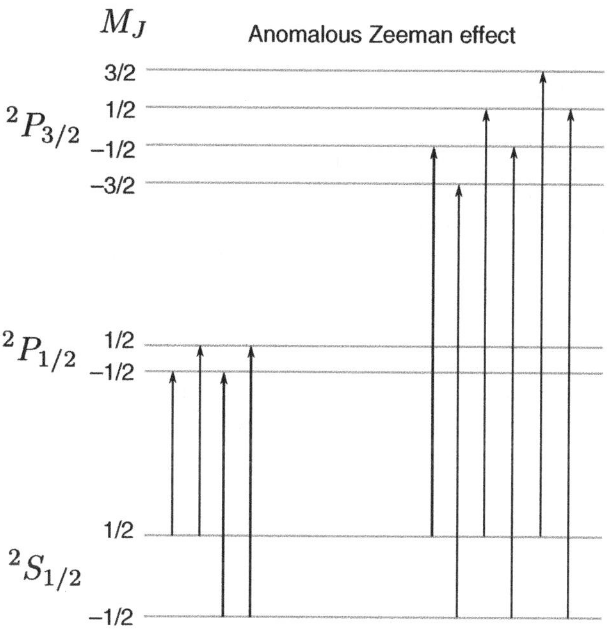

Figure 8.5 Zeeman splitting of the D-lines in sodium.

In the special case $S = 0$, one will have $J = L$ and $g_J = 1$ and the energy of a given level is then

$$E(J, M_J; L, S) = E(^{2S+1}L_J) + \mu_B BM_J, \quad (8.100)$$

resulting in transition energies

$$h\nu = h\nu_0 + \mu_B B\Delta M_J. \quad (8.101)$$

As an example, we may consider the red line of cadmium at 643.847 nm. It corresponds to the transition $5s5p(^1P_1) \rightarrow 5s5d(^1D_2)$. In this case the line splits into a triplet, as seen in Figure 8.6, corresponding to the normal Zeeman effect.

In the absence of spin–orbit coupling, the energy of a given level is

$$E(L, M_L, S, M_S) = E(^{2S+1}L) + \mu_B B(M_L + g_e M_S), \quad (8.102)$$

and transition energies are given by

$$h\nu = h\nu_0 + \mu_B B\Delta M_L. \quad (8.103)$$

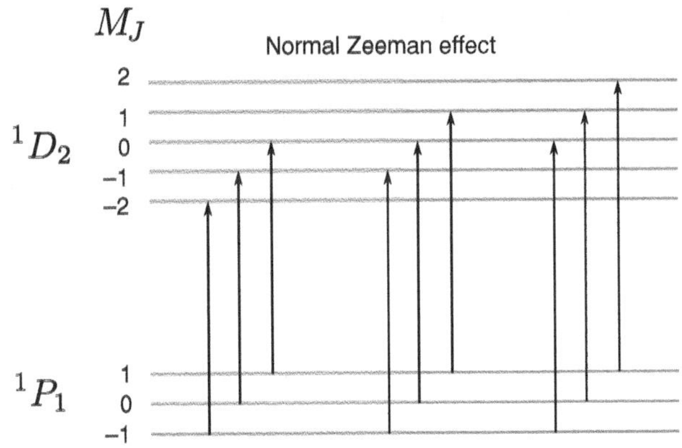

Figure 8.6 Zeeman splitting of the red line in cadmium.

The ΔM_S term is absent since $\Delta M_S = 0$ in electric dipole transitions. To lowest order, this situation also corresponds to the *Paschen–Back effect* where the applied magnetic field is stronger than the effective field associated with spin–orbit coupling.

Zeeman splittings in molecules or solids are often described through the introduction of a phenomenological Hamiltonian

$$\hat{H}^Z = \mu_B \mathbf{B} \cdot g \cdot \hat{\mathbb{S}}, \tag{8.104}$$

where the magnitude of the pseudo-spin operator $\hat{\mathbb{S}}$ is chosen such that its multiplicity corresponds to the degeneracy of the level under consideration, just as we saw for the phenomenological Hamiltonian for hyperfine interactions, Eq. (8.54). The 3×3 matrix g is often referred to as the *g-tensor*, but does not generally transform as one with a change of basis due to the presence of the pseudo-spin operator in the phenomenological Hamiltonian.[9] We shall therefore, in the following, refer to it as the g-matrix. In the absence of spin–orbit coupling and orbital degeneracy, pseudo-spin becomes the regular electronic spin and the g-matrix becomes isotropic and equal to $g = g_e I_3$.

Let us look at the simplest, but chemically relevant case of a doublet. In the presence of spin–orbit coupling spin symmetry is lost, but it can to some extent be replaced by time-reversal symmetry (see Section 4.2). We shall now specifically look at a Kramers doublet, that is, two states $\{|0\rangle, |\bar{0}\rangle\}$ that are degenerate and related by time-reversal symmetry in the absence of the external magnetic field. In this case, the magnitude of the pseudo-spin is $\mathbb{S} = 1/2$ and can be represented by the Pauli spin matrices, Eq. (2.49), and we may write the phenomenological Hamiltonian, Eq. (8.104), as

$$\hat{H}^Z = \mathbf{V} \cdot \boldsymbol{\sigma} = V_x\sigma_x + V_y\sigma_y + V_z\sigma_z = \begin{bmatrix} V_z & V_- \\ V_+ & -V_z \end{bmatrix}; \quad V_\alpha = \frac{1}{2}\hbar\mu_B B_\beta g_{\beta\alpha}. \tag{8.105}$$

The resulting 2×2 matrix can be seen to take the form of a time-antisymmetric matrix [Eq. (4.46) with $\theta_t = -1$]. This may be compared with the 2×2 matrix representation of the true, relativistic interaction operator, Eq. (8.86),

$$\begin{bmatrix} \langle 0|\hat{H}^Z_{rel}|0\rangle & \langle 0|\hat{H}^Z_{rel}|\bar{0}\rangle \\ \langle \bar{0}|\hat{H}^Z_{rel}|0\rangle & \langle \bar{0}|\hat{H}^Z_{rel}|\bar{0}\rangle \end{bmatrix}. \tag{8.106}$$

9 See for instance L. F. Chibotaru, A. Ceulemans, and H. Bolvin. *Phys. Rev. Lett.*, 101:033003, 2008.

Because the relativistic Zeeman operator, Eq. (8.86), is antisymmetric under time reversal, the matrix will have the same time-antisymmetric structure as in Eq. (8.105). This allows for a straightforward identification of the elements of the g-matrix

$$g_{\alpha x} = \frac{2}{\mu_B}\frac{\partial}{\partial B_\alpha}\mathrm{Re}\langle 0|\hat{H}^Z_{rel}|\bar{0}\rangle; \quad g_{\alpha y} = -\frac{2}{\mu_B}\frac{\partial}{\partial B_\alpha}\mathrm{Im}\langle 0|\hat{H}^Z_{rel}|\bar{0}\rangle; \quad g_{\alpha z} = \frac{2}{\mu_B}\frac{\partial}{\partial B_\alpha}\langle 0|\hat{H}^Z_{rel}|0\rangle, \tag{8.107}$$

and shows that at the relativistic level of theory, the elements of the g-matrix can be calculated as first-order properties.

Let us compare this with the nonrelativistic approach. We replace the relativistic operator of Eq. (8.106) by

$$\hat{H}^Z = \frac{e}{2m}(\hat{\mathbf{L}}_G + g_e\hat{\mathbf{S}})\cdot\mathbf{B}, \tag{8.108}$$

which is valid in the weak-field case (we have dropped the diamagnetic contribution to the nonrelativistic Zeeman operator). We represent the zeroth-order Kramers doublet (no spin–orbit) by a spin doublet

$$|0^{(0)}\rangle = |\Phi_0^{(0)};+1/2\rangle; \quad |\bar{0}^{(0)}\rangle = |\Phi_0^{(0)};-1/2\rangle, \tag{8.109}$$

where $\Phi_0^{(0)}$ is the real, spatial part of the zeroth-order wave function. Here we have associated the $M_S = +1/2$ ($M_S = -1/2$) component with the unbarred (barred) Kramers partner, but it is important to note that this is arbitrary; one can equally well do the opposite, or any intermediate, assignment. When forming the 2×2 representation of the nonrelativistic Zeeman operator in this basis, the contribution from the orbital angular momentum vanishes. Diagonal elements such as

$$\langle\Phi_0^{(0)};\ 1/2|\hat{\mathbf{L}}\cdot\mathbf{B}|\Phi_0^{(0)};+1/2\rangle = \langle\Phi_0^{(0)}|\hat{\mathbf{L}}\cdot\mathbf{B}|\Phi_0^{(0)}\rangle\langle+1/2|+1/2\rangle = \langle\Phi_0^{(0)}|\hat{\mathbf{L}}\cdot\mathbf{B}|\Phi_0^{(0)}\rangle = 0, \tag{8.110}$$

vanish because it reduces to the expectation value of a purely imaginary operator with respect to the purely real spatial wave function. Off-diagonal elements vanish due to orthonormality of the spin functions, for instance,

$$\langle\Phi_0^{(0)};+1/2|\hat{\mathbf{L}}\cdot\mathbf{B}|\Phi_0^{(0)};-1/2\rangle = \langle\Phi_0^{(0)}|\hat{\mathbf{L}}\cdot\mathbf{B}|\Phi_0^{(0)}\rangle\langle+1/2|-1/2\rangle = 0. \tag{8.111}$$

From the spin contribution we get

$$\frac{1}{2}\mu_B g_e\begin{bmatrix} B_z & B_- \\ B_+ & -B_z \end{bmatrix}, \tag{8.112}$$

and from a comparison with Eq. (8.105), we find that the zeroth-order g-matrix is diagonal and isotropic, that is, $g^{(0)} = g_e I_3$. To get a nontrivial result, we have to go beyond zeroth order. The first-order contribution to the 2×2 interaction matrix reads

$$\begin{bmatrix} \langle 0^{(0)}|\hat{H}^Z|0^{(1)}\rangle & \langle 0^{(0)}|\hat{H}^Z|\bar{0}^{(1)}\rangle \\ \langle\bar{0}^{(0)}|\hat{H}^Z|0^{(1)}\rangle & \langle\bar{0}^{(0)}|\hat{H}^Z|\bar{0}^{(1)}\rangle \end{bmatrix} + \begin{bmatrix} \langle 0^{(1)}|\hat{H}^Z|0^{(0)}\rangle & \langle 0^{(1)}|\hat{H}^Z|\bar{0}^{(0)}\rangle \\ \langle\bar{0}^{(1)}|\hat{H}^Z|0^{(0)}\rangle & \langle\bar{0}^{(1)}|\hat{H}^Z|\bar{0}^{(0)}\rangle \end{bmatrix}. \tag{8.113}$$

From the time-antisymmetric nature of the operator $\hat{H}_Z$, we can compress the above expression to

$$2\begin{bmatrix} \mathrm{Re}\langle 0^{(0)}|\hat{H}^Z|0^{(1)}\rangle & \langle 0^{(0)}|\hat{H}^Z|\bar{0}^{(1)}\rangle \\ \langle\bar{0}^{(0)}|\hat{H}^Z|0^{(1)}\rangle & -\mathrm{Re}\langle 0^{(0)}|\hat{H}^Z|0^{(1)}\rangle \end{bmatrix}. \tag{8.114}$$

The first-order functions read

$$|0^{(1)}\rangle = \sum_{i\neq 0}(c_{\alpha i}|\Phi_i^{(0)};+1/2\rangle + c_{\beta i}|\Phi_i^{(0)};-1/2\rangle), \tag{8.115}$$

$$|\overline{0}^{(1)}\rangle = \sum_{i\neq 0}(-c_{\beta i}^*|\Phi_i^{(0)};+1/2\rangle + c_{\alpha i}^*|\Phi_i^{(0)};-1/2\rangle), \tag{8.116}$$

where the relation between expansion coefficients for $|0^{(1)}\rangle$ and $|\overline{0}^{(1)}\rangle$ follows from time-reversal symmetry [compare with Eq. (4.41)]. The first-order functions are chosen to be orthonormal to the zeroth-order one (intermediate normalization). A consequence of this is that the spin contribution to the first-order matrix vanishes. However, the angular momentum part of the interaction will generally introduce off-diagonal contributions (anisotropy) to the g-matrix. What we learn from this exercise is that in a nonrelativistic approach, the g-matrix is calculated as a second-order property.

We can also relate our theoretical considerations to the splitting actually observed in experiment. Upon diagonalization of the time-antisymmetric matrix in Eq. (8.105), we obtain the eigenvalues

$$\lambda_{\pm} = \pm\sqrt{|V_x|^2 + |V_y|^2 + |V_z|^2}, \tag{8.117}$$

which means that the energy splitting is

$$\Delta E = 2\sqrt{|V_x|^2 + |V_y|^2 + |V_z|^2}. \tag{8.118}$$

Starting from the 2×2 matrix of the phenomenological Hamiltonian $\hat{H}_Z$, we obtain

$$\Delta E = \mu_B[B_\alpha B_\beta G_{\alpha\beta}]^{1/2}, \tag{8.119}$$

where appears the G-tensor defined by $G = \mathbf{g}\mathbf{g}^T$. Starting from the matrix of Eq. (8.106), we obtain

$$\Delta E = 2\left[\sum_{m=0,\overline{0}}|\langle 0|\hat{H}^Z|m\rangle|^2\right]^{1/2} = 2\left[B_\alpha B_\beta \sum_{m=0,\overline{0}}\left\langle 0\left|\frac{\partial \hat{H}^Z}{\partial B_\alpha}\right|m\right\rangle\left\langle m\left|\frac{\partial \hat{H}^Z}{\partial B_\beta}\right|0\right\rangle\right]^{1/2}. \tag{8.120}$$

Since these results are independent of the actual orientation of the magnetic field, we conclude that the G-tensor extracted from experiment can be connected to theory as

$$G_{\alpha\beta} = \frac{2}{\mu_B}\sum_{m=0,\overline{0}}\left\langle 0\left|\frac{\partial \hat{H}^Z}{\partial B_\alpha}\right|m\right\rangle\left\langle m\left|\frac{\partial \hat{H}^Z}{\partial B_\beta}\right|0\right\rangle. \tag{8.121}$$

A nice feature of the g-matrix is that it provides a very compact 3×3 representation of possibly complex interactions. In the case of $\mathbb{S} = 1/2$ there is no loss of information, in the sense that the 2×2 matrix representation of the true interaction operator can be expanded in the corresponding spin matrices (the number of degrees of freedom matches). However, for $\mathbb{S} > 1/2$ this is no longer guaranteed and higher orders of the pseudo-spin operator may have to be introduced into the phenomenological Hamiltonian in order to account for the actual splittings.

Zeeman splittings may be observed in optical spectra, along with hyperfine structure, but a more direct and precise method for their observation is provided by electron paramagnetic resonance (EPR), which has become a tool for the investigation of open-shell systems, such as organic radicals or transition metal complexes. In EPR the sample is placed in a magnetic field and Zeeman splitting, as well as additional structure, is probed by electromagnetic radiation, typically in the microwave range. The underlying principles of EPR are very much akin to NMR, which will be discussed at length in Section 8.8.

A phenomenological spin Hamiltonian is used to account for the spectral features of EPR. In its simplest form it is given by

$$\hat{\mathcal{H}}^{\mathrm{EPR}} = \mu_{\mathrm{B}}\mathbf{B}\cdot g\cdot \hat{\mathbb{S}} + \hat{\mathbf{I}}_K\cdot A_K\cdot \hat{\mathbb{S}} + \hat{\mathbb{S}}\cdot D\cdot \hat{\mathbb{S}}. \tag{8.122}$$

The first two terms have already been encountered in the form of Eqs. (8.104) and (8.54). They represent the Zeeman interaction and hyperfine interaction due to nuclear magnetic dipole moments, respectively. The third term describes *zero-field splitting* (ZFS). It has the form of a spin–spin interaction and evidently requires pseudo-spin $\mathbb{S} \geq 1$ (beyond Kramers doublets) in order to occur. It leads to splittings of ${}^{2S+1}\Gamma$ electronic terms even in the absence of external fields, hence the name. ZFS arises from direct spin–spin coupling between electrons and/or from spin–orbit interaction. The former interaction dominates in systems with light atoms only, such as organic radicals, whereas spin–orbit interaction takes over when relativistic effects become more predominant, for instance in transition metal complexes. Both effects are accounted for by the four-component Dirac–Coulomb–Breit Hamiltonian,[10] and as such come "for free" in calculations based on this Hamiltonian.

8.3 Polarizabilities

We discussed in Section 5.2 the microscopic molecular electric properties, and in particular the effects of a time-dependent electric field on the dipole moment of a molecule, that is, the induced polarization in the molecule. It was shown that the induced polarization could be written as a power series in the strength of the applied electric field, relating molecular response properties such as the electric-dipole polarizability $\alpha_{\alpha\beta}(-\omega;\omega)$ to the linear response function involving two dipole operators $\langle\langle\hat{\mu}_\alpha;\hat{\mu}_\beta\rangle\rangle_\omega$, and similarly for higher order responses such as the first-order hyperpolarizability $\beta_{\alpha\beta\gamma}(-\omega_\sigma;\omega_1,\omega_2)$ and the second-order hyperpolarizability $\gamma_{\alpha\beta\gamma\delta}(-\omega_\sigma;\omega_1,\omega_2,\omega_3)$. We also discussed their origin by analyzing their expression for a two-level system. We will here briefly revisit some of these aspects and also consider some more qualitative properties of polarizabilities. Our main emphasis will be on the electric-dipole polarizability, being the dominating response of a material to an external electric field. One particularly important aspect of the polarizability in relation to macroscopic observables is the relation between the polarizability and the refractive index of the material. This has already been addressed in Section 3.5.2, so we will not repeat this very important area of application of the polarizability here. We will also briefly consider higher order responses, as these are becoming increasingly important with the development of nonlinear optical materials with unique optical properties. The coupling with the magnetic field component of the frequency-dependent light, Eq. (3.186), will be deferred to Section 8.6 where we will be discussing birefringences and dichroisms.

8.3.1 Linear Polarizability

Polarizability is often introduced already in the first course of general chemistry, where it is used to explain the aggregation state of the halogens through the ease by which the electron

10 The Dirac–Coulomb Hamiltonian contains one-electron spin–orbit interaction as well as two-electron spin–same orbit interaction, whereas the Breit term, Eq. (3.161), provides two-electron spin–other orbit interaction and direct spin–spin coupling. The distinction between spin–same and spin–other orbit interactions arises from the fact that the electronic Hamiltonian within the Born–Oppenheimer approximation is formulated in the nuclear frame and refers to relative motion of the reference and "other" electron with respect to this frame. A detailed discussion can be found in T. Saue. *ChemPhysChem*, 12:3077, 2011.

density can be shifted as a consequence of charge fluctuations in nearby halogen molecules. These charge fluctuations may lead to a nonisotropic electron density that will give rise to a local electric field. This electric field, in turn, can induce a dipole moment in atoms and molecules, as illustrated in Figure 1.2. From our discussion in Chapter 5, we know that for nonpolar molecules, the leading-order contribution to the induced dipole moment is the molecular polarizability, and we can therefore expect that these weak intermolecular forces, the so-called *London* or *dispersion* forces, can be related to the molecular polarizability.

However, before discussing weak intermolecular forces, we will first look in somewhat qualitative terms at the polarizability itself. Polarizability is related to the linear response function through Eq. (5.169) and is repeated here for ease of reference

$$\alpha_{\alpha\beta}(-\omega;\omega) = \frac{1}{\hbar}\sum_n{}' \left[\frac{\langle 0|\hat{\mu}_\alpha|n\rangle\langle n|\hat{\mu}_\beta|0\rangle}{\omega_{n0}-\omega} + \frac{\langle 0|\hat{\mu}_\beta|n\rangle\langle n|\hat{\mu}_\alpha|0\rangle}{\omega_{n0}+\omega}\right]. \tag{8.123}$$

Let us first assume a static electric field perturbation, and secondly that we can approximate the excitation energies occurring in the denominator with an average excitation energy ΔE. This is sometimes referred to as the Unsöld approximation, and for the isotropic polarizability gives

$$\begin{aligned}\overline{\alpha} &\approx \frac{2}{3\Delta E}\sum_n{}' \langle 0|\hat{\mu}_\alpha|n\rangle\langle n|\hat{\mu}_\alpha|0\rangle \\ &= \frac{2}{3\Delta E}\left[\sum_n \langle 0|\hat{\mu}_\alpha|n\rangle\langle n|\hat{\mu}_\alpha|0\rangle - \langle 0|\hat{\mu}_\alpha|0\rangle\langle 0|\hat{\mu}_\alpha|0\rangle\right] \\ &= \frac{2[\langle\hat{\mu}^2\rangle - \langle\hat{\mu}\rangle^2]}{3\Delta E},\end{aligned} \tag{8.124}$$

where we have completed the sum-over-states expression to allow for identification of the resolution of the identity. The static polarizability can thus be considered an approximation to the variance of the dipole moment from its equilibrium value. Our qualitative understanding of the weak forces between nonpolar molecules as due to the instantaneous variations in the charge density can now be seen to be closely related to the molecular polarizability. Even for molecules without a permanent dipole moment, the dipole fluctuations will be nonzero and lead to a nonvanishing molecular polarizability.

If we now limit ourselves to nonpolar system, and use the expression $\hat{\mu}_\alpha = -er_\alpha$, the isotropic polarizability in Eq. (8.124) can be written as

$$\overline{\alpha} \approx \frac{2e^2\langle r^2\rangle}{3\Delta E}, \tag{8.125}$$

and is thus approximately proportional to the second moment of electronic charge of the system. The larger the system, the larger the polarizability, and thus we can, through this relation, rationalize why fluorine and chlorine are gases, bromine a liquid at room temperature, and iodine a solid material.

The notion that the polarizability can be related to the size of the atomic or molecular system suggests that the polarizability of a molecular system can be written as a sum of atomic, bond, or group polarizabilities. Although approximate, such approaches often give surprisingly good results. The simplest of these approaches arise from assuming that we can write the polarizability as the sum of *free-atom polarizabilities* (FAP)

$$\alpha_{\alpha\beta} \approx \alpha_{\alpha\beta}^{\mathrm{FAP}} = \sum_K \alpha_{\alpha\beta}^K, \tag{8.126}$$

where the summation runs over all atoms in the molecule. This approximation gives in practice rather poor results as the atoms are significantly perturbed when forming molecules. A better result is obtained by fitting a large data set of molecular polarizabilities into a set of *dressed-atom polarizabilities* (DAP)

$$\alpha_{\alpha\beta} \approx \alpha_{\alpha\beta}^{\mathrm{DAP}} = \sum_K \tilde{\alpha}_{\alpha\beta}^K. \tag{8.127}$$

This dressed-atom model can give median errors below 3%, and is thus a surprisingly accurate approximation to the polarizability of the molecule. In general, it is found that the DAP are smaller than those of the free atom. This is perhaps not too surprising, considering that the polarizability is related to the size of the system and that bonding in general leads to a contraction of the electron density outside the bonding region. This is sometimes formulated as the *principle of minimum polarizability,* that is, molecules seek to minimize the polarizability.

We illustrate the additivity of the polarizability for the case of normal alkanes C_nH_{2n+2} in Figure 8.7, where we have plotted the in-chain component of the polarizability α_{xx} together with the two components perpendicular to the chain length, with α_{yy} being the in-plane component and α_{zz} the component perpendicular to the plane of the alkane chain, as a function of the number of methylene (CH_2) units in the chain. We note that for the components perpendicular to the chain length, the polarizability per methylene unit is slowly decreasing and seems to be converging to a stable value for each methylene unit. The xx-component, on the other hand, shows a weak increase with increasing chain length, which does not appear to saturate.

This increase with increasing chain length becomes much more pronounced if instead we consider the conjugated *trans*-polyenes. As seen in Figure 8.8, the long in-plane component of the polarizability displays a very strong dependence on the chain length. We can rationalize this observed nonlinear increase with increasing chain length if we approximate the electrons in the carbon chain as independent particles in a one-dimensional box. Inserting the eigenfunctions and eigenvalues for a particle in a box

$$\Psi_n(x) = \sqrt{\frac{2}{L}} \sin\left(\frac{n\pi x}{L}\right); \quad E_n = \frac{n^2 h^2}{8 m_e L^2}; \quad n = 1, 2, 3, \ldots \tag{8.128}$$

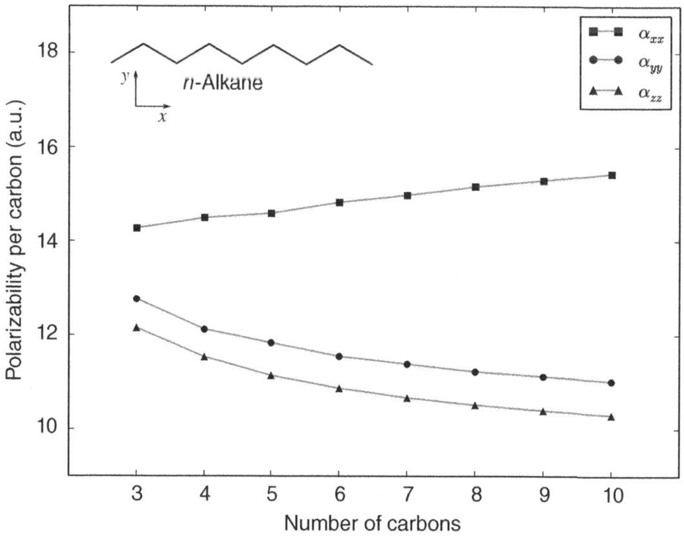

Figure 8.7 Polarizability per carbon atom for the n-alkanes C_nH_{2n+2}. Results are obtained at the time-dependent Hartree–Fock level of theory with use of the aug-cc-pVDZ basis set.

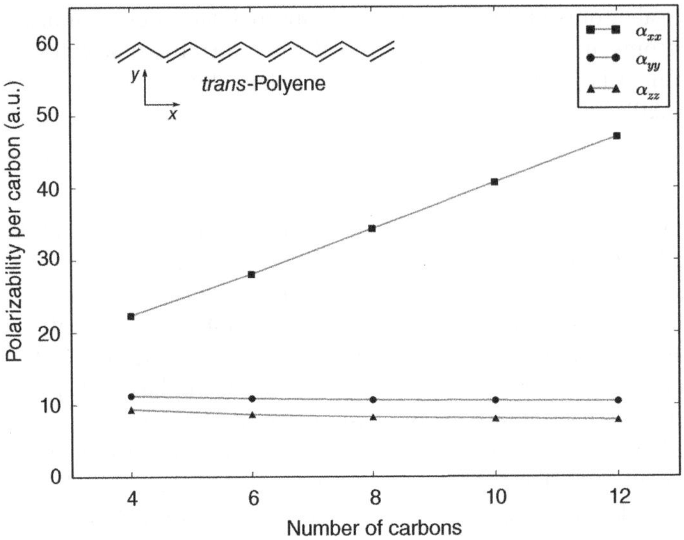

Figure 8.8 Polarizability per carbon atom for the *trans*-polyenes $C_{2n}H_{2n+2}$. Results are obtained at the time-dependent Hartree–Fock level of theory with use of the aug-cc-pVDZ basis set.

into Eq. (8.123), it can be shown that the polarizability for the one-dimensional particle-in-a-box model within the dipole approximation for the interaction between the electric field and the electron density is given as

$$\alpha_{xx} = \frac{256e^2 m_e L^4}{\pi^4 \hbar^2} \sum_{n \neq k} \frac{n^2 k^2 [(-1)^{n+k} - 1]^2}{(n^2 - k^2)^5}, \tag{8.129}$$

where n and k are the quantum numbers of the reference and excited states, respectively, and L is the length of the box. The first thing to notice, from the numerator, is that only every second state contributes in the sum-over-states expression for the polarizability, which is due to the symmetry of the system in the absence of a permanent dipole moment. For a particular reference state n, we also note that the polarizability increases with the fourth power of the length of the box. As the chain length grows, the quantum number of the highest occupied orbital and lowest unoccupied orbitals will increase, and the denominator will thus increase and partially cancel the polynomial growth with box length. We would therefore expect that the quartic growth with box length would be dominating for short lengths, and eventually be damped by the increasing denominator.

Neither the polarizability for the poly-alkanes nor that for the poly-alkenes grows as the fourth power with respect to the number of units in the chain. For the poly-alkanes, the growth is close to linear with respect to the number of methylene units, whereas a fit of the polyene results shows a growth closer to 1.67. This reflects the fact that the nuclei create local potentials that deviate from the potential-free particle-in-a-box model. However, the much stronger growth of the polarizability in the polyene chain reflects the fact that the dominating contribution in these systems arises from the π-electrons, which occupy a conjugated band of π orbitals, allowing these electrons to move more freely than the σ-electrons in the poly-alkane chain.

There has been much effort devoted to identifying compounds that display large polarizabilities and hyperpolarizabilities. Our analysis of the particle-in-a-box model shows that increasing the effective "length" that the electrons can move in leads to larger polarizabilities. To gain further insight into how to increase the polarizability, let us return to the two-level system discussed in Section 5.1, now in the context of a two-state model, that is, assume that there is only one excited state that contributes significantly to the observed polarizability. Starting from Eq. (8.123) and assuming only one excited state, we arrive at a generalization of Eq. (5.7) to a frequency-dependent polarizability in a two-state model

$$\alpha_{\alpha\beta} = \frac{1}{\hbar}\left[\frac{\mu_\alpha^{01}\mu_\beta^{10}}{\omega_{10}-\omega} + \frac{\mu_\beta^{01}\mu_\alpha^{10}}{\omega_{10}+\omega}\right] = \frac{2\omega_{10}}{\hbar(\omega_{10}^2-\omega^2)}\mu_\alpha^{01}\mu_\beta^{10}. \tag{8.130}$$

This equation shows that there are three ways of increasing the polarizability of a system, namely, by (i) reducing the band gap, that is, reducing the difference in energy between the ground state and the lowest excited state; (ii) increasing the transition dipole moment; or (iii) allowing the frequency of the incident light to approach that of the excitation energy, thus approaching resonance conditions. However, an exact matching of the frequency will in general lead to absorption instead of scattering of the incident light, and thus a balance must be found. In conjugated systems, the band gap decreases with increasing chain length, and thus this effect supports extending the length of molecules to increase the polarizability. Increasing the transition dipole moment requires more trial and error, as this can be achieved in multiple ways. Often, introducing donor or acceptor groups on one or both sides of the conjugated backbone will lead to the desired increase in the polarizability of the system.

8.3.1.1 Weak Intermolecular Forces

Our analysis of the polarizability showed that it could be related to the fluctuation of the dipole moment. Such charge fluctuations can be induced by the electric field created by a nearby molecule, and the polarizability thus plays an important role in the theory of weak intermolecular forces. It is not our purpose to discuss this theory in detail, but as it is an important area of application for the polarizability tensor, we will briefly introduce here the theory of long-range intermolecular forces, focusing our attention primarily on how polarizability governs these forces.

Let us consider two nonoverlapping charge distributions $\rho_A(\mathbf{r}_a)$ and $\rho_B(\mathbf{r}_b)$, each described by their own Hamiltonians $\hat{H}_A$ and $\hat{H}_B$ that only depend on the coordinates of molecules A and B, respectively. It is in practice not possible to ensure that there is no overlap between the two charge densities. This approximation is nevertheless a useful starting point, although we should keep in mind that there will be contributions missing, which we will not discuss further here. If we assume as a first approximation that there is no interaction between the two charge densities, then the Hamiltonian for this unperturbed system is $\hat{H}_A + \hat{H}_B$, and it follows that the energy of the composite system is the sum of the energies of the two subsystems $E_A + E_B$, and the wave function is thus a product of the wave functions of the two subsystems $\Psi_{mn} = \Psi_m^A\Psi_n^B = |mn\rangle$, where we also have introduced a shorthand notation for the state of the noninteracting systems.

The interaction between the two charge densities will be small, and we can thus use perturbation theory to describe the interaction between them. The Hamiltonian describing this interaction is then given by

$$\hat{H}^{\text{int}} = \int\int \frac{\hat{\rho}_A(\mathbf{r}_a)\hat{\rho}_B(\mathbf{r}_b)}{4\pi\varepsilon_0|\mathbf{r}_a - \mathbf{r}_b|} d^3\mathbf{r}_a d^3\mathbf{r}_b, \tag{8.131}$$

where appears the charge density operators $\hat{\rho}_A$ and $\hat{\rho}_B$ of each subsystem, showing that the Hamiltonian is limited to interactions between the two charge densities and not interactions within each subsystem.

From perturbation theory, we find that first- and second-order energy corrections are given by

$$E_{00}^{(1)} = \langle 00|\hat{H}^{\text{int}}|00\rangle, \tag{8.132}$$

$$E_{00}^{(2)} = -\sum_{(m,n)\neq(0,0)} \frac{\langle 00|\hat{H}^{\text{int}}|mn\rangle\langle mn|\hat{H}^{\text{int}}|00\rangle}{E_{mn}^{(0)} - E_{00}^{(0)}}. \tag{8.133}$$

The first-order energy correction corresponds to the electrostatic interaction between two charge distributions. It can either be calculated directly from the interaction operator, being in structure very identical to the Coulomb interaction between electrons and nuclei in molecules, or restricted to interactions between the two subsystems. More commonly, this contribution is calculated by expanding the electron density of one of the subsystems in a multipolar expansion as described in Section 3.3.2 and evaluating the interaction of these multipoles with the charge density of the other subsystem. However, in relation to our interest in the polarizability, we will not discuss this contribution any further here.

We recall from our discussion of the vibronic polarizabilities in Section 6.2 that it was convenient to separate the sum over all excited vibronic states into excited electronic and vibrational states, respectively. Following this strategy, we partition the second-order energy correction in Eq. (8.133) into three contributions, two involving excitations only within one subsystem, and one contribution involving simultaneous excitations in both subsystems, that is

$$E_{00}^{(2)} = E_{\text{ind}}^{A} + E_{\text{ind}}^{B} + E_{\text{disp}}, \tag{8.134}$$

where we have defined

$$E_{\text{ind}}^{A} = -\sum_{(m,0)\neq(0,0)} \frac{\langle 00|\hat{H}^{\text{int}}|m0\rangle\langle m0|\hat{H}^{\text{int}}|00\rangle}{E_{m0}^{(0)} - E_{00}^{(0)}}, \tag{8.135}$$

$$E_{\text{ind}}^{B} = -\sum_{(0,n)\neq(0,0)} \frac{\langle 00|\hat{H}^{\text{int}}|0n\rangle\langle 0n|\hat{H}^{\text{int}}|00\rangle}{E_{0n}^{(0)} - E_{00}^{(0)}}, \tag{8.136}$$

$$E_{\text{disp}} = -\sum_{m\neq0,n\neq0} \frac{\langle 00|\hat{H}^{\text{int}}|mn\rangle\langle mn|\hat{H}^{\text{int}}|00\rangle}{E_{mn}^{(0)} - E_{00}^{(0)}}. \tag{8.137}$$

The first two of these contributions are the *induction energies* arising from the induced polarization of the electron density of one subsystem due to the permanent electric moments of the other subsystem. Although in principle they are always present, in practice the induction energies will only be significant if a subsystem has a permanent charge or dipole moment, as higher order corrections will in general be smaller than the contribution from the last term in Eq. (8.134). The last term in Eq. (8.134) is the *dispersion energy* and will be present also in the case of nonpolar molecules.

Let us first consider the induction energy. We first perform a multipole expansion of the interaction operator Eq. (8.131), including only charge and dipole-moment terms

$$\hat{H}^{\text{int}} = \frac{q^A q^B}{4\pi\epsilon_0 R} + \frac{(q^B \hat{\mu}_\alpha^A - q^A \hat{\mu}_\alpha^B)R_\alpha}{4\pi\epsilon_0 R^3} - \frac{(3R_\alpha R_\beta - R^2\delta_{\alpha\beta})\hat{\mu}_\alpha^A \hat{\mu}_\beta^B}{4\pi\epsilon_0 R^5} + \cdots, \tag{8.138}$$

where $R = |\mathbf{R}_B - \mathbf{R}_A|$ is the distance between the multipole expansion points of subsystems A and B. We insert this multipolar expansion of the interaction operator into the expression for the induction energy, Eq. (8.135), keeping in mind that the charge of the subsystem is independent of the electronic coordinates and that the ground and excited states of a given subsystem are orthogonal. Integrating over the two subsystems, we find that the induction energy of subsystem B is given by

$$E^B_{\text{ind}} = -\left(-\frac{q^A R_\alpha}{4\pi\epsilon_0 R^3} - \frac{(3R_\alpha R_\beta - R^2\delta_{\alpha\beta})\mu^A_\beta}{4\pi\epsilon_0 R^5} - \cdots \right) \times \sum_{m\neq 0} \frac{\langle 0|\hat{\mu}^B_\alpha|m\rangle\langle m|\hat{\mu}^B_{\alpha'}|0\rangle}{E^B_m - E^B_0} \times \left(-\frac{q^A R_{\alpha'}}{4\pi\epsilon_0 R^3} - \frac{(3R_{\alpha'} R_{\beta'} - R^2\delta_{\alpha'\beta'})\mu^A_{\beta'}}{4\pi\epsilon_0 R^5} - \cdots \right) = -\frac{1}{2}F^A_\alpha(\mathbf{B})\alpha^B_{\alpha\alpha'}F^A_{\alpha'}(\mathbf{B}), \tag{8.139}$$

where we have inserted the definition of the static polarizability of subsystem B and inserted the definition of the electric field due to a charge distribution as given by Eq. (3.240).

We recall that the induced dipole moment to first order in the electric field can be expressed as being proportional to the polarizability multiplied with the external electric field, Eq. (5.10), and we can thus write the induction energy as $-1/2F^A_\alpha(\mathbf{A})\mu^{\text{ind}}_B$, with μ^{ind}_A being the dipole moment induced in subsystem B by the charge distribution of subsystem A. The induction energy can thus be seen to arise from the interaction of the dipole moment induced in one subsystem with the electric field due to the charge distribution of the other subsystem. The induction energy thus corresponds to the classical energy of a polarizable charge distribution, and only the magnitude of the polarization is being dictated by the quantum-mechanical polarizability.

In the asymptotic limit of large separations between the subsystems, the induction energy will remain the dominating interaction. From Eq. (8.139), we note that if the interacting subsystems are charged, the first term in the multipolar expansion will dominate and the interaction will go asymptotically as R^{-4} with the distance separating the two subsystems. If, on the other hand, the subsystems are neutral but have a dipole moment, the induction energy will go asymptotically as R^{-6}. For nondipolar systems, we could in principle include quadrupole contributions in the expansion of the charge density of the subsystem. However, these are in general much smaller than the last term in our second-order energy expression for the interaction energy Eq. (8.134), the dispersion energy, and we now turn our attention to this contribution.

To calculate the dispersion energy, we return to the expression in Eq. (8.137), using the interaction Hamiltonian in Eq. (8.138). As we will now be considering nondipolar systems, there are no permanent charges in the subsystems; we thus focus on the dipole–dipole interaction contribution in the interaction Hamiltonian. Inserting this term into Eq. (8.137) and extracting the separation distance between the two subsystems, we get

$$E_{\text{disp}} = -\frac{(3R_\alpha R_\beta - R^2\delta_{\alpha\beta})}{4\pi\epsilon_0 R^5}\frac{(3R_\gamma R_\delta - R^2\delta_{\gamma\delta})}{4\pi\epsilon_0 R^5} \times \sum_{m\neq 0, n\neq 0} \frac{\langle 0|\hat{\mu}^A_\alpha|m\rangle\langle m|\hat{\mu}^A_\gamma|0\rangle\langle 0|\hat{\mu}^B_\beta|n\rangle\langle n|\hat{\mu}^B_\delta|0\rangle}{E^A_{m0} + E^B_{n0}}. \tag{8.140}$$

The denominator in the sum-over-states expressions prevents a separation of this term into the polarizabilities of the individual subsystems. Unlike the case of separating the electronic and vibrational excited states when deriving the pure vibrational contributions to the polarizability, the excited-state energies of the two subsystems are of the same order, and we thus cannot assume one of them to be smaller than the other.

There are two approaches to proceed with. One approach, first used by Fritz London in 1930, is to introduce the Unsöld approximation, approximating the excitation energies of the individual subsystems by an average excitation energy. By choosing these average energies to be close to the excited states that dominate the polarizability of the individual subsystems, the error in this approximation can be minimized. Inserting our expression for the polarizability in Eq. (8.140), rewriting the energy denominator and introducing an average excitation energy for each subsystem, we arrive at the following expression for the dispersion energy

$$E_{\text{disp}} \approx -\frac{1}{4}\frac{(3R_\alpha R_\beta - R^2\delta_{\alpha\beta})}{4\pi\epsilon_0 R^5}\frac{(3R_\gamma R_\delta - R^2\delta_{\gamma\delta})}{4\pi\epsilon_0 R^5}\frac{\Delta\tilde{E}_A\Delta\tilde{E}_B}{\Delta\tilde{E}_A + \Delta\tilde{E}_B}\alpha^A_{\alpha\gamma}\alpha^B_{\beta\delta}. \tag{8.141}$$

For the special case of atoms, for which the polarizability is diagonal and has the same value for all diagonal elements, for example, $\alpha^A_{\alpha\gamma} = \overline{\alpha}^A\delta_{\alpha\gamma}$, the dispersion energy can be written as

$$E_{\text{disp}} \approx -\frac{3}{2}\frac{\Delta\tilde{E}_A\Delta\tilde{E}_B}{\Delta\tilde{E}_A + \Delta\tilde{E}_B}\frac{\overline{\alpha}^A\overline{\alpha}^B}{(4\pi\epsilon_0)^2R^6} = -C_6R^{-6}. \tag{8.142}$$

This equation can also be used for freely rotating molecules. The dispersion energy thus has an asymptotic R^{-6} dependence on the separation distance of the systems, the strength of the interaction being described by the C_6 coefficient of the system. Higher order corrections to this expression can be obtained, involving C_8, C_{10} and higher order expansion coefficients.

Keeping in mind our analysis of polarizability as a measure of the fluctuation of the dipole moment, we can interpret the dispersion energy in Eq. (8.142) as arising from the interaction of the fluctuating dipole moments of the two subsystems. Unlike the induction energy, the dispersion energy is a true quantum effect that would not be present if a classical description of the electrostatic interactions were used.

The problem with the dispersion energy expression in Eq. (8.142) is the choice of average excitation energy in the two subsystems. Without a detailed knowledge of the excited-state spectrum of the two subsystems, the approximation may end up being rather poor. In most cases the ionization energy is selected, but this is likely to be a rather poor estimate.

A more rigorous approach was made by Casimir and Polder when studying retardation effects in the dispersion energy due to the finite speed of light. We will not dive into QED here. Rather, we will use the identity

$$\frac{1}{A+B} = \frac{2}{\pi}\int_0^\infty \frac{AB}{(A^2+\gamma^2)(B^2+\gamma^2)}d\gamma, \tag{8.143}$$

to rewrite the energy denominators in Eq. (8.140), allowing the terms in the sum-over-states expression in Eq. (8.140) to be written as a product of two terms, each only depending on the properties of one of the two subsystems

$$E_{\text{disp}} = -\frac{2\hbar}{\pi}\frac{(3R_\alpha R_\beta - R^2\delta_{\alpha\beta})}{4\pi\epsilon_0 R^5}\frac{(3R_\gamma R_\delta - R^2\delta_{\gamma\delta})}{4\pi\epsilon_0 R^5} \tag{8.144}$$
$$\times\int_0^\infty\left[\frac{1}{\hbar}\sum_{m\neq 0}\frac{\langle 0|\hat{\mu}^A_\alpha|m\rangle\langle m|\hat{\mu}^A_\gamma|0\rangle\omega^A_m}{(\omega^A_m)^2+\gamma^2}\right]\left[\frac{1}{\hbar}\sum_{n\neq 0}\frac{\langle 0|\hat{\mu}^B_\beta|n\rangle\langle n|\hat{\mu}^B_\delta|0\rangle\omega^B_n}{(\omega^B_n)^2+\gamma^2}\right]d\gamma,$$

where we should not confuse the subscript γ indicating a Cartesian direction with the damping parameter γ. Comparing Eq. (8.144) with Eq. (5.292), we recognize in this expression the real (dispersive) part of the complex polarizability at $\omega = 0$. We can thus evaluate the dispersion energy by integrating the polarizability of the two subsystems over the complex frequency range

$$E_{\text{disp}} = -\frac{\hbar}{2\pi}\frac{(3R_\alpha R_\beta - R^2\delta_{\alpha\beta})}{4\pi\epsilon_0 R^5}\frac{(3R_\gamma R_\delta - R^2\delta_{\gamma\delta})}{4\pi\epsilon_0 R^5}\int_0^\infty \alpha^A_{\alpha\gamma}(i\gamma)\alpha^B_{\beta\delta}(i\gamma)d\gamma. \tag{8.145}$$

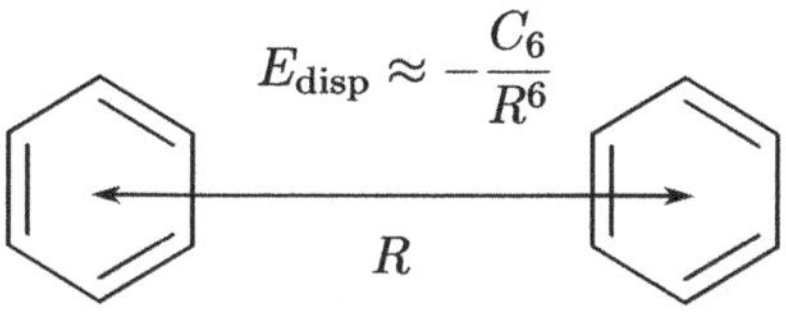

N	C_6
4	1741.1
6	1736.9
8	1736.5
10	1736.5

Figure 8.9 Convergence of the Gauss–Legendre quadrature scheme with respect to the number of quadrature points N used for obtaining the C_6 coefficients (in a.u.) of benzene from Eq. (8.145). Results are obtained at the Kohn–Sham/B3LYP level of theory using the polarized basis set of triple zeta quality by Sadlej.

Although it may appear somewhat strange to evaluate the real (dispersive) part of the complex polarizability, Eq. (5.292), over a set of imaginary frequencies (damping factors), the mathematical object is well-behaved, as illustrated by the results in Figure 5.25. Indeed, using the Gauss–Legendre quadrature scheme and well-selected imaginary frequencies, convergence can be achieved with less than 10 frequency points, as illustrated for the C_6 coefficients of benzene in Figure 8.9.

8.3.2 Nonlinear Polarizabilities

We discussed in Section 5.2 some of the many nonlinear polarizabilities that can be observed, all depending on the experimental setup, and these were summarized in Table 5.1. Together with the discussion found in Section 5.1 of the induced polarization in a two-level system, we gained a qualitative understanding of how nonlinear polarization effects occur and manifest themselves in molecular systems. We will not repeat this discussion here, nor will we discuss individual experimental setups or approaches to the determination of different nonlinear polarizabilities, despite the large experimental activity in the field of nonlinear optics. We will merely conclude that from a computational and theoretical point of view, response theory provides the necessary framework to calculate the microscopic molecular properties that govern the observable quantities.

However, we will address one particular case here, namely, nonlinear processes that involve one or more static electric fields. From Table 5.1, examples of such processes are the electro-optic Pockels effect, the electro-optic Kerr effect as well as electric-field-induced second-harmonic generation. From our discussion of polarization in linear media in Section 3.5.2, we recall that for dipolar systems in the presence of a static electric field, the distribution of the dipoles in the sample will no longer be isotropic due to partial orientation of the dipolar molecules relative to the direction of the applied electric field. This leads to an orientational contribution to the polarization that is proportional to the electric field and governed by the Boltzmann distribution of the molecular dipoles in the sample. This orientational effect influences the nonlinear polarization, and, following the procedure adopted for the linear polarization in Section 3.5.2, we find for the electric-field-induced second-harmonic generation process a nonlinear susceptibility given by

$$\chi_{\mathrm{NL}} = \frac{1}{15}\mathcal{N}\left\{\frac{\mu_\alpha\overline{\beta}_\alpha(-2\omega;\omega,\omega)}{kT} + [2\gamma_{\alpha\alpha\beta\beta}(-2\omega;\omega,\omega,0) + \gamma_{\alpha\beta\beta\alpha}(-2\omega;\omega,\omega,0)]\right\}, \tag{8.146}$$

where $\mathcal{N}$ is the number density of molecules. Here $\overline{\beta_\alpha}(-2\omega;\omega,\omega)$ refers to the quantity that we encountered in Eq. (5.217). With a similar analysis of the electro-optic Kerr effect, we get a linear as well as a quadratic dependence on the inverse temperature. More specifically, the molar Kerr constant is given by

$$\begin{aligned} {}_mK = \frac{2\pi\mathcal{N}}{405}\{&2\gamma_{\alpha\alpha\beta\beta}(-\omega;\omega,0,0) \\ &+\frac{1}{kT}[4\mu_\alpha\beta_{\alpha\beta\beta}(-\omega;\omega,0)+3(\alpha_{\alpha\beta}(-\omega;\omega)\alpha_{\alpha\beta}(0;0)-3\alpha_{\alpha\alpha}(-\omega;\omega)\alpha_{\beta\beta}(0;0))] \\ &+\frac{3}{k^2T^2}[\alpha_{\alpha\beta}(-\omega;\omega)\mu_\alpha\mu_\beta-\alpha_{\alpha\alpha}(-\omega;\omega)\mu_\beta^2]\}. \end{aligned} \tag{8.147}$$

We note that each order of the static field may lead to an ordering contribution arising from the permanent dipole moments in the molecules, as discussed in Section 3.5.2.

The temperature dependence of many nonlinear optical processes provides a means of experimentally determining the different contributions to a nonlinear optical process. By measuring the nonlinear optical response at different temperatures and plotting these against, for example, $1/kT$ in the case of the electric-field-induced second-harmonic generation, the slope of the line will provide us with the value of $\mu_\alpha\overline{\beta}_\alpha(-2\omega;\omega,\omega)$, whereas the intercept of this line with the x-axis (corresponding to an infinite temperature) provides us with the contribution from the second-order hyperpolarizability γ, see Eq. (8.146).

8.4 Magnetizability

In the previous section, we considered the interactions of a static or frequency-dependent electric field with a molecular system. We now turn to the interactions between a static magnetic field and a molecular system. We will at this point, however, restrict our discussion. We will not consider the interactions of the external field with the nuclear magnetic moments, nor the interactions with the magnetic moments arising from the spin of the unpaired electrons, which were dealt with in Sections 8.1 and 8.2, respectively. Within these limits, the set of measurable interactions between closed-shell molecules and magnetic fields is much smaller than that of the corresponding electric properties. One reason for this is that the magnetic field strengths are smaller than those of the electric fields by a factor of $1/c$ as can be seen from Eq. 3.76. Secondly, the magnetic field strengths that can be created in a laboratory are in general very small on a molecular scale, as discussed in Section 3.5.6. Finally, only even-order magnetic interactions will survive for closed-shell molecules. This follows from time reversal symmetry, that we discussed in Section 4.2: Odd-order magnetic interactions will involve time antisymmetric combinations of operators and are therefore not totally symmetric. Alternatively, in the non-relativistic domain, we can understand this from the fact that the angular momentum operator is purely imaginary, whereas the non-relativistic wave function for closed-shell molecules can always be chosen to be real. For molecules with unpaired electrons, one may have first-order magnetic interactions, but for molecules with a nonzero total spin, the interaction with the magnetic field will in general be dominated by the spin-magnetic moment rather than the orbital magnetic moment.

The first nonvanishing contribution to an expansion of the energy of a closed-shell molecule in the presence of a static external magnetic field is thus of second order in the magnetic field

$$\Delta E(\mathbf{B}) = -\frac{1}{2}B_\alpha\xi_{\alpha\beta}B_\beta. \tag{8.148}$$

The quantity ξ is known as the molecular magnetizability. Considering the form of the non-relativistic, electronic Hamiltonian in the presence of a magnetic vector potential Eq. (3.153), we can see that there will be two contributions to the magnetizability: One contribution is quadratic in the external magnetic field and can be calculated as the second derivative of the expectation value of the operator in Eq. (8.83); the second contribution will in analogy with the polarizability arise from the term in the Hamiltonian linear in the external magnetic field, Eq. (8.81), leading to a contribution involving a linear response function. The expression for the magnetizability can therefore be written as

$$\xi_{\alpha\beta} = -\frac{d^2}{dB_\alpha dB_\beta}[\langle \hat{H}^{\text{dia}}_{B\,B;\alpha\beta}\rangle + \langle\langle \hat{H}^{\text{OZ}}_\alpha; \hat{H}^{\text{OZ}}_\beta\rangle\rangle_0]. \tag{8.149}$$

A negative magnetizability corresponds to an induced magnetic moment in the molecule, opposing the applied magnetic field, and is experimentally manifested by the molecule trying to get away from the poles of a magnet and thus aligns itself across the applied magnetic field, an effect referred to as *diamagnetism*. The opposite effect, where the induced magnetic moment enforces the magnetic field is referred to as *paramagnetism*, and in this case the molecule will seek to orient itself in the direction of the poles of the magnet and therefore along the magnetic field. An extreme case of paramagnetism is the collective phenomenon of ferromagnetism, where the electron spins in a bulk system are coupled in such a manner that they give the largest possible total spin-magnetic moment for the system.

The expectation value and the linear response terms contributing to the magnetizability in Eq. (8.149) are conventionally referred to as the *diamagnetic* and *paramagnetic* contributions, respectively. This is historically motivated by the observation that only the expectation value contribution to the magnetizability is nonzero for closed-shell atoms in the nonrelativistic picture. This is due to the spherical symmetry of the atoms, which allows us to choose the z direction as the axis of quantization for the z-component of the magnetic dipole operator. Since the atomic states are eigenfunctions of the z component of the angular momentum operator appearing in $\hat{H}^{\text{OZ}}$, the transition moment appearing in of the response function becomes

$$\langle L, M_L \mid \hat{L}_z \mid L', M'_L\rangle = \hbar M'_L \langle L, M_L \mid L', M'_L\rangle = 0 \quad \{L, M_L\} \neq \{L', M'_L\}, \tag{8.150}$$

using the atom as the natural gauge origin for this system. Only the diamagnetic contribution to the magnetizability is thus nonzero, and the diagonal elements of the diamagnetic part of the magnetizability, and thus also the isotropic magnetizability, are negative since

$$\xi^{\text{dia}}_{\alpha\beta} = -\frac{d^2}{dB_\alpha dB_\beta}\langle \hat{H}^{\text{dia}}_{BB;\alpha\beta}\rangle = -\frac{e^2}{4m_{\text{e}}}\langle r^2_G \delta_{\alpha\beta} - r_{G,\alpha} r_{G,\beta}\rangle. \tag{8.151}$$

We note that also for linear molecules, the reference state is an eigenstate of the magnetic moment operator along the molecular axis, and thus this component of the magnetizability has only a diamagnetic contribution. We recall from Section 8.2 that the analysis in terms of diamagnetic and paramagnetic contributions will not apply in the relativistic domain due to the fact that the magnetic dipole operator is proportional to the kinetic angular momentum. Indeed, in the relativistic domain, the magnetizability is given as a linear response function only, involving the Zeeman interaction in the relativistic picture, Eq. (8.86), leading to the following relativistic expression for the magnetizability

$$\begin{aligned}\xi_{\alpha\beta} &= -\frac{d^2}{dB_\alpha dB_\beta}\langle\langle ec\alpha_\gamma A_\gamma; ec\alpha_\delta A_\delta\rangle\rangle_0 \\ &= -\frac{e^2}{2}\epsilon_{\alpha\gamma\delta}\epsilon_{\beta\mu\nu}\langle\langle r_{G,\gamma} c\alpha_\delta; r_{G,\mu} c\alpha_\nu\rangle\rangle_0.\end{aligned} \tag{8.152}$$

For closed-shell molecules, the expectation value contribution almost always dominates, and almost all closed-shell molecules are therefore diamagnetic. Exceptions can occur for molecules for which there are near degeneracies in the electronic ground state, where the denominator of the linear response function Eq. (5.168) may become small and therefore increase the magnitude of the paramagnetic term sufficiently to make the overall magnetizability positive.

Although this has been predicted through computation, an experimental verification of the existence of closed-shell paramagnetic molecules is still lacking.

It is worth comparing the expression for the diamagnetic magnetizability Eq. (8.151) with that of the electronic contribution to the traceless electric quadrupole moment in Eq. (3.227). We see that the molecular quadrupole moment is very closely related to the diamagnetic part of the magnetizability. In microwave Zeeman experiments it is possible to extract the diamagnetic contribution to the magnetizability (at least in the nonrelativistic limit), and microwave Zeeman spectroscopy can thus serve as a route to the experimental determination of the molecular quadrupole moment.[11] Microwave Zeeman experiments also allow the anisotropy of the magnetizability $\Delta\xi$ to be determined, although in general the accuracy of the measured quantities are rather low. Nevertheless, this is one of the very few ways of determining molecular quadrupole moments and magnetizability anisotropies experimentally. In this particular case, the origin of the quadrupole moment calculated will be the center of mass of the molecule, since this is the position about which the molecule rotates. An alternative approach to estimate molecular quadrupole moments is to extract these from the temperature dependence of the external-field-induced birefringences, which we will briefly discuss in Section 8.6.3.

8.4.1 The Origin Dependence of the Magnetizability

In our discussion of multipole expansions in Section 3.3.4, it was noted that only the lowest order, nonvanishing multipole moment would be origin independent, whereas any experimentally observable quantity by necessity will be origin independent. The issue of origin dependence of the magnetic properties in general, and the magnetizability in particular, in approximate calculations is important, and we will therefore discuss this point again here for the case of the magnetizability itself.

The interaction of the molecule with the external magnetic field is determined by the introduction of a magnetic vector potential $\mathbf{A}_G(\mathbf{r})$ which has an arbitrariness in the definition of the gauge origin, as discussed in Section 3.3.4. For this reason, the operators in Eqs. (8.81) and (8.83) are not uniquely defined. Indeed, by changing the gauge origin appearing in the definition of the magnetic vector potential Eq. (3.212), we can change the values of the diamagnetic and the paramagnetic contributions *without changing the total magnetizability*. This arbitrariness makes it clear that focusing discussions of magnetic properties too strongly on the diamagnetic or paramagnetic contributions—or derived properties such as dia- and paramagnetic ring currents—is likely to be counterproductive since the analysis can be very different if a change of gauge origin is made. In particular, this makes the comparison of dia- and paramagnetic contributions in different molecules difficult, since no unique gauge origin can be chosen that ensures an equivalent partitioning of the magnetizability in the two molecules.

Let us start by showing that the magnetizability is origin independent for exact states (or any approximate state fulfilling certain operator equivalences following from the Ehrenfest theorem as discussed in Section 7.2.1). With the shift of the gauge origin from $\mathbf{G}$ to $\mathbf{G}' = \mathbf{G} + \mathbf{D}$, the diamagnetic contribution changes as (using here an off-diagonal element as an example)

$$\begin{aligned}\xi^{\text{dia}}_{\alpha\beta}(\mathbf{r}_{G+D}) &= \frac{e^2}{4m_\text{e}}\langle 0|(r_{G,\alpha} - D_\alpha)(r_{G,\beta} - D_\beta)|0\rangle \\ &= \xi^{\text{dia}}_{\alpha\beta}(\mathbf{r}_G) - \frac{e^2}{4m_\text{e}}(D_\alpha\langle 0|r_\beta|0\rangle + D_\beta\langle 0|r_\alpha|0\rangle) + \frac{e^2}{4m_\text{e}}N_\text{e}D_\alpha D_\beta,\end{aligned} \tag{8.153}$$

where N_e is the total number of electrons in the system. It now remains to show that the changes occurring in the paramagnetic contribution are equal to the changes occuring in the

11 W. H. Flygare. *Chem. Rev.*, 79:653 (1974).

diamagnetic contribution, but of opposite sign. We have

$$
\begin{aligned}
\xi^{\text{para}}_{\alpha\beta}(\mathbf{r}_{G+D}) &= -\frac{e^2}{4m_e^2}\langle\langle \hat{l}_{(\mathbf{G}+\mathbf{D}),\alpha};\hat{l}_{(\mathbf{G}+\mathbf{D}),\beta}\rangle\rangle_0 \\
&= -\frac{e^2}{4m_e^2}\langle\langle \epsilon_{\alpha\gamma\delta}(r_{G,\gamma}-D_\gamma)\hat{p}_\delta;\epsilon_{\beta\mu\nu}(r_{G,\mu}-D_\mu)\hat{p}_\nu\rangle\rangle_0 \\
&= \xi^{\text{para}}_{\alpha\beta}(\mathbf{r}_G) + \frac{e^2}{4m_e^2}(\epsilon_{\alpha\gamma\delta}D_\gamma\langle\langle\hat{p}_\delta;\hat{l}_\beta\rangle\rangle_0 + \epsilon_{\beta\mu\nu}D_\mu\langle\langle\hat{l}_\alpha;\hat{p}_\nu\rangle\rangle_0 \\
&\quad -\epsilon_{\alpha\gamma\delta}\epsilon_{\beta\mu\nu}D_\gamma D_\mu\langle\langle\hat{p}_\delta;\hat{p}_\nu\rangle\rangle_0).
\end{aligned}
\tag{8.154}
$$

Recalling Eq. (7.69), we can rewrite the terms linear in the displacement vector $\mathbf{D}$ as

$$
\begin{aligned}
&\frac{e^2}{4m_e^2}(\epsilon_{\alpha\gamma\delta}D_\gamma\langle\langle\hat{p}_\delta;\hat{l}_\beta\rangle\rangle_0 + \epsilon_{\beta\mu\nu}D_\mu\langle\langle\hat{l}_\alpha;\hat{p}_\nu\rangle\rangle_0) \\
&\quad = \frac{e}{4m_e}(\epsilon_{\alpha\gamma\delta}D_\gamma\epsilon_{\delta\beta\epsilon}\langle 0|\hat{\mu}_\epsilon|0\rangle + \epsilon_{\beta\mu\nu}D_\mu\epsilon_{\nu\alpha\epsilon}\langle 0|\hat{\mu}_\epsilon|0\rangle) \\
&\quad = \frac{e^2}{4m_e}(D_\beta\langle 0|r_\alpha|0\rangle + D_\alpha\langle 0|\hat{r}_\beta|0\rangle),
\end{aligned}
\tag{8.155}
$$

where we have used the properties of products of Levi–Civita symbols, and remembering that we are considering an off-diagonal element. We note that the terms that are linear in the shift of the gauge origin are of equal magnitude but opposite sign to the terms linear in the shift of the gauge origin for the diamagnetic magnetizability, as required.

The term quadratic in the displacement of the origin can be rewritten using Eqs. (5.231) and (7.64), in the latter case noting that our magnetic field is static,

$$
-\frac{e^2}{4m_e^2}\epsilon_{\alpha\gamma\delta}\epsilon_{\beta\mu\nu}D_\gamma D_\mu\langle\langle\hat{p}_\delta;\hat{p}_\nu\rangle\rangle_0 = -\frac{e^2}{4m_e}\epsilon_{\alpha\gamma\delta}\epsilon_{\beta\mu\nu}N_e D_\gamma D_\mu\delta_{\delta\nu} = -\frac{e^2}{4m_e}N_e D_\alpha D_\beta,
\tag{8.156}
$$

which proves the origin independence of the total magnetizability tensor.

However, we note an important point regarding this derivation, namely, that it relies on the fulfillment of sum rules. We recall from Section 7.2 that these relations are only fulfilled for exact states or for variational wave functions in the limit of a complete basis set. In practice, the calculated magnetizability (as well as any other property that depends on the external magnetic field) will therefore not be origin independent.

There are many ways of achieving a *constrained* origin independence for calculated magnetic properties. We will here focus on one of the oldest and most versatile approaches to overcome the gauge-origin problem in approximate calculations, the so-called *London atomic orbitals* (LAOs), also known as gauge-including atomic orbitals (GIAOs).

In 1937, London was studying ring-current effects in aromatic systems using Hückel theory. In order to obtain results that were independent of the gauge origin, he suggested introducing an imaginary phase factor attached to the basis functions

$$
\chi^{\text{LAO}}_{M,\mu}(\mathbf{r},\mathbf{B}) = \exp\left(-\frac{i}{\hbar}e\mathbf{A}_G(\mathbf{R}_M)\cdot\mathbf{r}\right)\chi_{M,\mu}(\mathbf{r}),
\tag{8.157}
$$

where $\chi_{M,\mu}$ is an ordinary basis function of any kind, such as a Gaussian basis function, located at center $\mathbf{R}_M$. Consider now the action of the operator of mechanical momentum on the London orbital

$$
[\hat{\mathbf{p}} + e\mathbf{A}_G(\mathbf{r})]\chi^{\text{LAO}}_{M,\mu} = \exp\left(-\frac{i}{\hbar}e\mathbf{A}_G(\mathbf{R}_M)\cdot\mathbf{r}\right)[\hat{\mathbf{p}} + e(\mathbf{A}_G(\mathbf{r}) - \mathbf{A}_G(\mathbf{R}_M))]\chi_{M,\mu}
\tag{8.158}
$$

Since $\mathbf{A}_G(\mathbf{r}) - \mathbf{A}_G(\mathbf{R}_M) = \mathbf{A}_M(\mathbf{r})$, the effect of this complex phase factor is to move the global gauge origin $\mathbf{G}$ to a local gauge origin for each individual basis function. For a basis function $\chi_{M,\mu}$ the local gauge origin will accordingly be its center $\mathbf{R}_M$.

By moving the global gauge origin to the local gauge origins of each basis function, the global gauge origin will disappear from the molecular integrals occurring in the expressions for the magnetizability, thus making the calculated magnetizabilities gauge-origin independent.

However, this origin independence is by design. This means that the use of London orbitals does not lead to a wave functional that has the variational freedom to satisfy general gauge invariance as defined in Section 3.3.4, and thus the use of London orbitals will, for instance, not ensure current conservation in approximate calculations.[12]

The use of LAOs not only leads to formally origin-independent results, they also lead to significantly faster basis set convergence, as illustrated for PF_3 in Table 8.5. One may wonder why this is the case, as there is no apparent reason why the use of the center to which the basis function is attached should be a better gauge origin than the global gauge origin. However, this can be rationalized by considering the case of a one-electron atom in an external magnetic field with a single London orbital, in terms of the solution φ_{nlm} of the unperturbed Hamiltonian $\hat{H}_0$, and assuming a magnetic field oriented in the z direction

$$\begin{aligned} \hat{H}\varphi_{nlm}^{\text{LAO}} &= \exp\left(-\frac{i}{\hbar}e\mathbf{A}_G(\mathbf{R}_M)\cdot\mathbf{r}\right)\left\{\frac{1}{2m_e}[\mathbf{p}+e\mathbf{A}_M(\mathbf{r})]^2-\frac{e^2}{4\pi\varepsilon_0}\frac{Z_M}{r_M}\right\}\varphi_{nlm} \\ &= \exp\left(-\frac{i}{\hbar}e\mathbf{A}_M\cdot\mathbf{r}\right)\left[\hat{H}_0+\frac{e}{2m_e}B_z\hat{l}_z\right]\chi_{nlm}+\mathcal{O}(\mathbf{B}^2) \\ &= \varphi_{nlm}^{\text{LAO}}\left(E_n+\frac{e\hbar}{2m_e}mB_z\right)+\mathcal{O}(\mathbf{B}^2). \end{aligned} \tag{8.159}$$

The London phase factor has ensured that the origin of the magnetic dipole operator Eq. (8.81) is the atomic nucleus, making the orbital an eigenfunction of the corresponding angular momentum operator. For a standard basis function, on the other hand, this will not be the case and the orbital is not an eigenfunction of the angular momentum operator. This basis function will therefore only be an eigenfunction to zeroth order in the magnetic field for an arbitrary gauge origin. There is of course no guarantee that the choice of the atomic centers will remain an optimal choice of local gauge origins in the case of molecules. However, computational experience has demonstrated that this choice is usually a very good one.

The discussion so far has overlooked one important point: for the exact wave function, the diamagnetic and paramagnetic contributions to the magnetizability have a dependence on the gauge origin, which are of the same magnitude but opposite sign, thus ensuring the origin independence of the total magnetizability. When London orbitals are used, both the "diamagnetic" and the "paramagnetic" contributions can be made origin independent since all the atomic integrals appearing in the calculation of the magnetizability are origin independent as we discussed above, at odds with the behavior of the exact wave function. This can also be traced to the

Table 8.5 Basis set convergence of the Hartree–Fock magnetizability of PF_3 calculated with and without London atomic orbitals and Dunning's augmented correlation-consistent basis sets.

	aug-cc-pVDZ		aug-cc-pVTZ		aug-cc-pVQZ		aug-cc-pV5Z	
	No-Lon	Lon	No-Lon	Lon	No-Lon	Lon	No-Lon	Lon
ξ^{iso}	−16.55	−6.80	−11.50	−6.67	−9.06	−6.61	−7.66	−6.57
$\Delta\xi$	−3.66	−0.25	−0.77	−0.32	0.48	−0.33	−0.28	−0.34

All results are given in atomic units.

12 S.T. Epstein. *J. Chem. Phys.*, 58:1592 (1973). See also J. Juselius, D. Sundholm and J. Gauss. *J. Chem. Phys.*, 121:3952 (2004).

lack of current conservation in the London orbital approach. Although of little importance for calculations of magnetizabilities, since the diamagnetic and paramagnetic contributions are not observable quantities, it does indicate a problem, namely, that the formalism so far presented does not yield a correct magnetic dipole operator. As we will observe when we consider oscillating electromagnetic fields, the origin independence of some molecular properties is ensured by the mutual cancelation of the origin dependence between the magnetic dipole operator and the quadrupole operator. This requires a redefinition of the magnetic dipole operator when LAOs are used.[13]

Once LAOs are used, the calculation of magnetizabilities is in general fairly straightforward, displaying in general fast basis set convergence if a set of diffuse polarization functions are included in the basis set. The aug-cc-pVDZ basis set thus in general gives near-basis-set-limit results as seen from Table 8.5. As already noted and as evident from Table 8.5, basis set convergence is much slower when LAOs are not used. The effect of the external magnetic field is to induce a current in the electron density, that is, a rotation of the electron density around the axis of the applied magnetic field. This induced current is difficult to model using conventional Cartesian basis sets, and we would expect that both diffuse orbitals and orbitals with high angular momentum will be required to accurately represent the magnetic-field-induced current. This is in contrast to the case of electric perturbations, which to a large extent induce linear polarization in the electron density, changes that can easily be represented by Cartesian basis functions. The LAOs respond well to magnetic-field-induced currents in the case of one-electron systems, as shown in Eq. (8.159), and these properties are transferred into multielectronic systems. Furthermore, the fact that London orbitals ensure the use of local gauge origins also means that the basis set primarily depends on a good description of the electron density close to each nuclear center.

8.4.2 Magnetizabilities from Magnetically Induced Currents

We have so far considered the magnetizability primarily as originating as an energy correction in the presence of an external magnetic field. Alternatively, we can consider the magnetizability as originating, in analogy with the polarizability, as the interaction of the current induced in the electron density by the external magnetic field with the magnetic field itself. In such a picture, the magnetizability for a static magnetic field is given by

$$\xi_{\alpha\beta} = \frac{d^2}{dB_\alpha dB_\beta} \int \mathbf{A}_G \cdot \mathbf{j}^{NR}(\mathbf{r}, \mathbf{B}) d^3\mathbf{r}, \tag{8.160}$$

where we have introduced the current density in the presence of a magnetic field that appeared in Eq. (3.168), explicitly indicating its dependence on the external magnetic field, and noting that the contribution from the spin will vanish because we consider closed-shell molecules. Just as we could understand, in the case of polarizability, the origin of large polarizabilities in terms of how easily the electron density could be shifted in the presence of an external electric field, we can rationalize the magnitude of the magnetizability by considering the strength of the currents induced in a molecule by an applied external magnetic field.

In Figure 8.10(a), we have plotted the magnetic-field-induced currents in the hydrogen molecule at an interatomic separation of 1.4 a_0, for a magnetic field applied perpendicular to the plane of the book, pointing toward the reader. The dominating effect of the induced current is a clockwise rotation of the electron density, which is in line with the diamagnetic character of the molecule. Such a clockwise rotation of the induced current is thus referred to as a *diatropic* current. Had the hydrogen molecule been paramagnetic, we would instead

13 J. Olsen, K. L. Bak, K. Ruud, T. Helgaker, and P. Jørgensen. *Theor. Chim. Acta*, 90:421, 1995.

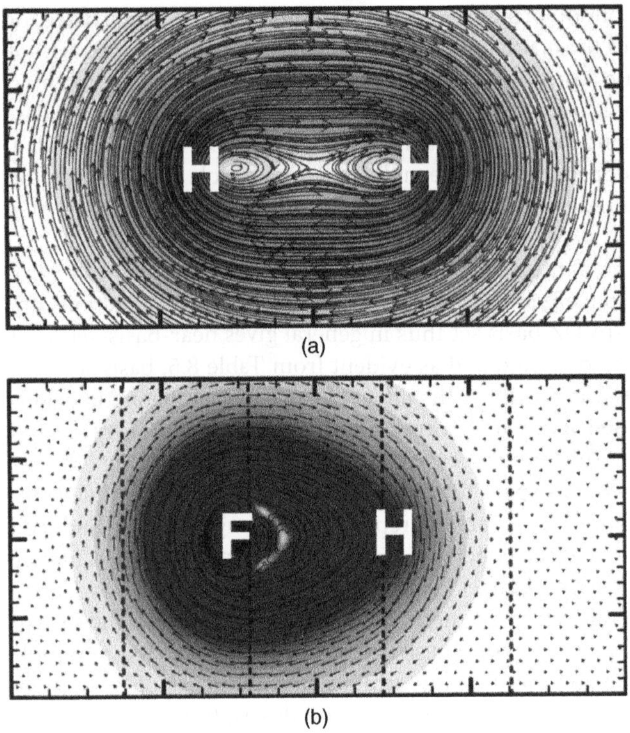

Figure 8.10 Magnetic-field-induced currents in the (a) hydrogen molecule and (b) hydrogen fluoride.

have seen a counterclockwise rotation of the electron density, and we would refer to this as a *paratropic* current. It is also worth noting that in this simple case, the induced current encompasses the entire molecular system, and this to some extent illustrates very clearly the concept of electron-pair sharing that is a characteristic of covalent bonds.

The induced currents are more complex once more electrons are added. In Figure 8.10(b), we have plotted the induced current in the HF molecule at a bond length of 1.94 a_0. There are a couple of things to note compared to the case of the hydrogen molecule. First of all, we now observe strong diatropic currents primarily circulating around the fluorine atom. These currents originate from the electrons not involved in bonding. Secondly, we also note a weak paratropic current in the immediate vicinity of the fluorine atom. This shows that although the molecule is diamagnetic at the global level, locally there can be regions that sustain either dia- or paramagnetic currents.

The magnetic-field-induced currents thus give us a novel insight into what happens to the electron density when an external magnetic field is applied. These induced currents can in turn be considered to generate local magnetic fields due to the circulation of the charged electrons. These local fields are key to the determination of chemical shielding constants observed in NMR spectroscopy. We will return to this in Section 8.8.

8.4.3 Isotropic Magnetizabilities and Pascal's Rule

The conventional way of measuring isotropic magnetizabilities is through the use of a Gouy balance. In the Gouy balance, the material of interest is placed in a uniform cylinder of uniform cross section, which is suspended from a magnetic balance in an inhomogeneous magnetic field. The inhomogeneous magnetic field will give rise to a force acting on the sample, leading to an apparent change in the mass of the sample. This apparent change in sample mass Δm can be related to the magnetizability of the sample ξ through

$$(\Delta m)g = \frac{1}{2}(\xi - \xi_{\text{air}})AB^2, \tag{8.161}$$

where g is the gravitational constant, A the area of the sample tube, and B the strength of the magnetic field at the lowest part of the tube. As it is in general difficult to measure the absolute magnetic field, most experiments are carried out relative to a sample with known magnetizability. It should also be noted that the effects are in general small for diamagnetic molecules, leading in many cases to fairly large error bars for the experimentally determined magnetizability of diamagnetic samples. We have collected some experimental gas-phase results in Table 8.6, the difference between the two being whether experimental or theoretical reference data have been used to calibrate the reference curve. In particular, the data depends strongly on the value for the compounds with high magnetizability, and in particular CH_4 and Ar. As Ar can be calculated very accurately, it seems reasonable to assume that in this case the data obtained by calibrating against theoretical reference data is of higher quality than those obtained when calibrating against older experimental reference values.

Considering that the effect of an external magnetic field is to induce currents in the molecule, a somewhat surprising property of the magnetizability is the so-called Pascal's rule,[14] which states that the molecular magnetizability can be expressed as a sum of "atomic" magnetizabilities

$$\xi = \sum_K c_K \xi^K, \tag{8.162}$$

where the summation runs over the different elements in the molecule, c_K is the number of atoms of a given element K, and ξ^K the "atomic" magnetizability of this element. We emphasize that these "atomic" magnetizabilities are not the magnetizability of the atom itself, but rather the

Table 8.6 The magnetizability of fluorine determined using, respectively, experimental data (Experiment estimated) and theoretical data (Theory estimated) for calibration.

System	Experiment	Experiment estimated	Theory	Theory estimated
He	−31.2	−32.4	−31.31	−30.6
H_2	−66.6	−64.8	−67.14	−65.1
Ne	−127	−125	−125.6	−129
F_2		−160		−166.6
CO	−196	−199		−208.9
CH_4	−289	−306	−318.2	−321.3
Ar	−325	−324	−343.1	−342.2
CO_2	−349	−347		−365.7

Results taken from K. Ruud, P. R. Taylor, and M. Jaszuński. *J. Phys. Chem. A*, 104:168, 2000, where more details about the experiment and the theoretical calculations can be found. All results reported in 10^{-30} JT^{-2}.

14 It should be noted that even though the additivity of the magnetizability bears the name of Pascal [*Ann. Chim. Phys.*, 19:5, 1910] due to his efforts in establishing this additivity together with Pacault [*Rev. Sci. Instrum.*, 86:38, 1948], the effect was first observed almost 20 years earlier by Henrichsen [*Wied. Ann.*, 34:180, 1888] at what was then called the University of Christiania, now the University of Oslo (Norway).

result of fitting experimental (or theoretical) observations to Eq. (8.162). This thus corresponds to the DAP model for polarizabilities, Eq. (8.127). However, this suggests that the majority of the induced current is located in the vicinity of the nuclei and thus where the majority of the electrons are. A parallel can be made with the energy as the majority of the electronic energy resides in the atomic energies, and energy stabilization arising from chemical bonding is a mere fraction of the total energy. This could also be an explanation as to why the LAOs improve basis set convergence, since it ensures that the gauge origin is always close to the region of the largest induced currents.

Deviations from Pascal's rule would thus suggest a large magnetic-field-induced current arising in areas of the molecule other than in the vicinity of nuclei. In Figure 8.11, we have collected the magnetizability of different hydrocarbons, obtained both using Hartree–Fock theory and calculated from atomic magnetizabilities derived by fitting the calculated data in Figure 8.12 to Eq. (8.162), giving $\xi^C = -81.4 \times 10^{-30}$ J T^{-2} and $\xi^H = -56.4 \times 10^{-30}$ J T^{-2}. For the normal hydrocarbons, the agreement between the Pascal rule and calculated magnetizabilities is good, although larger differences occur for molecules with significant strain in their molecular structure. However, there are two notable deviations from the additivity rule, namely, cyclobutadiene and benzene, giving respectively a much lower and a much higher magnetizability (in absolute value) than that predicted by Pascal's rule. These molecules are prototypical examples of antiaromatic and aromatic molecules, respectively. A feature of these molecules is the conjugated π-electron network. Indeed, considering the magnetic-field-induced current densities of cyclobutadiene and benzene measured 1 bohr above the molecular plane, shown in Figure 8.12, we clearly see the current induced in the π-electrons of these two molecules.

From Figure 8.12, we also note that the induced π-ring current in panel (a) circulates counterclockwise, that is, it is a paratropic ring current, whereas that in benzene circulates clockwise and is thus a diatropic ring current. This is in accordance with the deviations from the Pascal rule seen in Figure 8.11, where benzene is more diamagnetic than you would expect from Pascal's rule, whereas cyclobutadiene in contrast is more paramagnetic. The fact that these archetypical antiaromatic and aromatic molecules display magnetizabilities that deviate

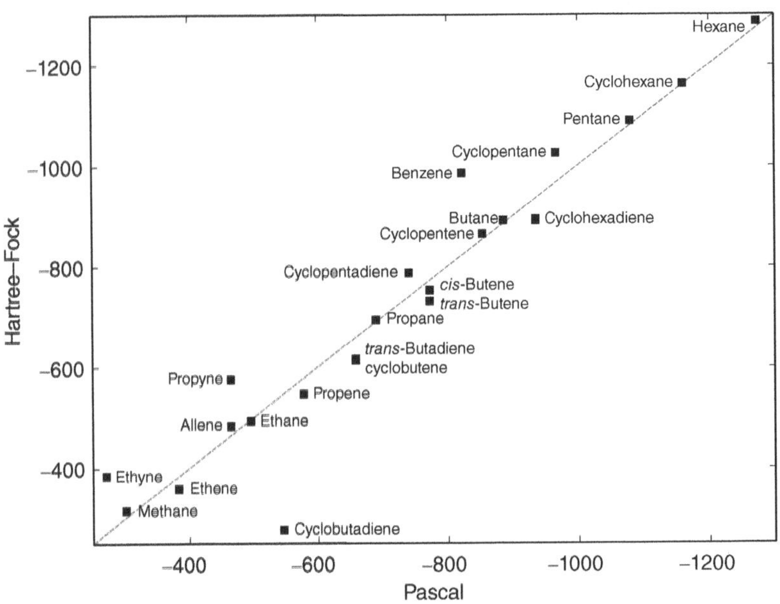

Figure 8.11 Magnetizabilities for a number of hydrocarbons as obtained at the Hartree–Fock/aug-cc-pVDZ level of theory and with the use of Pascal's rule. All results are reported in units of 10^{-30} J T^{-2}.

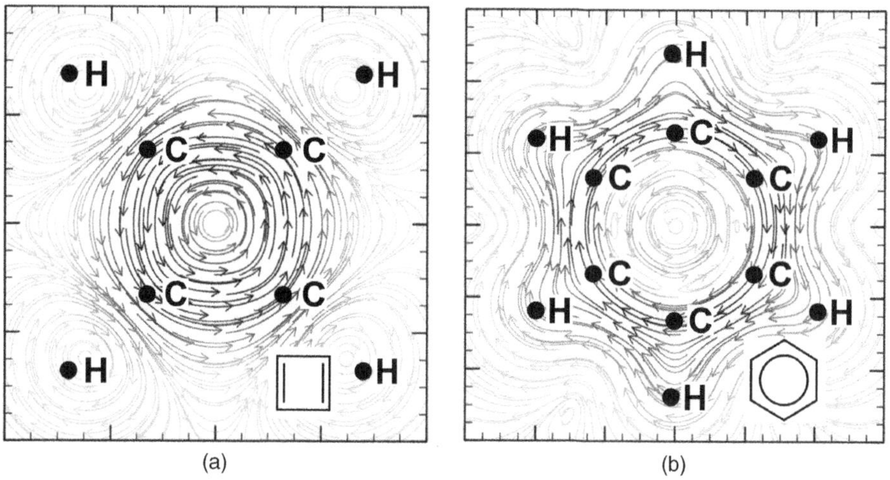

Figure 8.12 Magnetic-field-induced ring currents in cyclobutadiene (a) and benzene (b) as due to a magnetic field applied perpendicular to the molecular plane. The currents shown are at a height of 1.0 a_0 above the molecular plane.

from the Pascal rule has led to the proposal of defining aromatic character from the magnetic properties of the molecules. The use of a more diamagnetic value for the magnetizabilty than that predicted from the Pascal rule, often referred to as diamagnetic exaltation, is made somewhat complicated due to its dependence on atomic magnetizabilities, and an alternative approach is to define the aromatic character in terms of the strength of the induced currents as discussed above. The literature on aromaticity is however vast, and this magnetic criterion is only one of many different definitions of aromaticity.

8.5 Electronic Absorption and Emission Spectroscopies

The absorption and emission of electromagnetic radiation in the visible or ultraviolet region of the electromagnetic spectrum are among the earliest forms of spectroscopy. They are often referred to as UV/vis spectroscopies, and are in their simplest form directly detectable by the human eye. The absorption or emission of radiation in this frequency range is due to transitions between different electronic states of a molecule. For most molecules, only a few of the lowest electronically excited states are accessible using radiation in this frequency range. These excited states largely involve the highest occupied and lowest unoccupied molecular orbitals, and the nature of the orbitals can thus be deduced from the observed UV/vis spectrum and can give detailed insight into the electronic structure of a molecule.

This information can in turn be invaluable in understanding chemical reactivity. The first principle of photochemical activation, the so-called Grotthuss–Draper law, states that in order to induce a photochemical reaction, light has to be absorbed by a molecule. The energy of this light can excite a molecule into reactive excited states and thus be used to initiate photochemical reactions. Such reactions are essential to many industrial and biological processes, such as photodynamic therapy (in which drugs often referred to as photosensitizers are activated by shining light on them), solar cells (also known as photovoltaic cells) that convert light into electricity, or the process of human vision, in which the first step is the photoisomerization of the 11-*cis*-retinal into the 11-*trans*-retinal isomer through the absorption of light as illustrated in Figure 8.13.

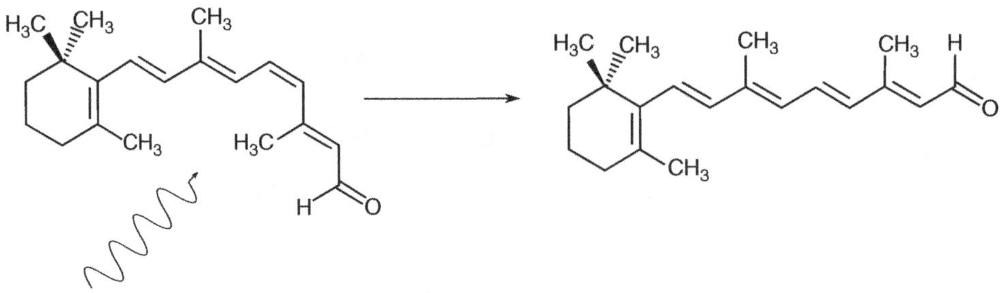

Figure 8.13 Photoisomerization of retinal.

The excitation of a molecule from its ground state to an excited electronic state is due to the absorption of a photon, and is for this reason often referred to as *one-photon absorption* (OPA). We will return to this process in Section 8.5.1. In very intense laser fields, the excited states can also be reached by simultaneous absorption of two photons with half the energy of the excitation energy, in what is known as a *two-photon absorption* (TPA) process. There are multiple benefits arising from the use of TPA or multiphoton absorption processes, which we will discuss in Section 8.5.4.

With the discovery of X-ray radiation by Wilhelm Röntgen in 1895, higher energy regions of the electromagnetic spectrum also became accessible. The absorption of X-ray radiation can lead to the excitation of an electron from one of the innermost orbitals of a molecule. As these orbitals in most cases retain their atomic character, X-ray spectroscopy provides information that is more atom specific than UV/vis spectroscopy, as the highest occupied and lowest unoccupied molecular orbitals in general are delocalized over a larger part of the molecular structure. Nevertheless, the core electrons are in X-ray absorption spectroscopy (XAS) excited into low-lying unoccupied orbitals, often in the valence band, and they can therefore provide information about the molecular electronic structure in the vicinity of the atom from which the core electron has been excited.

The hole created by the excitation of the electron from a core orbital will in general have a rather short lifetime, and it will be rapidly filled by an electron from occupied orbitals that are higher in energy, accompanied by the emission of a photon or a so-called Auger electron (see Section 8.5.5). This process is called X-ray fluorescence, and because only orbitals that are largely atomic in nature are involved, the photon emitted will have a wavelength that is characteristic for the atom, allowing X-ray fluorescence to be used for determining the elemental composition of a compound or material, and in particular for identifying the presence of metals. We will not discuss this application of X-ray spectroscopy in this book, but rather in Section 8.5.5 focus on XAS.

The energies involved when exciting an electron to a higher electronic state are, on a molecular scale, quite large, being far larger than the energies involved in excitations in the vibrational or rotational manifolds of the molecule. Molecules in electronic excited states eventually want to get rid of the excess energy acquired in the absorption process, and there are a number of processes by which this can be achieved. These processes are normally summarized in a *Jablonski diagram*, as shown in Figure 5.21. In this particular diagram, the processes that may take place when one or two photons are absorbed by the molecule, bringing the molecule from the electronic ground state S_0 to an excited electronic state S_n, are illustrated. Let us now briefly discuss these different processes.

We recall from the discussion of vibronic transitions in Section 6.7 that, within the Franck–Condon approximation, the excitation will in general occur to a number of different vibronic states in the electronically excited state, and the probability for excitation to a specific vibronic state is given by the Franck–Condon factor. In an ensemble of molecules, electronic excitations will lead to the population of a number of vibronic states of the particular excited electronic state. The molecules will in general seek to reach the lowest vibrational level of the electronically excited state, and will do so through vibrational relaxation processes in which the vibrational energy is dissipated through collisions with surrounding molecules, converting the excess vibrational energy into kinetic energy in the surrounding molecules. These collisions and the intramolecular vibrations can also induce radiationless transitions to a vibronic state of a different electronically excited state. This is referred to as *internal conversion (IC)*, and is an important mechanism for the depopulation of higher electronic excited states, in most cases eventually down to the lowest singlet-excited state.

In cases where vibronic states with different spin multiplicities have significant overlap, the molecular collisions may also induce changes from one electronically excited state to an electronically excited state of different spin multiplicity. This is referred to as *intersystem crossing* (ISC) and is an important process by which excited states with spin multiplicities different from that of the electronic ground state can be reached. Excited states with a spin multiplicity different from that of the electronic ground state can also be reached through spin-forbidden excitations, although the absorption cross section for such processes are much smaller than electric dipole-allowed transitions between states of the same multiplicity. We note, however, that for compounds containing heavy elements, relativistic effects may lead to sizeable absorption cross sections for transitions that are spin forbidden in the nonrelativistic framework.

Molecular collisions occur frequently in solution, and closed-shell molecules excited to an electronically excited state will, through rapid IC and intersystem crossings, relax down to their lowest singlet or triplet excited state. Assuming a sufficiently long radiative lifetime for these lowest excited states, the molecular structure will relax to the corresponding equilibrium geometry. These nonradiative processes leading to the population largely of the lowest excited states of different spin multiplicities forms the basis for Kasha's rule, which states that photoemission, be it fluorescence or phosphorescence, only occurs with an appreciable yield from the lowest excited state of a given spin multiplicity.

One may wonder why these nonradiative processes do not always bring the molecules all the way down to the electronic ground state. This can be understood by noting that IC and intersystem crossings are most likely to occur when there is strong coupling between the vibronic wave functions in the two electronic states. These couplings will in most cases be the largest when there is a small energy difference between the two vibronic states (and thus the electronic energies). While the separation between different electronically excited states in general is small, leading to efficient IC and ISC processes, the electronic ground state is in most cases well isolated from the electronically excited states, and thus does not have significant overlap with excited states except at significantly distorted, and thus unlikely, geometries. The efficiency of the IC and ISC processes depends to a large extent on the density of excited electronic states. These considerations also indicate that exceptions to Kasha's rule may occur: If a particular excited state is well separated from all other electronic states, it may obtain a sizeable population despite not being the lowest excited state of its spin multiplicity. Azulene is an example of a molecule with an isolated S_2 state that shows substantial radiative emission.

If the lifetime of the excited state is sufficiently long, the molecules may also absorb additional photons. Following our discussion of IC and ISC, such excited-state absorption (ESA) is most likely to occur from the lowest excited state of a given spin multiplicity. There is no principal difference between this ESA process and absorption of radiation from the electronic ground state. However, from a computational point of view, modeling ESA will in general require that we optimize the molecular structure in the excited electronic state. This requires in principle the use of a multistate method that has an explicit reference to the excited-state wave function.

However, we recall from Section 5.9 that we from the double residue of the quadratic response function can obtain electric dipole transition moments between different excited states as well as expectation value properties of an excited state; see Table 5.3. This approach thus allows us to determine both the excited-state equilibrium geometry from the excited-state molecular gradient and the transition moments between different excited states, even though we only explicitly optimize the ground-state wave function or density.

A molecule in an electronically excited state will eventually return to the electronic ground state and this can occur either by means of nonradiative pathways as described above or by means of emission of radiation. It follows from Kasha's rule that photoemission largely occurs from the lowest excited state of a given spin multiplicity and this emission can be spontaneous or stimulated with a probability that is proportional to the transition moment between ground and excited states. This emission process is referred to as photoluminescence and it is further separated into the two subcategories of *fluorescence* and *phosphorescence*.[15] The emission of radiation upon de-excitation from an excited state of singlet spin symmetry (or more precisely, an excited state of the same spin symmetry as the electronic ground state) is referred to as fluorescence, and will be discussed in Section 8.5.2. The emission of light from an excited triplet state (or more precisely, an excited state with a different spin multiplicity than the electronic ground state) is referred to as phosphorescence and is often associated with rather long lifetimes for the electronically excited state, in some cases as long as minutes or hours. Phosphorescence will be discussed in more detail in Section 8.5.3.

This brief discussion of the Jablonski diagram in Figure 5.21 illustrates the complex photophysical processes that can follow upon molecular photoactivation. It also illustrates the insight that can be obtained from the analysis of spectra. We will, in the next sections, consider some of the most central concepts in absorption and emission spectroscopies.

8.5.1 Visible and Ultraviolet Absorption

In UV/vis absorption spectroscopy, we use electromagnetic radiation in the visible and ultraviolet energy range to induce transitions between different electronic states. The light is absorbed by the molecule and used to bring the molecule to an excited electronic state. Experimentally, the absorbance is given by the Beer–Lambert law, Eq. (3.411), where the absorption cross section in the electric dipole approximation is given by Eq. (5.321). We recall that excitation energies can be determined from the poles of the linear response function. The electric dipole transition moments that appear in the expression for the absorption cross section, Eq. (5.321), can be determined from the residue of the linear response function as discussed in Section 5.9. The electronic absorption spectrum consists of all possible excitations from the electronic ground state S_0 to the excited electronic states S_n. Although in principle covering all excited states, current UV/vis detectors limit the excitation wavelengths to about 180 nm, corresponding to an excitation energy of approximately 7 eV.

Before proceeding, let us consider the implications of Eq. (5.321). First, we note that because there is no spin component in the interaction Hamiltonian in the electric-dipole approximation, there can in the nonrelativistic case be no transitions between states of different spin multiplicity. Excitation and de-excitation processes will thus, in general, preserve the spin symmetry of the reference state. However, we recall that changes in spin multiplicity can occur through

15 Fluorescence is used to describe a process in which a molecule that has been excited to an excited electronic state through the absorption of light returns to the electronic ground state through the emission of light. Luminescence describes any process in which light is emitted irrespective of how the molecule reached the excited state from which it decays (except heat). This can involve biological processes (bioluminescence), chemical processes (chemiluminescence), or excitations by photon (photoluminescence).

intersystem crossings or through mixing of states with different spin multiplicity through the spin–orbit interaction. In the relativistic picture, however, spin is no longer a good quantum number due to the spin–orbit interaction and it is possible to induce transitions between states that in a nonrelativistic framework would be considered to have different spin multiplicities, although such transitions tend to be weak.

Second, the expression in Eq. (5.321) shows that the transition moment will only be nonvanishing if the transition is electric dipole allowed, as discussed in Chapter 4. If the electronic ground state is totally symmetric, which is always the case for molecules with a closed-shell electronic structure, this implies that only electronic excited states that transform according to the same irreducible representations as the components of the electric dipole operator will be nonzero. As a special case of this symmetry selection rule, we note that for molecules with inversion symmetry, the parity of the molecule will have to change upon electronic excitation (Laporte's rule). States that cannot be reached by electric dipole excitations, or for which the oscillator strength accidentally is small even though the transition is electric dipole allowed, are referred to as *dark* states.

If in the multipolar expansion of the interaction Hamiltonian, Eq. (3.186), we include contributions beyond the leading-order electric dipole term, we can enable transitions through the electric quadrupole and magnetic dipole moments. The electric quadrupole and magnetic dipole interactions are in general two to three orders of magnitude smaller than the electric dipole interactions, and such states will at best be seen as very weak signals in the absorption spectrum. In practice, such transitions are only likely to be observed in highly symmetric molecules, where the limited number of electric dipole-allowed transitions will lead to a spectrum with well-separated absorption bands. In less symmetric molecules, the broadness of the absorption bands will lead to significant overlap of bands, making the decomposition of the observed spectrum into individual electronic transitions difficult and furthermore hiding any weak nondipolar transitions. A more effective way to reach electric dipole-forbidden excited states is instead to use multiphoton absorption processes, which we will return to in Section 8.5.4. Nevertheless, the electric quadrupole and magnetic dipole interactions are important in determining the interaction of circularly polarized light with the electronic structure of a molecule, as observed, for instance, in circular birefringence experiments. We will return to these interaction mechanisms in Section 8.6.

In its simplest form, an absorption spectrum will consist of a set of absorption lines occurring at the energies matching dipole-allowed excited states, the intensity of the lines being proportional to the oscillator strength of the transitions. This ideal form of an absorption spectrum is in practice only experimentally realized for low-pressure, gas-phase spectra of atoms, where the absence of a surrounding medium and rotational and vibrational degrees of freedom allows the electronic transitions to be determined with very high accuracy. The wavelengths and transitions observed in the absorption spectrum of the hydrogen atom are illustrated in Figure 8.14.

In molecules, the rotational and vibrational motions of the molecule lead to a broadening of the absorption bands, which we have seen can arise from vibronic effects as discussed in Section 6.7. The zero-point vibrational motion of the atoms in a molecule also implies that even though excitations are most likely to occur at the ground-state equilibrium geometry, absorption may also occur at nonequilibrium geometries, leading to additional broadening effects. Temperature effects, molecular collisions, and the finite lifetime of the excited electronic state can all lead to a broadening of the absorption band, as can the presence of a solvent through changes in the local electric fields.

From Figure 6.12, we note that the absorption maximum does not correspond to the transition from the lowest vibrational state of the electronic ground state to the lowest vibrational state of the excited state, but rather to the excited vibronic state with the largest Franck–Condon

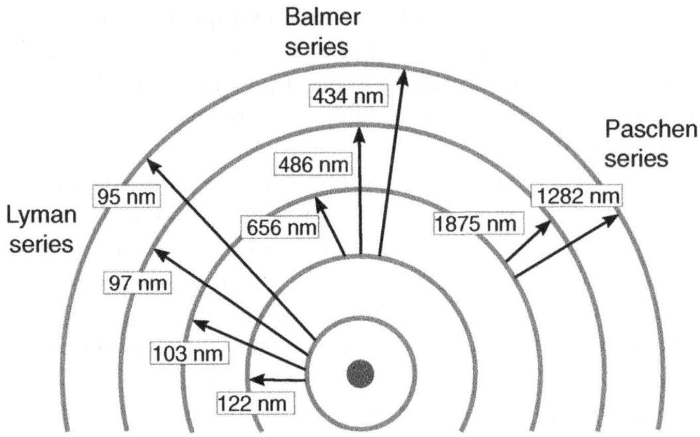

Figure 8.14 The Lyman, Balmer, and Paschen series of hydrogen with transitions from levels with principal quantum numbers $n = \{1, 2, 3\}$, respectively.

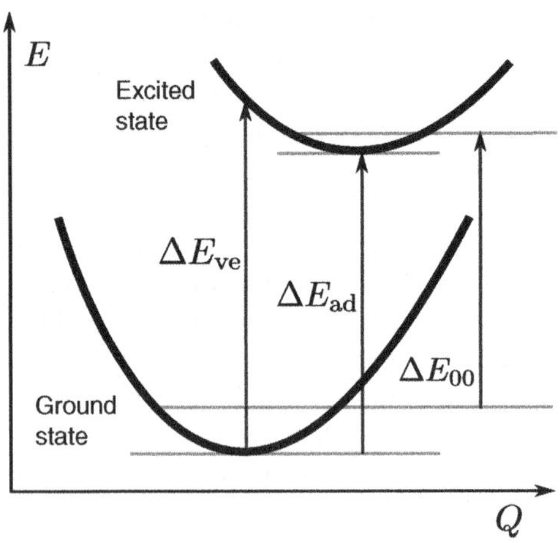

Figure 8.15 Illustration of vertical, adiabatic, and 0–0 excitation energies.

factor. The absorption band maximum can thus only be rigorously modeled through a full vibronic analysis.

In place of a full vibronic analysis, different computationally accessible excitation energies are often used, and these are illustrated in Figure 8.15. If we ignore the nuclear motions altogether and assume that excitations are most likely to occur at the equilibrium geometry, we obtain vertical transition energies $\Delta E_{\rm ve}$. These are the excitation energies obtained from the poles of the linear response function when we only consider the electronic wave function or density. However, the vertical excitation energy does not correspond to any experimentally observable excitation energy.

In contrast, in well-resolved gas-phase absorption spectra, the 0–0 transition can be directly observed as the lowest energy vibronic transition; see Figure 6.12. This transition energy corresponds to the excitation from the lowest vibrational state of the electronic ground state to the lowest vibrational state of the excited state, and we will denote this energy difference as ΔE_{00}. From Figure 6.12, we also note that the 0–0 transition is the only overlapping band between the absorption and fluorescence spectra. For more complex molecules in solution, the crossing point of the fluorescence and absorption bands can be expected to be close to the 0–0 transition if geometry relaxation is small in the excited state. The fluorescence–absorption crossing point can therefore be used to derive approximate experimental 0–0 transition energies.

The calculation of ΔE_{00} requires the determination of the excited-state Hessian matrix. This can be computationally challenging, and as an approximation to the experimental 0–0 transition energy we may instead use the adiabatic excitation energy $\Delta E_{\mathrm{ad}} = \Delta E_{00} - \Delta E_{\mathrm{zpv}}$, which is the energy difference between the energy of the ground and excited electronic states at their respective equilibrium geometries. The validity of using the adiabatic excitation energy for comparison with experimental 0–0 transition energies relies on the force fields of the ground and excited electronic states being rather similar. We recall from Section 6.7 that this is a commonly used approximation when calculating vibronic spectra.

An example of a calculated molecular absorption spectrum is given in Figure 8.16, with the oscillator strengths for the vertical electronic transitions represented by sticks. Such a stick spectrum bears little direct resemblance to an experimental absorption or emission spectrum as it does not include broadening effects. While broadening effects can be introduced through the use of complex response theory as discussed in Section 5.10, this phenomenological approach cannot account for all possible sources of broadening.

This fact, in combination with the computational efficiency, has made it customary to convolute each vertical electronic transition with a lineshape function defined such that when integrated over the entire frequency range, it integrates to unity

$$\int_{\mathrm{band}} L(\omega)d\omega = 1. \tag{8.163}$$

One such broadening function is the Lorentzian lineshape that arises from the use of the Cauchy distribution given in Eq. (3.455); see also Eq. (5.296). The Cauchy distribution corresponds to the situation in which all molecules interact in a similar manner with the electromagnetic radiation in the frequency range contained by the line shape, and for this reason the Lorentzian

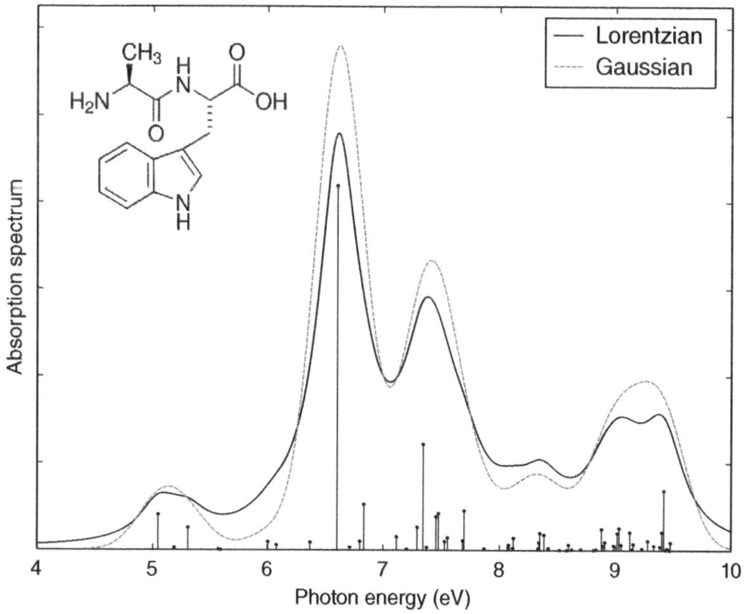

Figure 8.16 Linear absorption spectrum of L-alanyl-L-tryptophan obtained at the CAM-B3LYP/6-31G level of theory. Oscillator strengths of electronic transitions are represented as bars and the corresponding absorption spectra are obtained with Lorentzian and Gaussian line broadenings employing $\gamma = 0.2$ and 0.4 eV, respectively.

lineshape describes *homogeneous broadening*, arising, for instance, from the finite lifetime of the excited electronic state or from collisional broadening.

In solid samples, individual molecules may experience different interactions with the electromagnetic fields, due, for instance, to the presence of dopants in the material. This gives rise to *inhomogeneous broadening*. Inhomogeneous broadening may also occur in polar solvents, where solvent reorganization effects following the excitation of a solvated molecule to an electronically excited state may lead to different local environments for the different solvated molecules. Inhomogeneous broadening is described by a Gaussian lineshape

$$g(\omega; \omega_{n0}, \sigma) = \frac{\sqrt{2}}{\sigma\sqrt{\pi}} \exp\left(-\frac{2(\omega - \omega_{n0})^2}{\sigma^2}\right), \tag{8.164}$$

for which the maximum value is $\sqrt{2}/\sigma\sqrt{\pi}$ and the half width at half maximum (HWHM) is $\sigma\sqrt{\ln 2}/\sqrt{2}$.

In solution, effects giving rise to both homogeneous and inhomogeneous broadening are present. This situation can be modeled using a Voigt lineshape that is defined as a convolution of the Gaussian and Lorentzian lineshape functions

$$V(\omega; \omega_{n0}, \sigma, \gamma) = \int_{-\infty}^{\infty} g(\omega'; \omega_{n0}, \sigma) f(\omega - \omega'; \omega_{n0}, \gamma) d\omega'. \tag{8.165}$$

These different forms of the lineshape functions are compared for a single absorption band in Figure 8.17.

In Figure 8.16, we report the theoretical spectrum of an organic molecule as obtained after applying lineshape broadening using a Lorentzian band-shape function with a value for γ of 0.2 eV. After broadening the theoretical spectrum, the agreement between theory and experiment can be assessed through a visual comparison. This is particularly useful if the density of

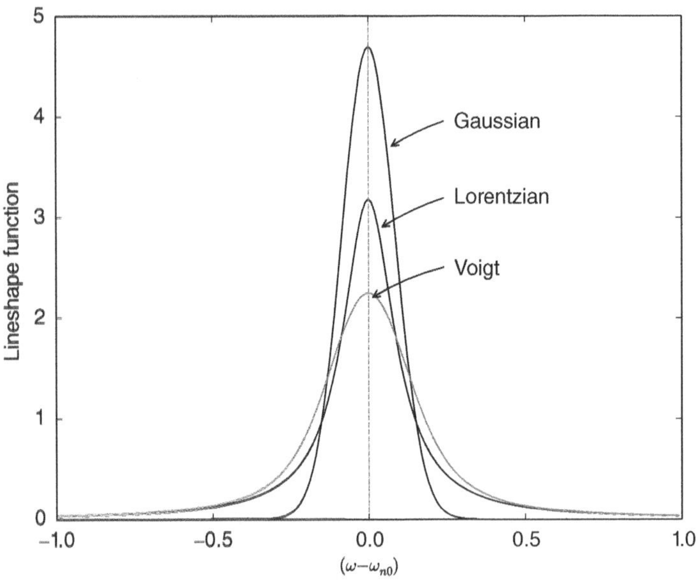

Figure 8.17 Illustration of Gaussian, Lorentzian, and Voigt lineshape functions. The σ and γ broadening parameters are so chosen as to give a common HWHM value of 0.1 for the Gaussian and Lorentzian functions, respectively.

states is high, and where it is not necessarily a single state that contributes to an absorption band, but rather a collection of excited states.

We have discussed the difficulty of extracting a computable excitation energy from an experimentally observed absorption spectrum unless a full vibronic analysis if performed. The vertical, adiabatic, and the 0–0 excitation energies all rely on the quality of the vertical excitation energy. In contrast to many other molecular properties, a given computational approach can have rather different errors in the vertical excitation energies, all depending on the nature of the excited state. As a consequence, the relative separation and even the ordering of the computed excited states may depend strongly on the choice of computational method. This is particularly important for density-functional theory (DFT) calculations, as, for instance, charge-transfer excitations and Rydberg excitations are poorly described by many exchange–correlation functionals, the former often being severely underestimated, in some cases by as much 1–2 eV. Fortunately, charge-transfer and Rydberg transitions are often associated with small orbital overlaps between initial and final state orbitals and the corresponding oscillator strengths are therefore also small so that the transitions do not contribute significantly to the observed absorption spectrum.

Vibronic effects give rise to band shapes that are nonsymmetric. We recall from Section 6.7 that vibronic effects may shift the absorption maximum to higher energies due in part to the higher adiabatic excitation energy relative to the vertical excitation energy, and due to the fact that the Franck–Condon factors may favor excitations from the vibrational ground state of the electronic ground state into a vibrationally excited state of the electronically excited state. This was illustrated in Figure 6.12 for the simple case of a diatomic molecule with a well-resolved absorption spectrum. A high-resolution absorption spectrum of a molecule in the gas phase will have a structure similar to that of the diatomic molecule, but will be more complicated as in principle each vibrational mode can give rise to their own vibronic progressions that will overlap in the observed spectrum. However, in many cases only a few selected vibrational modes dominate the vibronic fine structure, and these vibrational modes often involve atoms that are included in the chromophore of the electronic excitation process. By identifying the vibrational modes giving rise to the vibronic progressions, insight into the nature of the excited state can be obtained. Such analysis can also be used to our advantage in resonance Raman experiments to investigate the local structural dynamics of an excitation process, as we will return to in Section 8.7.3.

For polyatomic molecules in solution, the vibronic fine structure is in most cases lost. In addition to the complexity due to multiple vibronic progressions, additional loss of detail is due to collisional broadening in the solvent that broadens each individual vibronic band and makes them overlap. Nevertheless, assuming rather isolated electronic states, the vibronic details will lead to rather characteristic asymmetric absorption profiles, having an extended tail into the high-energy part of the spectrum. However, it is difficult a priori from experimental data alone to determine whether an electronic excited state is isolated and thus that the absorption profile is due to vibronic progression, or whether an asymmetric absorption profile is due to overlapping absorption bands arising from two or more different electronic states of different intensity.

8.5.2 Fluorescence Spectroscopy

In fluorescence spectroscopy we measure emitted photons due to the relaxation from an excited state to the ground state. The quantum yield for this process is defined as

$$\Phi = \frac{\text{Number of emitted photons}}{\text{Number of absorbed photons}},$$

and it is governed by the competition between radiative and nonradiative relaxation channels. The transition rate of stimulated emission is equal to that of stimulated absorption and, within the electric dipole approximation, it is given by the Einstein B coefficient found in Eq. (5.338). Following the pumping of excited states, a molecular system can also return to its ground state by means of spontaneous emission (or fluorescence) at a rate given instead by the Einstein A coefficient given in Eq. (5.347).

Let us first consider an excitation into the lowest electronic excited state, and let us assume that this state is isolated from both the electronic ground state and other excited states, as is, for instance, the case for the excited states of the diatomic molecule in Figure 6.10. Upon excitation, an electron is moved from a bonding to a nonbonding molecular orbital. Assuming a sufficiently long radiative lifetime, the molecule will relax its geometry to the equilibrium geometry of the excited electronic state. Owing to the changes in the electronic structure relative to the ground state, this will in most cases lead to an elongation of the bonds closest to the chromophore, as illustrated in Figure 8.15. From Figure 6.10, we see that, under these circumstances, the wavelength of the emitted light will be longer than the absorption wavelength, and the fluorescence spectrum will therefore be redshifted relative to the corresponding absorption spectrum, as shown in Figure 6.12 for the special case of a diatomic molecule. Fluorescence and absorption spectra will in general not display identical vibronic profiles, as the Franck–Condon factors will differ due to the difference in equilibrium geometries as well as force fields for the ground and excited states. Nevertheless, unless the structural rearrangements are significant, the fluorescence spectrum will often show resemblance to the absorption spectrum, but redshifted and with a vibronic profile that is the inverse of the absorption profile, having the vibronic tail extending to the lower energy part of the spectrum, as follows from the discussion of the special case of a diatomic molecule in Section 6.7.2.

If we consider more general absorption processes, in which also higher excited electronic states are accessed in the absorption process, we recall that Kasha's rule states that the only significant emission yield will come from the lowest excited state of a given spin multiplicity. It follows as a corollary that the luminescence quantum yield is independent of the excitation wavelength. This is sometimes referred to as the Vavilov rule. Kasha's rule also implies that while an absorption spectrum in general will have a large number of absorption bands, the emission spectrum will be dominated by a single emission band, unless a given excited state is well isolated and also accessible through IC processes, as previously discussed. As the lowest excited state often displays the highest absorption, the difference in energy between the maximum of the largest absorption bands and the emission band is called the Stokes shift. This shift is related to the structural relaxation in the lowest excited electronic state as well as the structural reorganization of the solvent, which leads to a difference in the local electric fields in the ground and excited electronic states.

8.5.3 Phosphorescence

At the nonrelativistic level of theory, spin is a good quantum number and electronic transitions due to the interactions with electromagnetic fields will conserve the spin angular momentum of a quantum mechanical system. When relativity is taken into account this is no longer the case, and selection rules based on spin are not present. However, for valence transitions, relativistic effects can often be treated by perturbation theory, and excited states in different spin manifolds are in these cases coupled by the perturbation. By phosphorescence one denotes light emission following the electronic relaxation from a triplet to a singlet state induced by electronic spin–orbit coupling. In a typical situation, phosphorescence occurs from the lowest triplet state as a consequence of IC in the triplet manifold; see Figure 5.21. The three sublevels of the triplet state are degenerate in the nonrelativistic realm, whereas ZFSs are present in the relativistic realm.

The phosphorescence lifetime τ_k from one of the three sublevels (indexed by k) of the lowest triplet state $|T_1^k\rangle$ is given by the inverse of the rate of spontaneous emission in Eq. (5.347). We thereby get

$$\frac{1}{\tau_k} = \frac{\omega_{f0}^3}{3\pi\varepsilon_0\hbar c^3} \sum_{\alpha\in\{x,y,z\}} |M_\alpha^k|^2, \tag{8.166}$$

where M_α^k is the electric dipole transition moment between the ground state and the triplet state. The radiative lifetime τ of the triplet state in the high-temperature limit is obtained by averaging over the three sublevels according to

$$\frac{1}{\tau} = \frac{1}{3}\sum_{k=1}^{3}\frac{1}{\tau_k}. \tag{8.167}$$

Excitation energies are given as poles of the linear response function as discussed in Section 5.9, but when it comes to the determination of transition moments there is a fundamental distinction to be made between the relativistic and nonrelativistic realms. In the former case, the electric dipole transition moments are identified as residues of the linear response function as described in Eq. (5.255), and there is, in that respect, no separation made between the fluorescence and the phosphorescence processes in the treatment. ZFSs of the spin sublevels are in principle also included in this case.

In the nonrelativistic realm, on the other hand, we need to invoke a perturbational treatment that requires the determination of the first-order response in the wave functions in order to calculate the transition moments (but not the excitation energy, if ZFSs are ignored). The first nonvanishing contribution to the transition moment comes from the first-order corrected wave functions. The sublevels of the triplet state are energetically degenerate, so the final expression reduces to

$$M_\alpha^k = \sum_{n=0}^{\infty} \frac{\langle S_0|\hat{\mu}_\alpha|S_n\rangle\langle S_n|\hat{H}_{\rm SO}|T_1^k\rangle}{E(S_n)-E(T_1)} + \sum_{n=1}^{\infty} \frac{\langle S_0|\hat{H}_{\rm SO}|T_n\rangle\langle T_n|\hat{\mu}_\alpha|T_1^k\rangle}{E(T_n)-E(S_0)}, \tag{8.168}$$

where the summation over intermediate triplet states includes all three sublevels of each triplet state and $\hat{H}_{\rm SO}$ is the electronic spin–orbit operator. The Breit–Pauli form of the spin–orbit operator is given by

$$\hat{H}_{\rm SO} = \frac{\alpha^2}{2}\left[\sum_{i,A} Z_A \frac{\hat{\mathbf{l}}_A(i)\cdot\hat{\mathbf{s}}(i)}{r_{iA}^3} - \sum_{i,j}\frac{\hat{\mathbf{l}}_j(i)\cdot[\hat{\mathbf{s}}(i)+2\hat{\mathbf{s}}(j)]}{r_{ij}^3}\right], \tag{8.169}$$

where Z_A is the atomic number of nucleus A, r_{iA} and r_{ij} are distances between electron i and nucleus A and electrons i and j, respectively, $\mathbf{s}(i)$ is the spin operator for electron i, and $\hat{\mathbf{l}}_A(i)$ and $\hat{\mathbf{l}}_j(i)$ are the orbital angular momentum operators for electron i with respect to the position of nucleus A and electron j, respectively. The Breit–Pauli operator, Eq. (8.169), includes spin–same orbit as well as spin–other orbit interactions, and the two-electron terms screen the spin–orbit interactions due to electronic motions about the nuclei. More approximate treatments of the spin–orbit interactions try to describe the screening effect in more computationally efficient ways. The mean-field spin–orbit interaction method introduces spin–orbit corrections for the valence electrons that are due to the exact one-electron term in Eq. (8.169) but with a two-electron screening in a mean-field manner. This scheme can be further approximated by neglect of all multicenter atomic orbital contributions so that the screening of the nuclear charge is performed for each atom separately. The resulting approach is known as the atomic mean-field interaction (AMFI) method.[16]

The representation of the triplet sublevels in Eq. (8.168) can be done in terms of the spin eigenfunctions with M_S as a good quantum number, but since only the averaged lifetime is of interest, see Eq. (8.167), any orthonormal set will do and the Cartesian components are often used in this respect; see Eq. (4.100). It is also customary to introduce a separation of the terms in the spin–orbit Hamiltonian according to

$$\hat{H}_{\mathrm{SO}} = \hat{H}^{x}_{\mathrm{SO}} + \hat{H}^{y}_{\mathrm{SO}} + \hat{H}^{z}_{\mathrm{SO}}, \tag{8.170}$$

where $\hat{H}^{x}_{\mathrm{SO}}$ contains terms of the type $\hat{l}_{K,x}\hat{s}_x$ and so on. The spin-forbidden electric-dipole transition matrix element in Eq. (8.168) can now be identified from the first-order residue of a quadratic response function in accordance with Eq. (5.256), so that we get a final expression that reads as

$$M^{k}_{\alpha} = \lim_{\omega\to\omega_{f0}} \hbar(\omega_{f0} - \omega)\langle\langle \hat{\mu}_{\alpha}; \hat{H}^{k}_{\mathrm{SO}}, \hat{\Omega}\rangle\rangle_{0,\omega} / \langle f|\hat{\Omega}|0\rangle. \tag{8.171}$$

We note that the reference to the operator $\hat{\Omega}$ is only a formal one, and after the division of the response function residue is made with the matrix element $\langle f|\hat{\Omega}|0\rangle$, there is no dependence on $\hat{\Omega}$ left in the equation.

Let us consider the calculation of the phosphorescence parameters for the $n \to \pi^*$ excited triplet state in formaldehyde as an example. To a first approximation the absorption and phosphorescence transition processes are treated as vertical adiabatic electronic transitions, as depicted in Figure 8.18. The ground-state optimized geometry is of C_{2v} symmetry, and the absorption energy is determined to be 3.46 eV in the time-dependent Hartree–Fock approximation. In the excited state, there will, due to its long lifetime, occur a nuclear relaxation in the triplet-state potential—bonds may be elongated and the molecular symmetry may be lowered as in the illustrated case. From the bottom of the excited-state potential, a phosphorescence decay occurs, resulting in a photon with an energy that is substantially lower compared to the

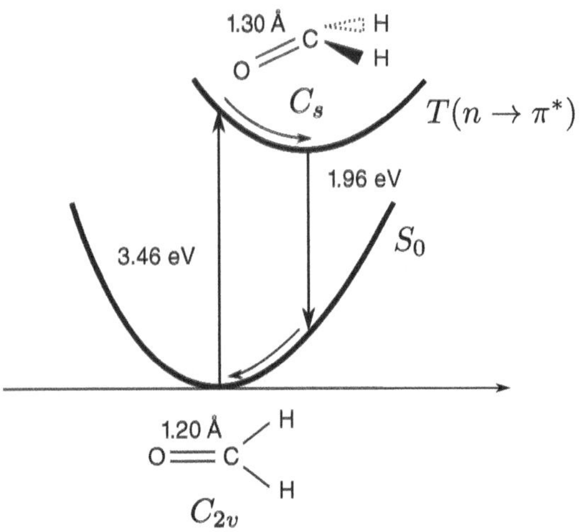

ΔE (eV)	f	τ
3.46	5.1×10^{-7}	—
1.96	7.3×10^{-7}	24.7 ms

Figure 8.18 Phosphorescence parameters for the $n \to \pi^*$ excited triplet state in formaldehyde. Molecular structures are optimized at the B3LYP/cc-pVTZ level of theory and property calculations are carried out at the four-component Hartree–Fock/taug-cc-pVTZ level of theory using the Dirac–Coulomb Hamiltonian.

16 B. A. Hess, C. M. Marian, U. Wahlgren, and O. Gropen. *Chem. Phys. Lett.*, 251:365, 1996.

absorption energy, in this case 1.96 eV. The oscillator strengths associated with the photon absorption and emission processes are on the order of 10^{-6}, which is many orders of magnitude smaller than typical oscillator strengths in absorption and fluorescence spectroscopy. The theoretically predicted radiative lifetime of the triplet state is 24.7 ms based on Eq. (8.167).

8.5.4 Multiphoton Absorption

It is interesting to note that the first theoretical prediction of multiphoton absorption processes was made rather soon after the establishment of the principles of quantum theory. In 1931, Maria Göppert–Mayer in her PhD thesis noted that not only one, but also two photons, could be absorbed simultaneously in a molecule and excite the molecule. Excited states could in this way be reached using photons with only half the energy of that of the excited state in question. However, the probability that two photons will be absorbed simultaneously—that is, the TPA cross section—is low. On the other hand, the TPA cross section grows nonlinearly with the power of the light source, and thus TPA can be realized by increasing the intensity of the light source. Indeed, with the advent of powerful light sources such as lasers,[17] TPA was experimentally realized in the early 1960s.

Today, TPA has become a commonly used experimental technique that shows great promise for many applications. Higher-order multiphoton processes have also been observed. In addition to providing valuable insight into the electronic structure of molecules, several novel areas of applications can be envisaged. Nevertheless, the low absorption cross sections and the challenges in the experimental setup make it unclear as to what extent the routine application of more than TPA processes can be realized in a routine manner.

In multiphoton processes, we can reach excited electronic states using infrared or near-infrared photons. This allows for deeper penetration into biological tissue with less damage to cellular components while still being able to excite molecules into higher electronic excited states. Through spontaneous emission, such multiphoton processes can be used, for instance, for the imaging of cellular tissue, an additional benefit being that multiphoton processes in general display higher focality. In addition to imaging, TPA processes can also be a potential tool for photodynamic therapy by exciting molecules into reactive excited states using long-wavelength radiation, inducing medically relevant chemical reactions.

The photons can also be in the visible energy range, in which case higher excited states, for instance, in the ultraviolet region, can be reached. However, such processes are in general of lesser interest since such states often will be reactive. More interesting in this respect is the combined use of light sources both in the visible and the infrared region. As we in such processes can induce excitations both in the electronic and vibrational manifold, a proper treatment of vibronic coupling between the electronic and vibrational states is required when the photons are absorbed simultaneously.

From the point of view of electronic structure theory and physical chemistry, one of the exciting opportunities offered by multiphoton absorption processes is the possibility to reach states that are one-photon forbidden because of symmetry. We recall from Section 8.5.1 that only dipole-allowed transitions in general can be observed in absorption spectroscopy. For the case of a centrosymmetric molecule such as ethylene, which has D_{2h} point group symmetry, we see from the group table in Table 4.2 that only the B_{1u}, B_{2u}, and B_{3u} states are dipole allowed. However, if instead we absorb two photons simultaneously, we can through intermediate states that are dipole allowed reach states that transform as the x^2 (y^2 and z^2), xy, yz, and xz symmetry elements—that is, also states of A_g, B_{1g}, B_{2g}, and B_{3g} symmetry. As such, two-photon processes are an important source of information about low-lying excited states that are dark in OPA.

---

17 Light Amplification through Stimulated Emission of Radiation.

From Table 4.2, we note that A_u states can be reached neither by OPA nor by TPA processes. However, by using three photons simultaneously of respectively x, y, and z polarization, such states can also be probed experimentally.

In the case of multiphoton-induced vibrational excitations, we recall that within the harmonic approximation, only excitations between neighboring states are allowed; see Eq. (6.37). Since in general only the lowest vibrational state is occupied at normal temperatures, only the first vibrationally excited states are reachable in regular infrared and Raman spectroscopy. Overtones and combination bands can be reached through either electric or mechanical overtones, but such effects will be weak and cannot easily be controlled. Instead, we may use two infrared laser sources and in this manner directly probe overtones and combination bands. In practice, most of such experiments are time resolved. As such, although the experiments in many cases can be modeled using response theory, the processes are not formally multiphoton absorption processes in the sense of simultaneous absorption of several photons, as the molecules first absorb one photon, bringing the molecule up to a specific vibrationally excited state, before a second photon brings the molecule into a second excited state. We will therefore not discuss those processes any further here.

8.5.4.1 Multiphoton Absorption Cross Sections

We have in Section 5.9 seen that two-photon transition matrix elements $M^{(2)}_{0f,\alpha\beta}$ can be obtained from the single residue of the second-order (quadratic) response function, Eq. (5.250). Correspondingly, j-photon transition matrix elements $M^{(j)}_{0f,\alpha\beta,\dots}$ can be obtained from the single residue of the jth-order response function. In the following, we will drop the states involved in the transition for ease of notation. Transition probabilities are given by expressions involving squares of these transition matrix elements. Alternatively, the (multiphoton) absorption spectra can be determined from the imaginary parts of complex response functions. Using this approach, the jth order absorption cross section is obtained from the imaginary part of the $2j-1$-order response function. In this section, we will focus on the residue approach. We will focus on the isotropic value of these multiphoton absorption cross sections, thus corresponding to the case of randomly oriented samples, such as liquids or gases. In Section 3.5.5, the general approach for obtaining rotational averages of tensors of different ranks were presented. Following this strategy, isotropic averages also for higher order multiphoton absorption cross sections can be obtained, and we have collected the explicit expression up to five-photon absorption in Table 8.7.[18]

The generalization in nonlinear optics of the absorption cross section, introduced in the Beer–Lambert law, Eq. (3.411), and in terms of the imaginary part of the polarizability in Eq. (5.335), is given by

$$\sigma(\omega) = \sigma^{(1)}(\omega) + \sigma^{(2)}(\omega)I + \sigma^{(3)}(\omega)I^2 + \cdots, \tag{8.172}$$

where I denotes the intensity of radiation. Here $\sigma^{(j)}(\omega)$ denotes the cross section for absorption of j photons in a single quantum mechanical process, and, since the SI units of cross section and intensity are m^2 and $J\ s^{-1}\ m^{-2}$, respectively, the SI unit of j-photon absorption cross sections is given by $m^{2j}\ s^{j-1}/J^{j-1}$. Regarding units, we note that the conventional unit for TPA cross section is Göppert-Mayer (GM),[19] where 1 GM corresponds to 10^{-50} cm^{4}s photon^{-1}.

Light absorption does not occur at discrete energy levels and spectral bands can be represented by the adoption of lineshape functions as introduced in Eq. (8.163). Using this approach,

18 The general approach to determine isotropic averages of multiphoton absorption cross sections can be found in D. H. Friese, M. T. P. Beerepoot, and K. Ruud. *J. Chem. Phys.*, 141:204103, 2014.

19 In honor of Maria Göppert–Mayer (1906–1972). Originally, it was proposed that the unit be called *Maria*.

Table 8.7 Isotropic averages for the lowest order multiphoton absorption processes.

$\langle \delta^{\text{OPA}} \rangle$	$\frac{1}{3}\sum_{\alpha} M_{\alpha}^{(1)}\overline{M}_{\alpha}^{(1)}$
$\langle \delta^{\text{TPA}} \rangle$	$\frac{1}{15}\sum_{\alpha\beta}(2M_{\alpha\beta}^{(2)}\overline{M}_{\alpha\beta}^{(2)} + M_{\alpha\alpha}^{(2)}\overline{M}_{\beta\beta}^{(2)})$
$\langle \delta^{\text{3PA}} \rangle$	$\frac{1}{35}\sum_{\alpha\beta\gamma}(2M_{\alpha\beta\gamma}^{(3)}\overline{M}_{\alpha\beta\gamma}^{(3)} + 3M_{\alpha\alpha\beta}^{(3)}\overline{M}_{\beta\gamma\gamma}^{(3)})$
$\langle \delta^{\text{4PA}} \rangle$	$\frac{1}{315}\sum_{\alpha\beta\gamma\delta}(8M_{\alpha\beta\gamma\delta}^{(4)}\overline{M}_{\alpha\beta\gamma\delta}^{(4)} + 24M_{\alpha\alpha\beta\gamma}^{(4)}\overline{M}_{\beta\gamma\delta\delta}^{(4)} + 3M_{\alpha\alpha\beta\beta}^{(4)}\overline{M}_{\gamma\gamma\delta\delta}^{(4)})$
$\langle \delta^{\text{5PA}} \rangle$	$\frac{1}{693}\sum_{\alpha\beta\gamma\delta\epsilon}(8M_{\alpha\beta\gamma\delta\epsilon}^{(5)}\overline{M}_{\alpha\beta\gamma\delta\epsilon}^{(5)} + 40M_{\alpha\alpha\beta\gamma\delta}^{(5)}\overline{M}_{\beta\gamma\delta\epsilon\epsilon}^{(5)} + 15M_{\alpha\alpha\beta\beta\gamma}^{(5)}\overline{M}_{\gamma\delta\delta\epsilon\epsilon}^{(5)})$

Note that for variational wave functions $\overline{M}_{\alpha\beta\ldots}^{(j)}$ is the complex conjugate of $M_{\alpha\beta\ldots}^{(j)}$, whereas for nonvariational wave functions, M^j and $\overline{M}^{(j)}$, correspond to right and left transition moments, respectively.

the general multiphoton absorption cross section can be written as

$$\sigma^{j\text{PA}} = (2\pi)^{j+1}\left(\frac{e^2}{\hbar c}\right)^j \cdot \left[\prod_{i=1}^{j} \omega_i\right] \cdot g(\omega_\sigma) \cdot \langle \delta^{j\text{PA}} \rangle, \tag{8.173}$$

where we remind the reader that ω_σ is used to denote the sum of optical frequencies. Specifically, the absorption cross sections for TPA, three- and four-photon absorption becomes

$$\sigma^{\text{TPA}} = 8\pi^3\left(\frac{e^2}{\hbar c}\right)^2 \cdot \omega_1\omega_2 \cdot g(\omega_\sigma) \cdot \langle \delta^{\text{TPA}} \rangle, \tag{8.174}$$

$$\sigma^{\text{3PA}} = 16\pi^4\left(\frac{e^2}{\hbar c}\right)^3 \cdot \omega_1\omega_2\omega_3 \cdot g(\omega_\sigma) \cdot \langle \delta^{\text{3PA}} \rangle, \tag{8.175}$$

$$\sigma^{\text{4PA}} = 32\pi^5\left(\frac{e^2}{\hbar c}\right)^4 \cdot \omega_1\omega_2\omega_3\omega_4 \cdot g(\omega_\sigma) \cdot \langle \delta^{\text{4PA}} \rangle, \tag{8.176}$$

where $\langle \delta^{\text{TPA}} \rangle$ is given in Table 8.7.

The most common experimental multiphoton experiments correspond to what is often referred to as the degenerate case, in which all photons have the same energy, being the excitation energy of the final state divided by the number of absorbed photons. The cross sections given above have also been derived for experiments in which the absorption cross sections match the photon dissipation rate. This is the case when two different light sources are used. It is, however, important to note that most modern multiphoton absorption setups use a single beam, in which the photon dissipation rate is j times that of the absorption cross section. In single-beam experiments, the absorption cross sections should therefore be multiplied by a factor j.[20] For a single-beam experiment, the TPA cross section in cgs units is given as

$$\sigma^{\text{TPA}} = \frac{4\pi^3\alpha a_0^5\omega^2}{c}\langle \delta^{\text{TPA}} \rangle g(\omega, \omega_0, \Gamma), \tag{8.177}$$

where α is the fine-structure constant and a_0 the Bohr radius.

20 A concise account of appropriate conversion factors in the case of TPA, taking into account both the experimental setup and the definition of the theoretical quantities, can be found in M. T. P. Beerepoot, D. H. Friese, N. H. List, J. Kongsted, and K. Ruud. *Phys. Chem. Chem. Phys.*, 17:19306, 2015.

8.5.4.2 Few-State Models for Two-Photon Absorption Cross Section

In order to get a better qualitative understanding of the molecular factors that govern multiphoton absorption properties in general, and TPA in particular, it is instructive to consider few-state models. The simplest model would of course be a two-state model, but for many systems such a simple model would not be able to capture salient features of the TPA process, and for this reason we will also consider a three-state model.

As discussed in Section 5.9, the single residue of the quadratic response function contains the information needed to obtain the two-photon transition moment, Eq. (5.250). Assuming for simplicity the degenerate case where the frequency of the incoming photons both have the same energy, corresponding to half the excitation energy of the final state ($\omega_1 = \omega_2 = \omega_{f0}/2$), the two-photon transition moment for a two-state model is given by

$$M^{(2)}_{0f,\alpha\beta} = \frac{1}{\hbar}\left[\frac{\langle 0|\hat{\mu}_\alpha|f\rangle\langle f|\bar{\hat{\mu}}_\beta|f\rangle}{\omega_{f0} - \omega_{f0/2}} + \frac{\langle 0|\hat{\mu}_\beta|f\rangle\langle f|\bar{\hat{\mu}}_\alpha|f\rangle}{\omega_{f0} - \omega_{f0/2}}\right]$$
$$= \frac{2}{\hbar\omega_{f0}}[\langle 0|\hat{\mu}_\alpha|f\rangle\langle f|\bar{\hat{\mu}}_\beta|f\rangle + \langle 0|\hat{\mu}_\beta|f\rangle\langle f|\bar{\hat{\mu}}_\alpha|f\rangle]. \quad (8.178)$$

Following Eq. (8.177), the TPA cross section for the two-state model can in the nonrelativistic framework, assuming real wave functions, be written as

$$\sigma^{\text{TPA}}_{0f} = \frac{16\pi^3\alpha a_0^5}{c}\frac{1}{\pi\Gamma}|\mu_{0f}|^2|\Delta\mu_{f0}|^2(2\cos^2\theta + 1), \quad (8.179)$$

where we have used the notation μ_{0f} to denote the dipole transition moment between the ground state 0 and the final state f, $\Delta\mu_{f0}$ the difference in the dipole moment of the final state and the ground state, and θ the angle between the dipole transition moment vector and the vector describing the change in the dipole moment upon excitation $\Delta\boldsymbol{\mu}_{f0} = \boldsymbol{\mu}_f - \boldsymbol{\mu}_0$.

The first thing to note is that in order for Eq. (8.179) to be nonzero, there must be a nonvanishing dipole transition moment. As such, the two-state model is only applicable for dipole-allowed states. Secondly, there must be a change in the dipole moment between the ground and the excited states, and the larger the difference in the dipole moment between the states, the larger the TPA cross section. For molecules with inversion symmetry, and thus for which there is no ground- or excited-state dipole moment, the two-state model cannot be applied. However, the two-state model suggests that excited states with strong charge-transfer character will display large TPA cross sections.

Finally, the largest TPA cross section can be obtained when $\cos\theta = \pm 1$, that is, when the transition dipole vector and the vector for the change in the dipole moment upon excitation are parallel or antiparallel. One way to achieve this is to make a dipole axis that aligns well with the molecular axis. The two-state model thus suggests that long-axis molecules with a prominent dipole axis in which strong, one-photon allowed charge–transfer excitations can be induced should be promising candidates as efficient two-photon absorbing molecules. This can often be achieved by attaching donor and acceptor groups on opposite ends of a conjugated backbone. A balance has to be found on the length of the backbone. A long backbone can increase the change in the dipole moment upon excitation, but too long a backbone will reduce the overlap between the orbitals occupied in the ground and the excited states and thus reduce the dipole transition moment.

Another somewhat surprising result arising from the two-state model is the fact that the TPA cross section is independent of the frequency of the incident light. We have seen that, in general, response functions have a dispersion, and we would expect this also to be the case for the two-photon transition moment considering that it is derived from the single residue of the quadratic response function. However, as there are only two states involved, the situation becomes similar to that observed in the case of OPA, which is also frequency independent.

The two-state model is only applicable for states that are one-photon active. As we discussed, one of the strengths of TPA is the possibility to explore excited states that are dipole forbidden. In order to rationalize such transitions, we need to also consider intermediate states. In the simplest approximation, we assume that there is only one intermediate state involved, giving us a three-state model. The three-state model for the two-photon transition moments follows from Eq. (5.250)

$$M^{(2)}_{0f,\alpha\beta} = \frac{1}{\hbar}\left[\frac{\langle 0|\hat{\mu}_\alpha|i\rangle\langle i|\hat{\mu}_\beta|f\rangle}{\omega_{i0}-\omega_{f0}/2} + \frac{\langle 0|\hat{\mu}_\beta|i\rangle\langle i|\hat{\mu}_\alpha|f\rangle}{\omega_{i0}-\omega_{f0}/2}\right]. \tag{8.180}$$

We can now insert this expression for the two-photon transition moment into the expression for the TPA cross section Eq. (8.177). Introducing as in the case of the two-state model the norm of the transition dipole moments and the angle between these two vectors, the TPA cross section for a three-state model becomes

$$\sigma^{2\mathrm{PA}}_{0f} = \frac{16\pi^3\alpha a_0^5}{c}\,\frac{1}{\pi\Gamma}\,\frac{|\mu_{0i}|^2|\mu_{if}|^2(2\cos^2\vartheta+1)}{\omega_{i0}-\omega_{f0}/2}. \tag{8.181}$$

An important difference in the three-state model compared to the two-state model is that dipole-forbidden states now can be addressed via excitations involving a (virtual) dipole-allowed state. The TPA cross section is maximized if the two transition dipole moments are (anti)parallel, and the intermediate state is strongly one-photon active both from the ground and the final excited states.

Another important difference compared to the two-state model is that the TPA cross section can be enhanced by allowing the denominator to approach zero, that is, when the intermediate state is located at half the frequency of the final excited state. A two-photon active state for which there exists such a resonance condition with an intermediate state is often referred to as being *resonance enhanced*. We emphasize that the standard response theory approach in which the excited states have infinite lifetimes will not be able to accurately model these resonance-enhanced states and that instead a complex polarization propagator approach must be used.

8.5.4.3 General Multiphoton Absorption Processes

As already noted, the technical complexity of experiments involving the simultaneous absorption of more than two photons makes these rather rare. Nevertheless, there is rapid progress, and several four- as well as five-photon absorption experiments have been reported. The theoretical and computational complexity of the expressions for the multiphoton absorption cross sections makes it difficult to rationalize the molecular structural characteristics that would give rise to large specific multiphoton absorption cross sections. However, from a qualitative analysis, we can get insight into when large multiphoton absorption cross sections can be expected.

In order to more clearly illustrate the inherent relations between multiphoton absorption cross sections of different orders, we will for simplicity assume a molecule with inversion symmetry. Many of the principles can be expected to hold also in the case of molecules without such symmetry, but this assumption enhances the effects and simplifies the notation and discussion of the multiphoton transition moments.

For molecules with inversion symmetry, all states can be classified as being of either *gerade* or *ungerade* symmetry. Furthermore, for closed-shell molecules, the ground state is always of *gerade* symmetry. For dipole-driven excitation processes, we can thus write the sum-over-states expression for the three-photon absorption process of a closed-shell molecule as

$$M^{(3)}_{0\mathcal{U}_f,\alpha\beta\gamma} = \frac{1}{\hbar^2}\sum \mathcal{P}_{\alpha\beta\gamma}\sum_k^{n_\mathcal{U}}\sum_l^{n_\mathcal{G}} \frac{\langle 0|\hat{\mu}_\alpha|\mathcal{U}_k\rangle\langle\mathcal{U}_k|\hat{\mu}_\beta|\mathcal{G}_l\rangle\langle\mathcal{G}_l|\hat{\mu}_\gamma|\mathcal{U}_f\rangle}{(\omega_\alpha-\omega_{\mathcal{U}_k})(\omega_\alpha+\omega_\beta-\omega_{\mathcal{G}_l})}, \tag{8.182}$$

where $|\mathcal{U}_k\rangle$ and $|\mathcal{G}_k\rangle$ denote ungerade and gerade states, respectively, and $n_\mathcal{U}$ and $n_\mathcal{G}$ the number of states of ungerade and gerade symmetry. For a molecule with a strong, isolated OPA state, the dominating excited state of *ungerade* symmetry is the final state of interest. In the spirit of the few-state models, the 3PA matrix element can then be reduced to

$$M^{\text{3PA}}_{0\mathcal{U}_f,\alpha\beta\gamma} \approx \frac{1}{\hbar^2}\sum \mathcal{P}_{\alpha\beta\gamma}\sum_{l=1}^{n_\mathcal{G}} \frac{\langle 0|\hat{\mu}_\alpha|\mathcal{U}_f\rangle\langle\mathcal{U}_f|\hat{\mu}_\beta|\mathcal{G}_l\rangle\langle\mathcal{G}_l|\hat{\mu}_\gamma|\mathcal{U}_f\rangle}{(\omega_\alpha-\omega_{\mathcal{U}_f})(\omega_\alpha+\omega_\beta-\omega_{\mathcal{G}_l})} \tag{8.183}$$

$$\approx \frac{1}{\hbar^2}\sum \mathcal{P}_{\alpha\beta\gamma}\frac{M^{\text{OPA}}_{0f,\alpha}}{(\omega_\alpha-\omega_{\mathcal{U}_f})}\cdot\sum_{l=1}^{n_\mathcal{G}} \frac{\langle\mathcal{U}_f|\hat{\mu}_\beta|\mathcal{G}_l\rangle\langle\mathcal{G}_l|\hat{\mu}_\gamma|\mathcal{U}_f\rangle}{(\omega_\alpha+\omega_\beta-\omega_{\mathcal{G}_l})}. \tag{8.184}$$

By assuming that the final state of the three-photon absorption process is isolated and strong, we see that the three-photon absorption cross section is proportional to the OPA cross section, the factor of proportionality being the polarizability of the excited state. We emphasize that in cases where the state of interest is one-photon forbidden, as is the case for the A_u state in the ethylene molecule, Eq. (8.184) would give zero and not be an appropriate approximation to the three-photon absorption cross section. In general, states that are one-photon forbidden will display weaker three-photon absorption than states that are one-photon allowed.

From Eq. (8.184) it follows that because the three-photon absorption cross section is modulated by the polarizability of the final excited state, we can enhance the three-photon absorption cross section by designing molecules in which there is a near-resonance condition in the excited-state polarizability. In particular in the degenerate case, if there is a state of gerade symmetry at two-thirds of the energy of the final state $\mathcal{U}_f$, the excited-state polarizability will have a pole and, following Eq. (8.184), so will the three-photon absorption cross section.

Let us in a similar manner consider the four-photon absorption cross section. Assuming a dominating, isolated two-photon active state of *gerade* symmetry, we can write

$$M^{(4)}_{0\mathcal{G}_f,\alpha\beta\gamma\delta} = \frac{1}{\hbar^3}\sum \mathcal{P}_{\alpha\beta\gamma\delta}\sum_k^{n_\mathcal{U}}\sum_l^{n_\mathcal{G}}\sum_m^{n_\mathcal{U}} \frac{\langle 0|\hat{\mu}_\alpha|\mathcal{U}_k\rangle\langle\mathcal{U}_k|\hat{\mu}_\beta|\mathcal{G}_l\rangle\langle\mathcal{G}_l|\hat{\mu}_\gamma|\mathcal{U}_m\rangle\langle\mathcal{U}_m|\hat{\mu}_\delta|\mathcal{G}_f\rangle}{(\omega_\alpha-\omega_{\mathcal{U}_k})(\omega_\alpha+\omega_\beta-\omega_{\mathcal{G}_l})(\omega_\alpha+\omega_\beta+\omega_\gamma-\omega_{\mathcal{U}_m})} \tag{8.185}$$

$$\approx \frac{1}{\hbar^3}\sum \mathcal{P}_{\alpha\beta\gamma\delta}\sum_k^{n_\mathcal{U}}\sum_m^{n_\mathcal{U}} \frac{\langle 0|\hat{\mu}_\alpha|\mathcal{U}_k\rangle\langle\mathcal{U}_k|\hat{\mu}_\beta|\mathcal{G}_f\rangle\langle\mathcal{G}_f|\hat{\mu}_\gamma|\mathcal{U}_m\rangle\langle\mathcal{U}_m|\hat{\mu}_\gamma|\mathcal{G}_f\rangle}{(\omega_\alpha-\omega_{\mathcal{U}_k})(\omega_\alpha+\omega_\beta-\omega_{\mathcal{G}_f})(\omega_\alpha+\omega_\beta+\omega_\gamma-\omega_{\mathcal{U}_m})} \tag{8.186}$$

$$\approx \frac{1}{2\hbar^2}\sum \mathcal{P}_{\alpha\beta\gamma\delta}\left(\frac{M^{\text{TPA}}_{0f,\alpha\beta}}{(\omega_\alpha+\omega_\beta-\omega_{\mathcal{G}_f})}\cdot\sum_m^{n_\mathcal{U}} \frac{\langle\mathcal{G}_f|\hat{\mu}_\gamma|\mathcal{U}_m\rangle\langle\mathcal{U}_m|\hat{\mu}_\delta|\mathcal{G}_f\rangle}{(\omega_\alpha+\omega_\beta+\omega_\gamma-\omega_{\mathcal{U}_m})}\right). \tag{8.187}$$

and thus we see that the four-photon absorption cross section is proportional to the TPA cross section. As in the case of the three-photon absorption process, the excited-state polarizability modulates the four-photon absorption cross section relative to the TPA cross section. By considering the poles of the excited-state polarizability, we see that we can selectively enhance the four-photon absorption process relative to the TPA process in molecules that have an intermediate state of *ungerade* symmetry at three fourths of the excitation energy of the final state.

For completeness and because we will see some new features, we will also consider a five-photon absorption process. Following what we have done for three- and four-photon absorption, we get

$$M^{(5)}_{0\mathcal{U}_f,\alpha\beta\gamma\delta\epsilon} = \frac{1}{\hbar^4}\sum \mathcal{P}_{\alpha\beta\gamma\delta\epsilon}\sum_{k,m}^{n_\mathcal{U}}\sum_{l,n}^{n_\mathcal{G}} \frac{\langle 0|\hat{\mu}_\alpha|\mathcal{U}_k\rangle\langle\mathcal{U}_k|\hat{\mu}_\beta|\mathcal{G}_l\rangle\langle\mathcal{G}_l|\hat{\mu}_\gamma|\mathcal{U}_m\rangle}{(\omega_\alpha-\omega_{\mathcal{U}_k})(\omega_\alpha+\omega_\beta-\omega_{\mathcal{G}_l})(\omega_\alpha+\omega_\beta+\omega_\gamma-\omega_{\mathcal{U}_m})} \times\frac{\langle\mathcal{U}_m|\hat{\mu}_\delta|\mathcal{G}_n\rangle\langle\mathcal{G}_n|\hat{\mu}_\epsilon|\mathcal{U}_f\rangle}{(\omega_\alpha+\omega_\beta+\omega_\gamma+\omega_\delta-\omega_{\mathcal{G}_n})}, \tag{8.188}$$

noting that we now have two *ungerade* intermediate states that can be substituted by the $\mathcal{U}_f$ state. If we only introduce the final state for the second *ungerade* state ($\mathcal{U}_m$ in the equation) we obtain

$$M^{(5)}_{0\mathcal{U}_f,\alpha\beta\gamma\delta\epsilon} \approx \frac{1}{6\hbar^2}\sum \mathcal{P}_{\alpha\beta\gamma\delta\gamma\epsilon}\left(\frac{M^{3\text{PA}}_{0f,\alpha\beta\gamma}}{(\omega_\alpha+\omega_\beta+\omega_\gamma-\omega_{\mathcal{U}_f})}\cdot\sum_{n=1}^{n_\mathcal{G}}\frac{\langle\mathcal{U}_f|\hat{\mu}_\delta|\mathcal{G}_n\rangle\langle\mathcal{G}_n|\hat{\mu}_\epsilon|\mathcal{U}_f\rangle}{(\omega_\alpha+\omega_\beta+\omega_\gamma+\omega_\delta-\omega_{\mathcal{G}_n})}\right), \tag{8.189}$$

which can be considered as a *three-photon-dominated five-photon absorption cross section*. This would correspond, for instance, to states of A_u symmetry in molecules with D_{2h} symmetry. In principle, there should also be a term arising from introducing the final state for the first *ungerade* intermediate state. However, this would lead to a term which would correspond to the OPA transition moment scaled by the excited-state second hyperpolarizability, and thus can be expected to be smaller. If the final state is also strongly OPA active, we can simplify Eq. (8.189) further by substituting all *ungerade* intermediate states by the final state, giving

$$\begin{aligned} M^{5\text{PA}}_{0\mathcal{U}_f,\alpha\beta\gamma\delta\epsilon} &\approx \frac{1}{\hbar^4}\sum \mathcal{P}_{\alpha\beta\gamma\delta\epsilon} \qquad (8.190)\\ &\sum_{l,n=1}^{n_\mathcal{G}}\frac{\langle 0|\hat{\mu}_\alpha|\mathcal{U}_f\rangle\langle\mathcal{U}_f|\hat{\mu}_\beta|\mathcal{G}_l\rangle\langle\mathcal{G}_l|\hat{\mu}_\gamma|\mathcal{U}_f\rangle\langle\mathcal{U}_f|\hat{\mu}_\delta|\mathcal{G}_n\rangle\langle\mathcal{G}_n|\hat{\mu}_\epsilon|\mathcal{U}_f\rangle}{(\omega_\alpha-\omega_{\mathcal{U}_f})(\omega_\alpha+\omega_\beta-\omega_{\mathcal{G}_l})(\omega_\alpha+\omega_\beta+\omega_\gamma-\omega_{\mathcal{U}_f})(\omega_\alpha+\omega_\beta+\omega_\gamma+\omega_\delta-\omega_{\mathcal{G}_n})}\\ &= \frac{1}{\hbar^4}\sum \mathcal{P}_{\alpha\beta\gamma\delta\epsilon}\frac{M^{\text{OPA}}_\alpha}{(\omega_\alpha-\omega_{\mathcal{U}_f})(\omega_\alpha+\omega_\beta+\omega_\gamma-\omega_{\mathcal{U}_f})}\\ &\times\sum_{l=1}^{n_\mathcal{G}}\frac{\langle\mathcal{U}_f|\hat{\mu}_\beta|\mathcal{G}_l\rangle\langle\mathcal{G}_l|\hat{\mu}_\gamma|\mathcal{U}_f\rangle}{(\omega_\alpha+\omega_\beta-\omega_{\mathcal{G}_l})}\sum_{n=1}^{n_\mathcal{G}}\frac{\langle\mathcal{U}_f|\hat{\mu}_\delta|\mathcal{G}_n\rangle\langle\mathcal{G}_n|\hat{\mu}_\epsilon|\mathcal{U}_f\rangle}{(\omega_\alpha+\omega_\beta+\omega_\gamma+\omega_\delta-\omega_{\mathcal{G}_n})}, \end{aligned}$$

which shows that the 5PA absorption cross section can be expected to be proportional to the OPA transition moment scaled by the excited-state polarizability squared.

To summarize, we have in this section seen that in the special case of molecules with inversion symmetry and for molecules with strong and isolated final states, there exist strong similarities between odd- and even-order multiphoton absorption processes. Odd-order processes have been shown to be proportional to the lowest order, nonvanishing odd-order process, be this either OPA or three-photon absorption processes depending on whether the state of interest is one-photon active or not. In a similar manner, even-order processes have in these systems been shown to be proportional to the TPA cross section. In all cases, the factor of proportionality is the polarizability of the final, excited state, but evaluated using different sums of the incoming photon energies. This provides a route for designing molecules in which specific higher order multiphoton processes can be selectively enhanced compared to the lower order processes.

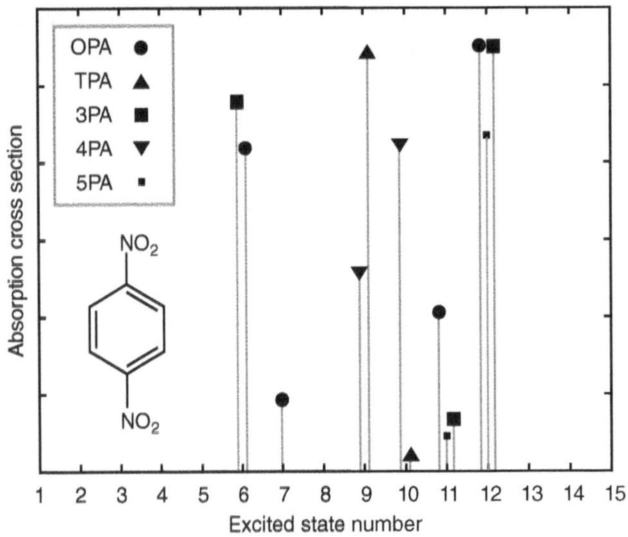

Figure 8.19 Dominant OPA, TPA, three-, four-, and five-photon absorption cross sections for the lowest 15 singlet states in *para*-dinitrobenzene. Results are obtained at the CAM-B3LYP/aug-cc-pVDZ level of theory.

To illustrate these principles, we show in Figure 8.19 the one- through five-photon absorption cross sections of the fifteen lowest states of *para*-dinitrobenzene using the CAM-B3LYP functional and the aug-cc-pVDZ basis set. Units have been selected so that the multiphoton absorption cross sections are comparable. We see that the excited states are divided into three groups—one set of states that show strong odd-order absorption (states 6–7 and 11–12 of symmetry B_{2u} and B_{3u}), one set of states that show strong even-order absorption (states 9–10 of symmetry B_{1g} and A_g), and a set of states that neither display large odd- nor even-order absorption.[21] Since the multiphoton absorption cross section at different orders have different units, a direct comparison is not possible. Nevertheless, looking at these relative intensities, there is one state that clearly stands out from the rest, namely state 10, which has a rather moderate TPA cross section, but in relative terms a significantly enhanced four-photon absorption cross section. This is the second excited state of B_{1g} symmetry that has an excitation energy of 6.20 eV. We recall that we could selectively enhance four-photon absorption compared to the TPA if there are intermediate states of *ungerade* symmetry at three fourths of the energy of the final state. Indeed, at the computational level chosen, excited state number 5 of B_{2u} symmetry of *para*-dinitrobenzene has an excitation energy of 4.62 eV. This will lead to a near-resonance condition in the excited-state polarizability that governs the four-photon absorption cross section, and the molecular characteristics thus explain the observed enhancement of the four-photon absorption cross section for the $2B_{1g}$ state relative to the TPA cross section for this state.

8.5.5 X-ray Absorption

Molecular excitation processes due to the absorption of photons in the soft X-ray region (< 2000 eV) involve initial electronic states that are localized and atomic-like in nature, and final states that are typically more delocalized, or even free-electron like. The initial electronic state in an X-ray absorption process belongs to a certain atomic shell *K*, *L*, and so on, with a specific ionization potential. The great majority of experiments are carried out at the *K*-edges of second-row elements in the periodic table but the *K*- and *L*-edges of third-row elements are

21 The one-, three-, and five-absorption cross sections for the $1B_{2u}$ state (state 5) are too small to be visible on the scale in this figure.

frequently addressed as well.[22] Upon closer inspection, it is noted that the L-edge consists of three individual components namely the L_1-, L_2-, and L_3-edges as due to electronic excitations from the atomic $2s$, $2p_{1/2}$, and $2p_{3/2}$ subshells, respectively, the splitting of the $2p$ orbitals being due to the spin–orbit interaction as discussed in Section 8.2. The spectral regions that are most rich in details are found close to the ionization edges and they are studied in near-edge X-ray absorption fine structure (NEXAFS) spectroscopy. The name X-ray absorption near-edge structure (XANES) is synonymous but we will use the terminology NEXAFS as it is more commonly adopted in surface and molecular physics. Our discussion in this section will be focused on this near-edge spectroscopy but note that XAS also includes the study of the extended X-ray absorption fine structure (EXAFS) regions found at higher energies, above the ionization edges.

The source of radiation in a NEXAFS experiment is a synchrotron in which charged particles (typically electrons) are orbiting at speeds near the speed of light. Accelerating electrons by means of electric fields to high kinetic energies (~GeV) requires magnetic fields in the bending magnets to be increased in synchronicity (hence the term synchrotron). The radiation that is generated in the bending magnets is transported via beamlines to the user's experimental chamber, and each synchrotron thus has multiple available beamlines set up for various experimental activities. In the third generation of synchrotrons, which are in use today, the radiation beam is characterized by high brightness, intensity, and collimation and, most importantly, also high degrees of wavelength tunability and field vector polarization. The next generation of radiation sources, that is, the fourth-generation synchrotrons, differ from their predecessors by delivering an extremely high peak brightness during pulse lengths of 100 fs or shorter. One of the several fundamental consequences of this development is that multiphoton X-ray interactions will become noticeable. At present, however, the vast majority of materials research is concerned with the linear photon processes and we will therefore focus on that.

8.5.5.1 Core-Excited States

In XAS, the observable is the variation of the photon absorption cross section with respect to the photon wavelength (or energy). This cross section is detected in an indirect manner by studying the decay of the core hole. The core hole is filled by an electron from an orbital higher in energy, which, as a consequence, will lead to the emission of either a fluorescence photon or an Auger electron, as shown in Figure 8.20. Since both decay channels are connected to past events of photon absorption, detection can be made of either the fluorescence photon in a fluorescence detector or the Auger electron in a channeltron.[23] For low-Z elements (such as carbon, nitrogen, and oxygen), the latter approach is advantageous since the Auger electron yield is then higher than the fluorescence photon yield as illustrated in Figure 8.21.

The most striking feature of the core-electron excited state is that the hole orbital is localized and atomic in nature, and as a consequence there is a strong reduction in the screening of the corresponding atomic nucleus. From the point of view of valence electrons, this reduction in nuclear screening amounts to almost an entire elementary charge, but, for fellow core electrons, the reduction is of course smaller. It is also clear that for a given molecule, the $1s$-orbital density in the manifold of core-excited states will closely resemble that found in the corresponding core-ionized species since what separates the core-excited and core-ionized states is the electron occupation of the virtual valence or Rydberg orbital in the former case, but this difference has a negligible effect on the core orbital. However, the difference in core-electron densities between the ground and core-ionized states is expected to be significant since the remaining

22 K-edges correspond to excitations of electrons occupying the $n = 1$ energy levels of atoms (the $1s$-orbitals), L-edges refer to the $n = 2$ levels ($2s$ and $2p$), and so on. One may wonder why the $1s$-level is denoted as by letter K in X-ray spectroscopies. The origin of this nomenclature can be traced to Charles G. Barkla, who when characterizing X-ray radiation in 1909 wanted to leave alphabetic room for even higher energy excitations than those arising from the K-shell.

23 The channeltron electron multiplier enhances the signal (i.e., the number of Auger electrons per time unit) by several orders of magnitude.

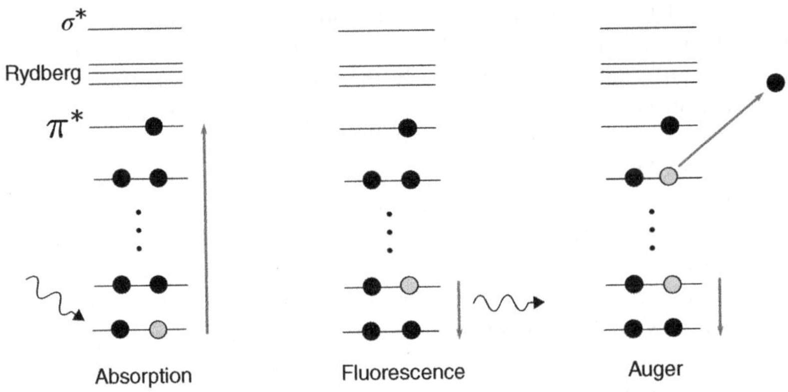

Figure 8.20 X-ray absorption and radiative and nonradiative core hole decay.

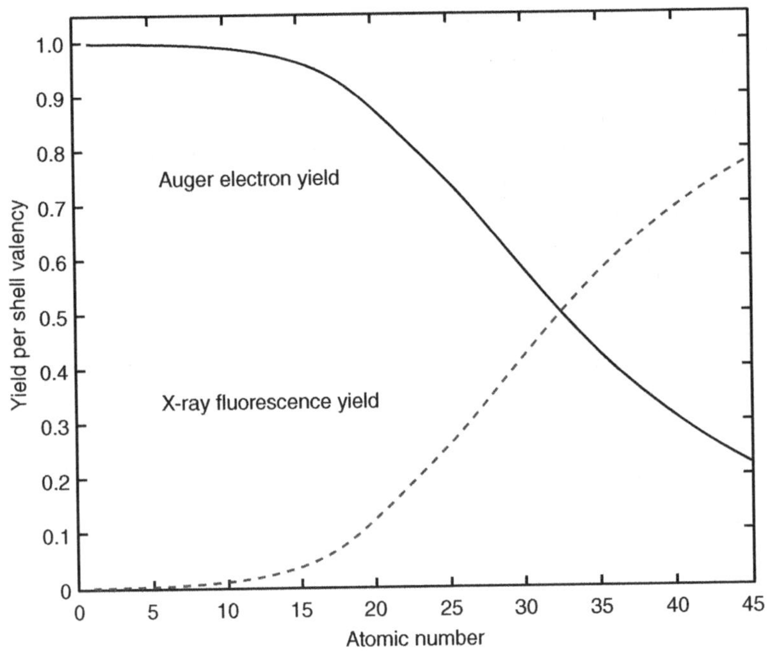

Figure 8.21 Fluorescence and Auger electron yields in K-edge NEXAFS as a function of atomic number.

core electron feels the full attraction of the atomic nucleus in the core-ionized species, without the screening from its Kramers (or spin) partner.

As an example, let us study the 1s-orbital of a carbon atom in ethylene and see how it is affected by core ionization. In Figure 8.22, we depict the radial densities of the core orbitals in the ground- and core-ionized states. In the ground state, the radial density finds a maximum value at a distance of 0.176 a_0, whereas, in the core-ionized state, this maximum is located at 0.168 a_0. The latter result is close to what is found for a hydrogenic atom with $Z = 6$, where the maximum occurs at 0.167 a_0, and which illustrates the minor influence of the valence electron density on the core electrons. But the contraction of the 1s-orbital that occurs upon core-electron excitation is significant and signals a need for flexibility in the basis set in the core region.

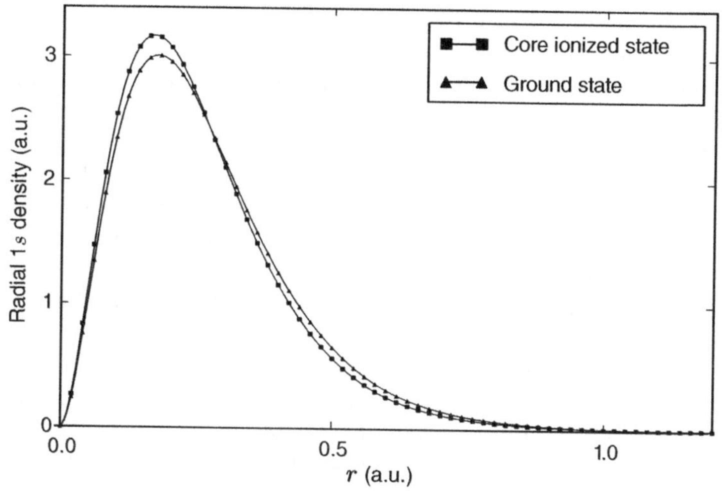

Figure 8.22 Radial densities for the carbon core 1s-orbitals in C_2H_4 in the ground- and core-ionized states at a frozen molecular geometry. The results refer to the nonrelativistic Hartree–Fock level of theory using an uncontracted basis set.

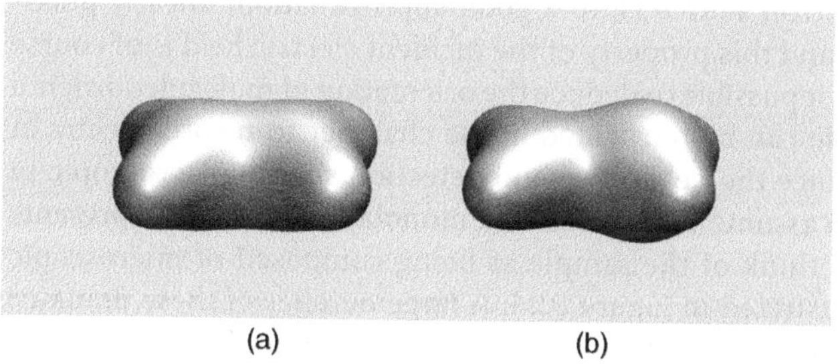

Figure 8.23 Valence electron density for C_2H_4 in (a) the ground state and (b) a core-ionized state with the core hole localized to the rightmost carbon atom. The plots correspond to isodensity surfaces of the valence electron densities as determined at the Hartree–Fock level of theory.

Returning to a discussion of the effects of core excitations on the valence electron density, we understand that due to the core–valence separation (CVS), it can be reasonable to approximate the removal of an electron from the K-shell by instead adding one unit of elementary charge to the nucleus. This approximation is known as the $Z + 1$-approximation. In our example, it amounts to replacing the carbon atom to which the core hole is localized by a nitrogen atom in its ground state. It becomes clear that inversion symmetry is lost in this process, as a consequence of the localization of the core hole to one of the two carbons, and the core hole appears highly attractive for the valence electrons. In the core-excited, or core-ionized, state, we will therefore find an excess of valence electron charge on the atomic center of the core hole. In the case of our ethylene example, this effect, which is known as the electron relaxation effect, is clearly seen in Figure 8.23.

In this example, we have explicitly taken into account the electron relaxation in the core-hole state by performing an open-shell SCF optimization of the excited-state wave function. This procedure is however not in line with the philosophy of response-theoretical approaches, where one avoids any explicit reference to the excited states and instead deduces excitation spectra from singularities in response matrices (see Section 7.4). We immediately anticipate two main obstacles in the application of response theory to calculations of X-ray absorption spectra, namely (i) to identify the matrix singularities that correspond to the highly excited core-excited states, and (ii) to deal with the electron relaxation effect, or the polarization of the valence electron density, which is depicted in Figure 8.23. The first issue is mainly related to the design of numerical recipes other than bottom-up techniques for finding matrix eigenvalues. Bottom-up approaches are conventionally used in the visible and ultraviolet regions of the spectrum, but they are bound to fail in searches for semibound states in the X-ray region that in reality are embedded in a continuum of valence-ionized states. From a response theory point of view, the second issue means that we see the effect of electron relaxation as an effect of strong electron correlation in the calculation of the ground-state response function. If we think of the core-excited wave function in terms of a CI expansion in the canonical Hartree–Fock orbitals of the ground state, it becomes clear that the polarization of the valence electron density requires, as a minimum, the invocation of double-excited determinants. The double-excited determinants of importance must include a combination of the core-hole excitation of interest and a valence-electron excitation, out of which the latter is responsible for the polarization effect. We shall see further on that the two issues discussed here have driven the design of methods in a direction where one, in some way or another, does make explicit reference to the core-hole state at the expense of the generality that is the hallmark of response theory.

8.5.5.2 Field Polarization

The radiation provided by a synchrotron source is, to a good approximation, linearly polarized in the plane of the storage ring, and this property of the incident electric field is of course exploited in the experiment. It makes it possible to deduce the orientation of molecules ordered by bonding to surfaces, as the sample can be reoriented in the chamber to record spectra at varying angles of beam incidence. Since the beam area and detection slit are macroscopic in dimension, it becomes reasonable to assume that a recorded monolayer spectrum represents an orientational average, that is, we think of the sample as being composed of microscopic domains of ordered molecules as illustrated in Figure 8.24. A large number of these domains are exposed to the X-ray radiation and the spectrum will correspond to an average over all possible directions in the xy-plane, that is, the plane of the substrate.

In the electric dipole approximation, the linear absorption cross section is proportional to the imaginary part of the electric-dipole polarizability, Eq. (5.335), and the integrated cross section

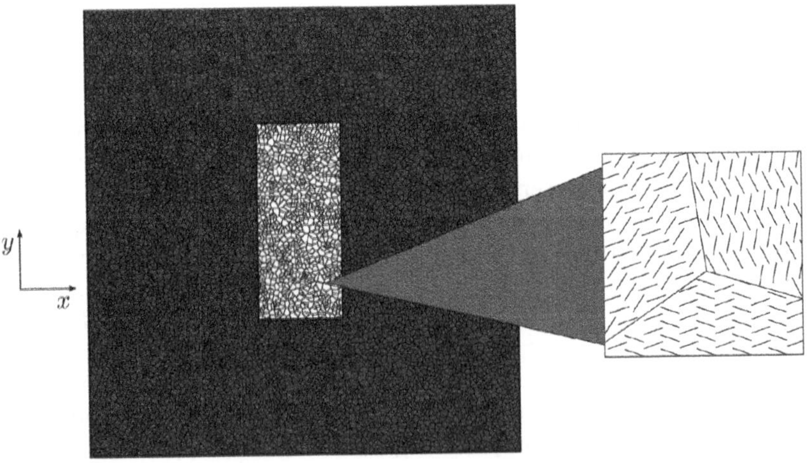

Figure 8.24 X-ray radiation beam illuminating domains of self-assembled molecules in a monolayer sample. Detection is made of particles (fluorescence photons or Auger electrons) passing through a slit. The z axis is aligned with the surface normal.

is proportional to the oscillator strength, Eq. (5.228), so the strength of a given transition due to linearly polarized light is governed by the square of the one-photon transition matrix element; see Eq. (5.248)

$$I \propto |\langle f|\hat{\boldsymbol{\mu}}|0\rangle \cdot \mathbf{F}|^2, \tag{8.191}$$

where **F** denotes the electric field vector. In the nonrelativistic realm, when the eigenstates of the zeroth-order molecular Hamiltonian can be assumed real, we introduce an absorption dipole vector $\mathbf{D} = \langle f|\hat{\boldsymbol{\mu}}|0\rangle$, which leaves us with the following expression for the transition intensity

$$I \propto \cos^2(D, F), \tag{8.192}$$

where (D, F) denotes the angle between the absorption dipole vector and the incident electric field. Transitions from spherically symmetric s orbitals to σ^*- and π^*-orbitals with distinct spatial orientations are of particular interest, since in this case the absorption dipole vector will be collinear with the orientation of the final state orbital, and this is the reason why a study of the dependence of the transition intensity with respect to the angle (D, F) will reveal information about the molecular orientation. For conjugated systems with extended and/or multiple π^*-orbitals, this procedure can become troublesome due to the fact that the orbitals may map out several varying spatial orientations. It is then important to attribute isolated peaks in the NEXAFS spectrum to specific moieties, which is a task well suited for calculations.

Equation (8.192) refers to a specific molecular orientation and in order to arrive at a final formula for the measured intensity, we need to average over the molecular orientations in the sample. For a self-assembled monolayer as illustrated in Figure 8.24, the operation of orientational averaging amounts to averaging the unit absorption dipole vector $\mathbf{e}_D$ over a cone, as illustrated in Figure 8.25. In doing so, we choose to orient the coordinate system so that the incident electric field vector lies in the xz-plane, that is, $\phi_F = 0$, and we thereby get

$$\begin{aligned}\cos^2(D, F) &= |\mathbf{e}_D \cdot \mathbf{e}_F|^2 \\ &= \sin^2\theta_D \sin^2\theta_F \cos^2\phi_D + \tfrac{1}{2}\sin 2\theta_D \sin 2\theta_F \cos\phi_D + \cos^2\theta_D \cos^2\theta_F,\end{aligned} \tag{8.193}$$

where θ_D and θ_F denote the polar angles for the absorption dipole and the electric field vectors, respectively. We briefly note that under conditions of normal incidence, that is, a direction of the X-ray beam propagation along the surface normal, we have $\theta_F = 90°$, whereas at grazing incidence we have $\theta_F < 90°$. Based on Eq. (8.193), the orientational average will become

$$I \propto \frac{1}{2\pi} \int_0^{2\pi} \cos^2(D, F)\, d\phi_D = \left(\frac{1}{2}\sin^2\theta_D \sin^2\theta_F + \cos^2\theta_D \cos^2\theta_F\right). \tag{8.194}$$

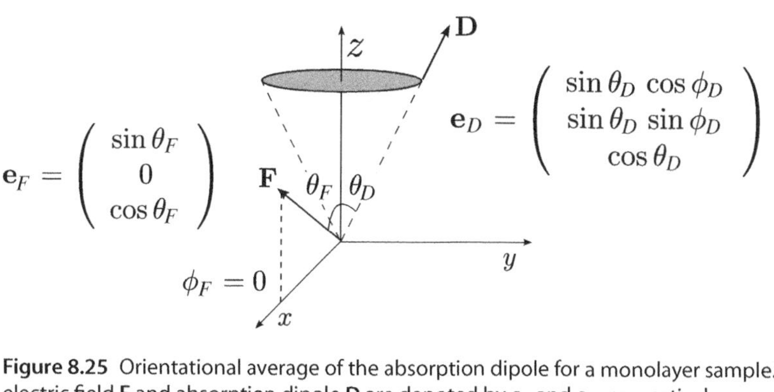

Figure 8.25 Orientational average of the absorption dipole for a monolayer sample. The unit vectors of the electric field **F** and absorption dipole **D** are denoted by $\mathbf{e}_F$ and $\mathbf{e}_D$, respectively.

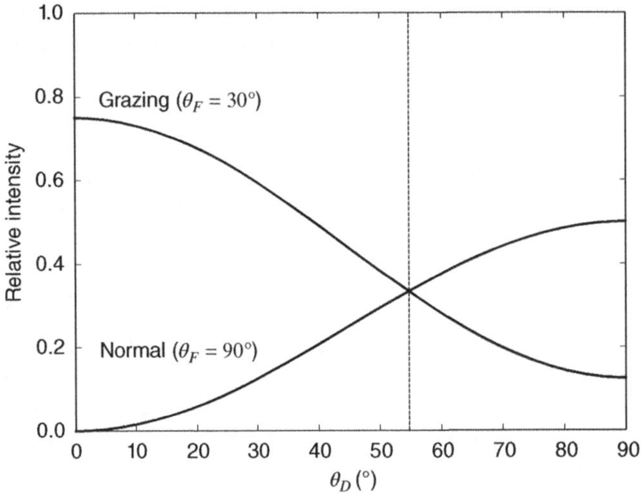

Figure 8.26 Relative intensities of a near K-edge transition at normal and grazing incidence of the X-ray beam. The vertical line is located at the magic angle.

We note that for a polar angle of the electric field equal to $\theta_F = \arctan \sqrt{2} \approx 54.7°$ (known as the *magic* angle), the intensity in Eq. (8.194) will become independent of the orientation of the absorption dipole vector or, equivalently, the molecular orientation. The converse is also true, namely that if the molecular orientation in the monolayer is such that θ_D equals the magic angle, then the measured intensity is independent of the orientation of the electric field. This situation is somewhat cumbersome since a transition intensity that is independent of the angle of incidence of the X-ray beam is also the signal of a randomly oriented multilayer sample, and it may therefore cast doubts on whether the sample preparation really resulted in a self-assembled monolayer.

Owing to this issue of the magic angle, it is customary to choose the angle of grazing incidence separate from the magic angle. With a choice of $\theta_F = 30°$ at grazing incidence, the dependencies of the transition intensity on the polar angle of the final-state orbital will take the form shown in Figure 8.26—the crossing point of the two curves occurs at the magic angle. The integrated absorption cross section of a specific transition is compared at the two angles of incidence, and given their ratio, the angle θ_D, that is, the molecular orientation, can be deduced from the figure.

Crucial to the analysis of the experimental spectra is that assignment of absorption bands to specific molecular moieties can be made. Such assignments are made easier in cases of clear band separations, whereas in situations of overlapping bands the analysis is more difficult. In this work of assignment, theoretical calculations constitute an indispensable tool, and below we will discuss two separate response theoretical approaches to determine X-ray absorption spectra.

8.5.5.3 Static Exchange Approximation

We discussed the issue of electron relaxation in the vicinity of a core hole, and Figure 8.23 illustrates the electron density obtained by an explicit optimization of a wave function for the core-ionized state. This orbital relaxation energy can be large, amounting to some 17 eV for the oxygen K-edge in water, and this is also reflected in the Hartree–Fock core-orbital eigenvalues that, via Koopmans' theorem, strongly overestimate core-ionization energies. Despite this strong effect of electron correlation (in the form of orbital relaxation) on the core excitation energies, it is possible to approximate both the ground- and core-excited states using state-specific methods such as the open-shell Hartree–Fock or the multiconfiguration SCF approximation. But in order to model the near-edge fine structure measured in a NEXAFS experiment, we need to take a large number of states into account, and this makes the application of state-specific methods computationally expensive and cumbersome. An additional problem is that it may be difficult to create a balanced model for all excited states in the sense that for different excited states, different active spaces may be needed in order to properly account for the electron correlation effects when using multiconfiguration state-specific methods.

From our discussion, it is clear that a conventional configuration interaction singles (CIS) approach is bound to be inaccurate due to a failure to describe valence orbital relaxation, but we may consider an approach that is somewhat modified. If we describe the excited state $|\epsilon\rangle$ as a linear combination of singly excited configurations

$$|\epsilon\rangle = \sum_{s}^{\text{virt}} \sum_{i}^{\text{core}} c_i^s \, \hat{a}_s^\dagger \hat{a}_i |\text{ref}\rangle, \tag{8.195}$$

but with $|\text{ref}\rangle$ being a closed-shell reference determinant constructed from the ground-state orbitals of the *core-ionized* system, then the orbital relaxation is approximately captured in the state $|\epsilon\rangle$.

The original formulation of the idea expressed in Eq. (8.195) was based on an explicit effective one-particle Hamiltonian for the excited electron in the static field of the core-ionized molecule, and the method was consequently coined as the *static exchange* (STEX) approach. However, when extending the STEX method to treat excitations from p and higher angular momenta shells, that is, including more than one core orbital in the model, this description is inappropriate, and STEX should then rather be thought of as a CIS performed with orbitals optimized for the core-ionized molecule. We also note that since the orbital relaxation is specific to the atom where the core hole is being created, a single STEX calculation can only properly describe excitations from core orbitals on a single atom. In order to obtain the final NEXAFS spectrum in an STEX calculation, one needs to perform separate calculations with, in turn, a core hole localized on each of the chemically inequivalent atoms and in the end add these separate contributions to a complete spectrum at the absorption edge of interest. The clear benefit, in comparison to separate state approaches, is that the explicit diagonalization of the CIS Hamiltonian provides us with a representation of somewhere between a few hundred to a few thousand excited states while needing only to perform one single SCF optimization (namely, that of the core ionized state).

Once the CIS Hamiltonian has been diagonalized and thus provided excitation energies (in form of the eigenvalues) as well as a representation of the excited states, then the transition strengths are evaluated as squares of transition electric dipole moments

$$\langle\epsilon|\hat{\mu}|0\rangle = \langle\epsilon|0\rangle\langle 0|\hat{\mu}|0\rangle + \langle\epsilon|\perp\rangle\langle\perp|\hat{\mu}|0\rangle, \tag{8.196}$$

where $\hat{\mu}$ is the electric-dipole operator, $|0\rangle$ denotes the Hartree–Fock ground state, and $|\perp\rangle$ its orthogonal complement. In the length gauge, the electronic part of the ground-state dipole moment is origin dependent, which introduces an origin dependence also in the transition dipole moments. In practice, however, the overlap $\langle\epsilon \mid 0\rangle$ is small, and this is of minor importance. A more important practical problem is that the transition moments have to be computed between nonorthogonal orbitals. Here a cofactor expansion is used, but this expansion may be equally computationally costly as the calculation of the excitation energies themselves. It scales as one matrix inversion per excited state, where the matrix dimension equals the number of

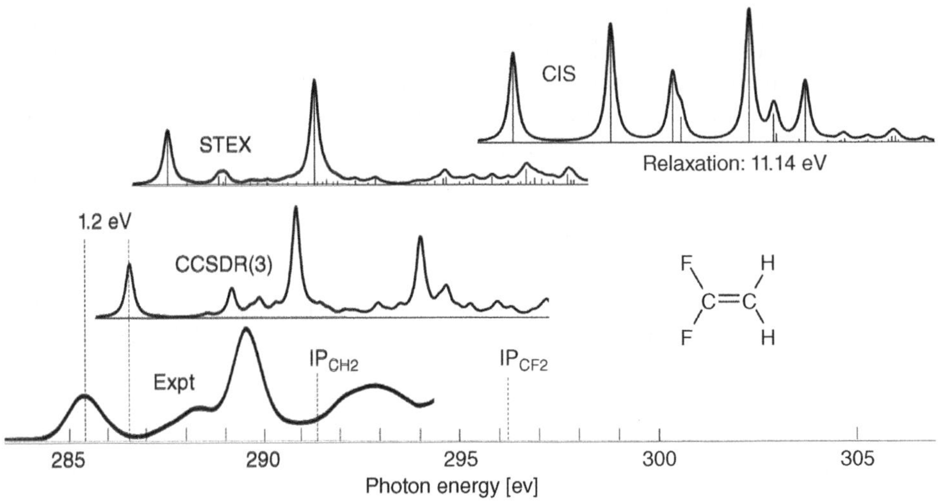

Figure 8.27 Near carbon *K*-edge X-ray absorption spectra for 1,1-difluoroethylene. Theoretical spectra are obtained with configuration interaction singles (CIS), STEX, and coupled cluster singles and doubles with perturbative treatment of triples excitations [CCSDR(3)]. A HWHM broadening parameter $\hbar\gamma = 0.124$ eV is used.

occupied orbitals, and so the computational cost is basis set independent. This should be contrasted with theories that work with a single set of orthonormal orbitals, where the calculation of transition moments is trivial.

The advantages of STEX are thus that it is a fast method that quickly produces the core-excitation spectrum over a large energy range (thousands of states) and the excited-state wave functions are immediately available for interpretation. The disadvantages are that STEX neglects effects of electron correlation and it also introduces issues of nonorthogonality. However, in contrast to pure state-specific methods, it maintains orthogonality between the excited states.

In Figure 8.27, the NEXAFS spectra of 1,1-difluoroethylene as obtained with a selection of different methods are shown as an illustration. The CIS spectrum based on the ground-state orbitals is seen to be nowhere near the experimental spectrum. If we consider the discrepancy in excitation energy for the first peak in the spectrum as a measure of the relaxation energy, it is seen to amount to 11.14 eV. Moreover, the CIS spectral profile is of very poor quality. The STEX spectrum is in far better agreement with experiment, displaying two distinct π^*-resonances due to the two inequivalent carbons. The peak lower in energy is associated with CH_2 and is less intense compared to the one higher in energy and associated with CF_2, a spectral feature that is well captured in the STEX calculation.

8.5.5.4 Complex or Damped Response Theory

A quite different view on the calculation of X-ray absorption spectra is obtained if one considers as a starting point the physical observable, namely, the linear absorption cross section given in Eq. (5.335). The molecular property of interest is in this case the imaginary part of the electric-dipole polarizability of the electronic *ground state*, and in this view, we thus completely circumvent a direct and explicit reference to the final state in the absorption process. The issue of electron relaxation in the excited state will not be dealt with by means of explicit SCF optimization of a core-ionized species, but rather by means of a treatment of electron correlation in the calculation of the ground-state response function

$$\alpha^I_{\alpha\beta}(\omega) = -\mathrm{Im}\langle\langle \hat{\mu}_\alpha; \hat{\mu}_\beta \rangle\rangle_\omega. \tag{8.197}$$

Such a complex response theory approach has several formal and practical advantages that have been addressed in this book. It explicitly optimizes the ground-state wave function (or density) only, it ensures orthogonality among states, it preserves sum rules, and it is applicable to all standard electronic structure methods (wave function and density based). In other words, it puts the calculation of an X-ray absorption spectrum on equal footing with the calculation of the corresponding UV/vis spectrum, where these methodological advantages have been recognized for a long time. In terms of practical calculations, it becomes highly advantageous to adopt a standard response theory approach for the treatment of molecules with several chemically inequivalent atoms, which is frequently encountered in addressing the carbon K-edge of biochemical molecules. In contrast to the STEX method discussed above, or any state-specific method, a full channel interaction is accounted for in the response theory approach and the contributions from all atomic centers is thus included given a complete manifold of excitation operators in the evaluation of the propagator. Full channel interaction here refers to the inclusion of *all* operators transferring electrons from occupied core to unoccupied virtual orbitals. Due to the large separation in energy between core and valence excitations, one can in linear XAS without any great loss of accuracy introduce the approximation of CVS, where electron excitations from the occupied valence orbitals are excluded in the calculation of the linear response function in Eq. (8.197). We would still refer to this case as providing full channel interaction and it in fact enables the calculation of a NEXAFS spectrum by also solving the generalized eigenvalue equation in the same manner as done in UV/vis absorption spectroscopy.

So much for pros, but what are the cons? In a wave function-based formalism, we have understood that a minimum requirement in a response theory treatment of XAS is the inclusion of doubly excited determinants in order to capture the electronic relaxation and polarization in the transition process. The correlation effect is dynamic in nature so it seems that the coupled cluster singles and doubles approach is a reasonable candidate. However, since the main advantages of the damped response theory technique for addressing XAS becomes apparent when treating medium- and large-sized molecules, there is a need for coupled cluster methodologies with reduced computational scaling before it can be applied on a routine basis. Currently, the main alternative for treating dynamic correlation in large systems is DFT, but since the operator manifold in this case is built from single-electron excitations, it is not the ideal candidate to capture the valence-electron relaxation in core excitations. Specific requirements must also be imposed on the exchange–correlation functional for it to be able to describe core excitations. The electronic transition takes place from a local atomic core orbital to a delocalized molecular orbital with a relatively small overlap between the initial- and final-state orbitals, and this makes it, in a sense, similar in nature to a valence charge-transfer transition that represents a well-known challenge for time-dependent DFT. The root of the problem of charge transfer in TD-DFT lies of course in the exchange–correlation functional which, due to an inexact exchange interaction, has an incorrect long-range asymptotic Coulomb interaction. One remedy to correct this problem is to invoke the technique of long-range corrections by means of range separation, which, in essence, turns on more exact exchange at long interelectron distances. A second concern using standard DFT in XAS is the self-interaction error that is particularly pronounced for the localized core orbitals. If this error is more or less constant for the core orbitals of interest, then it can be accounted for by an overall energy shift of the final spectrum. There also exist schemes for calculating orbital-specific self-interaction corrections. All things taken into account, the spectral profile as obtained at the DFT level of theory can reach a quality that is close to what one finds at the CCSD level of theory as illustrated in Figure 8.28. Both spectra are shifted as to align the first peak with the corresponding peak in the experiment, but, with regard to the spectral profile, it is seen that the DFT spectrum is in close agreement with the CCSD one. The DFT

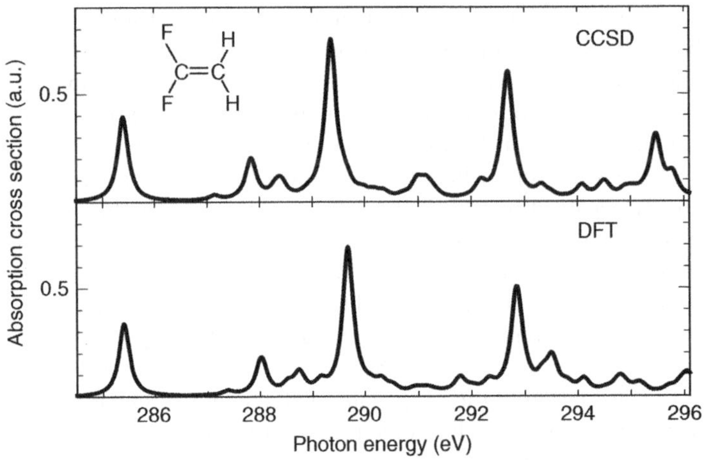

Figure 8.28 Near carbon *K*-edge X-ray absorption spectra for 1,1-difluoroethylene obtained with the complex polarization propagator method in conjunction with a description of the electronic structure at the levels of DFT and CCSD. The CCSD and DFT spectra are shifted −3.0 and −10.2 eV, respectively, so as to align the first peak with the corresponding experimental transition energy. A HWHM broadening parameter $\hbar\gamma = 0.124$ eV is used.

spectrum is here obtained with use of the CAM-B3LYP functional with a parametrization that provides a correct asymptotic limit of the long-range hole–electron Coulomb interaction. In Figure 8.27, the CCSDR(3) spectrum is presented. At this level of theory, CCSD transition energies are corrected by a perturbative treatment of triple excitations, whereas transition moments and thus intensities are not. It is noted that about 90% of the relaxation energy is captured by the treatment of electron correlation at this level of theory. Not only the π^*-resonances but also other resonances are well described in the theoretical spectrum, allowing an unambiguous assignment of the spectrum to be made.

8.6 Birefringences and Dichroisms

In this section, we will consider in more detail the molecular properties that arise from the fact that a dipole moment can be induced in a molecule by a time-dependent magnetic field such as that associated with a propagating electromagnetic wave, as discussed in Section 3.5.5. The polarization of light is closely related to the propagation of the magnetic component of the electromagnetic wave, as discussed in Section 3.1.5, and thus we will be concerned with the dependence of the induced dipole moment on the time variation of the magnetic component of the electromagnetic wave. This induced dipole moment may then differ for different polarizations of the light, giving rise to two important phenomena: *birefringences* (from Latin: *bi*, "two"+*refringere*, "to break up") and their absorptive counterparts, *dichroisms* (from Greek: *di*, "two"+*chroma*, "color"). Some birefringences, such as the Kerr effect introduced in Section 8.3.2, are defined by the electric field component of the light only, but focus will be on birefringences arising from the magnetic component of the electromagnetic wave.

There are many different kinds of birefringences and dichroisms, and they are all characterized by the dispersive or absorptive components of the refractive index depending on the polarization and propagation direction of the light. We will then measure the difference between different components of the refractive index, which will have a well-defined magnitude and sign. Birefringences and dichroisms can provide information about molecular structure that can complement information obtained in the conventional refraction and absorption of light.

Birefringences and dichroisms provide in particular two sets of information about the molecule and the experimental setup that often cannot be obtained in any other way. The first, and most important, is that the observables can distinguish between enantiomers of *chiral molecules*, that is, molecules whose mirror image is not superimposable on the molecule itself, such as the CHFClBr molecule shown in Figure 3.29. The two enantiomers of the molecule will have the same magnitude of the observed birefringence or dichroism, but they will differ in their sign. In the case of nonchiral molecules, there will in general not be a difference in the observables in the two components of the refractive index, and thus there would be no observed dichroism or birefringence. However, for such molecules, we can instead induce an asymmetry in the experimental setup, creating a setup which is not superimposable on its own mirror image. In this case, birefringences and dichroisms can be observed even for nonchiral molecules, but the sign itself is no longer directly connected to the molecular structure.

The second set of information that can be obtained from, in particular, nonlinear birefringence measurements, are lower order electric multipole moments or polarizabilities, as already discussed for the case of polarization of linear media in Section 3.5.2. The orientation of molecules relative to the applied electric field leads to temperature-dependent contributions to the observable properties involving lower order molecular properties.

Birefringences are normally divided into three main classes based on which difference in the components of the refractive index we measure. The birefringence that is most frequently encountered is circular birefringence, in which one measures the difference in refraction of light that is left n_+ and right n_- circularly polarized. In general, we can write a circular birefringence as

$$\Delta n^{\mathrm{circ}} = n_+ - n_-. \tag{8.198}$$

One important application of circular birefringence is the observation of the rotation of plane-polarized light as it passes through a sample of chiral molecules, which is known as *optical rotation*. We will discuss this property in more detail in Section 8.6.1. Its absorptive counterpart is analogously referred to as *(electronic) circular dichroism*, which we will discuss in Section 8.6.2. These are among the earliest observations of birefringence, and are still important for determining the absolute stereochemistry of chiral molecules or the enantiomeric purity of a sample. For achiral systems, a circular birefringence can be induced by applying an external magnetic field along the direction of light propagation, as illustrated in Figure 8.29. This phenomenon is known as the Faraday effect and it has an absorptive counterpart called magnetic circular dichroism (MCD).

A second class of birefringences is known as *linear birefringence*. Linear birefringence is the difference in refraction for linearly polarized light beams with orthogonal polarization. In general, a linear birefringence can be written as

$$\Delta n^{\mathrm{lin}} = n_{\parallel} - n_{\perp}, \tag{8.199}$$

and it can, for instance, be measured parallel and perpendicular to a direction given by an applied static electric field, as illustrated in Figure 8.29. This is known as the Kerr effect and depends quadratically on the strength of the externally applied static electric field. The polarization of the incident beam is at a 45° angle with respect to the applied electric field, and it can be decomposed into components parallel and perpendicular to the applied field. Since the two components of the light travel at different speeds, a phase shift will arise that causes the polarization of the total field to become elliptically polarized.

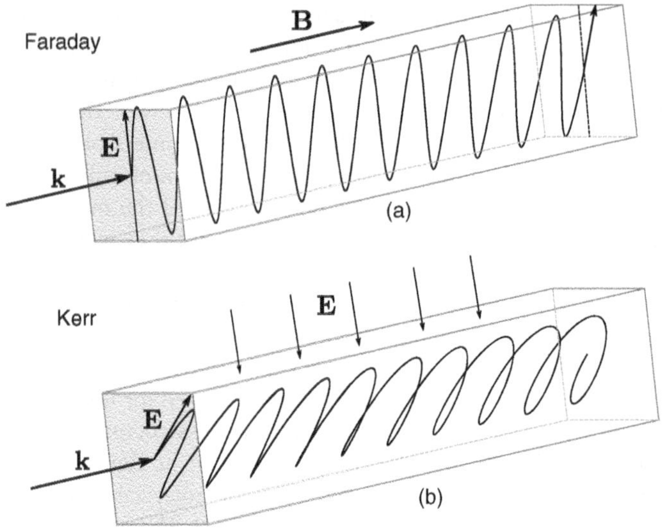

Figure 8.29 Illustration of classes of birefringences: (a) A circular birefringence known as the Faraday effect causes a rotation of the plane of polarization that is linearly proportional to the component of the uniform magnetic field in the direction of light propagation. (b) A linear birefringence known as the Kerr effect causes a linearly polarized field to become elliptically polarized in quadratic proportion to the externally applied static electric field.

The third and final class of birefringences is known as *axial birefringence*. In this case, the effect can be observed even with the use of unpolarized light, but only for chiral molecules. An example of such an axial birefringence is the magnetochiral effect, where the difference in the refractive index is measured when the light wave is propagating parallel $n_{\uparrow\uparrow}$ or antiparallel $n_{\uparrow\downarrow}$ to an applied external magnetic field. The birefringence is thus given by

$$\Delta n^{\mathrm{ax}} = n_{\uparrow\uparrow} - n_{\uparrow\downarrow}. \tag{8.200}$$

We will now first consider the simplest of the birefringences, namely, natural optical activity.

8.6.1 Natural Optical Activity

We recall from Section 3.5.2 that the polarizability of a molecule determines the interaction between a propagating light wave and a molecular material through the refractive index of the material. In this process, we did not have to consider the polarization state of the light. We will here consider the phase of the wave vector, and in particular consider light that is circularly polarized. We will see that for chiral molecules there is a difference in the refractive index when left or right circularly polarized light is passed through the sample, and this is an example of *natural optical activity*.

It is interesting to note that the first observations of the circular polarization of light were done by Arago in 1811 and by Biot in 1812, and it is a historical curiosity that Biot studied circular polarization as a means of providing support for the corpuscular theory of light. The connection between molecular asymmetry and the difference in absorption of the left and right circular polarization states of light was made by Pasteur in 1847 in his studies of the asymmetry of crystals of tartaric acid, which had been motivated by the observation by Biot in 1832 that tartaric acid rotated polarized light.[24]

24 In a crime novel by Dorothy L. Sayers, *The Documents in the Case*, optical activity plays a key role.

In our discussion of electromagnetic waves and their polarization in Section 3.1.5, we noted that we could write plane-polarized light as a sum of left- and right-circularly polarized light, Eq. (3.89). We found in Section 3.5.3 that the speed of propagation of light in a medium depends on the frequency-dependent refractive index $n(\omega)$ of the medium, Eq. (3.396). This means that if the refractive index differs for the left- and right-circularly polarized components of the light, the speed of propagation of the right- and left-circularly polarized components will differ. Considering a wave propagating along the z direction in a medium with refractive indexes $n_{\pm}(\omega)$, the wave vector will be given by $k_{\pm} = \omega/v_{\pm}(\omega)$, and using Eq. (3.396) we find that the phase ϕ of the circularly polarized waves, Eq. (3.83), can be written as

$$\phi_{\pm} = k_{\pm}z - \omega t + \delta = \frac{\omega n_{\pm}(\omega)z}{c} - \omega t + \delta. \tag{8.201}$$

It is now convenient to introduce an average refractive index $\overline{n} = (n_{+} + n_{-})/2$ and the difference in the refractive index for the left and right components $\Delta n = n_{+} - n_{-}$. The phase factor can then be written as

$$\phi_{\pm} = \phi \pm \frac{\omega z \Delta n}{2c}; \quad \phi = \frac{\overline{n}\omega z}{c} - \omega t + \delta. \tag{8.202}$$

When the medium is not circularly birefringent, $\Delta n = 0$ and we recover the relevant modification of Eq. (3.89) for a wave propagating in the z direction in a medium of refractive index $\overline{n}$

$$\begin{aligned}\mathbf{E}_{\mathrm{lp}} &= \frac{1}{2}E_0\mathrm{Re}\left\{\mathbf{e}_{-}\exp\left[i\left(\frac{\omega\overline{n}z}{c} - \omega t + \delta + \theta\right)\right]\right\} \\ &+ \frac{1}{2}E_0\mathrm{Re}\left\{\mathbf{e}_{+}\exp\left[i\left(\frac{\omega\overline{n}z}{c} - \omega t + \delta - \theta\right)\right]\right\}.\end{aligned} \tag{8.203}$$

If instead we have a nonzero difference in the refractive index for the given medium, the angle θ giving the orientation of the plane of polarization with respect to the x-axis is replaced by

$$\theta'(z) = \theta - \frac{\omega \Delta n z}{2c}. \tag{8.204}$$

The resulting equation still describes a plane-polarized wave in the medium, but, after moving a distance Δz, the plane of polarization has been rotated an angle $\Delta\theta$

$$\Delta\theta = -\frac{\omega \Delta z \Delta n}{2c}. \tag{8.205}$$

Thus, if the medium possesses a difference in the refractive index for left- and right-circularly polarized light, the speed of the transversal rotation for the left- and right-circularly polarized components of the light will differ, leading to a rotation of the plane of polarization. Since the refractive index is an extensive property, the amount of rotation of the plane of polarization will depend on the path length, the concentration of the sample, and of course on how large the difference in the refractive index is for the left- and right-circularly components of the light. It is customary to refer to samples that rotate the plane clockwise, when viewed against the direction of the source as dextrorotatory ($\Delta\theta < 0$), whereas compounds that rotate the plane counterclockwise are levorotatory ($\Delta\theta < 0$). This nomenclature is also the origin of the historical classification of chiral molecules as being in a D or an L form. Unfortunately, there is no link between the handedness of the specific enantiomer and the sign of its optical rotation, making this nomenclature largely of historical interest.

The refractive index was shown to be related, in the optical region of interest to us, to the molecular polarizability through the Lorentz–Lorenz relation, Eq. (3.461). The polarizability, in turn, could be calculated from the linear response function involving two electric dipole operators Eq. (5.168). However, this would not allow us to account for the circular polarization of the light. In order to account for the polarization state, we recall that there is also a magnetic component associated with the electric field present in the plane-polarized wave, see Eq. (3.74). Thus, in order to explain the difference in the refractive index for the left- and right-circularly polarized light, we have to go beyond the electric dipole approximation and also consider the dipole moment induced by the magnetic dipole and electric quadrupole interactions, see Eq. (8.1).

We know that the magnetic dipole operator is of the same order in the wave vector as the electric quadrupole operator and that these are coupled through Eq. (3.209). In general, the optical rotation depends on both the electric dipole–magnetic dipole polarizability and the electric dipole–electric quadrupole polarizability, and this will have to be kept in mind when oriented samples are considered. However, most optical rotation measurements are carried out in liquid samples, and we will therefore be concerned primarily with the isotropic average of the two polarizability tensors. However, for isotropic samples, the contribution from the mixed electric dipole–electric quadrupole polarizability vanishes and it therefore suffices to consider only the electric dipole–magnetic dipole polarizability. In terms of response functions and sum-over-states expressions, it can be written as

$$G'_{\alpha\beta}(-\omega;\omega) = -\mathrm{Im}\langle\langle\hat{\mu}_\alpha;\hat{m}_\beta\rangle\rangle_\omega = -\frac{2\omega}{\hbar}\mathrm{Im}\sum_{n\neq 0}\frac{\langle 0\mid\hat{\mu}_\alpha\mid n\rangle\langle n\mid\hat{m}_\beta\mid 0\rangle}{(\omega_{n0}-\omega)^2}. \tag{8.206}$$

From the sum-over-states expression, it is clear that this mixed polarizability will vanish for static fields where $\omega = 0$.

The most common experimental setup uses light with a frequency corresponding to the sodium D-line (589.29 nm).[25] The optical rotation is reported as the number of degrees of rotation of the plane of polarization per mole of sample for a sample cell of length 1 dm, and at a temperature of 25 °C. Assuming only one enantiomer in the sample, this specific optical rotation $[\alpha]_\mathrm{D}^{25}$ is defined as

$$[\alpha]_\mathrm{D}^{25} = 288\times 10^{-30}\frac{\pi^2\mathcal{N}a_0^4\omega^2}{M}\,{}^{\mathrm{ORD}}\overline{\beta}, \tag{8.207}$$

where M is the molecular mass in g mol^{-1} and ${}^{\mathrm{ORD}}\overline{\beta}$ is given in atomic units, and where we have introduced

$${}^{\mathrm{ORD}}\overline{\beta} = -\frac{1}{3\omega}G'_{\alpha\alpha}. \tag{8.208}$$

The tensor ${}^{\mathrm{ORD}}\beta_{\alpha\beta}$ is often referred to as the Rosenfeld tensor.

Even though the majority of optical rotation measurements are conducted at a single wavelength, and in most cases at the sodium D-line, historically the dispersion of the optical rotation was also important as it gave additional insight into the excited electronic states of chiral molecules. In Figure 8.30 we show the calculated *optical rotatory dispersion* (ORD) spectrum of S-methyloxirane up to about 10 eV (124 nm). As was discussed in Section 5.10, the introduction of relaxation effects in our response functions as obtained from the Liouville equation allows us to calculate response functions also as we pass its poles, making it possible for us to create a complete ORD spectrum. For reference, we have in Figure 8.30 also indicated the excitation energies and the corresponding oscillator strengths in order to facilitate the understanding of how the structure of the ORD spectrum can be used to extract information about the excited electronic states of molecules. Another potential benefit of using a frequency different from that

25 This is the average of the D_1 and D_2 lines; see Section 8.2.

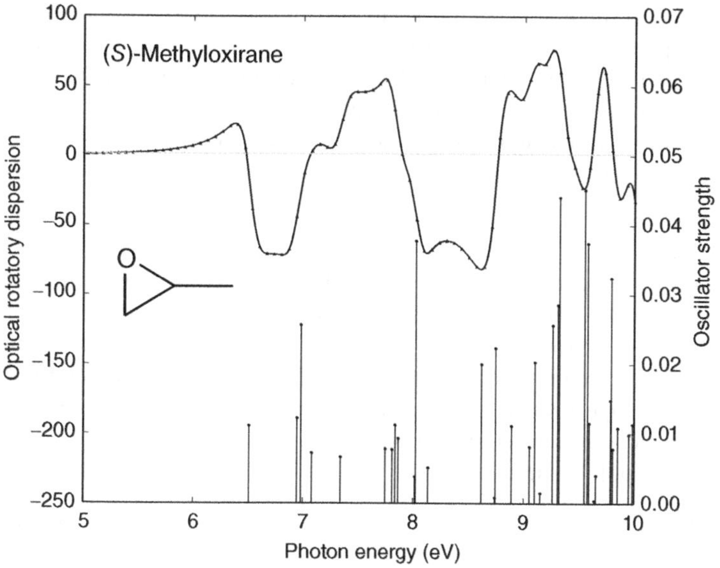

Figure 8.30 Optical rotatory dispersion [in units of 10^3 deg/(dm g cm^{-3})] and oscillator strengths in the region of the lowest singlet excited states of (S)-methyloxirane. Results are obtained at the B3LYP/aug-cc-pVDZ level of theory and with use of a damping parameter $\gamma = 0.124$ eV.

of the sodium D-line, at least from a computational point of view, is that by approaching a pole, the magnitude of the optical rotation rises significantly, reducing the concern that errors in the calculations may lead to wrong sign of the optical rotation, and thus the wrong assignment of the absolute stereochemistry of the molecule.

Since the mixed electric dipole–magnetic dipole polarizability involves the magnetic dipole operator, Eq. (8.84), we recall from the discussion in Section 8.4 that G' will carry an origin dependence in approximate calculations. Indeed, each of the individual tensor elements of G' is origin dependent and, for a gauge origin $\mathbf{O}' = \mathbf{O} + \mathbf{D}$, we have

$$\begin{aligned} G'_{\alpha\beta}(\mathbf{O}') &= -\frac{e^2}{2m_e}\mathrm{Im}\langle\langle r_\alpha; \epsilon_{\beta\gamma\delta}(r_\gamma - D_\gamma)\hat{p}_\delta\rangle\rangle_\omega \\ &= G'_{\alpha\beta}(\mathbf{O}) + \frac{e^2}{2m_e}\epsilon_{\beta\gamma\delta}D_\gamma \mathrm{Im}\langle\langle r_\alpha; \hat{p}_\delta\rangle\rangle_\omega. \end{aligned} \tag{8.209}$$

Using Eq. (7.66), the origin dependence of the optical rotation can be written as

$$G'_{\alpha\beta}(\mathbf{O}') = G'_{\alpha\beta}(\mathbf{O}) + \frac{\omega e^2}{2}\epsilon_{\beta\gamma\delta}D_\gamma\alpha_{\alpha\delta}. \tag{8.210}$$

We thus notice that an individual tensor element is origin dependent. For isotropic samples, we are concerned with the trace of the mixed electric dipole–magnetic dipole polarizability, which is origin independent

$$\epsilon_{\alpha\gamma\delta}D_\gamma\alpha_{\alpha\delta} = 0, \tag{8.211}$$

which follows from the fact that $\epsilon_{\alpha\gamma\delta} = -\epsilon_{\delta\gamma\alpha}$ and that the polarizability is a symmetric tensor for real wavefunctions.

For approximate variational wave functions, the origin independence of the trace of G' is only achieved in the limit of complete basis sets since Eq. (7.66) only is fulfilled in this limit, as discussed in Section 7.2. One way to overcome this problem is to introduce local gauge origins by using LAOs, as we discussed for the magnetizabilities in Section 8.4. However, to define a proper LAO magnetic dipole operator that has the correct origin dependence for finite basis sets and which can be used in frequency-dependent calculations is not straightforward. It would take us too far astray to discuss this here, and we refer instead to the literature.[26]

A more straightforward way to achieve origin-independent results is to rewrite the electric dipole operator in the expression for G' using Eq. (7.65)

$$G'_{\alpha\beta} = \frac{e}{m_e\omega}\langle\langle \hat{p}_\alpha ; \hat{m}_\beta \rangle\rangle_\omega, \tag{8.212}$$

which can be shown to be origin independent irrespective of the choice of wave function and basis set. However, if we compare this equation with the sum-over-states expression in Eq. (8.206), we can see that there is a caveat. While G' as defined in Eq. (8.206) vanishes for static fields, this is not the case for G' in the velocity gauge, Eq. (8.212). This can be understood as a consequence of the lack of gauge invariance in approximate-state theory. A pragmatic approach to resolving this conundrum is to simply subtract the static-limit value from the mixed dipole velocity–magnetic dipole polarizability

$$G'_{\alpha\beta} = \frac{e}{m_e\omega}(\langle\langle \hat{p}_\alpha ; \hat{m}_\beta \rangle\rangle_\omega - \langle\langle \hat{p}_\alpha ; \hat{m}_\beta \rangle\rangle_0), \tag{8.213}$$

as this can be shown to lead to better basis set convergence and a correct limiting behavior with increasing wavelength of the incoming light.

From Figure 8.30, we note that the optical rotation is small before reaching the first pole. Indeed, at the sodium *D*-line, which corresponds to an energy of 2.1 eV, the optical rotation is rather small for most molecules. However, even though the trace of the mixed electric dipole–magnetic dipole polarizability is fairly small, the individual diagonal elements are in many cases fairly large, almost cancelling each other in the trace. This is shown for a few selected molecules in Table 8.8. A consequence of this is that the calculated optical rotation is sensitive to the molecular geometry as well as to effects that are normally considered small, such as zero-point vibrational effects or solvent effects, as even small changes in the diagonal elements due to changes in the electronic structure of the molecule can lead to rather large changes in ${}^{\mathrm{ORD}}\overline{\beta}$.

The sensitivity of the optical rotation is perhaps most clearly seen for conformationally flexible molecules—that is, molecules that have a significant population of multiple stable minima. As exemplified for paraconic acid (a substituted γ-butyrolactone) in Figure 8.31, different molecular conformations can have optical rotations that differ by orders of magnitude and even in sign. The optical rotation of such a flexible molecule can then be determined by a Boltzmann averaging over the dominant molecular conformations.

Table 8.8 Diagonal elements of the mixed electric dipole–magnetic dipole polarizability and the trace of the tensor for a few selected molecules.

Molecule	$\omega^{-1}G'_{xx}$	$\omega^{-1}G'_{yy}$	$\omega^{-1}G'_{zz}$	${}^{\mathrm{ORD}}\overline{\beta}$	$[\alpha]_D^{25}$
H_2O_2 ($\phi = 120°$)	−1.201	0.387	0.960	−0.049	−55.3
(+)-Methyloxirane	−0.820	−0.812	1.587	0.015	10.2
(+)-Camphor	0.240	4.891	−5.847	0.239	60.5

Results (atomic units) obtained at the DFT/B3LYP level of theory using the polarized basis set of triple zeta quality by Sadlej. $[\alpha]_D^{25}$ in units of deg/(dm g cm^{-3}, other quantities reported in atomic units.

26 J. Olsen, K. L. Bak, K. Ruud, T. Helgaker, and P. Jørgensen. *Theor. Chim. Acta*, 90:421, 1990.

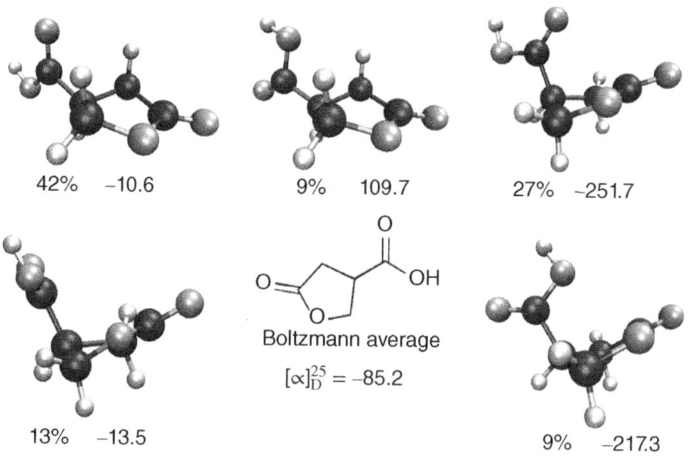

Figure 8.31 Specific optical rotation [deg cm^3 dm^{-1} g^{-1}] for different conformers of (*S*)-(−)-paraconic acid together with the Boltzmann-weighted average amounting to −85.2 (same units).

We recall from Section 8.4 that the use of LAOs leads not only to gauge origin-independent results, but also to much improved basis set convergence. This is not the case for the optical rotation. The magnetic dipole operator defined using London orbitals inherits the slow basis set convergence of the standard magnetic dipole operator. There is therefore little difference in basis set convergence between the optical rotation calculated directly either using the definition in Eq. (8.206), the dipole-velocity form Eq. (8.212) or when using LAOs. To ensure converged results for optical rotation, basis sets of polarized valence double-zeta quality are required for SCF-based methods such as Hartree–Fock and DFT, and it is particularly important that the basis set includes diffuse, polarizing functions. More specifically, diffuse *p* functions in the outer regions of the electron density, which in most cases is described by the electron density of hydrogen atoms, are required to ensure qualitatively correct results. In case of correlated methods, the basis set needs, in addition to the diffuse polarizing functions, to have the flexibility to accurately describe the effects of electron correlation. An extreme case of the importance of diffuse polarizing functions is shown for propane in Figure 8.32, where the optical rotation for different dihedral angles θ(H–C–C–C) is plotted. While both the aug-cc-pV(D/T)Z and the 6-31++G**/6-311++G** basis sets contain diffuse functions, only the aug-cc-pV(D/T)Z basis sets contain diffuse, polarizing functions, which can be seen to be key for a correct description of the dependence of the optical rotation on the dihedral angle.

There are no empirical rules relating the stereochemistry of a molecule to the observed sign of the optical rotation. An important area of application for optical rotation calculations is therefore to combine theory and experiment and through this determine the absolute configuration of a chiral molecule. Such a joint study can also be used to determine the relative population of different conformations of a flexible molecule. However, this is a difficult task due in part to the sensitivity of the optical rotation to the structure of the molecule, in part because the magnitude of $^{\mathrm{ORD}}\beta$ is small, and in part because we only have one number with which to compare theory and experiment. ORD provides a way of extracting more information from the molecule, theoretically and experimentally. However, rather than using ORD, it is more common to use electronic circular dichroism (ECD) to obtain more confidence in the comparison between calculated results and experimental observation and at the same time gain insight into the excited electronic states of a molecule, and we turn to this topic next.

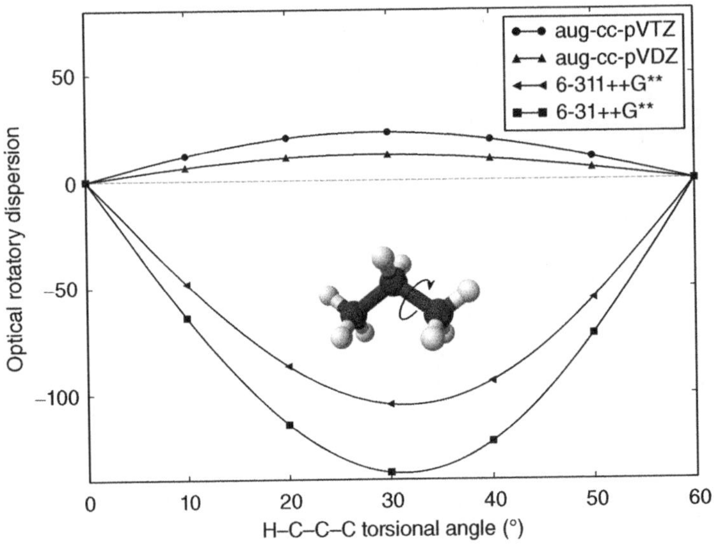

Figure 8.32 Optical rotation at 589 nm of propane as a function of the dihedral H-C-C-C angle. Results are obtained at the DFT/B3LYP level of theory with use of different basis sets and London orbitals. Optical rotatory dispersion in units of deg/(dm g cm^{-3}).

8.6.2 Electronic Circular Dichroism

Optical rotation was in the previous section seen to be a measure of the difference in the refractive index of left- and right-circularly polarized light. It was seen to arise from contributions to the refractive index of the molecule arising from terms beyond the electric dipole in the multipolar expansion of the electromagnetic wave. In a similar manner, the absorption of light, bringing a molecule into an excited electronic state, has a chiroptical analog, known as ECD, or more commonly just circular dichroism (CD). In ECD, we measure the differential absorption of left- and right-circularly polarized light. This will give us a spectrum similar in structure to a regular absorption spectrum, but because the differential absorption may be both positive and negative, the ECD spectrum will display both positive and negative bands, where the intensity and sign of the different bands are related to the so-called rotatory strengths of the molecule. As for optical rotation, ECD spectra will in isotropic media only be observed for chiral molecules, and the ECD spectra of two enantiomers of a chiral molecule will be each other's mirror image.

Following Section 8.5.1, the effects of broadening as well as vibronic effects apply also to ECD and will not be repeated here. We only note that just as we could have emission from an electronic excited state, we can also measure the differential emission of left- and right-circularly polarized light from an excited electronic state, in a process commonly referred to as circularly polarized luminescence (CPL). A particularly attractive application of CPL is the study of molecules that are nonchiral in their electronic ground state, but for which the equilibrium molecular structure of a given excited state is chiral. An example of this is shown in Figure 8.33, where the excited-state structure of a ketone leads to pyramidalization of the ketone carbon atom, making the molecule chiral, and thus giving rise to CPL if the lifetime of the excited state is sufficiently long to allow the structure of the molecule to relax in the given excited state. In such a pyramidalization, there are often two pathways, and as the CPL in most cases would have opposite signs for the two chiral excited-state structures, CPL can be a very powerful method to determine the excited-state structures of such molecules.

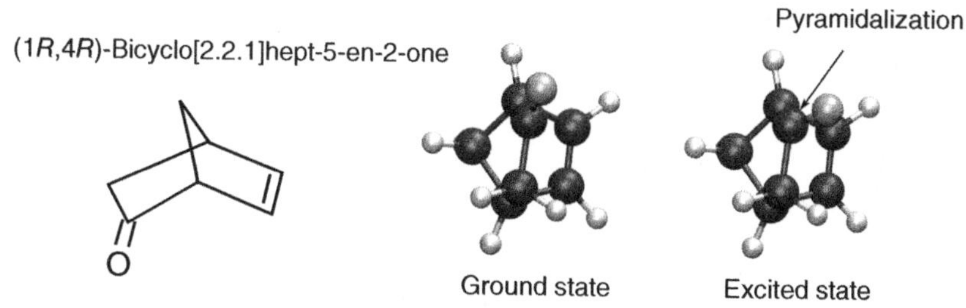

Figure 8.33 Ground- and excited-state structures of (1*R*,4*R*)-bicyclo[2.2.1]hept-5-en-2-one, where chirality may be induced in the excited state through pyramidalization of the ketone carbon, thus giving rise to circularly polarized luminescence.

We recall that the oscillator strength of an absorption process, Eq. (5.228), is related to the single residue of the polarizability tensor, Eq. (5.255). By analogy, we would thus expect that the rotatory strength of a given absorption band is related to the single residue of the optical rotation tensor, Eq. (8.206). Indeed, the strength of an ECD band is given by the anisotropy of the decadic molar extinction coefficient

$$\Delta\epsilon(\omega) = \epsilon_+(\omega) - \epsilon_-(\omega) = \frac{16\pi^2 \mathcal{N}\omega}{3 \times 1000 \ln(10)(4\pi\epsilon_0)\hbar c^2} \sum_n f(\omega;\omega_{n0},\gamma)\, R_{n0}. \tag{8.214}$$

We have here introduced the rotatory strength R_{n0} defined as

$$R_{n0} = \lim_{\omega\to\omega_{n0}} (\omega_{n0} - \omega)\mathrm{Im}\langle\langle \hat{\mu}_\alpha; \hat{m}_\alpha \rangle\rangle_\omega, \tag{8.215}$$

where $\mathcal{N}$ is the number density of the sample. The rotatory strength will vanish if either the electric or the magnetic transition dipole moment is zero, or if the two vectors are orthogonal to each other. The latter will always be the case in molecules with an improper rotation axis, and thus in isotropic samples, ECD is only observed for chiral molecules, as was the case for optical rotation.

We recall from our discussion of absorption spectroscopy in Section 8.5.1 that Eq. (8.215) will give rise to a stick spectrum, whereas an experimental spectrum in general will be broadened due to various effects, which can be taken into account when representing the theoretical results using a lineshape function $f(\omega;\omega_{n0},\gamma)$, as done in Eq. (8.214). The Lorentzian lineshape used in this equation could be replaced with either Gaussian, Eq. (8.164), or Voigt, Eq. (8.165), lineshapes depending on the experimental situation.

We recall that an alternate approach to the single residue of the linear response function is to use damped response theory as discussed in Section 5.10, where the absorptive properties can be extracted from the absorptive part of the linear damped response function in Eq. (5.290), using the electric dipole and magnetic dipole operators. In Figure 8.34, we show the ECD spectrum of (*S*)-methyloxirane obtained using the damped response function of the mixed electric dipole–magnetic dipole polarizability together with the corresponding stick absorption spectrum. The ECD spectrum can also be compared to the ORD spectrum in Figure 8.30, which corresponds to the dispersive part of the damped response function, Eq. (5.290). When comparing the ORD and ECD spectra, we note that while the dispersive part of the response function is zero at the excitation energies, the absorptive part reaches its maximum absolute value at these excitation energies. Both quantities contain a wealth of information about the electronic excited states in the molecule as well as clearly identify the molecular stereochemistry. However, we emphasize that it is not possible to a priori deduce the molecular stereochemistry from the sign of the specific optical rotation or from the shape of the ECD spectrum. This can only

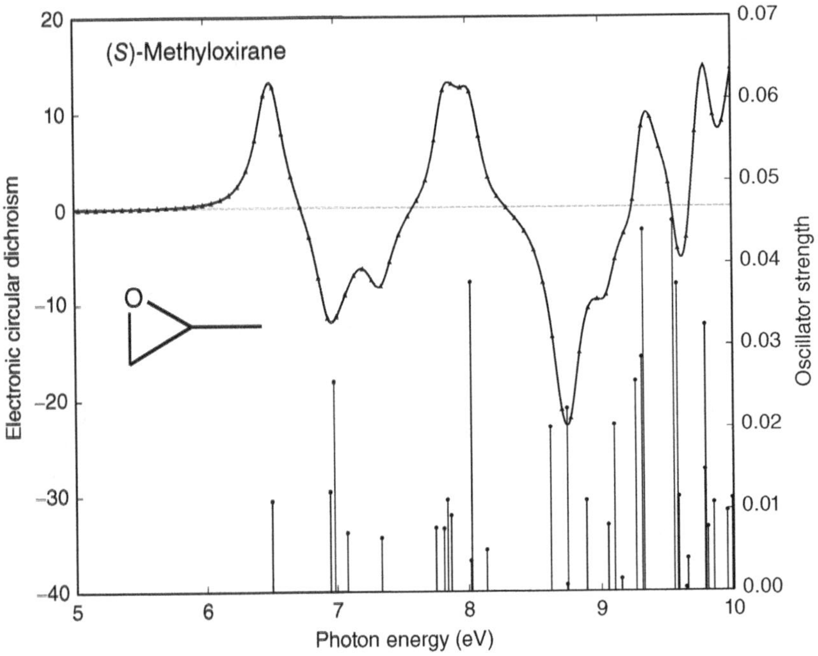

Figure 8.34 Electronic circular dichroism [in units of L mol^{-1} cm^{-1}] and oscillator strengths in the region of the lowest singlet-excited states of (S)-methyloxirane. Results are obtained at the B3LYP/aug-cc-pVDZ level of theory and with use of a damping parameter $\gamma = 0.124$ eV.

be achieved by comparing the experimental results with results obtained from an *ab initio* calculation. In the case of ECD, a number of phenomenological rules-of-thumb have been devised as a way of linking the observed ECD spectrum to the absolute stereochemistry, and although in many cases they are working well, incorrect assignments have been reported based on the failure of these simple rules.[27]

Comparing the absorption spectrum in Figure 8.16 and the ECD spectrum in Figure 8.34, we note that the fact that the ECD spectrum can have both positive and negative rotatory strengths may make it easier to separate different electronic excited states in what would otherwise only be a broad set of overlapping absorption bands in the case of UV/vis spectroscopy. Nevertheless, a detailed comparison of experimental ECD spectra with calculated ones is made difficult by experimental broadening, as in absorption spectroscopy, by the quality of the calculated excited states, as well as by the strong sensitivity of the calculated rotatory strengths on the molecular conformation, comparable to what is seen for optical rotation in Figure 8.31.

Since the rotatory strength of ECD is related to the imaginary part of the mixed electric dipole–magnetic dipole polarizability, we would expect that the lack of gauge-origin independence observed in approximate calculations also would apply for calculations of ECD using Eq. (8.215), and this is indeed the case. By using LAOs or by switching to the velocity gauge through the use of Eq. (7.65), we can obtain gauge-origin-independent expressions also for the rotatory strength. In the velocity gauge, the rotatory strength is given by

$$R_{n0} = -\frac{e}{m_e \omega_{n0}} \lim_{\omega \to \omega_{n0}} (\omega_{n0} - \omega) \langle\langle \hat{p}_\alpha ; \hat{m}_\alpha \rangle\rangle_\omega . \tag{8.216}$$

27 An example of such a misassignment based on ECD is Tröger's base, where the exciton coupling model was used to deduce the absolute configuration of this compound. X-ray analysis and vibrational circular dichroism (VCD) were later used to demonstrate the incorrect assignment (see, e.g., A. Aamouche, F. J. Devlin and P. J. Stephens. *Chem. Commun.*, 361–362, 1999).

We recall from Section 5.7 that the Thomas–Reiche–Kuhn sum rule identifies the sum over all states of the oscillator strengths to the total number of electrons in the system. A similar sum rule exists for the rotatory strength, namely, that the sum over all states of the rotatory strength should be zero

$$\sum_{n\neq 0} R_{n0} = \mathrm{Im} \sum_{n\neq 0} \langle 0|\hat{\mu}_\alpha|n\rangle\langle n|\hat{m}_\alpha|0\rangle = \mathrm{Im}\langle 0|\hat{\mu}_\alpha\hat{m}_\alpha|0\rangle = 0, \tag{8.217}$$

where we have completed the sum to also include the electronic ground state in order to invoke the resolution of the identity, noting that $\langle 0|\hat{m}_\alpha|0\rangle$ is zero for nondegenerate electronic ground states. A consequence of this sum rule is that the optical rotation will approach zero for high-frequency radiation ($\omega \gg \omega_{n0}$) since we can write

$$[\alpha]^{25} = 288 \times 10^{-30} \frac{\pi^2 \mathcal{N} a_0^4 \omega^2}{3M} \sum_{n\neq 0} \frac{R_{n0}}{\omega^2} = 288 \times 10^{-30} \frac{\pi^2 \mathcal{N} a_0^4}{3M} \sum_{n\neq 0} R_{n0} = 0. \tag{8.218}$$

8.6.3 Nonlinear Birefringences

Nonlinear birefringences historically played an important role in the development of the theory of electromagnetism. While it was known in 1845 that materials could modify the direction of polarization of light when properly aligned, the origin of this effect was not clear. Maxwell showed that when light passed through a material parallel to an external magnetic field, the polarized light was rotated, the angle of rotation being proportional to the strength of the magnetic field. This gave strong support for Maxwell's belief that light was an electromagnetic phenomenon.[28]

The quantitative study of the Faraday effect was made by Verdet, who showed that the angle of rotation α of the plane of polarization per unit path length could be written as

$$\alpha = VB\cos\theta, \tag{8.219}$$

where θ is the angle between the direction of the magnetic field **B** and the propagation of the light beam, and where V is the Verdet constant of the material for a given temperature and wavelength.

Considering that birefringences such as that observed for the Faraday effect are related to the refractive index of the material, the fact that for a general material the effect occurs only in the presence of a magnetic field and that the dependence on the strength of the magnetic field is linear, it would appear that the Verdet constant would be related to the linear term in a perturbation expansion of the refractive index with respect to an external magnetic field

$$\alpha_{\alpha\beta}(-\omega;\omega,\mathbf{B}) = \alpha_{\alpha\beta}(-\omega;\omega) + \frac{d\alpha_{\alpha\beta}(-\omega;\omega)}{dB_\gamma} B_\gamma + \cdots , \tag{8.220}$$

which will then be related to a quadratic response function of the form $\langle\langle \hat{\mu}_\alpha; \hat{\mu}_\beta, \hat{m}_\gamma \rangle\rangle_{\omega,0}$.

Equation (8.220) indicates that we can obtain a large number of birefringences by expanding the polarizability in a power series with respect to different sources of external electric or magnetic fields as well as electric field gradients. We recall from Section 5.2 the richness in the number of nonlinear processes that could be devised when considering only electric field perturbations. Including also higher order terms in the multipolar expansion of light, external perturbations, and the corresponding absorption processes, it is clear that the field of nonlinear

28 Maxwell also tried to demonstrate, unsuccessfully, the effect of electric fields on the polarization of light. This birefringence was demonstrated about 30 years later by John Kerr, in the effect that now carries his name, the electro-optical Kerr effect as discussed in Section 8.3.2.

birefringences and dichroisms is very rich in observable effects, although they are in many cases small and difficult to measure accurately.

Instead of covering a large number of such nonlinear birefringences and dichroisms, we will instead only consider two of these processes in some detail here: MCD and electric field gradient-induced birefringence (Buckingham birefringence), but in neither case will we try to provide a full account of the effects, for which we instead refer to suggested further reading. We will discuss MCD because it is one of the most common experimental nonlinear dichroisms, and the absorptive counterpart of the Faraday effect. It can provide unique insight into the electronic structure of a molecule, in particular molecules of high symmetry. Buckingham birefringence will be discussed primarily as it is one of the very few means of determining the molecular quadrupole moment. Our discussions will be brief, and we will in particular not dwell on the issue of gauge-origin dependence. Although relevant for approximate calculations of most birefringences and dichroisms, this can be solved through the use of LAOs.

8.6.3.1 Magnetic Circular Dichroism

In MCD, the differential absorption of left- and right-circularly polarized light is measured for a molecule in the presence of an external magnetic field. It is the absorptive counterpart of the Faraday effect discussed above, and as such we can expect that the property is determined from the single residue of the electric dipole–electric dipole–magnetic dipole polarizability. As for the Faraday effect, the application of the magnetic field ensures that an MCD spectrum can be observed for all molecules, also if they are nonchiral. MCD therefore has in principle a much wider range of application than does regular ECD.

The most important areas of applications for MCD are not related to the magnetic-field-perturbed polarizability *per se*, but rather the effects of the magnetic field on highly symmetric molecules for which there can be orbital degeneracies, either in the electronic ground state or in the excited electronic states. For highly symmetric molecules, many excited electronic states are not dipole allowed, and the applied magnetic field will lower the symmetry of the total system and make more of these excited states dipole allowed. For orbitally degenerate ground or excited states, the magnetic field will split the orbital degeneracy, and the two states will display different polarization, which give rise to very characteristic bisignate signals, as we will see shortly. For orbitally degenerate ground states with unpaired electrons, the magnetic field will also split the degenerate orbitals and the electronic population of these states will be determined by the magnitude of the splittings and the temperature at which the measurement is done. This provides a powerful route to determine the symmetry and occupancy of the degenerate orbitals in highly symmetric, open-shell molecules.

Based on the considerations above, we can expect that the molar decadic extinction coefficients associated with an MCD experiment conducted in the presence of an external magnetic field **B** will consist of three different contributions. It is indeed given by

$$\Delta\epsilon(\omega) = -\frac{8\pi^2 N_A \omega}{3 \times 1000 \times \ln(10)(4\pi\varepsilon_0)\hbar c} B\sum_n \left\{ \frac{1}{\hbar}\frac{\partial f(\omega;\omega_{n0},\gamma)}{\partial\omega} A_{0n} + f(\omega;\omega_{n0},\gamma)(\omega)\left[B_{0n} + \frac{C_{0n}}{kT}\right]\right\}, \tag{8.221}$$

where we have introduced the A_{0n}, B_{0n}, and C_{0n} terms for an excitation from an electronic ground state to the excited state n. The three terms are defined as

$$A_{0n} = \frac{1}{2g}\sum_{0_\alpha n_\beta}\sum_{(0'_\alpha, n'_\beta)\neq 0} \mathrm{Im}[(\langle n'_\beta|\hat{\mathbf{m}}|n_\beta\rangle\delta_{0'_\alpha 0_\alpha} - \langle 0_\alpha|\hat{\mathbf{m}}|0'_\alpha\rangle)\cdot\langle 0'_\alpha|\hat{\boldsymbol{\mu}}|n'_\beta\rangle \times \langle n_\beta|\hat{\boldsymbol{\mu}}|0_\alpha\rangle], \tag{8.222}$$

$$B_{0n} = \frac{1}{g} \sum_{0_\alpha n_\beta} \mathrm{Im}\left(\sum_{k \neq 0} \frac{\langle k|\hat{\mathbf{m}}|0_\alpha\rangle \cdot \langle n_\beta|\hat{\boldsymbol{\mu}}|k\rangle \times \langle 0_\alpha|\hat{\boldsymbol{\mu}}|n_\beta\rangle}{\omega_{k0}} + \sum_{k \neq n} \frac{\langle n_\beta|\hat{\mathbf{m}}|k\rangle \cdot \langle k|\hat{\boldsymbol{\mu}}|0_\alpha\rangle \times \langle 0_\alpha|\hat{\boldsymbol{\mu}}|n_\beta\rangle}{\omega_{kn}} \right), \tag{8.223}$$

$$C_{0n} = \frac{1}{2g} \sum_{0_\alpha n_\beta} \sum_{0'_\alpha} \mathrm{Im}[\langle 0_\alpha|\hat{\mathbf{m}}|0'_\alpha\rangle \cdot (\langle 0'_\alpha|\hat{\boldsymbol{\mu}}|n_\beta\rangle \times \langle n_\beta|\hat{\boldsymbol{\mu}}|0_\alpha\rangle)]. \tag{8.224}$$

In these equations, g denotes the degeneracy of the ground electronic state and the subscripts α and β denote the different magnetic components of the different degenerate states.

From the expression of the A term in Eq. (8.222), we note that it measures the difference in the induced magnetic moment in the ground and the excited states, weighted by the dipole transition moments coupling the respective ground and excited states of a given magnetic component. As a magnetic moment can be induced by an external magnetic field only in case we have orbital degeneracy, the A term will be zero if neither the ground nor the excited electronic state are orbitally degenerate. The A term of MCD can for nondegenerate electronic ground states be calculated as the double residue of the quadratic response function, see Eq. (5.257)

$$A_{0n} = \frac{1}{2}\epsilon_{\alpha\beta\gamma}[\lim_{\omega_\sigma \to \omega_{n0}} (\omega_{n0} - \omega_\sigma)(\lim_{\omega_2 \to \omega_{n0}} (\omega_{n0} - \omega_2)\mathrm{Im}\langle\langle \hat{\mu}_\gamma; \hat{m}_\beta, \hat{\mu}_\alpha \rangle\rangle_{0,\omega_2})]. \tag{8.225}$$

In the case where the electronic ground state is nondegenerate, a somewhat more straightforward approach is to obtain the A term directly from the damped expression for the quadratic response function.

The fact that the A term only appears if there are orbital degeneracies in either the electronic ground state or, more commonly, an excited electronic state, is one important reason why MCD is an important technique for the study of highly symmetric molecules. The high symmetry allows for multidimensional irreps that will support degenerate states. The static magnetic field splits the orbitally degenerate states into a multiplet of states with a permanent magnetic moment. These magnetic-field split states have different polarization. The plane-polarized light may induce transitions to or from these states, and this will lead to a very characteristic bisignate MCD response that makes it possible to determine degenerate excited states, as illustrated in Figure 8.35.

The B term in Eq. (8.223) bears strong similarities to the expression for the residue of the quadratic response function Eq. (5.256) for a static field in the case where the ground electronic state is orbitally nondegenerate. However, we note that there is a difference in that the first sum-over-states expression in Eq. (8.223) excludes the electronic ground state, whereas this is included in the residue of the quadratic response function. However, by projecting out the

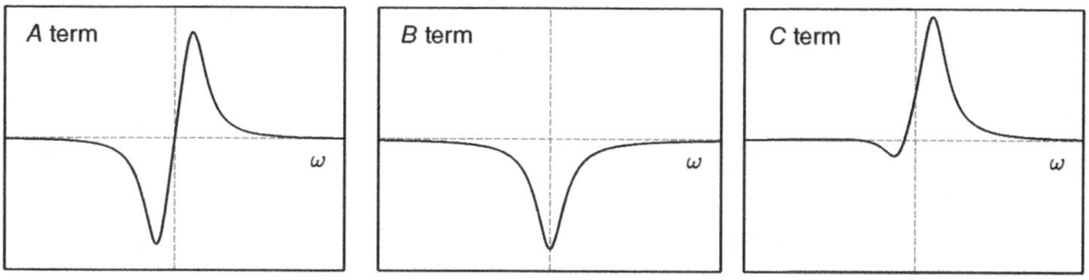

Figure 8.35 Schematic representation of the *A*, *B*, and *C* bands of MCD spectra. The *C* term is temperature dependent and will become progressively more absorptive in character the higher the temperature, as this would suggest a more equal population of the different sublevels.

reference state, we can obtain the B term from the linear residue of the quadratic response function (assuming a nondegenerate electronic ground state)

$$B_{ij} = \mathrm{Im}[\epsilon_{\alpha\beta\gamma} \lim_{\omega_2 \to \omega_{n0}} (\omega_{n0} - \omega_2)\langle\langle \hat{\mu}_\gamma; \hat{m}_\beta, \hat{\mu}_\alpha \rangle\rangle_{0,\omega_2}]. \tag{8.226}$$

In this expression, we can consider the external magnetic field to perturb each individual electronic state, leading to perturbed states that may have different polarization. These bands are referred to as *B bands*; see Figure 8.35 for a negative B band. An isolated band will have a regular absorption profile, but can be of either positive or negative sign, and an MCD spectrum for a molecule with a symmetry that does not support any degenerate irreducible representations will have a shape very similar to that of a regular ECD spectrum. However, such an MCD spectrum can be recorded for any molecule, irrespective of whether it is chiral or not.

One thing to note from Eq. (8.223), which in principle applies to any quadratic response function when static fields are involved, is that if there are excited states close in energy to the final state, we may have a divergence in the second sum-over-states expression in Eq. (8.223). As these states can be shown to have different signs for their MCD response, these near-degeneracies will also give rise to a bisignate signal and are for this reason often referred to as pseudo-A terms. Indeed, for the special case of state k being identical to the final state n, it can be shown that the B term reduces to the A term. From a physical and computational point of view, the A and B terms thus have the same origin (ignoring the case of a degenerate electronic ground state), and the separation into different contributions is motivated primarily because of their difference in absorption band structure as well as difference in computational approach.

If we use the damped expression for the quadratic response function, we will be able to handle these divergences and to recover the A, pseudo-A, and B terms of MCD for a molecule with an orbitally nondegenerate ground state through the equation

$$\Delta\epsilon(\omega) = \frac{8\pi^2 N_A \omega}{3 \times 1000 \times \ln(10)(4\pi\varepsilon_0)\hbar c} B\epsilon_{\alpha\beta\gamma} Re\langle\langle \hat{\mu}_\alpha; \hat{m}_\beta, \hat{\mu}_\gamma \rangle\rangle_{0,\omega}, \tag{8.227}$$

where we use the absorptive component of the damped quadratic response function in Eq. (5.310), and B is the strength of the magnetic field.

This is illustrated in Figure 8.36 for the porphyrin molecule which has D_{4h} symmetry, where we also have bent the porphyrin structure out of its planar geometry. In the planar form, the dominating feature of the MCD spectrum is a bisignate band corresponding to an excitation into the $2E_u$ excited state that matches the Soret band of conventional porphyrin absorption spectra. Bending the molecule, the symmetry is lowered, and in particular orbital degeneracy is lost. Nevertheless, the MCD spectrum calculated with the damped expression for the quadratic response function looks more or less identical to the MCD spectrum for the planar structure, despite the formal lack of an A term.

A particular class of highly symmetric molecules for which MCD is an important technique is open-shell systems. In this case, the electronic ground state may be orbitally degenerate. The splitting of the electronic ground state induced by the external magnetic field will mean that the unpaired electron, assuming for simplicity a doublet state, can occupy two different states of different energy, the population fraction being determined by the Boltzmann ensemble averages, with the lowest state having the greatest population. The absorption of light from these two states will also lead to a bisignate signal, but because of the difference in the population of the states, the signal will now be asymmetric, as can be seen from Figure 8.35. This band shape is often referred to as *C band* in the MCD literature. It would take us too far astray to discuss the origin and mechanisms contributing to the C term of MCD, in particular for molecules with multiple unpaired electrons and with heavy elements, as spin–orbit effects quickly become

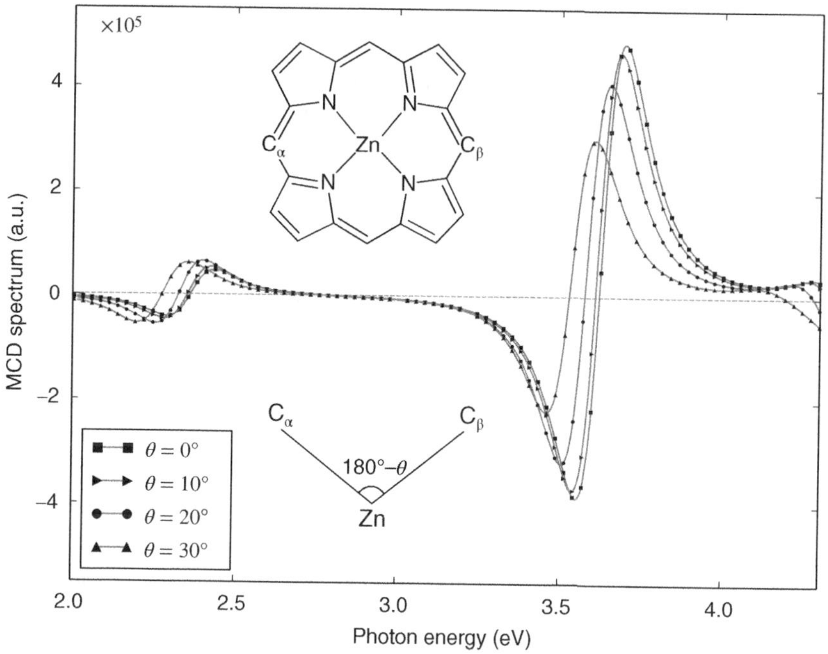

Figure 8.36 MCD spectra of Zn-porphyrin for various degrees of bending about the mesocarbon -metal–mesocarbon axis. Results are obtained at the CAM-B3LYP/aug-cc-pVDZ level of theory. A lifetime broadening of 1000 cm^{-1} has been used.

important. The interested reader is instead referred to more specialized books and research papers on this topic.[29]

8.6.3.2 Electric Field Gradient-Induced Birefringence

In electric field gradient-induced birefringence, an external field gradient is applied, which, as we recall from Section 3.3.2, can interact with the molecular quadrupole moment. A differential refractive index for the two directions perpendicular to the direction of propagation of the light can then be observed. For an ideal gas at constant pressure and for light along the z axis, the expression for this difference in the refractive index can be written as

$$\Delta n = n_x - n_y = \frac{3E^{[1]}}{2V_{\mathrm{m}}} {}_mQ(\omega, T); \quad {}_mQ(\omega, T) = \frac{2N_{\mathrm{A}}}{45\epsilon_0} s. \tag{8.228}$$

In this equation, N_{A} is Avogadro's number, V_{m} is the molar volume, and $\mathbf{E}^{[1]}$ is the external electric field gradient, where we have created a setup such that $E^{[1]}_{xx} = -E^{[1]}_{yy}$ and $E^{[1]}_{zz} = 0$.

In Eq. (8.228), we have introduced the molecular quantity s, which for nondipolar molecules is defined as

$$s = b(\omega) + \frac{1}{kT}\Theta^{[2]}_{\alpha\beta}\alpha_{\alpha\beta}(-\omega; \omega). \tag{8.229}$$

The quantity thus has, as discussed in Section 3.5.2, both a temperature-independent contribution $b(\omega)$ and a temperature-dependent part, which involves the molecular polarizability as well as the molecular quadrupole moment, the latter contribution arising from the alignment of the

29 See, for instance, the book by Mason given in further reading, or F. Neese and E. I. Solomon. *Inorg. Chem.*, 38:1847, 1999.

molecules to the applied external field gradient. It is therefore in principle possible to determine the molecular quadrupole moment assuming we can determine the temperature-dependent part of the electric field gradient-induced birefringence, for instance, through measurements at multiple temperatures, and we know the value of the molecular polarizability. From a theoretical point of view, the quadrupole moment is of course easily obtained as the expectation value of the traceless quadrupole moment operator, Eq. (3.227).

The temperature-independent part of the electric field gradient-induced birefringence is given as a sum of different quadratic response functions

$$b(\omega) = B_{\alpha\beta,\alpha\beta}(-\omega;\omega,0) - \mathcal{B}_{\alpha,\alpha\beta,\beta}(-\omega;\omega,0) - \frac{5}{\omega}\epsilon_{\alpha\beta\gamma}J'_{\alpha,\beta,\gamma}(-\omega;\omega,0) \tag{8.230}$$

and is a combination of three mixed hyperpolarizabilities

$$B_{\alpha\beta,\gamma\delta}(-\omega;\omega,0) = \langle\langle\hat{\mu}_\alpha;\hat{\mu}_\beta,\hat{\Theta}^{[2]}_{\gamma\delta}\rangle\rangle_{\omega,0}, \tag{8.231}$$

$$\mathcal{B}_{\alpha,\beta\gamma,\delta}(-\omega;\omega,0) = \langle\langle\hat{\mu}_\alpha;\hat{\Theta}^{[2]}_{\beta\gamma},\hat{\mu}_\delta\rangle\rangle_{\omega,0}, \tag{8.232}$$

$$J'_{\alpha,\beta,\gamma}(-\omega;\omega,0) = i\langle\langle\hat{\mu}_\alpha;\hat{m}_\beta,\hat{\mu}_\gamma\rangle\rangle_{\omega,0}. \tag{8.233}$$

In experimental measurements of electric field gradient-induced birefringence, varying the temperature has in practice proved difficult, and if at all possible, the accuracy in the measurements is not high enough to determine reliably the temperature-independent part of the induced birefringence. It is therefore often assumed that the temperature-independent part is small and negligible. In many cases this is a fairly good approximation, as can be seen from a consideration of the contributions to the electric field gradient-induced birefringence of some nonpolar molecules in Table 8.9.

Let us finally note that the quantity s in Eq. (8.229) will be more complicated for dipolar molecules, being now given by the expression

$$s = b(\omega) + \frac{1}{kT}\left(\Theta^{[2]}_{\alpha\beta}\alpha_{\alpha\beta}(-\omega;\omega) - \mu_\alpha\left[A_{\beta,\alpha\beta}(-\omega;\omega) + \frac{5}{\omega}\epsilon_{\alpha\beta\gamma}G'_{\beta\gamma}(-\omega;\omega)\right]\right). \tag{8.234}$$

However, for dipolar molecules we can define a so-called *effective quadrupole center* as the point in space where the last term in Eq. 8.234, involving the molecular dipole moment μ_α and the mixed polarizabilities $A_{\alpha,\beta\gamma}(-\omega;\omega)$ of Eq. (8.265) and $G'_{\alpha\beta}(-\omega;\omega)$ of Eq. (8.206), vanishes. Eq. (8.229) is then still valid, but only at one point in space.

Table 8.9 Electric field gradient-induced birefringence of H_2, N_2, and C_2H_2 at temperatures 297.15 K (H_2), 298.15 K (N_2), and 294.0 K (C_2H_2).

	H_2		N_2		C_2H_2	
Property	Theory	Experiment	Theory	Experiment	Theory	Experiment
b	−36.75	−46±15	−36.13		−193.3	
F[a)]	120.9	145	−719.3	−700	8481	
$_mQ(T)$	528.7		−4747		52070	49700±1500
Θ	0.4561	0.484	−1.119	−1.09±0.07	4.869	4.55±0.2
$\Delta\alpha$	1.8710	2.12	4.554	4.749	12.09	12.13±0.24

a) $F = 2\Theta\Delta\alpha/15k$.
Theoretical results are obtained at the CCSD level of theory with use of the d-aug-cc-pV5Z (H_2) and d-aug-cc-pVQZ (N_2 and C_2H_2) basis sets. All results reported in atomic units.

8.7 Vibrational Spectroscopies

8.7.1 Infrared Absorption

We recall from Chapter 5 that a linear absorption spectrum can be determined either by the calculation of oscillator strengths from the residue of a linear response function, Eq. (5.228), or from the imaginary part of the electric-dipole polarizability, Eq. (5.335). The discussion at that point was very generic and made in the context of exact states. These states could in principle be electronic, vibrational, or vibronic states. In our discussion of UV/vis spectroscopy, we noted that the absorption cross section was given by Eq. (5.321), where the dipole transition strength could be calculated from the single residue of the electronic polarizability, or alternatively from the imaginary part of the damped linear response function. In this chapter, we will look in more detail into spectroscopies that involve the vibrational degrees of freedom of a molecule, building on our analysis of the vibrational problem in Chapter 6.

In infrared (IR) absorption spectroscopy, we are concerned with the absorption of light in the wavelength region from about 800 nm to 1000 μm, of which we are predominantly concerned with the mid-IR region from 2.5 to 25 μm and corresponding to wave numbers between 4000 and 400 cm^{-1}.[30] This selected wavelength region is suited for the study of fundamental vibrations of covalent molecular bonds, and the photon energy is assumed to be significantly smaller than the electronic excitation energies of the system. This assumption holds true for most situations, but may need to be examined in cases with near degeneracies among the frontier orbitals.

When we selectively consider vibrational excitations, the relevant molecular property to determine in IR absorption spectroscopy is the pure vibrational polarizability that was presented in Eq. (6.16). Following the discussion in Chapter 5, the absorption cross section for infrared radiation is still given by Eq. (5.321), where the transition strength can be obtained from the single residue of the *pure vibrational polarizability*, Eq. (6.16), or from the corresponding damped pure vibrational linear response function. The absorption cross section can then be written as

$$\begin{aligned}\sigma(\omega) &= \frac{\pi\omega}{3\varepsilon_0 c\hbar}\langle 0|\mu_\alpha|k\rangle\langle k|\mu_\alpha|0\rangle f(\omega;\omega_{k0},\gamma_{k0}) \\ &= \frac{\pi\omega}{3\varepsilon_0 c}\lim_{\omega\to\omega_{k0}}(\omega_{k0}-\omega)\langle\langle\mu_\alpha;\mu_\alpha\rangle\rangle_\omega f(\omega;\omega_{k0},\gamma_{k0}),\end{aligned} \tag{8.235}$$

where γ_{k0} is the relaxation parameter due to the finite lifetime of the excited vibrational state $|k\rangle$ in the electronic ground state, and μ_α is the molecular (total) dipole moment of the electronic ground state at the equilibrium geometry.

What remains is then to evaluate these vibrational dipole transition moments, and we will follow the approach outlined in Chapter 6, and use vibrational perturbation theory. We will then represent the vibrational wave functions as (perturbed) harmonic oscillator wave functions, and we will express the geometry dependence of the dipole moment in a Taylor expansion with respect to displacements along the different normal coordinates, Eq. (6.57).

8.7.1.1 Double-Harmonic Approximation

The by far most common approximation applied in order to evaluate Eq. (8.235) is to adopt the double-harmonic approximation, as already discussed in Section 6.6. We recall that in this approximation the vibrational states are products of harmonic oscillator functions for all the normal modes in the molecule, and we only include the linear dependence of the dipole moment

30 We will follow the conventions used in vibrational spectroscopy for the vibrational frequencies and use wave numbers, as introduced in Section 3.1.5.

on the nuclear displacements. As a consequence, only normal coordinates that lead to a change in the dipole moment upon geometric distortions will contribute to the IR absorption spectrum. These vibrational modes are said to be *infrared active*.

Using symmetry, we can predict the IR activity of the different vibrational modes. A normal coordinate Q_a and a component of the electric dipole moment operator $\hat{\mu}_\alpha$ span, respectively, irreducible representations $\Gamma(Q_a)$ and $\Gamma(\hat{\mu}_\alpha)$ of the molecular point group. Following the symmetry analysis for transition matrix elements in Section 4.1.2, we can predict that vibrational modes will only be IR active if

$$\Gamma(Q_a) \otimes \Gamma(\hat{\mu}_\alpha) = \Gamma(\hat{I}), \tag{8.236}$$

as the vibrational ground state is always totally symmetric.

As an example, let us consider the water molecule in the C_{2v} point group and placed in the yz-plane with z as the principal axis. From the group table in Table 4.1, we see that the three electric dipole moment components $\hat{\mu}_x$, $\hat{\mu}_y$, and $\hat{\mu}_z$ span the irreducible representations B_1, B_2, and A_1, respectively, whereas the bending, symmetric stretch, and asymmetric stretch normal modes illustrated in Figure 6.3 span the irreducible representations A_1, A_1, and B_2, respectively. From the discussion above, it is then clear that all three modes are IR active with excitations of the bending and symmetric stretch modes being induced by an electric field polarized along the molecular z-axis and excitations of the asymmetric stretch mode being induced by an electric field polarized along the y-axis.

For the IR-active modes, it remains to determine the absorption frequencies and absorption cross section. In the double-harmonic approximation, the transition energies $\hbar\omega_a$ are given in terms of the harmonic oscillator frequencies for a given normal coordinate Q_a, recalling that the properties of the harmonic oscillator wave functions ensure that only excitations into the first vibrationally excited state of each normal mode are possible. From Eq. (6.37), it follows that the vibrational transition moment for an IR-active vibrational mode in the double-harmonic approximation is given by Eq. (6.74), giving the following expression for the transition strength of a vibrational transition of normal coordinate Q_a

$$\sigma_a(\omega) = \frac{\pi\omega}{3\varepsilon_0 c\hbar}\langle 0_a|\mu_\alpha|1_a\rangle\langle 1_a|\mu_\alpha|0_a\rangle = \frac{\pi}{3\varepsilon_0 c}\frac{\partial\mu_\alpha}{\partial Q_a}\frac{\partial\mu_\alpha}{\partial Q_a}f(\omega;\omega_{k0},\gamma_{k0}). \tag{8.237}$$

We note, that in most cases, IR spectra are reported in terms of the absorbance, which is described by the Beer–Lambert law, Eq. (3.411).

The numerical values of the vibrational frequencies and dipole moment gradients for water, as obtained at the B3LYP/cc-pVTZ level of theory are given in Table 8.10. The sign of the permanent dipole moment is negative with a coordinate system chosen such that the z-axis points from the hydrogens toward the oxygen, and displacements along the normal coordinates (with positive directions indicated by the arrows in Figure 6.3) will in all cases provide negative

Table 8.10 Infrared absorption properties of H_2O in terms of vibrational wave numbers (cm^{-1}) and dipole moment gradients with respect to normal coordinates (10^{-3} a.u.).

a	Sym	ω_a		$\partial\mu_x/\partial Q_a$	$\partial\mu_y/\partial Q_a$	$\partial\mu_z/\partial Q_a$
1	A_1	1640	2.8%	0	0	−6.25
2	A_1	3800	3.9%	0	0	−1.35
3	B_2	3900	3.8%	0	−4.78	0

Theoretical values calculated at the B3LYP/cc-pVTZ level of theory with percentage corrections of ω_a referring to comparisons made against experimental fundamental frequencies.

contributions to the dipole moment. The vibrational frequencies of the harmonic force field will in general provide an overestimation of the fundamental ones, as anharmonic effects in general will lower the vibrational frequencies, as discussed in Section 6.5.

Alternatively, we could start from the expression for the damped polarizability in Eq. (5.290), the transition strength being obtained as the absorptive part of the polarizability, Eq. (5.297), where we recall that the expression will, depending on the value of the damping parameter, be nonzero only in regions close to vibrational transitions. It follows from Eq. (5.335) that the absorption cross section can be written as

$$\sigma_a(\omega) = \frac{\omega}{3\varepsilon_0 c}\mathrm{Im}[\alpha^{\mathrm{v}}_{\alpha\alpha}(-\omega;\omega)], \tag{8.238}$$

which can be shown to be identical to Eq. (8.237) when we evaluate the sum-over-states expression in the imaginary part of the vibrational polarizability at resonance, Eq. (5.293).

8.7.1.2 Anharmonic Corrections

Electrical anharmonicity is induced by quadratic and higher order terms in the expansion of the dipole moment with respect to normal coordinates as given by Eq. (6.57). The quadric term in this expansion gives rise to the possibility of observing *overtones* and *combination bands* in the IR absorption spectrum—the former as a consequence of transition moments of the form $\langle 2_a|Q_a^2|0_a\rangle$ being nonzero and the latter as a consequence of $\langle 1_a|Q_a|0_a\rangle\langle 1_b|Q_b|0_b\rangle$ being nonzero.

Mechanical anharmonicity is induced by cubic and higher order terms in the expansion of the potential as given in Eq. (6.27). The cubic term may give rise to *Fermi resonances*. Fermi resonances arise because of an interaction between an overtone or a combination band on the one hand and a fundamental transition on the other hand. These interactions are due to matrix elements of the cubic term of the potential of the form $\langle 0_a 2_b|Q_a Q_b^2|1_a 0_b\rangle$ (overtone resonance) or $\langle 0_a 1_b 1_c|Q_a Q_b Q_c|1_a 0_b 0_c\rangle$ (resonance with combination band) being nonzero. As the name indicates, this occurs only if the vibrational energy of the overtone or combination band is close to the energy of the fundamental vibrational state, as illustrated in Figure 8.37. The otherwise weak overtone or combination band can borrow intensity from the strong fundamental transition. As a consequence, instead of one strong and one very weak band very close in energy, two rather strong bands shifted in energy relative to the fundamental band are observed in the spectrum.

The anharmonic energy corrections are straightforwardly determined by means of perturbation theory as described in Section 6.4. For our example in terms of the water molecule, we get the second-order correction to the vibrational energy from Eq. (6.50) and the results are presented in Table 8.11. As seen in the table, the inclusion of anharmonic corrections leads to theoretical estimates for vibrational frequencies that fall below the experimental values. The

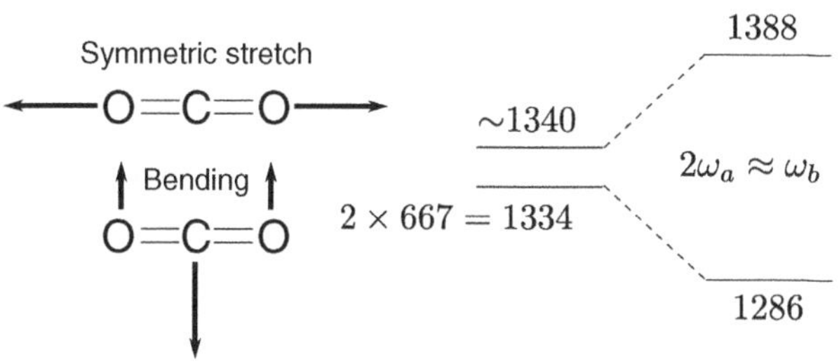

Figure 8.37 Illustration of the Fermi resonance between the first overtone of the two degenerate bending modes in CO_2 and the fundamental mode arising from the symmetric stretching motion. In Raman spectroscopy, the bending mode is symmetry forbidden, but Fermi observed in 1931 that the overtone, which is Raman allowed, could be observed due to intensity-borrowing from the fundamental mode.

Table 8.11 Harmonic and anharmonic vibrational frequencies (cm^{-1}) for H_2O.[a)]

a	Sym	$\omega_a^{(0)}$	$\Delta_{exp}^{(0)}$	$\omega_a^{(2)}$	$\Delta_{exp}^{(2)}$	ω_a^{expt}
1	A_1	1640	45	1575	−20	1595
2	A_1	3800	143	3630	−27	3657
3	B_2	3900	144	3703	−53	3756

a) $\Delta^{(i)} = \omega_a^{(i)} - \omega_a^{expt}$.
Theoretical values are obtained at the B3LYP/cc-pVTZ level of theory and compared against experimental fundamental frequencies.

remaining discrepancy between theoretical and experimental results is predominantly due to the limited accuracy in the potential energy surface at the adopted level of theory.

8.7.2 Vibrational Circular Dichroism

VCD is the vibrational analogue of ECD. It thus measures the difference in the absorption of left- and right-circularly polarized light when exciting the molecule from the vibrational ground to a vibrational excited state. In analogy with ECD, we would therefore expect the differential absorption of left- and right-circularly polarized light to be proportional to the product of the electric dipole and magnetic dipole operators. For a transition between the ground vibrational state and an excited vibrational state k of vibrational mode Q_a, we can write

$$R_{0k} = \text{Im}\langle 0|\mu_\alpha|k\rangle\langle k|m_\alpha|0\rangle, \tag{8.239}$$

where μ_α and m_α are components of the total electric and magnetic dipole moments given as functions of normal coordinates. In analogy with ECD, the VCD rotational strength will only be nonvanishing for molecules that are vibrationally chiral. While all molecules that are electronically chiral are also vibrationally chiral, it may be that even though a molecule is electronically achiral (within the Born–Oppenheimer approximation), the molecule may be vibrationally chiral due to, for instance, isotopic substitution. An example of such a molecule would be mono-deuterated oxirane, where the deuterium gives rise to a vibrationally chiral structure. Thus, whereas methyloxirane would give rise to both an electronic and a VCD spectrum, only VCD would give a nonzero spectrum in the case of deutero-oxirane.

Following the by now well-known strategy to evaluate vibrational matrix elements, we expand the dependence of the electric and magnetic dipole moments with respect to their dependence on the vibrational normal coordinates as in Eq. (6.57), and the leading-order double-harmonic contribution to the rotational strength of a vibrational transition of normal coordinate Q_a becomes

$$R_a = \frac{\hbar}{2\omega_a}\text{Im}\frac{\partial\mu_\alpha}{\partial Q_a}\frac{\partial m_\alpha}{\partial Q_a}. \tag{8.240}$$

As for infrared spectroscopy, excitations in VCD are only possible from the vibrational ground state to the first vibrationally excited state of each vibrational mode. We recall that vectors changing their sign during inversion are referred to as polar tensors, whereas vectors keeping their sign are referred to as axial tensors; see Eqs. (4.51) and (4.52). In Eq. (8.240), the electric dipole gradient is therefore commonly referred to as the atomic polar tensor contribution to VCD, whereas the magnetic dipole gradient is referred to as the atomic axial tensor.

In Eq. (8.240), we have indicated that we take the derivatives of the electric and magnetic dipole moments with respect to displacements along the normal coordinates, and this would correspond to the vibrational modes observed experimentally. In practice, derivatives of the properties are calculated with respect to Cartesian nuclear displacements and then transformed to the normal coordinates using the transformation matrix L in Eq. (6.31). In the following, we will therefore indicate the geometry dependence of the properties through the derivatives with respect to Cartesian displacements of the nuclei, keeping in mind that the final tensors obtained must be transformed to the normal-coordinate basis.

There is, however, a problem with the expression for the rotational strength in Eq. (8.240): As the vibrational wave function is real and the geometric derivative of the magnetic dipole operator is purely imaginary, the vibrational transition moment that defines the atomic axial tensor vanishes, and thus the rotational strength vanishes. In order to resolve this apparent conundrum, we note that a small magnetic moment can arise from the fact that the electrons do not instantaneously adjust to a change in the nuclear positions, that is, that the Born–Oppenheimer approximation is inexact. As the electrons do not follow instantaneously the changes in the positions of the nuclei during the vibrational motion, this will lead to a small vibrationally induced magnetic moment.

Before we address in more detail how we can calculate the vibrationally induced magnetic moment that appears in the atomic axial tensor, let us first consider the evaluation of the dipole gradient. As the dipole moment can be evaluated as the first derivative of the molecular energy with respect to an applied electric field, it follows that the dipole gradient is the second derivative of the energy, once with respect to the electric field and once with respect to nuclear distortions. The dipole gradient can therefore be expressed as a linear response function, evaluated at zero frequency and assuming real wave functions

$$\begin{aligned}\frac{d\mu_\alpha}{dR_{\mathrm{N},\beta}} &= Z_{\mathrm{N}}\delta_{\alpha\beta} + \left\langle\left\langle \hat{\mu}^{\mathrm{e}}_\alpha ; \frac{\partial \hat{H}^{\mathrm{tot}}}{\partial R_{\mathrm{N},\beta}} \right\rangle\right\rangle_0 \\ &= Z_{\mathrm{N}}\delta_{\alpha\beta} + \frac{2}{\hbar}\sum_{K\neq 0}(\omega_{K0})^{-1}\langle 0|\hat{\mu}^{\mathrm{e}}_\alpha|K\rangle \left\langle K\left|\frac{\partial \hat{H}^{\mathrm{tot}}}{\partial R_{\mathrm{N},\beta}}\right|0\right\rangle .\end{aligned} \quad (8.241)$$

We will assume that we are working with exact states, and thus that there is no explicit dependence of the electronic states on the nuclear positions within the Born–Oppenheimer approximation. In this picture, the gradient of the molecular Hamiltonian with respect to the distortion of the nuclei is the electric field at the nucleus, that is, the derivative of the electron–nuclear potential with respect to the nuclear coordinate.

To take into account the breakdown of the Born–Oppenheimer approximation, we can perturb the ground-state vibronic wave function to first order in the nonadiabatic coupling matrix elements, following our discussion in Section 6.1. Assuming for simplicity that the reference wave function is the vibronic ground-state wave function, we can write the wave function that includes first-order vibronic corrections as

$$|\Psi^{\mathrm{pert}}\rangle = |\Psi^{(0)}\rangle + |\Psi^{(1)}\rangle = |0,0\rangle - \frac{1}{\hbar}\sum_{(K,k)\neq(0,0)} \frac{\langle k,K|\hat{H}^{\mathrm{tot}}|0,0\rangle}{\omega_{Kk,00}}|K,k\rangle, \quad (8.242)$$

using the full vibronic Hamiltonian defined in Eq. (6.1). We recall from the discussion in Section 6.1 that we in the Born–Oppenheimer approximation ignored the first- and second-order variations of the electronic wave function with respect to the nuclear coordinates. In order to go beyond the Born–Oppenheimer, these corrections must now be retained in the evaluation of the matrix elements in Eq. (8.242) following Eq. (6.5). Inserting this perturbed wave function into the expression for the magnetic dipole operator, it can be shown that the final expression for the nonvanishing, nonadiabatic magnetic dipole operator is

$$m_{\mathrm{N},\alpha\beta} = \frac{i}{4}\epsilon_{\alpha\beta\gamma}Z_{\mathrm{N}}R_{\mathrm{N},\gamma}\delta_{\beta\gamma} - \frac{1}{\hbar^2}\sum_{K\neq 0}(\omega_{KO})^{-2}\left\langle 0\left|\frac{\partial\hat{H}^{\mathrm{OZ}}}{\partial B_\alpha}\right|K\right\rangle\left\langle K\left|\frac{\partial\hat{H}^{\mathrm{tot}}}{\partial R_{\mathrm{N},\beta}}\right|0\right\rangle, \tag{8.243}$$

where the first term is a purely nuclear contribution, $\hat{H}^{\mathrm{OZ}}$ is the orbital Zeeman operator defined in Eq. (8.81), and the summation in the electronic part runs over all excited electronic states of the molecule.

We note that the expression for the atomic axial tensor in Eq. (8.243) does not correspond directly to any particular response function, as the expression for the frequency-dependent response function corresponding to the perturbation operators $\partial\hat{H}/\partial R_{K,\alpha}$ and $\partial\hat{H}^{\mathrm{OZ}}/\partial B_\beta$ reads

$$\left\langle\left\langle\frac{\partial\hat{H}^{\mathrm{OZ}}}{\partial B_\alpha};\frac{\partial\hat{H}^{\mathrm{tot}}}{\partial R_{\mathrm{N},\beta}}\right\rangle\right\rangle_\omega = \frac{2}{\hbar}\sum_{K\neq 0}(\omega_{K0}-\omega)^{-1}\left\langle 0\left|\frac{\partial\hat{H}^{\mathrm{OZ}}}{\partial B_\alpha}\right|K\right\rangle\left\langle K\left|\frac{\partial\hat{H}^{\mathrm{tot}}}{\partial R_{\mathrm{N},\beta}}\right|0\right\rangle. \tag{8.244}$$

However, from this equation, it can be realized that we may express the electronic contribution to the atomic axial tensor as the frequency derivative of the linear response function defined with respect to nuclear distortions and the magnetic moment of the molecule, that is,

$$m_{N,\alpha\beta} = \frac{i}{4}\epsilon_{\alpha\beta\gamma}Z_N R_{N\gamma}\delta_{\beta\gamma} - \frac{1}{2}\frac{d}{d\omega}\left\langle\left\langle\frac{\partial\hat{H}^{\mathrm{tot}}}{\partial B_\beta};\frac{\partial\hat{H}^{\mathrm{tot}}}{\partial R_{N,\alpha}}\right\rangle\right\rangle_\omega\Bigg|_{\omega=0}. \tag{8.245}$$

We recall that a key difference between electronic absorption and ECD is that the absorption cross section only has a magnitude, whereas the rotational strength has both a magnitude and a sign. ECD thus provides more information about the molecular structure than does electronic absorption. In particular, ECD provides a direct connection to the stereochemistry of a chiral molecule. This is also the case for VCD. Compared to a regular infrared vibrational spectrum, in VCD each vibrational normal mode will have a rotatory strength that, in addition to a magnitude, will also have a sign. VCD thus provides the same information about normal modes as regular infrared vibrational spectroscopy, but will provide additional insight into the stereochemistry and absolute configuration of chiral molecules. Furthermore, while ECD is only applicable to molecules that have a chiral electronic structure, VCD can also be measured for molecules that are vibrationally chiral, due to, for instance, isotopic substitution, as noted above.

A challenge in the use of ECD for determining the absolute stereochemistry of molecules, in particular if there are multiple chiral centers, is that for most molecules there is only a limited number of electronic excited states in the frequency range accessible to modern spectrometers (down to around 180–200 nm), and different states may be difficult to identify due to broad, overlapping bands. Even for chiral molecules with only a single chiral center, a reliable assignment of the absolute stereochemistry is made difficult when comparing an experimental spectrum to that obtained from *ab initio* calculations due to the limited accuracy of the calculated excited states, the few spectral features observed experimentally, and the complexity that arises from conformational flexibility.

These limitations in the use of ECD for determining the absolute configurations of chiral molecules are largely lifted in VCD. The vibrational spectrum is determined by the ground-state electronic density only, which can be much more reliably determined computationally than electronic excited states. Furthermore, as the number of vibrational modes is $3N-6$ for a molecule with N atoms, there are in a given VCD spectrum a large number of modes for which a comparison can be made. Even though there may be disagreements between a calculated spectrum and that observed experimentally for selected VCD bands, in particular for weak vibrational modes, there may be a number of strong VCD bands that are robust, showing little dependence on molecular conformation or computational level.

Although strong VCD bands in general are rather insensitive to conformational and structural changes and thus allow them to be used as reliable marker bands for the absolute configuration of a chiral center, there is an important caveat that must be kept in mind. In order to understand this, we will revisit the equation for the rotational strength in Eq. (8.239). For a given normal mode, we can write the rotational strength as a scalar product of the vibrational electric dipole transition vector and the corresponding vibrational magnetic dipole transition vector. The rotational strength for a given normal mode Q_a can thus be written as

$$R_a = \boldsymbol{\mu}_{01}^a \cdot \mathbf{m}_{10}^a = |\mu_{01}^a||m_{10}^a| \cos\theta, \tag{8.246}$$

where θ now defines the angle between the electric dipole and magnetic dipole transition vectors of the vibrational motion being considered.

This expression gives some interesting perspectives of the origin of VCD. In general, the transition electric and magnetic dipole moments are nonzero also for nonchiral molecules. This means that for achiral molecules, the electric and magnetic dipole transition vectors are orthogonal to each other, that is, $\cos\theta = 0$.

Because two enantiomers have the opposite sign but the same magnitude for the rotational strengths, it also follows that

$$\cos\theta_R^i = -\cos\theta_S^i \Rightarrow \theta_R^i = 180^\circ - \theta_S^i, \tag{8.247}$$

and thus the relative orientation of the vibrational electric dipole and magnetic dipole transition vectors are the opposite for the two enantiomers. We also observe that the rotatory strength is positive for angles in the range $0^\circ \leq |\theta| < 90^\circ$, and similarly that the rotatory strength will be negative for $90^\circ < |\theta| \leq 180^\circ$. One can extend this analysis to chiral molecules with symmetry. For such molecules, only selected relative orientations are possible for $\boldsymbol{\mu}_{01}^a$ and $\mathbf{m}_{01}^a$ for a given vibrational mode. For instance, for totally symmetric vibrations of molecules with C_2 point group symmetry, the vectors can only be aligned parallel or antiparallel to each other. This can be realized by noting that only the z component of the electric and magnetic dipole transition moments are nonzero for totally symmetric vibrations. In contrast, for vibrations transforming according to the B representation, any orientation of the vectors in the xy plane is possible, thus imposing no constraints on the angle between the two vectors. For the general and most dominating case of a chiral molecule without any point-group symmetry, any relative orientation of the vibrational electric and magnetic dipole transition moments are possible.

We would in general expect that modes that have an angle $\theta \approx 90^\circ$ would have small rotational strengths, as $\cos\theta \approx 0$. We would also expect that vibrational modes whose angles between the electric and magnetic dipole transition vectors are close to 90° would be easily susceptible to sign changes when there are small changes in the molecular structure, either due to small differences in the description of the electronic structure of the molecule, in the vibrational normal modes or in conformational differences. In contrast, modes for which $|\theta|$ is rather far away from 90°, often 60° or less, can be expected to be rather robust toward the choice of computational level as well as toward conformational flexibility. Such modes are to be preferred when comparing experimental and theoretical VCD spectra to determine the absolute configuration of a chiral molecule.

To illustrate the information richness of a VCD spectrum, we present in Figure 8.38 the VCD spectrum of methyloxirane, including the spectrum both for S and R enantiomers. We immediately note that the spectra are each other's mirror images, and that it is easy from a comparison with experiment to determine which enantiomer (or enantiomeric excess) we have experimentally. In particular, although there will be differences between experiment and the calculated spectra, the bisignate structure around 1000 cm^{-1}, composed of multiple strong bands, will be a signature for the comparison with experiment. IR and VCD spectra can often be obtained

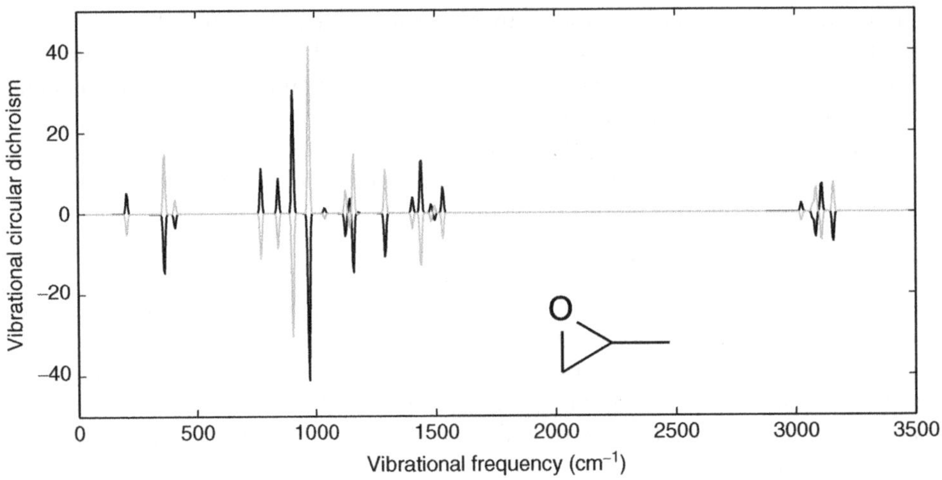

Figure 8.38 Vibrational circular dichroism of (*S*)-methyloxirane. Rotational strengths are given in units of 10^{-44} esu^2 cm^2 and broadened by Gaussian line profiles. A spectral line profile is also provided for the (*R*)-enantiomer in grey color.

with rather high resolution, allowing many vibrational modes to be identified. In the spectrum in Figure 8.38, we have used a broadening of 8 cm^{-1} to give a spectrum that is most directly comparable to the resolution that can be obtained in experiment.

8.7.3 Raman Scattering

8.7.3.1 Raman Scattering from a Classical Point of View

Inelastic light scattering was first observed by Raman and Krishnan in 1928 and is today known as Raman scattering. It is theoretically founded on the Kramers–Heisenberg–Dirac expression, Eq. (8.250), relating transition moments and excitation energies to the scattering cross section. From a classical point of view, vibrational Raman scattering together with the concomitant Rayleigh scattering are understood as the consequences of electronic charge oscillations in molecules as we saw in Section 3.3.7. The charge oscillations are *mechanical*, due to the atomic vibrational motions on the potential energy surface, and *electrical*, due to the forces extorted by the applied radiation, and they give rise to a time-dependent induced dipole moment. The different frequency components of the induced dipole moment will act as individual sources of scattering. To linear order, the time-dependent dipole moment couples to the external electric field of frequency ω and strength F^ω via the polarizability tensor, and we may express $\mu(t)$ as

$$\mu_\alpha(t) = \sum_\omega \alpha^e_{\alpha\beta}[\omega; Q(t)]F^\omega_\beta e^{-i\omega t}, \tag{8.248}$$

where it has been explicitly noted that the electronic polarizability not only varies with the frequency of the field (dispersion) but also with the molecular configuration given in terms of the set of vibrational normal mode coordinates $Q(t)$.

Let us illustrate this principle by considering molecular nitrogen, for which the single vibrational coordinate is the internuclear distance $R(t)$ and the potential energy surface is depicted in Figure 8.39. We address the situation of a dilute gas for which the relevant induced dipole moment is equal to the product of the isotropic average of the polarizability and the external

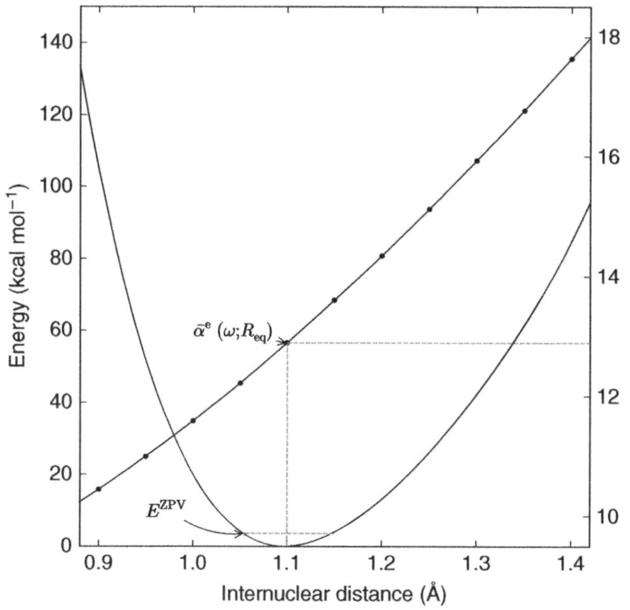

Figure 8.39 Potential energy curve and isotropic polarizability $\bar{\alpha}$ for N_2. The polarizability $\bar{\alpha}(\omega; R)$ is determined at the B3LYP/taug-cc-pVTZ level of theory for $\omega = 2\pi \times 10^{15}$ rad s^{-1}. The zero-point vibrational energy E^{ZPV} as determined at the B3LYP/cc-pVTZ level of theory is indicated in the figure.

electric field, that is,

$$\mu(t) = \bar{\alpha}^e[\omega; R(t)]F(t).$$

Zero-point vibrational motion corresponds to time-dependent variations in $R(t)$ in the close vicinity of the equilibrium value R_{eq}—the classical turning points of the nuclear motion are the crossings of the zero-point vibrational energy E^{ZPV} with the potential curve. This vibrational motion will not be fully harmonic as the mechanical anharmonicity is apparent in the figure plot, but the electrical anharmonicity, on the other hand, is small as reflected in the near linear dependence of the isotropic polarizability $\bar{\alpha}$ on R in the region of vibration. The dynamic polarizability data are obtained for an angular frequency of $\omega = 2\pi \times 10^{15}$ rad s^{-1} and refer to calculations performed at the level of time-dependent Kohn–Sham DFT for different bond distances. It is clear that during the course of molecular vibrations with a bond distance that varies with time, the molecular property (here polarizability) will also vary with time. We will now turn to the determination of this time-dependent bond distance $R(t)$.

With the given force field, we perform a molecular dynamics (MD) simulation with conditions of constant energy and an initial value of R equal to the shorter of the two classical turning points. The oscillatory motion of the atoms is depicted in the lower panel of Figure 8.40, showing a fundamental period time of 14.5 fs (corresponding to a frequency of 69.0 THz) whereas the harmonic period time of 13.6 fs is the one corresponding to the B3LYP/cc-pVTZ harmonic wave number of 2450 cm^{-1} (or E^{ZPV} of 3.50 kcal mol^{-1}). The vibrational motion gives rise to a time-dependent isotropic polarizability $\bar{\alpha}[\omega; R(t)]$ as depicted in the lower-mid panel of the figure, where also $\bar{\alpha}(\omega; R_{eq})$ is shown by means of the horizontal line.

In this state of vibration, the nitrogen molecule is subjected to a monochromatic external electric field as depicted in the upper-mid panel of Figure 8.40 with frequency 1.0 PHz (corresponding to a photon energy of 4.14 eV) and strength 5.0×10^{-5} a.u. (corresponding to that

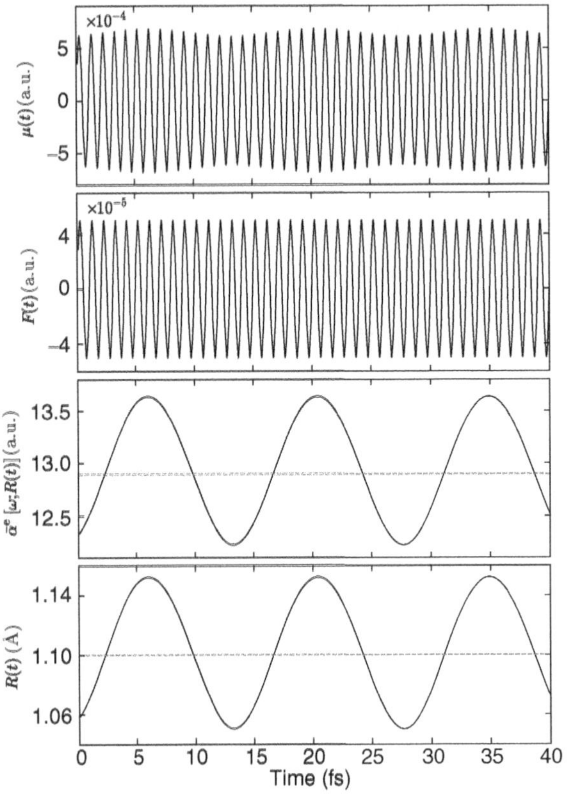

Figure 8.40 The molecular dynamics of N_2 as given in terms of the internuclear distance $R(t)$ and which gives rise to the time-dependent polarizability $\overline{\alpha}[\omega; R(t)]$ (here ω is equal to $2\pi \times 10^{15}$ rad). The application of an electric field with strength $F(t)$ and frequency 1.0 PHz results in a time-dependent induced dipole moment $\mu(t) = \overline{\alpha}F$.

of a pulsed laser of modest energy). The resulting dipole moment is plotted in the upper panel of Figure 8.40. In analogy with amplitude modulation in radio communication, we can say that the "carrier wave" in $\mu(t)$, that is, $\overline{\alpha}(\omega; R_{\mathrm{eq}})F(t)$, is the source of Rayleigh scattering whereas the "information signal" is the source of vibrational Raman scattering. In Figure 8.41(c), we have plotted this source of Raman scattering, that is,

$$r(t) = \mu(t) - \overline{\alpha}^{e}(\omega; R_{\mathrm{eq}})F(t).$$

The signal $r(t)$ is out of phase with respect to the external field due to the nonzero phase of the vibrational coordinate $R(t)$ at $t = 0$. Since the phases of vibrations differ in a random manner from one molecule to another, it is clear that Raman scattering is inherently incoherent. A comparison of the envelope function of $r(t)$ in the half-period intervals (say from 2–9 to 9–16 fs) and the fact that they differ is a direct sign of mechanical anharmonicity in the system.

The frequency components in $r(t)$ reveal themselves best by consideration of the autocorrelation function, which is a mathematical technique designed to discover repeat patterns in signals. With access to $r(t)$ in a time window $t \in [0, T]$ (in our case $T = 40$ fs), we determine the autocorrelation according to

$$R_{rr}(\tau) = \int_{-\infty}^{\infty} r(t)[\Theta(t) - \Theta(t-T)]r(t-\tau)[\Theta(t-\tau) - \Theta(t-\tau-T)]dt, \tag{8.249}$$

where Θ is the Heaviside step function that here serves the purpose of limiting the time integration to regions where the Raman signal is available. The maximum value of the autocorrelation function occurs for $\tau = 0$ and, when normalized against this maximum value, the autocorrelation function of $r(t)$ takes the form presented in Figure 8.41(b). As the Fourier transform

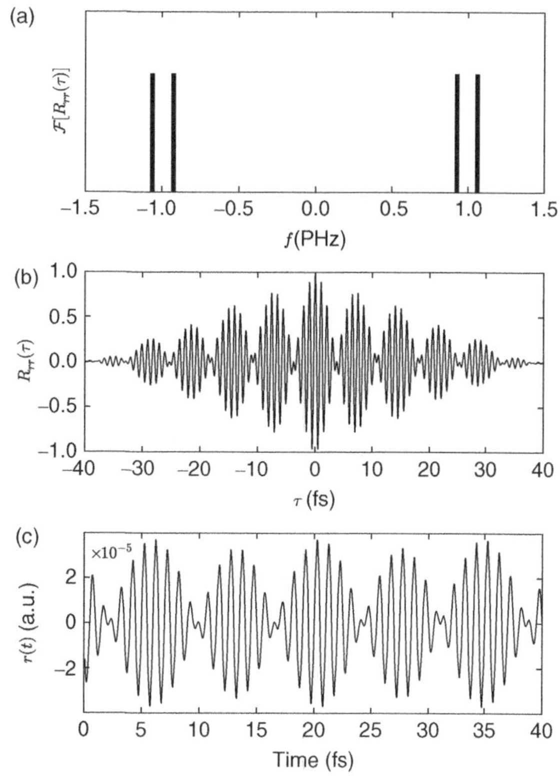

Figure 8.41 The Raman signal of N_2 as given by $r(t) = \mu(t) - \overline{\alpha}(\omega; R_{\text{eq}})F(t)$ together with its normalized autocorrelation function $R_{rr}(\tau)$ and the Fourier transform $\mathcal{F}[R_{rr}(\tau)](f)$. The Fourier transform shows the Stokes and anti-Stokes Raman scattering signals.

presented in panel (a) in Figure 8.41 shows, there are two frequency components in the Raman signal, namely, $\omega \pm 2\pi\nu$, where ν is the fundamental frequency of vibration. The minus and plus signs correspond to Stokes and anti-Stokes Raman scattering, respectively, or, in other words, the Stokes line is associated with energy loss in the scattering process whereas the reverse is true for the anti-Stokes line. In both cases, the difference in energy between incoming and outgoing photons is associated with a change in the vibrational state before and after the scattering event. In principle, since there is a significant mechanical anharmonicity in this system, there are also overtones in the Raman signal $r(t)$. Owing to their relative smallness, the overtone signals do not, however, become visible in our plot of the Fourier spectrum.

From a principal point of view, we have demonstrated that combined use of molecular dynamics (here based on a classical molecular mechanics force field) and quantum chemical response theory to treat atomic vibrational motions and electronic field responses, respectively, is a viable approach to study vibrational normal Raman scattering. This approach comes with an appealing feature in that it includes mechanical and electrical anharmonicites. It also possesses a favorable scaling with system size since the computational effort of relevance is restricted to the determination of the polarizability tensor in the time window $t \in [0, T]$, regardless of system size. This technique has, however, not been much employed in the literature where one has instead favored perturbational expansions of the potential and property surfaces, so that the computational burden is shifted toward a determination of force constants and property gradients. In the subsequent section, we will adopt to this approach and view the atomic vibrations as the motion of quantum-mechanical, rather than classical, oscillators.

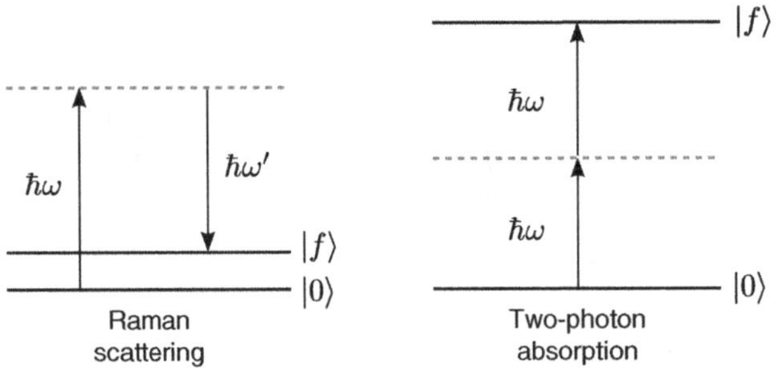

Figure 8.42 Illustration of the processes of nonresonant Raman scattering and two-photon absorption. States $|0\rangle$ and $|f\rangle$ refer to the initial (ground) and final states, respectively, and the dashed line represents an intermediate virtual state.

8.7.3.2 Raman Scattering from a Quantum Mechanical Point of View

Raman scattering is a second-order process involving an incoming photon of energy $\hbar\omega$ and an outgoing photon of energy $\hbar\omega'$ in accordance with Figure 8.42. It bears strong resemblance to the process of TPA, which is also illustrated in the figure. The relevant molecular property for Raman scattering is the transition polarizability tensor for which the Kramers–Heisenberg–Dirac expression reads as

$$[\alpha_{\alpha\beta}]_{f0} = \frac{1}{\hbar}\sum_n \left[\frac{\langle f|\hat{\mu}_\alpha|n\rangle\langle n|\hat{\mu}_\beta|0\rangle}{\omega_{n0} - \omega - i\gamma_n} + \frac{\langle f|\hat{\mu}_\beta|n\rangle\langle n|\hat{\mu}_\alpha|0\rangle}{\omega_{n0} + \omega' + i\gamma_n} \right], \tag{8.250}$$

where $|0\rangle$ and $|f\rangle$ are the initial (here assumed to be the ground state) and final vibronic states in the scattering process. By comparison, it is clear that in nonresonant situations, when the lifetime broadening can be ignored, the calculation of the inelastic scattering parameter is analogous to the calculation of the two-photon transition amplitude, see Eq. (5.250). But a difference arises because, in the former case the frequency of the applied field largely exceeds the transition frequency ω_{f0} and in the latter case it is equal to $\omega_{f0}/2$, which means that standard response theory calculations are applicable to electronic Raman scattering as resulting from nonresonant inelastic scattering of photons in the X-ray region of the spectrum. When it comes to vibrational Raman scattering, on the other hand, the final state $|f\rangle$ is an excited *vibronic* state, which is a situation ill suited for quantum chemical response theory methods that introduce parameterizations for either electronic or vibrational wave functions, but not for vibronic wave functions.

As closely related as Raman scattering and TPA may be, there are also fundamental differences. A nonresonant scattering process is to be regarded as instantaneous where one photon is annihilated and another one is created, whereas in the TPA process, the two involved light quanta are both annihilated. The observable in Raman scattering, that is, the intensity of scattered radiation I, is proportional to the square magnitude of the transition dipole moment

$$I \propto |\boldsymbol{\mu}_{f0}|^2 \propto |\boldsymbol{\alpha}_{f0} \cdot \mathbf{F}|^2 \propto I_0, \tag{8.251}$$

where $\mathbf{F}$ and I_0 are the electric field and intensity associated with the incident radiation. So Raman scattering is a linear optical process. In contrast, the observable in TPA, which is the loss of energy in the applied field per time unit as defined by Poynting's theorem, Eq. (3.131), is proportional to I_0^2 as this is a nonlinear optical process. As we have shown above, however, both phenomena can, from a computational aspect, be addressed in a common framework, where the computational quantity of interest can be identified from the first-order residue of the second-order (nonlinear) response function.

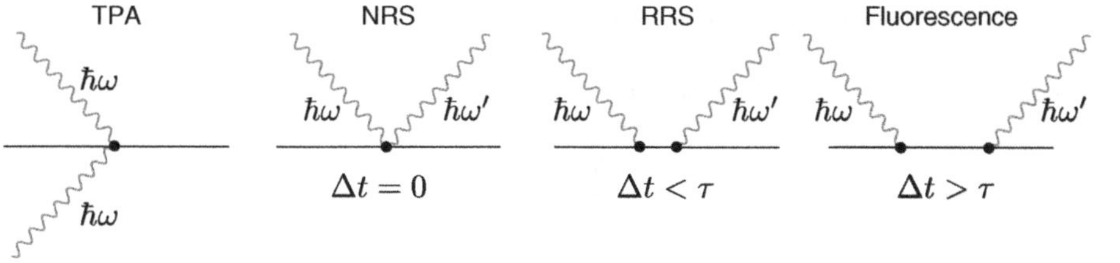

Figure 8.43 Illustration of optical processes: (i) coherent two-photon absorption, (ii) normal instantaneous Raman scattering, (iii) resonant Raman scattering with interaction time Δt smaller than excited-state lifetime τ, and (iv) fluorescence with $\Delta t > \tau$ and nonradiative energy transfer to the environment in the excited state.

The aspect of interaction time separates the resonance Raman process from that of fluorescence; see Figure 8.43. In the former case, the system resides in the excited state for a time Δt shorter than the lifetime τ before emission of a photon occurs, whereas in the latter case, the interaction time is larger than the vibronic state lifetime so that nonradiative relaxation occurs in the electronically excited-state potential. This means that the only photons that ought to be attributed to Raman scattering are those that are emitted within the excited-state lifetime, which is a factor that complicates the experiment. From a theoretical perspective, it means that different approximations may be called for depending on the excited-state lifetimes since these indirectly enter in the form of the damping terms in Eq. (8.250).

Placzek Theory of Normal Raman Scattering In a situation where the frequency of the external field is intermediate to vibrational and electronic transition frequencies, a series of approximations can be adopted in the evaluation of the transition polarizability tensor, Eq. (8.250), without significant loss in accuracy. First, the excited-state lifetimes are typically in the picoseconds region, which is far longer than the period time of the nonresonant external field, so it is safe to ignore the imaginary relaxation parameters. Second, since the vibrational polarizability is known to be small in the optical region, one is motivated to exclude from the summation in Eq. (8.250) all states involving the ground electronic state. Third, for electronically excited states, the vibronic transition frequencies ω_{nf} and ω_{n0} can be replaced with their vertical electronic counterparts. Within the Born–Oppenheimer approximation, these three approximations lead to an expression for the transition polarizability that reads

$$[\alpha_{\alpha\beta}]_{f0} = \langle f|\alpha^{e}_{\alpha\beta}(\omega; Q)|0\rangle, \tag{8.252}$$

where α^{e} is the electronic polarizability and $|0\rangle$ and $|f\rangle$ have come to denote the initial and final vibrational states, both belonging to the electronic ground state. The intermediate steps in this derivation are analogous to those made to arrive at Eq. (6.17). In accordance with Eq. (6.57), it is a standard practice for polyatomic molecules to introduce a Taylor expansion of molecular properties in the normal mode coordinates Q. If we consider an expansion of α^{e} to linear order and introduce harmonic oscillator functions, Eq. (6.35), as approximations of $|0\rangle$ and $|f\rangle$, then only transitions with $f = 1$ will have nonzero intensity. The notation $f = 1$ here implies that a single vibrational mode (say a) is excited, that is, $f_a = 1$ and $f_b = 0$ for all $b \neq a$. Each of the normal modes will act as an individual source of scattering with an intensity governed by a transition polarizability tensor of the form

$$[\alpha_{\alpha\beta}]_{10} = \sqrt{\frac{\hbar}{2\omega_a}} \left(\frac{\partial \alpha^{e}_{\alpha\beta}}{\partial Q_a} \right)_{Q=0}. \tag{8.253}$$

The evaluation of harmonic oscillator matrix elements that is performed in the intermediate steps is described in Section 6.3. From Eq. (8.253), it is clear that Raman scattering will occur for

Table 8.12 Isotropic transition polarizabilities $[\overline{\alpha}]_{f0}$ (in a.u.) for N_2.

	Harmonic		Fundamental	
f	$E_f/(hc)$	$[\overline{\alpha}^e]_{f0}$	$E_f/(hc)$	$[\overline{\alpha}^e]_{f0}$
1	3,675.3	0.4338[a]	3,649.2	0.4335
2	6,125.5	0	6,049.6	−0.0128
3	8,575.7	0	8,425.6	0.0005
4	11,025.9	0	10,769.1	0.0000

a) Obtained from Eq. (8.253) based on numerical results: $\partial\overline{\alpha}^e/\partial Q = 64.8 \times 10^{-3}$ a.u.

The initial vibrational state is taken as the ground state. For the final vibrational states we report energies E_f divided by hc in units of cm^{-1}

normal modes along which the polarizability changes significantly for small-amplitude motion around the equilibrium structure. Such modes are referred to as being *Raman active*. This includes, for instance, the vibrational mode of homonuclear diatomic molecules such as our example of N_2. In view of Eq. (8.236), it is also clear that symmetry selection rules for IR and Raman activity will be very different. If we, for instance, consider a molecule with inversion symmetry, then it is noted that none of the modes of vibration will be both Raman and IR active.

In view of this quantum-mechanical treatment of Raman scattering, let us return to the illustrative example of molecular nitrogen for which the electronic polarizability as a function of the single vibrational coordinate is illustrated in Figure 8.39. For diatomics, the fundamental frequencies and associated vibrational wave functions can be found from the standard Numerov–Cooley procedure with transition integrals given by means of numerical integration. Based on the B3LYP/cc-pVTZ potential energy curve, this procedure gives the transition polarizabilities presented in Table 8.12. The $0 \rightarrow 1$ transition dominates strongly with a value of $[\overline{\alpha}]_{0f}$ that is 0.4335 a.u. There is also a noticeable value of −0.0128 a.u. for the first overtone, whereas values for higher overtones are vanishingly small. In the harmonic approximation, on the other hand, the quantity needed is the derivative of the isotropic electronic polarizability with respect to the mass-weighted vibrational normal mode coordinate. At our level of theory, the value for this derivative is equal to 64.8×10^{-3} a.u., which results in a transition polarizability of 0.4338 a.u. In the harmonic approximation, there is no contribution to Raman scattering from overtones.

For normal vibrational Raman scattering, overtones and combination bands are much weaker than the fundamentals. The harmonic approximation, as illustrated here by the example of N_2, is quite accurate and electron correlation effects are relatively small when it comes to calculations of polarizabilities (far smaller than for hyperpolarizabilities). So from a computational perspective, simulations of vibrational normal Raman scattering can be a relatively straightforward matter, even for large molecules. One complicating issue that we have not mentioned till now is the existence of low-frequency modes, such as torsional motion, and the effects of the environment. With respect to low-frequency modes it is clear that a description in terms of expansions about an equilibrium point on the potential energy surface is quite unrealistic, and techniques of configuration averaging are instead called for to obtain the broadening of the targeted high-frequency modes.

Resonance Raman Scattering The resonance Raman signal depends not only on the properties of the electronic ground state but also on the properties of one, or a few, of the excited states. This fact is a photophysical blessing as well as a computational curse since it, on the one hand, opens up for a characterization of the excited state but, on the other hand, makes the greatly simplifying, yet accurate, approximations of the Placzek theory in general invalid. Several alternative approximation schemes have been developed in order to make the evaluation of Eq. (8.250) computationally tractable for the application of discrete resonance Raman scattering.

The most obvious approximations to be made in the resonance case are to disregard the nonresonant (second) term in Eq. (8.250) and to restrict the summation over intermediate states $|n\rangle = |K\rangle|k\rangle$ to include vibronic states belonging to a single, or possibly a few, electronic bands. Both these approximations are quantitatively accurate and need not be further discussed. The next approximation in line is to Taylor expand potential energy surfaces and the transition dipole moment surface $\mu^{0K}(Q)$ about the ground-state equilibrium geometry, which immediately gives rise to series of approximation levels depending on the choice of truncations. For medium- and large-scale molecules, the Franck–Condon (FC) approximation is normally invoked, that is, $\mu^{0K}(Q)$ is regarded as constant over the region of vibrations, which gives a simplified formula (here including a single electronic band) for the transition polarizability according to

$$[\alpha_{\alpha\beta}]_{f0} = (\mu_\alpha^{0K})_0(\mu_\beta^{K0})_0\frac{1}{\hbar}\sum_k \frac{\langle f|k\rangle\langle k|0\rangle}{\omega_{Kk,00} - \omega - i\gamma_k}. \tag{8.254}$$

The FC approximation cannot be expected to be quantitatively correct in general, often due to both strong and nonlinear dependencies of transition moments on nuclear displacements around the equilibrium geometry. But it is an approximation that leads the way toward computationally tractable schemes for polyatomic molecules and which we do expect to be qualitatively correct.

In the independent mode displaced harmonic oscillator (IMDHO) model, the evaluation of the overlap of vibrational wave functions is greatly simplified. The assumptions made are essentially the same as in the linear coupling model presented in Section 6.7.3. In addition, one assumes that overtones are not of concern ($f = 1$). The notation $f = 1$ again implies that a single vibrational mode (say a) is excited, that is, $f_a = 1$ and $f_b = 0$ for all $b \neq a$. Given this scenario, we have

$$\langle f|k\rangle\langle k|0\rangle = \langle 1_a|k_a\rangle\langle k_a|0_a\rangle \prod_{b\neq a}^{3N-6} |\langle 0_b|k_b\rangle|^2, \tag{8.255}$$

where we again refer to Section 6.3 for a description of the evaluation of matrix elements for harmonic oscillator functions.

Let us illustrate the theory of resonance Raman scattering by a simple model system in terms of the ethylene molecule. The vibrational motion in our model system is restricted to the C—C bond stretching coordinate. We focus on the strong $\pi - \pi^*$ band in the absorption spectrum, which is coupled to the ground state by the long in-plane dipole moment operator. The potential energy curves for the ground and excited states with respect to the C—C bond distance are presented in Figure 8.44 together with the electronic transition dipole moment—all results are obtained at the B3LYP/taug-cc-pVTZ level of theory. The ground-state equilibrium geometry is depicted by the dashed vertical line in the figure and it is clear that, around this point, the transition dipole moment has a strong and nonlinear dependence on $R_{\text{C—C}}$. The ground-state harmonic potential is shown in the figure as an approximation not only of the ground-state potential but also of the excited-state potential. In the latter case, it has been shifted so that the gradients of the harmonic and true potential are in agreement at $R_{\text{eq}} = 1.3241$ Å—the numerical value of the gradient $\partial E/\partial R$ at this point is -0.126 a.u. The ground-state harmonic potential is

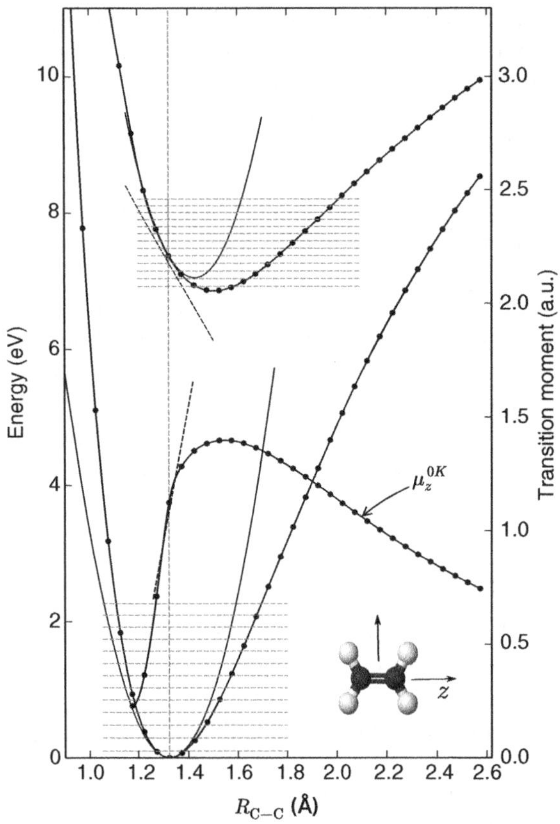

Figure 8.44 Potential energy curves (PECs) for electronic ground and $\pi - \pi^*$ excited states and electronic transition moment for C_2H_4. Data are determined at the B3LYP/taug-cc-pVTZ level of theory. Fundamental vibrational energies are indicated by horizontal dashed lines. The harmonic ground-state PEC is drawn both in the ground and excited states; in the excited state, the gradient (slanted dashed line) of the harmonic potential matches that of the true PEC at the ground-state equilibrium distance. The linear approximation of the transition moment is shown as a slanted dashed line.

clearly a poor approximation of the excited-state potential in general. However, by virtue of error cancelation, the strong curvature of the harmonic potential gives a relatively close fit to the true excited-state potential for $R \lessapprox R_{\text{eq}}$, which means that the displacement Δ from the equilibrium structure will be too small in the IMDHO model.

The vibrational motion of our model system is one-dimensional and can be correctly treated using the Numerov–Cooley procedure by treating the CH_2 groups as the particles of vibration, each with a mass of 14.0156 amu. The calculated fundamental and harmonic frequencies for this motion are 1538 and 1564 cm^{-1}, respectively, which is to be compared with the harmonic frequency of 1693 cm^{-1} for the symmetric carbon stretching mode in true ethylene. The fundamental frequency of the $\pi - \pi^*$ excited state shown in Figure 8.44 is 948 cm^{-1}, which quantifies the inaccuracy in the IMDHO model that approximates the excited state PEC with the harmonic potential of the ground state.

With access to the vibrational wave functions of the electronic ground and excited states, the vibrationally resolved absorption spectrum is determined in terms of the oscillator strengths for the vibronic states $|n\rangle = |K\rangle|k\rangle$:

$$f_{0n} = \frac{2m\omega_{n0}}{3\hbar e^2} \sum_{\alpha=x,y,z} |\langle 0|\hat{\mu}_\alpha^{0K}(Q)|k\rangle|^2. \tag{8.256}$$

As mentioned above, only the z-component of the transition dipole moment contributes in the current case. The oscillator strengths for excitations from the vibrational ground state of the electronic ground state are reported in Table 8.13, and it is noted that the absorption maximum occurs for $k = 3$ with an excitation energy that is equal to 7.165 eV, as compared to the vertical

Table 8.13 Vibronic transition energies ΔE_n (eV), oscillator strengths f_{0n}, Stokes Raman shifts $\tilde{\nu}$ (cm^{-1}), and absolute values of complex transition polarizabilities $|[\alpha_{zz}(\omega)]_{f0}|$ (a.u.) for vibronic transitions in C_2H_4.

	$\pi - \pi^*$ state			$\lvert[\alpha_{zz}(\omega)]_{f0}\rvert$[a)]				
Level	ΔE_n[b)]	f_{0n}	$\tilde{\nu}$	A[c)]	B[d)]	C[e)]	D[f)]	E[g)]
0	6.822	0.0050	0	201.1	200.7	189.4	216.9	179.5
1	6.938	0.0175	1,526.6	154.9	155.3	137.1	184.1	125.9
2	7.052	0.0322	3,038.8	118.4	118.2	69.6	145.4	90.3
3	7.165	0.0416	4,536.5	147.3	147.3	149.9	196.7	111.1
4	7.276	0.0412	6,019.3	119.6	119.5	115.5	172.7	89.4
5	7.384	0.0337	7,495.6	85.2	85.2	76.3	133.8	64.2
6	7.492	0.0229	8,946.4	120.0	120.0	126.3	203.1	90.2
7	7.598	0.0131	10,391.0	125.7	125.7	132.0	224.5	93.1
8	7.702	0.0063	11,835.5	84.0	84.0	95.2	159.5	62.1
9	7.805	0.0023	13,260.5	68.7	68.7	91.1	140.1	51.0
10	7.907	0.0006	14,667.8	110.0	110.1	123.5	241.1	83.1
11	8.007	0.0001	16,071.4	132.0	132.0	135.5	307.2	100.1
12	8.106	0.0000	17,453.4	131.7	131.7	131.1	324.5	100.3

The C—C bond length is considered as the sole vibrational coordinate, in accordance with Figure 8.44. The frequency ω is in resonance with the fourth vibrational state ($k = 3$) of the $\pi - \pi^*$ excited state. A common lifetime of 0.1ps has been adopted for excited vibronic states.

a) Between the vibrational ground ($|0\rangle$) and excited ($|f\rangle$) states of the electronic ground state.
b) Between the vibrational ground state of the electronic ground state and vibrational states of the $\pi - \pi^*$ excited state $|n\rangle = |K\rangle|k\rangle$.
c) From Eq. (8.250) using vibronic transition moments.
d) Disregarding the nonresonant term in Eq. (8.250).
e) From Eq. (8.250) but with inclusion only of $k = \{2, 3, 4\}$.
f) From Eq. (8.250) with linear transition dipole moment, the Herzberg–Teller approximation.
g) From Eq. (8.254) based on the Franck–Condon approximation.

electronic transition energy of 7.372 eV. The discrepancy between these two energies is the root of the problem when comparing experimental λ_{max}-values to vertical electronic transition wavelengths.

For the study of the resonant Raman scattering response, we focus on an optical laser frequency in resonance with the vibronic absorption maximum ($k = 3$) and we will assume a common lifetime of the intermediate vibronic states that amounts to 0.1 ps ($\gamma_n = 0.04136$ eV). The results obtained from the full Kramers–Heisenberg–Dirac formula, Eq. (8.250), are presented in column A in Table 8.13. The most prominent difference in comparison with the nonresonant case is that the values of the transition polarizability are about two orders of magnitude larger in the resonance case (cf. Table 8.12), meaning that Raman intensities are about 10^4–10^5 times higher. The second key difference between nonresonant and resonant Raman scattering is that, in the latter case, overtones are of about equal strength as fundamentals—compare the results for $f = 1$ to values for $f > 1$. As we have seen above, more approximate theoretical models focus on the fundamentals, which means that, in comparing with experiments, overtones of lower frequency modes can make the comparison complicated.

In column B of Table 8.13, we present the results obtained by excluding the second (nonresonant) term in Eq. (8.250). As concluded above, this approximation must be highly accurate in the case of resonant Raman scattering, and the comparison of the results in columns A and B gives numerical evidence for this statement. Our results are of course based on the inclusion of a quite limited number of vibrational states in the excited state; as a matter of fact, we only include the 16 lowest states in the summations. This approximation is well motivated by the absorption spectrum discussed earlier and, in fact, there is reason to believe that one could be even more restrictive without inflicting severe inaccuracies. In column C, we present the results obtained by inclusion of states $k = \{2, 3, 4\}$ and it is clear that results are still qualitatively correct.

Starting points for more approximate treatments applicable to large molecules are based on the Herzberg–Teller or Franck–Condon approximations, which consider the dependence of the transition dipole moment on nuclear distortions to be linear or constant, respectively; see Section 6.7. We have illustrated the Herzberg–Teller approximation in Figure 8.44 with a dashed slanted line, approximating the transition dipole moment in the region of the equilibrium geometry. For C—C bond contractions, the linear approximation is quite accurate, but for bond elongations it leads to a severe overestimation of the true transition dipole moment (which should decrease upon bond dissociation). The results presented in column D are based on the Herzberg–Teller approximation or, to be more specific, on Eq. (8.250) using a linear approximation for the transition dipole moment. It is clear that results are no longer in quantitative agreement with the reference data (column A) but a qualitative agreement is preserved for the lower modes ($f = \{0, 1, 2, 3\}$). For the higher modes, the Herzberg–Teller approximation becomes qualitatively incorrect since there are important contributions to the transition polarizabilities from the integration of nuclear geometries far away from the equilibrium. In general, the Herzberg–Teller approximation leads to overestimated transition polarizabilities due to the overestimation of the transition dipole moment at longer C—C bond distances discussed above. In column E, finally, we present results based on the Franck–Condon approximation, given in Eq. (8.254). At a first glance of Figure 8.44, it appears less than clear that this approximation makes much sense. However, in the region near the equilibrium geometry (where the linear approximation is relatively accurate) there will be a favorable error cancelation in the integration from the overestimation of the transition moment at short internuclear distances and the underestimation at long distances. Outside this region, at longer internuclear separations, the Franck–Condon approximation is a better choice than the Herzberg–Teller approximation, which is clearly seen by comparing results in columns D and E for the higher modes against the benchmark data in column A.

Obviously, models introducing additional approximations to those discussed above cannot be expected to perform any better, but a qualitatively correct and computationally inexpensive model is still very valuable for the work of assigning spectra of large molecules. The need to make accurate predictions of absolute cross sections is less urgent. In the IMDHO model, the key molecular property to be determined in the quantum chemical calculation is the excited-state gradient. In our model system, the single vibrational coordinate is the interatomic separation and we get

$$\left(\frac{\partial E}{\partial \xi_a}\right)_0 = \sqrt{\frac{\hbar}{\omega_a}}\left(\frac{\partial E}{\partial Q_a}\right)_0 = \sqrt{\frac{2\hbar}{m\omega_a}}\left(\frac{\partial E}{\partial R}\right)_{R_{\mathrm{eq}}} = -0.0136 \text{ a.u.} \tag{8.257}$$

Given this result and a value of the transition moment at the equilibrium geometry that equals 1.124 a.u., it is a straightforward task to determine the transition polarizability. The reference value for the fundamental transition is presented in column A of Table 8.13 and is $|[\alpha_{zz}(\omega)]_{01}| =$ 154.9 a.u. for a photon energy $\hbar\omega$ in vibronic resonance with $k = 3$ at 7.165 eV. We compare this reference value to the result obtained at the vibronic resonance at 7.053 eV in the IMDHO model and which is 166.9 a.u. As discussed above, the agreement cannot be quantitative due to the nature of the underlying approximations but, at the same time, it is clear that the IMDHO result is in good qualitative agreement with the reference data in the present case.

A Raman spectrum is typically thought of as a one-dimensional presentation of the scattering cross section as a function of frequency of emitted photons for a given constant wavelength of the incident laser field. Further information about the molecular system can be obtained by presenting the dependence of the cross sections not only with respect to the frequency of the emitted photons but also with respect to the frequency (or wavelength) of photons in the incident field. This gives rise to two-dimensional Raman spectroscopy for which we have till now only discussed a single plane characterized by a constant laser wavelength. Data representation becomes an issue in multidimensional spectroscopies in general, but in Raman spectroscopy there exists another clear and illustrative choice in terms of the tracking of the scattering cross section of a specific vibrational mode as a function of the laser wavelength. This refers to the visualization of a plane in our two-dimensional Raman data set characterized by a constant wavelength of emitted photons. In Figure 8.45 we give such a presentation for our model system. The three panels in the figure refer to different levels of approximation in the calculation with the top panel referring to Kramers–Heisenberg–Dirac level of theory, Eq. (8.250); the mid panel also refers to Eq. (8.250) but supplemented with Franck–Condon transition dipole moments; and the lower panel refers to the IMDHO approximation, Eq. (8.255). The general trend for the Franck–Condon approach is that it leads to an underestimated spectral intensity (as discussed above) in addition to a spectral broadening due to the underestimated transition dipole moment in the region of long bond lengths. The vibronic resonances in the excited electronic state are clearly seen in all three spectra and the inaccurate positioning of vibrational levels in the IMDHO model becomes apparent when comparing the lower panel with the others. In our example based on a system with a single vibrational mode one can, as we did above, compare transition polarizabilities obtained under conditions of vibronic resonance. This leads to a comparison of results at different laser frequencies; the comparisons discussed above were made at the vibronic resonance $k = 3$ for Kramers–Heisenberg–Dirac and Franck–Condon levels of theory with resonance $k = 1$ for the IMDHO model, depicted by the fourth and second dashed vertical lines in the respective panels of Figure 8.45. For a true polyatomic system with multiple vibrational modes, on the other hand, caution is called for in the comparison between theory and experiment. The detailed conditions of vibronic resonances may not be met in the calculation due to the errors made in the calculation of vibronic energy levels.

The vibrational progression is distinct and clear in Figure 8.45 as a consequence of the relatively long lifetime (0.1 ps) of the vibronic states; as we noted in Figure 8.40, the nuclear motion for a second-row element may complete about 5–10 periods of oscillation during this time. If Raman scattering is restricted to a shorter timescale due to either a shorter vibronic lifetime or anhamonicities causing the wavepacket not to complete several oscillations, then the imaginary damping term γ_n in Eq. (8.250) become larger. Such a situation is referred to as the *short-time approximation* since it suggests another route than the Franck–Condon approach to turn the Kramers–Heisenberg–Dirac expression into a computationally tractable form. When $\hbar\gamma_n$ becomes significantly larger than the energy separation of vibrational levels in the excited state, then it becomes reasonable to replace $\hbar\omega_{n0}$ in the denominator of Eq. (8.250) by the vertical electronic transition energy. After this step is taken (and with $\gamma_n = \gamma$), one can identify the resolution of the identity operator and remove the intermediate summation over vibrational states to arrive at Eq. (8.252), which is a technique also adopted in Placzek theory. From a computational point of view, the key quantity that needs to be determined is in this case the derivative of the electronic polarizability with respect to normal mode coordinates for the evaluation of Eq. (8.253). The fact that the optical frequency is in resonance with electronic transitions in the molecular system means that quantum chemical calculations of the polarizability must be based on a theory that includes damping. The key advantage of this Placzek-type approach to resonance Raman scattering is the (implicit) inclusion of all electronic states in the

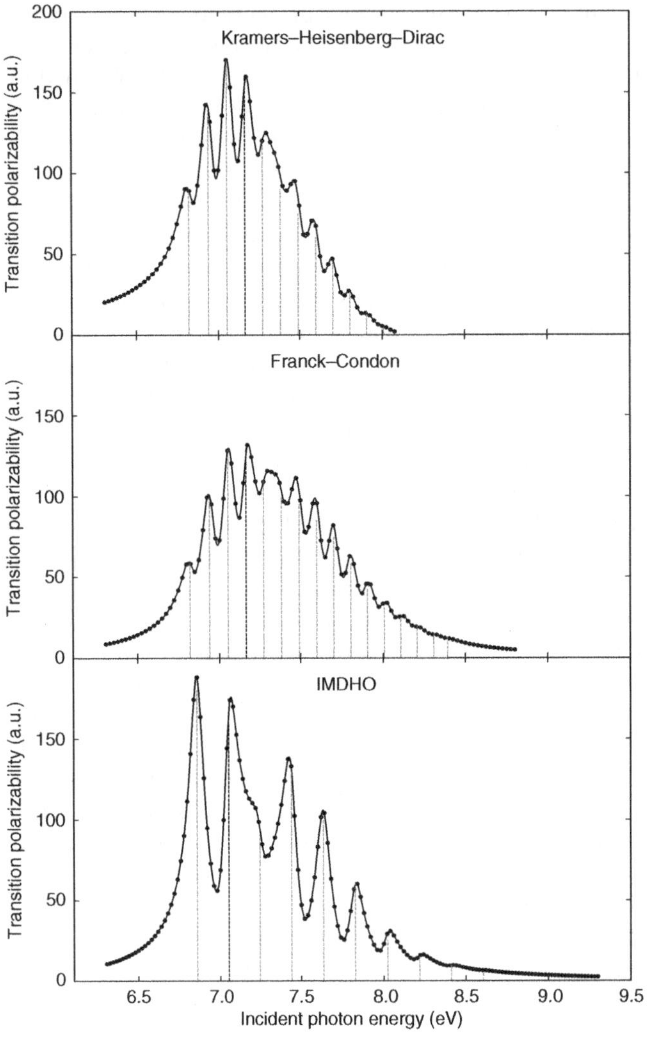

Figure 8.45 Absolute value of transition polarizability $|[\alpha_{zz}(\omega)]_{10}|$ for C_2H_4 in the region of electronic $\pi - \pi^*$ resonance. Results are obtained with use of Kramers–Heisenberg–Dirac theory, Eq. (8.250) (upper panel), the Franck–Condon approximation, Eq. (8.254) (mid panel), and the IMDHO model, Eq. (8.255) (lower panel) assuming a lifetime of 0.1 ps for the excited vibronic states. Vertical dashed lines indicate the vibronic resonances.

calculations. For small molecules of low symmetry, electronic states are typically well separated in the UV/vis region of the spectrum and this may not be an issue, but for large molecules, or molecules with three-fold (or higher) rotation symmetry, it becomes prohibitively difficult to use state-specific approaches.

With an excited-state lifetime that is an order of magnitude smaller than that previously considered, that is, $\tau = 10$ fs, there is no time to complete repeated oscillations in the excited-state potential before relaxation occurs, and the short-time approximation becomes well motivated. For our model system, transition polarizabilities for the fundamental transition $[\alpha_{zz}(\omega)]_{01}$ are determined in the harmonic approximation from Eq. (8.253) based on numerical derivatives of the complex electronic polarizability. These results are shown in Figure 8.46

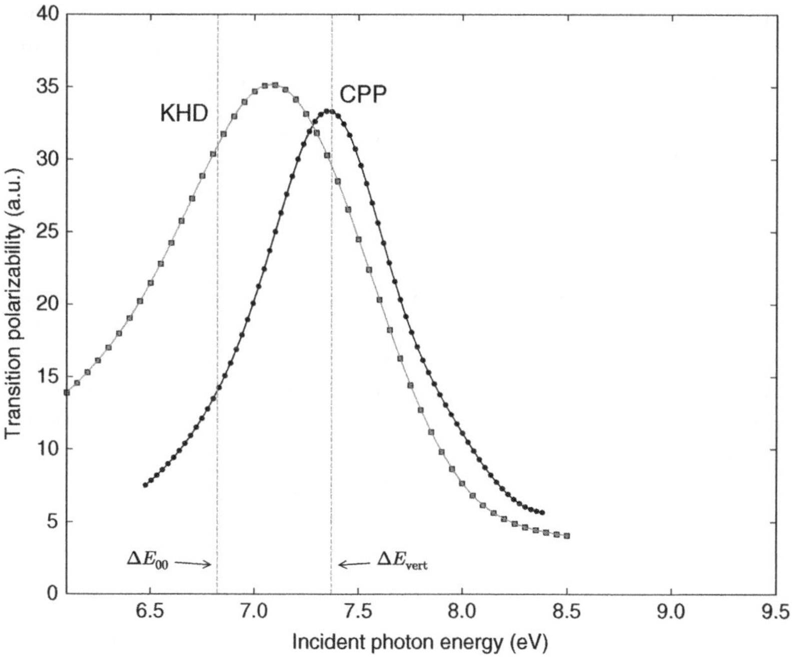

Figure 8.46 Absolute value of transition polarizability $|[\alpha_{zz}(\omega)]_{10}|$ for C_2H_4 in the region of electronic $\pi - \pi^*$ resonance. Result curves labeled by KHD and CPP are obtained with use of Kramers–Heisenberg–Dirac theory, Eq. (8.250), and the short-time approximation, Eq. (8.252), respectively, assuming a lifetime of 10 fs for the excited vibronic states. The vertical dashed lines indicate the vertical and 0–0 electronic transition energies.

(the curve labeled by "CPP") and refer to the B3LYP/taug-cc-pVTZ level of theory adopting a symmetric differentiation with a step length of $\Delta Q = 0.08543$ a.u. and a damping parameter of $\gamma = 1/\tau = 0.4136$ eV. The Raman scattering response, by construction, displays a maximum at the vertical electronic transition energy depicted by the vertical dashed line in the figure ($\Delta E_{vert} = 7.372$ eV), and the absolute value of the transition polarizability reaches a maximum of 33.3 a.u. With the same damping parameter, the transition polarizability has also been determined from the Kramers–Heisenberg–Dirac expression, Eq. (8.250), and these reference results are shown in the curve labeled by "KHD" in Figure 8.46. The KHD transition polarizability displays a maximum between the vertical and the 0–0 transition energy ($\Delta E_{00} = 6.822$ eV), reflecting the vibronic absorption spectrum presented in Table 8.13. The maximum value of the absolute transition polarizability amounts to 35.1 a.u., which serves as a benchmark for the CPP result presented above and we can conclude that the CPP result is in almost quantitative agreement in the present case with the adopted short lifetime of the excited vibronic states. The most distinct discrepancy between the CPP and KHD excitation profiles is that the profile in the CPP case is too narrow due to the absence of the nonvertical low vibronic resonances.

Time-Dependent Formulation of Raman Scattering Also after adopting the Franck–Condon approximation to arrive at Eq. (8.254) one is left with a summation over all vibrational levels of the excited electronic state, which can be an insurmountable task for polyatomic molecules, even if the IMDHO approximation is invoked. In the limit of short interaction times (Δt in illustration (iii) of Figure 8.43), we could argue that we could ignore the dependence on the vibrational levels in the denominator and thereby be able to identify a resolution of the identity operator in the numerator—this leads us to the CPP approximation. It turns out, however, that one can remove the summation over intermediate vibrational levels even without invoking this argument but by instead turning to a time-domain formulation of Eq. (8.254)

$$\begin{aligned}[\alpha_{\alpha\beta}]_{f0} &= (\mu_\alpha^{0K})_0(\mu_\beta^{K0})_0\frac{i}{\hbar}\sum_k\int_0^\infty \langle f|e^{-i\hat{H}_{\mathrm{n}}^K t/\hbar}|k\rangle\langle k|0\rangle e^{-i([\Delta E_{\mathrm{ad}}-E_0]/\hbar-\omega)t}e^{-\gamma t}dt\\ &= (\mu_\alpha^{0K})_0(\mu_\beta^{K0})_0\frac{i}{\hbar}\int_0^\infty \langle f|0(t)\rangle e^{-i([\Delta E_{\mathrm{ad}}-E_0]/\hbar-\omega)t}e^{-\gamma t}dt,\end{aligned} \tag{8.258}$$

where $\hat{H}_{\mathrm{n}}^K$ is the Hamiltonian of nuclear motion in the excited electronic state K and for which $|k\rangle$ is an eigenstate with eigenvalue E_k. The transition energy $\hbar\omega_{Kk,00}$ has been written as the sum of the adiabatic excitation energy ΔE_{ad} (see Figure 8.15) and the difference in vibrational energies $(E_k - E_0)$. After the resolution of identity has been identified as the sum over intermediate vibrational states, the time propagation operator is left to act on the ground vibrational state $|0\rangle$. As a result, the time-dependent state $|0(t)\rangle$ is obtained, which thus refers to the time propagation of the initial wave packet on the excited-state potential energy surface.

8.7.4 Vibrational Raman Optical Activity

We recall that VCD is the chiroptical analog of infrared spectroscopy, providing valuable structural information about chiral molecules and, in particular, the possibility of determining the absolute stereochemistry of a chiral molecule when comparing experimental spectra with spectra calculated using *ab initio* methods. Also for Raman spectroscopy there exists a chiroptical analog, known as Raman optical activity (ROA). In ROA, one measures the differential scattering of right- and left-circularly polarized light.[31] Considering the rather weak Raman effect, and the fact that we can expect the ROA differential scattering cross sections to be much weaker than the Raman scattering, this a property that is complicated to measure experimentally. However, in contrast to infrared spectroscopy, Raman and ROA spectra can be measured in an aqueous solution, and combined with its stereochemical information, ROA is a powerful technique to study biomolecular systems in their native environment. While large-amplitude, low-frequency vibrations associated with dynamic conformational changes may cancel over the time used to record the ROA spectrum, more static secondary structures such as protein α-helices and β-sheets give unique spectroscopic features. This allows for secondary structures of biomolecules in their native state to be studied and identified. As such, ROA is a spectroscopy that holds great promise for unraveling relationships between biomolecular structure, function, and reactivity.

It is beyond the scope of this book to discuss in detail all the possible experimental realizations of the phenomenon of ROA. Different experimental setups will lead to different observable quantities and it is therefore important that the experimental setup is considered when doing calculations of scattering cross sections. However, in all cases, the observables reduce to different linear combinations of geometry derivatives of three different linear response tensors.

In the Raman process, we have both incident and scattered light. This gives us multiple ways in which we can measure the difference in scattering of right- and left-circularly polarized

31 Note that the sign convention used for ROA differs from that used for absorption spectroscopies, where one considers the difference between left- and right-circularly polarized light. This difference in convention does not have much practical application, except in the case of resonant ROA spectra. Such spectra will have the appearance of a regular Raman spectrum but with an overall positive or negative differential scattering intensity whose sign is determined by the rotatory strength of the given excited state. Owing to the different conventions, the CD band and the resonance ROA circular intensity differences will have opposite sign.

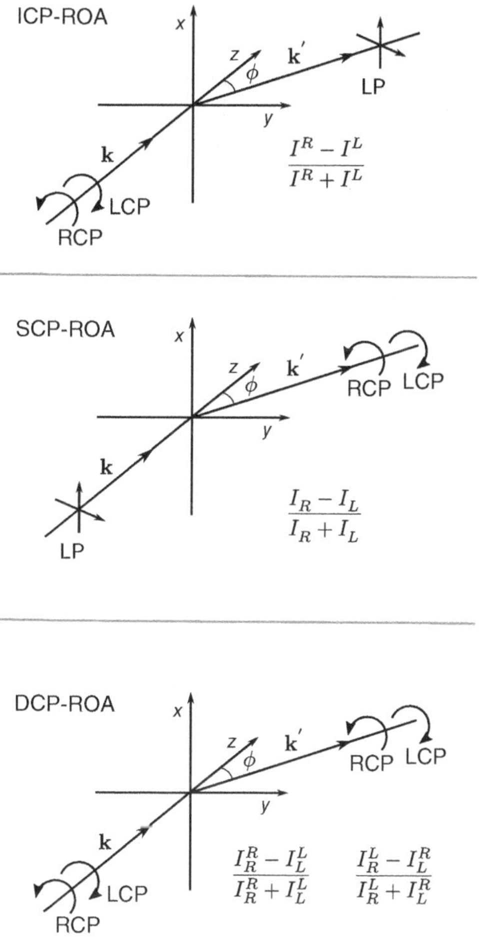

Figure 8.47 Setups for ROA measurements based on incident, scattered, and dual circular polarization (ICP, SCP, and DCP) with light incident along the *z*-axis and scattering intensities detected at an angle ϕ in the *yz*-plane. Scattering angles ϕ of 0°, 90°, and 180° correspond to forward, right-angle, and backward scattering geometries, respectively.

light, as shown in Figure 8.47. One possibility, and the one that was the first to be experimentally realized, is to modulate between right- and left-circular polarization of the incident light, and to measure the scattered light at a fixed polarization. The benefit of this approach, the so-called *incident circular polarization* (ICP) approach, is that it does not require any changes to the detectors of the scattered light and that a charge-coupled device (CCD) detector can be used. An alternative approach is to use incident linearly polarized light and to measure the difference in the right- and left-circularly polarized scattered light. This *scattered circular polarization* (SCP) approach faced for a long time experimental difficulties, but is currently the *de facto* standard for modern, commercial ROA instrumentation due to the superior quality of the resulting spectrum. Finally, we can combine the two approaches, modulating between left- and right-circularly polarized incident light, while also recording the left- and right-circularly polarized light. This is referred to as a *dual circular polarization* (DCP) setup, but it has not found much practical use.

After scattering off the electronic structure of the molecule, photons will leave in any direction. There is thus a choice to be made at what angle to place the detector relative to the direction of the incident light. The measured cross section will depend on the choice of this scattering angle, and by making a careful selection of angles, it is in principle possible to single out individual ROA invariants in the experiment. The explicit expressions for the scattering cross sections can be derived using scattering theory, but this goes beyond the scope of this book, as, from a computational point of view, it can be considered as defined by the experimental setup. In principle, any scattering angle can be used, but the three most commonly used experimental setups are for 0°, 90°, and 180°. These three choices are referred to as forward, right-angle, and back-scattering setups, respectively.

By now, we are well familiar with the general strategy that must be adopted in order to identify the contributions to the response of a molecule to right- and left-circularly polarized light, namely that we must consider contributions beyond the electric dipole approximation, including in general both the magnetic dipole and the electric quadrupole contributions, and otherwise follow the general approach described in Section 8.7.3 to derive the relevant vibrational scattering cross sections. Limiting ourselves to the dominating SCP experiments, the difference in scattering intensity of right- and left-circularly polarized light for a given vibrational mode with frequency ω_a at a scattering angle ϕ, is given by the expression

$$\begin{aligned} I_R^{\rm SCP}(\phi) - I_L^{\rm SCP}(\phi) = \frac{1}{90c}\left(\frac{\omega_a^2\mu_0}{4\pi}\right)^2 \frac{\hbar}{2\omega_a}[90aG'_a + 26\beta_{G_a}^2 - 2\beta_{A_a}^2 \\ + \cos(\phi)(180aG'_a - 20\beta_{G_a}^2 - 12\beta_{A_a}^2) \\ + \cos^2(\phi)(90aG'_a + 2\beta_{G_a}^2 + 6\beta_{A_a}^2)]. \end{aligned} \tag{8.259}$$

The resulting expressions for the most common experimental setups, both in terms of the Raman and the ROA observables, are collected in Table 8.14.

In Eq. (8.259), we have introduced a number of Raman and ROA invariants that are defined in terms of the corresponding first-order geometrical derivatives of the electric dipole polarizability, mixed electric dipole–magnetic dipole polarizability and the electric dipole–electric quadrupole polarizability

$$a_a^2 = \frac{1}{9}\left(\frac{\partial \alpha_{\alpha\alpha}}{\partial Q_a}\right)\left(\frac{\partial \alpha_{\beta\beta}}{\partial Q_a}\right), \tag{8.260}$$

$$\beta_a^2 = \frac{1}{2}\left[3\left(\frac{\partial \alpha_{\alpha\beta}}{\partial Q_a}\right)\left(\frac{\partial \alpha_{\alpha\beta}}{\partial Q_a}\right) - \left(\frac{\partial \alpha_{\alpha\alpha}}{\partial Q_a}\right)\left(\frac{\partial \alpha_{\beta\beta}}{\partial Q_a}\right)\right], \tag{8.261}$$

$$aG'_a = \frac{1}{9}\left(\frac{\partial \alpha_{\alpha\alpha}}{\partial Q_a}\right)\left(\frac{\partial G'_{\beta\beta}}{\partial Q_a}\right), \tag{8.262}$$

Table 8.14 Explicit expressions for the Raman scattering intensity and the difference in scattering of right and left circularly polarized light observed in Raman optical activity for the most commonly used experimental setups of backward-, right-angle- and forward-scattering in scattered circular polarization experiments.

Experimental setup	Raman scattering	ROA scattering differential
Forward scattering	$90a_a^2 + 14\beta_a^2$	$\frac{2}{c}[180aG'_a + 4\beta_{G_a}^2 - 4\beta_{A_a}^2]$
Backward scattering	$90a_a^2 + 14\beta_a^2$	$\frac{2}{c}[24\beta_{G_a}^2 + 8\beta_{A_a}^2]$
Polarized right-angle scattering	$90a_a^2 + 14\beta_a^2$	$\frac{2}{c}[90aG'_a + 14\beta_{G_a}^2 + 2\beta_{A_a}^2]$
Depolarized right-angle scattering	$12\beta_a^2$	$\frac{2}{c}[12\beta_{G_a}^2 - 4\beta_{A_a}^2]$

$$\beta^2_{G_a} = \frac{1}{2}\left[3\left(\frac{\partial \alpha_{\alpha\beta}}{\partial Q_a}\right)\left(\frac{\partial G'_{\alpha\beta}}{\partial Q_a}\right) - \left(\frac{\partial \alpha_{\alpha\alpha}}{\partial Q_a}\right)\left(\frac{\partial G'_{\beta\beta}}{\partial Q_a}\right)\right], \tag{8.263}$$

$$\beta^2_{A_a} = \frac{\omega}{2}\left[\left(\frac{\partial \alpha_{\alpha\beta}}{\partial Q_a}\right)\left(\frac{\varepsilon_{\alpha\gamma\delta}\partial A_{\gamma,\delta\beta}}{\partial Q_a}\right)\right], \tag{8.264}$$

where ω is the frequency of the incident laser source, G' the mixed electric dipole–magnetic dipole polarizability, Eq. (8.206), and the electric dipole–electric quadrupole polarizability is defined as

$$A_{\alpha,\beta\gamma} = \langle\langle \hat{\mu}_\alpha; \Theta^{[2]}_{\beta\gamma}\rangle\rangle_\omega, \tag{8.265}$$

where $\Theta^{[2]}_{\alpha\beta}$ is the traceless quadrupole moment, Eq. (3.227).

We note that it is also common to report the circular intensity differential (CID), which is the difference of the scattering of right- and left-circularly polarized light relative to the total scattering intensity

$$\Delta_{\text{CID}} = \frac{I_R^{\text{SCP}} - I_L^{\text{SCP}}}{I_R^{\text{SCP}} + I_L^{\text{SCP}}}, \tag{8.266}$$

which then becomes a dimensionless quantity and would in general be of the order of 10^{-3} to 10^{-4}, in accordance with the prefactor that accompanies the magnetic field and electric-field gradient components of the electromagnetic wave relative to its electric field strength. The benefit of the CID over the scattering differential, Eq. (8.259), is that it is independent of sample size and concentration, as it becomes a relative measure compared to the total scattering cross section. This is particularly useful when comparing theory and experiment, as absolute Raman scattering cross sections in general are very difficult to obtain experimentally. Nevertheless, from a computational point of view, either the difference scattering or the CID is easily obtained from the Raman and ROA invariants, and a visual comparison of experimentally observed and calculated scattering differentials is unproblematic even though the total scattering intensity is not known.

Each of the ROA invariants in Eqs. (8.260)–(8.264) is origin independent for exact states, as can be verified by shifting the gauge origin in the invariants that involve the electric dipole–magnetic dipole and electric dipole–electric quadrupole polarizabilities. In approximate state theories, gauge-origin independent calculations of the G' tensor invariants can be achieved by use of LAOs. While all ROA invariants in principle should be included in the estimation of the total difference in the scattering of right- and left-circularly polarized light, a number of studies have shown that the ROA invariant involving the electric dipole–electric quadrupole component $\beta^2_{A_a}$ often is small and thus in many cases can be ignored. This is illustrated for the ROA invariants of methyloxirane in Table 8.15 for a backscattering setup, where we see that the $\beta^2_{A_a}$ contribution is significantly smaller than the total differential scattering cross section for most modes. Exceptions do exist, however, such as the vibrational mode with a vibrational frequency of 411 cm^{-1} in methyloxirane. But since this is a rather weak band it will not be critical for the determination of the stereochemistry of the molecule.

In view of the fact that the electric dipole–electric quadrupole invariant, Eq. (8.264), in general only gives a minor contribution to the differential scattering of right- and left-circularly polarized light, the key to an accurate calculation of ROA spectra is, in addition to the quality of the vibrational force field itself, the quality of the mixed electric dipole–magnetic dipole polarizability tensor, which contributes to the two invariants in Eqs. (8.262) and (8.263). We recall from our discussion of the optical rotation tensor in Section 8.6.1 that a key to the reliable calculation of G' is the inclusion of diffuse polarizing functions in order to model the chirality

Table 8.15 Contributions of individual ROA tensor invariants in (S)-methyloxirane for a backscattering SCP experiment.

ω_a	$45(a_a^2)(aG_a')$	$7\beta^2_{G_a'}$	$\beta^2_{A_a}$	Contribution (%)
3,155	1,127	2,234	352	9
3,107	−1,114	4,063	439	13
3,084	525	455	39	4
3,082	−1,680	−11,148	−1,037	7
3,070	1,306	5,408	23	0
3,024	−316	−443	27	−4
1,531	−8	−618	−41	6
1,497	−23	2,519	212	8
1,482	0	−989	−32	3
1,439	−244	−710	−38	4
1,404	11	156	−24	−17
1,293	−181	293	−13	−13
1,183	161	−1,049	−71	7
1,161	−32	838	65	7
1,146	7	193	−3	−2
1,128	−41	207	1	1
1,039	17	206	45	17
972	117	184	17	5
905	5	−1,114	−137	11
844	−37	−1,259	−31	2
772	128	848	43	4
411	−75	40	10	−40
366	−11	−211	−7	3
207	−6	32	2	7

The last column indicates the percentage contribution from the mixed electric dipole–electric quadrupole invariant. All results in atomic units.

induced in the electronic density by the circular polarization of the light. As ROA is in most cases used on organic or biomolecular systems for which hydrogens often reside at the perimeter of the molecular structure, one can often get surprisingly good ROA spectra using minimal basis sets, as long as diffuse *p* functions are included on the hydrogens, of course transforming the polarizability derivatives from a Cartesian form to normal coordinates using a force field of sufficiently high quality.

We present in Figure 8.48 the Raman scattering and optical activity spectrum of (*S*)-methyloxirane. As for VCD, the ROA spectrum provides additional information about the stereochemistry of the molecule, and although not explicitly reported in the figure, the ROA spectrum of (*R*)-methyloxirane would of course be the mirror image of the ROA spectrum in Figure 8.48, whereas the Raman spectrum would remain unchanged. In many respects, when comparing the VCD spectrum of *S*-methyloxirane in Figure 8.38 with the corresponding ROA

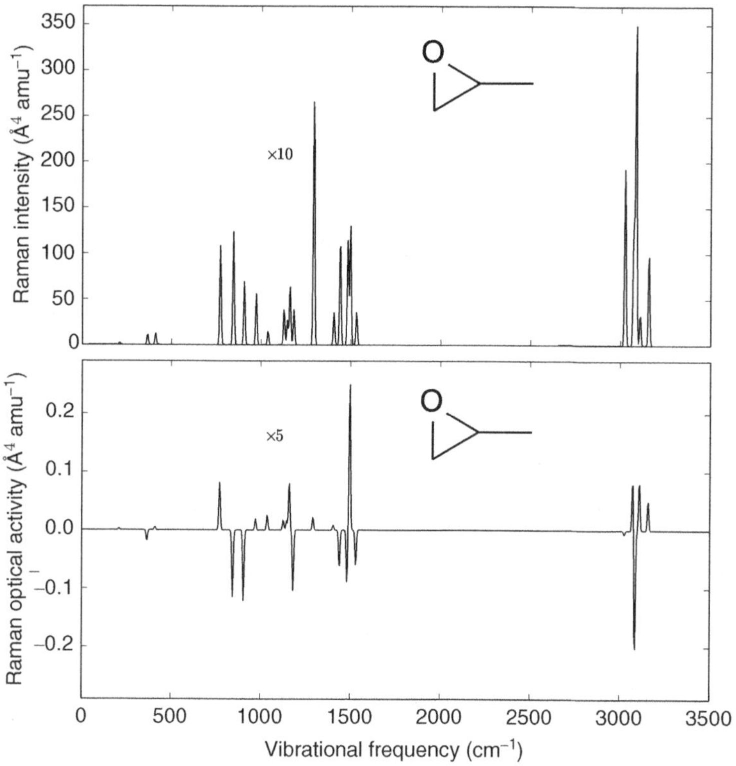

Figure 8.48 Raman scattering intensities and differential Raman optical activity scattering intensities for (*S*)-methyloxirane for a backscattering experiment. The intensities and optical activities in the low-frequency region are scaled by factors of 10.0 and 5.0, respectively.

spectrum in Figure 8.48, it is clear that the information content of VCD and ROA spectroscopy is much the same: They both provide insight into the vibrational modes of a molecule and provide in addition a signed rotational strength or differential scattering cross section for each of these vibrational transitions. The main difference between the two experimental techniques in terms of the information we can extract about the molecular structure is primarily a consequence of differences in the intensity of individual vibrational bands, including here possible differences in relative sign between different modes. Indeed, there is no correlation between the signs or the relative signs of VCD and ROA bands; this may help decongest either spectrum, and together the two spectroscopies may provide the necessary information to disentangle structural information from the experimental spectra. This is particularly important for flexible molecules with a large number of accessible conformers and potentially multiple stereocenters, for which the VCD or ROA spectrum alone may not be sufficient to determine the absolute stereochemistry of the molecule. An additional experimental difference lies in the fact that ROA in general can be recorded from the very low-frequency modes that are characteristic of weak inter- and intramolecular bonds, such as hydrogen bonds, to the high-frequency modes corresponding to aliphatic CH stretching modes. In contrast, VCD spectroscopy is in general limited to the frequency range 200–1600 cm^{-1}, at least within regular experimental setups. Beyond these experimental differences, the analysis of ROA spectra follow largely the same principles as those that apply to VCD.

8.8 Nuclear Magnetic Resonance

8.8.1 The NMR Experiment

I remember, in the winter of our first experiments,[...],looking on snow with new eyes. There the snow lay around my doorstep — great heaps of protons quietly precessing in the earth's magnetic field.

Edward Mills Purcell Nobel lecture (1952)

In Section 8.1.2 we learned that some nuclei have spins and that corresponding nuclear magnetic dipole moments manifest themselves through hyperfine splittings of optical spectra. However, a more direct and accurate way of probing nuclear magnetic dipole moments became available with the advent of the NMR experiment. In the NMR experiment, the sample is placed in a magnetic field and transitions between levels split by the nuclear spin Zeeman interaction are induced by radiation in the radiofrequency domain. The NMR experiment was conceived by physicists to determine nuclear magnetic dipole moments, but soon it was realized that the NMR signal showed great sensitivity to the chemical environment of the nuclei. Today NMR is one of the most powerful tools of chemistry, with a wide range of applications, as illustrated in Figure 8.49. The by far most important nucleus in NMR spectroscopy is the proton ^1H.

In the following, the NMR experiment is described from a time-dependent perspective. It provides a nice illustration of concepts introduced in previous chapters of the book. We start by considering the effect of the applied magnetic field. Specifically, we place a nucleus with magnetic moment $\mathbf{m} = \gamma \mathbf{I}$ in a static magnetic field $\mathbf{B}_0 = B_0 \mathbf{n}$. The interaction spin Hamiltonian

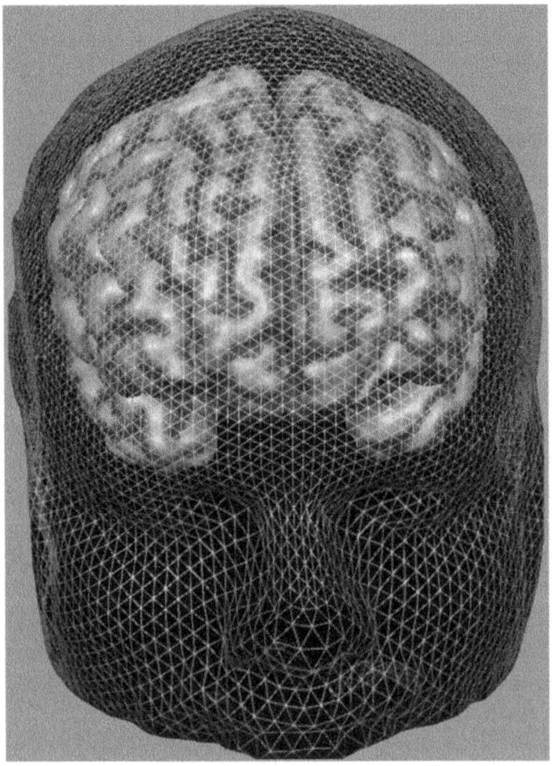

Figure 8.49 The power of nuclear magnetic resonance is illustrated by magnetic resonance imaging, here of the brain of one of the authors (TS).

is given by

$$\hat{H}_0 = -\hat{\mathbf{m}} \cdot \mathbf{B}_0 = -\gamma \hat{\mathbf{I}} \cdot \mathbf{B}_0. \tag{8.267}$$

From Eqs. (2.68) and (2.75), it follows that the time evolution of the nuclear spin function is

$$\psi(t) = e^{-i\hat{H}_0 t/\hbar}\psi(0) = e^{+i\gamma B_0 t(\mathbf{n}\cdot\hat{\mathbf{I}})}\psi(0). \tag{8.268}$$

From Eq. (4.71) it follows that we can rewrite the nuclear spin function as

$$\psi(t) = \hat{\mathcal{R}}^I(-\gamma B_0 t, \mathbf{n})\psi(0), \tag{8.269}$$

where appears the rotation operator $\hat{\mathcal{R}}^I(\phi, \mathbf{n})$ associated with nuclear spin I corresponding to a rotation of the nuclear spin about the magnetic field vector $\mathbf{n}$ with angular frequency $\omega_0 = -\gamma B_0$.

Another way to arrive at the same conclusion is to employ the Ehrenfest theorem, Eq. (5.117), according to which the time-evolution of component i of the nuclear spin $\mathbf{I}$ is

$$\frac{d\langle \hat{I}_i \rangle}{dt} = -\frac{i}{\hbar}\langle [\hat{I}_i, H_0] \rangle = +i\frac{\gamma}{\hbar}\langle [\hat{I}_i, \hat{I}_j] \rangle B_{0;j} = -\gamma \epsilon_{ijk} \langle \hat{I}_k \rangle B_{0;j}. \tag{8.270}$$

We can rearrange the above expression to

$$\frac{d\mathbf{I}}{dt} = \mathbf{I} \times \mathbf{B}_0, \tag{8.271}$$

which again shows the nuclear spin precessing about the magnetic field vector. We refer to this as *Larmor precession*, and the Larmor frequency is $\omega_0 = -\gamma B_0$. Most nuclei have *positive* γ such that the Larmor precession is *clockwise*, as shown in Figure 8.50. Without loss of generality, we can choose our axis system such that the magnetic field is along the z-axis $\mathbf{B}_0 = B_0 \mathbf{e}_z$. We can then write the time evolution of the nuclear spin function as

$$\psi(t) = \hat{\mathcal{R}}^I_z(\omega_0 t)\psi(0). \tag{8.272}$$

The energy of each nuclear level is

$$E(I, m_I) = \langle I, m_I | -\gamma \hat{I}_z B | I, m_I \rangle = -\gamma \hbar m_I B. \tag{8.273}$$

A positive gyromagnetic ratio γ implies that the nuclear levels are split in the energy order $\{+m_I, \ldots, -m_I\}$ with $+m_I$ starting from below, whereas a negative γ will invert this ordering.

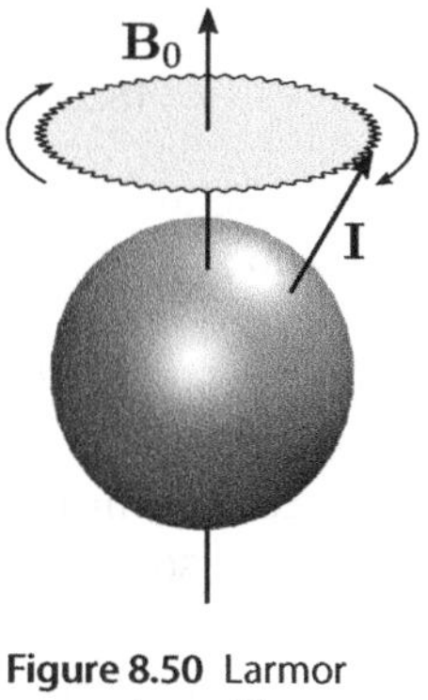

Figure 8.50 Larmor precession with a radiofrequency probe adding nutation.

The system is now probed by a weak radiofrequency pulse perpendicular to the magnetic field (we choose the x-axis)

$$\mathbf{B}_1 = B_1 \mathbf{e}_x \cos(\Phi_p); \quad \Phi_p(t) = \omega_{\mathrm{ref}} t. \tag{8.274}$$

In Section 3.1.5 we learned that a linearly polarized wave can be decomposed into two circularly polarized ones, Eq. (3.89), and this will be useful in the present case. We therefore rewrite Eq. (8.274) as

$$\begin{aligned}\mathbf{B}_1 = \mathbf{B}_1^+ + \mathbf{B}_1^-; \quad \mathbf{B}_1^\pm &= \frac{1}{2}B_1 \mathrm{Re}\{\mathbf{e}_\pm \exp(i\Phi_p)\} \\ &= \frac{1}{4}B_1\{\mathbf{e}_\pm \exp(i\Phi_p) + \mathbf{e}_\mp \exp(-i\Phi_p)\},\end{aligned} \tag{8.275}$$

in terms of the complex unit vectors of Eq. (3.84). The right-handed component $\mathbf{B}_1^-$ describes circular polarization following the Larmor precession, whereas the left-handed component $\mathbf{B}_1^+$ describes circular polarization in the opposite sense and can in most situations be discarded, as we will do in the following.[32] The Hamiltonian describing the radio pulse is therefore approximated as

$$\hat{H}_1^- = -\gamma\hat{\mathbf{I}}\cdot\mathbf{B}_1^- = -\frac{1}{4}\gamma B_1(\hat{I}_-e^{+i\Phi_p} + \hat{I}_+e^{-i\Phi_p});\quad \hat{I}_\pm = \hat{I}_x \pm i\hat{I}_y. \tag{8.276}$$

In order to simplify the math, we shall switch to a rotating frame that follows $\mathbf{B}_1^-$. In this frame the nuclear spin function is

$$\psi_\mathrm{R}(t) = \hat{\mathcal{R}}_z^I(-\Phi)\psi(t);\quad \Phi(t) = \omega_\mathrm{ref}t + \phi_\mathrm{ref}, \tag{8.277}$$

where ϕ_ref is a phase constant. In the original static frame, the time-dependent Schrödinger equation is

$$\hat{H}\hat{\mathcal{R}}_z^I(\Phi)\psi_\mathrm{R}(t) = i\hbar\frac{d}{dt}\hat{\mathcal{R}}_z^I(\Phi)\psi_\mathrm{R}(t) = \hat{\mathcal{R}}_z^I(\Phi)\left(\omega_\mathrm{ref}\hat{I}_z + i\hbar\frac{d}{dt}\right)\psi_\mathrm{R}(t). \tag{8.278}$$

Premultiplication with $\hat{\mathcal{R}}_z^I(-\Phi)$ gives the corresponding equation in the rotating frame

$$\hat{H}_R\psi_\mathrm{R}(t) = i\hbar\frac{d}{dt}\psi_\mathrm{R}(t), \tag{8.279}$$

where we have introduced the rotating-frame spin Hamiltonian

$$\hat{H}_R = \hat{\mathcal{R}}_z^I(-\Phi)\hat{H}\hat{\mathcal{R}}_z^I(\Phi) - \omega_\mathrm{ref}\hat{I}_z. \tag{8.280}$$

To complete our derivation we need to find

$$\hat{\mathcal{R}}_z^I(-\Phi)(\hat{H}_0 + \hat{H}_1^-)\hat{\mathcal{R}}_z^I(\Phi) = \hat{\mathcal{R}}_z^I(-\Phi)\left(\omega_0\hat{I}_z - \frac{1}{4}\gamma B_1(\hat{I}_-e^{+i\Phi_p} + \hat{I}_+e^{-i\Phi_p})\right)\hat{\mathcal{R}}_z^I(\Phi), \tag{8.281}$$

that is, we need to transform the spin operators to the rotating frame. This can be done using the Baker–Campbell–Hausdorff (BCH) expansion, see Eq. (E.6), and the commutation rules for spin operators, in particular

$$[\hat{I}_z, \hat{I}_\pm] = \pm\hbar\hat{I}_\pm. \tag{8.282}$$

We then get

$$\hat{\mathcal{R}}_z^I(-\Phi)\hat{I}_z\hat{\mathcal{R}}_z^I(\Phi) = \hat{I}_z, \tag{8.283}$$

$$\hat{\mathcal{R}}_z^I(-\Phi)\hat{I}_\pm\hat{\mathcal{R}}_z^I(\Phi) = \sum_{n=0}^{\infty}\frac{1}{n!}(\pm i\Phi)^n\hat{I}_\pm = \exp[\pm i(\omega_\mathrm{ref}t + \phi_\mathrm{ref})]\hat{I}_\pm. \tag{8.284}$$

This allows us to express the rotating-frame spin Hamiltonian as

$$\hat{H}_R = (\omega_0 - \omega_\mathrm{ref})\hat{I}_z - \frac{1}{4}\gamma B_1(\hat{I}_-e^{-i\phi_\mathrm{ref}} + \hat{I}_+e^{+i\phi_\mathrm{ref}}). \tag{8.285}$$

We can now employ our freedom in choosing the phase constant ϕ_ref to bring the total spin Hamiltonian into identical form for both positive and negative gyromagnetic ratio γ:

$$\hat{H}_R = (\omega_0 - \omega_\mathrm{ref})\hat{I}_z + \omega_\mathrm{nut}\hat{I}_x;\quad \phi_\mathrm{ref} = \begin{cases}\pi \text{ for } \gamma > 0\\ 0 \text{ for } \gamma < 0\end{cases}, \tag{8.286}$$

where we have introduced the *nutation frequency* $\omega_\mathrm{nut} = \frac{1}{2}|\gamma|B_1$. The presence of $\hat{I}_x$ in the Hamiltonian adds a *nutation* to the precession generated by $\hat{I}_z$, as shown in Figure 8.50.

32 The validity of this so-called *rotating wave approximation* has the interesting consequence that the sense of the Larmor precession and thus the sign of the gyromagnetic factor can be determined experimentally using a circularly polarized oscillating magnetic field. This method can also be applied in EPR spectroscopy to determine the sign of the determinant of the g-matrix, which in a principal axis system corresponds to the triple product $g_{xx}g_{yy}g_{zz}$.

Let us now see the rotating-frame spin Hamiltonian, Eq. (8.286), in action. To do so, we will introduce some notation. The usual spin functions α and β refer to spin quantization along the z-axis, so we shall emphasize this through the notation

$$|+\tilde{z}\rangle = |\alpha\rangle = \begin{bmatrix} 1 \\ 0 \end{bmatrix} \quad \text{and} \quad |-\tilde{z}\rangle = |\beta\rangle = \begin{bmatrix} 0 \\ 1 \end{bmatrix}. \tag{8.287}$$

We can now generate eigenfunctions of the spin operators $\hat{s}_x$ and $\hat{s}_y$, corresponding to spin quantization axes x and y, respectively, by intuitive geometric operations using the spin-$\frac{1}{2}$ rotation operator, Eq. (4.76), which we encountered in Chapter 4, giving

$$\begin{aligned}
|+\tilde{x}\rangle &= \hat{\mathcal{R}}_y^{\frac{1}{2}}\left(\tfrac{\pi}{2}\right)|+\tilde{z}\rangle = \tfrac{1}{\sqrt{2}}\begin{bmatrix} 1 & -1 \\ 1 & 1 \end{bmatrix}\begin{bmatrix} 1 \\ 0 \end{bmatrix} = \tfrac{1}{\sqrt{2}}\begin{bmatrix} 1 \\ 1 \end{bmatrix}, \\
|-\tilde{x}\rangle &= \hat{\mathcal{R}}_y^{\frac{1}{2}}\left(-\tfrac{\pi}{2}\right)|+\tilde{z}\rangle = \tfrac{1}{\sqrt{2}}\begin{bmatrix} 1 & 1 \\ -1 & 1 \end{bmatrix}\begin{bmatrix} 1 \\ 0 \end{bmatrix} = \tfrac{1}{\sqrt{2}}\begin{bmatrix} 1 \\ -1 \end{bmatrix}, \\
|+\tilde{y}\rangle &= \hat{\mathcal{R}}_x^{\frac{1}{2}}\left(-\tfrac{\pi}{2}\right)|+\tilde{z}\rangle = \tfrac{1}{\sqrt{2}}\begin{bmatrix} 1 & i \\ i & 1 \end{bmatrix}\begin{bmatrix} 1 \\ 0 \end{bmatrix} = \tfrac{1}{\sqrt{2}}\begin{bmatrix} 1 \\ i \end{bmatrix}, \\
|-\tilde{y}\rangle &= \hat{\mathcal{R}}_x^{\frac{1}{2}}\left(\tfrac{\pi}{2}\right)|+\tilde{z}\rangle = \tfrac{1}{\sqrt{2}}\begin{bmatrix} 1 & -i \\ -i & 1 \end{bmatrix}\begin{bmatrix} 1 \\ 0 \end{bmatrix} = \tfrac{1}{\sqrt{2}}\begin{bmatrix} 1 \\ -i \end{bmatrix}.
\end{aligned} \tag{8.288}$$

We leave it to the reader to verify that $|\pm\tilde{x}\rangle$ and $|\pm\tilde{y}\rangle$ are indeed eigenfunctions of $\hat{s}_x$ and $\hat{s}_y$, respectively.

Consider now the application of a radiofrequency pulse at resonance, that is, when $\omega_{\text{ref}} = \omega_0$. The spin Hamiltonian in the rotating frame, Eq. (8.286), then simplifies to

$$\hat{H}_R = \omega_{\text{nut}}\hat{I}_x = \frac{1}{2}\hbar\omega_{\text{nut}}\sigma_x, \tag{8.289}$$

and the time development of the spin function in the rotating frame is

$$\psi_{\text{R}}(t) = e^{-i\hat{H}_{\text{R}}t/\hbar}\psi_{\text{R}}(0) = e^{-i\frac{1}{2}\omega_{\text{nut}}\sigma_x t}\psi_{\text{R}}(0) = \hat{\mathcal{R}}_x^{\frac{1}{2}}(\omega_{\text{nut}}t)\psi_{\text{R}}(0). \tag{8.290}$$

Let us choose the spin function at $t = 0$ to be $\psi_{\text{R}}(0) = |+\tilde{z}\rangle$. We then get

$$\psi_{\text{R}}(t) = \hat{\mathcal{R}}_x^{\frac{1}{2}}(\omega_{\text{nut}}t)|+\tilde{z}\rangle = \cos\left(\frac{1}{2}\omega_{\text{nut}}t\right)|+\tilde{z}\rangle - i\sin\left(\frac{1}{2}\omega_{\text{nut}}t\right)|-\tilde{z}\rangle. \tag{8.291}$$

We see that the spin function cycles between states $|\pm\tilde{z}\rangle$, and so these are Rabi oscillations of frequency $\frac{1}{2}\omega_{\text{nut}} = \frac{1}{4}B_1|\gamma|$. We encountered Rabi oscillations already in Section 5.8; see Figure 5.16 for a two-level atom subject to an electric field with frequency in the vicinity of resonance. A nonperturbative solution for that system is given in Appendix I, starting from Eq. (I.1) with $F(t) = F^{\omega}\sin\omega t$. In the present case, the time-dependent Schrödinger equation in the initial frame reads

$$\begin{bmatrix} +\frac{1}{2}\hbar\omega_0 & -\frac{1}{2}B_1\gamma\cos(\Phi_p) \\ -\frac{1}{2}B_1\gamma\cos(\Phi_p) & -\frac{1}{2}\hbar\omega_0 \end{bmatrix}\begin{bmatrix} c_\alpha(t) \\ c_\beta(t) \end{bmatrix} = i\hbar\begin{bmatrix} \dot{c}_\alpha(t) \\ \dot{c}_\beta(t) \end{bmatrix}, \tag{8.292}$$

and so we see that our derivation so far provides another route to the solution of this problem, highlighting different aspects of the underlying physics.

The Rabi cycles occur upon continuous radiation at resonance. In general, however, the radiofrequency pulse is chosen to have a specific duration τ_p, which allows the selection of the final state

$$\psi_R(\tau_p) = \hat{\mathcal{R}}_x^{\frac{1}{2}}(\beta_p)|+\tilde{z}\rangle = \cos\frac{1}{2}\beta_p|+\tilde{z}\rangle - i\sin\frac{1}{2}\beta_p|-\tilde{z}\rangle, \tag{8.293}$$

where we have introduced the *flip angle* $\beta_p = \omega_{\text{nut}}\tau_p$. Two examples are

$$\begin{array}{lll} \beta_p = \frac{\pi}{2}: & \Rightarrow \psi_R(\tau_p) = & \frac{1}{\sqrt{2}}|+\tilde{z}\rangle - \frac{i}{\sqrt{2}}|-\tilde{z}\rangle = |-\tilde{y}\rangle \\ \beta_p = \pi: & \Rightarrow \psi_R(\tau_p) = & -i|-\tilde{z}\rangle \end{array}. \tag{8.294}$$

These are referred to as $(\pi/2)_x$- and $(\pi)_x$- pulses, respectively, and are seen to have a straightforward geometrical interpretation.

So far our equations involve a single nuclear spin, whereas the NMR experiment operates on a manifold of spins. We can go to a macroscopic picture by replacing nuclear spin **I** in Eq. (8.271) by nuclear magnetization **M**. We then get

$$\frac{d\mathbf{M}(t)}{dt} = \gamma\mathbf{M}(t) \times \mathbf{B}_0 \tag{8.295}$$

The phenomenological *Bloch equations* are obtained by including relaxation

$$\begin{aligned} \frac{d\mathbf{M}_x(t)}{dt} &= \gamma(\mathbf{M}(t) \times \mathbf{B}_0)_x - \frac{M_x(t)}{T_2}, \\ \frac{d\mathbf{M}_y(t)}{dt} &= \gamma(\mathbf{M}(t) \times \mathbf{B}_0)_y - \frac{M_y(t)}{T_2}, \\ \frac{d\mathbf{M}_z(t)}{dt} &= \gamma(\mathbf{M}(t) \times \mathbf{B}_0)_z - \frac{M_z(t) - M_{\text{eq}}}{T_1}. \end{aligned}$$

They describe an experiment in which the sample is placed in a strong magnetic field, which induces a *longitudinal* magnetization

$$M_z(t) = M_{\text{eq}}(1 - \exp\{-t/T_1\}), \tag{8.296}$$

where T_1 is the longitudinal (or spin-lattice) relaxation time constant. *Transverse* nuclear magnetization is induced by a weak radiofrequency pulse at resonance ω_0

$$\begin{aligned} M_x(t) &= M_{\text{eq}}\sin(\omega_0 t)\exp\{-t/T_2\}, \\ M_y(t) &= -M_{\text{eq}}\cos(\omega_0 t)\exp\{-t/T_2\}, \end{aligned}$$

where T_2 is the transverse relaxation time constant.

Originally, the NMR experiment was operated under continuous-wave mode, in which either the magnetic field $\mathbf{B}_0$ was held constant and the radiofrequency signal swept to determine resonances (frequency-sweep method), or the radiofrequency signal held constant and the magnetic field $\mathbf{B}_0$ adjusted to bring the energy splittings into resonance (field-sweep method). Nowadays one rather employs Fourier-transform NMR, where the magnetic field B_0 is held fixed and a short (1–10 μs) square radiofrequency pulse is applied. The NMR signal that is detected is the oscillating electric current, the *free-induction decay* (FID), induced by the precessing nuclear transverse magnetization, which is Fourier transformed to give the final spectrum.

8.8.2 NMR Parameters

The NMR spectrum can be described by a phenomenological Hamiltonian

$$\hat{\mathcal{H}}^{\text{NMR}} = -\sum_K \hat{\mathbf{m}}_K \cdot (I_3 - \sigma_K)\mathbf{B}_0 + \frac{1}{2}\sum_{K \neq L} \hat{\mathbf{m}}_K \cdot (D_{KL} + K_{KL})\hat{\mathbf{m}}_L, \tag{8.297}$$

where summations run over nuclei. The chemical interest of the NMR experiment arises from the *shielding tensor* σ_K as well as the reduced *indirect nuclear spin–spin coupling tensor* K_{KL}. The first term of the above Hamiltonian describes the nuclear Zeeman interaction

$$\hat{\mathcal{H}}_Z^{\text{NMR}} = -\sum_K \hat{\mathbf{m}}_K \cdot \mathbf{B}_0^{\text{loc};K}, \tag{8.298}$$

but in terms of local magnetic fields $\mathbf{B}_0^{\text{loc};K}$ rather than the externally applied one. The local magnetic field can to a very good approximation be assumed linear in the applied external field

$$\mathbf{B}_0^{\text{loc};K} = (I_3 - \sigma_K)\mathbf{B}_0; \quad \sigma_K = \begin{bmatrix} \sigma_{K;xx} & \sigma_{K;xy} & \sigma_{K;xz} \\ \sigma_{K;yx} & \sigma_{K;yy} & \sigma_{K;yz} \\ \sigma_{K;zx} & \sigma_{K;zy} & \sigma_{K;zz} \end{bmatrix}, \tag{8.299}$$

where the shielding tensor captures the modification of the applied field due to the electronic environment as experienced by the nuclear magnetic moment. This modification is small and the shielding is generally reported in parts-per-million (ppm). The second term of Eq. (8.297) describes the interaction of the nuclear magnetic dipole moments. The direct coupling assumes the form

$$\hat{\mathcal{H}}_{\text{DD}}^{\text{NMR}} = \frac{1}{2}\sum_{K \neq L} \hat{\mathbf{m}}_K \cdot D_{KL}\hat{\mathbf{m}}_L = -\frac{1}{2}\sum_{K \neq L} \left\{ \frac{3(\hat{\mathbf{m}}_K \cdot \mathbf{n}_{KL})(\hat{\mathbf{m}}_L \cdot \mathbf{n}_{KL}) - \hat{\mathbf{m}}_K \cdot \hat{\mathbf{m}}_L}{4\pi\varepsilon_0 c^2 R_{KL}^3} \right\}, \tag{8.300}$$

where $\mathbf{R}_{KL} = R_{KL}\mathbf{n}_{KL}$ and we have coupled the nuclear magnetic dipole $\mathbf{m}_K$ with the magnetic field, Eq. (3.204), associated with a (point) magnetic dipole $\mathbf{m}_L$. The contact term can safely be ignored since $R_{KL} > 0$.

The indirect nuclear magnetic dipole–dipole coupling describes the modification of the "bare" coupling due to the electronic environment. What is observed is the indirect spin–spin coupling tensor J_{KL}, defined by

$$2\pi J_{KL} = \gamma_K \gamma_L K_{KL}. \tag{8.301}$$

For the study of theoretical trends it is better to use the reduced tensor K_{KL}, though, since it does not contain the nuclear gyromagnetic ratios γ_K that varies in a somewhat unpredictable manner amongst nuclear isotopes.

In an isotropic liquid, which will be the situation considered in the following, the direct dipole–dipole interaction averages to zero, whereas only the isotropic part of the shielding and indirect nuclear spin–spin coupling tensors are retained

$$X^{\text{iso}} = \frac{1}{3}(X_{xx} + X_{yy} + X_{zz}); \quad X = \sigma_K, J_{KL}, \tag{8.302}$$

according to Eq. (3.474), and are referred to as the shielding and indirect nuclear spin–spin coupling constants, respectively. The phenomenological Hamiltonian then reduces to

$$\hat{\mathcal{H}}_{\text{iso}}^{\text{NMR}} = -\sum_K (1 - \sigma_K^{\text{iso}})\hat{m}_{K;z}B_0 + \frac{1}{2}\sum_{K \neq L} K_{KL}^{\text{iso}}(\hat{\mathbf{m}}_K \cdot \hat{\mathbf{m}}_L). \tag{8.303}$$

To understand the form of the first term, one should keep in mind that we are now dealing with an isotropic system and so we let the magnetic field define the z-axis of our system, just as we do in the atomic case. If we now restrict attention to a single nucleus in an isotropic environment, the energy of a single nuclear level is

$$E_K(I,M_I) = -\gamma_K\langle I,M_I|(1-\sigma_K^{\text{iso}})\hat{I}_z B_0|I,M_I\rangle = -\hbar\gamma_K(1-\sigma_K^{\text{iso}})M_I B_0. \tag{8.304}$$

Within the electric dipole approximation, only transitions between neighboring levels are allowed, and the transition energy will be

$$\Delta E_K = h\nu_K = \hbar\gamma_K(1-\sigma_K^{\text{iso}})B_0. \tag{8.305}$$

We already pointed out that physicists devised the NMR experiment for the direct measurement of nuclear magnetic dipole moment, and to them the shielding constant σ_K^{iso} is clearly a nuisance, because it precludes such direct measurement. One option to determine nuclear magnetic dipole moments is to combine the NMR experiment with accurate calculation of the shielding constants. For the chemist, NMR shielding is rather a blessing, but the direct experimental measurement of shielding constants is only possible to the extent that accurate values of nuclear magnetic dipole moments are available, and also to the extent that the magnetic field applied to the sample can be determined accurately. What is usually reported is the *chemical shift* δ_K, which is defined as the relative resonance frequency shift with respect to a reference compound, that is,

$$\delta_K = \frac{\nu_K - \nu_{K;\text{ref}}}{\nu_{K;\text{ref}}} = \frac{\sigma_{K;\text{ref}} - \sigma_K}{1-\sigma_{K;\text{ref}}} \approx \sigma_{K;\text{ref}} - \sigma_K, \tag{8.306}$$

where we use Eq. (8.305) to go from frequencies ν_K to absolute shieldings σ_K. From the above formula one can see that the sign of the chemical shift and the shielding constant differ, and an increase in the chemical shift suggests the nucleus of interest is deshielded relative to the reference nucleus and experiences a stronger local field. In general, the more electronegative the element is, the smaller its nuclear shielding, but other factors, such as ring currents and bond strain, may also play a role.

Of prime interest to the chemist is the shielding tensor σ_K and the reduced indirect spin–spin coupling tensor K_{KL}. These NMR parameters can be calculated by mapping the phenomenological NMR Hamiltonian, Eq. (8.297), onto the expression for the electronic energy in the presence of nuclear spins and uniform magnetic fields, that is,

$$\sigma_{K;\mu\nu} = \left.\frac{d^2E}{dm_{K;\mu}dB_{0;\nu}}\right|_0 + \delta_{\mu\nu},$$

$$K_{KL:\mu\nu} = \left.\frac{d^2E}{dm_{K;\mu}dm_{L;\nu}}\right|_0,$$

where the derivatives are taken at $\mathbf{B}_0 = \{\mathbf{m}_K\} = \mathbf{0}$. The vector potentials associated with the nuclear spins and the external magnetic field are given by Eqs. (3.202) and (3.212), respectively. For closed-shell molecules, the nonrelativistic expression for elements of the shielding tensor is given by

$$\sigma_{K;\mu\nu} = \frac{\partial^2}{\partial m_{K;\mu}\partial B_{0;\nu}}[\langle\langle \hat{h}^{\text{OZ}}; \hat{h}_K^{\text{pso}}\rangle\rangle_0 + \langle 0|\hat{h}_{BK}^{\text{dia}}|0\rangle]. \tag{8.307}$$

The paramagnetic contribution is given by a linear response function coupling the orbital Zeeman operator, Eq. (8.81), with the paramagnetic hyperfine spin–orbit operator, Eq. (8.42), whereas the diamagnetic contribution is the expectation value of the operator

$$\hat{h}_{BK}^{\text{dia}} = \frac{e}{4\pi\varepsilon_0}\frac{1}{2mc^2}\left[\frac{(\mathbf{B}_0\cdot\mathbf{m}_K)(\mathbf{r}_{iG}\cdot\mathbf{r}_{iK}) - (\mathbf{B}_0\cdot\mathbf{r}_{iK})(\mathbf{r}_{iG}\cdot\mathbf{m}_K)}{r_{iK}^3}\right]. \tag{8.308}$$

The diamagnetic contribution is found to dominate in many cases, but this is less systematic than for magnetizabilities. Thus, in certain instances the shielding constant may be negative corresponding to deshielding, and the nuclear magnetic moment experiences a local field that is stronger than the externally applied magnetic field. For nonrelativistic atoms, however, the paramagnetic contribution vanishes.

The corresponding relativistic expression for elements of the shielding tensor is

$$\sigma^{\text{rel}}_{K;\mu\nu} = \frac{\partial^2}{\partial m_{K;\mu}\partial B_{0;\nu}}\langle\langle \hat{h}^{\text{hfs}}_{K;\text{rel}}; \hat{h}^{Z}_{\nu;\text{rel}}\rangle\rangle_0, \tag{8.309}$$

and contains only a single linear response function coupling the relativistic hyperfine operator, Eq. (8.47), with the relativistic Zeeman operator, Eq. (8.86). There is accordingly no separation of para- and diamagnetic contributions, as in the nonrelativistic expression, Eq. (8.307). It is in fact possible to separate para- and diamagnetic contributions in the relativistic case, and one can at the SCF level show that the diamagnetic contributions arise from the negative-energy orbitals of the unperturbed system. This sounds very exotic, but it should be kept in mind that a complete basis, including both positive- and negative-energy orbitals, is needed to build the first-order corrected wave function. An important consequence is that the paramagnetic contribution does not vanish for relativistic atoms, contrary to the nonrelativistic case.

The nonrelativistic expression for elements of the reduced indirect spin–spin coupling tensor for closed-shell molecules is

$$K_{KL:\mu\nu} = \frac{\partial^2}{\partial m_{K;\mu}\partial m_{L;\nu}}\left[\left\langle 0\left|\hat{h}^{\text{dso}}_{KL}\right|0\right\rangle + \left\langle\left\langle \hat{h}^{\text{pso}}_{K}; \hat{h}^{\text{pso}}_{L}\right\rangle\right\rangle_0 + \left\langle\left\langle \hat{h}^{\text{fc}}_{K} + \hat{h}^{\text{sd}}_{K}; \hat{h}^{\text{fc}}_{L} + \hat{h}^{\text{sd}}_{L}\right\rangle\right\rangle_0\right], \tag{8.310}$$

and consists of one diamagnetic and two paramagnetic contributions. The diamagnetic contribution is the expectation value of the diamagnetic hyperfine spin–orbit operator, Eq. (8.45). The paramagnetic contribution splits into a term coupling singlet (spin-free) operators and a term coupling triplet (spin-dependent) operators. The first term couples the paramagnetic hyperfine spin–orbit operators, Eq. (8.42), associated with the two nuclei. The second term couples the hyperfine Fermi-contact and spin-dipole operators, Eqs. (8.43) and (8.44), of the two centers. It is in general found that K^{dso}_{KL} is very small and effectively cancels the paramagnetic spin-orbit contribution. The latter contribution may become substantial, for instance, for couplings between nuclei with multiple bonds. K^{fc}_{KL} dominates in most cases the isotropic coupling. The mixed term $K^{\text{fc/sd}}_{KL}$ vanishes for isotropic couplings, but often dominates anisotropic couplings. From a computational point of view, it should be noted that the dominance of K^{fc}_{KL} implies that only methods that are stable toward triplet perturbations can be used. The corresponding relativistic expression for elements of the reduced indirect spin–spin coupling tensor is

$$K^{\text{rel}}_{KL:\mu\nu} = \frac{\partial^2}{\partial m_{K;\mu}\partial m_{L;\nu}}\langle\langle \hat{h}^{\text{hfs}}_{K,\text{rel};\mu}; \hat{h}^{\text{hfs}}_{L,\text{rel};\nu}\rangle\rangle_0, \tag{8.311}$$

which is a single linear response function coupling the relativistic hyperfine operators, Eq. (8.46), of the two centers K and L.

The computational expressions for the NMR parameters contain operators that probe the electron density in the vicinity of nuclei and their calculation therefore requires use of basis sets with sufficient flexibility in the core region. It also means that NMR parameters are quite sensitive to relativistic effects. Absolute shieldings are more sensitive to relativistic effects than chemical shifts since there is the possibility of cancelation of errors in the latter quantity. The

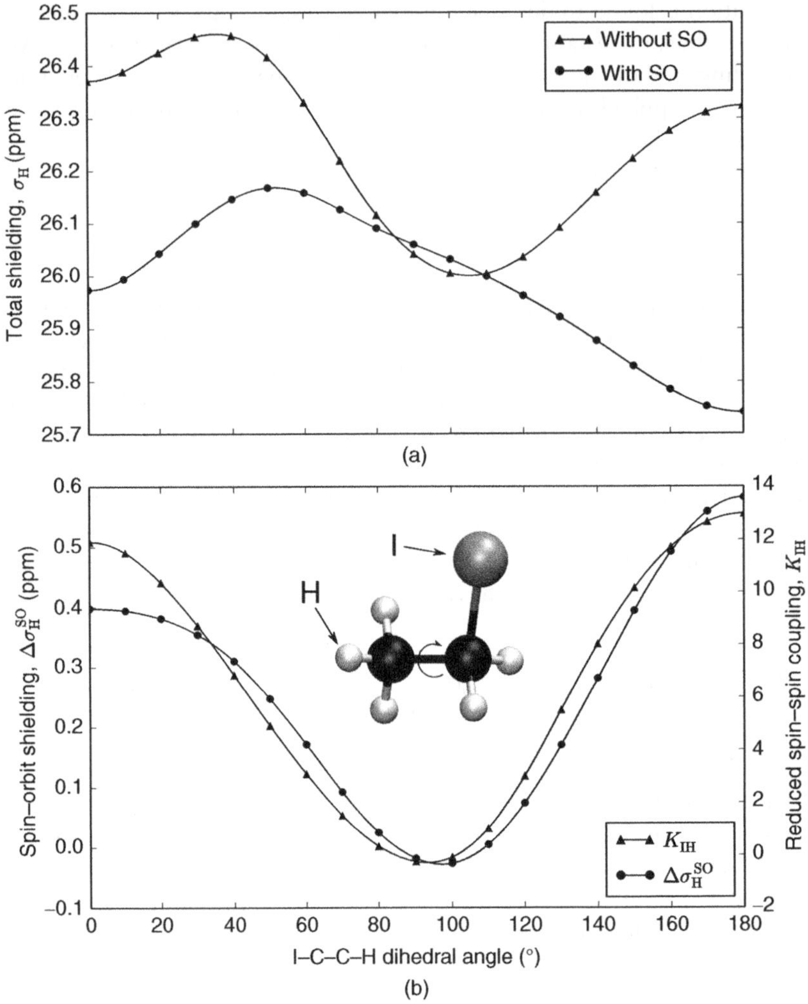

Figure 8.51 Hydrogen shielding constant σ_H (ppm) and indirect iodine–hydrogen spin–spin coupling constant K_{IH} (10^{-19} kg m^{-2} s^{-2} A^{-2}) for iodoethane as a function of dihedral angle (the hydrogen and iodine atoms are pointed out with arrows): (a) σ_H calculated with and without inclusion of spin–orbit interactions and (b) comparison of K_{IH} with the spin–orbit contribution to the total shielding $\Delta\sigma_H^{SO}$, that is, the difference of curve plots in panel (a).

effect of spin–orbit interaction can be quite substantial for shieldings. An interesting observation is that the spin–orbit contribution to shieldings bears some similarity to indirect spin–spin couplings. A striking example[33] is seen when plotting the spin–orbit contribution to the shielding constant $\sigma(H_\beta)$ of iodoethane alongside the indirect spin–spin coupling constant $K(I, H_\beta)$ as a function of dihedral angle ϕ (see Figure 8.51). The spin–spin coupling curve $K(I, H_\beta)$ fits very well a Karplus curve[34] $K(I, H_\beta) = A + B\cos\phi + C\cos(2\phi)$ with constants $A = 6.18$, $B = -0.57$ and $C = 6.35$, all constants in the unit 10^{-19} kg m^{-2} s^{-2} A^{-2} of the reduced spin–spin coupling constant. The spin–orbit contribution to $\sigma(H_\beta)$ follows this curve quite closely, albeit with a somewhat reduced fit and scaled down by a numerical factor of about 25. It is also

33 M. Kaupp, O. L. Malkina, V. G. Malkin, and P. Pyykkö. *Chem. Eur. J.*, 4:118, 1998.
34 M. Karplus. *J. Am. Chem. Soc.*, 85:2870, 1963.

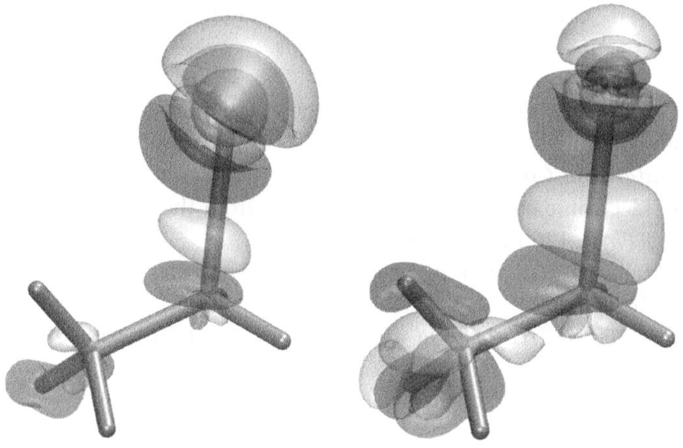

Figure 8.52 Property surfaces of the indirect spin–spin coupling constant K_{IH} (a) and the shielding constant σ_{H} (b) of iodoethane at an dihedral angle of 120°. Dark and light gray surfaces refer to positive and negative property values, respectively.

quite remarkable to compare corresponding property surfaces at the dihedral angle 120° in Figure 8.52.[35] The Karplus curve for the spin–spin coupling is dominated by the Fermi-contact operator, Eq. (8.43), which, for positive gyromagnetic ratio γ, favors an antiparallel configuration of nuclear spin and electron spin density in the vicinity of the nucleus. This introduces spin polarization, which can propagate through bonds. In the case of the spin–orbit contribution to shielding, the role of the nuclear spin appears to be taken by the electron spin density on the heavy nucleus due to the very local, core-like nature of the spin–orbit interaction, as can be seen from the Breit–Pauli form of the spin–orbit operator in Eq. (8.169). The downside to this similarity is that the shielding constant becomes susceptible to triplet instabilities when the spin–orbit interaction is strong.

Further Reading

1 P. W. Atkins and R. S. Friedman. *Molecular Quantum Mechanics*. Oxford University Press, 2011.
2 L. D. Barron. *Molecular Light Scattering and Optical Activity*. Cambridge University Press, 2004.
3 D. M. Bishop and P. Norman. Calculations of dynamic hyperpolarizabilities for small and medium sized molecules. In H. S. Nalwa, editor, *Handbook of Advanced Electronic and Photonic Materials*. Academic Press, 2001.
4 S. Coriani, O. Christiansen, T. Fransson, and P. Norman. Coupled-cluster response theory for near-edge X-ray-absorption fine structure of atoms and molecules. *Phys. Rev. A*, 85:022507, 2012.
5 U. Ekström, P. Norman, V. Carravetta, and H. Ågren. Polarization propagator for X-ray spectra. *Phys. Rev. Lett.*, 97:143001, 2006.
6 T. Fahleson, H. Ågren, and P. Norman. A polarization propagator for nonlinear X-ray spectroscopies. *J. Phys. Chem. Lett.*, 7:1991–1995, 2016.

35 Property surfaces are scalar functions which, integrated over all space, gives the value of the property.

7 T. Helgaker, S. Coriani, P. Jørgensen, K. Kristensen, J. Olsen, and K. Ruud. Recent advances in wave-function-based methods of molecular-property calculations. *Chem. Rev.*, 112:543–631, 2012.

8 T. Helgaker, M. Jaszuński, and K. Ruud. Ab initio methods for the calculation of NMR shielding and indirect spin–spin coupling constants. *Chem. Rev.*, 99:293, 1999.

9 F. Jensen. *Introduction to Computational Chemistry*. Wiley, 2013.

10 A. Jiemchooroj and P. Norman. Electronic circular dichroism spectra from the complex polarization propagator. *J. Chem. Phys.*, 126:134102, 2007.

11 M. Kaupp, M. Bühl, and V. G. Malkin. *Calculation of NMR and EPR Parameters*. Wiley-VCH Verlag GmbH, Weinheim, 2004.

12 D. A. Long. *The Raman Effect: A Unified Treatment of the Theory of Raman Scattering by Molecules*. Wiley, 2002.

13 W. R. Mason. *A Practical Guide to Magnetic Circular Dichroism Spectroscopy*. Wiley, Hoboken, NJ, 2007.

14 A. B. Myers. Resonance Raman intensities and charge-transfer reorganization energies. *Chem. Rev.*, 96:911–926, 1996.

15 L. A. Nafie. *Vibrational Optical Activity: Principles and Applications*. Wiley, Chichester, 2011.

16 J. Neugebauer, M. Reiher, C. Kind, and B. A. Hess. Quantum chemical calculation of vibrational spectra of large molecules–Raman and IR spectra for buckminsterfullerene. *J. Comput. Chem.*, 23:895–910, 2002.

17 P. Norman, A. Jiemchooroj, and B. E. Sernelius. Polarization propagator calculations of the polarizability tensor at imaginary frequencies and long-range interactions for the noble gases and n-alkanes. *J. Chem. Phys.*, 118:9167–9174, 2003.

18 P. Norman, K. Ruud, and T. Helgaker. Density-functional theory calculations of optical rotatory dispersion in the nonresonant and resonant frequency regions. *J. Chem. Phys.*, 120:5027–5035, 2004.

19 W. A. Saidi and P. Norman. Probing single-walled carbon nanotube defect chemistry using resonance Raman spectroscopy. *Carbon*, 67:17–26, 2014.

20 H. Solheim, K. Ruud, S. Coriani, and P. Norman. Complex polarization propagator calculations of magnetic circular dichroism spectra. *J. Chem. Phys.*, 128:094103, 2008.

21 J. Stöhr. *NEXAFS spectroscopy*, volume 25. Springer, 2013.

22 A. J. Stone. *The Theory of Intermolecular Forces*. Oxford University Press, Oxford, 1996.

23 D. Sundholm, H. Fliegl, and R. J. F. Berger. Calculations of magnetically induced current densities: theory and applications. *WIREs Comput. Mol. Sci.*, 6:639–678, 2016.

24 A. X. Trautwein, P. Gütlich, and E. Bill. *Mössbauer Spectroscopy and Transition Metal Chemistry: Fundamentals and Applications*. Springer, 2011.

Appendicies

A

Abbreviations

Table A.1 Abbreviations in theoretical chemistry used in this book.

Abbreviation	Meaning
AMFI	atomic mean-field interaction
BCH	Baker–Campbell–Hausdorff
BLYP	Becke–Lee–Yang–Parr
CAM	Coulomb attenuated method
CAS	complete active space
CC	coupled cluster
CCSD	coupled cluster with singly and doubly excited configurations
CCSD(T)	coupled cluster with singly, doubly, and selected triply excited configurations
CI	configuration interaction
CIS	configuration interaction with singly excited configurations
CISD	configuration with singly and doubly excited configurations
CPP	complex polarization propagator
CSF	configuration state function
CVS	core-valence separation
DFT	density functional theory
DKH	Douglas–Kroll–Hess
FCI	full configuration interaction
GIAO	gauge-including atomic orbital
HF	Hartree–Fock
IMDHO	independent-mode displaced harmonic oscillator
KHD	Kramers–Heisenberg–Dirac
KS	Kohn–Sham
LAO	London atomic orbital
LDA	local density approximation
MCSCF	multiconfigurational self-consistent field
MD	molecular dynamics
MM	molecular mechanics
MP*n*	*n*th order Møller–Plesset perturbation theory

Principles and Practices of Molecular Properties: Theory, Modeling and Simulations, First Edition.
Patrick Norman, Kenneth Ruud, and Trond Saue.

Table A.1 (Continued)

Abbreviation	Meaning
PCM	polarizable continuum model
RPA	random phase approximation
SCF	self-consistent field
STEX	static exchange
TD	time-dependent
X2C	exact two-component
XC	exchange-correlation
ZORA	zeroth-order regular approximation
ZPV	zero-point vibration
ZPVA	zero-point vibrational averaging

B

Units

Table B.1 Atomic unit conversion factors.

	Symbol	Atomic unit	SI equivalent	
Energy	E	$1\,E_h$	$4.359\,744 \times 10^{-18}$	J
Reduced Planck constant	$\hbar$	$1\,\hbar$	$1.054\,572 \times 10^{-34}$	J s
Time	t	$1\,\hbar E_h^{-1}$	$2.418\,884 \times 10^{-17}$	s
Length	l	$1\,a_0$	$5.291\,772 \times 10^{-11}$	m
Speed of light	c	$137.036\,a_0 E_h \hbar^{-1}$	$2.997\,925 \times 10^{8}$	$\mathrm{m\,s^{-1}}$
Electric constant	ε_0	$e^2(4\pi a_0 E_h)^{-1}$	$8.854\,188 \times 10^{-12}$	$\mathrm{F\,m^{-1}}$
Fine structure constant	α	$1/137.036\,e^2$ $(a_0 E_h 4\pi\varepsilon_0)^{-1}$	$7.297\,353 \times 10^{3}$	
Charge	q	$1\,e$	$1.602\,176 \times 10^{-19}$	C
Electric field	F	$1\,E_h(ea_0)^{-1}$	$5.142\,207 \times 10^{11}$	$\mathrm{V\,m^{-1}}$
Dipole moment	μ	$1\,ea_0$	$8.478\,353 \times 10^{-30}$	C m
Polarizabilities	α	$1\,e^2 a_0^2 E_h^{-1}$	$1.648\,777 \times 10^{-41}$	$\mathrm{C^2\,m^2\,J^{-1}}$
	β	$1\,e^3 a_0^3 E_h^{-2}$	$3.206\,361 \times 10^{-53}$	$\mathrm{C^3\,m^3\,J^{-2}}$
	γ	$1\,e^4 a_0^4 E_h^{-3}$	$6.235\,380 \times 10^{-65}$	$\mathrm{C^4\,m^4\,J^{-3}}$
Mass	m	$1\,m_e$	$9.109\,383 \times 10^{-31}$	kg
Bohr magneton	μ_B	$1/2\,ehm_e{}^{1}$	$9.274\,009 \times 10^{24}$	$\mathrm{J\,T^{1}}$

Principles and Practices of Molecular Properties: Theory, Modeling and Simulations, First Edition.
Patrick Norman, Kenneth Ruud, and Trond Saue.

C

Second Quantization

C.1 Creation and Annihilation Operators

The quantum mechanical Hamiltonian $\hat{H}$ is obtained from the classical Hamiltonian function $H \equiv H(\boldsymbol{r}, \boldsymbol{p})$, by replacing the dynamical variables, position $\boldsymbol{r}$ and momentum $\boldsymbol{p}$, by operators in accordance with the fundamental commutator relation

$$[r_i, p_j] = i\hbar\delta_{ij}. \tag{C.1}$$

In the coordinate representation this leads to the choice

$$\boldsymbol{r} \rightarrow \hat{\boldsymbol{r}} = \boldsymbol{r}; \quad \boldsymbol{p} \rightarrow \hat{\boldsymbol{p}} = -i\hbar\frac{\partial}{\partial \boldsymbol{r}} = -i\hbar\boldsymbol{\nabla}, \tag{C.2}$$

whereas in the momentum representation one chooses

$$\boldsymbol{r} \rightarrow \hat{\boldsymbol{r}} = i\hbar\frac{\partial}{\partial \boldsymbol{p}}; \quad \boldsymbol{p} \rightarrow \hat{\boldsymbol{p}} = \boldsymbol{p}. \tag{C.3}$$

For localized systems, such as molecules, the coordinate representation is preferred.

Once the Hamiltonian is defined, the wave equation may (in principle) be solved to provide the wave function from which information about the system can be extracted. In the time-independent case the equation to be solved is

$$\hat{H}\Psi = E\Psi. \tag{C.4}$$

Quantization of the energy E is achieved from consideration of boundary conditions, in particular that the wave function should go to zero at boundaries or at infinity. Another constraint on the wave function follows from the Pauli principle, namely, that the total wave function should be antisymmetric under the exchange of identical fermions and symmetric under the exchange of identical bosons. An important observation is that the Pauli principle is not built into the Hamiltonian. However, this is possible upon the use of second quantization.

One way to introduce second quantization is to start from the probabilistic interpretation of the wave function. For a system of N electrons, the quantity

$$\Psi^{\dagger}(1, 2, \ldots, N)\Psi(1, 2, \ldots, N)d1d2\cdots dN \tag{C.5}$$

represents the probability of finding the electrons in the infinitesimal volumes $d1,\ d2, \ldots, dN$ about coordinates $1,\ 2, \ldots, N$, where each coordinate collects both spatial and spin coordinates of a given electron. This leads to the normalization condition

$$\int \Psi^{\dagger}(1, 2, \ldots, N)\Psi(1, 2, \ldots, N)d1d2\cdots dN = 1. \tag{C.6}$$

Principles and Practices of Molecular Properties: Theory, Modeling and Simulations, First Edition.
Patrick Norman, Kenneth Ruud, and Trond Saue.

The electron density can be obtained from the quantity

$$\rho(1) = N \int \Psi^{\dagger}(1,2,\ldots,N)\Psi(1,2,\ldots,N)\, d2\cdots dN, \tag{C.7}$$

where we have integrated over all coordinates except one. The electron density integrates to the number of electrons

$$\int \rho(1)\, d1 = N. \tag{C.8}$$

Now suppose that we replace the above relation by the following one

$$\hat{N} = \int \hat{\psi}^{\dagger}(1)\hat{\psi}(1)\, d1. \tag{C.9}$$

Following Dirac, we think of the quantities $\hat{\psi}^{\dagger}(1)$ and $\hat{\psi}(1)$ as *operators*: $\hat{\psi}(1)$ annihilates the total electron density amplitude at position 1 and $\hat{\psi}^{\dagger}(1)$ creates it back. We want the total operator $\hat{N}$, when acting on an object representative for an N-electron system, to return the particle number N.

The quantities $\hat{\psi}^{\dagger}(1)$ and $\hat{\psi}(1)$ are collectively denoted field operators. In order to represent electrons (fermions), they have to obey the following anticommutation relations:

$$\begin{aligned}
[\hat{\psi}^{\dagger}(1), \hat{\psi}^{\dagger}(2)]_+ &= \hat{\psi}^{\dagger}(1)\hat{\psi}^{\dagger}(2) + \hat{\psi}^{\dagger}(2)\hat{\psi}^{\dagger}(1) = 0,\\
[\hat{\psi}(1), \hat{\psi}(2)]_+ &= \hat{\psi}(1)\hat{\psi}(2) + \hat{\psi}(2)\hat{\psi}(1) = 0,\\
[\hat{\psi}(1), \hat{\psi}^{\dagger}(2)]_+ &= \hat{\psi}(1)\hat{\psi}^{\dagger}(2) + \hat{\psi}^{\dagger}(2)\hat{\psi}(1) = \delta(1-2),
\end{aligned} \tag{C.10}$$

whereas bosonic field operators obey the corresponding commutator relations.

Suppose now that we have some orthonormal basis

$$\{\varphi_p(1)\}_{p=1}^{M}; \quad \int \varphi_p^{\dagger}(1)\varphi_q(1)d1 = \delta_{pq}. \tag{C.11}$$

We may expand the field operators in this basis

$$\hat{\psi}(1) = \sum_q \varphi_q(1)\hat{a}_q; \quad \hat{\psi}^{\dagger}(1) = \sum_q \varphi_q^{\dagger}(1)\hat{a}_q^{\dagger}. \tag{C.12}$$

We find the expansion coefficients $\hat{a}_p$ and $\hat{a}_p^{\dagger}$ by projection

$$\hat{a}_p = \int \varphi_p^{\dagger}(1)\hat{\psi}(1)d1; \quad \hat{a}_p^{\dagger} = \int \hat{\psi}^{\dagger}(1)\varphi_p(1)\mathrm{d}1. \tag{C.13}$$

The expansion coefficients $\hat{a}_p^{\dagger}$ and $\hat{a}_p$ are operators as well, and are denoted creation and annihilation operators, respectively. We may deduce their algebra from the algebra of the field operators. For instance,

$$\begin{aligned}
[\hat{a}_p, \hat{a}_q^{\dagger}]_+ &= \left[\int \varphi_p^{\dagger}(1)\hat{\psi}(1)d1, \int \hat{\psi}^{\dagger}(2)\varphi_q(2)d2\right]_+\\
&= \int\int \varphi_p^{\dagger}(1)\varphi_q(2)[\hat{\psi}(1), \hat{\psi}^{\dagger}(2)]_+ d1d2\\
&= \int\int \varphi_p^{\dagger}(1)\varphi_q(2)\delta(1-2)d1d2 = \int \varphi_p^{\dagger}(1)\varphi_q(1)d1 = \delta_{pq}.
\end{aligned}$$

In passing we may note that in a nonorthogonal basis the Kronecker delta δ_{pq} will be replaced by an element S_{pq} of the overlap matrix. Proceeding in the same manner we also find the relations

$$[\hat{a}_p^{\dagger}, \hat{a}_q^{\dagger}]_+ = [\hat{a}_p, \hat{a}_q]_+ = 0. \tag{C.14}$$

C.2 Fock Space

An approximate N-electron wave function is typically expanded in Slater determinants. We now show how Slater determinants can be mapped into occupation number vectors residing in Fock space. Let us consider a simple example of four orbitals $\{\varphi_1, \varphi_2, \varphi_3, \varphi_4\}$ $(M = 4)$. With two electrons $(N = 2)$, we can build $\binom{4}{2} = 6$ determinants. One example is

$$\Phi(1,2) = \frac{1}{\sqrt{2!}} \begin{vmatrix} \varphi_1(1) & \varphi_3(1) \\ \varphi_1(2) & \varphi_3(2) \end{vmatrix}. \tag{C.15}$$

Very often one uses a shorthand notation

$$\Phi(1,2) = |\varphi_1\varphi_3|, \tag{C.16}$$

since the Slater determinant is uniquely determined once one specifies the orbitals included and their order (the latter is required to fix the phase). We can map this into an occupation number vector

$$\Phi_k(1,2) = |\varphi_1\varphi_3| \quad \rightarrow \quad |\boldsymbol{k}\rangle = |k_1, k_2, k_3, k_4\rangle = |1,0,1,0\rangle, \tag{C.17}$$

where occupation numbers k_p are either 1 or 0, since electrons are fermions. Another example is

$$\Phi_m(1,2,3) = \frac{1}{\sqrt{3!}} \begin{vmatrix} \varphi_1(1) & \varphi_2(1) & \varphi_4(1) \\ \varphi_1(2) & \varphi_2(2) & \varphi_4(2) \\ \varphi_1(3) & \varphi_2(3) & \varphi_4(3) \end{vmatrix} = |\varphi_1\varphi_2\varphi_4| \quad \rightarrow \quad |\boldsymbol{m}\rangle = |1,0,1,1\rangle, \tag{C.18}$$

and we can already anticipate that occupation number vectors handle variable particle number in a seamless manner. A special occupation number vector is the vacuum state

$$|\text{vac}\rangle = |0,0,0,0\rangle. \tag{C.19}$$

Annihilation operators reduce occupation numbers by one and therefore all give zero when acting on $|\text{vac}\rangle$

$$\hat{a}_p|\text{vac}\rangle = 0; \quad \forall \hat{a}_p. \tag{C.20}$$

This even serves as a definition of the vacuum state.

Creation operators, on the other hand, increase occupation numbers by one

$$\hat{a}_1^\dagger|\text{vac}\rangle = |1,0,0,0\rangle$$
$$\hat{a}_2^\dagger|\text{vac}\rangle = |0,1,0,0\rangle$$
$$\hat{a}_3^\dagger|\text{vac}\rangle = |0,0,1,0\rangle$$
$$\hat{a}_4^\dagger|\text{vac}\rangle = |0,0,0,1\rangle,$$

but, since they refer to fermions, occupation numbers cannot be greater than one

$$\hat{a}_1^\dagger|1,0,0,0\rangle = \hat{a}_1^\dagger\hat{a}_1^\dagger|\text{vac}\rangle = 0. \tag{C.21}$$

This follows directly from the special case

$$[\hat{a}_p^\dagger, \hat{a}_p^\dagger]_+ = \hat{a}_p^\dagger\hat{a}_p^\dagger + \hat{a}_p^\dagger\hat{a}_p^\dagger = 2\hat{a}_p^\dagger\hat{a}_p^\dagger = 0 \tag{C.22}$$

We can build occupation number vectors corresponding to $N = 2$

$$\hat{a}_1^\dagger|0,1,0,0\rangle = |1,1,0,0\rangle = \hat{a}_1^\dagger\hat{a}_2^\dagger|\text{vac}\rangle. \tag{C.23}$$

Using the algebra of creation operators, we find that

$$\hat{a}_2^\dagger|1,0,0,0\rangle = \hat{a}_2^\dagger\hat{a}_1^\dagger|\text{vac}\rangle = -\hat{a}_1^\dagger\hat{a}_2^\dagger|\text{vac}\rangle = -|1,1,0,0\rangle. \tag{C.24}$$

This demonstrates how the fermion antisymmetry is built into the operators.

Occupation number vectors have the general form

$$|\boldsymbol{k}\rangle = |k_1, k_2, \ldots, k_M\rangle; \quad k_p = 0 \text{ or } 1. \tag{C.25}$$

and reside in a Fock space of dimension 2^M. There is no restriction on particle number N in Fock space, except that $N \leq M$. Any occupation number vector can be generated from the vacuum state

$$|\boldsymbol{k}\rangle = \prod_{p=1}^{M} (\hat{a}_p^\dagger)^{k_p}|\text{vac}\rangle \tag{C.26}$$

An inner-product in Fock space is defined by

$$\langle\boldsymbol{k}|\boldsymbol{m}\rangle = \prod_{p=1}^{M} \delta_{k_p m_p} \tag{C.27}$$

and is one if all occupation numbers are identical, zero otherwise. A special case is

$$\langle\text{vac}|\text{vac}\rangle = 1. \tag{C.28}$$

The dual vector $\langle\boldsymbol{k}|$ appearing in the inner product is obtained by conjugation, for example, starting from

$$|\boldsymbol{k}\rangle = |1,0,1,1\rangle = \hat{a}_1^\dagger\hat{a}_3^\dagger\hat{a}_4^\dagger|\text{vac}\rangle, \tag{C.29}$$

we have

$$\langle\boldsymbol{k}| = \langle 1,0,1,1| = \langle\text{vac}|(\hat{a}_1^\dagger\hat{a}_3^\dagger\hat{a}_4^\dagger)^\dagger = \langle\text{vac}|\hat{a}_4\hat{a}_3\hat{a}_1. \tag{C.30}$$

One should note that the order of operators changes under conjugation and that annihilation operators become creators when operating to the left. In fact, we can define the dual vacuum state by

$$\langle\text{vac}|\hat{a}_p^\dagger = 0; \quad \forall \hat{a}_p^\dagger. \tag{C.31}$$

C.3 The Number Operator

Let us now return to Eq. (C.9). By expanding the field operators in an orthonormal basis, we obtain

$$\hat{N} = \int \hat{\psi}^\dagger(\boldsymbol{r})\hat{\psi}(\boldsymbol{r})\mathrm{d}^3\boldsymbol{r} = \sum_{p,q}\left\{\int \varphi_p^\dagger(\boldsymbol{r})\varphi_q(\boldsymbol{r})\mathrm{d}^3\boldsymbol{r}\right\}\hat{a}_p^\dagger\hat{a}_q = \sum_{p,q}\delta_{pq}\hat{a}_p^\dagger\hat{a}_q = \sum_p \hat{a}_p^\dagger\hat{a}_p, \tag{C.32}$$

which defines the number operator $\hat{N}$. For instance, we have

$$\hat{N}|1,0,1,1\rangle = 3|1,0,1,1\rangle, \tag{C.33}$$

which shows that the occupation number vectors are eigenvectors of the number operator. In passing, we may note that the electron field operators do not relate to specific electrons. Rather, they sample contributions to the electron quantum field. Quantum field theory in fact explains why electrons are the same everywhere; they all belong to the same field !

We have previously stated that creation (annihilation) operators increase (reduce) particle number, respectively. We can now demonstrate this using the number operator. Suppose that we have

$$\hat{N}|\boldsymbol{k}\rangle = N|\boldsymbol{k}\rangle. \tag{C.34}$$

We now consider $\hat{N}a_p^\dagger|\boldsymbol{k}\rangle$. We can rewrite this as

$$\hat{N}\hat{a}_p^\dagger|\boldsymbol{k}\rangle = (\hat{a}_p^\dagger\hat{N} + [\hat{N},\hat{a}_p^\dagger])|\boldsymbol{k}\rangle. \tag{C.35}$$

To proceed, we need to solve the commutator

$$[\hat{N},\hat{a}_p^\dagger] = \sum_q [\hat{a}_q^\dagger\hat{a}_q,\hat{a}_p^\dagger]. \tag{C.36}$$

We may use a commutator rule such as

$$[\hat{A}\hat{B},\hat{C}] = \hat{A}[\hat{B},\hat{C}] + [\hat{A},\hat{C}]\hat{B}. \tag{C.37}$$

However, since the algebra of creation and annihilation operators is expressed in terms of anti-commutators, it is better to use

$$[\hat{A}\hat{B},\hat{C}] = \hat{A}[\hat{B},\hat{C}]_+ - [\hat{A},\hat{C}]_+\hat{B}, \tag{C.38}$$

which directly gives

$$[\hat{N},\hat{a}_p^\dagger] = \sum_q \left(\hat{a}_q^\dagger \underbrace{[\hat{a}_q,\hat{a}_p^\dagger]_+}_{=\delta_{pq}} - \underbrace{[\hat{a}_q^\dagger,\hat{a}_p^\dagger]_+}_{=0}\hat{a}_q \right) = \hat{a}_p^\dagger, \tag{C.39}$$

such that

$$\hat{N}\hat{a}_p^\dagger|\boldsymbol{k}\rangle = \hat{a}_p^\dagger(\hat{N} + 1)|\boldsymbol{k}\rangle = (N+1)\hat{a}_p^\dagger|\boldsymbol{k}\rangle. \tag{C.40}$$

In the same manner we find that

$$\hat{N}\hat{a}_p|\boldsymbol{k}\rangle = (\hat{a}_p\hat{N} + [\hat{N},\hat{a}_p])|\boldsymbol{k}\rangle = (\hat{a}_p\hat{N} - [\hat{N},\hat{a}_p^\dagger]^\dagger)|\boldsymbol{k}\rangle = (N-1)\hat{a}_p|\boldsymbol{k}\rangle. \tag{C.41}$$

An extension of the number operator is

$$\hat{N}_{pair} = \frac{1}{2}\int \hat{\psi}^\dagger(1)\hat{\psi}^\dagger(2)\hat{\psi}(2)\hat{\psi}(1)\,\mathrm{d}1\mathrm{d}2 = \frac{1}{2}\sum_{p,q}\hat{a}_p^\dagger\hat{a}_q^\dagger\hat{a}_q\hat{a}_p. \tag{C.42}$$

We do some further operator algebra

$$\hat{a}_p^\dagger\hat{a}_q^\dagger\hat{a}_q\hat{a}_p = -\hat{a}_p^\dagger\hat{a}_q^\dagger\hat{a}_p\hat{a}_q = -\hat{a}_p^\dagger(\delta_{pq} - \hat{a}_p\hat{a}_q^\dagger)\hat{a}_q = \hat{a}_p^\dagger\hat{a}_p\hat{a}_q^\dagger\hat{a}_q - \delta_{pq}\hat{a}_p^\dagger\hat{a}_q, \tag{C.43}$$

which shows that the pair operator can be expressed as

$$\hat{N}_{\mathrm{pair}} = \frac{1}{2}\hat{N}(\hat{N}+1), \tag{C.44}$$

and counts electron pairs.

C.4 The Electronic Hamiltonian on Second-Quantized Form

The generic form of the first-quantized electronic Hamiltonian is

$$\hat{H} = V_{NN} + \sum_{i=1}^{N}\hat{h}(i) + \frac{1}{2}\sum_{i\neq j}^{N}\hat{g}(i,j). \tag{C.45}$$

Without further ado, we write the electronic Hamiltonian on second-quantized form

$$\hat{H} = \int \hat{\psi}^{\dagger}(1)\hat{h}(1)\hat{\psi}(1)\mathrm{d}1 + \frac{1}{2}\int \hat{\psi}^{\dagger}(1)\hat{\psi}^{\dagger}(2)\hat{g}(1,2)\hat{\psi}(2)\hat{\psi}(1)\mathrm{d}1\mathrm{d}2 + V_{NN}, \tag{C.46}$$

but one may note that the one- and two-electron parts contain generalized forms of the number and pair operators, respectively. One should also carefully note the order of electron coordinates in the two-electron operator, showing that the left-hand product of creation operators is obtained by conjugation of the right-hand product of annihilation operators.

Expansion of the field operators in an orthonormal basis gives the final form

$$\hat{H} = \sum_{p,q} h_{pq}\hat{a}_p^{\dagger}\hat{a}_q + \frac{1}{2}\sum_{p,q,r,s} V_{pq,rs}\hat{a}_p^{\dagger}\hat{a}_q^{\dagger}\hat{a}_s\hat{a}_r + V_{NN}, \tag{C.47}$$

where appear one- and two-electron integrals

$$\begin{aligned} h_{pq} &= \int \varphi_p^{\dagger}(1)\hat{h}(1)\varphi_q(1)\mathrm{d}1 &&= \langle\varphi_p|\hat{h}|\varphi_q\rangle, \\ V_{pq,rs} &= \int \varphi_p^{\dagger}(1)\varphi_q^{\dagger}(2)\hat{g}(1,2)\varphi_r(1)\varphi_s(2)\mathrm{d}1\mathrm{d}2 &&= \langle\varphi_p\varphi_q|\hat{g}|\varphi_r\varphi_s\rangle. \end{aligned} \tag{C.48}$$

The "sandwiching" of the first-quantized two-electron operator by the field operators leads to the particular ordering of indices seen in the final second-quantized form.

The above forms can be extended to any one- and two-electron operators. For several reasons, it represents a very convenient operator form: (i) as already mentioned, the fermion antisymmetry is built into the operator; (ii) the operator is expressed in terms of one- and two-electron integrals, which are the basic ingredients of quantum chemistry codes; and (iii) the form is universal in the sense that there is no reference made to the number of electrons. However, as clearly shown by Eq. (C.12), these are *projected* operators that only "live" in the space defined by the chosen set of orbitals $\{\varphi_p\}_{p=1}^{M}$. This means that operator relations in first quantization are generally only valid in the corresponding second-quantized operators when the underlying basis is complete. As an example, we may consider the operator equality Eq. (7.65). In terms of second-quantized operators, the left-hand side of the relation becomes

$$\sum_{p,q,r,s}[\langle\varphi_p|\hat{r}_\alpha|\varphi_q\rangle\hat{a}_p^{\dagger}\hat{a}_q, \langle\varphi_r|\hat{h}|\varphi_s\rangle\hat{a}_r^{\dagger}\hat{a}_s] = \sum_{p,q,r,s}\langle\varphi_p|\hat{r}_\alpha|\varphi_q\rangle\langle\varphi_r|\hat{h}|\varphi_s\rangle[\hat{a}_p^{\dagger}\hat{a}_q, \hat{a}_r^{\dagger}\hat{a}_s], \tag{C.49}$$

where we for simplicity consider a one-electron Hamiltonian. Evaluation of the commutator gives

$$[\hat{a}_p^{\dagger}\hat{a}_q, \hat{a}_r^{\dagger}\hat{a}_s] = \delta_{rq}\hat{a}_p^{\dagger}\hat{a}_s - \delta_{ps}\hat{a}_r^{\dagger}\hat{a}_q. \tag{C.50}$$

Since we have the summation over all indices in the starting expression, we use the freedom to rename summation indices to simplify the resulting expression to

$$\sum_{p,q}[\langle\varphi_p|\hat{r}_\alpha|\varphi_t\rangle\langle\varphi_t|\hat{h}|\varphi_q\rangle - \langle\varphi_p|\hat{h}|\varphi_t\rangle\langle\varphi_t|\hat{r}_\alpha|\varphi_q\rangle]\hat{a}_p^{\dagger}\hat{a}_q. \tag{C.51}$$

This may be compared to the right-hand side of Eq. (7.65), which in second-quantized form becomes

$$\frac{i\hbar}{m_e}\sum_{p,q}\langle\varphi_p|\hat{p}_\alpha|\varphi_q\rangle\hat{a}_p^{\dagger}\hat{a}_q. \tag{C.52}$$

The above two expressions are only equal in a complete basis in which the implied sum over index t becomes a resolution of identity.[1]

1 A truly complete basis is hard, if not impossible, to obtain. However, it is possible to satisfy the operator equality at the matrix level by a suitable choice of basis, as explained in K. G. Dyall, I. P. Grant, and S. Wilson. *J. Phys. B: At. Mol. Phys.*, 17:493, 1984.

C.5 Spin in Second Quantization

In a nonrelativistic formalism, the spatial and spin parts of orbitals are separated, that is,

$$\varphi(1) = \phi(\mathbf{r})\chi(\sigma). \tag{C.53}$$

This separation can be carried over to operators on second-quantized form. For instance, we may separate out spin from spatial parts of the creation and annihilation operators, giving

$$[\hat{a}^{\dagger}_{p\sigma}, \hat{a}^{\dagger}_{q\sigma'}]_{+} = 0; \quad [\hat{a}_{p\sigma}, \hat{a}_{q\sigma'}]_{+} = 0; \quad [\hat{a}_{p\sigma}, \hat{a}^{\dagger}_{q\sigma'}]_{+} = \delta_{pq}\delta_{\sigma\sigma'}; \quad \sigma, \sigma' = \alpha \text{ or } \beta. \tag{C.54}$$

Proceeding to the electronic Hamiltonian, we obtain for the (nonrelativistic) one-electron part

$$\begin{aligned}
\hat{H}_1 &= \sum_{p,q}\sum_{\sigma,\sigma'}\langle\phi_p\sigma|\hat{h}|\phi_q\sigma'\rangle\hat{a}^{\dagger}_{p\sigma}\hat{a}_{q\sigma'}\\
&= \sum_{p,q}\sum_{\sigma,\sigma'}\langle\phi_p|\hat{h}|\phi_q\rangle\langle\sigma|\sigma'\rangle\hat{a}^{\dagger}_{p\sigma}\hat{a}_{q\sigma'}\\
&= \sum_{p,q}\langle\phi_p|\hat{h}|\phi_q\rangle\hat{E}_{pq}; \quad \hat{E}_{pq} = \sum_{\sigma}\hat{a}^{\dagger}_{p\sigma}\hat{a}_{q\sigma'}.
\end{aligned}$$

Likewise, for the (nonrelativistic) two-electron part we obtain

$$\begin{aligned}
\hat{H}_2 &= \frac{1}{2}\sum_{p,q,r,s}\sum_{\sigma,\tau,\sigma',\tau'}\langle\phi_p\sigma\phi_q\tau|\hat{g}|\phi_r\sigma'\phi_s\tau'\rangle\hat{a}^{\dagger}_{p\sigma}\hat{a}^{\dagger}_{q\tau}\hat{a}_{s\tau'}\hat{a}_{r\sigma'}\\
&= \frac{1}{2}\sum_{p,q,r,s}\sum_{\sigma,\tau,\sigma',\tau'}\langle\phi_p\phi_q|\hat{g}|\phi_r\phi_s\rangle\hat{e}_{pq,rs}; \quad \hat{e}_{pq,rs} = \sum_{\sigma,\tau}\hat{a}^{\dagger}_{p\sigma}\hat{a}^{\dagger}_{q\tau}\hat{a}_{s\tau}\hat{a}_{r\sigma}.
\end{aligned}$$

Using operator algebra, we find that

$$\hat{e}_{pq,rs} = \hat{E}_{pr}\hat{E}_{qs} - \delta_{qr}\hat{E}_{ps}. \tag{C.55}$$

D

Fourier Transforms

A Fourier transform connects functions in the time and frequency domain (or, in position and momentum space) in a nonlocal way. Starting with some function $f(t)$ in the time domain, its Fourier transform is defined by

$$\hat{f}(\omega) = \mathcal{F}[f(t)](\omega) = \frac{1}{2\pi}\int_{-\infty}^{\infty} f(t)e^{+i\omega t}dt \tag{D.1}$$

The inverse transformation reads

$$f(t) = \mathcal{F}^{-1}[\hat{f}(\omega)](t) = \int_{-\infty}^{+\infty} \hat{f}(\omega)e^{-i\omega t}d\omega \tag{D.2}$$

and the functions $\{f(t),\hat{f}(\omega)\}$ constitute a Fourier transform pair. We distinguish the Fourier transformed function from the original function with a hat, but elsewhere in the book we will not make this distinction, and rather let the distinction be implied from the argument to the function. An unfortunate aspect of Fourier transform is that there is no universal agreement on whether the negative sign in the exponent should appear in the Fourier transform or its inverse or how to distribute the prefactor $1/2\pi$ between the Fourier transform and its inverse. We have chosen a convention in agreement with the seminal paper of J. Olsen and P. Jørgensen. *J. Chem. Phys.*, 82:3235, 1985.

A representation of the Dirac delta function is obtained from a combination of forward and backward Fourier transforms

$$f(t) = \frac{1}{2\pi}\int_{-\infty}^{+\infty} d\omega e^{-i\omega t}\int_{-\infty}^{+\infty} dt' f(t')e^{+i\omega t'} = \int_{-\infty}^{+\infty} dt' f(t')\left\{\frac{1}{2\pi}\int_{-\infty}^{+\infty} e^{-i\omega(t-t')}d\omega\right\}, \tag{D.3}$$

where the last term is obtained from a reversal of the order of integration. If we now compare with the expression

$$f(t) = \int_{-\infty}^{+\infty} f(t')\delta(t-t')dt', \tag{D.4}$$

we conclude that

$$\delta(t-t') = \frac{1}{2\pi}\int_{-\infty}^{+\infty} e^{-i\omega(t-t')}d\omega. \tag{D.5}$$

Setting $t' = 0$ and comparing with Eq. (D.2) we see that $\{2\pi\delta(t), 1\}$ is a Fourier transform pair. In a similar manner we find that

$$\delta(\omega-\omega') = \frac{1}{2\pi}\int_{-\infty}^{+\infty} e^{+i(\omega-\omega')t}dt, \tag{D.6}$$

so that $\{1, \delta(\omega)\}$ is another Fourier transform pair.

Principles and Practices of Molecular Properties: Theory, Modeling and Simulations, First Edition.
Patrick Norman, Kenneth Ruud, and Trond Saue.

Taking the time derivative on both sides of the inverse transform gives the relation

$$\frac{df(t)}{dt} = \int_{-\infty}^{+\infty} \{-i\omega\hat{f}(\omega)\}e^{-i\omega t}d\omega \tag{D.7}$$

showing that

$$\{f(t),\hat{f}(\omega)\} \quad \Rightarrow \quad \left\{\frac{df(t)}{dt}, -i\omega\hat{f}(\omega)\right\}. \tag{D.8}$$

This relation does not generally hold in the opposite sense. As an example we may consider the *Heaviside step function*, defined by

$$H(t) = \begin{cases} 0, & t < 0 \\ 1/2, & t = 0 \\ 1, & t > 0 \end{cases} \tag{D.9}$$

Its derivative is the Dirac delta function

$$\frac{dH(t)}{dt} = \delta(t) \tag{D.10}$$

and if we go the "wrong" way in Eq. (D.8) we would conclude that its Fourier transform is $\mathcal{F}[H(t)](\omega) = i/2\pi\omega$. However, let us consider the backward transformation of this function

$$\mathcal{F}^{-1}[i/2\pi\omega](t) = \frac{i}{2\pi}\int_{-\infty}^{+\infty} \frac{e^{-i\omega t}}{\omega}d\omega. \tag{D.11}$$

After splitting the integral and a change of variable we obtain

$$\mathcal{F}^{-1}[i/2\pi\omega](t) = \frac{i}{2\pi}\int_{0}^{+\infty} \frac{e^{-i\omega t} - e^{+i\omega t}}{\omega}d\omega = \frac{1}{\pi}\int_{0}^{+\infty} \frac{\sin\omega t}{\omega}d\omega = \frac{1}{2}\text{sgn}(t). \tag{D.12}$$

The sign function is defined as

$$\text{sgn}(t) = \begin{cases} -1, & t < 0 \\ 0, & t = 0 \\ 1, & t > 0 \end{cases}. \tag{D.13}$$

It is related to the Heaviside step function as

$$\text{sgn}(t) = 2H(t) - 1 \tag{D.14}$$

and its derivative is accordingly

$$\frac{d\text{sgn}(t)}{dt} = 2\delta(t). \tag{D.15}$$

Our above derivation shows that the Fourier transform of the sign function is

$$\mathcal{F}[\text{sgn}(t)](\omega) = \frac{i}{\pi\omega}, \tag{D.16}$$

from which we further deduce

$$\mathcal{F}[H(t)](\omega) = \frac{i}{2\pi\omega} + \frac{1}{2}\delta(\omega). \tag{D.17}$$

To conclude this section let us consider the Fourier transform of a product of functions

$$\mathcal{F}[f(t)g(t)](\omega) = \frac{1}{2\pi}\int_{-\infty}^{+\infty} f(t)g(t)e^{+i\omega t}dt. \tag{D.18}$$

We would like to express the final results in terms of the Fourier transform of each function. To do so, we replace $g(t)$ by its inverse Fourier transform

$$\mathcal{F}[f(t)g(t)](\omega) = \frac{1}{2\pi}\int_{-\infty}^{+\infty} dt f(t)e^{+i\omega t}\int_{-\infty}^{+\infty} d\omega' \hat{g}(\omega')e^{-i\omega' t}. \tag{D.19}$$

By rearrangement, we get

$$\mathcal{F}[f(t)g(t)](\omega) = \frac{1}{2\pi}\int_{-\infty}^{+\infty} d\omega' \hat{g}(\omega')\int_{-\infty}^{+\infty} dt f(t)e^{+i(\omega-\omega')t} = \int_{-\infty}^{+\infty} \hat{f}(\omega-\omega')\hat{g}(\omega')d\omega', \tag{D.20}$$

which shows that the Fourier transform of a product of two functions is given by the *convolution* of the Fourier partners. In the same manner we find that

$$\mathcal{F}^{-1}[\hat{f}(\omega)\hat{g}(\omega)](t) = \frac{1}{2\pi}\int_{-\infty}^{+\infty} dt' f(t-t')g(t')dt'. \tag{D.21}$$

E

Operator Algebra

Lemma E.1 Let A and B be two matrices, or operators, that in general do not commute. Then

$$BA^n = \sum_{k=0}^{n} (-1)^k \binom{n}{k} A^{n-k} \underline{\underline{A}}^k B, \tag{E.1}$$

where the superoperator $\underline{\underline{A}}$ is defined by its action on operator B according to $\underline{\underline{A}}B = [A, B]$.

Proof: The result can be shown by induction. It holds true for $n = 0$. Given that it holds for $n = p$, we get for $n = p + 1$

$$\begin{aligned}
BA^{p+1} &= \left(\sum_{k=0}^{p} (-1)^k \binom{p}{k} A^{p-k} \underline{\underline{A}}^k B \right) A = /\text{using} (\underline{\underline{A}}^k B) A = A \underline{\underline{A}}^k B - \underline{\underline{A}}^{k+1} B / \\
&= \sum_{k=0}^{p} (-1)^k \binom{p}{k} A^{p+1-k} \underline{\underline{A}}^k B + \sum_{k=1}^{p+1} (-1)^k \binom{p}{k-1} A^{p+1-k} \underline{\underline{A}}^k B \\
&= \sum_{k=0}^{p+1} (-1)^k \binom{p+1}{k} A^{p+1-k} \underline{\underline{A}}^k B.
\end{aligned}$$

Lemma E.2 Let A and B be two matrices, or operators, that in general do not commute. Then

$$A^n B = \sum_{k=0}^{n} \binom{n}{k} \left[\underline{\underline{A}}^k B \right] A^{n-k}. \tag{E.2}$$

Proof: Analogous to proof of Lemma E.1.

Lemma E.3 Let A and B be two matrices, or operators, that in general do not commute. Then

$$\underline{\underline{A}}^n B = \sum_{k=0}^{n} (-1)^k \binom{n}{k} A^{n-k} B A^k. \tag{E.3}$$

Principles and Practices of Molecular Properties: Theory, Modeling and Simulations, First Edition.
Patrick Norman, Kenneth Ruud, and Trond Saue.

Proof: The result can be shown by induction. It holds true for $n = 0$. Given that it holds for $n = p$, we get for $n = p + 1$

$$
\begin{aligned}
\underline{\underline{A}}^{p+1}B &= \underline{\underline{A}}\left(\sum_{k=0}^{p}(-1)^k\binom{p}{k}A^{p-k}BA^k\right)\\
&= \sum_{k=0}^{p}(-1)^k\binom{p}{k}A^{p+1-k}BA^k + \sum_{k=1}^{p+1}(-1)^k\binom{p}{k-1}A^{p+1-k}BA^k\\
&= \sum_{k=0}^{p+1}(-1)^k\binom{p+1}{k}A^{p+1-k}BA^k
\end{aligned}
$$

Lemma E.4 Let A and B be two matrices, or operators, that in general do not commute. Then

$$
\sum_{k=0}^{n}A^{n-k}BA^k = \sum_{k=0}^{n}(-1)^k\binom{n+1}{k+1}A^{n-k}\underline{\underline{A}}^kB. \tag{E.4}
$$

Proof: With use of Lemma E.1, we get

$$
\begin{aligned}
\sum_{k=0}^{n}A^{n-k}BA^k &= \sum_{k=0}^{n}A^{n-k}\sum_{l=0}^{k}(-1)^l\binom{k}{l}A^{k-l}\underline{\underline{A}}^lB\\
&= \sum_{k=0}^{n}\sum_{l=0}^{k}(-1)^l\binom{k}{l}A^{n-l}\underline{\underline{A}}^lB\\
&= \sum_{l=0}^{n}(-1)^l\left[\sum_{k=l}^{n}\binom{k}{l}\right]A^{n-l}\underline{\underline{A}}^lB\\
&= \sum_{l=0}^{n}(-1)^l\binom{n+1}{l+1}A^{n-l}\underline{\underline{A}}^lB,
\end{aligned}
$$

where we from the third to the fourth line made use of the hockey-stick identity.

Lemma E.5 Let A and B be two matrices, or operators, that in general do not commute. Then

$$
\sum_{k=0}^{n}A^kBA^{n-k} = \sum_{k=0}^{n}\binom{n+1}{k+1}[\underline{\underline{A}}^kB]A^{n-k}. \tag{E.5}
$$

Proof: With use of Lemma E.2, we get

$$
\begin{aligned}
\sum_{k=0}^{n}A^kBA^{n-k} &= \sum_{k=0}^{n}\left[\sum_{l=0}^{k}\binom{k}{l}\underline{\underline{A}}^lBA^{k-l}\right]A^{n-k}\\
&= \sum_{k=0}^{n}\sum_{l=0}^{k}\binom{k}{l}[\underline{\underline{A}}^lB]A^{n-l}\\
&= \sum_{l=0}^{n}\left[\sum_{k=l}^{n}\binom{k}{l}\right][\underline{\underline{A}}^lB]A^{n-l}\\
&= \sum_{l=0}^{n}\binom{n+1}{l+1}[\underline{\underline{A}}^lB]A^{n-l}
\end{aligned}
$$

Lemma E.6 Let A and B be two matrices, or operators, that in general do not commute. Then

$$e^A B e^{-A} = \sum_{n=0}^{\infty} \frac{1}{n!} \underline{\underline{A}}^n B. \tag{E.6}$$

This result is sometimes referred to as the Baker–Campbell–Hausdorff (BCH) expansion.

Proof: By expanding the exponentials, we get

$$\begin{aligned} e^A B e^{-A} &= 1 + (AB - BA) + (\frac{A^2}{2} B - ABA + B\frac{A^2}{2}) + \cdots \\ &= \sum_{n=0}^{\infty} \frac{1}{n!} \sum_{k=0}^{n} (-1)^k \binom{n}{k} A^{n-k} B A^k = /\text{Lemma E.3}/ \\ &= \sum_{n=0}^{\infty} \frac{1}{n!} \underline{\underline{A}}^n B. \end{aligned}$$

This expansion can also be demonstrated by introducing the function

$$f(\lambda) = e^{\lambda A} B e^{-\lambda A}$$

such that

$$f(0) = B; \qquad f(1) = e^A B e^{-A}.$$

We now express $f(1)$ by means of a Taylor (McLaurin) expansion of $f(\lambda)$ around $\lambda = 0$:

$$f(1) = \sum_{n=0}^{\infty} \frac{f^{(n)}(0)}{n!},$$

where the first derivative with respect to λ is equal to

$$f^{(1)}(\lambda) = e^{\lambda A}[A, B]e^{-\lambda A} \Rightarrow f^{(1)}(0) = [A, B].$$

Higher order derivatives are evaluated accordingly, which completes the proof.

Lemma E.7 Let A be a time-dependent matrix, or operator. Then

$$\frac{d}{dt} e^A = e^A \sum_{n=0}^{\infty} \frac{(-1)^n}{(n+1)!} \underline{\underline{A}}^n \dot{A}, \tag{E.7}$$

where $\dot{A}$ denotes the time derivative of A.

Proof: By expanding the exponential, we get

$$\begin{aligned} \frac{d}{dt} e^A &= \dot{A} + \frac{1}{2}(A\dot{A} + \dot{A}A) + \cdots \\ &= \sum_{n=0}^{\infty} \frac{1}{(n+1)!} \sum_{k=0}^{n} A^{n-k} \dot{A} A^k = /\text{Lemma E.4}/ \\ &= \sum_{n=0}^{\infty} \frac{1}{(n+1)!} \sum_{k=0}^{n} (-1)^k \binom{n+1}{k+1} A^{n-k} \underline{\underline{A}}^k \dot{A} = /\text{interchange order of summations}/ \\ &= \sum_{k=0}^{\infty} \frac{(-1)^k}{(k+1)!} \left[\sum_{n=k}^{\infty} \frac{1}{(n-k)!} A^{n-k} \right] \underline{\underline{A}}^k \dot{A} \end{aligned}$$

$$= \sum_{k=0}^{\infty} \frac{(-1)^k}{(k+1)!} \left[\sum_{n=0}^{\infty} \frac{1}{n!} A^n \right] \underline{\underline{A}}^k \dot{A}$$

$$= \left[\sum_{n=0}^{\infty} \frac{1}{n!} A^n \right] \sum_{k=0}^{\infty} \frac{(-1)^k}{(k+1)!} \underline{\underline{A}}^k \dot{A}$$

Lemma E.8 Let A be a time-dependent matrix, or operator. Then

$$\frac{d}{dt} e^A = \sum_{n=0}^{\infty} \frac{1}{(n+1)!} [\underline{\underline{A}}^n \dot{A}] e^A. \tag{E.8}$$

Proof: In accordance with the proof of Lemma E.7, we get

$$\frac{d}{dt} e^A = \sum_{n=0}^{\infty} \frac{1}{(n+1)!} \sum_{k=0}^{n} A^k \dot{A} A^{n-k} = /\text{Lemma E.5}/$$

$$= \sum_{n=0}^{\infty} \frac{1}{(n+1)!} \sum_{k=0}^{n} \binom{n+1}{k+1} [\underline{\underline{A}}^k \dot{A}] A^{n-k}$$

$$= \sum_{k=0}^{\infty} \frac{1}{(k+1)!} [\underline{\underline{A}}^k \dot{A}] \sum_{n=k}^{\infty} \frac{1}{(n-k)!} A^{n-k}$$

$$= \sum_{k=0}^{\infty} \frac{1}{(k+1)!} [\underline{\underline{A}}^k \dot{A}] \left[\sum_{n=0}^{\infty} \frac{1}{n!} A^n \right]$$

Lemma E.9 Let A be a time-dependent matrix, or operator, and B a time-independent matrix, or operator. Then

$$\frac{d}{dt} e^{-A} B e^A = e^{-A} \left[\sum_{n=0}^{\infty} \frac{1}{(n+1)!} \underline{\underline{B}}\, \underline{\underline{A}}^n \dot{A} \right] e^A. \tag{E.9}$$

Proof: By time differentiation, we get

$$\frac{d}{dt} e^{-A} B e^A = \left(\frac{d}{dt} e^{-A} \right) B e^A + e^{-A} B \left(\frac{d}{dt} e^A \right) = \text{Lemmas E.7 and E.8}$$

$$= -\left(e^{-A} \sum_{n=0}^{\infty} \frac{1}{(n+1)!} \underline{\underline{A}}^n \dot{A} \right) B e^A + e^{-A} B \left(\sum_{n=0}^{\infty} \frac{1}{(n+1)!} [\underline{\underline{A}}^n \dot{A}] e^A \right)$$

$$= -e^{-A} \left[\sum_{n=0}^{\infty} \frac{1}{(n+1)!} \underline{\underline{A}}^n \dot{A}, B \right] e^A$$

F

Spin Matrix Algebra

The basic building blocks for constructing spin functions are the functions α and β, corresponding to spin-up and spin-down wave functions of spin $s = 1/2$. The matrix representation of the spin operator $\hat{\mathbf{s}} = (\hat{s}_x, \hat{s}_y, \hat{s}_z)$ is $\frac{\hbar}{2}\boldsymbol{\sigma}$, where $\boldsymbol{\sigma}$ is the vector of Pauli spin matrices

$$\sigma_x = \begin{bmatrix} 0 & 1 \\ 1 & 0 \end{bmatrix}; \quad \sigma_y = \begin{bmatrix} 0 & -i \\ i & 0 \end{bmatrix}; \quad \sigma_z = \begin{bmatrix} 1 & 0 \\ 0 & -1 \end{bmatrix}. \tag{F.1}$$

A useful identity is the Dirac identity

$$(\boldsymbol{\sigma} \cdot \mathbf{A})(\boldsymbol{\sigma} \cdot \mathbf{B}) = \mathbf{A} \cdot \mathbf{B} + i\boldsymbol{\sigma} \cdot (\mathbf{A} \times \mathbf{B}), \tag{F.2}$$

where **A** and **B** are vector operators. Using the Dirac identity, one can easily derive the commutator relations

$$[(\boldsymbol{\sigma} \cdot \mathbf{A}), (\boldsymbol{\sigma} \cdot \mathbf{B})] = 2i\boldsymbol{\sigma} \cdot (\mathbf{A} \times \mathbf{B}) \tag{F.3}$$

and

$$[(\boldsymbol{\sigma} \cdot \mathbf{A}), (\boldsymbol{\sigma} \cdot \mathbf{B})]_+ = 2\mathbf{A} \cdot \mathbf{B}, \tag{F.4}$$

valid only when **A** and **B** commute.

These two relations show that the operators $(\boldsymbol{\sigma} \cdot \mathbf{A})$ and $(\boldsymbol{\sigma} \cdot \mathbf{B})$ *commute* when vectors **A** and **B** are *parallel*, that is, $(\mathbf{A} \parallel \mathbf{B})$, and the operators *anticommute* when vectors **A** and **B** are *perpendicular*, that is, $(\mathbf{A} \perp \mathbf{B})$. As a consequence, we also have that

$$(\boldsymbol{\sigma} \cdot \mathbf{A}_{\parallel}) \exp[(\boldsymbol{\sigma} \cdot \mathbf{B})] = \exp[(\boldsymbol{\sigma} \cdot \mathbf{B})](\boldsymbol{\sigma} \cdot \mathbf{A}_{\parallel}) \tag{F.5}$$

and

$$(\boldsymbol{\sigma} \cdot \mathbf{A}_{\perp}) \exp[(\boldsymbol{\sigma} \cdot \mathbf{B})] = \exp[-(\boldsymbol{\sigma} \cdot \mathbf{B})](\boldsymbol{\sigma} \cdot \mathbf{A}_{\perp}). \tag{F.6}$$

Principles and Practices of Molecular Properties: Theory, Modeling and Simulations, First Edition.
Patrick Norman, Kenneth Ruud, and Trond Saue.

G

Angular Momentum Algebra

The high symmetry of atoms allows us to learn a lot about their properties using the powerful machinery of angular momentum algebra. The generalized angular momentum $\mathbf{j}$ is a vector, but the three components of the corresponding operator $\hat{\mathbf{j}}$ do not commute, which means that they cannot be measured simultaneously. All three components commute with the squared operator $\hat{j}^2$, though, and so we may have experimental access to the length of the vector and one of its components, usually chosen to be the z-component. The corresponding eigenfunctions $|jm_j\rangle$ satisfy

$$\hat{j}^2|jm_j\rangle = \hbar^2 j(j+1)|jm_j\rangle$$
$$\hat{j}_z|jm_j\rangle = \hbar m_j|jm_j\rangle,$$

so the length of $\mathbf{j}$ is $\hbar\sqrt{j(j+1)}$ and its projection on the z-axis is $\hbar m_j$. It is furthermore useful to introduce ladder operators that satisfy

$$\hat{j}_\pm|jm_j\rangle = \hbar\sqrt{j(j+1) - m_j(m_j \pm 1)}|j, m_j \pm 1\rangle; \quad \hat{j}_\pm = \hat{j}_x \pm i\hat{j}_y. \tag{G.1}$$

The possible values of total angular momentum J, obtained by the coupling of two angular momenta j_1 and j_2, are given by the Clebsch–Gordan series

$$\mathbf{J} = \mathbf{j}_1 + \mathbf{j}_2; \quad J = |j_1 - j_2|, \ldots, j_1 + j_2, \tag{G.2}$$

and the resultant eigenfunctions $|JM_J\rangle$ can be expressed as linear combinations of microstates $|j_1 m_{j1}; j_2 m_{j2}\rangle$ in terms of Clebsch–Gordan coefficients

$$|JM_J\rangle = \sum_{m_{j1}}\sum_{m_{j2}} |j_1 m_{j1}; j_2 m_{j2}\rangle\langle j_1 m_{j1}; j_2 m_{j2}|JM_J\rangle. \tag{G.3}$$

Written in this manner, we see that the Clebsch–Gordan coefficients are simply obtained as the projection of the microstates onto the resultant state. The above summation is restricted to combinations such that $m_{j1} + m_{j2} = M_J$, but this restriction as well as the triangle condition (G.2) are built into the Clebsch–Gordan coefficients.

An eigenfunction $|jm_j\rangle$ spans row m_j of irreducible representation j of the rotation group SO(3), which completely specifies its transformation properties under rotations. In the same manner, one may introduce *spherical tensor operators* $T^{(j)}_{m_j}$ that satisfy

$$\left[\hat{J}_\pm, T^{(j)}_{m_j}\right] = \hbar\sqrt{j(j+1) - m_j(m_j \pm 1)}T^{(j)}_{m_j \pm 1},$$
$$\left[\hat{J}_z, T^{(j)}_{m_j}\right] = \hbar m_j T^{(j)}_{m_j},$$

and have similar well-defined behavior under rotations.

Principles and Practices of Molecular Properties: Theory, Modeling and Simulations, First Edition.
Patrick Norman, Kenneth Ruud, and Trond Saue.

An extremely powerful result is the *Wigner–Eckart theorem* that states that

$$\langle \alpha j m_j | T_q^{(k)} | \alpha' j' m_j' \rangle = \langle j' m_j' k q | j m_j \rangle \langle \alpha j \| T^{(k)} \| \alpha' j' \rangle. \quad \text{(G.4)}$$

In this expression, α and α' refer to quantum numbers that are invariant under rotations. The reduced matrix element $\langle \alpha j \| T^{(k)} \| \alpha' j' \rangle$ is independent of the azimuthal quantum numbers m_j and m_j' such that all oriental dependence is encoded in the Clebsch–Gordan coefficient $\langle j' m_j' k q | j m_j \rangle$. In order to determine the reduced matrix element $\langle \alpha j \| T^{(k)} \| \alpha' j' \rangle$, it suffices to be able to evaluate one of the integrals $\langle \alpha j m_j | T_q^{(k)} | \alpha' j' m_j' \rangle$ on the left-hand side (there are $(2j+1)(2q+1)(2j'+1)$ such integrals). Another approach is to find another spherical tensor operator $\tilde{T}_q^{(k)}$, which has the same transformation properties and therefore satisfies

$$\langle \beta j m_j | \tilde{T}_q^{(k)} | \beta' j' m_j' \rangle = \langle j' m_j' k q | j m_j \rangle \langle \beta j \| \tilde{T}^{(k)} \| \beta' j' \rangle, \quad \text{(G.5)}$$

but which is easier to handle. A simple combination of Eqs. (G.4) and (G.5) shows the very useful result that corresponding matrix elements of the two spherical tensor operators are proportional

$$\langle \alpha j m | \tilde{T}_q^{(k)} | \alpha' j' m' \rangle = \langle \beta j m | T_q^{(k)} | \beta' j' m' \rangle \frac{\langle \alpha j \| \tilde{T}^{(k)} \| \alpha' j' \rangle}{\langle \beta j \| T^{(k)} \| \beta' j' \rangle}.$$

Consider, for instance, the special case of spherical *vector* operators

$$\langle \alpha J M | \hat{\mathbf{A}} | \alpha' J M' \rangle = \frac{\langle \alpha J \| \hat{\mathbf{A}} \| \alpha' J \rangle}{\langle \alpha J \| \hat{\mathbf{J}} \| \alpha J \rangle} \langle \alpha J M | \hat{\mathbf{J}} | \alpha J M' \rangle.$$

Notice that we set $\alpha' = \alpha$ in the matrix elements involving the angular momentum operator, since they otherwise would be zero. For the same reason, we also restrict attention to the case $J' = J$. For matrix elements diagonal in J we effectively have

$$\hat{\mathbf{A}} = c\hat{\mathbf{J}}.$$

To find the proportionality constant we multiply with $\hat{\mathbf{J}}$ and consider matrix elements diagonal in both J and M

$$\langle \alpha J M | \hat{\mathbf{A}} \cdot \hat{\mathbf{J}} | \alpha' J M \rangle = c \langle \alpha J M | \hat{J}^2 | \alpha J M \rangle = c\hbar^2 J(J+1),$$

allowing us to write

$$\langle \alpha J M | \hat{\mathbf{A}} | \alpha' J M' \rangle = \frac{\langle \alpha J M | \hat{\mathbf{A}} \cdot \hat{\mathbf{J}} | \alpha' J M \rangle}{\hbar^2 J(J+1)} \langle \alpha J M | \hat{\mathbf{J}} | \alpha J M' \rangle \quad \text{(G.6)}$$

This result is known as the *projection theorem* since the matrix element $\langle \alpha J M | \hat{\mathbf{A}} | \alpha' J M' \rangle$ is proportional to $\langle \alpha J M | \hat{\mathbf{J}} | \alpha J M' \rangle$, with a proportionality constant determined by the projection of $\hat{\mathbf{A}}$ onto $\hat{\mathbf{J}}$.

An example of the use of the projection theorem concerns the spin–orbit splitting of an LS-term. In approximate relativistic theories, the spin–orbit operator has the form

$$\hat{H}_{SO} = \sum_i^N \zeta_i \hat{\mathbf{l}}_i \cdot \hat{\mathbf{s}}_i. \quad \text{(G.7)}$$

and is expressed in terms of the orbital angular momentum operator $\hat{\mathbf{l}}_i$ and spin operator $\hat{\mathbf{s}}_i$ of each individual electron. However, these operators transform under rotations as the total operators for orbital angular momentum and spin, that is, $\hat{\mathbf{L}}$ and $\hat{\mathbf{S}}$, respectively. If we limit attention to spin–orbit splitting within a single LS-term, the operators are effectively proportional

$$\hat{\mathbf{l}}_i = c_i(L)\hat{\mathbf{L}}; \quad \hat{\mathbf{s}}_i = c_i(S)\hat{\mathbf{S}}. \quad \text{(G.8)}$$

This allows us to introduce an effective spin–orbit operator

$$\hat{H}_{\text{SO}}^{\text{eff}} = \sum_i \zeta_i \hat{\mathbf{l}}_i \cdot \hat{\mathbf{s}}_i = \sum_i \zeta_i c_i(L) c_i(S) \hat{\mathbf{L}} \cdot \hat{\mathbf{S}} = \zeta(^{2S+1}L) \hat{\mathbf{L}} \cdot \hat{\mathbf{S}} \tag{G.9}$$

expressed in terms of total operators $\hat{\mathbf{L}}$ and $\hat{\mathbf{S}}$, and where appears a spin–orbit parameter $\zeta(^{2S+1}L)$ specific to the chosen LS term. The angular momenta $\mathbf{L}$ and $\mathbf{S}$ couple to total angular momentum $\mathbf{J} = \mathbf{L} + \mathbf{S}$ with possible magnitudes $J = |L - S|, \ldots, L + S$. Eigenfunctions of $\hat{J}^2$ and $\hat{J}_z$ within this space are given as

$$|LSJM_J\rangle = \sum_{M_L} \sum_{M_S} |LM_L; SM_S\rangle \langle LM_L; SM_S | JM_J \rangle,$$

where the sums over M_L and M_S are restricted to combinations such that $M_L + M_S = M_J$. The energy of each term is given by

$$E(^{2S+1}L_J) = E_0 + \zeta(^{2S+1}L) \langle \alpha LSJM_J | \hat{\mathbf{L}} \cdot \hat{\mathbf{S}} | \alpha LSJM_J \rangle,$$

where E_0 is the energy of the term ^{2S+1}L in the absence of spin–orbit coupling. To proceed we note that

$$\hat{J}^2 = (\hat{\mathbf{L}} + \hat{\mathbf{S}})^2 = \hat{L}^2 + \hat{S}^2 + 2\hat{\mathbf{L}} \cdot \hat{\mathbf{S}} \quad \Rightarrow \quad \hat{\mathbf{L}} \cdot \hat{\mathbf{S}} = \frac{1}{2}(\hat{J}^2 - \hat{L}^2 - \hat{S}^2), \tag{G.10}$$

such that

$$E(^{2S+1}L_J) = E_0 + \frac{1}{2}\hbar^2 \zeta(^{2S+1}L)[J(J+1) - L(L+1) - S(S+1)].$$

The spacing of energy levels is given by

$$E(^{2S+1}L_J) - E(^{2S+1}L_{J-1}) = \hbar^2 \zeta(^{2S+1}L) J,$$

which is known as *Landé's interval rule*.

For completeness we note that another way of writing the effective spin–orbit operator, suitable for the original $|LM_L; SM_S\rangle$ basis, is

$$\hat{H}_{SO}^{\text{eff}} = \zeta(^{2S+1}L) \left[\frac{1}{2}(L_- S_+ + L_+ S_-) + L_z S_z \right]. \tag{G.11}$$

This means that diagonal elements in the original basis are

$$E_0 + \langle \alpha LM_L; SM_S | \zeta(^{2S+1}L) \mathbf{L} \cdot \mathbf{S} | \alpha LM_L; SM_S \rangle = E_0 + \hbar^2 \zeta(^{2S+1}L) M_L M_S. \tag{G.12}$$

The spin–orbit contributions cancel out when we take the trace of the Hamiltonian matrix in the original basis. Since the trace is conserved upon the unitary transformation to the $\langle \alpha LSJM_J \rangle$, we arrive at a formula that allows us to extract the unperturbed energies from the weighted sum over J–levels

$$E_0 = \sum_J \frac{E(^{2S+1}L_J)}{(2L+1)(2S+1)}. \tag{G.13}$$

In order to see the validity and limitations of Landé's interval rule we can have a look at the first five energy levels arising from the np^2 valence configuration of the group 14 elements, given in Table G.1. The energy levels are given in the first five lines of the table. We then calculate the unperturbed energy of the first three J-levels according to Eq. (G.13) as well as the spin–orbit constant $\zeta(^3P)$. Using this data we recalculate the first three energy levels and indicate the mean absolute error (MAE). We see that Landé's interval rule holds quite well for carbon and silicon, but then starts breaking down. This is because the rule was derived assuming that spin–orbit interaction only caused splittings within a single LS-term, in other words,

Table G.1 The first five energy levels (in cm^{-1}) arising from the np^2 valence configuration of the group 14 elements.

LS	J	C	Si	Ge	Sn	Pb	*jj*
^3P	0	0.000	0.000	0.000	0.000	0.000	$6p_{1/2}^2$
	1	16.417	77.112	557.134	1,691.806	7,819.263	$6p_{1/2}^1 6p_{3/2}^1$
	2	43.413	223.157	1,409.961	3,427.673	10,650.327	
^1D	2	10,192.660	6,298.847	7,125.299	8,612.955	21,457.798	$6p_{3/2}^2$
^1S	0	21,648.020	15,394.362	16,367.333	17,162.499	29,466.830	
$E_0(^3P)$		29.591	149.680	969.023	2,468.198	8,523.269	
$\zeta(^3P)$		14.958	75.067	491.774	1,279.870	4,617.398	
MAE_ζ		1.081	1.515	48.415	305.138	2,371.752	
E_{av}		4,858.509	3,215.715	4,047.669	5,496.070	14,231.016	

Further explanation is given in the text.

that spin–orbit interaction is much weaker than the splitting between *LS*-terms. To see the extent to which this is a reasonable assumption we have also calculated the average energy E_{av}, weighted with respect to multiplicity, for the first five energy levels. Comparing those values with the spin–orbit constants $\zeta(^3P)$, we see that for carbon the latter is only 0.3% of the former, but grows to 33% for lead and can no longer be considered a perturbation. As we proceed down the series, we go from *LS*- to *jj*-coupling as indicated by the far left and right column headings.

H

Variational Perturbation Theory

In Chapter 5 expressions for molecular properties to various orders in the perturbation strength(s) were derived in the framework of Rayleigh–Schrödinger perturbation theory, assuming that the exact solutions of the unperturbed Hamiltonian are available. The more typical situation is discussed in Chapter 7, namely, that we have access only to approximate solutions of the time-independent Schrödinger equation and most often only for a single targeted reference state (typically the ground state). Most methods for obtaining such solutions start from some *ansatz* for the wave function where appears a certain number of variational parameters. In methods such as Hartree–Fock, Kohn–Sham DFT, MCSCF, or CI, the approximate ground state is found variationally, that is, by minimizing the energy with respect to the set of variational parameters. A powerful framework for deriving expressions for molecular properties in such cases is *variational perturbation theory*. In this appendix, we give a general outline of how it works.

We start from an energy function

$$E \equiv E(\lambda, \varepsilon) \tag{H.1}$$

where vectors λ and ε collect variational parameters and perturbation field strengths, respectively. In what follows it will be important to distinguish total and partial derivatives. A nice way to do so is to consider the differential

$$E \equiv E(\lambda, \varepsilon) \quad \Rightarrow \quad dE = \left(\frac{\partial E}{\partial \varepsilon_X}\right) d\varepsilon_X + \left(\frac{\partial E}{\partial \lambda_i}\right) d\lambda_i, \tag{H.2}$$

which expresses how infinitesimal changes in the variables (λ, ε) propagate into an infinitesimal change in the energy. The reader is reminded that we use the Einstein summation convention. When any variable is varied the others are held fixed, which is why partial derivatives appear.

In Chapter 2, we pointed out that the Hellmann–Feynman theorem, Eq. (2.107), provided a link between energy derivatives and molecular properties, but considered the exact state case there. Let us therefore look at the total derivative of the approximate state energy with respect to some field strength. Starting from Eq. (H.2), we obtain

$$\frac{dE}{d\varepsilon_A} = \left(\frac{\partial E}{\partial \varepsilon_X}\right) \frac{d\varepsilon_X}{d\varepsilon_A} + \left(\frac{\partial E}{\partial \lambda_i}\right) \frac{d\lambda_i}{d\varepsilon_A}. \tag{H.3}$$

The field strengths are independent variables, but we open up the possibility that the variational parameters λ depend on the perturbation strengths. We therefore obtain the final expression

$$\frac{dE}{d\varepsilon_A} = \frac{\partial E}{\partial \varepsilon_A} + \frac{\partial E}{\partial \lambda_i} \frac{d\lambda_i}{d\varepsilon_A}. \tag{H.4}$$

Principles and Practices of Molecular Properties: Theory, Modeling and Simulations, First Edition.
Patrick Norman, Kenneth Ruud, and Trond Saue.

The right-hand side expression contains the explicit derivative of the energy with respect to the field strength as well as the implicit derivative through the variational parameters. For further clarification, let us consider the differential of a variational parameter

$$\lambda_i \equiv \lambda_i(\varepsilon) \quad \Rightarrow \quad d\lambda_i = \left(\frac{\partial \lambda_i}{\partial \varepsilon_X}\right) d\varepsilon_X. \tag{H.5}$$

The total derivative of the variational parameter with respect to a perturbation strength then becomes

$$\frac{d\lambda_i}{d\varepsilon_A} = \left(\frac{\partial \lambda_i}{\partial \varepsilon_X}\right)\frac{d\varepsilon_X}{d\varepsilon_A} = \frac{\partial \lambda_i}{\partial \varepsilon_A}. \tag{H.6}$$

We find in fact that the total derivative is equal to the partial derivative since the perturbation strengths are independent variables. In practice, it turns out to be more convenient to work with partial derivatives of variational parameters with respect to perturbation strengths.

Terms in the sum over i in the second term of Eq. (H.4) may disappear for any of two reasons:

1) The energy is optimized with respect to variational parameter λ_i, such that $\dfrac{\partial E}{\partial \lambda_i} = 0$.
2) The variational parameter λ_i is not affected by the perturbation, such that $\dfrac{d\lambda_i}{d\varepsilon_A} = 0$.

If it so happens that the second term of Eq. (H.4) drops out, the total derivative reduces to the explicit derivative

$$\frac{dE}{d\varepsilon_A} = \frac{\partial E}{\partial \varepsilon_A}. \tag{H.7}$$

Furthermore, if the approximate energy is formulated as the expectation value of the trial function $|0\rangle$ with respect to the Hamiltonian *and* the trial function has no *explicit* dependence on the perturbation, we recover the Hellmann–Feynman theorem, that is,

$$\frac{dE}{d\varepsilon_A} = \frac{\partial E}{\partial \varepsilon_A} = \langle 0|\frac{\partial H}{\partial \varepsilon_A}|0\rangle. \tag{H.8}$$

To see what applies to a particular method and a particular perturbation it is useful to keep in mind that energy derivatives can also be obtained by finite-field methods, that is, we calculate the derivatives numerically from energy calculations at selected perturbation strengths. One then has to think about what parameters define the trial function and to what extent they vary when the perturbation strength is modified during such calculations. In Chapter 7, we saw that the Hartree–Fock trial function can be conveniently parametrized using orbital rotations, Eq. (7.1). However, in order to fully specify the Hartree–Fock trial function, we need to specify the complete set of unperturbed orbitals, which implies specifying the underlying basis functions and their contractions into orbitals.[1] Assume now that we carry out Hartree–Fock calculations at different electric field strengths. Upon optimization of the wave function for a given field strength, the orbital rotation parameters (or contraction coefficients) are optimized whereas the underlying basis is not modified. This means that all parameters defining the Hartree–Fock trial function are either optimized or not affected by the perturbation, meaning that the Hellmann–Feynman theorem is valid in this case. If we instead perform a series of Hartree–Fock calculations at different nuclear configurations, then the molecular gradient, that is, the derivative of the Hartree–Fock energy with respect to nuclear displacements, does not satisfy the Hellmann–Feynman theorem. This is because the basis functions typically follow nuclear centers, but their positions are not optimized.

1 We can also use the contraction coefficients of the orbitals as variational parameters, but we then need to include the constraint of orbital orthonormality through Lagrange multipliers.

A key tenet of variational perturbation theory is to assume that the energy is stationary at any field strength

$$\left.\frac{\partial E}{\partial \lambda_i}\right|_{\varepsilon} = 0, \quad \forall \lambda_i. \tag{H.9}$$

The variational parameters λ then become functions of the perturbation strengths ε, that is, $\lambda \equiv \lambda(\varepsilon)$. Furthermore, total derivatives of the variational condition, Eq. (H.9), to any order, are zero, giving us *response equations* that allow us to determine perturbed variational parameters to any order. Specifically, up to third order we have

$$\frac{d}{d\varepsilon_A}\left(\frac{\partial E}{\partial \lambda_i}\right) = \frac{\partial^2 E}{\partial \varepsilon_A \partial \lambda_i} + \frac{\partial^2 E}{\partial \lambda_i \partial \lambda_j}\frac{\partial \lambda_j}{\partial \varepsilon_A} = 0 \tag{H.10}$$

$$\begin{aligned}\frac{d^2}{d\varepsilon_A d\varepsilon_B}\left(\frac{\partial E}{\partial \lambda_i}\right) &= \frac{\partial^3 E}{\partial \varepsilon_A \partial \varepsilon_B \partial \lambda_i} + \frac{\partial^3 E}{\partial \varepsilon_A \partial \lambda_i \partial \lambda_j}\frac{\partial \lambda_j}{\partial \varepsilon_B} + \frac{\partial^3 E}{\partial \varepsilon_B \partial \lambda_i \partial \lambda_j}\frac{\partial \lambda_j}{\partial \varepsilon_A}\\ &\quad + \frac{\partial^3 E}{\partial \lambda_i \partial \lambda_j \partial \lambda_k}\frac{\partial \lambda_j}{\partial \varepsilon_A}\frac{\partial \lambda_k}{\partial \varepsilon_B} + \frac{\partial^2 E}{\partial \lambda_i \partial \lambda_j}\frac{\partial^2 \lambda_j}{\partial \varepsilon_A \partial \varepsilon_B} = 0\end{aligned} \tag{H.11}$$

$$\begin{aligned}&\frac{d^3}{d\varepsilon_A d\varepsilon_B d\varepsilon_C}\left(\frac{\partial E}{\partial \lambda_i}\right)\\ &= \frac{\partial^4 E}{\partial \varepsilon_A \partial \varepsilon_B \partial \varepsilon_C \partial \lambda_i} + \frac{\partial^4 E}{\partial \varepsilon_A \partial \varepsilon_B \partial \lambda_i \partial \lambda_j}\frac{\partial \lambda_j}{\partial \varepsilon_C} + \frac{\partial^4 E}{\partial \varepsilon_A \partial \varepsilon_C \partial \lambda_i \partial \lambda_j}\frac{\partial \lambda_j}{\partial \varepsilon_B}\\ &+ \frac{\partial^4 E}{\partial \varepsilon_B \partial \varepsilon_C \partial \lambda_i \partial \lambda_j}\frac{\partial \lambda_j}{\partial \varepsilon_A} + \frac{\partial^4 E}{\partial \varepsilon_A \partial \lambda_i \partial \lambda_j \partial \lambda_k}\frac{\partial \lambda_j}{\partial \varepsilon_B}\frac{\partial \lambda_k}{\partial \varepsilon_C} + \frac{\partial^4 E}{\partial \varepsilon_B \partial \lambda_i \partial \lambda_j \partial \lambda_k}\frac{\partial \lambda_j}{\partial \varepsilon_A}\frac{\partial \lambda_k}{\partial \varepsilon_C}\\ &+ \frac{\partial^4 E}{\partial \varepsilon_C \partial \lambda_i \partial \lambda_j \partial \lambda_k}\frac{\partial \lambda_j}{\partial \varepsilon_A}\frac{\partial \lambda_k}{\partial \varepsilon_B} + \frac{\partial^3 E}{\partial \varepsilon_A \partial \lambda_i \partial \lambda_j}\frac{\partial^2 \lambda_j}{\partial \varepsilon_B \partial \varepsilon_C} + \frac{\partial^3 E}{\partial \varepsilon_B \partial \lambda_i \partial \lambda_j}\frac{\partial^2 \lambda_j}{\partial \varepsilon_A \partial \varepsilon_C}\\ &+ \frac{\partial^3 E}{\partial \varepsilon_C \partial \lambda_i \partial \lambda_j}\frac{\partial^2 \lambda_j}{\partial \varepsilon_A \partial \varepsilon_B} + \frac{\partial^4 E}{\partial \lambda_i \partial \lambda_j \partial \lambda_k \partial \lambda_l}\frac{\partial \lambda_j}{\partial \varepsilon_A}\frac{\partial \lambda_k}{\partial \varepsilon_B}\frac{\partial \lambda_l}{\partial \varepsilon_C} + \frac{\partial^3 E}{\partial \lambda_i \partial \lambda_j \partial \lambda_k}\frac{\partial^2 \lambda_j}{\partial \varepsilon_A \partial \varepsilon_C}\frac{\partial \lambda_k}{\partial \varepsilon_B}\\ &+ \frac{\partial^3 E}{\partial \lambda_i \partial \lambda_j \partial \lambda_k}\frac{\partial \lambda_j}{\partial \varepsilon_A}\frac{\partial^2 \lambda_k}{\partial \varepsilon_B \partial \varepsilon_C} + \frac{\partial^3 E}{\partial \lambda_i \partial \lambda_j \partial \lambda_k}\frac{\partial \lambda_k}{\partial \varepsilon_C}\frac{\partial^2 \lambda_j}{\partial \varepsilon_A \partial \varepsilon_B} + \frac{\partial^2 E}{\partial \lambda_i \partial \lambda_j}\frac{\partial^3 \lambda_j}{\partial \varepsilon_A \partial \varepsilon_B \partial \varepsilon_C} = 0.\end{aligned} \tag{H.12}$$

Under the variational condition in Eq. (H.9), the first-order energy derivative is given by Eq. (H.7), whereas the second-order derivative becomes

$$\frac{d^2 E}{d\varepsilon_A d\varepsilon_B} = \frac{\partial^2 E}{\partial \varepsilon_A \partial \varepsilon_B} + \frac{\partial^2 E}{\partial \varepsilon_A \partial \lambda_i}\frac{\partial \lambda_i}{\partial \varepsilon_B}. \tag{H.13}$$

It requires first-order variational parameters, which we can obtain from the first-order response equation, Eq. (H.10). In passing we may note that the second derivative is not symmetric in perturbation indices A and B. This is because we start from Eq. (H.7). If we instead start from Eq. (H.4) we obtain a symmetric expression

$$\frac{d^2 E}{d\varepsilon_A d\varepsilon_B} = \frac{\partial^2 E}{\partial \varepsilon_A \partial \varepsilon_B} + \frac{\partial^2 E}{\partial \varepsilon_A \partial \lambda_i}\frac{\partial \lambda_i}{\partial \varepsilon_B} + \frac{\partial^2 E}{\partial \varepsilon_B \partial \lambda_i}\frac{\partial \lambda_i}{\partial \varepsilon_A}. \tag{H.14}$$

It requires the first-order response of variational parameters with respect to both perturbation A and B, but, in practical calculations, it has the merit of providing higher accuracy in the evaluation of the linear response function.[2]

2 See H. Sellers. *Int. J. Quantum Chem.*, 30:433, 1986.

Continuing from Eq. (H.13) to third derivatives, we obtain

$$\frac{d^3E}{d\varepsilon_A d\varepsilon_B d\varepsilon_C} = \frac{\partial^3 E}{\partial\varepsilon_A\partial\varepsilon_B\partial\varepsilon_C} + \frac{\partial^3 E}{\partial\varepsilon_A\partial\varepsilon_B\partial\lambda_i}\frac{\partial\lambda_i}{\partial\varepsilon_C} + \frac{\partial^3 E}{\partial\varepsilon_A\varepsilon_C\partial\lambda_i}\frac{\partial\lambda_i}{\partial\varepsilon_B} + \frac{\partial^3 E}{\partial\varepsilon_A\partial\lambda_i\partial\lambda_j}\frac{\partial\lambda_i}{\partial\varepsilon_B}\frac{\partial\lambda_j}{\partial\varepsilon_C} + \frac{\partial^2 E}{\partial\varepsilon_A\partial\lambda_i}\frac{\partial^2\lambda_i}{\partial\varepsilon_B\partial\varepsilon_C}. \tag{H.15}$$

The third-order energy derivative appears to require both first- and second-order parameters. However, we can use the first-order response equation, Eq. (H.10), to rewrite the final term as

$$\frac{d^3E}{d\varepsilon_A d\varepsilon_B d\varepsilon_C} = \frac{\partial^3 E}{\partial\varepsilon_A\partial\varepsilon_B\partial\varepsilon_C} + \frac{\partial^3 E}{\partial\varepsilon_A\partial\varepsilon_B\partial\lambda_i}\frac{\partial\lambda_i}{\partial\varepsilon_C} + \frac{\partial^3 E}{\partial\varepsilon_A\partial\varepsilon_C\partial\lambda_i}\frac{\partial\lambda_i}{\partial\varepsilon_B} + \frac{\partial^3 E}{\partial\varepsilon_A\partial\lambda_i\partial\lambda_j}\frac{\partial\lambda_i}{\partial\varepsilon_B}\frac{\partial\lambda_j}{\partial\varepsilon_C} - \frac{\partial^2 E}{\partial\lambda_i\partial\lambda_j}\frac{\partial^2\lambda_i}{\partial\varepsilon_B\partial\varepsilon_C}\frac{\partial\lambda_j}{\partial\varepsilon_A}. \tag{H.16}$$

Thereafter, we apply the second-order response equation, Eq. (H.11), to the final term to obtain

$$\begin{aligned}\frac{d^3E}{d\varepsilon_A d\varepsilon_B d\varepsilon_C} &= \frac{\partial^3 E}{\partial\varepsilon_A\partial\varepsilon_B\partial\varepsilon_C} + \frac{\partial^3 E}{\partial\varepsilon_A\partial\varepsilon_B\partial\lambda_i}\frac{\partial\lambda_i}{\partial\varepsilon_C} + \frac{\partial^3 E}{\partial\varepsilon_A\partial\varepsilon_C\partial\lambda_i}\frac{\partial\lambda_i}{\partial\varepsilon_B}\\ &+ \frac{\partial^3 E}{\partial\varepsilon_B\partial\varepsilon_C\partial\lambda_i}\frac{\partial\lambda_i}{\partial\varepsilon_A} + \frac{\partial^3 E}{\partial\varepsilon_A\partial\lambda_i\partial\lambda_j}\frac{\partial\lambda_i}{\partial\varepsilon_B}\frac{\partial\lambda_j}{\partial\varepsilon_C} + \frac{\partial^3 E}{\partial\varepsilon_B\partial\lambda_i\partial\lambda_j}\frac{\partial\lambda_i}{\partial\varepsilon_C}\frac{\partial\lambda_j}{\partial\varepsilon_A}\\ &+ \frac{\partial^3 E}{\partial\varepsilon_C\partial\lambda_i\partial\lambda_j}\frac{\partial\lambda_i}{\partial\varepsilon_B}\frac{\partial\lambda_j}{\partial\varepsilon_A} + \frac{\partial^3 E}{\partial\lambda_i\partial\lambda_j\partial\lambda_k}\frac{\partial\lambda_i}{\partial\varepsilon_B}\frac{\partial\lambda_j}{\partial\varepsilon_A}\frac{\partial\lambda_k}{\partial\varepsilon_C}.\end{aligned} \tag{H.17}$$

The expression becomes longer, but only contains first-order variational parameters. For the fourth-order energy derivative, we get

$$\begin{aligned}&\frac{d^4E}{d\varepsilon_A d\varepsilon_B d\varepsilon_C d\varepsilon_D}\\
&= \frac{\partial^4 E}{\partial\varepsilon_A\partial\varepsilon_B\partial\varepsilon_C\partial\varepsilon_D} + \frac{\partial^4 E}{\partial\varepsilon_A\partial\varepsilon_B\partial\varepsilon_C\partial\lambda_i}\frac{\partial\lambda_i}{\partial\varepsilon_D} + \frac{\partial^4 E}{\partial\varepsilon_A\partial\varepsilon_B\partial\varepsilon_D\partial\lambda_i}\frac{\partial\lambda_i}{\partial\varepsilon_C}\\
&+ \frac{\partial^4 E}{\partial\varepsilon_A\partial\varepsilon_C\partial\varepsilon_D\partial\lambda_i}\frac{\partial\lambda_i}{\partial\varepsilon_B} + \frac{\partial^4 E}{\partial\varepsilon_B\partial\varepsilon_C\partial\varepsilon_D\partial\lambda_i}\frac{\partial\lambda_i}{\partial\varepsilon_A} + \frac{\partial^4 E}{\partial\varepsilon_A\partial\varepsilon_B\partial\lambda_i\partial\lambda_j}\frac{\partial\lambda_i}{\partial\varepsilon_C}\frac{\partial\lambda_j}{\partial\varepsilon_D}\\
&+ \frac{\partial^4 E}{\partial\varepsilon_A\varepsilon_C\partial\lambda_i\partial\lambda_j}\frac{\partial\lambda_i}{\partial\varepsilon_B}\frac{\partial\lambda_j}{\partial\varepsilon_D} + \frac{\partial^3 E}{\partial\varepsilon_B\partial\varepsilon_C\partial\lambda_i\partial\lambda_j}\frac{\partial\lambda_i}{\partial\varepsilon_A}\frac{\partial\lambda_j}{\partial\varepsilon_D} + \frac{\partial^4 E}{\partial\varepsilon_A\partial\varepsilon_D\partial\lambda_i\partial\lambda_j}\frac{\partial\lambda_i}{\partial\varepsilon_B}\frac{\partial\lambda_j}{\partial\varepsilon_C}\\
&+ \frac{\partial^4 E}{\partial\varepsilon_B\partial\varepsilon_D\partial\lambda_i\partial\lambda_j}\frac{\partial\lambda_i}{\partial\varepsilon_C}\frac{\partial\lambda_j}{\partial\varepsilon_A} + \frac{\partial^4 E}{\partial\varepsilon_C\partial\varepsilon_D\partial\lambda_i\partial\lambda_j}\frac{\partial\lambda_i}{\partial\varepsilon_B}\frac{\partial\lambda_j}{\partial\varepsilon_A} + \frac{\partial^3 E}{\partial\varepsilon_A\partial\varepsilon_B\partial\lambda_i}\frac{\partial^2\lambda_i}{\partial\varepsilon_C\partial\varepsilon_D}\\
&+ \frac{\partial^3 E}{\partial\varepsilon_A\varepsilon_C\partial\lambda_i}\frac{\partial^2\lambda_i}{\partial\varepsilon_B\partial\varepsilon_D} + \frac{\partial^3 E}{\partial\varepsilon_B\partial\varepsilon_C\partial\lambda_i}\frac{\partial^2\lambda_i}{\partial\varepsilon_A\partial\varepsilon_D} + \frac{\partial^4 E}{\partial\varepsilon_A\partial\lambda_i\partial\lambda_j\partial\lambda_k}\frac{\partial\lambda_i}{\partial\varepsilon_B}\frac{\partial\lambda_j}{\partial\varepsilon_C}\frac{\partial\lambda_k}{\partial\varepsilon_D}\\
&+ \frac{\partial^4 E}{\partial\varepsilon_B\partial\lambda_i\partial\lambda_j\partial\lambda_k}\frac{\partial\lambda_i}{\partial\varepsilon_C}\frac{\partial\lambda_j}{\partial\varepsilon_A}\frac{\partial\lambda_k}{\partial\varepsilon_D} + \frac{\partial^4 E}{\partial\varepsilon_C\partial\lambda_i\partial\lambda_j\partial\lambda_k}\frac{\partial\lambda_i}{\partial\varepsilon_B}\frac{\partial\lambda_j}{\partial\varepsilon_A}\frac{\partial\lambda_k}{\partial\varepsilon_D}\\
&+ \frac{\partial^4 E}{\partial\varepsilon_D\partial\lambda_i\partial\lambda_j\partial\lambda_k}\frac{\partial\lambda_i}{\partial\varepsilon_B}\frac{\partial\lambda_j}{\partial\varepsilon_A}\frac{\partial\lambda_k}{\partial\varepsilon_C} + \frac{\partial^3 E}{\partial\varepsilon_A\partial\lambda_i\partial\lambda_j}\frac{\partial^2\lambda_i}{\partial\varepsilon_B\partial\varepsilon_D}\frac{\partial\lambda_j}{\partial\varepsilon_C}\\
&+ \frac{\partial^3 E}{\partial\varepsilon_A\partial\lambda_i\partial\lambda_j}\frac{\partial\lambda_i}{\partial\varepsilon_B}\frac{\partial^2\lambda_j}{\partial\varepsilon_C\partial\varepsilon_D} + \frac{\partial^3 E}{\partial\varepsilon_B\partial\lambda_i\partial\lambda_j}\frac{\partial^2\lambda_i}{\partial\varepsilon_C\partial\varepsilon_D}\frac{\partial\lambda_j}{\partial\varepsilon_A} + \frac{\partial^3 E}{\partial\varepsilon_B\partial\lambda_i\partial\lambda_j}\frac{\partial\lambda_i}{\partial\varepsilon_C}\frac{\partial^2\lambda_j}{\partial\varepsilon_A\partial\varepsilon_D}\\
&+ \frac{\partial^3 E}{\partial\varepsilon_C\partial\lambda_i\partial\lambda_j}\frac{\partial^2\lambda_i}{\partial\varepsilon_B\partial\varepsilon_D}\frac{\partial\lambda_j}{\partial\varepsilon_A} + \frac{\partial^3 E}{\partial\varepsilon_C\partial\lambda_i\partial\lambda_j}\frac{\partial\lambda_i}{\partial\varepsilon_B}\frac{\partial^2\lambda_j}{\partial\varepsilon_A\partial\varepsilon_D}\end{aligned}$$

$$+\frac{\partial^4 E}{\partial\lambda_i\partial\lambda_j\partial\lambda_k\partial\lambda_l}\frac{\partial\lambda_i}{\partial\varepsilon_B}\frac{\partial\lambda_j}{\partial\varepsilon_A}\frac{\partial\lambda_k}{\partial\varepsilon_C}\frac{\partial\lambda_l}{\partial\varepsilon_D}+\frac{\partial^3 E}{\partial\lambda_i\partial\lambda_j\partial\lambda_k}\frac{\partial^2\lambda_i}{\partial\varepsilon_B\partial\varepsilon_D}\frac{\partial\lambda_j}{\partial\varepsilon_A}\frac{\partial\lambda_k}{\partial\varepsilon_C}$$
$$+\frac{\partial^3 E}{\partial\lambda_i\partial\lambda_j\partial\lambda_k}\frac{\partial\lambda_i}{\partial\varepsilon_B}\frac{\partial^2\lambda_j}{\partial\varepsilon_A\partial\varepsilon_D}\frac{\partial\lambda_k}{\partial\varepsilon_C}+\frac{\partial^3 E}{\partial\lambda_i\partial\lambda_j\partial\lambda_k}\frac{\partial\lambda_i}{\partial\varepsilon_B}\frac{\partial\lambda_j}{\partial\varepsilon_A}\frac{\partial^2\lambda_k}{\partial\varepsilon_C\partial\varepsilon_D}. \tag{H.18}$$

In this case, no reductions are possible and we need both first- and second-order responses in the parameters. Generally, the variational parameters comply with the Wigner $(2n+1)$-rule:[3] With variational parameters to order n we can calculate the energy to order $2n+1$.

The expressions so far assumed that the second term in the right-hand side of Eq. (H.4) is zero, in particular that the energy is fully optimized with respect to all variational parameters, at all field strengths, as expressed by Eq. (H.9). However, it is easy to find exceptions to this rule. For instance, the CI energy is stationary with respect to CI coefficients, but not with respect to orbitals, which are optimized at the Hartree–Fock level. Worse, coupled cluster amplitudes are normally found by means of projection and not by optimization. Fortunately, in such cases it is possible to recover the machinery of variational perturbation theory by the introduction of Lagrange multipliers. For instance, for the calculation of molecular properties at the CI level, we may start from the Lagrangian[4]

$$L^{\mathrm{CI}}(R,\kappa,\overline{\kappa})=E^{\mathrm{CI}}(R,\kappa)+\overline{\kappa}_{pq}\frac{\partial E^{\mathrm{HF}}(\kappa)}{\partial\kappa_{pq}}, \tag{H.19}$$

using the CI and HF parametrization of Eqs. (7.38) and (7.1), respectively. The HF variational conditions, multiplied by Lagrange multipliers $\overline{\kappa}_{pq}$, have been added to the CI energy (here with an implied summation over the pair of indices pq). We now insist on the Lagrangian being stationary with respect to all parameters. This gives three kinds of equations

$$\frac{\partial L^{\mathrm{CI}}}{\partial R_n}=\frac{\partial E^{\mathrm{CI}}}{\partial R_n}=0, \tag{H.20}$$
$$\frac{\partial L^{\mathrm{CI}}}{\partial\overline{\kappa}_{rs}}=\frac{\partial E^{\mathrm{HF}}}{\partial\kappa_{rs}}=0, \tag{H.21}$$
$$\frac{\partial L^{\mathrm{CI}}}{\partial\kappa_{rs}}=\frac{\partial E^{\mathrm{CI}}}{\partial\kappa_{rs}}+\overline{\kappa}_{pq}\frac{\partial^2 E^{\mathrm{HF}}}{\partial\kappa_{pq}\partial\kappa_{rs}}=0. \tag{H.22}$$

The first two equations are simply the CI and HF variational conditions, whereas the third equation has the form of a first-order response equation involving the HF Hessian [see Eq. (H.10)] to be solved for the Lagrange multipliers.

A first-order molecular property is obtained from the first-order derivative of the Lagrangian with respect to the perturbation strength. This derivative becomes equal to

$$\frac{dL^{\mathrm{CI}}}{d\varepsilon_A}=\frac{\partial L^{\mathrm{CI}}}{\partial\varepsilon_A}+\frac{\partial L^{\mathrm{CI}}}{\partial R_n}\frac{\partial R_n}{\partial\varepsilon_A}+\frac{\partial L^{\mathrm{CI}}}{\partial\kappa_{rs}}\frac{\partial\kappa_{rs}}{\partial\varepsilon_A}+\frac{\partial L^{\mathrm{CI}}}{\partial\overline{\kappa}_{rs}}\frac{\partial\overline{\kappa}_{rs}}{\partial\varepsilon_A}. \tag{H.23}$$

Using the variational conditions Eqs. (H.20)–(H.22), we get

$$\begin{aligned}\frac{dL^{\mathrm{CI}}}{d\varepsilon_A}=\frac{\partial L^{\mathrm{CI}}}{\partial\varepsilon_A}&=\frac{\partial E^{\mathrm{CI}}}{\partial\varepsilon_A}+\frac{\partial\overline{\kappa}_{pq}}{\partial\varepsilon_A}\frac{\partial E^{\mathrm{HF}}}{\partial\kappa_{pq}}+\overline{\kappa}_{pq}\frac{\partial^2 E^{\mathrm{HF}}}{\partial\varepsilon_A\partial\kappa_{pq}}\\&=\frac{\partial E^{\mathrm{CI}}}{\partial\varepsilon_A}+\overline{\kappa}_{pq}\frac{\partial^2 E^{\mathrm{HF}}}{\partial\varepsilon_A\partial\kappa_{pq}},\end{aligned} \tag{H.24}$$

---

3 The classical reference to this proof is E. Wigner, *Math. Natur. Anz. (Budapest)*, 53:477, 1930, but the result is already found in E. A. Hylleraas, *Z. Phys.*, 65:209, 1930. A nice proof is given in I. Mayer. *Simple Theorems, Proofs and Derivations in Quantum Chemistry*. Kluwer, 2003, as well as in T. Helgaker *et al.*, *Chem. Rev.*, 112:543, 2012.

4 See also T. Helgaker and P. Jørgensen, *Theor. Chim. Acta*, 75:111, 1989.

where the HF variational condition, Eq. (H.21), has been used in the final step. Since the derivatives are normally evaluated at zero field strength, we need only the unperturbed Lagrange multipliers that can be calculated once and for all from Eq. (H.22), irrespective of the molecular property at hand.

Proceeding to the second-order derivative with respect to perturbation strength, we obtain, starting from Eq. (H.23),

$$\begin{aligned}\frac{d^2L^{\mathrm{CI}}}{d\varepsilon_A d\varepsilon_B} &= \frac{\partial^2 L^{\mathrm{CI}}}{\partial\varepsilon_A\partial\varepsilon_B} + \frac{\partial^2 L^{\mathrm{CI}}}{\partial R_m \partial R_n}\frac{\partial R_n}{\partial\varepsilon_A}\frac{\partial R_m}{\partial\varepsilon_B} + \frac{\partial^2 L^{\mathrm{CI}}}{\partial\kappa_{rs}\partial\kappa_{tu}}\frac{\partial\kappa_{rs}}{\partial\varepsilon_A}\frac{\partial\kappa_{tu}}{\partial\varepsilon_B}\\
&+\sum \mathcal{P}_{AB}\left[\frac{\partial^2 L^{\mathrm{CI}}}{\partial\varepsilon_A\partial R_n}\frac{\partial R_n}{\partial\varepsilon_B} + \frac{\partial^2 L^{\mathrm{CI}}}{\partial\varepsilon_A\partial\kappa_{rs}}\frac{\partial\kappa_{rs}}{\partial\varepsilon_B} + \frac{\partial^2 L^{\mathrm{CI}}}{\partial\varepsilon_A\partial\bar{\kappa}_{rs}}\frac{\partial\bar{\kappa}_{rs}}{\partial\varepsilon_B}\right.\\
&\left.+\frac{\partial^2 L^{\mathrm{CI}}}{\partial R_n\partial\kappa_{rs}}\frac{\partial R_n}{\partial\varepsilon_A}\frac{\partial\kappa_{rs}}{\partial\varepsilon_B} + \frac{\partial^2 L^{\mathrm{CI}}}{\partial R_n\partial\bar{\kappa}_{rs}}\frac{\partial\bar{\kappa}_{rs}}{\partial\varepsilon_A}\frac{\partial R_n}{\partial\varepsilon_B} + \frac{\partial^2 L^{\mathrm{CI}}}{\partial\bar{\kappa}_{rs}\partial\kappa_{tu}}\frac{\partial\bar{\kappa}_{rs}}{\partial\varepsilon_A}\frac{\partial\kappa_{tu}}{\partial\varepsilon_B}\right]\\
&+\frac{\partial L^{\mathrm{CI}}}{\partial R_n}\frac{\partial^2 R_n}{\partial\varepsilon_A\partial\varepsilon_B} + \frac{\partial L^{\mathrm{CI}}}{\partial\kappa_{rs}}\frac{\partial^2\kappa_{rs}}{\partial\varepsilon_A\partial\varepsilon_B} + \frac{\partial L^{\mathrm{CI}}}{\partial\bar{\kappa}_{rs}}\frac{\partial^2\bar{\kappa}_{rs}}{\partial\varepsilon_A\partial\varepsilon_B} + \frac{\partial^2 L^{\mathrm{CI}}}{\partial\bar{\kappa}_{rs}\partial\bar{\kappa}_{tu}}\frac{\partial\bar{\kappa}_{rs}}{\partial\varepsilon_A}\frac{\partial\bar{\kappa}_{tu}}{\partial\varepsilon_B},\end{aligned}$$

where $\sum \mathcal{P}_{AB}$ indicates the sum obtained from the permutation of labels A and B. In this expression both first- and second-order derivatives of the variational parameters and Lagrange multipliers appear. However, using the variational conditions all second-order parameters (the first three terms on the final line) may be eliminated. In addition, the second-order partial derivative of the Lagrangian with respect to Lagrange multipliers (final term) will be zero since the multipliers only appear linearly.

It appears as if all first-order parameters are needed. However, by writing out separately the contributions from the two terms of the Lagrangian, we obtain

$$\begin{aligned}\frac{d^2L^{\mathrm{CI}}}{d\varepsilon_A d\varepsilon_B}\\
&= \frac{\partial^2 E^{\mathrm{CI}}}{\partial\varepsilon_A\partial\varepsilon_B} + \frac{\partial^2 E^{\mathrm{CI}}}{\partial R_m\partial R_n}\frac{\partial R_n}{\partial\varepsilon_A}\frac{\partial R_m}{\partial\varepsilon_B} + \frac{\partial^2 E^{\mathrm{CI}}}{\partial\kappa_{rs}\partial\kappa_{tu}}\frac{\partial\kappa_{rs}}{\partial\varepsilon_A}\frac{\partial\kappa_{tu}}{\partial\varepsilon_B}\\
&+\sum \mathcal{P}_{AB}\left[\frac{\partial^2 E^{\mathrm{CI}}}{\partial\varepsilon_A\partial R_n}\frac{\partial R_n}{\partial\varepsilon_B} + \frac{\partial^2 E^{\mathrm{CI}}}{\partial\varepsilon_A\partial\kappa_{rs}}\frac{\partial\kappa_{rs}}{\partial\varepsilon_B} + \frac{\partial^2 E^{\mathrm{CI}}}{\partial R_n\partial\kappa_{rs}}\frac{\partial R_n}{\partial\varepsilon_A}\frac{\partial\kappa_{rs}}{\partial\varepsilon_B}\right]\\
&+\bar{\kappa}_{pq}\left[\frac{\partial^3 E^{\mathrm{HF}}}{\partial\varepsilon_A\partial\varepsilon_B\partial\kappa_{pq}} + \frac{\partial^3 E^{\mathrm{HF}}}{\partial\kappa_{pq}\partial\kappa_{rs}\partial\kappa_{tu}}\frac{\partial\kappa_{rs}}{\partial\varepsilon_A}\frac{\partial\kappa_{tu}}{\partial\varepsilon_B} + \sum \mathcal{P}_{AB}\left(\frac{\partial^3 E^{\mathrm{HF}}}{\partial\varepsilon_A\partial\kappa_{pq}\partial\kappa_{rs}}\frac{\partial\kappa_{rs}}{\partial\varepsilon_B}\right)\right]\\
&+2\sum \mathcal{P}_{AB}\left[\frac{\partial\bar{\kappa}_{pq}}{\partial\varepsilon_A}\left(\frac{\partial^2 E^{\mathrm{HF}}}{\partial\varepsilon_B\partial\kappa_{pq}} + \frac{\partial^2 E^{\mathrm{HF}}}{\partial\kappa_{pq}\partial\kappa_{rs}}\frac{\partial\kappa_{rs}}{\partial\varepsilon_B}\right)\right].\end{aligned} \tag{H.25}$$

We then see that the first-order Lagrange multipliers are eliminated since they multiply the left-hand side of first-order HF response equation [compare with Eq. (H.10)], which is zero. This is an instance of the observation that the Lagrange multipliers, compared to the variational parameters, obey a stricter $(2n+2)$-rule.[5]

5 For proof, see T. Helgaker *et al.*, *Chem. Rev.*, 112:543, 2012.

I

Two-Level Atom

In Chapter 5, we presented numerical solutions to the Schrödinger equation as obtained by means of time propagations of a two-level atom subjected to a monochromatic, time-oscillating electric field. This allowed us to grasp a number of concepts without having to go through the mathematics needed to find analytic solutions. In contrast, this appendix will provide several of the formulas for the time evolution of the two-level system that were left out.

A matrix representation of the time-dependent wave equation of the two-level atom is given by

$$\begin{bmatrix} E_a & -\mu_{ab}F(t) \\ -\mu_{ab}F(t) & E_b \end{bmatrix} \begin{bmatrix} c_a(t) \\ c_b(t) \end{bmatrix} = i\hbar \begin{bmatrix} \dot{c}_a(t) \\ \dot{c}_b(t) \end{bmatrix}, \tag{I.1}$$

in agreement with the illustration given in Figure 5.1. In the absence of an external field $F(t)$, the time evolution operator, Eq. (2.75), takes the form

$$\mathcal{U}(t, t_0) = e^{-i\mathcal{H}_0(t-t_0)/\hbar} = \begin{bmatrix} e^{-iE_a(t-t_0)/\hbar} & 0 \\ 0 & e^{-iE_b(t-t_0)/\hbar} \end{bmatrix}, \tag{I.2}$$

and the wave function evolves according to

$$\begin{bmatrix} c_a(t) \\ c_b(t) \end{bmatrix} = \mathcal{U}(t, 0) \begin{bmatrix} c_a(0) \\ c_b(0) \end{bmatrix}, \tag{I.3}$$

that is,

$$\psi(t) = c_a(0)e^{-i\omega_a t}\psi_a + c_b(0)e^{-i\omega_b t}\psi_b, \tag{I.4}$$

where, for instance, $E_a = \hbar\omega_a$. We see that the expansion coefficients become equal to their initial values multiplied with simple time-dependent phase factors.

In the subsequent sections, we will introduce the monochromatic field given in Eq. (5.15). First, we consider the nonperturbative solution near resonance that leads to the Rabi oscillations seen in Figure 5.16. Next, we consider the time evolution of the two-level system in the perturbative regime, to begin with from the time evolution of the projection amplitudes of Eq. (5.106) and thereafter within the quasi-energy approach using the parametrization of the phase-isolated wave function given in Eq. (5.81).

We note that in Figure 5.16, we used a very large electric field amplitude in order to obtain a Rabi frequency that fit into the adopted time window of 500 a.u. If we had used an electric field amplitude of 5.0×10^{-5} a.u. as adopted for most parts of Chapter 5, the principal behavior would have been the same but with a lower Rabi frequency. The Rabi oscillations in the excited state population for this situation are shown in Figure 5.17.

Principles and Practices of Molecular Properties: Theory, Modeling and Simulations, First Edition.
Patrick Norman, Kenneth Ruud, and Trond Saue.

I.1 Rabi Oscillations

In this section, we consider a monochromatic field given by

$$F(t) = F^{\omega} \sin \omega t. \tag{I.5}$$

Contrary to Eq. (5.15), we do not introduce switch-on function. Rather, we shall consider times sufficiently far from the initial time such that any effect of the switch-on is erased.

It will be convenient to introduce a new set of expansion coefficients for the wave function according to

$$c_k(t) = d_k(t) e^{-i\phi_k t}. \tag{I.6}$$

Note that these are different from the parametrization in Eq. (5.106) in that the phases ϕ_k are not set equal to state energies. The time-dependent wave equation is then given by

$$\begin{bmatrix} E_a - \hbar\phi_a & -\mu_{ab}F^{\omega}e^{-i(\phi_b-\phi_a)t}\sin\omega t \\ -\mu_{ab}F^{\omega}e^{i(\phi_b-\phi_a)t}\sin\omega t & E_b - \hbar\phi_b \end{bmatrix} \begin{bmatrix} d_a(t) \\ d_b(t) \end{bmatrix} = i\hbar \begin{bmatrix} \dot{d}_a(t) \\ \dot{d}_b(t) \end{bmatrix} \tag{I.7}$$

and the corresponding density matrix, Eq. (5.264), reads

$$\rho = \begin{bmatrix} |c_a|^2 & c_a c_b^* \\ c_b c_a^* & |c_b|^2 \end{bmatrix} = \begin{bmatrix} |d_a|^2 & d_a d_b^* e^{i(\phi_b-\phi_a)t} \\ d_b d_a^* e^{-i(\phi_b-\phi_a)t} & |d_b|^2 \end{bmatrix}. \tag{I.8}$$

It is seen that, reformulated in terms of the coefficients d_a and d_b, only the phase difference $(\phi_b - \phi_a)$ comes into play. We will now choose to set the phases to values according to

$$\phi_b + \phi_a = (E_b + E_a)/\hbar, \tag{I.9}$$

$$\phi_b - \phi_a = \omega, \tag{I.10}$$

where Eq. (I.9) makes the Hamiltonian traceless. By furthermore decomposing the sine functions into complex exponentials the equation is transformed into

$$[\tilde{H}_0 + \tilde{H}_1] \begin{bmatrix} d_a(t) \\ d_b(t) \end{bmatrix} = i\hbar \begin{bmatrix} \dot{d}_a(t) \\ \dot{d}_b(t) \end{bmatrix}, \tag{I.11}$$

where

$$\tilde{H}_0 = \begin{bmatrix} -\frac{1}{2}\hbar(\omega_{ba} - \omega) & \frac{i}{2}\mu_{ab}F^{\omega} \\ -\frac{i}{2}\mu_{ab}F^{\omega} & \frac{1}{2}\hbar(\omega_{ba} - \omega) \end{bmatrix}; \quad \tilde{H}_1 = \begin{bmatrix} 0 & -\frac{i}{2}\mu_{ab}F^{\omega}e^{-i2\omega t} \\ \frac{i}{2}\mu_{ab}F^{\omega}e^{i2\omega t} & 0 \end{bmatrix} \tag{I.12}$$

and ω_{ba} is the angular transition frequency. A simple analytic solution is not available for the full equation. However, for ω close to resonance ($\omega \approx \omega_{ba}$), we may ignore the time-dependent part $\tilde{H}_1$ of the Hamiltonian. In Section 3.1.5, we learned that a linearly polarized electromagnetic wave can be decomposed into two oppositely circularly polarized waves. Discarding $\tilde{H}_1$ is known as the *rotating wave approximation* and corresponds to ignoring one circular component of the incoming linearly polarized field.

Under near resonance conditions, it is convenient to introduce the *detuning* Δ and the *Rabi frequency* Ω:

$$\Delta = \omega_{ba} - \omega, \tag{I.13}$$

$$\Omega = \frac{\mu_{ab}F^{\omega}}{\hbar}. \tag{I.14}$$

We then have

$$\tilde{H}_0 = \frac{\hbar}{2} \begin{bmatrix} -\Delta & i\Omega \\ -i\Omega & \Delta \end{bmatrix}. \tag{I.15}$$

Since $\tilde{H}_0$ is time independent, we may obtain our approximate solution in terms of the time-evolution operator, that is,

$$\begin{bmatrix} d_a(t) \\ d_b(t) \end{bmatrix} = e^{-i\tilde{H}_0 t/\hbar} \begin{bmatrix} d_a(0) \\ d_b(0) \end{bmatrix}. \tag{I.16}$$

A great simplification is obtained by noting that we can rewrite the Hamiltonian matrix in terms of the Pauli spin matrices according to

$$\tilde{H}_0/\hbar = -\frac{1}{2}[\Delta\sigma_z + \Omega\sigma_y] = -\frac{1}{2}\Omega'(\boldsymbol{\sigma} \cdot \boldsymbol{n}), \tag{I.17}$$

where

$$\Omega' = \sqrt{\Omega^2 + \Delta^2} \quad \text{and} \quad \boldsymbol{n} = (0, \Omega, \Delta)/\Omega' \tag{I.18}$$

The time evolution operator accordingly assumes the form of a spin-1/2 rotation operator [cf. Eq. (4.76)]:

$$U(t,0) = e^{-i\tilde{H}_0 t/\hbar} = \exp\left[\frac{i}{2}\Omega'(\boldsymbol{\sigma} \cdot \boldsymbol{n})t\right] = \cos(\Omega' t/2) + i(\boldsymbol{\sigma} \cdot \boldsymbol{n})\sin(\Omega' t/2), \tag{I.19}$$

or, in full,

$$U(t,0) = \begin{bmatrix} \cos(\Omega' t/2) + \frac{\Delta}{\Omega'}\sin(\Omega' t/2) & -i\frac{\Omega}{\Omega'}\sin(\Omega' t/2) \\ i\frac{\Omega}{\Omega'}\sin(\Omega' t/2) & \cos(\Omega' t/2) - \frac{\Delta}{\Omega'}\sin(\Omega' t/2) \end{bmatrix}. \tag{I.20}$$

If we impose that at $t = 0$ only the lower level a is populated then we obtain the specific solution

$$\begin{bmatrix} d_a(t) \\ d_b(t) \end{bmatrix} = e^{-i\tilde{H}_0 t/\hbar} \begin{bmatrix} 1 \\ 0 \end{bmatrix} = \begin{bmatrix} \cos(\Omega' t/2) + \frac{\Delta}{\Omega'}\sin(\Omega' t/2) \\ i\frac{\Omega}{\Omega'}\sin(\Omega' t/2) \end{bmatrix} \tag{I.21}$$

and the population of the upper level is given by

$$\rho_b(t) = |c_b(t)|^2 = |d_b(t)|^2 = \frac{\Omega^2}{\Omega^2 + \Delta^2}\sin^2(\Omega' t/2). \tag{I.22}$$

At resonance ($\omega = \omega_{ba}$), the time evolution operator in Eq. (I.20) reduces to

$$U(t,0) - \begin{bmatrix} \cos(\Omega t/2) & -i\sin(\Omega t/2) \\ i\sin(\Omega t/2) & \cos(\Omega t/2) \end{bmatrix}, \tag{I.23}$$

and

$$\rho_b(t) = |c_b(t)|^2 = |d_b(t)|^2 = \sin^2(\Omega t/2) = \frac{1}{2}[1 - \cos\Omega t]. \tag{I.24}$$

We may also note that at resonance, the wave function can be written as

$$\psi(t) = \cos(\Omega t/2)e^{-iE_a t/\hbar}\psi_a - i\sin(\Omega t/2)e^{-iE_b t/\hbar}\psi_b \tag{I.25}$$

$$= \frac{1}{2}[e^{-i(\omega_a+\Omega/2)t} + e^{-i(\omega_a-\Omega/2)t}]\psi_a + \frac{1}{2}[e^{-i(\omega_b+\Omega/2)t} - e^{-i(\omega_b-\Omega/2)t}]\psi_b. \tag{I.26}$$

Comparing with Eq. (I.4), we see that the wave function is a superposition of the two levels of our system, but each level is split by the Rabi frequency Ω. In 1955, Autler and Townes suggested that if Rabi oscillations were set up between isolated levels a and b of some molecular system, the resulting superposition could be probed by a weak oscillating field inducing transitions to a third level c. Each level (a and b) would then appear split by the Rabi frequency, which they could show experimentally. This is known as the dynamical Stark effect or simply the Autler–Townes effect.

I.2 Time-Dependent Perturbation Theory

We now apply time-dependent perturbation theory based on a perturbation expansion of the projection amplitudes appearing in Eq. (5.106). Proceeding as in Section 5.4.1, we find that the first-order projection amplitude is given by

$$d_m^{(1)}(t) = \frac{1}{i\hbar}\int_0^t V_{m0}(t')e^{i\omega_{m0}t'}dt'. \tag{I.27}$$

For a two-level atom in the presence of the electric field of Eq. (5.15), we obtain

$$d_a^{(1)}(t) = 0 \tag{I.28}$$

$$d_b^{(1)}(t) = -\frac{\mu_{ab}F^\omega}{i\hbar}\int_0^t \sin\omega t' \times \text{erf}\,(at')e^{i\omega_{m0}t'}dt' \tag{I.29}$$

$$= \frac{\mu_{ab}F^\omega}{2\hbar}\int_0^t (e^{i(\omega_{ba}+\omega)t'} - e^{i(\omega_{ba}-\omega)t'}) \times \text{erf}\,(at')dt'. \tag{I.30}$$

We shall use the formula

$$\int \text{erf}\,(at)e^{i\delta t}dt = \frac{1}{i\delta}\left[e^{i\delta t}\text{erf}\,(at) - \exp\left(\frac{-\delta^2}{4a^2}\right)\text{erf}\,\left(at - \frac{i\delta}{2a}\right)\right]. \tag{I.31}$$

In our nonresonant example, we used $a = 0.01$ a.u., $\omega = 0.1$ a.u., and $\omega_{ba} = 0.5$ a.u. This means that the second of the above terms is strongly damped by the exponential and will be ignored in the following. Our final expression for the first-order projection amplitude becomes

$$d_b^{(1)}(t) = \frac{\mu_{ab}F^\omega}{\hbar}e^{i\omega_{ba}t}\left[\frac{\omega_{ba}\sin\omega t + i\omega\cos\omega t}{(\omega_{ba}^2 - \omega^2)}\right]\text{erf}\,(at), \tag{I.32}$$

and one may note that it is proportional to the Rabi frequency. The corresponding excited state population is

$$\rho_b(t) = \left(\frac{\mu_{ab}F^\omega}{\hbar}\right)^2\left[\frac{\omega^2 + (\omega_{ba}^2 - \omega^2)\sin^2\omega t}{(\omega_{ba}^2 - \omega^2)^2}\right]\text{erf}^2(at), \tag{I.33}$$

in perfect agreement with the population shown in Figure 5.2 and which is the result of a numerical solution of the Schrödinger equation (without invoking perturbation theory). The excited state population cannot exceed a numerical value of 1 and so our perturbation approach is only valid for

$$\left|\left(\frac{\mu_{ab}F^\omega}{\hbar}\right)\frac{\omega_{ba}}{(\omega_{ba}^2 - \omega^2)}\right| > 1. \tag{I.34}$$

Let us consider what happens at resonance. Going back to Eq. (I.30) and using only the resonant part, we get

$$d_b^{(1)}(t) = -\frac{\mu_{ab}F^\omega}{2\hbar}\int_0^t \text{erf}\,(at')dt' = -\frac{\mu_{ab}F^\omega}{2\hbar}\left[t'\,\text{erf}\,(at') + \frac{e^{-(at')^2}}{a\sqrt{\pi}}\right]_0^t$$

$$= -\frac{\mu_{ab}F^\omega}{2\hbar}\left(t\,\text{erf}\,(at) + \frac{e^{-(at)^2} - 1}{a\sqrt{\pi}}\right), \tag{I.35}$$

which is depicted in Figure 5.15.

such that the excited state population becomes

$$\rho_b(t) = \frac{1}{4}\left(\frac{\mu_{ab}F^{\omega}}{\hbar}\right)^2\left(t\ \mathrm{erf}\,(at) + \frac{e^{-(at)^2}-1}{a\sqrt{\pi}}\right)^2. \tag{I.36}$$

I.3 The Quasi-energy Approach

In this section, we explore the quasi-energy approach introduced in Section 5.1. In order to conserve normalization of the phase-isolated wave function we shall employ the exponential parametrization of Eq. (5.81). Upon perturbation expansion we find the first- and second-order phase-isolated wave functions of Eqs. (5.131) and (5.132), respectively. Proceeding as in Section 5.4.2, we find that the first-order rotation amplitude is given by

$$P_n^{(1)}(t) = e^{-i\omega_{n0}t}\frac{1}{\hbar}\int_0^t \hat{V}_{n0}e^{i\omega_{n0}t'}dt'. \tag{I.37}$$

We may compare this with the expression for the total amplitude

$$d_n^{(1)}e^{-iE_nt/\hbar} = -ie^{-iE_0t/\hbar}P_n^{(1)}(t), \tag{I.38}$$

which shows that the phase associated with the ground-state energy has been taken out of the phase-isolated wave function. We may also note that there is no contribution from the initial state to the first-order phase-isolated wave function, whereas the corresponding projection amplitude is

$$d_0^{(1)}e^{-iE_0t/\hbar} = e^{-iE_0t/\hbar}\frac{1}{i\hbar}\int_0^t V_{00}(t')dt'. \tag{I.39}$$

Summing up the zeroth- and first-order contributions to the ground state projection amplitude

$$\begin{aligned}(d_0^{(0)} + d_0^{(1)})e^{-iE_0t/\hbar} &= e^{-iE_0t/\hbar}\left(1 + \frac{1}{i\hbar}\int_0^t V_{00}(t')dt'\right)\\ &\approx e^{-iE_0t/\hbar}\exp\left[-i\int_0^t V_{00}(t')dt'/\hbar\right],\end{aligned} \tag{I.40}$$

we find that it corresponds to a perturbation expansion of the phase in Eq. (5.25).

In the case of our two-level atom, we find that the first-order rotation amplitude of the excited state is

$$P_b^{(1)} = id_b^{(1)}e^{-i\omega_{ba}t} = i\frac{\mu_{ab}F^{\omega}}{\hbar}\left[\frac{\omega_{ba}\sin\omega t + i\omega\cos\omega t}{(\omega_{ba}^2-\omega^2)}\right]\mathrm{erf}\,(at). \tag{I.41}$$

The first-order phase-isolated wave function accordingly reads

$$\overline{\psi}^{(1)}(t) = \frac{\mu_{ab}F^{\omega}}{\hbar}\left[\frac{\omega_{ba}\sin\omega t + i\omega\cos\omega t}{(\omega_{ba}^2-\omega^2)}\right]\mathrm{erf}\,(at)\ \psi_b. \tag{I.42}$$

The zeroth-order contribution to the quasi-energy is equal to the ground-state energy:

$$Q^{(0)}(t) = \langle\overline{\psi}^{(0)}(t)|\left(\hat{H}_0 - i\hbar\frac{\partial}{\partial t}\right)|\overline{\psi}^{(0)}(t)\rangle = \langle\psi_a|\left(\hat{H}_0 - i\hbar\frac{\partial}{\partial t}\right)|\psi_a\rangle = E_a. \tag{I.43}$$

The first-order contribution is zero:

$$
\begin{aligned}
Q^{(1)}(t) &= \langle \overline{\psi}^{(0)}(t)|\hat{V}(t)|\overline{\psi}^{(0)}(t)\rangle + \langle \overline{\psi}^{(1)}(t)|\hat{H}_0 - i\hbar\frac{\partial}{\partial t}|\overline{\psi}^{(0)}(t)\rangle + \langle \overline{\psi}^{(0)}(t)|\hat{H}_0 - i\hbar\frac{\partial}{\partial t}|\overline{\psi}^{(1)}(t)\rangle \\
&= 2\mathrm{Re}\left[iP_b^{(1)}\langle \psi_b|\hat{H}_0 - i\hbar\frac{\partial}{\partial t}|\psi_a\rangle\right] - i\hbar\frac{\partial}{\partial t}[-iP_b^{(1)}\langle \psi_a|\psi_b\rangle] = 0,
\end{aligned}
$$

where we have used Eq. (5.35). We split the second-order contribution into one part that only contains the Hamiltonian,

$$
\begin{aligned}
Q_1^{(2)}(t) &= \langle \overline{\psi}^{(1)}(t)|\hat{H}_0|\overline{\psi}^{(1)}(t)\rangle + 2\mathrm{Re}[\langle \overline{\psi}^{(2)}(t)|\hat{H}_0|\overline{\psi}^{(0)}(t)\rangle] + 2\mathrm{Re}[\langle \overline{\psi}^{(1)}(t)|\hat{V}(t)|\overline{\psi}^{(0)}(t)\rangle] \\
&= |P_b^{(1)}|^2 E_b - |P_b^{(1)}|^2 E_a + 2\mathrm{Re}[iP_b^{(1)*}\langle \psi_b|\hat{V}(t)|\psi_a\rangle] \\
&= \hbar\omega_{ba}\left(\frac{\mu_{ab}F^{\omega}}{\hbar}\right)^2 \left[\frac{\omega^2 - (\omega_{ba}^2 - \omega^2)\sin^2\omega t}{(\omega_{ba}^2 - \omega^2)^2}\right] \mathrm{erf}^2(at),
\end{aligned}
$$

and a second part involving the time derivative,

$$
\begin{aligned}
Q_2^{(2)}(t) &= -i\hbar\langle \overline{\psi}^{(1)}(t)|\frac{\partial}{\partial t}|\overline{\psi}^{(1)}(t)\rangle - i\hbar\frac{\partial}{\partial t}[\langle \overline{\psi}^{(0)}(t)|\overline{\psi}^{(2)}(t)\rangle] \\
&= \hbar\ \mathrm{Im}\left[(P_b^{(1)*})\frac{\partial}{\partial t}(P_b^{(1)})\right] \\
&= -\hbar\omega_{ba}\left(\frac{\mu_{ab}F^{\omega}}{\hbar}\right)^2 \frac{\omega^2}{(\omega_{ba}^2 - \omega^2)^2}\mathrm{erf}^2(at),
\end{aligned}
$$

where we have omitted a term involving the time derivative of the error function, since it will be very small. The total second-order quasi-energy accordingly reads

$$
Q^{(2)}(t) = -\hbar\omega_{ba}\left(\frac{\mu_{ab}F^{\omega}}{\hbar}\right)^2 \frac{\sin^2\omega t}{(\omega_{ba}^2 - \omega^2)}\mathrm{erf}^2(at), \tag{I.44}
$$

which is in excellent agreement with the result presented in Figure 5.3 and obtained without invoking perturbation theory.

Index

Note: page numbers in italics refer to figures; page numbers in bold refer to tables; a page number followed by 'n' indicates a footnote.

a

abbreviations 419–20
Abelian groups 117–18, 227
absorption 185–90
absorption and emission spectroscopies 335–64
  fluorescence spectroscopy 343–4
  multiphoton absorption 347–54
  phosphorescence 344–7
  visible and ultraviolet absorption 335, 338–43
  X-ray absorption 354–64
absorption coefficient 94, *105*
absorption cross sections 95, 381
  band-integrated 290–1
  and electric dipole polarizability *292*
  exact-state response theory 207–10
  multiphoton 210, 347–54
  vibrational spectroscopies 381, 383
  X-ray absorption 355, 358–9, *364*
active space 264
adiabatic vibronic theory 244–53
  Franck–Condon integrals 248–50
  linear coupling model 252
  vibronic effects in a diatomic system 250–2
AMFI (atomic mean-field interaction) method 346
Ampère–Maxwell equation 33, 76
Ampère's law 33, 76
angular momentum algebra 441–4
anharmonic corrections 383–4
anisotropy 82
antilinear operators 115, 117
anti-Stokes Raman scattering 391
approximate electronic state response theory 255–93
  equations of motion 271–6
  reference state parameterizations 255–70
  relaxation 291–3
  residue analysis 288–91
  response functions 276–88
asymmetry parameter 307
atomic axial tensor 384–6
atomic mean-field interaction (AMFI) method 346
atomic polar tensor contribution to VCD 384
atomic unit conversion factors 421
Auger electron 336, 355, *356*
axial birefringence 366
axial vector 126

b

Baker–Campbell–Hausdorff (BCH) expansion 154, 256, 410
Beer–Lambert law 95–6, 105, 338, 348, 381
Biot–Savart law 36, 39, 61, 304
birefringences 364–80
  natural optical activity 366–71
  nonlinear birefringences 375–80
Bloch equations 412
Bohr–Weisskopf effect 303
Born–Oppenheimer approximation 52, 213–15, 216, 218, 245, 385
boundary conditions 34, 82–6
Breit operator 56

Principles and Practices of Molecular Properties: Theory, Modeling and Simulations, First Edition.
Patrick Norman, Kenneth Ruud, and Trond Saue.

c

CAS (complete active space) 264, 266–7
Cauchy distribution 104, 188, 202, 341
Cauchy moments 180
CC (coupled cluster) *see* coupled cluster (CC) models
character tables 118–19
charge density 2–3, 32–3, 56–7, 87
  bound 77, 78, 85
  free 77, 82
charge distribution 2
charge-transfer excitations 343
chemical shift 414
chiral molecules 366, 384
  birefringences and dichroisms 365, 366
  electronic circular dichroism 372, 373
  isotope substitutions 233
  natural optical activity 367, 368, 371
  Raman optical activity (ROA) 402–7
  rotational averages 109
  vibrational circular dichroism (VCD) 384, 386–7
CI (configuration interaction) *see* configuration interaction (CI) models
CID (circular intensity differential) 405
circular birefringence 365
circular dichroism (CD) *see* electronic circular dichroism (ECD)
circular intensity differential (CID) 405
circular polarization (cp) 44, 367–8
circularly polarized luminescence (CPL) 372
Clausius–Mossotti relation 88, 90, 106, 107
Clebsch–Gordan coefficients 308, 441
complete active space (CAS) 264, 266–7
computational cost 3
configuration interaction (CI) models 3–5, 259, 263–6
  full configuration interaction (FCI) model 267
  matrix structure 281–5
  residue analysis 289
  response functions 281
  singles (CIS) approximation 261–3, 361–2
  singles and doubles (CISD) method 4–5, 263, 269
configuration state functions (CSFs) 264–5
conservation of energy 26, 52, 121
constitutive relations 80, 81
contact density 301
continuity equation 33, 46, 79
continuum dielectric medium models 8
conversion factors 421
core-excited states in X-ray absorption spectroscopy 355–8
core–valence separation (CVS) 357
Coulomb gauge 40, 41, 47, 55, 59
Coulomb operator 21–2, 56, 262
Coulomb's law 36, 38
coupled cluster (CC) models 3–5, 268–70, 275–6
  response functions 285–8
  with singly, doubly, and selected triply excited configuration (CCSD(T)) 276
  with singly and doubly excited configurations (CCSD) 4
CPL (circularly polarized luminescence) 372
creation and annihilation operators 256, 423–4
CSFs (configuration state functions) 264–5
cubic response function 177–9
curl theorem 33, 34, 36–7, 76, 82
current density 32–3, 41, 49, 56–7, 78–9
curvature contribution to polarizability 242, 244
CVS (core–valence separation) 357

d

d'Alembertian operator 38, 46
damped linear response function 202, 381, 383
damped response theory 201–2, 362–4, 373
damping terms 102, 104, 197, 201
DAP (dressed-atom polarizabilities) 319
dark states 339
dc optical rectification **151**
Debye model 98–101
density matrix 195–6, 200–1, 204
density operator 195–6
density-functional theory (DFT) *see* Kohn–Sham density functional theory
diamagnetic exaltation 335
diamagnetic magnetizability 326–35, 414–6
diamagnetic spin–orbit (dso) 304
diatropic current 331

dichroisms 364–5
electronic circular dichroism (ECD) 372–5
magnetic circular dichroism (MCD) 376–9
difference-frequency generation 149, *150*
dimensional analysis 110
Dirac constant 5
Dirac delta function 21, 32, 49, 104, 188–9, 202, 296
Dirac equation 19, 53–4, 123
Dirac Hamiltonian 53–4
Dirac identity 55, 439
Dirac–Coulomb–Breit Hamiltonian 317
disconnected clusters 269
dispersion 81, 140
dispersion energy 322–4
dispersion forces *see* London forces
divergence theorem 33–4, 36, 82
double-harmonic approximation 237, 238, 241, 381–3
Douglas–Kroll–Hess (DKH) Hamiltonian 53
dressed-atom polarizabilities (DAP) 319
dso (diamagnetic spin–orbit) 304
dual circular polarization (DCP) 403
Duschinsky rotation 247, 248

e

ECD *see* electronic circular dichroism (ECD)
effective electronic potential 214
effective molecular geometry 243
effective quadrupole center 380
Ehrenfest theorem 26, 120–1, 122, 159, 205, 207, 271–5, 409
eigenstates 27
eigenvalue equation 2
eigenvalues 27
Einstein coefficient of absorption 210–11
elastic scattering 146
electric and magnetic multipoles 57–72
electric dipole approximation and beyond 63–4
electric dipole radiation 70–2
electric multipoles 59–60, 65–9
general versus traceless forms 65–8
symmetry considerations 68–9
magnetic multipoles 60, 69–70
symmetry considerations 70
multipolar gauge 57–9
multipole expansions 59–63
origin dependence 64–5
electric dipole approximation 63–4
electric dipole moment 2, 3, 60, 295
see also induced dipole moments
electric dipole polarizability *see* polarizability
electric dipole radiation 70–2
electric dipole–electric quadrupole polarizability 295, 368, 404, 405
electric dipole–magnetic dipole polarizability 109, 295, 369, *370*, 404
electric displacement 75
electric quadrupole coupling 66–8
electric quadrupole moments 60, 295, 305–10
see also nuclear electric quadrupole moments
electric quadrupole–magnetic dipole approximation 64
electrical anharmonicity 237
electric-field-induced second-harmonic generation 325, 326
electrodynamics 31
electromagnetic fields
force acting on a point charge 50
Lagrangian density 48
principle of minimal electromagnetic coupling 49, 55
electromagnetic potentials 39–41
electromagnetic tensor 48
electromagnetic wave equation 41–2
inhomogeneous 38
linear media 93
electromagnetic waves
absorbance 94–6
absorption coefficient 94, *105*
absorption cross-section 95
handedness convention 44–5
intensity 52, 94
in linear media 92–6
penetration depth 105
phase, period, wavelength and frequency 42
phase velocity 93
polarization 43–5, 367
Poynting vector 52, 72, 94
signal velocity 93
wavefront 42
wavelength 42
wavenumber 42
see also Maxwell's equations
electron paramagnetic resonance (EPR) 6, 316–17

electron spin 18–19, 54, 56
see also Pauli spin matrices; spin–orbit interactions
electronic charge distribution 2
electronic circular dichroism (ECD) 365, 372–5
see also vibrational circular dichroism (VCD)
electronic energy of a molecule 238, 240, 241
electronic Hamiltonian 52–3, 213–14, 427–8
electronic polarizability 218, 240, 241
electronic structure methods 4
electronic wave function 1, 18–19, 21
approximate-state theory 28
energy basis expansion 27–8
electron–nucleus interaction 296, 299
elliptic polarization (ep) 43
emission spectroscopies *see* absorption and emission spectroscopies
energy conservation 26, 52
energy eigenstates 28
energy smearing 207
ep (elliptic polarization) 43
EPR (electron paramagnetic resonance) 6, 316–17
equations of motion **12**, 14, 156–63
in applied external electromagnetic fields 49
approximate electronic state response theory 271–6
Ehrenfest theorem 271–5
quasi-energy derivatives 275–6
time evolution of projection amplitudes 157–9
time evolution of rotation amplitudes 159–63
ESA (excited-state absorption) *195*, 337
Euler–Lagrange equations 14, 48
exact two-component (X2C) Hamiltonian 53
exact-state response theory 135–212
absorption 185–90
dispersion 179–83
equations of motion 156–63
molecular electric properties 145–51
oscillator strength and sum rules 183–5
reference-state parameterizations 151–5
relaxation 194–211
residue analysis 190–4
response functions 163–79
first-order properties 166
fourth-order properties 174–9
higher-order properties 179
second-order properties 166–9
third-order properties 169–74
responses in two-level system 135–45
EXAFS (extended X-ray absorption fine structure) 355
exchange-correlation (XC) functional 4
excited-state absorption (ESA) *195*, 337
expectation value 1, 2, 19–20, 25–7
exponential parametrization 153, 255–6, 455
extended X-ray absorption fine structure (EXAFS) 355

f

Faraday effect 365, *366*, 375
Faraday's law 32
FC (Franck–Condon) *see* Franck–Condon (FC) approximation
Fermi 296
Fermi distribution 297
Fermi resonances 383
fermions 129
Fermi's golden rule 189, 207
ferromagnetism 327
few-state models of multiphoton absorption 350–1
field polarization in X-ray absorption spectroscopy 358–60
first-order correction to the wave function 161, 166–7, 205
first-order hyperpolarizability 137, 147, 171–2, *181*, 182, 191
first-order nonlinear response function 169–71, *191*, 204
five-photon absorption 351–3
flip angle 412
fluorescence 338
fluorescence spectroscopy 343–4, 355, *356*
Fock space 425–6
form factor 296
Fourier transforms 431–3
four-photon absorption 349, 351, 352
Franck–Condon (FC) approximation 245–6, 247, 336–7, 395, 398, 399, *400*
Franck–Condon (FC) integrals 248–50, 251–2
Franck–Condon principle 245, *246*, 339–40, 343

free-atom polarizabilities (FAP) 318
free-induction decay (FID) 412
frequency decomposition of molecular polarization 148–50
full configuration interaction (FCI) model 267
functional derivatives 13–14
functionals 12–13

g

gas-phase absorption spectra 339, 340
gauge freedom 39–40
gauge function 40
gauge origin 311, 328–30, 369–70, 371
gauge transformations 40, 47, 58
gauge-including atomic orbitals (GIAOs) *see* London atomic orbitals (LAOs)
Gauss' law 32
Gaussian distribution 297, *298*
Gaussian lineshape *341*, 342, 373
Gaussian surfaces 73–4, 87, 297
Gauss–Legendre quadrature scheme 325
gerade symmetry 351, 352
GIAOs (gauge-including atomic orbitals) *see* London atomic orbitals (LAOs)
g-matrix 314–316
Göppert-Mayer (GM) 348
Gouy balance 332
Grotthuss–Draper law 335
group theory 118
g-tensor *see* g-matrix
gyromagnetic ratio 302

h

half-integer spin systems (fermions) 129
Hamiltonian 7–9, 15–18, 22
  classical mechanical 48–52
  conservation of energy 26, 52, 121
  coordinate representation 17
  Douglas–Kroll–Hess (DKH) 53
  energy eigenstates 28
  exact two-component (X2C) 53
  generic form 52
  Hamilton's equations **12**, 16
  minimal substitutions 49
  momentum representation 17
  quantum-mechanical 17, 52–7
  relativistic 17–18, 56–7
  for a system of particles and fields 31
  zeroth-order regular approximation (ZORA) 54
harmonic oscillator approximation 222–5, 381, 394
harmonic oscillator wave function 24–5, 221, 222, 248
Hartree–Fock (HF) method 4, 181, 261–2, *264*, 265, 269, 291, 446
Heaviside step function 297
Heisenberg equation of motion 26, 185
Hellmann–Feynman theorem 29, 144, 163–4, 446
Helmholtz theorem 35–6, 39
Hermitian operators **12**, 20, 27, 115–16
Herzberg–Teller corrections 245, 252–3, 398
HF (Hartree–Fock) method 4, 181, 261–2, *264*, 265, 269, 291, 446
Hilbert space 18
Hückel theory 329
hyperfine structure (nuclear interactions) 301–10
hyperpolarizability 3n, 147
  first-order 137, 147, 171–2, *181*, 182, 191
  second-order 3, 137, 147, 176, 182–3, *184*, 192–3
hypopolarizability 3n

i

IC (internal conversion) *195*, 337
ICP (incident circular polarization) 403
identity operator 20
incident circular polarization (ICP) 403
independent mode displaced harmonic oscillator (IMDHO) 395, 396, 398, 399, *400*
indirect nuclear spin–spin coupling tensor 412–17
induced dipole moments 3, 80, 86, 102, 107, 109, 139, 295
induced polarization 80, 136, 137, 317
induction energy 322–3
infrared (IR) absorption
  anharmonic corrections 383–4
  double-harmonic approximation 381–3
infrared (IR) spectroscopy 347, 348, 381
inhomogeneous electromagnetic wave equation 38
intensity-dependent refractive index **151**
internal conversion (IC) *195*, 337

internal field 86, 88
International System of Units (SI) 110
inter-system crossing (ISC) *195*, 337
IR absorption *see* infrared (IR) absorption
irreducible representation 118
ISC (inter-system crossing) *195*, 337
isotope shifts 231–2
isotropic tensors 108–9

j
Jablonski diagram *195*, 336
Jefimenko's equations 39

k
Kasha's rule 337, 338, 344
Kerr effect **151**, 325–6, 365, *366*
Kohn–Sham density functional theory (KS-DFT) 4, 279–80
Kramers basis 124
Kramers degeneracy 124
Kramers partners 124
Kramers–Heisenberg–Dirac expression 392, 397, 399, *400*, 401
Kramers–Kronig relations 97–8, 201
Kronecker delta 108, 109, 157, 164
KS-DFT (Kohn–Sham density functional theory) 4, 279–80
Kubo relation 163

l
Lagrangian 48–9
Lagrangian density 48
Lagrangian mechanics 12–15
Landé's interval 443
Langevin function 91, *92*
LAOs (London atomic orbitals) 329–30, 370, 371
Laplace's equation 35
Laporte's rule 339
Larmor precession 409
LDA (local density approximation) 280
Legendre transformation 15–16
Leibniz integral rule 102
Levi–Civita symbol 59, 108, 109, 329
Lévy-Leblond equation 54
linear birefringence 365
linear coupling model 247, 252
linear media 81–111, 167–8
  boundary conditions 82–6
  dimensions, units, and magnitudes 110–11
  electromagnetic waves in 92–6
  frequency dependence of the permittivity 96–107
  polarization in 86–92
  rotational averages 107–10
linear momentum **20**
linear operators 115
linear polarizability 140, 168, 179, 317–25
linear polarization (lp) 43
linear polarization propagator 168
linear response functions 163, 167–9, 201–4, 206
Liouville equation 196–200, 205
local density approximation (LDA) 280
London atomic orbitals (LAOs) 329–30, 370, 371
London forces 318, 324
London phase factor 329–30
Lorentz factor 5
Lorentz force 50, 110, 121, 125
Lorentz gauge 40, 41
Lorentz model 101–4, 184
Lorentz transformation 46–7
Lorentzian lineshape *341*, 342, 373
Lorentz–Lorenz relation 106, 367
lp (linear polarization) *see* linear polarization (lp)

m
macroscopic magnetization 79
macroscopic polarization 78
magic angle 360
magnetic circular dichroism (MCD) 376–9
magnetic dipole moment 60–2, *61*, 295
  see also induced dipole moments; nuclear magnetic dipole moments
magnetic hysteresis 80
magnetic multipoles 60, 69–70
  origin dependence 64–5
  symmetry considerations 70
magnetic quadrupole interaction 70
magnetic susceptibility 76–7
magnetically induced currents 331–2
magnetizability 326–35
  isotropic magnetizabilities and Pascal's rule 332–5
  magnetically induced currents 331–2

origin dependence 328–31
magnetization 75–7, 79
magnetochiral effect 366
mass-velocity correction 18
mass-weighted coordinates 221–2
material constants 81
Maxwell's equations 32–47
constitutive relations 79–80
dimensional analysis 110
electrodynamics 45–7
electromagnetic potentials and gauge freedom 39–41
general case 38–9
macroscopic 72–80
in matter 77–9
microscopic 32–47
nonrelativistic limit 47
polarization and magnetization 73–7
spatial averaging 72–3
stationary case 34–8
MCD (magnetic circular dichroism) 376–9
MCSCF *see* multiconfiguration self-consistent field (MCSCF)
MD (molecular dynamics) 8, 389
mechanical anharmonicity 228, 230, 237
mechanical momentum 15, 49, 50, 110
microwave Zeeman spectroscopy 328
minimal electromagnetic coupling 49, 55
MM (molecular mechanics) 8
molecular dynamics (MD) 8, 389
molecular electric properties 145–51
molecular Hamiltonian 213–14
molecular magnetizability 326, 333
molecular mechanics (MM) 8
molecular orbital coefficients 255, 266
molecular polarization 148–50
molecular quadrupole moment 328
molecular vibrations 216–21
and normal coordinates 221–5
perturbation theory (PT) for vibrational wave functions 225–7
pure vibrational contributions to properties **229**, 235–44
temperature effects 234, *235*
zero-point vibrational contributions to properties 227–34
Møller–Plesset perturbation theory 276
Morse oscillator 230
Mössbauer isomer shift 300–1
Mössbauer spectroscopy 295, 300–1
MP2 second-order (Møller–Plesset) 276
multiconfiguration self-consistent field (MCSCF) 266–8, 274–5
matrix structure 281–5
residue analysis 289
response functions 281
multiphoton absorption 336, 347–54
cross sections 210, 347–9
few-state models for two-photon absorption 350–1
general processes 351–4
multiphoton interactions 146, 149–50
multipolar gauge 57–9
see also electric and magnetic multipoles

n

natural optical activity 366–71
near-edge X-ray absorption fine structure (NEXAFS) 355–64
NMR *see* nuclear magnetic resonance (NMR)
nonlinear birefringences 375–80
electric field gradient-induced 379–80
magnetic circular dichroism (MCD) 376–9
nonlinear optical processes **151**
nonlinear polarizabilities 147–51, 325–6
nonlinear response function 171–4, 202, 206
first-order response function 169–71, *191*, 204
nonrelativistic velocity **20**
NQCC (nuclear quadrupole coupling constant) 307–8
NQR (nuclear quadrupole resonance) spectroscopy 307
nuclear charge distribution 296–301
nuclear electric quadrupole moments 295, 305–10
nuclear interactions 295–310
hyperfine structure 301–10
nuclear electric quadrupole moments 295, 305–10
nuclear magnetic dipole moment 295, 301–5
nuclear charge distribution 296–301
nuclear kinetic energy operator 213–15
nuclear magnetic dipole moments 295, 301–5

nuclear magnetic resonance (NMR) 233, 295, 407–17
NMR experiment 407–12
parameters 412–17
nuclear magneton 302
nuclear mass density 296
nuclear quadrupole coupling constant (NQCC) 307–8
nuclear quadrupole resonance (NQR) spectroscopy 307
nuclear relaxation polarizability 240, 241, 242
nuclear Schrödinger equation 215, 221
nuclear size 300
nuclear spin 53, 301, 306, 408–10
nuclear spin–spin coupling 231
see also indirect nuclear spin–spin coupling tensor
number operator 426–7
Numerov–Cooley procedure 394, 396
nutation frequency 410

o

one-photon absorption (OPA) 189, 192, **194**, *195*, 336
operator algebra 435–8
optical activity 366–71
optical permittivity 101
optical rectification **151**
optical rotation 365, 368–71
optical rotatory dispersion (ORD) spectrum 368–9, 371
orbital angular momentum **20**, 123, 127, 129
see also angular momentum algebra; spin–orbit interactions
ORD (optical rotatory dispersion) spectrum 368–9, 371
origin dependence of electric and magnetic multipoles 64–5
origin dependence of magnetizability 328–31
oscillator strength and sum rules 183–5

p

paramagnetic magnetizability 328, 330
see also electron paramagnetic resonance (EPR)
paramagnetic spin–orbit (pso) 304
paramagnetism 327
parameterization 151–5
based on projections 152–3, 157–9
based on rotations 153–5, 159–63
see also reference state parameterizations
paratropic current 331
particles and fields 31
particles in electromagnetic fields 48–57
classical mechanical Hamiltonian 48–52
quantum-mechanical Hamiltonian 17, 52–7
Pascal rule 333–4
Paschen–Back effect 314
Pauli exclusion principle 21
Pauli spin matrices 18, 53, 123, 314, 349
PCM (polarizable continuum model) 8
PECs (potential energy curves) 231, *232*, *389*, 394, 395, *396*
permeability 32, 76
permittivity 32, 75
frequency dependence in linear media 96–107
Kramers–Kronig relations 97–8, 201
refraction and absorption 105–7
relaxation in the Debye model 98–101
resonances in the Lorentz model 101–4
perturbation theory (PT) 138–40, 142
density matrix from 200–1
equations of motion from 156–63
pure vibrational contributions to properties 235–8
for vibrational wave functions 225–7
see also Ehrenfest theorem; quasi-energy approach
phase-isolated wave function 140, 142, 153, 159, 161
phosphorescence 338, 344–7
photochemical reactions 335
photoisomerization of retinal 335, *336*
photoluminescence 338
Placzek theory 393–4, 395
Planck's radiation law 211
Pockels effect **151**, 325
point electric dipole 60, 62
point magnetic dipole 60–2
Poisson equation 35, 40
polar vector 126
polarizability 3, 86–8, 102, 107, 109, 295
curvature contribution 240, 242, 244
electronic polarizability 218, 240, 241
linear 140, 168, 179, 317–25

nonlinear 147–51, 325–6
nuclear relaxation polarizability 240, 241, 242, 244
principle of minimum polarizability 319
pure vibrational 217–20, 235–8
vibronic 216–21, 322
zero-point vibrationally averaged 218, 239–40, 242–3
see also hyperpolarizability
polarizable continuum model (PCM) 8
polarization (electric) 73–5
in linear media 86–92
macroscopic 78
time-dependent 97–9, 147
polarization (electromagnetic waves) 43–5, 367
poles and residues 190–4
positrons 53
potential energy curves (PECs) 231, *232*, *389*, 394, 395, *396*
potential energy surfaces 215
surface curvature 240, 242, 244
Poynting vector 52, 72, 94
Poynting's theorem 51–2, 110, 392
principle of minimal electromagnetic coupling 49, 55
principle of minimum polarizability 319
principle of stationary action 13
probability density function 11
pseudo-spin operator 305
pseudovector *see* axial vector
pso (paramagnetic spin–orbit) 304
PT (perturbation theory) *see* perturbation theory (PT)
pure vibrational contributions to properties 235–44
electric-field dependence of themolecular geometry 238–44
perturbation theory approach 235–8
pure vibrational polarizability 217–20, 235–8, 383

q

quantum mechanics 11–29
quantum-mechanical operators 20–1, **20**
time evolution of expectation values 25–7
time evolution of wave functions 22–5
time-inversion symmetry 122
variational principle 27–9
quasi-energy approach 135, 140–5, 163–4, 275–6, 455–6

r

Rabi cycles 186–7, 410, 453
Raman active modes 394
Raman optical activity (ROA) 402–7
Raman scattering 388–402
classical point of view 388–91
quantum mechanical point of view 392–402
resonance Raman scattering 394–401
Raman spectroscopy 348, 399
random phase approximation (RPA) 279, 284, 292
Rayleigh scattering 388, 390
reaction field 88–9
reducible representation 118
reference state parameterizations 151–5, 255–70
see also configuration interaction (CI) models; coupled cluster (CC) models; Hartree–Fock (HF) method; multiconfiguration self-consistent field (MCSCF)
refractive index 93, 105–6, 366–8
relative permeability 76
relative permittivity 75, 96–107
relativistic effects 5–7, 17–18, 38, 45–7, 49
relativistic velocity **20**
relaxation 194–211
absorption cross section 207–10
approximate electronic state response theory 291–3
in the Debye model 98–101
density matrix from perturbation theory 200–1
density operator 195–6
Einstein coefficient of absorption 210–11
linear response functions from the density matrix 201–4
Liouville equation 196–200
nonlinear response functions from the density matrix 204
in wave function theory 204–7
relaxation time 99, 102
residue analysis 190–4, 288–91

resonance enhanced states 351
resonances in the Lorentz model 101–4
response functions **194**, 216–21, 276–88
  configuration interaction (CI) models 281
  coupled cluster (CC) models 285–8
  matrix structure in the SCF, CI, and MCSCF approximations 281
  multiconfiguration self-consistent field (MCSCF) 281
  single determinant approaches 276–80
  see also linear response functions; nonlinear response function
ROA (Raman optical activity) 402–7
Röntgen current 79
Rosenfeld tensor 368
rotating wave approximation 409n, 451
rotation matrices 129–31, 132–3
rotation operator 127, 128
rotations 127–33
  parameterization by 153–5, 159–63
Rowland ring *75*, 76
RPA (random phase approximation) 279, 284, 292
Rydberg excitations 343

s

scalar relativistic effects 6, 7
scattered circular polarization (SCP) 403
Schrödinger equation 54–5
  nuclear 215, 221
  parameterization of the wave function 151–5
  time inversion of the wave function 122–3
  time-dependent (TDSE) 22, 26
  time-independent (TISE) 29
SCP (scattered circular polarization) 403
second quantization 255–6, 423–9
second-harmonic generation 146, **151**, 325, 326
second-order correction to the wave function 161, 206
second-order hyperpolarizability 3, 137, 147, 176, 182–3, *184*, 192–3
second-order nonlinear response function *192*
second-order polarization 169
selection rules in spectroscopy 118–19
self-adjoint operators 19–20
self-consistent field (SCF) theory 261, 273
  matrix structure 281–5
  residue analysis 289
  see also multiconfiguration self-consistent field (MCSCF)
shielding tensor 412–13, 414
short-time approximation 399, 400, *401*
simple linear dielectrics 82
single and double (SD) excitations 4
size consistent methods 268
size extensive methods 268
Slater determinants 4, 21, 256–7, 258, 425
spatial dispersion 81
spatial inversion/parity 125–7
spectral line broadening 195, 207, 250–1, 310, 341–2
spectroscopies
  electronic absorption and emission 335–64
  fluorescence 343–4
  multiphoton absorption 347–54
  nuclear magnetic resonance (NMR) 407–17
  phosphorescence 344–7
  selection rules 118–19
  visible and ultraviolet absorption 335, 338–43
  X-ray absorption 354–64
spherical tensor operators 441
spin *see* electron spin; nuclear spin
spin angular momentum **20**, 126, 128
  see also angular momentum algebra
spin in second quantization 429
spin matrix algebra 439
spin multiplicities 337–9
spin–magnetic field interaction 55
spin-magnetic moment 131–2
spin–orbit interactions 6–7, 133, 317n, 345, 442–3
  paramagnetic and diamagnetic 304, 305, 311, 414
spontaneous emission 195, 211
state-transfer amplitudes 153
static exchange (STEX) approximation 361–2
Stokes Raman scattering 391
Stokes' theorem 33n
sum-frequency generation 149
symmetry 68–9, 113–34
  operations and groups 113–17
    direct product groups 119

group representation 117–20
point group 115
principal axis 114
subgroups 119
symmetry element 114
rotations 127–33
spatial inversion/parity 125–7
time symmetries 120–4
symmetry operator 116–17
symmetry-adapted basis function 118

t

Tamm–Dancoff approximation 261
TDHF (time-dependent Hartree–Fock) approximation 181, 263, 272, 289–90
temperature effects 92, 234, *235*, 326
birefringences 379–80
magnetic circular dichroism (MCD) 376, *377*
temporal dispersion *see* dispersion
third-harmonic generation **151**
third-order polarization 174–6
Thomas–Reiche–Kuhn sum rule 185, 291, 329, 375
three-photon absorption (3PA) 348, 349, 352
time-averaged quasi-energy 144–5, 163–4
time-dependent coupled cluster theory 275–6
time-dependent Hartree–Fock (TDHF) approximation 181, 263, 272, 289–90
time-dependent Hellmann–Feynman theorem 144, 163–4
time-dependent Kohn–Sham theory 279–80
time-dependent perturbation theory 454–5
see also Ehrenfest theorem; time-averaged quasi-energy
time-dependent polarization 97, 147
time-dependent reference states 255
time-dependent Schrödinger equation (TDSE) 22, 26
time-dependent variational principle 144, 159, 276, 389
time-energy uncertainty relation 146
time-independent Schrödinger equation (TISE) 29, 53
time-inversion symmetry 121
time-reversal operator 123
time-reversal symmetry 121–4
time-translational symmetry 121
two-level atom 135–45, 185–7, 451–6
two-photon absorption (TPA) 149, 189, 190, 192–3, **194**, *195*, 336, 347, 349–1

u

ungerade symmetry 351–3
units 421
Unsöld approximation 318, 324
Unsöld's theorem 308
UV/vis spectroscopies 335, 338–43

v

variational perturbation theory 240, 445–50
conversion of formulas **236**
variational wave function theories *see* configuration interaction (CI) models; multiconfiguration self-consistent field (MCSCF); self-consistent field (SCF) theory
Vavilov rule 344
Verdet constant 375
vibrational circular dichroism (VCD) 384–8, 406–7
vibrational excitations 105, 216–21
vibrational frequencies 224, 225, 227, 232
vibrational perturbation theory 225–7, 235–8
vibrational Raman optical activity 402–7
vibrational spectroscopies 381–407
infrared (IR) absorption spectroscopy 381–4
Raman scattering 388–402
vibrational circular dichroism (VCD) 384–8
vibrational Raman optical activity 402–7
vibrational wave functions 218, 222, 223, 226–7
vibronic absorption spectra 244, 252, 397
vibronic polarizabilities 216–21, 322
vibronically induced transitions 245–53, 336–7, 343
diatomic system 250–2
Franck–Condon integrals 248–50
Herzberg–Teller corrections 252–3
multiphoton absorption 348

w

weak intermolecular forces 321–5
Wigner rule 243
Wigner–Eckart theorem 442

X

X2C (exact two-component) Hamiltonian 53
XANES (X-ray absorption near-edge structure) 355–64
XAS (X-ray absorption spectroscopy) 336
X-ray absorption 354–64
  complex or damped response theory 362–4
  core-excited states 355–8
  field polarization 358–60
  response theory 357–8, 362–4
  static exchange approximation 360–2
X-ray absorption near-edge structure (XANES) 355–64
X-ray absorption spectroscopy (XAS) 336
X-ray fluorescence 336

Z

Zeeman interaction 310–17, 327, 328, 412
zero-field splitting (ZFS) 317, 345
zero-point vibrational averaging (ZPVA) 227, 228, 230–4
zero-point vibrational contributions to properties 227–34
zero-point vibrational corrections 228, 230–4, 239
zero-point vibrationally averaged polarizability 218, 239–40, 242–3
zeroth-order regular approximation (ZORA) Hamiltonian 54
ZFS (zero-field splitting) 317, 345
Zitterbewegung 27
ZPVA (zero-point vibrational averaging) 227, 228, 230–4

Printed and bound by CPI Group (UK) Ltd, Croydon, CR0 4YY
16/09/2026
14963473-0008